Symmetry Groups and Their Applications

Pure and Applied Mathematics

A Series of Monographs and Textbooks

Editors **Samuel Eilenberg and Hyman Bass**

Columbia University, New York

RECENT TITLES

XIA DAO-XING. Measure and Integration Theory of Infinite-Dimensional Spaces: Abstract Harmonic Analysis

RONALD G. DOUGLAS. Banach Algebra Techniques in Operator Theory

WILLARD MILLER, JR. Symmetry Groups and Their Applications

ARTHUR A. SAGLE AND RALPH E. WALDE. Introduction to Lie Groups and Lie Algebras

T. BENNY RUSHING. Topological Embeddings

JAMES W. VICK. Homology Theory: An Introduction to Algebraic Topology

E. R. KOLCHIN. Differential Algebra and Algebraic Groups

GERALD J. JANUSZ. Algebraic Number Fields

A. S. B. HOLLAND. Introduction to the Theory of Entire Functions

WAYNE ROBERTS AND DALE VARBERG. Convex Functions

A. M. OSTROWSKI. Solution of Equations in Euclidean and Banach Spaces, Third Edition of Solution of Equations and Systems of Equations

H. M. EDWARDS. Riemann's Zeta Function

SAMUEL EILENBERG. Automata, Languages, and Machines: Volume A. *In preparation:* Volume B

MORRIS HIRSCH AND STEPHEN SMALE. Differential Equations, Dynamical Systems, and Linear Algebra

WILHELM MAGNUS. Noneuclidean Tesselations and Their Groups

FRANÇOIS TREVES. Basic Linear Partial Differential Equations

WILLIAM M. BOOTHBY. An Introduction to Differentiable Manifolds and Riemannian Geometry

BRAYTON GRAY. Homotopy Theory: An Introduction to Algebraic Topology

ROBERT A. ADAMS. Sobolev Spaces

JOHN J. BENEDETTO. Spectral Synthesis

D. V. WIDDER. The Heat Equation

IRVING EZRA SEGAL. Mathematical Cosmology and Extragalactic Astronomy

J. DIEUDONNÉ. Treatise on Analysis: Volume II, enlarged and corrected printing; Volume IV

WERNER GREUB, STEPHEN HALPERIN, AND RAY VANSTONE. Connections, Curvature, and Cohomology: Volume III, Cohomology of Principal Bundles and Homogeneous Spaces

In preparation

I. MARTIN ISAACS. Character Theory of Finite Groups

SYMMETRY GROUPS AND THEIR APPLICATIONS

WILLARD MILLER, Jr.

School of Mathematics
University of Minnesota
Minneapolis, Minnesota

ACADEMIC PRESS New York San Francisco London 1972

A Subsidiary of Harcourt Brace Jovanovich, Publishers

ACADEMIC PRESS, INC.
111 Fifth Avenue, New York, New York 10003

United Kingdom Edition published by
ACADEMIC PRESS, INC. (LONDON) LTD.
24/28 Oval Road, London NW1

LIBRARY OF CONGRESS CATALOG CARD NUMBER: 72-82628

AMS(MOS) 1970 Subject Classifications: 20C30, 20G05, 22E05, 81A78

PRINTED IN THE UNITED STATES OF AMERICA

Contents

Preface ix

Chapter 1 **Elementary Group Theory**

1.1 Abstract Groups 1
1.2 Subgroups and Cosets 4
1.3 Homomorphisms, Isomorphisms, and Automorphisms 6
1.4 Transformation Groups 8
1.5 New Groups from Old Ones 13
Problems 15

Chapter 2 **The Crystallographic Groups**

2.1 The Orthogonal Group in Three-Space 16
2.2 The Euclidean Group 20
2.3 Symmetry and the Discrete Subgroups of $E(3)$ 23
2.4 Point Groups of the First Kind 27
2.5 Point Groups of the Second Kind 32
2.6 Lattice Groups 34
2.7 Crystallographic Point Groups 38
2.8 The Bravais Lattices 42
2.9 Crystal Structure 52
2.10 Space Groups 55
Problems 59

Chapter 3 Group Representation Theory

3.1 A Group Representation 61
3.2 Reducible Representations 67
3.3 Irreducible Representations 69
3.4 Group Characters 74
3.5 New Representations from Old Ones 78
3.6 Character Tables 85
3.7 The Method of Projection Operators 92
3.8 Applications 100
Problems 114

Chapter 4 Representations of the Symmetric Groups

4.1 Conjugacy Classes in S_n 116
4.2 Young Tableaux 119
4.3 Symmetry Classes of Tensors 128
4.4 The Simple Characters of S_α 147
Problems 151

Chapter 5 Lie Groups and Lie Algebras

5.1 The Exponential of a Matrix 152
5.2 Local Lie Groups 162
5.3 Lie Algebras 166
5.4 The Classical Groups 171
5.5 The Exponential Map of a Lie Algebra 175
5.6 Local Homomorphisms and Isomorphisms 178
5.7 Subgroups and Subalgebras 183
5.8 Representations of Lie Groups 186
5.9 Local Transformation Groups 188
5.10 Examples of Transformation Groups 199
Problems 205

Chapter 6 Compact Lie Groups

6.1 Invariant Measures on Lie Groups 206
6.2 Compact Linear Lie Groups 210
6.3 Group Characters and Representations 217
Problems 221

Chapter 7 The Rotation Group and Its Representations

7.1 The Groups $SO(3)$ and $SU(2)$ 222
7.2 Irreducible Representations of $SU(2)$ 228
7.3 Irreducible Representations of $sl(2)$ 233
7.4 Expansion Theorems for Functions on $SU(2)$ 236

7.5 New Realizations of the Irreducible Representations 239
7.6 Applications to Physics 247
7.7 The Clebsch–Gordan Coefficients 256
7.8 Applications of the Clebsch–Gordan Series 263
7.9 Double-Valued Representations of the Crystallographic Groups 271
7.10 The Wigner–Eckart Theorem and Its Applications 273
7.11 Spinor Fields and Invariant Equations 278
Problems 284

Chapter 8 **The Lorentz Group and Its Representations**

8.1 The Homogeneous Lorentz Group 285
8.2 The Physical Significance of Lorentz Invariance 293
8.3 Representations of the Lorentz Group 297
8.4 Models of the Representations 307
8.5 Lorentz-Invariant Equations 313
Problems 320

Chapter 9 **Representations of the Classical Groups**

9.1 Representations of the General Linear Groups 321
9.2 Character Formulas 334
9.3 The Irreducible Representations of $GL(m, R)$, $SL(m, \mathfrak{C})$, and $SU(m)$ 340
9.4 The Symplectic Groups and Their Representations 344
9.5 The Orthogonal Groups and Their Representations 351
9.6 Dirac Matrices and the Spin Representations of the Orthogonal Groups 362
9.7 Examples and Applications 368
9.8 The Pauli Exclusion Principle and the Periodic Table 377
9.9 The Group Ring Revisited 387
9.10 Semisimple Lie Algebras 391
Problems 395

Chapter 10 **The Harmonic Oscillator Group**

10.1 The Harmonic Oscillator 397
10.2 Representations of the Harmonic Oscillator Group 402
Problems 405

Appendix **Hilbert Space** 407

References 415

Symbol Index 421
Index 425

Preface

This is a beginning graduate level textbook on applied group theory. Only those aspects of group theory are treated which are useful in the physical sciences, but the mathematical apparatus underlying the applications is presented with a high degree of rigor.

The principal characters in this book are symmetry groups of mathematical physics. The first four chapters are primarily concerned with finite or discrete symmetry groups, e.g., the point, space, and permutation groups. The last six chapters are devoted to Lie groups.

The theory presented here is largely algebraic in nature; the more complicated global topological problems are avoided. Thus topics such as the representation theory of Euclidean, Poincaré, and space groups are omitted. (These topics will be included in a projected second volume by the author which will be primarily devoted to topological aspects of applied group theory.) It is assumed that the reader is proficient in linear algebra and advanced calculus. Such concepts as finite-dimensional vector spaces, linear operators, and Jacobians are used without prior definition. An appendix on Hilbert space lists all the information the reader needs on that topic. There are a few places where greater mathematical sophistication is needed. In Chapter 5 the existence and uniqueness theorem for solutions of ordinary differential equations and some simple properties of power series are employed. In Chapter 6 the Peter–Weyl theorem is stated but not proved.

Most of the theory presented here is applied to quantum mechanics. Thus, it is desirable, though not essential, for the reader to be familiar with the basic concepts of quantum theory, particularly the probabilistic and physical interpretations. To make the applications clearer and to avoid unnecessary detail, the version of quantum theory presented here is slightly oversimplified. (In particular, only a qualitative perturbation theory of energy eigenvalues of the Hamiltonian is presented, and the physical interpretation in terms of spectral lines is omitted.) The author hopes in this way to explain some of the beautiful applications of group theory in atomic and nuclear physics to mathematics students unfamiliar with the physical literature.

There are several features which together differentiate this book from prior works on applications of group theory: (1) A rigorous derivation of point and space groups including a derivation of the fourteen Bravais lattices. (2) A simplified but rigorous presentation of the theory of local linear Lie groups. (3) A construction of the representations of the classical groups using both weights and Young diagrams. (4) An integrated theory which includes applications not only to classical and quantum physics but also to geometry and special function theory.

Finally, the author wishes to acknowledge his debt to those mathematicians and physicists whose writings form the main content of this volume, especially H. Boerner, I. Gel'fand, S. Lie, G. Liubarskii, M. Naimark, N. Vilenkin, H. Weyl, and E. Wigner. (In particular, Chapter 9 is adapted from Weyl's Princeton lecture notes [2].)

Chapter 1

Elementary Group Theory

1.1 Abstract Groups

A group is an abstract mathematical entity which expresses the intuitive concept of symmetry.

Defintion. A **group** G is a set of objects $\{g, h, k, \ldots\}$ (not necessarily countable) together with a binary operation which associates with any ordered pair of elements g, h in G a third element gh. The binary operation (called **group multiplication**) is subject to the following requirements:

(1) There exists an element e in G called the **identity element** such that $ge = eg = g$ for all $g \in G$.
(2) For every $g \in G$ there exists in G an **inverse element** g^{-1} such that $gg^{-1} = g^{-1}g = e$.
(3) **Associative law.** The identity $(gh)k = g(hk)$ is satisfied for all $g, h, k \in G$.

Thus, any set together with a binary operation which satisfies conditions (1)–(3) is called a group. If $gh = hg$ we say that the elements g and h **commute.** If all elements of G commute then G is a **commutative** or **abelian** group. If G has a finite number of elements it has **finite order** $n(G)$, where $n(G)$ is the number of elements. Otherwise, G has **infinite order.**

A **subgroup** H of G is a subset which is itself a group under the group multiplication defined in G. The subgroups G and $\{e\}$ are called **improper** subgroups of G. All other subgroups are **proper.**

Theorem 1.1. A nonempty subset H of a group G is a subgroup if and only if the following two conditions hold:

(1) If $h, k \in H$ then $hk \in H$.
(2) If $h \in H$ then $h^{-1} \in H$.

Proof. If H is a subgroup then (1) and (2) clearly hold. Conversely, suppose these conditions hold. We show that H satisfies the requirements for a group. The associative law holds since it holds for G. There exists some $h \in H$ since H is nonempty and by (2) we have $h^{-1} \in H$. By (1), $hh^{-1} = e \in H$, so H has an identity. Q.E.D.

The identity element e of a group is unique: Suppose $e' \in G$ such that $e'g = ge' = g$ for all $g \in G$. Setting $g = e$, we find $ee' = e'e = e$. But $e'e = e'$ since e is an identity element. Therefore, $e' = e$.

A similar proof shows that the inverse element g^{-1} of g is unique. Suppose $g' \in G$ such that $gg' = e$. Multiplying on the left by g^{-1} and using the associative law, we get $g^{-1} = g^{-1}e = g^{-1}(gg') = (g^{-1}g)g' = eg' = g'$.

The following examples indicate the variety of mathematical objects which have the structure of groups. (For most groups in this book the associative law will be trivial to verify. We shall specifically verify the law only in those cases where it is not obvious.)

Example 1. The real numbers R with addition as the group product. The product of two elements r_1, r_2 is their sum $r_1 + r_2$. The identity is 0 and the inverse of an element is its negative. R is an infinite abelian group. Among the subgroups of R are the integers, the even integers, and the group consisting of the element zero alone.

Example 2. The nonzero real numbers in R with multiplication of real numbers as the group product. The identity is 1 and the inverse of $r \in R$ is $1/r$. Group multiplication is again commutative. One of the subgroups is the group of positive numbers.

Example 3. The group containing two elements $\{0, 1\}$ with group multiplication given by $0 \cdot 0 = 0$, $0 \cdot 1 = 1 \cdot 0 = 1$, $1 \cdot 1 = 0$. The identity element is 0. This is an abelian group of order two. It has only two subgroups, $\{0\}$ and $\{0, 1\}$.

Example 4. The **complex general linear group** $GL(n, \mathfrak{C})$. Here n is a positive integer. The group elements A are nonsingular $n \times n$ matrices with complex coefficients:

$$GL(n, \mathfrak{C}) = \{A = (A_{ij}), \quad 1 \le i, j \le n : A_{ij} \in \mathfrak{C} \quad \text{and} \quad \det A \neq 0\}. \tag{1.1}$$

Group multiplication is ordinary matrix multiplication. The identity element is the identity matrix $E = (\delta_{ij})$, where δ_{ij} is the **Kronecker delta,** (see the Symbol Index). The inverse of an element A is its matrix inverse, which exists since A is nonsingular. Clearly $GL(n, \mathfrak{C})$ is infinite and nonabelian. Among its subgroups are the **real general linear group** $GL(n, R)$ which consists of the **real** $n \times n$ nonsingular matrices, the **complex special linear group**

(1.2) $$SL(n, \mathfrak{C}) = \{A \in GL(n, \mathfrak{C}) : \det A = 1\},$$

and the **real special linear group**

(1.3) $$SL(n, R) = \{A \in GL(n, R) : \det A = 1\}.$$

Example 5. The **symmetric group** S_n. Let n be a positive integer. A **permutation of n objects** (say the set $X = \{1, 2, \ldots, n\}$) is a 1–1 mapping of X onto itself. Such a permutation s is written

(1.4) $$s = \begin{pmatrix} 1 & 2 & \cdots & n \\ p_1 & p_2 & \cdots & p_n \end{pmatrix}$$

and we say: 1 is mapped into p_1, 2 into $p_2, \ldots, n$ into p_n. The numbers $p_1, \ldots, p_n$ are a reordering of $1, 2, \ldots, n$ and no two of the p_j are the same. The order in which the columns of (1.4) are written is unimportant. The **inverse** permutation s^{-1} is given by

$$s^{-1} = \begin{pmatrix} p_1 & p_2 & \cdots & p_n \\ 1 & 2 & \cdots & n \end{pmatrix}.$$

The **product** of two permutations s and t,

$$t = \begin{pmatrix} q_1 & q_2 & \cdots & q_n \\ 1 & 2 & \cdots & n \end{pmatrix},$$

is given by the permutation

$$st = \begin{pmatrix} q_1 & q_2 & \cdots & q_n \\ p_1 & p_2 & \cdots & p_n \end{pmatrix},$$

where the product is read from right to left. That is, the integer q_i is mapped to i by t and i is mapped to p_i by s, so q_i is mapped to p_i by st. The **identity** permutation is

$$e = \begin{pmatrix} 1 & 2 & \cdots & n \\ 1 & 2 & \cdots & n \end{pmatrix}.$$

With these definitions it is easy to show that the permutations of n objects form a group S_n called the symmetric group. S_n has order $n!$.

Instead of (1.4) we will often use the convenient **cycle notation,** which is best explained by an example. Consider the permutation

$$s = \begin{pmatrix} 1 & 2 & 3 & 4 & 5 & 6 & 7 & 8 \\ 5 & 1 & 6 & 7 & 4 & 3 & 2 & 8 \end{pmatrix}.$$

Starting with the symbol 1, we see that s maps 1 into 5, 5 into 4, 4 into 7, 7 into 2, and 2 into 1, closing a **cycle**. We write (15472). We now chose a symbol in the top line which is not in the first cycle, say 3. The permutation generates a second cycle (36). The only remaining symbol in the top row is 8, which is mapped into itself and generates the cycle (8). Finally we write

$$s = (15472)(36)(8) = (15472)(36),$$

where in the second expression we have omitted the unpermuted symbol. (This last simplification can only be used if we keep in mind the number of elements permuted.) In writing an individual cycle it makes no difference where we start. Thus, $(36) = (63)$ and $(15472) = (21547) = (72154) = (47215) = (54721)$. Furthermore, it makes no difference in which order we write the cycles in a given permutation as long as the cycles contain no common elements, e.g., $(15472)(36) = (36)(15472)$. We present a final example showing the computation of a product of permutations in S_8 with the cycle notation: $(872)(34)(432) = (2387)$.

1.2 Subgroups and Cosets

Let H be a subgroup of the group G and $g \in G$. The set

$$gH = \{gh\colon h \in H\}$$

is called a **left coset** of H. There is a similar definition for right cosets. Every element g in G is contained in some left coset of H. In particular, $g = ge \in gH$. Furthermore, two left cosets are either identical or have no element in common. To see this, assume the cosets gH and kH have at least one element a in common. Thus, $a = gh_1 = kh_2$ with $h_1, h_2 \in H$, which implies $g = kh_2h_1^{-1} \in kH$ and $gH \subseteq kH$. Similarly, $k = gh_1h_2^{-1} \in gH$ and $kH \subseteq gH$. We conclude that $gH = kH$, i.e., the sets gH and kH have the same elements.

Suppose G is a finite group of order $n(G)$. Then the subgroup H is also finite and it is easy to show that each left coset gH contains exactly $n(H)$ distinct elements. As we have seen, it is possible to partition the elements of G into a finite number of disjoint left cosets $g_1H, g_2H, \ldots, g_mH$. That is, every element of G lies in exactly one of the cosets g_iH. Since there are m cosets and each coset contains $n(H)$ elements, it follows that $n(G) = m \cdot n(H)$. The integer $m = n(G)/n(H)$ is called the **index** of H in G. We have proved the following theorem due to Lagrange.

Theorem 1.2. The order of a subgroup of a finite group divides the order of the group.

Lagrange's theorem severely restricts the possible orders of subgroups. Thus, a group G of order 15 can have at most subgroups of order 1, 3, 5, or 15. The subgroup of order 15 is G itself, the group of order 1 is $\{e\}$, while the other possibilities lead to proper subgroups of G. A group of order p, where p is prime, has no proper subgroups.

By using left (or right) cosets we have partitioned the elements of G into disjoint sets. Another way to partition G is by means of conjugacy classes. A group element h is said to be **conjugate** to the group element k, $h \sim k$, if there exists a $g \in G$ such that $k = ghg^{-1}$. It is easy to show that conjugacy is an equivalence relation, i.e., (1) $h \sim h$ (reflexive), (2) $h \sim k$ implies $k \sim h$ (symmetric), and (3) $h \sim k$, $k \sim j$ implies $h \sim j$ (transitive). Thus, the elements of G can be divided into **conjugacy classes** of mutually conjugate elements. The class containing e consists of just one element since $geg^{-1} = e$ for all $g \in G$. Different conjugacy classes do not necessarily contain the same number of elements. We will study specific examples of conjugacy classes later where it will become apparent that such classes have simple geometrical interpretations.

If G is finite the number of elements in each conjugacy class is a factor of $n(G)$. To see this, choose some $g \in G$ and consider the set

$$H^g = \{h \in G: hgh^{-1} = g\}.$$

H^g is clearly a subgroup of G. The number of elements conjugate to g is equal to the number of distinct elements kgk^{-1} which can be formed by letting k run over G. We show that this is just the number of left cosets of H^g, a factor of $n(G)$. Indeed, if $k_1 g k_1^{-1} = k_2 g k_2^{-1}$, then $(k_1^{-1}k_2)g(k_1^{-1}k_2)^{-1} = g$, so $k_1^{-1}k_2 \in H^g$ or $k_2 \in k_1 H^g$. Conversely, if $k_2 \in k_1 H^g$ then $k_1 g k_1^{-1} = k_2 g k_2^{-1}$. Q.E.D.

The subgroup H of G is said to be **conjugate** to the subgroup K if there is a $g \in G$ such that $K = gHg^{-1}$ as sets, i.e., $Kg = gH$. Note that gHg^{-1} is a subgroup of G for any $g \in G$. Just as above, we can use this notion to partition the subgroups of G into conjugacy classes. A subgroup N is **normal** (**invariant, self-conjugate**) if $gNg^{-1} = N$ for all $g \in G$. Equivalently, N is normal if and only if $gN = Ng$ for all $g \in G$.

If N is a normal subgroup we can construct a group from the cosets of N, called the **factor group** G/N. The elements of G/N are the cosets gN, $g \in G$. Of course two cosets gN, $g'N$ containing the same elements of G define the same element of G/N: $gN = g'N$. Since N is normal it follows that $(g_1N)(g_2N) = (g_1N)(Ng_2) = g_1Ng_2 = g_1g_2N$ as sets. (Note that $NN = N$ as sets.) Therefore, we define group multiplication in G/N by

$$(g_1N)(g_2N) = g_1g_2N.$$

If G is finite, the order of G/N is clearly the index of N in G.

Corresponding to any element g of G we define the group element g^n, n an integer, by

$$g^n = \begin{cases} e & \text{if} \quad n = 0 \\ gg \cdots g \quad (n \text{ times}) & \text{if} \quad n > 0 \\ g^{-1}g^{-1} \cdots g^{-1} \quad (-n \text{ times}) & \text{if} \quad n < 0. \end{cases}$$

The reader can easily verify that $g^{n+m} = g^n g^m$ and $g^n g^{-n} = e$.

Suppose $S = \{g, h, \ldots\}$ is an arbitrary subset of G. Consider the set G_S consisting of all finite products of the form $g_1^{n_1} g_2^{n_2} \cdots g_j^{n_j}$, where $g_1, \ldots, g_j \in S$, $n_1, \ldots, n_j$ run over the integers, and j runs over the positive integers. Under the group product inherited from G, G_S is a subgroup called the subgroup **generated** by the set S. Here G_S can be characterized as follows: If H is a subgroup of G and $S \subseteq H$ then $G_S \subseteq H$. That is, G_S is the smallest subgroup of G containing S. If a group H is generated by $S = \{g\}$, i.e., if every $h \in H$ can be written in the form $h = g^n$, then H is **cyclic.**

The **order of an element** $g \in G$ is the order of the cyclic subgroup generated by $\{g\}$, i.e., the smallest positive integer m such that $g^m = e$. By Theorem 1.2, m divides the order of G.

Theorem 1.3. If G is a finite group of order $2n$ and N is a subgroup of order n then N is normal and the factor group G/N is cyclic of order two.

Proof. Since $2n(N) = n(G)$ there are only two left cosets in G: $eN = N$ and gN, where $g \notin N$. Similarly, there are only two right cosets N and Ng. Since every element of G is contained in exactly one left coset and exactly one right coset, we must have $gN = Ng$ for all $g \in G$, $g \notin N$. This last relation is also true if $g \in N$. Therefore, N is normal. The relations $NN = N$, $N(gN) = (gN)N = gN$, and $(gN)(gN) = N$, $g \notin N$, imply G/N is cyclic of order two The last relation follows from the fact that $g^2 \in N$. For, if $g^2 \in gN$ then $g \in N$, a contradiction. Q.E.D.

1.3 Homomorphisms, Isomorphisms, and Automorphisms

A **homomorphism** μ is a mapping from a group G into a group G' which transforms products into products. Thus, to every $g \in G$ there is associated $\mu(g) \in G'$ such that $\mu(g_1g_2) = \mu(g_1)\mu(g_2)$ for all $g_1, g_2 \in G$. Let e, e' be the identity elements of G, G', respectively. Then $\mu(e) = \mu(ee) = \mu(e)\mu(e)$, which implies $\mu(e) = e'$ by multiplication on the right with $\mu(e)^{-1} \in G'$.

Thus, μ maps the identity element of G into the identity element of G'. A similar argument shows $\mu(g^{-1}) = \mu(g)^{-1}$, i.e., μ maps inverses into inverses.

Homomorphisms are important because they are exactly the maps from one group to another that preserve group structure. They are the group analogy of linear transformations on vector spaces. Here we discuss homomorphisms from an abstract viewpoint, but in the following sections we will return to this topic and stress its geometrical aspects.

A homomorphism μ from G to G' is often designated by $\mu: G \longrightarrow G'$. The **domain** of μ is G, the **range** of μ is $\mu(G) = \{\mu(g) \in G' : g \in G\}$. Clearly, $\mu(G)$ is a subgroup of G'. If $\mu(G) = G'$ then μ is said to be **onto**. In case $\mu(g_1) \neq \mu(g_2)$ whenever $g_1 \neq g_2$ we say μ is 1–1. A homomorphism which is 1–1 and onto is an **isomorphism**. If μ is an isomorphism then it can be inverted in an obvious manner to define an isomorphism μ^{-1} of G' onto G. From the point of view of abstract group theory, isomorphic groups can be identified. In particular, isomorphic groups have identical multiplication tables. However, for the purposes of physical and geometrical applications it is frequently useful to distinguish between groups which are abstractly isomorphic. We shall return to this point in Section 1.4.

The above concepts are obvious analogies for groups of concepts related to a linear mapping of one vector space into another. We continue this analogy by defining the **kernel** K of μ as the set

$$K = \{g \in G: \mu(g) = e'\}.$$

The kernel of μ is the analogy of the null space of a linear transformation.

Theorem 1.4. K is a normal subgroup of G.

Proof. If $k_1, k_2 \in K$ then $\mu(k_1k_2) = \mu(k_1)\mu(k_2) = e'e' = e'$, so $k_1k_2 \in K$. Furthermore, if $k \in K$ then $\mu(k^{-1}) = \mu(k)^{-1} = (e')^{-1} = e'$, so $k^{-1} \in K$. By Theorem 1.1, K is a subgroup of G. To prove that K is normal it is enough to show $gkg^{-1} \in K$ for all $k \in K$, $g \in G$. This follows from $\mu(gkg^{-1}) = \mu(g)\mu(k)\mu(g)^{-1} = \mu(g)e'\mu(g)^{-1} = e'$. Q.E.D.

All elements in a left coset gK are mapped into the same element $\mu(g)$ in G' since $\mu(gk) = \mu(g)\mu(k) = \mu(g)$ for all $k \in K$. Furthermore, two elements with the same image under μ lie in the same left coset. Indeed, if $\mu(g_1) = \mu(g_2)$ then $\mu(g_1^{-1}g_2) = e'$, which implies $g_1^{-1}g_2 \in K$ or $g_2 \in g_1K$. This argument leads to several important results. First of all, μ is 1–1 if and only if the kernel consists of the identity element alone. Second, the fact that μ is constant on left cosets of K means that we can define a transformation μ': $G/K \longrightarrow \mu(G)$ mapping the factor space G/K (which makes sense since K is normal) onto the subgroup $\mu(G)$ of G'. This map is defined by $\mu'(gK) =$

$\mu(g)$, $g \in G$. The above argument shows μ' is 1–1 and onto. It is a homomorphism since $\mu'[(g_1K)(g_2K)] = \mu'[g_1g_2K] = \mu(g_1g_2) = \mu(g_1)\mu(g_2) = \mu'(g_1K)\mu'(g_2K)$.

Theorem 1.5. Let K be the kernel of the homomorphism $\mu: G \longrightarrow G'$. Then $\mu(G)$ is isomorphic to the factor group G/K.

An isomorphism $\nu: G \longrightarrow G$ of a group G onto itself is called an **automorphism.** For fixed $h \in G$ the map $\nu_h(g) = hgh^{-1}$ is an automorphism, since $\nu_h(g_1g_2) = hg_1g_2h^{-1} = (hg_1h^{-1})(hg_2h^{-1}) = \nu_h(g_1)\nu_h(g_2)$ and ν_h is clearly 1–1 and onto. The mappings ν_h, $h \in G$, are called **inner automorphisms.** It is not necessarily true that all automorphisms of a group are inner. The set of all automorphisms of G itself forms a group $A(G)$, the **automorphism group**. The product $\nu_1\nu_2$ of two automorphisms is defined by $\nu_1\nu_2(g) = \nu_1(\nu_2(g))$, $g \in G$, and the identity automorphism is the identity map of G onto itself. The set $I(G)$ of inner automorphisms of G forms a subgroup of $A(G)$.

1.4 Transformation Groups

Up to now our presentation of group theory has been entirely abstract and there has been little apparent connection with the study of symmetry. The missing link between abstract group theory and the notion of symmetry is the transformation group.

Definition. A **permutation** of a nonempty set X is a 1–1 mapping of X onto itself.

Thus, if the elements of X are denoted $x, y, z, \ldots$ a permutation $\boldsymbol{\sigma}$ is a map from X to X such that (1) $\boldsymbol{\sigma}(x) = \boldsymbol{\sigma}(y)$ if and only if $x = y$ and (2) for every $z \in X$ there exists an $x \in X$ such that $\boldsymbol{\sigma}(x) = z$. One such permutation is the **identity permutation** $\mathbf{1}(x) = x$ for all $x \in X$. The set S_X of all permutations of X forms a group, the **full symmetric group on** X. The product $\boldsymbol{\sigma\tau}$ of two permutations $\boldsymbol{\sigma}, \boldsymbol{\tau} \in S_X$ is given by $\boldsymbol{\sigma\tau}(x) = \boldsymbol{\sigma}(\boldsymbol{\tau}(x))$ for all $x \in X$. clearly $\boldsymbol{\sigma\tau}$ is again a permutation of X. The identity element of S_X is $\mathbf{1}$ and the inverse $\boldsymbol{\sigma}^{-1}$ of $\boldsymbol{\sigma}$ is defined by the requirement $\boldsymbol{\sigma}^{-1}(x) = y$ if and only if $\boldsymbol{\sigma}(y) = x$. Elements of S_X are said to **act** or **operate** on elements of X.

The set X may have an infinite number of elements, e.g., it may consist of all the points in the plane. If X is infinite then S_X is an infinite group. If X has a finite number of elements, say n, then we can identify S_X with the symmetric group S_n defined in Section 1.1. In other words, the groups S_X and S_n are isomorphic in case X has n elements. Recall that S_n has order $n!$.

Definition. A **transformation (permutation) group** on X is a subgroup of S_X.

If G is a transformation group on X then the elements $\mathbf{g}$ of G define permutations $\mathbf{g}(x)$ of X. (Henceforth we will drop the parentheses and write $\mathbf{g}x$ for these mappings. This should not result in any confusion.) These permutations can be used to decompose X into mutually disjoint subsets. Let $x, y \in X$.

Definition. We say x is ***G*-equivalent** to y $(x \sim y)$ if $\mathbf{g}x = y$ for some $g \in G$.

Let us show that $\sim$ is an equivalence relation. Now $\mathbf{1}x = x$ implies $x \sim x$. Furthermore, if $\mathbf{g}x = y$ then $x = \mathbf{g}^{-1}\mathbf{g}x = \mathbf{g}^{-1}y$, so $x \sim y$ implies $y \sim x$. Suppose $x \sim y$ and $y \sim z$. Then there exist elements $\mathbf{g}_1, \mathbf{g}_2$ in G such that $\mathbf{g}_1 x = y$, $\mathbf{g}_2 y = z$. But $(\mathbf{g}_2\mathbf{g}_1)x = \mathbf{g}_2 y = z$, so $x \sim z$ and $\sim$ is an equivalence relation.

Definition. The equivalence classes of X under the equivalence relation $\sim$ are called ***G*-orbits** or just **orbits**.

Thus x and y belong to the same orbit if and only if $y = \mathbf{g}x$ for some $\mathbf{g} \in G$. The orbit containing x is the set $\{\mathbf{g}x: \mathbf{g} \in G\}$. If there is only one G-orbit in X we say G is **transitive**. In this case for every pair of points x, y in X there is a $\mathbf{g} \in G$ such that $y = \mathbf{g}x$.

Example. Introduce a rectangular coordinate system (x_1, x_2) in the Euclidean plane X and let G be the set of all rotations about the origin. The elements $\mathbf{g}_\varphi$ of G are labeled by the continuous parameter φ, which is the angle of rotation in radians measured from the positive x_1-axis. If $x \in X$ has coordinates (x_1, x_2) then $y = \mathbf{g}_\varphi x$ has coordinates

$$\begin{pmatrix} y_1 \\ y_2 \end{pmatrix} = \begin{pmatrix} \cos\varphi & -\sin\varphi \\ \sin\varphi & \cos\varphi \end{pmatrix} \begin{pmatrix} x_1 \\ x_2 \end{pmatrix}.$$

Note that $\mathbf{g}_\varphi = \mathbf{g}_{\varphi+2\pi}$, since both group elements lead to the same transformation of the plane. The elements $\mathbf{g}_\varphi$ clearly form a group since $\mathbf{g}_\varphi \mathbf{g}_\theta = \mathbf{g}_{\varphi+\theta}$. The orbits are concentric circles about the origin. The rotation group in two-space is clearly isomorphic to the matrix group $SO(2, R)$,

$$\begin{pmatrix} \cos\varphi & -\sin\varphi \\ \sin\varphi & \cos\varphi \end{pmatrix}, \qquad 0 \le \varphi < 2\pi$$

called the **real special orthogonal group in two-space**.

If Y is a subset of X and $\mathbf{g} \in G$, we denote by $\mathbf{g}(Y)$ the set $\{\mathbf{g}y: y \in Y\}$.

Definition. A subset Y of X is ***G*-invariant** or just **invariant** if $\mathbf{g}(Y) \subseteq Y$ for all $\mathbf{g} \in G$.

In particular, the subset $\{x\}$ of X is invariant if and only if $\mathbf{g}x = x$ for all $\mathbf{g} \in G$, i.e., if and only if the G-orbit containing x consists of x alone. In the above example the only invariant point is the origin. A general invariant set is formed by taking arbitrary unions of concentric circles about the origin.

We are now in a position to state a major theme of this book. Given a transformation group G we can look for all G-invariant subsets Y of X. The group G is an invariance or symmetry group of the objects Y. As in the example given above, such subsets often have geometrical significance. They can always be expressed as unions of orbits. Similarly, given an arbitrary subset Y of X we can find a subgroup

$$K = \{\mathbf{g} \in G\colon \mathbf{g}(Y) \subseteq Y\}.$$

It is easy to show that K is itself a transformation group and Y is a K-invariant subset of X. Frequently we shall refer to K as the **G-symmetry** or **symmetry** group of the object Y. This simple relationship between objects and their symmetry groups provides us with a means of applying group-theoretic concepts to geometrical problems.

Example. The symmetries of the square.

Let X be the Euclidean plane and G the group $O(2)$ of all rotations and reflections in the plane which leave a fixed point p invariant. [We will explicitly define the orthogonal group $O(2)$ later. Its exact definition is not important for our example.]

Consider the square $ABCD$ with center p as pictured in Fig. 1.1. We look for all rotations and reflections in $O(2)$ which map the square onto itself. There are eight such symmetries of the square: the identity permutation $\mathbf{1}$, the 90° clockwise rotation $\mathbf{r}$, clockwise rotations $\mathbf{r}^2$ and $\mathbf{r}^3$ through 180° and

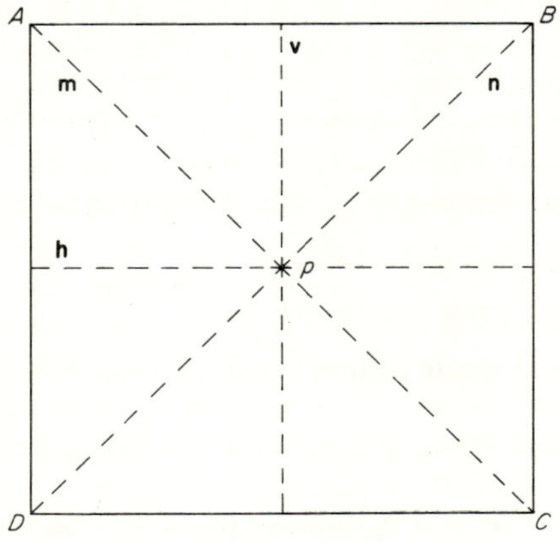

FIGURE 1.1

270°, respectively, and four reflections $\mathbf{h}, \mathbf{v}, \mathbf{m}$, and $\mathbf{n}$ about horizontal, vertical, major diagonal, and minor diagonal axes, respectively. A convenient way to list these symmetries is by means of the cycle notation for the permutation of the vertices induced by each symmetry. Thus, we can write $\mathbf{r} = (ABCD)$, $\mathbf{r}^2 = (AC)(BD)$, $\mathbf{r}^3 = (ADCB)$, $\mathbf{h} = (AD)(BC)$, $\mathbf{v} = (AB)(CD)$, $\mathbf{m} = (BD)$, and $\mathbf{n} = (AC)$. These eight symmetries form the group D_4, the dihedral group of order eight. Indeed $\mathbf{1} \in D_4$ and the inverse of each $\mathbf{g} \in D_4$ is in D_4, e.g., $\mathbf{h}^{-1} = \mathbf{h}$, $\mathbf{r}^{-1} = \mathbf{r}^3$. Furthermore, the product of two symmetries is again a symmetry. Thus $\mathbf{nh} = (AC)(AD)(BC) = (ADCB) = \mathbf{r}^3$, or the result of reflecting the square about the horizontal axis followed by a reflection about the minor axis is equivalent to a clockwise rotation of 270°. Recall that the group operations are performed from right to left. (The reader would do well to work out the complete multiplication table for D_4 to be sure he understands this important example.) Note that D_4 is isomorphic to a subgroup of S_4 and our realization of D_4 by permutations constitutes a 1–1 homomorphism of D_4 into S_4.

The conjugacy classes of D_4 can contain possibly one, two, or four elements since those are the factors of eight. (No conjugacy class can contain eight elements since $\{\mathbf{1}\}$ is always a class by itself.) A simple computation shows that there are five classes, $\{\mathbf{1}\}$, $\{\mathbf{r}, \mathbf{r}^3\}$, $\{\mathbf{r}^2\}$, $\{\mathbf{h}, \mathbf{v}\}$, and $\{\mathbf{m}, \mathbf{n}\}$. Note that $\mathbf{1}$ and $\mathbf{r}^2$ commute with all $\mathbf{g} \in D_4$ and thus lie in classes containing only one element. The conjugacy classes have a simple geometrical interpretation. They correspond to rotations through 0°, 90°, and 180°, respectively, and reflections about an axis through either the midpoints of opposite sides or two opposite vertices. Note that a clockwise rotation of 270° leads to the same result as a counterclockwise rotation through 90°. A conjugacy relation such as $\mathbf{rhr}^{-1} = \mathbf{v}$ can be interpreted as follows: To perform a reflection about the vertical axis, rotate the square counterclockwise 90°, reflect in the horizontal axis, and then rotate back 90° in the clockwise direction.

We now return to a general discussion of the transformation group G on X. For any $x \in X$ the group

$$G^x = \{\mathbf{g} \in G: \mathbf{g}x = x\}$$

is called the **isotropy subgroup** of G at x. It contains those elements of G that leave x invariant.

Theorem 1.6. Each left coset of G^x consists of all elements of G that map x to a specific point y. Thus there is a 1–1 relationship between the points in the G-orbit containing x and the left cosets of G^x. If G is finite, the G-orbit containing x consists of $n(G)/n(G^x)$ points, a factor of $n(G)$.

Proof. The last statement is immediate once we establish the 1–1 relationship between points in the G-orbit through x and left cosets of G^x. Let $y \in X$

such that $x \sim y$, i.e., there is a $\mathbf{g} \in G$ such that $y = \mathbf{g}x$. Then $\mathbf{gh}x = \mathbf{g}x = y$ for all $\mathbf{h} \in G^x$, so all elements in the coset $\mathbf{g}G^x$ map x onto y. Conversely, if $y = \mathbf{k}x$ for some $k \in G$ then $\mathbf{g}x = \mathbf{k}x$ or $x = \mathbf{g}^{-1}\mathbf{k}x$, so $\mathbf{g}^{-1}\mathbf{k} \in G^x$, which implies $\mathbf{k} \in \mathbf{g}G^x$. Q.E.D.

This theorem provides an important connection between the algebraic notion of coset and the geometrical notion of orbit.

It is sometimes helpful to view a transformation group G as an abstract group $\tilde{G}$ together with a 1–1 homomorphism μ of $\tilde{G}$ into the group S_X of permutations of X. Then, $\mu(\tilde{G}) = G$ is isomorphic to $\tilde{G}$. Here we are distinguishing between the abstract multiplicative structure $\tilde{G}$ and the transformation group G on X. Clearly the same abstract group can have many different realizations as a transformation group. We will usually distinguish between two abstractly isomorphic transformation groups if they correspond to physically distinct types of transformations. For example, the cyclic group of order two consists of the elements $\{e, g\}$ with $g^2 = e$. This group can be realized as a transformation group in the plane where g corresponds to a 180° rotation about a point p, e.g., $g = \mathbf{r}^2$ in our last example. Another realization is obtained by letting g correspond to a reflection about a line in the plane, e.g., $g = \mathbf{v}$ in the last example. These groups are isomorphic, but for the purposes of applications to physics and geometry we usually distinguish between them. Nevertheless, we will often use the same symbol G to describe both an abstract group and any transformation group obtained from it.

Any abstract group G can be realized as a transformation group acting on itself. Indeed the mapping $L: G \longrightarrow G$, called the **left regular representation** of G and defined by $\mathbf{L}(a)g = ag$, $a, g \in G$, is easily shown to be a 1–1 homomorphism of G into S_G. That is, $\mathbf{L}(ab) = \mathbf{L}(a)\mathbf{L}(b)$ for $a, b \in G$, each $\mathbf{L}(a)$ is a permutation of G, and $\mathbf{L}(a) = \mathbf{1}$ if and only if $a = e$. This proves Cayley's theorem, which is as follows.

Theorem 1.7. Any group G is isomorphic to a subgroup of the full permutation group S_G. In particular, any finite group of order n is isomorphic to a subgroup of S_n.

Note that the left regular representation is transitive. However, if we restrict L to a proper subgroup H of G, thus defining H as a transformation group on G, the space $G = X$ splits up into orbits which are exactly the right cosets Hg of H.

As a final remark we clarify the meaning of conjugacy in a transformation group. Let G be a transformation group acting on X, $\mathbf{g}, \mathbf{h} \in G$ and $x, y \in X$.

Theorem 1.8. (1) The permutation $\mathbf{g}$ sends x into y if and only if $\mathbf{hgh}^{-1}$ sends $\mathbf{h}x$ into $\mathbf{h}y$.

(2) The point x is invariant under $\mathbf{g}$ if and only if $\mathbf{h}x$ is invariant under $\mathbf{hgh}^{-1}$.

(3) If G^x is the isotropy group of x and $\mathbf{h}$ sends x into y, then $\mathbf{h}G^x\mathbf{h}^{-1} = G^y$, i.e., points in the same G-orbit have conjuate, hence isomorphic, isotropy groups.

Proof. (1) $\mathbf{g}x = y$ if and only if $(\mathbf{hgh}^{-1})\mathbf{h}x = y$. (2) $\mathbf{g}x = x$ if and only if $(\mathbf{hgh}^{-1})\mathbf{h}x = \mathbf{h}x$. (3) Follows from (2). Q.E.D.

1.5 New Groups from Old Ones

Given the groups G and G', we discuss two different ways to construct new groups which contain subgroups isomorphic to G and G'.

Definition. The **direct product** $G \times G'$ is the group consisting of all ordered pairs (g, g') with $g \in G$ and $g' \in G'$. The product of two group elements is given by $(g_1, g_1')(g_2, g_2') = (g_1g_2, g_1'g_2')$.

It is easy to show that $G \times G'$ is a group with identity element (e, e') where e, e' are the identity elements of G, G', respectively. Indeed $(g, g')^{-1} = (g^{-1}, g'^{-1})$ and the associative law is trivial to verify. The subgroup $G \times \{e'\} = \{(g, e') \colon g \in G\}$ of $G \times G'$ is isomorphic to G with the ismorphism given by $(g, e') \longleftrightarrow g$. Similarly the subgroup $\{e\} \times G'$ is isomorphic to G'. Since $(g, e')(e, g') = (e, g')(g, e') = (g, g')$ it follows that (1) the elements of $G \times \{e'\}$ commute with the elements of $\{e\} \times G'$ and (2) every element of $G \times G'$ can be written uniquely as a product of an element in $G \times \{e'\}$ and an element in $\{e\} \times G'$. Frequently one identifies the isomorphic groups $G \times \{e'\}$ and G as well as $\{e\} \times G'$ and G', and writes $(g, e') = g$, $(e, g') = g'$, $(g, g') = gg' = g'g$ for $g \in G$, $g' \in G'$. This point of view leads to the following definition.

Definition. A group G is the **direct product** of its subgroups H and K ($G = H \times K$) if (1) $hk = kh$ for all $h \in H$, $k \in K$, and (2) every $g \in G$ can be expressed uniquely in the form $g = hk$, $h \in H$, $k \in K$. The subgroups H and K are said to be **direct factors** of G.

It follows from (2) that H and K have only the identity element in common. For, if $g \in H \cap K$ then $g = ge = eg$ and by uniqueness we must have $g = e$. Furthermore, the reader can show that H and K are normal subgroups of G.

The above definitions can easily be extended to define direct products $G_1 \times G_2 \times \cdots \times G_n$ of n groups. Furthermore, if the G_i have finite order it is clear that the order of the direct product is the product of the orders of the direct factors.

As an example we use the first definition to construct the direct product $G \times G'$ of the cyclic group of order two, $G = \{e, a\}$, $a^2 = e$, and the cyclic group of order three, $G' = \{e', b, b^2\}$, $b^3 = e'$. The group $G \times G'$ has order six and contains the element $j = (a, b)$ of order six. Thus, $G \times G'$ is the cyclic group of order six generated by j.

A more general, but more complicated, method of building a new group from two old ones is the semidirect product.

Definition. Let H and K be groups and let the map $k \longrightarrow \nu_k$ be a homomorphism of K into the automorphism group $A(H)$ of H. Then the set of all ordered pairs $\langle h, k\rangle$, $h \in H$, $k \in K$, forms a group, the **semidirect product** of H and K, with group multiplication

$$\langle h, k\rangle\langle h', k'\rangle = \langle h\nu_k(h'), kk'\rangle. \tag{5.1}$$

It is necessary to verify that this definition makes sense. First of all, the map ν_k is an automorphism of H for each $k \in K$. Furthermore, $\nu_e = 1$, the identity automorphism, and $\nu_{kk'}(h) = \nu_k[\nu_{k'}(h)]$ for all $k, k' \in K$, $h \in H$. To show that the binary relation (5.1) defines a group, we check the standard group definition in Section 1.1. The associative law follows from

$$\begin{aligned}(\langle h_1, k_1\rangle\langle h_2, k_2\rangle)\langle h_3, k_3\rangle &= \langle h_1\nu_{k_1}(h_2), k_1k_2\rangle\langle h_3, k_3\rangle \\ &= \langle h_1\nu_{k_1}(h_2)\nu_{k_1k_2}(h_3), k_1k_2k_3\rangle\end{aligned} \tag{5.2}$$

and

$$\begin{aligned}\langle h_1, k_1\rangle(\langle h_2, k_2\rangle\langle h_3, k_3\rangle) &= \langle h_1, k_1\rangle\langle h_2\nu_{k_2}(h_3), k_2k_3\rangle \\ &= \langle h_1\nu_{k_1}(h_2\nu_{k_2}(h_3)), k_1k_2k_3\rangle \\ &= \langle h_1\nu_{k_1}(h_2)\nu_{k_1k_2}(h_3), k_1k_2k_3\rangle.\end{aligned} \tag{5.3}$$

The identity element is $\langle e, e\rangle$ since

$$\begin{aligned}\langle h, k\rangle\langle e, e\rangle &= \langle h\nu_k(e), k\rangle = \langle h, k\rangle \\ \langle e, e\rangle\langle h, k\rangle &= \langle \nu_e(h), k\rangle = \langle h, k\rangle.\end{aligned} \tag{5.4}$$

It is left as an exercise to verify that the element inverse to $\langle h, k\rangle$ is $\langle \nu_{k^{-1}}(h^{-1}), k^{-1}\rangle$.

If $\nu_k \equiv 1$ for all $k \in K$ then the semidirect product reduces to the direct product. Just as with the direct product, we can identify the groups $\langle H, e\rangle$ and H using the map $\langle h, e\rangle \longleftrightarrow h$ as well as the groups $\langle e, K\rangle$ and K. Thus we can write any element g of the semidirect product G uniquely in the form $g = \langle h, k\rangle = hk$ and group multiplication becomes

$$(h_1k_1)(h_2k_2) = (h_1\nu_{k_1}(h_2))(k_1k_2). \tag{5.5}$$

From this identification it follows that $H \cap K = \{e\}$, K is a subgroup of G, and H is a normal subgroup of G. Indeed, if $k \in K$ and $h \in H$ then

$$khk^{-1} = \langle e, k\rangle\langle h, k^{-1}\rangle = \nu_k(h), \tag{5.6}$$

so for $g = hk \in G$, $h' \in H$, we have

$$gh'g^{-1} = (hk)h'(k^{-1}h^{-1}) = h(kh'k^{-1})h^{-1} \in H, \tag{5.7}$$

and H is normal.

As an example we note that the dihedral group D_4 is isomorphic to a semidirect product of the cyclic group H of order four and the cyclic group K of order two. If h, k are the generators of H, K respectively, then the automorphism ν_k of H is defined by $\nu_k(h) = h^{-1} = h^3$, i.e., $khk^{-1} = h^{-1}$. We shall see later that the Euclidean and Poincaré groups can also be expressed as semidirect products of simpler groups.

Problems

1.1 Prove: A group G has no proper subgroups if and only if the order of G is finite and prime.

1.2 Let G be a finite group and S a nonempty subset of G such that $gh \in S$ for all g, $h \in S$. Prove that S is a subgroup. What if G is an infinite group?

1.3 Prove: If the group G has exactly one element h of order two then $gh = hg$ for every $g \in G$.

1.4 Show that there are exactly two groups of order four, one of which is cyclic. Find all groups of order six.

1.5 Construct a homomorphism of D_4 onto the cyclic group of order two.

1.6 Determine all subgroups of S_4 and sort them into classes of conjugate subgroups.

1.7 Show that the symmetry group of a regular hexagon consists of 12 elements and determine the conjugacy classes.

1.8 The **commutator subgroup** G_c is the subgroup of G generated by all elements of the form $ghg^{-1}h^{-1}$, $g, h \in G$. Prove that G_c is a normal subgroup of G and G/G_c is commutative.

1.9 Let g, h, k be elements of the group G. Prove that ghk, hkg, and kgh have the same order.

Chapter 2

The Crystallographic Groups

2.1 The Orthogonal Group in Three-Space

Let R_3 be three-dimensional real Euclidean space. We erect a cartesian coordinate system with origin C in this space and associate with each point P in R_3 a unique triple of real numbers (x_1, x_2, x_3), the projections of P on the three mutually perpendicular coordinate axes. It is useful to think of R_3 as a three-dimensional vector space with elements $\mathbf{x} = (x_1, x_2, x_3) = \sum_{i=1}^{3} x_i \mathbf{e}_i$, where $\mathbf{e}_1$, $\mathbf{e}_2$, and $\mathbf{e}_3$ are unit vectors along the coordinate axes. As is well known, the bilinear form

(1.1) $$\langle \mathbf{x}, \mathbf{y} \rangle = \sum_{i=1}^{3} x_i y_i, \qquad \mathbf{x}, \mathbf{y} \in R_3$$

defines an inner product on this space. The norm $\|\mathbf{x}\| = \langle \mathbf{x}, \mathbf{x} \rangle^{1/2}$ is the Euclidean length of the vector $\mathbf{x}$ and

(1.2) $$\cos \varphi = \langle \mathbf{x}, \mathbf{y} \rangle / \|\mathbf{x}\| \|\mathbf{y}\|$$

is the cosine of the angle φ between the vectors $\mathbf{x}$ and $\mathbf{y}$.

We look for all linear transformations $\mathbf{O}: R_3 \to R_3$ which preserve length, i.e., all linear transformations $\mathbf{O}$ such that $\langle \mathbf{Ox}, \mathbf{Ox} \rangle = \langle \mathbf{x}, \mathbf{x} \rangle$ for all $\mathbf{x} \in R_3$. Because of the identity

(1.3) $$4\langle \mathbf{x}, \mathbf{y} \rangle = \langle \mathbf{x} + \mathbf{y}, \mathbf{x} + \mathbf{y} \rangle - \langle \mathbf{x} - \mathbf{y}, \mathbf{x} - \mathbf{y} \rangle$$

it follows that

(1.4) $$\langle \mathbf{Ox}, \mathbf{Oy} \rangle = \langle \mathbf{x}, \mathbf{y} \rangle$$

and the transformation **O** also preserves angle. That is, the angle between the vectors **x** and **y** is equal to the angle between vectors **Ox** and **Oy**.

To compute the possible length-preserving transformations **O** we pass to matrices. Recall that the matrix T of a linear transformation $\mathbf{T}: R_3 \longrightarrow R_3$ with respect to the basis $\mathbf{e}_1, \mathbf{e}_2, \mathbf{e}_3$ is defined by $T = (T_{ij})$, $1 \leq i, j \leq 3$, where

$$\mathbf{T}\mathbf{e}_j = \sum_{i=1}^{3} T_{ij}\mathbf{e}_i. \tag{1.5}$$

The identity operator $\mathbf{Ex} = \mathbf{x}$ has the matrix $E_3 = (\delta_{ij})$, where δ_{ij} is the Kronecker delta. The product **TQ** of two transformations defined by $\mathbf{TQx} = \mathbf{T}(\mathbf{Qx})$ corresponds to the matrix product TQ where $(TQ)_{ij} = \sum_k T_{ik}Q_{kj}$. Furthermore, the inverse $\mathbf{T}^{-1}$ of an invertible operator **T** corresponds to the inverse matrix T^{-1} of the nonsingular matrix T.

Using the fact that $\mathbf{Ox} = \sum_{i,j} O_{ij}x_j\mathbf{e}_i$ and writing the equation $\langle \mathbf{Ox}, \mathbf{Oy} \rangle = \langle \mathbf{x}, \mathbf{y} \rangle$ in component form, we obtain the result

$$\sum_{i=1}^{3} O_{ij}O_{ik} = \delta_{jk} \tag{1.6}$$

or $O^tO = E_3$ in terms of matrix multiplication. Here, O^t is the transpose of the matrix O, $O^t_{ij} = O_{ji}$. Thus $O^t = O^{-1}$ is a necessary and sufficient condition that the operators **O** preserve inner product. Let $O(3) = \{3 \times 3$ matrices $O: O^tO = E_3\}$. Clearly the matrices in $O(3)$ are all nonsingular. Now, $(O_1O_2)^t(O_1O_2) = O_2^tO_1^tO_1O_2 = O_2^tO_2 = E_3$ if $O_1, O_2 \in O(3)$, so $O_1O_2 \in O(3)$. Furthermore, $E_3 \in O(3)$ and $(O_1^{-1})^tO_1^{-1} = O_1O_1^{-1} = E_3$, so $O_1^{-1} \in O(3)$. Thus, $O(3)$ is a group, the **real orthogonal group** in three-space. The operators **O** also form a group and the correspondence $\mathbf{O} \longleftrightarrow O$ defines an isomorphism between the two groups. Both groups are usually called $O(3)$. Any abstract group-theoretic property which holds for one realization of $O(3)$ automatically holds for the other. We shall sometimes use the operator form of the group and at other times use the matrix form.

Lemma 2.1. $\det O = \pm 1$ if $O \in O(3)$.

Proof. Since $O^tO = E_3$ it follows that $\det(O^tO) = 1$. But, $\det(O^tO) = (\det O^t)\cdot(\det O) = (\det O)^2$.

Both signs of the determinant occur. Indeed, E_3 and $I_3 = -E_3$ are elements of $O(3)$ with $\det E_3 = 1$ and $\det I_3 = -1$. The operator **I** with matrix I_3 is defined by $\mathbf{Ix} = -\mathbf{x}$, all $\mathbf{x} \in R_3$, and called the **inversion** operator. Note that $\mathbf{I}^2 = \mathbf{E}$. Since $\det(O_1O_2) = (\det O_1)\cdot(\det O_2)$ it follows that the set

$$SO(3) = \{O \in O(3): \det O = +1\} \tag{1.7}$$

forms a subgroup of $O(3)$, called the **special orthogonal group (proper orthogonal group) in three-space** or just the **rotation group.** The map $O \longrightarrow \det O$ defines a homomorphism of $O(3)$ onto the cyclic group of order two with elements $1 = e$ and -1. The kernel of this homomorphism is $SO(3)$, which implies that $SO(3)$ is a normal subgroup of $O(3)$. Furthermore, by Theorem 1.5 there are exactly two $SO(3)$-cosets in $O(3)$: $SO(3)$ and $I_3 \cdot SO(3)$. The elements of the first coset are all proper orthogonal (rotation) matrices and the elements of the second coset are all **improper,** i.e., they have negative determinants. Thus, every improper element O' can be written uniquely in the form $O' = I_3 O$, a rotation followed by inversion.

The groups $O(3)$ and $SO(3)$ have now been realized as transformation groups on the set R_3. We will show that the elements of $SO(3)$ are exactly the possible geometrical rotations about all axes in R_3 passing through the origin, while $O(3)$ consists of all possible geometrical rotations and rotation-inversions in R_3 that fix the origin.

Theorem 2.1. Let $O \in SO(3)$. Then there is a vector $\mathbf{f}_3 \in R_3$, $\|\mathbf{f}_3\| = 1$, such that $\mathbf{Of}_3 = \mathbf{f}_3$. If $\mathbf{O} \neq \mathbf{E}$ the axis designated by $\pm\mathbf{f}_3$ is called the **axis of rotation.**

Proof. The theorem asserts that the operator $\mathbf{O}$ has a unit eigenvector $\mathbf{f}_3$ with eigenvalue $\lambda = 1$. This is equivalent to the assertion that $\lambda = 1$ is a solution of the characteristic equation $\det(O - \lambda E_3) = 0$. But $\det(O - E_3) = \det(O^t - E_3^t) = \det(O^{-1} - E_3) = (\det O^{-1}) \cdot [\det(-E_3)] \cdot \det(O - E_3) = -\det(O - E_3)$. Therefore, $\det(O - E_3) = 0$ and $\lambda = 1$ is an eigenvalue of $\mathbf{O}$. Thus, a desired vector $\mathbf{f}_3$ exists, though it is not unique.

Now choose unit vectors $\mathbf{f}_1$ and $\mathbf{f}_2$ so that $\{\mathbf{f}_1, \mathbf{f}_2, \mathbf{f}_3\}$ is an orthonormal basis for R_3, i.e., $\langle \mathbf{f}_j, \mathbf{f}_k \rangle = \delta_{jk}$. We will compute the matrix $\tilde{O}$ of $\mathbf{O}$ with respect to this basis. The relations $\langle \mathbf{Of}_j, \mathbf{Of}_k \rangle = \langle \mathbf{f}_j, \mathbf{f}_k \rangle = \delta_{jk}$ lead to

$$\begin{aligned} \mathbf{Of}_1 &= \alpha_1 \mathbf{f}_1 + \beta_1 \mathbf{f}_2, & \alpha_1\alpha_2 + \beta_1\beta_2 &= 0 \\ \mathbf{Of}_2 &= \alpha_2 \mathbf{f}_1 + \beta_2 \mathbf{f}_2, & \alpha_1^2 + \beta_1^2 &= 1 \\ \mathbf{Of}_3 &= \mathbf{f}_3, & \alpha_2^2 + \beta_2^2 &= 1. \end{aligned} \tag{1.8}$$

These equations have the unique solution ($\det \tilde{O} = 1$)

$$\alpha_1 = \beta_2 = \cos\theta, \qquad \beta_1 = -\alpha_2 = \sin\theta, \qquad 0 \leq \theta < 2\pi, \tag{1.9}$$

so that the matrix of $\mathbf{O}$ in the $\mathbf{f}$-basis is

$$\tilde{O} = \begin{pmatrix} \cos\theta & -\sin\theta & 0 \\ \sin\theta & \cos\theta & 0 \\ 0 & 0 & 1 \end{pmatrix}. \tag{1.10}$$

It follows from (1.10) and elementary analytic geometry that $\mathbf{O}$ can be interpreted as a counterclockwise rotation through the angle θ about the axis of rotation $\mathbf{f}_3$. We adopt the notation $O = C_{\mathbf{k}}(\theta)$, where $\mathbf{k} = \mathbf{f}_3$ is the axis of rotation and θ is the rotation angle. Note that

$$C_{\mathbf{k}}(\theta + \varphi) = C_{\mathbf{k}}(\theta)C_{\mathbf{k}}(\varphi), \tag{1.11}$$

where we assume $C_{\mathbf{k}}(\alpha + 2\pi) = C_{\mathbf{k}}(\alpha)$. Furthermore, both $\mathbf{k}$ and $-\mathbf{k}$ serve to define the same axis of rotation so

$$C_{\mathbf{k}}(\theta) = C_{-\mathbf{k}}(2\pi - \theta). \tag{1.12}$$

Since O and $\tilde{O}$ are matrices of the same transformation $\mathbf{O}$ viewed in different basis systems, these matrices must be similar, i.e., $\tilde{O} = QOQ^{-1}$, where Q is the orthogonal matrix denoting the change of basis. Thus, O and $\tilde{O}$ have the same determinant and trace. In particular,

$$\operatorname{tr} O = \sum_{i=1}^{3} O_{ii} = \operatorname{tr} \tilde{O} = 1 + 2\cos\theta. \tag{1.13}$$

The improper rotations also have a simple geometrical interpretation. An improper rotation O' can be written uniquely in the form

$$O' = I_3 C_{\mathbf{k}}(\pi + \theta) = I_3 C_{\mathbf{k}}(\pi) C_{\mathbf{k}}(\theta) = \sigma_{\mathbf{k}} C_{\mathbf{k}}(\theta), \tag{1.14}$$

where $\sigma_{\mathbf{k}} = I_3 C_{\mathbf{k}}(\pi)$ is the reflection in the plane through the origin of R_3 perpendicular to $\mathbf{k}$. Thus, any improper rotation (rotation-inversion) is equal to a rotation about some axis $\mathbf{k}$ followed by a reflection in the plane perpendicular to $\mathbf{k}$. We write $S_{\mathbf{k}}(\theta) = \sigma_{\mathbf{k}} C_{\mathbf{k}}(\theta)$.

The conjugacy classes in $O(3)$ and $SO(3)$ have a simple physical significance. The relation

$$OC_{\mathbf{k}}(\theta)O^{-1} = C_{O\mathbf{k}}(\theta), \qquad O \in SO(3) \tag{1.15}$$

shows that all rotations through the angle θ about any axis lie in the same conjugacy class of $SO(3)$. Thus the conjugacy classes can be labled by the rotation angle θ, $0 \leq \theta \leq \pi$. To prove (1.15) we chose an orthonormal basis $\{\mathbf{f}_j\}$ for R_3 corresponding to $C_{\mathbf{k}}(\theta)$, just as in (1.8). In particular, $\mathbf{f}_3 = \mathbf{k}$. Then $\{\mathbf{Of}_1, \mathbf{Of}_2, \mathbf{Of}_3\}$ is also an orthonormal basis and the matrix of $\mathbf{O}C_{\mathbf{k}}(\theta)\mathbf{O}^{-1}$ in this new basis is just (1.10) again. Thus $\mathbf{O}C_{\mathbf{k}}(\theta)\mathbf{O}^{-1}$ is a rotation of angle θ about the axis $\mathbf{Ok}$.

It is left to the reader to verify the following:

$$OC_{\mathbf{k}}(\theta)O^{-1} = C_{\epsilon O\mathbf{k}}(\theta), \qquad OS_{\mathbf{k}}(\theta)O^{-1} = S_{\epsilon O\mathbf{k}}(\theta), \qquad O \in O(3), \tag{1.16}$$

where $\epsilon = \det O$, which show that the conjugacy classes of $O(3)$ fall into two types. One type consists of all rotations through a fixed angle θ, $0 \leq \theta \leq \pi$, and the other consists of all rotation-inversions through a fixed angle θ', $0 \leq \theta' \leq \pi$.

2.2 The Euclidean Group

We now seek all transformations $\mathbf{T}$ of R_3 onto R_3 that preserve the distance between any pair of points, i.e.,

$$\|\mathbf{Tx} - \mathbf{Ty}\| = \|\mathbf{x} - \mathbf{y}\|, \qquad \mathbf{x}, \mathbf{y} \in R_3. \tag{2.1}$$

Such transformations are called *isometries*. We do not assume that the transformations $\mathbf{T}$ are linear. It may be helpful to think of $\mathbf{T}$ as a permutation of the elements of R_3 which also preserves distance. Let $E(3)$ be the set of all isometries.

Theorem 2.2. $E(3)$ is a group, the Euclidean group in three-space.

Proof. The identity mapping $\mathbf{E}$ is clearly in $E(3)$. If $\mathbf{T} \in E(3)$ and $\mathbf{Tx} = \mathbf{Ty}$ then by (2.1), $\|\mathbf{x} - \mathbf{y}\| = 0$ or $\mathbf{x} = \mathbf{y}$. Therefore, $\mathbf{T}$ is invertible. Corresponding to any two vectors $\mathbf{w}, \mathbf{z}$ in R_3 there exist unique vectors $\mathbf{x}, \mathbf{y}$ such that $\mathbf{Tx} = \mathbf{w}$ and $\mathbf{Ty} = \mathbf{z}$. Thus $\|\mathbf{T}^{-1}\mathbf{w} - \mathbf{T}^{-1}\mathbf{z}\| = \|\mathbf{x} - \mathbf{y}\| = \|\mathbf{Tx} - \mathbf{Ty}\| = \|\mathbf{w} - \mathbf{z}\|$, where we have made use of (2.1) again, and $\mathbf{T}^{-1} \in E(3)$. It is an elementary argument to show that the product $\mathbf{T}_1\mathbf{T}_2\mathbf{x} = \mathbf{T}_1(\mathbf{T}_2\mathbf{x})$ of $\mathbf{T}_1, \mathbf{T}_2 \in E(3)$ is again in $E(3)$. Q.E.D.

As its name suggests, the Euclidean group is basic to the study of Euclidean geometry. In Euclidean geometry, two subsets S, S' of R_3 are said to be **congruent** if there is a $\mathbf{T} \in E(3)$ such that $S' = \mathbf{T}S$, i.e., if the points of S can be made coincident with the points of S' by a distance-preserving transformation. Since congruent triangles have corresponding angles equal, it is easy to show that each $\mathbf{T} \in E(3)$ also preserves the angle between intersecting straight lines.

Among the elements of $E(3)$ the easiest to construct are the **translations** $\mathbf{T_a}$, $\mathbf{a} \in R_3$:

$$\mathbf{T_a x} = \mathbf{x} + \mathbf{a}, \qquad \mathbf{x} \in R_3. \tag{2.2}$$

Under $\mathbf{T_a}$ each point of R_3 is displaced by $\mathbf{a}$. The set $T(3)$ of all translations of three-space forms a subgroup of $E(3)$. This subgroup is abelian since

$$\mathbf{T_a T_b} = \mathbf{T_b T_a} = \mathbf{T_{(a+b)}}. \tag{2.3}$$

Let $\mathbf{T}$ be an arbitrary element of $E(3)$ and suppose $\mathbf{T\theta} = \mathbf{a}$, where $\boldsymbol{\theta} = (0, 0, 0)$ is the origin. Then $\mathbf{T_{-a}T\theta} = \boldsymbol{\theta}$, so $\mathbf{T_{-a}T} = \mathbf{O}$ is an element of $E(3)$ which leaves the origin invariant. Now it is clear that all $\mathbf{O} \in O(3)$, as constructed in the preceding section, are elements of $E(3)$ which leave the origin invariant. In fact $O(3)$ is a subgroup of $E(3)$. However, it is not so obvious that the elements of $O(3)$ are the *only* isometries that fix $\boldsymbol{\theta}$. In particular it is not obvious (but true) that every distance-preserving transformation of

R_3 that fixes $\mathbf{\theta}$ is necessarily a linear transformation, hence an element of $O(3)$. We assume this fact here and refer the reader to Yale [1] for a proof. Thus, every $\mathbf{T} \in E(3)$ can be written uniquely in the form

$$\mathbf{T} = \mathbf{T_a O} = \{\mathbf{a}, \mathbf{O}\}, \qquad \mathbf{O} \in \mathbf{O}(3). \tag{2.4}$$

Conversely, every product of the form (2.4) defines an element of $E(3)$. Note that $\mathbf{T_a} = \{\mathbf{a}, \mathbf{E}\}$ and $\mathbf{O} = \{\mathbf{\theta}, \mathbf{O}\}$. The action of the elements of $E(3)$ on R_3 is given by

$$\{\mathbf{a}, \mathbf{O}\}\mathbf{x} = \mathbf{Ox} + \mathbf{a}, \qquad x \in R_3, \tag{2.5}$$

and the product rule is

$$\{\mathbf{a}_1, \mathbf{O}_1\}\{\mathbf{a}_2, \mathbf{O}_2\} = \{\mathbf{a}_1 + \mathbf{O}_1\mathbf{a}_2, \mathbf{O}_1\mathbf{O}_2\}. \tag{2.6}$$

Comparing this expression with (5.1) of Chapter 1, we see that $E(3)$ is a semidirect product of $T(3)$ and $O(3)$. Indeed, the map $\mathbf{O} \longrightarrow \nu_{\mathbf{O}}$, where $\nu_{\mathbf{O}}(\mathbf{a}) = \mathbf{Oa}$, is a homomorphism of $O(3)$ into the automorphism group of $T(3)$. One consequence of this result is that $T(3)$ is a normal subgroup of $E(3)$. The factor group $E(3)/T(3)$ is isomorphic to $O(3)$.

Suppose $\mathbf{T} \in E(3)$ leaves a point $\mathbf{a}$ invariant, i.e., $\mathbf{T(a)} = \mathbf{a}$. Then $\mathbf{T_a^{-1}TT_a} = \mathbf{O}$ leaves $\mathbf{\theta}$ invariant, so

$$\mathbf{T} = \mathbf{T_a O T_a^{-1}}, \qquad \mathbf{O} \in O(3). \tag{2.7}$$

Conversely, any group element of the form $\mathbf{T_a O T_a^{-1}}$ leaves $\mathbf{a}$ invariant. The reader should have no trouble in verifying that the elements (2.7) are rotations or rotation-inversions about axes through $\mathbf{a}$. All such elements clearly form a subgroup $O_{\mathbf{a}}(3)$, the orthogonal group at $\mathbf{a}$. From (2.7) we have

$$O_{\mathbf{a}}(3) = \mathbf{T_a} O(3) \mathbf{T_a^{-1}}. \tag{2.8}$$

The subgroup of rotations and rotation-inversions about $\mathbf{a}$ is conjugate, hence isomorphic, to $O(3)$. A slight extension of this argument shows that all rotations by a fixed angle $\theta, 0 \le \theta \le \pi$, through any axis in R_3 form a single conjugacy class in $E(3)$. The same holds for all rotation-inversions by a fixed angle θ'.

We now give a geometrical interpretation of the elements of $E(3)$. First, consider the element $\{\mathbf{a}, \mathbf{C_k}(\theta)\}$, where $\theta \neq 0$ and $\langle \mathbf{a}, \mathbf{k}\rangle = 0$, i.e., $\mathbf{a}$ is perpendicular to the axis of rotation. This transformation has a fixed point $\mathbf{b}$. Indeed, the formula

$$\mathbf{T_b^{-1}}\{\mathbf{a}, \mathbf{O}\}\mathbf{T_b} = \{\mathbf{a} - \mathbf{b} + \mathbf{Ob}, \mathbf{O}\}$$

and the remarks preceding (2.7) show that $\{\mathbf{a}, \mathbf{C_k}(\theta)\}$ leaves $\mathbf{b}$ invariant if

$$\mathbf{b} - \mathbf{C_k}(\theta)\mathbf{b} = \mathbf{a}. \tag{2.9}$$

Looking at a plane through $\mathbf{a}$ and perpendicular to $\mathbf{k}$ we have the situation shown in Fig. 2.1. There are an infinite number of solutions $\mathbf{b}$ forming an axis

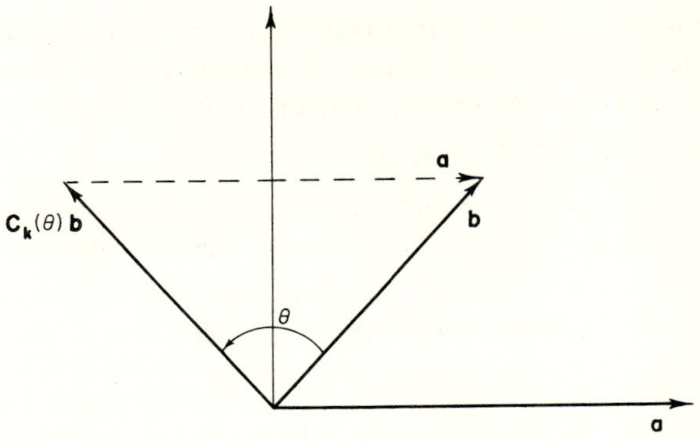

FIGURE 2.1

parallel to **k**. Thus, $\{\mathbf{a}, \mathbf{C}_\mathbf{k}(\theta)\}$ corresponds to a rotation through the angle θ about this invariant axis.

If **a** is an arbitrary vector, we can write $\mathbf{a} = \mathbf{a}_1 + \mathbf{a}_2$ uniquely, where $\mathbf{a}_1$ is parallel to **k** and $\mathbf{a}_2$ is perpendicular to **k**. Then

$$\{\mathbf{a}, \mathbf{C}_\mathbf{k}(\theta)\} = \mathbf{T}_{\mathbf{a}_1}\{\mathbf{a}_2, \mathbf{C}_\mathbf{k}(\theta)\}. \tag{2.10}$$

The transformation (2.10) is called a **screw displacement,** a rotation through an angle θ about the **screw axis,** followed by a translation along the axis by a distance $\|\mathbf{a}_1\|$. (If we think of the rotation and translation as being performed simultaneously we get a right-handed screwing motion, which justifies the name.)

A similar analysis shows that the isometry

$$\{\mathbf{a}, \mathbf{S}_\mathbf{k}(0)\} = \{\mathbf{a}, \boldsymbol{\sigma}_\mathbf{k}\} \tag{2.11}$$

is the product of the reflection in a plane perpendicular to **k**, the **glide plane,** and a translation in this plane. This transformation is called a **glide reflection.**

Finally, the isometry

$$\{\mathbf{a}, \mathbf{S}_\mathbf{k}(\theta)\}, \qquad \theta \neq 2\pi n, \tag{2.12}$$

represents a rotation-inversion about some point. To see this we introduce a new rectangular coordinate system for R_3 centered at the origin and such that the vector **k** points along the 3-axis. The transformation (2.12) maps the point $\mathbf{x} = (x_1, x_2, x_3)$ into $(x_1 \cos\theta - x_2 \sin\theta + a_1, x_1 \sin\theta + x_2 \cos\theta + a_2, a_3 - x_3)$. The reader can verify that this transformation has a unique fixed point $\mathbf{x}_0$ for $0 < \theta < 2\pi$. Thus, (2.12) must either be a rotation or a rotation-inversion about $\mathbf{x}_0$ and it is easy to show that it cannot be a rotation.

We have given geometrical interpretations for all elements of $E(3)$. The Euclidean group is made up of translations, rotations, rotation-inversions, screw displacements, and glide reflections. (There is a slight overlap in this classification since, for example, there are degenerate screw displacements which are also rotations.) From another point of view, the above constitutes a list of conjugacy classes of $E(3)$. The reader should have no difficulty in showing that all translations through a distance d form a single conjugacy class, all screw displacements with angle θ and translation distance d form a single conjugacy class, and so on,

The map $\{\mathbf{a}, \mathbf{O}\} \longrightarrow \det O$ defines a homomorphism of $E(3)$ onto the cyclic group of order two. The kernel of this homomorphism is $E^+(3)$, the **proper Euclidean group** in three-space or the **group of rigid motions.** Clearly, $E^+(3)$ is a normal subgroup of $E(3)$ and consists of all translations, rotations, and screw displacements. The elements of $E^+(3)$ are also called **direct isometries** or **direct symmetries.**

2.3 Symmetry and the Discrete Subgroups of $E(3)$

Let S be a subset of the space R_3 and define

$$G = \{\mathbf{T} \in E(3) : \mathbf{T}S = S\},$$

the group of all elements of $E(3)$ that map S onto itself. We call G the **complete symmetry group** of S. Any subgroup of G is called a **symmetry group** of S. It is not required that any point of S be fixed under $\mathbf{T}$, merely that $\mathbf{T}$ act as a permutation of the points of S. For example, the complete symmetry group of R_3 is $E(3)$. The complete symmetry group of a sphere with center at θ is $O(3)$. A right-pyramid-shaped figure with base given by Fig. 2.2 has C_2, the

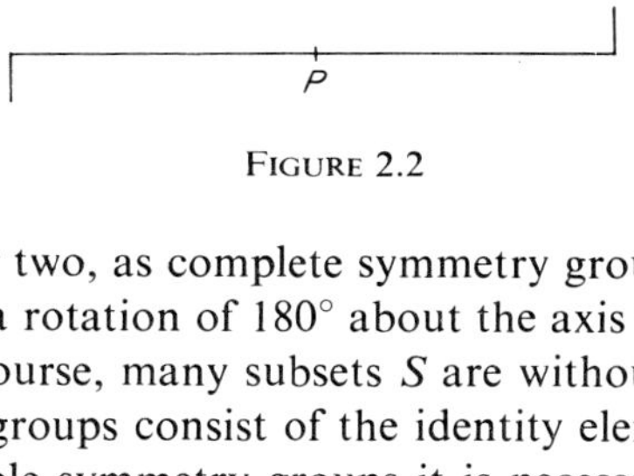

FIGURE 2.2

cyclic group of order two, as complete symmetry group. The elements of C_2 are the identity and a rotation of 180° about the axis through P perpendicular to the base. Of course, many subsets S are without symmetry, i.e., their complete symmetry groups consist of the identity element alone.

To find all possible symmetry groups it is necessary to classify all subgroups of $E(3)$. This is an extremely difficult problem! Fortunately, only two types of symmetry groups occur with much frequency in the physical sciences: discrete groups and Lie groups. These two types are easy to handle mathematically, which is one reason why they occur in applications. However, there are also good geometrical and physical reasons for limiting ourselves to such groups.

The groups $E(3)$, $O(3)$, $T(3)$, and C_∞ [the group of all rotations about a fixed axis, isomorphic to $SO(2)$] are examples of Lie groups. They will be studied in detail in subsequent chapters.

Definition. A **discrete group** G is a subgroup of the transformation group $E(3)$ such that for any $\mathbf{x} \in R_3$ and any sphere $B_r = \{\mathbf{y} \in R_3 : \|\mathbf{y}\| \leq r\}$ there are only a finite number of points in the G-orbit of $\mathbf{x}$ that are contained in B_r.

If G is a discrete group then the points $G\mathbf{x} = \{\mathbf{y} = \mathbf{gx} : \mathbf{g} \in G\}$ are distributed in R_3 so that only a finite number of them are contained in each bounded subset of R_3. Clearly, every finite subgroup of $E(3)$ is discrete since every G-orbit of a finite group contains only a finite number of elements. The group of all translations $\{\mathbf{T}_\mathbf{a}\}$ where

$$\mathbf{a} = a_1\mathbf{e}_1 + a_2\mathbf{e}_2 + a_3\mathbf{e}_3 \tag{3.1}$$

and a_1, a_2, and a_3 are integers is an infinite discrete group. On the other hand the group of all translations $\{T_\mathbf{a}\}$ where $\mathbf{a}$ takes the form (3.1) with a_1, a_2, and a_3 rational is neither a Lie group or a discrete group. Another infinite nondiscrete group is generated by a rotation through the angle $2\pi/a$ about an axis, where a is not a rational number. (Prove it!)

Let us search for the possible discrete symmetry groups of objects (or sets) of finite extent, i.e., objects which can be wholly contained inside some sufficiently large sphere B_r. Clearly such symmetry groups cannot contain nontrivial translations, screw displacements, or glide reflections since one of these transformations indefinitely repeated would map the object outside of the bounding sphere. Thus, the only allowable symmetry operations are rotations and rotation-inversions. It follows that every symmetry of a finite object has at least one fixed point. However, it is not so obvious that the symmetries have a common fixed point.

Theorem 2.3. Let S be a nonempty set of finite extent and let G be a discrete symmetry group of S. Then there is at least one point $\mathbf{y} \in R_3$ which is fixed by all $\mathbf{g} \in G$.

Proof. Since S is bounded it can be enclosed inside some sphere B_r. Let $\mathbf{x} \in S$ and consider the G-orbit containing $\mathbf{x}$. All points in this G-orbit must be in S, hence in B_r. Since G is discrete it follows that the orbit is finite:

$$G\mathbf{x} = \{\mathbf{x}_1 = \mathbf{x}, \mathbf{x}_2, \ldots, \mathbf{x}_n\}.$$

Let

$$\mathbf{y} = \Big(\sum_{i=1}^{n} \mathbf{x}_i\Big)\Big/ n \tag{3.2}$$

be the **centroid** of $G\mathbf{x}$. Since the action of an element of $E(3)$ on R_3 is given by (2.5) it follows easily that

$$\mathbf{gy} = \left(\sum_{i=1}^{n} \mathbf{gx}_i\right)\Big/ n, \qquad \mathbf{g} \in G, \tag{3.3}$$

i.e., the centroid of the finite set $\{\mathbf{x}_i\}$ is mapped onto the centroid of $\{\mathbf{gx}_i\}$ for any $\mathbf{g} \in G$. On the other hand, the transformation $\mathbf{g}$ merely permutes the elements of the G-orbit through $\mathbf{x}$. Thus, the terms on the right-hand sides of (3.2) and (3.3) are equal except for order, and the sums are the same. We conclude that $\mathbf{gy} = \mathbf{y}$, so the centroid $\mathbf{y}$ is a common fixed point for all symmetries of S.

Corollary 2.1. The elements of a finite subgroup of $E(3)$ have a common fixed point.

Theorem 2.4. Let G be a discrete subgroup of $E(3)$ whose elements have a common fixed point $\mathbf{y}$. Then G is a finite subgroup of the orthogonal group $O(3)$ of rotations and rotation-inversions about $\mathbf{y}$.

Proof. Let $\mathbf{x}_1, \ldots, \mathbf{x}_4$ be four noncoplanar points lying within a sphere B_r with center $\mathbf{y}$. Since $\|\mathbf{gx}_i - \mathbf{y}\| = \|\mathbf{gx}_i - \mathbf{gy}\| = \|\mathbf{x}_i - \mathbf{y}\|$, the orbits $\{G\mathbf{x}_i\}$ all lie inside B_r. Since G is discrete, there are only a finite number of points in these G-orbits. Now the transformation $\mathbf{g} \in G$ is uniquely determined by the four noncoplanar points $\{\mathbf{gx}_i\}$. For, if $\mathbf{gx}_i = \mathbf{g}'\mathbf{x}_i$, $1 \leq i \leq 4$, then the $\{\mathbf{x}_i\}$ are invariant under $\mathbf{g}^{-1}\mathbf{g}'$, so $\mathbf{g}^{-1}\mathbf{g}' = \mathbf{E}$, the identity operator. [An element of $E(3)$ is uniquely determined by its action on the $\{\mathbf{x}_i\}$.] Hence, $\mathbf{g}' = \mathbf{g}$ and our argument shows that G is a finite group. Q.E.D.

Thus, a discrete symmetry group of a body S of finite extent is always a finite group G of rotations and rotation-inversions about some fixed point $\mathbf{y}$. If we consider $O(3)$ as the orthogonal group with fixed point $\boldsymbol{\theta}$, it is clear that G is conjugate to the finite subgroup $\mathbf{T}_\mathbf{y}^{-1} G \mathbf{T}_\mathbf{y} = K$ of $O(3)$, and K is a symmetry group of $\mathbf{T}_\mathbf{y}^{-1} S$. Similarly, if $\mathbf{T} \in E(3)$ with $\mathbf{T}\boldsymbol{\theta} = \mathbf{y}$ then $\mathbf{T}K\mathbf{T}^{-1}$ is a finite group of rotations and rotation-inversions with fixed point $\mathbf{y}$. To simplify the classification of symmetry groups we will identify conjugate subgroups of $E(3)$. Conjugate symmetry groups are physically indistinguishable. Our listing of symmetry groups will really be a listing of equivalence classes of conjugate subgroups.

In abstract group theory, one identifies two groups if they are isomorphic, i.e., if they have the same multiplication table. This is not the same as the classification into conjugate subgroups of $E(3)$. Conjugate subgroups are isomorphic, but isomorphic subgroups of $E(3)$ need not be conjugate. For example, the cyclic groups of order two generated by a rotation of $180°$ about an axis and a reflection in a plane, respectively, are isomorphic but not conjugate.

To recapitulate, the problem of classifying all discrete symmetry groups of objects of finite extent reduces to the problem of listing all finite subgroups of $O(3)$. [It is obvious that each finite subgroup of $O(3)$ is the symmetry group of some object.] Subgroups of the transformation group $O(3)$ are called **point groups** since they always have a fixed point. (Without loss of generality we can assume that this fixed point is $\mathbf{0}$.) These groups are of two types: point groups of the **first kind,** which contain only rotations, and point groups of the **second kind,** which also contain rotation-inversions.

To a great extent the problem of classifying point groups can be reduced to the problem of classifying point groups of the first kind. Let G be a finite point group and consider the homomorphism $\mathbf{g} \longrightarrow \det g$, $\mathbf{g} \in G$ which maps G into the cyclic group of order two, i.e., $\mathbf{g}$ maps to $+1$ if it is a rotation and to -1 if it is a rotation-inversion. If the kernel of this homomorphism is G then G is a point group of the first kind. If the kernel is K, a proper subgroup of G, then by Theorems 1.4 and 1.5, K is a normal subgroup with half as many elements as G. Furthermore, the coset decomposition of G is $\{K, \mathbf{g}_0K = K\mathbf{g}_0\}$. The elements of K are rotations and the elements of $\mathbf{g}_0K$, including $\mathbf{g}_0$ itself, are rotation-inversions. There are two possibilities: either G contains the inversion $\mathbf{I} = -\mathbf{E}$ or it does not. If $\mathbf{I} \in G$ then $\mathbf{I} \in \mathbf{g}_0K$, so $\mathbf{I}K = \mathbf{g}_0K$ and we can take $\mathbf{g}_0 = \mathbf{I}$. Conversely, if K is a finite rotation group then the set $\{K, \mathbf{I}K\}$ forms a point group of the second kind. If $\mathbf{I} \notin G$ then the description of G becomes a little more complicated. Let

$$K^+ = \{\mathbf{Ig}: \mathbf{g} \in G, \quad \mathbf{g} \notin K\}.$$

It is easy to check that (1) the set K^+ consists of proper rotations, (2) $K^+ \cap K$ is empty, and (3) K^+ and K contain the same number of elements. In particular, (2) follows from the fact that $\mathbf{I} \notin G$. Now let $G^+ = K \cup K^+$. We will show that G^+ is a point group of the first kind isomorphic to G. The isomorphism is the identity on K and maps $\mathbf{g} \notin K$ into $\mathbf{Ig} \in K^+$. This map is a homomorphism because $\mathbf{I}$ commutes with all group elements. Indeed the elements of G^+ can be written in the form $\mathbf{I}^\varepsilon\mathbf{g}$ for $\mathbf{g} \in G$, where $\varepsilon = 0$ if $\mathbf{g} \in K$ and $\varepsilon = 1$ if $g \notin K$. Then

$$(\mathbf{I}^{\varepsilon_1}\mathbf{g}_1)(\mathbf{I}^{\varepsilon_2}\mathbf{g}_2) = \mathbf{I}^{\varepsilon_1+\varepsilon_2}\mathbf{g}_1\mathbf{g}_2,$$

where $\mathbf{I}^{\varepsilon_1+\varepsilon_2} = \mathbf{E}$ if $\mathbf{g}_1\mathbf{g}_2 \in K$, and is equal to $\mathbf{I}$ otherwise; which proves that the map is a homomorphism. Note that K is a normal subgroup of G^+ of index two. We have proved the following result.

Theorem 2.5. Let G be a finite subgroup of $O(3)$ and let $K = G \cap SO(3)$, the subgroup of rotations in G. There are exactly three possibilities: (1)

$G = K$, (2) $G = K \cup \mathbf{I}K$, (3) $G \neq K, \mathbf{I} \notin G$. In the last case G is isomorphic to the group of rotations $G^+ = K \cup K^+$.

Theorem 2.5 tells us how to construct all point groups of the second kind once we are given all point groups of the first kind. The only nontrivial constructions are in class (3), where we have to determine all point groups of the first kind G^+ that contain a normal subgroup K of index two. The point group G is then defined by means of the isomorphism discussed in the proof of the theorem.

2.4 Point Groups of the First Kind

Let G be a finite subgroup of $SO(3)$ with order $n(G) \geq 2$. Then G acts as a transformation group in Euclidean space whose elements have the origin $\boldsymbol{\theta}$ as a common fixed point. Let B_r be a sphere in R_3 with center at the origin and radius $r > 0$. The elements of G clearly map the surface S_r of the sphere onto itself. A point $\mathbf{x}$ on S_r is said to be a **pole** if $\mathbf{gx} = \mathbf{x}$ for some $\mathbf{g} \in G$, not the identity element. That is, a pole is a point of intersection of S_r and the axis of a nontrivial rotation in G. Clearly, each element of G except the identity is associated with two poles. The transformation group G maps poles into poles. Indeed, if $\mathbf{x}$ is a pole associated with $\mathbf{g}_1$ then $\mathbf{g}_2\mathbf{x}$ is a pole associated with $\mathbf{g}_2\mathbf{g}_1\mathbf{g}_2^{-1}$. It follows that the set of poles on S_r is partitioned into G-orbits. According to Theorem 1.6 the number of poles in the orbit containing $\mathbf{x}$ is $p = n(G)/n(G^{\mathbf{x}})$, where $G^{\mathbf{x}}$ is the isotropy subgroup of G corresponding to $\mathbf{x}$, i.e., $G^{\mathbf{x}}$ is the subgroup of all rotations with pole $\mathbf{x}$. Suppose there are k orbits. Choosing a point $\mathbf{x}_i$ in each orbit we see that the number of nontrivial rotations with pole $\mathbf{x}_i$ is $n_i - 1 = -1 + n/p_i$, where $n = n(G)$, $n_i = n(G^{\mathbf{x}_i})$, and p_i is the number of poles in the ith orbit. (Recall that n_i is the same for all points in the ith orbit.). We have subtracted 1 since the identity element in $G^{\mathbf{x}_i}$ is a trivial rotation. The total number of rotations leaving some pole in the ith orbit fixed is thus $p_i(n_i - 1)$. Summing over the orbits, we find that the total number of rotations leaving some pole fixed is $\sum_{i=1}^{k} p_i(n_i - 1)$. Since each rotation is associated with two poles this sum equals $2(n - 1)$, i.e., each nontrivial rotation is counted twice. Thus we have the identity

$$2(1 - 1/n) = \sum_{i=1}^{k} (1 - 1/n_i), \tag{4.1}$$

where $n \geq n_i \geq 2$. This equation can be solved only if $2 \leq k \leq 3$. If $k = 2$ then (4.1) becomes

$$2/n = 1/n_1 + 1/n_2. \tag{4.2}$$

Furthermore, (4.2) can be solved if and only if $n_1 = n_2 = n$, $n = 2, 3, \ldots$.

Thus, the finite rotation groups G with two orbits are associated with two poles, each pole fixed by all elements of G. There is only one axis of rotation.

The cyclic groups C_n of order n ($n = 2, 3, \ldots$) generated by a rotation through the angle $2\pi/n$ about a fixed axis clearly satisfy the above requirements. We shall show that these are the only point groups of the first kind whose poles can be partitioned into two orbits.

Lemma 2.2. Let G be a group of order $n \geq 2$ consisting of rotations about a fixed axis. Then $G \cong C_n$.

Proof. The n elements $\mathbf{e}, \mathbf{g}_1, \ldots, \mathbf{g}_{n-1}$ of G correspond to rotations through the angles $0, \theta_1, \ldots, \theta_{n-1}$ about the fixed axis, where 0 corresponds to the identity element. We can assume $0 < \theta_i < 2\pi$, $1 \leq i \leq n-1$, if the rotation angles are expressed in radians, and renumber the elements of G so that θ_1 is the smallest positive rotation angle. Using the Euclidean algorithm we see that for each θ_i, $2 \leq i \leq n-1$, there is an integer m_i such that

$$\theta_i = m_i\theta_1 + \varphi_i, \qquad 0 \leq \varphi_i < \theta_1.$$

But $\mathbf{g}_i\mathbf{g}_1^{-m_i} \in G$, so φ_1 is the rotation angle of some element of G. Since θ_1 is the smallest positive rotation angle, the only possibility is $\varphi_i = 0$. Thus G is a cyclic group generated by $\mathbf{g}_1$. Since G has order n it follows that $\theta_1 = 2\pi/n$. Q.E.D.

We now return to the solution of (4.1) for $k = 3$,

$$1 + 2/n = 1/n_1 + 1/n_2 + 1/n_3, \qquad n \geq n_i \geq 2. \tag{4.3}$$

It can be assumed that $n_1 \leq n_2 \leq n_3$. Clearly, there is no solution for $3 \leq n_1$ since in that case

$$1 + 2/n > 1 \geq 1/n_1 + 1/n_2 + 1/n_3.$$

Therefore $n_1 = 2$. If $n_2 = 2$ we get the unique solution

(a) $\quad n_1 = n_2 = 2, \qquad n_3 = n/2, \qquad n$ even, $\quad n \geq 4.$

If $n_2 \geq 4$ there is no solution since

$$1 + 2/n > 1 \geq \tfrac{1}{2} + \tfrac{1}{4} + 1/n_3, \qquad n_3 \geq 4.$$

Thus, the only remaining possibility is $n_2 = 3$:

$$1/6 + 2/n = 1/n_3, \qquad n \geq n_3 \geq 3, \qquad 6 > n_3.$$

The possible solutions are

(b) $\quad n_1 = 2, \qquad n_2 = n_3 = 3, \qquad n = 12,$

(c) $\quad n_1 = 2, \qquad n_2 = 3, \qquad n_3 = 4, \qquad n = 24,$

(d) $\quad n_1 = 2, \qquad n_2 = 3, \qquad n_3 = 5, \qquad n = 60.$

This exhausts the solutions of (4.1). We will show that each of the solutions (a)–(d) uniquely defines a point group of the first kind.

In solution (a) set $n = 2m$, $m \geq 2$ an integer. There is a rotation axis corresponding to a rotation subgroup of order m. It follows from Lemma 2.2 that this subgroup is C_m and is generated by the rotation through the angle $2\pi/m$ about the fixed axis L. We say L is an ***m*-fold** axis. The poles of L lie in the same orbit. We have now determined m elements of the point group. To get the remaining elements, note that there are m twofold axes of rotation, $l_1, \ldots, l_m$ whose poles are divided into two orbits of m poles each. Since the two poles of L form a single orbit, each of the rotations by π radians about a twofold axis l_i must interchange the poles. Thus the twofold axes are perpendicular to L. A rotation by $2\pi/m$ about L maps the l_i into themselves. By considering rotations about twofold axes we can easily show that the angle between two adjacent l_i in the plane perpendicular to L is a fixed constant. Thus, the angle between any two adjacent l_i must be π/m.

The abstract structure of the transformation group corresponding to (a) is now uniquely determined. Let $\mathbf{C}$ be the rotation of $2\pi/m$ about L and let $\boldsymbol{\tau}$ be a rotation by π about one of the twofold axes. Since the cyclic group C_m generated by $\mathbf{C}$ has order m, it follows that the elements of G can be divided into two cosets C_m and $\boldsymbol{\tau} C_m$. The m elements in the second coset are of order two since they interchange the poles of L. Thus, $\boldsymbol{\tau}^{-1} = \boldsymbol{\tau}$ and $(\boldsymbol{\tau} C)^2 = \mathbf{e}$, or

$$\boldsymbol{\tau} C = C^{-1}\boldsymbol{\tau}. \tag{4.4}$$

Any element $\mathbf{g}$ of G can be written uniquely in the form

$$\mathbf{g} = \boldsymbol{\tau}^{\varepsilon}\mathbf{C}^{k}, \qquad \varepsilon = 0, 1, \qquad k = 0, 1, \ldots, m-1. \tag{4.5}$$

The multiplication of two group elements is then uniquely determined by (4.4). For example,

$$(\boldsymbol{\tau}\mathbf{C}^{k_1})(\boldsymbol{\tau}\mathbf{C}^{k_2}) = \mathbf{C}^{k_2-k_1}, \qquad (\boldsymbol{\tau}\mathbf{C}^{k_1})(\mathbf{C}^{k_2}) = \boldsymbol{\tau}\mathbf{C}^{k_1+k_2}. \tag{4.6}$$

The abstract group defined by these rules is denoted D_m, the **dihedral group** of order $2m$.

We will list the conjugacy classes of D_m since they are of importance for representation theory. The details in the straightforward proofs will be left to the reader. Because of the presence of rotations which interchange the poles of L, the rotations $\mathbf{C}^k$ and $\mathbf{C}^{-k} = \mathbf{C}^{m-k}$, $k = 1, \ldots, m-1$, are conjugate. (The axis L is called **two-sided** because both poles of L lie in the same orbit.) Since C_m is a normal subgroup of D_m the conjugacy classes $\{\mathbf{C}^k, \mathbf{C}^{m-k}\}$ contain no elements not in C_m. There are $1 + (m/2)$ such classes if m is even and $(m+1)/2$ classes if m is odd. The m rotations $\boldsymbol{\tau}_1, \ldots, \boldsymbol{\tau}_m$ about the twofold axes form a single conjugacy class if m is odd and two classes if m is even. Thus D_m has a total of $(3 + m)/2$ classes for odd m and $3 + (m/2)$ classes for even m.

For each m we shall exhibit a solid with D_m as its largest point symmetry group of the first kind. An ***m*-prism** is a right cylinder with base a regular m-sided polygon and height not equal to one side of the polygon. In the case $m = 2$ we define a regular two-sided polygon as a plane figure looking as shown in Fig. 2.3. The reader can check that D_m is a symmetry group of the

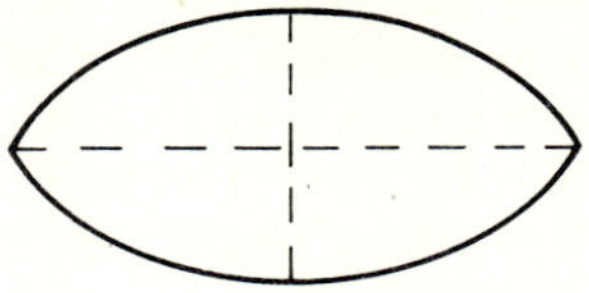

FIGURE 2.3

m-prism. In particular the axis L of the m-prism is the m-fold axis and the midpoint of L is invariant under all elements of D_m. Similarly, it is easy to show that C_m is the maximal direct symmetry group of an ***m*-pyramid.** An m-pyramid is a right pyramid with base a regular m-sided polygon such that the distance from the vertex of the pyramid to a vertex of the base is not equal to one side of the polygon.

Next we consider solution (b) of (4.1). There are four threefold axes $L_1, \ldots, L_4$ three twofold axes l_1, l_2, l_3. Let $\{\mathbf{x}_1, \ldots, \mathbf{x}_4\}$ be the poles in one of the orbits which contains four elements. The elements of G permute these poles transitively and effectively (since a nontrivial rotation can fix at most two poles). Any pole $\mathbf{x}_1$, say, is fixed by a cyclic subgroup C_3 of order three. It follows that for $2 \le i < j \le 4$ there exists $\mathbf{g} \in C_3$ such that $\mathbf{gx}_1 = \mathbf{x}_1$, $\mathbf{gx}_i = \mathbf{x}_j$. Therefore, for distinct i, j, k there is a $\mathbf{g} \in G$ such that $\mathbf{gx}_i = \mathbf{x}_i$, $\mathbf{gx}_j = \mathbf{x}_k$, i.e., the line segment $[\mathbf{x}_i, \mathbf{x}_j]$ is mapped onto $[\mathbf{x}_i, \mathbf{x}_k]$ by the rotation $\mathbf{g}$. Thus, the poles $\mathbf{x}_1, \ldots, \mathbf{x}_4$ are spaced equidistant from one another on the sphere S_r and the tetrahedron with these poles as vertices admits G as a symmetry group. Clearly, G is a subgroup of the **tetrahedral group** T of all direct symmetries of the tetrahedron. However, $n(T) = 12$, so $G = T$, Indeed, every symmetry of the tetrahedron can be represented uniquely as a permutation of the four vertices $\mathbf{x}_1, \ldots, \mathbf{x}_4$. Out of the 24 possible permutations in S_4, the reader can easily verify that only a subgroup of order 12, the even permutations, correspond to direct symmetries of the tetrahedron. The four threefold axes pass through the four vertices of the tetrahedron, while the three twofold axes join the midpoints of nonintersecting edges. There are four conjugacy classes: the identity, four rotations by 120°, four rotations by 240°, and three rotations by 180°.

Solution (c) of (4.1) corresponds to three fourfold axes L_1, L_2, and L_3, four threefold axes, and six twofold axes. All axes are two-sided. Consider the orbit with six poles (the poles of the L_i axes). G acts transitively and effectively on the poles of this orbit. Exactly two of the poles, say $\mathbf{x}_1$ and $\mathbf{x}_2$, are fixed under the subgroup C_4 of rotations about the axis L_1 containing $\mathbf{x}_1$ and $\mathbf{x}_2$. The four remaining poles are permuted by the elements of C_4. Since C_4 contains an element of order four it follows that for $3 \leq i \leq j \leq 6$ there is a $\mathbf{g} \in C_4$ such that $\mathbf{g}\mathbf{x}_k = \mathbf{x}_k$, $k = 1, 2$, and $\mathbf{g}\mathbf{x}_i = \mathbf{x}_j$. Thus $\mathbf{g}$ maps the line segments $[\mathbf{x}_1, \mathbf{x}_i]$ and $[\mathbf{x}_2, \mathbf{x}_i]$ onto $[\mathbf{x}_1, \mathbf{x}_j]$ and $[\mathbf{x}_2, \mathbf{x}_j]$, respectively. The distance between any two poles not on the same axis is a fixed constant, so the three L-axes are mutually orthogonal. Thus, we can construct a cube $\mathcal{C}$ such that the six poles form the midpoints of the six faces of $\mathcal{C}$. The group G is obviously a subgroup of the **octahedral group** O, the direct symmetry group of $\mathcal{C}$. The reader can verify that $n(O) = 24$, so $G = O$. As we have mentioned, the three fourfold axes pass through the midpoints of the faces of $\mathcal{C}$. The four threefold axes pass through the vertices of $\mathcal{C}$ and the six twofold axes pass through the midpoints of the edges of $\mathcal{C}$. The octahedral group contains five conjugacy classes. Since all axes are two-sided, the three rotations of 90° and the three rotations of 270° about the L_i form a single class, as do the four rotations of 120° and four rotations of 240° about the threefold axes. The remaining classes contain the identity, the three rotations of 180° about the L_i, and the six rotations of 180° about the twofold axes, respectively.

The octahedral group O is also the direct symmetry group of the octahedron, a figure formed by connecting the midpoints of adjacent faces of $\mathcal{C}$ with straight lines. The octahedron is a regular polyhedron with 8 triangular faces, 12 edges, and 6 vertices.

Solution (d) of (4.1) corresponds to six fivefold axes $L_1, \ldots, L_6$, ten threefold axes, and fifteen twofold axes. All axes are two-sided. The 12 poles of the L_i axes lie in a single orbit. The transformation group G permutes these poles transitively and effectively. Let us choose an axis L_1 with poles $\mathbf{x}_1, \mathbf{x}_2$. The subgroup of rotations that fix each of $\mathbf{x}_1, \mathbf{x}_2$ is isomorphic to C_5. In particular the rotation $\mathbf{g}$ through the angle 72° about L_1 is an element of order five. The action of $\mathbf{g}$ on the orbit can thus be represented by the permutation

$$(\mathbf{x}_3\mathbf{x}_4 \cdots \mathbf{x}_7)(\mathbf{x}_8\mathbf{x}_9 \cdots \mathbf{x}_{12}) \tag{4.7}$$

if the poles are suitably labeled. This is the only possibility since $\mathbf{g}$ must have order five and leave none of the poles $\mathbf{x}_3, \ldots, \mathbf{x}_{12}$ fixed. Therefore, under the action of C_5 the single G-orbit splits into four C_5-orbits: two fixed poles and two orbits containing five poles each.

Think of $\mathbf{x}_1$ and $\mathbf{x}_2$ as the north and south poles of the sphere B_r. The remaining poles $\mathbf{x}_3, \ldots, \mathbf{x}_{12}$ cannot all lie on the equator since then a rotation through 72° about one of the axes L_i, $2 \leq i \leq 6$, would map some of the

$\mathbf{x}_i$ into points which are neither on the equator or at the north or south poles. This is impossible because the G-orbit contains only 12 elements. Therefore, without loss of generality we can assume that $\mathbf{x}_3$ is in the northern hemisphere. There must then be five poles in the northern hemisphere since rotations by 72° about L_1 map $\mathbf{x}_3$ into $\mathbf{x}_4, \ldots, \mathbf{x}_7$, successively. The five remaining poles are in the southern hemisphere since they lie on the other ends of axes through the poles $\mathbf{x}_3, \ldots, \mathbf{x}_7$. Our original choice of the axis L_1 was arbitrary, so we have established that each pole has five nearest-neighbor poles, five distant poles, and its antipode. The distance between nearest-neighbor poles is a fixed constant.

Now draw straight lines connecting each pole to its five nearest neighbors. The figure thus formed, assuming it exists, is a regular polyhedron with 12 vertices (the poles), 30 edges, and 20 faces. (Prove it!) The faces are equilateral triangles.

Such a regular polyhedron does exist. It is called the icosahedron and the dubious reader can construct it by gluing 20 congruent equilateral triangles together along the edges. The direct symmetry group of the icosahedron is the **icosahedral group** Y. Clearly G is a subgroup of Y. It is easy to enumerate the possible direct symmetries of the icosahedron. The only possible axes are: 6 fivefold axes through pairs of opposite vertices, 10 threefold axes through the midpoints of opposite faces, and 15 twofold axes through the midpoints of opposite edges. Thus Y contains a total of 60 elements. Since $n(G) = 60$ it follows that $G = Y$.

There are five conjugacy classes in Y: the class of the identity element, the class containing 15 rotations of 180°, the class containing 10 rotations of 120° and 10 rotations of 240°, the class containing 6 rotations of 72° and 6 rotations of 288°, and the class containing 6 rotations of 144° and 6 rotations of 216°.

The icosahedral group is also the direct symmetry group of the dodecahedron. This regular polyhedron can be obtained by joining with straight lines the midpoints of adjacent faces of the icosahedron. The dodecahedron has 20 vertices and 30 edges. Its 12 faces are regular pentagons.

We have shown that a complete list of point groups of the first kind is given by the cyclic groups C_m, the dihedral groups D_m, $m \geq 2$, the tetrahedral group T, the octahedral group O, and the icosahedral group Y. No two groups in this list are isomorphic.

2.5 Point Groups of the Second Kind

A list of point groups of the second kind can be obtained from Theorem 2.5 and results of the last section. First we list all groups generated by the inversion $\mathbf{I}$ and a point group of the first kind K. Clearly $n(G) = 2n(K)$. As an abstract group G is isomorphic to the direct product $K \times H = K \cup \mathbf{I}K$, where H is the group with two elements $\{\mathbf{E}, \mathbf{I}\}$. Thus, the multiplication table for G can be obtained in an obvious way from the multiplication table for K. The number of conjugacy classes for G is just twice the number for K. The list is as follows:

(1) $C_n \cup \mathbf{I}C_n$. This is an abelian group of order $2n$ consisting of all rotations through multiples of the angle $2\pi/n$ about a fixed axis and all such rotations followed by an inversion. The group has $2n$ conjugacy classes, each class containing one element. For n odd there is an isomorphism $C_{2n} \cong C_n \cup \mathbf{I}C_n$. However, these two groups are not conjugate subgroups of $E(3)$. Also, $D_2 \cong C_2 \cup \mathbf{I}C_2$, but again the two subgroups are not conjugate.

(2) $D_n \cup \mathbf{I}D_n$, $n \geq 2$. This group of order $4n$ has $3 + n$ conjugacy classes if n is odd and $6 + n$ if n is even. For odd $n \geq 3$ there is an isomorphism $D_n \cup \mathbf{I}D_n \cong D_{2n}$, but the two subgroups are not conjugate.

(3) $T \cup \mathbf{I}T = T_h$. The group T_h is of order 24 and contains 8 conjugacy classes.

(4) $O \cup \mathbf{I}O = O_h$. The group O_h is the complete symmetry group of the cube. It has order 48 and contains 12 conjugacy classes.

(5) $Y \cup \mathbf{I}Y = Y_h$. This is the complete symmetry group of the icosahedron. It contains 120 elements divided into 10 conjugacy classes.

Next we construct the groups mentioned in part (3) of Theorem 2.6. We look for all point groups G^+ of the first kind such that G^+ contains a subgroup K of index two. With $G^+ = K \cup K^+$ it follows that $G = K \cup \mathbf{I}K^+$ is a point group of the second kind isomorphic (but not conjugate) to G^+. Examining our list of point groups of the first kind we find the possibilities given in Table 2.1. Perhaps the easiest way to obtain these results is to search

TABLE 2.1

	G^+	K	Order of G	Number of conjugacy classes
(6)	C_{2n}	C_n	$2n$	$2n$
(7)	D_n	$C_n,\ n \geq 2$	$2n$	$(3+n)/2$, n odd; $3 + (n/2)$, n even
(8)	D_{2n}	$D_n,\ n \geq 2$	$4n$	$3 + n$
(9)	O	T	24	5

for all homomorphisms of G^+ onto the cyclic group of order two. An element of odd order in G^+ is necessarily in the kernel K of each homomorphism. Only the elements of even order have to be examined with special care.

Since $G \cong G^+$, the multiplication table and the number of conjugacy classes for G are the same as for G^+. However, G and G^+ act differently as transformation groups because one group contains rotation-inversions and the other does not.

The group of type (9) is usually denoted as T_d in Schöenflies notation (Hamermesh [1]). T_d contains T as a normal subgroup and is the complete symmetry group of the tetrahedron. The groups of type (7) are denoted C_{nv}, $n = 2, 3, \ldots$. The C_{nv} group is the complete symmetry group of an n-pyramid. It contains the subgroup C_n of rotations about the vertical n-fold axis of the pyramid as well as reflections in n vertical planes passing through this axis.

The groups of types (1), (2), (6), and (8) are classified in a different manner by Schöenflies. The type (1) group for odd n and the type (6) group for even n are lumped together to form the cyclic group S_{2n} of order $2n$. A generator of S_{2n} is given by the rotation-inversion $S(\pi/n)$, i.e., a rotation of π/n about an axis followed by reflection in a plane perpendicular to the axis. The even powers of $S(\pi/n)$ form the subgroup C_n. The type (1) group for even n and the type (6) group for odd n are combined to form the abelian group C_{nh}, which consists of the $2n$ rotations and rotation-inversions about a fixed axis by all multiples of $2\pi/n$.

The type (2) group for even n and the type (8) group for odd n form D_{nh}, the complete symmetry group of the n-prism. This group of order $4n$ contains C_{nh} as a subgroup of order $2n$. The type (2) group for odd n and the type (8) group for even n form D_{nd} of order $4n$. The group D_{nd} is the complete symmetry group of a **twisted *n*-prism**, obtained by joining together two n-prisms at their bases in such a way that the prisms are rotated relative to one another by the angle π/n. Here, D_{nd} contains S_{2n} as a subgroup.

We have not listed solids whose complete symmetry groups are T_h, S_{2n}, and C_{nh}. Such solids are not difficult to construct, however, and we refer the interested reader to Yale [1].

2.6 Lattice Groups

A **lattice group** G is a nontrivial discrete subgroup of $T(3)$, the translation group in three-space. By nontrivial we mean that G is not just the identity element. Since the elements $\mathbf{T_a}$ of $T(3)$ are completely determined by the 3-vectors $\mathbf{a} = \alpha_1\mathbf{e}_1 + \alpha_2\mathbf{e}_2 + \alpha_3\mathbf{e}_3$, we can think of G as a group of 3-vectors $\mathbf{a}$ whose law of group multiplication is vector addition:

$$\mathbf{T}_{\mathbf{a}_1}\mathbf{T}_{\mathbf{a}_1} = \mathbf{T}_{\mathbf{a}_1+\mathbf{a}_2}.$$

If the vector group G contains three linearly independent vectors it is said to be **three-dimensional.** If G contains only two linearly independent vectors, i.e., if all the vectors lie in a plane through $\boldsymbol{\theta}$, then G is **two-dimensional.** If all the vectors lie on a line through $\boldsymbol{\theta}$, then G is **one-dimensional.** If $\mathbf{a}_1, \ldots, \mathbf{a}_k$

are linearly independent vectors in a k-dimensional lattice group then every $\mathbf{a} \in G$ can be written uniquely in the form

$$\mathbf{a} = \alpha_1 \mathbf{a}_1 + \cdots + \alpha_k \mathbf{a}_k, \tag{6.1}$$

where the α_i are real numbers. We shall be primarily concerned with three-dimensional lattice groups and we shall always consider G as a group of vectors under addition. In this way we obtain a geometrical model of each lattice group.

Two linearly independent vectors $\mathbf{a}_1$ and $\mathbf{a}_2$ in a lattice group determine a parallelogram with vertices $\boldsymbol{\theta}$, $\mathbf{a}_1$, $\mathbf{a}_2$, and $\mathbf{a}_1 + \mathbf{a}_2$ (all in G) (Fig. 2.4).

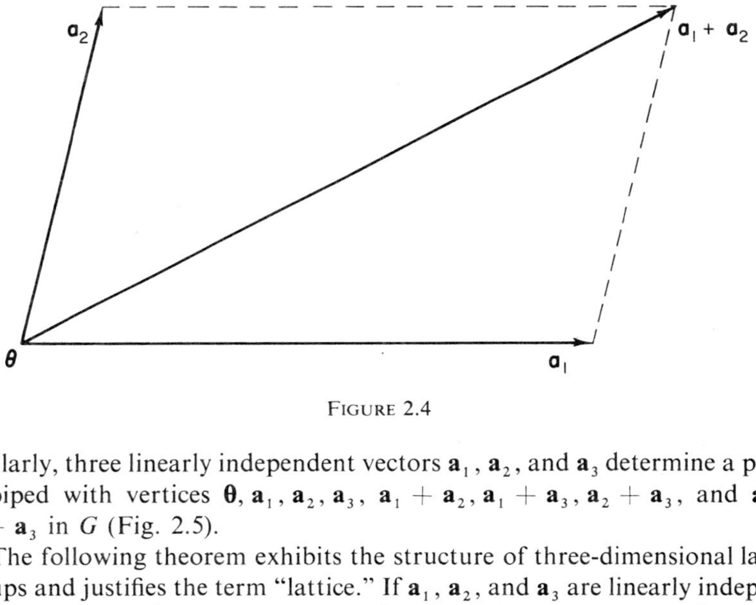

FIGURE 2.4

Similarly, three linearly independent vectors $\mathbf{a}_1$, $\mathbf{a}_2$, and $\mathbf{a}_3$ determine a parallelepiped with vertices $\boldsymbol{\theta}$, $\mathbf{a}_1$, $\mathbf{a}_2$, $\mathbf{a}_3$, $\mathbf{a}_1 + \mathbf{a}_2$, $\mathbf{a}_1 + \mathbf{a}_3$, $\mathbf{a}_2 + \mathbf{a}_3$, and $\mathbf{a}_1 + \mathbf{a}_2 + \mathbf{a}_3$ in G (Fig. 2.5).

The following theorem exhibits the structure of three-dimensional lattice groups and justifies the term "lattice." If $\mathbf{a}_1$, $\mathbf{a}_2$, and $\mathbf{a}_3$ are linearly independent then the set $\{\alpha_1 \mathbf{a}_1 + \alpha_2 \mathbf{a}_2 + \alpha_3 \mathbf{a}_3\}$, where the α_i run over all possible integers, is clearly a subgroup of G. We show it is possible to choose the $\mathbf{a}_i$ so this set is all of G.

Theorem 2.6. Let G be a three-dimensional lattice group. Then there exist linearly independent vectors $\mathbf{b}_1$, $\mathbf{b}_2$, $\mathbf{b}_3$ in G such that every $\mathbf{a} \in G$ can be written uniquely in the form

$$\mathbf{a} = n_1 \mathbf{b}_1 + n_2 \mathbf{b}_2 + n_3 \mathbf{b}_3 \tag{6.2}$$

where the n_i are integers.

Proof. Let $\mathbf{a}_1$, $\mathbf{a}_2$, $\mathbf{a}_3$ be linearly independent vectors in G and let P be the cell in R_2 determined by these vectors. (We think of P as consisting of the

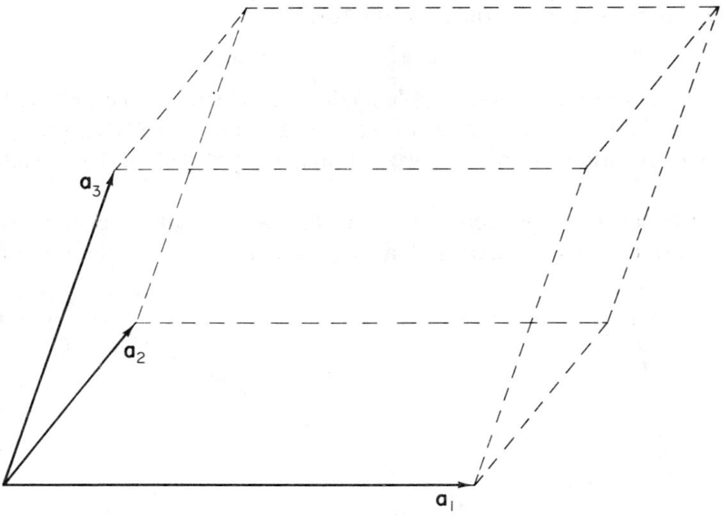

FIGURE 2.5

interior of the parallelepiped as well as its boundary faces, edges, and vertices.) There are only a finite number of elements of the discrete group G in P. Let $\mathbf{b}_1$ be the shortest nonzero vector in $G \cap P$ that is parallel to $\mathbf{a}_1$. That is, on the edge of P with endpoints $\boldsymbol{\theta}$ and $\mathbf{a}_1$ we choose the element $\mathbf{b}_1 \neq \boldsymbol{\theta}$ of G closest to $\boldsymbol{\theta}$. Now let $\mathbf{b}_2$ be an element of G in the parallelogram generated by $\mathbf{a}_1, \mathbf{a}_2$ such that the parallelogram generated by $\mathbf{b}_1, \mathbf{b}_2$ has the smallest possible nonzero area. Finally, choose $\mathbf{b}_3 \in G \cap P$ such that the parallelepiped Q generated by $\mathbf{b}_1, \mathbf{b}_2, \mathbf{b}_3$ has the smallest possible nonzero volume. We show that the $\mathbf{b}_i$ have property (6.2). Clearly, these vectors are linearly independent. Given any $\mathbf{a} \in G$ there exist unique real numbers α_i such that

$$\mathbf{a} = \alpha_1\mathbf{b}_1 + \alpha_2\mathbf{b}_2 + \alpha_3\mathbf{b}_3.$$

Let n_i be the largest integer in α_i. Then

$$\mathbf{a} - \sum_{i=1}^{3} n_i\mathbf{b}_i = \sum_{i=1}^{3} \beta_i\mathbf{b}_i = \mathbf{b} \tag{6.3}$$

with $0 \leq \beta_i < 1$. The vector $\mathbf{b}$ defined by (7.3) is clearly an element of $G \cap Q$. We will show that $\mathbf{b} = \boldsymbol{\theta}$.

Suppose $0 < \beta_3 < 1$. Then the volume $V(Q')$ of the parallelepiped Q' generated by $\mathbf{b}_1, \mathbf{b}_2, \mathbf{b}$ is strictly less than the volume $V(Q)$. In fact $V(Q') = \beta_3 V(Q)$, but if $\mathbf{b} \in P$ this is impossible since it contradicts our choice of $\mathbf{b}_3$. If $\mathbf{b} \notin P$ we can find integers m_1, m_2 such that $\mathbf{b}' = \mathbf{b} + m_1\mathbf{a}_1 + m_2\mathbf{a}_2 \in P$ and the parallelepiped Q'' generated by $\mathbf{b}_1, \mathbf{b}_2, \mathbf{b}'$ has volume $V(Q'') =$

$\beta_3 V(Q) < V(Q)$. This is impossible! Thus $\beta_3 = 0$ and $\mathbf{b}$ lies in the plane spanned by $\mathbf{a}_1$ and $\mathbf{a}_2$. If $0 < \beta_2 < 1$ then the area of the parallelogram generated by $\mathbf{b}_1, \mathbf{b}$ is β_2 times the area of the parallelogram generated by $\mathbf{b}_1, \mathbf{b}_2$, which contradicts the choice of $\mathbf{b}_2$ if $\mathbf{b} \in P$. If $\mathbf{b} \notin P$ then there is an integer m_1 such that $\mathbf{b}' = \mathbf{b} + m_1\mathbf{a}_1 \in P$ and the area of the parallelogram generated by $\mathbf{b}_1, \mathbf{b}'$ is β_2 times the area of the parallelogram generated by $\mathbf{b}_1, \mathbf{b}_2$. This is impossible! Thus $\beta_2 = 0$ and $\mathbf{b} = \beta_1\mathbf{b}_1$. If $0 < \beta_1 < 1$ then $\mathbf{b}$ is closer to $\boldsymbol{\theta}$ than is $\mathbf{b}_1$. This is impossible, so $\mathbf{b} = \boldsymbol{\theta}$. Q.E.D.

Now we return to the idea of the lattice group as a transformation group on R_3. Suppose the elements $\mathbf{b}_1, \mathbf{b}_2, \mathbf{b}_3$ in G satisfy property (6.2). We call such a triple **basic vectors.** Let $\mathbf{x} \in R_3$. Applying to $\mathbf{x}$ those transformations in G corresponding to $\mathbf{b}_1, \mathbf{b}_2, \mathbf{b}_3, \mathbf{b}_1 + \mathbf{b}_2, \mathbf{b}_1 + \mathbf{b}_3, \mathbf{b}_2 + \mathbf{b}_3$, and $\mathbf{b}_1 + \mathbf{b}_2 + \mathbf{b}_3$, we get a parallelepiped in R_3 called a **primitive cell** or **basic parallelepiped.** By applying all elements of G to $\mathbf{x}$, i.e., by constructing the G-orbit containing $\mathbf{x}$, we form a geometrical lattice of points in R_3. Indeed, it follows from Theorem 2.6 that this lattice is just what we would get by stacking together copies of the primitive cell so that they fill all of R_3. The lattice points are the vertices of the primitive cells. If the $\mathbf{b}_i$ were linearly independent but did not satisfy (6.2) we could still carry out the above construction and fill R_3 with cells constructed on $\mathbf{x}$. However, the vertices of these cells would not exhaust the points in the G-orbit containing $\mathbf{x}$.

We can construct a lattice containing any point $\mathbf{x}$. Two points lie on the same lattice if and only if they are in the same G-orbit. The totality of all lattices, i.e., all G-orbits, is called the **crystal lattice** or **space lattice.** Ordinarily it is most convenient to discuss lattices based on $\mathbf{x} = \boldsymbol{\theta}$.

By suitably eliminating certain faces, edges, and vertices from a primitive cell we can construct a **fundamental domain** for G, i.e., a subset D of R_3 such that any point $\mathbf{x} \in R_3$ lies in the same G-orbit as some $\mathbf{y} \in D$ and no two points in D lie in the same G-orbit. Thus, D consists of exactly one point from each G-orbit of R_3.

The proof of Theorem 2.6 shows that even though the crystal lattice of G is uniquely determined by G, the primitive cell is not. In fact, there are an infinite number of possible basic vectors $\mathbf{b}_i$.

Corollary 2.2. Corresponding to any two linearly independent vectors $\mathbf{a}_1, \mathbf{a}_2 \in G$ there exists a primitive cell Q with an edge directed along $\mathbf{a}_1$ and a face in the plane spanned by $\mathbf{a}_1$ and $\mathbf{a}_2$. (We assume $\mathbf{x} = \boldsymbol{\theta}$.)

The primitive cells of G can be characterized as the cells with smallest volume. Let $\mathbf{b}_1, \mathbf{b}_2, \mathbf{b}_3$ be basic vectors with primitive cell Q of volume $V(Q)$ and let $\mathbf{a}_1, \mathbf{a}_2, \mathbf{a}_3$ be three linearly independent vectors in G with cell P.

Theorem 2.7. $V(Q) \leq V(P)$.

Proof. In terms of the standard orthonormal basis vectors $\mathbf{e}_1, \mathbf{e}_2, \mathbf{e}_3$ the $\mathbf{a}_i$ and $\mathbf{b}_i$ can be represented as

$$\mathbf{a}_i = \sum_{j=1}^{3} a_{ji}\mathbf{e}_j, \qquad \mathbf{b}_k = \sum_{j=1}^{3} b_{jk}\mathbf{e}_j, \qquad i, k = 1, 2, 3, \tag{6.4}$$

where the 3×3 matrices

$$A = (a_{ij}), \qquad B = (b_{kj})$$

are nonsingular. Similarly,

$$\mathbf{a}_i = \sum_{k=1}^{3} c_{ki}\mathbf{b}_k, \qquad i = 1, 2, 3, \tag{6.5}$$

and the nonsingular matrix

$$C = (c_{ik})$$

has integer matrix elements, since the $\mathbf{b}_k$ are basic vectors. It follows from (6.4) and (6.5) that $A = BC$ in terms of matrix multiplication. Furthermore,

$$V(P) = |\mathbf{a}_1 \cdot (\mathbf{a}_2 \times \mathbf{a}_3)| = |\det A|. \tag{6.6}$$

Thus,

$$V(P) = |\det BC| = |\det C| \cdot |\det B| = |\det C| \cdot V(Q), \tag{6.7}$$

and $|\det C| \geq 1$ since C has integer matrix elements. Q.E.D.

In particular, $V(P)$ is an integral multiple of $V(Q)$.

Corollary 2.3. A primitive cell of G is a cell with minimum nonzero volume. The volumes of any two primitive cells are equal.

Corollary 2.4. If the vectors $\mathbf{a}_i$ in G are related to the basic vectors $\mathbf{b}_i$ by (6.5) then the $\mathbf{a}_i$ are basic vectors if and only if $\det C = \pm 1$.

2.7 Crystallographic Point Groups

Let H be a three-dimensional lattice group and consider the lattice L formed by the action of H on a given point $\mathbf{x} \in R_3$. For convenience we assume $\mathbf{x} = \boldsymbol{\theta}$. Since L is a (unbounded) point set in R_3, it has a complete symmetry group G. We will soon see that G is discrete. Clearly, H is a translation subgroup of G since the elements of H map L onto itself.

Suppose $\mathbf{t}$ is an element of $G \cap T(3)$, i.e., $\mathbf{t}$ is a translation in G. Then $\mathbf{t}\boldsymbol{\theta} = \mathbf{b}$ is a lattice point of L. If $\mathbf{b}_1, \mathbf{b}_2, \mathbf{b}_3$ are basic vectors for H there exist unique integers n_i such that

$$\mathbf{b} = n_1\mathbf{b}_1 + n_2\mathbf{b}_2 + n_3\mathbf{b}_3.$$

Since $\mathbf{t}$ is a translation it maps any $\mathbf{y} \in R_3$ onto $\mathbf{y} + n_1\mathbf{b}_1 + n_2\mathbf{b}_2 + n_3\mathbf{b}_3$. Thus, $\mathbf{t} \in H$ and $H = G \cap T(3)$.

Now if $\mathbf{g} \in G$ then $\mathbf{g}\boldsymbol{\theta} = \mathbf{b}$ is a lattice point of L. If $\mathbf{t} \in H$ is the lattice translation that maps $\boldsymbol{\theta}$ into $\mathbf{b}$ then $\mathbf{t}^{-1}\mathbf{g}\boldsymbol{\theta} = \boldsymbol{\theta}$, i.e., the transformation $\mathbf{f} = \mathbf{t}^{-1}\mathbf{g} \in G$ leaves the point $\boldsymbol{\theta}$ fixed. Thus, every element $\mathbf{g}$ of G can be written uniquely in the form $\mathbf{g} = \mathbf{tf}$, where $\mathbf{t} \in H$ and $\mathbf{f}$ leaves $\boldsymbol{\theta}$ fixed. Denoting by F the subgroup of G fixing $\boldsymbol{\theta}$, we see that $G = HF$ and G is the semidirect product of H and F. In particular the product of two elements $\mathbf{t}_1\mathbf{f}_1$ and $\mathbf{t}_2\mathbf{f}_2$ in G is given by

$$(\mathbf{t}_1\mathbf{f}_1)(\mathbf{t}_2\mathbf{f}_2) = \mathbf{t}_1(\mathbf{f}_1\mathbf{t}_2\mathbf{f}_1^{-1})(\mathbf{f}_1\mathbf{f}_2), \tag{7.1}$$

since $\mathbf{f}_1\mathbf{t}_2\mathbf{f}_1^{-1} \in G \cap T(3) = H$. Furthermore, the elements of G preserve distance and there are only a finite number of lattice points inside any sphere centered at $\boldsymbol{\theta}$, so F must be a finite point group. As a consequence, G is necessarily discrete. Any two three-dimensional lattice groups H_1, H_2 are clearly isomorphic. Thus to compute all complete symmetry groups of lattices up to isomorphism (hence to classify all lattices by symmetry type) it is enough to compute all possible point groups F. An arbitrary symmetry group of L, not necessarily the complete group of symmetries, is an arbitrary subgroup G' of G.

Definition. A subgroup of $E(3)$ which fixes a point $\mathbf{x}$ and maps a three-dimensional lattice L containing $\mathbf{x}$ into itself is called a **crystallographic point group.** The largest crystallographic point group F at $\mathbf{x}$ is called the **holohedry** of L at $\mathbf{x}$.

We have shown that a crystallographic point group is necessarily finite. Furthermore, if $\mathbf{x}$ and $\mathbf{y}$ are points contained in the same lattice L then the holohedries fixing $\mathbf{x}$ and $\mathbf{y}$, respectively, are conjugate subgroups of $E(3)$. The crystallographic point groups are just the subgroups of the holohedries.

Not all point groups are crystallographic point groups. The requirement that a point group leave a lattice invariant is a strong restriction on the elements of the group.

Theorem 2.8. (The crystallographic restriction). Let K be a crystallographic point group. If $\mathbf{g} \in K$ is a nontrivial rotation then $\mathbf{g}$ is of order two, three, four, or six. If $\mathbf{g} = \mathbf{Ik}$ is a rotation-inversion in K then the rotation $\mathbf{k}$ is of order one, two, three, four, or six.

Proof. Let $\mathbf{b}_1, \mathbf{b}_2, \mathbf{b}_3$ be basic vectors for the lattice L on which K acts. Writing

$$\mathbf{g}\mathbf{b}_i = \sum_{j=1}^{3} c_{ji}\mathbf{b}_j \tag{7.2}$$

we see that

$$C = (c_{ij})$$

is the matrix of the transformation **g** in the basis $\{\mathbf{b}_i\}$. Recall that the trace of a matrix is invariant under similarity transformations, i.e., the trace of **g** is independent of basis. Since the $\mathbf{b}_i$ are basic vectors of L it follows that the c_{ij} are integers, so tr C is an integer. However, from (1.10), we see that for an orthonormal basis with one basis vector along the axis of rotation, the trace is $\pm(1 + 2\cos\varphi)$, where φ is the rotation angle corresponding to **g**. The minus sign applies to rotation-inversions. Thus,

$$\text{tr}\, C = \pm(1 + 2\cos\varphi) \tag{7.3}$$

and the only way this can be an integer is for $\varphi = \pi/2, 3\pi/2, n\pi/3$, with $n = 0, 1, \ldots, 5$. (Note that necessarily $|\text{tr}\, C| \leq 3$.) Q.E.D.

This theorem shows that no point group which contains elements with rotational parts of order five or greater than six can be a crystallographic point group. It follows from our classification of point groups in Sections 2.4 and 2.5 that all but 32 point groups can be eliminated as candidates for crystallographic groups. The possible point groups of the first kind are the cyclic groups C_1, C_2, C_3, C_4, C_6, the dihedral groups D_2, D_3, D_4, D_6, the tetrahedral group T, and the octahedral group O. The possible point groups of the second kind are $S_2, S_4, S_6, C_{1h}, C_{2h}, C_{3h}, C_{4h}, C_{6h}, C_{2v}, C_{3v}, C_{4v}, C_{6v}, D_{2h}, D_{3h}, T_h, T_d, D_{4h}, D_{6h}, D_{2d}, D_{3d}$, and O_h. We will show that each member of this list is in fact a symmetry group of some lattice, and we will relate these groups to the study of crystal structure in physics.

We first classify the holohedries (or maximal crystallographic point groups) of lattices.

Definition. Two lattices L, L' are in the same **crystal system** if their holohedries F, F' are conjugate subgroups of $E(3)$.

We know that all lattices of a lattice group H lie in the same crystal system, so it also makes sense to speak of a classification of lattice groups into crystal systems.

The possible holohedries can be obtained from our list of the 32 possible crystallographic point groups. However, the following theorems show that there are at most seven holohedries. Let L be a lattice which for convenience we assume based at $\mathbf{x} = \boldsymbol{\theta}$ and let F be its holohedry at $\mathbf{x}$.

Theorem 2.9. The inversion **I** is an element of F.

Proof. If $\mathbf{b}_1, \mathbf{b}_2, \mathbf{b}_3$ are basic vectors for L, the lattice points of L are exactly the points

$$\mathbf{b} = n_1\mathbf{b}_1 + n_2\mathbf{b}_2 + n_3\mathbf{b}_3, \qquad n_1, n_2, n_3 \quad \text{integers.}$$

It follows from this representation that $-\mathbf{b} \in L$ whenever $\mathbf{b} \in L$. Thus $\mathbf{I} \in F$. Q.E.D.

We conclude that holohedries are necessarily point groups of the second kind containing $\mathbf{I}$.

Theorem 2.10. If F contains the cyclic subgroup C_n, $n = 3, 4, 6$, then F contains C_{nv}.

Proof. We have to show that if F contains an n-fold rotation axis l then it also contains a reflection plane P in which l lies. (The reflection and C_n generate C_{nv}.) Let $\mathbf{C} \in F$ be a rotation about l with rotation angle $2\pi/n$ and let Q be the plane through $\mathbf{x} = \mathbf{0}$ perpendicular to l. If $\mathbf{y}$ is a lattice point of L not on l then $\mathbf{Cy} - \mathbf{y}$ is a nonzero lattice point lying in Q. Therefore, $Q \cap L$ contains nonzero vectors. Let $\mathbf{b}_1$ be a nonzero vector of minimum length in $Q \cap L$. According to Theorem 2.6 and its corollary we can embed $\mathbf{b}_1$ in a system of basic vectors $\mathbf{b}_1, \mathbf{b}_2, \mathbf{b}_3$ for L such that $\mathbf{b}_2$ lies in Q. In fact we can set $\mathbf{b}_2 = \mathbf{Cb}_1$ (Fig 2.6), for if there is an $\mathbf{a} \in Q \cap L$ in the interior of the parallelogram generated by $\mathbf{b}_1$ and $\mathbf{Cb}_1$ then at least one of the lattice vectors $\mathbf{Cb}_1 - \mathbf{a}$, $\mathbf{Cb}_1 + \mathbf{b}_1 - \mathbf{a}$, $\mathbf{b}_1 - \mathbf{a}$, $\mathbf{a}$ is shorter than $\mathbf{b}_1$. This is impossible, so $\mathbf{b}_2 = \mathbf{Cb}_1$. All we know about $\mathbf{b}_3$ initially is that it is not in Q. We can write it uniquely in the form

$$\mathbf{b}_3 = \mathbf{u} + \mathbf{v} \tag{7.4}$$

where the vector $\mathbf{u}$ points along l and $\mathbf{v}$ lies in Q. (Here, $\mathbf{u}$ and $\mathbf{v}$ are just the projections of $\mathbf{b}_3$ on l and Q: they are not necessarily lattice vectors.) Since

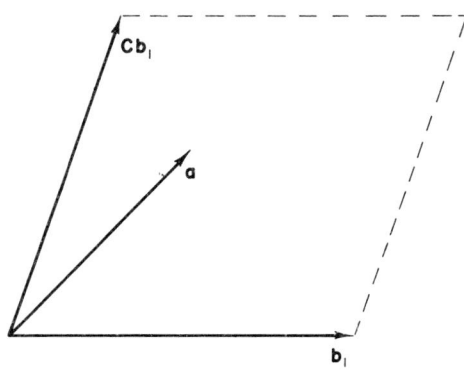

FIGURE 2.6

$\mathbf{Cb}_3 - \mathbf{b}_3 \in L \cap Q$ and $\mathbf{Cu} = \mathbf{u}$, there exist integers n_1, n_2 such that

$$\mathbf{Cb}_3 - \mathbf{b}_3 = \mathbf{Cv} - \mathbf{v} = n_1\mathbf{b}_1 + n_2\mathbf{Cb}_1. \tag{7.5}$$

Multiplying both sides of this expression by $\mathbf{C}^{-1}$ and then subtracting from (7.5) we get

$$\mathbf{Cv} + \mathbf{C}^{-1}\mathbf{v} - 2\mathbf{v} = n_2\mathbf{Cb}_1 - n_1\mathbf{C}^{-1}\mathbf{b}_1 + (n_1 - n_2)\mathbf{b}_1.$$

A little trigonometry yields (Fig. 2.7)

$$\mathbf{Cv} + \mathbf{C}^{-1}\mathbf{v} = 2\cos(2\pi/n)\mathbf{v},$$

$$\mathbf{Cb}_1 + \mathbf{C}^{-1}\mathbf{b}_1 = 2\cos(2\pi/n)\mathbf{b}_1.$$

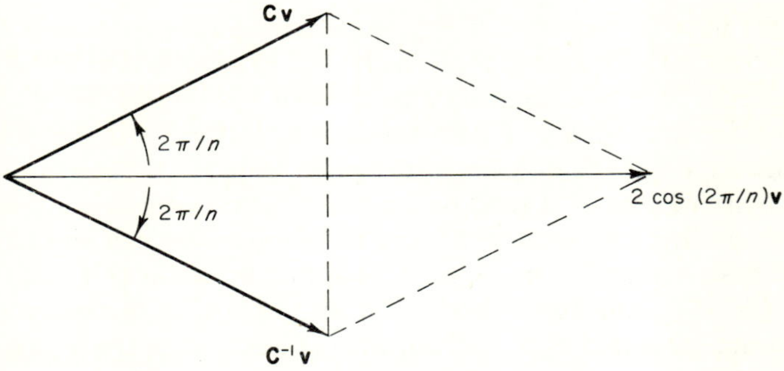

FIGURE 2.7

Thus,

$$2[\cos(2\pi/n) - 1]\mathbf{v} = (n_1 + n_2)\mathbf{b}_2 + [n_1 - n_2 - 2n\cos(2\pi/n)]\mathbf{b}_1. \tag{7.6}$$

Let $\boldsymbol{\sigma}$ be the reflection in the plane P containing l and perpendicular to $\mathbf{b}_1$. Clearly, $\boldsymbol{\sigma}\mathbf{b}_1 = -\mathbf{b}_1$, $\boldsymbol{\sigma}\mathbf{u} = \mathbf{u}$. If we can show that $\boldsymbol{\sigma}\mathbf{b}_2$ and $\boldsymbol{\sigma}\mathbf{b}_3$ are lattice points it will follow that $\boldsymbol{\sigma} \in F$.

For $n = 3$ we have $\boldsymbol{\sigma}\mathbf{b}_2 = \mathbf{b}_1 + \mathbf{b}_2 \in L$: for $n = 4$, $\boldsymbol{\sigma}\mathbf{b}_2 = \mathbf{b}_2 \in L$; and for $n = 6$, $\boldsymbol{\sigma}\mathbf{b}_2 = \mathbf{b}_2 - \mathbf{b}_1 \in L$. In each case, (7.4) and (7.6) yield

$$\boldsymbol{\sigma}\mathbf{b}_3 = \mathbf{u} + \boldsymbol{\sigma}\mathbf{v} = \mathbf{u} + \mathbf{v} + (n_1 - n_2)\mathbf{b}_1 = \mathbf{b}_3 + (n_1 - n_2)\mathbf{b}_1 \in L. \quad \text{Q.E.D.}$$

From our list of 32 possible crystallographic point groups only 7 satisfy the conditions imposed by Theorems 2.9 and 2.10. They are S_2, C_{2h}, D_{2h}, D_{3d}, D_{4h}, D_{6h}, and O_h. Thus there are at most 7 holohedries.

2.8 The Bravais Lattices

We shall verify that the seven groups S_2, C_{2h}, D_{2h}, D_{3d}, D_{4h}, D_{6h}, and O_h are holohedries by explicitly constructing all lattices L for which they are maximal crystallographic point groups. To avoid overly complicated calculations it is necessary to choose basic vectors for L in a convenient manner. For this, Theorem 2.6 and its corollary will prove useful.

Two lattices (or lattice groups) L_0, L_1 belonging to the same holohedry F are of the same **type** if one of them can be obtained from the other by a continuous lattice deformation L_t, $0 \leq t \leq 1$, in such a way that during the deformation process the holohedry F_t contains F. As we shall see, two lattices belonging to the same holohedry need not have isomorphic complete symmetry groups. However, two lattices belonging to the same type necessarily have isomorphic complete symmetry groups. The seven crystal systems will subdivide into 14 lattice types, the 14 Bravais lattices. Every lattice belongs to exactly one lattice type and this type determines the complete symmetry group of the lattice, up to isomorphism.

Our candidates for holohedries satisfy the subgroup relations given by Fig. 2.8. In particular, $S_2 = \{\mathbf{E}, \mathbf{I}\}$ is contained in all these groups. As we have seen in Theorem 2.9 every lattice admits S_2 as a crystallographic point group.

$$\begin{array}{ccccccccc} S_2 & \subset & C_{2h} & \subset & D_{2h} & \subset & D_{4h} & \subset & O_h \\ & & \cap & & \cap & & & & \\ & & D_{3d} & \subset & D_{6h} & & & & \end{array}$$

FIGURE 2.8

Except for S_2 these groups contain an n-fold rotation axis $l(n = 2, 4,$ or $6)$ and a reflection $\boldsymbol{\sigma}_h$ in the plane P through $\boldsymbol{\theta}$ perpendicular to this axis. Let $\mathbf{C}$ be a rotation through the angle $2\pi/n$ about l. We will determine the restrictions on a lattice L in order that it admit $\mathbf{C}$ and $\boldsymbol{\sigma}_h$ as symmetries. It is always possible to choose basic vectors $\mathbf{b}_1, \mathbf{b}_2, \mathbf{b}_3$ for L such that $\mathbf{b}_1$ and $\mathbf{b}_2$ lie in P. Indeed, if $\mathbf{a}_1$ and $\mathbf{a}_2$ are any two lattice points not lying in the same plane containing l then $\mathbf{a}_1 + \boldsymbol{\sigma}_h\mathbf{a}_1$ and $\mathbf{a}_2 + \boldsymbol{\sigma}_h\mathbf{a}_2$ are linearly independent lattice vectors in P. By Corollary 2.2 we can choose $\mathbf{b}_1$ and $\mathbf{b}_2$ in P. We write the third basic vector uniquely in the form

$$\mathbf{b}_3 = \mathbf{u} + \mathbf{v} \tag{8.1}$$

where $\mathbf{u}$ lies along l and $\mathbf{v}$ lies in P. Notice that the volume of the cell generated by $\mathbf{b}_1, \mathbf{b}_2, \mathbf{b}_3$ is the same as the volume of the cell generated by $\mathbf{b}_1, \mathbf{b}_2, \mathbf{b}_3'$, where

$$\mathbf{b}_3' = \mathbf{b}_3 + m_1\mathbf{b}_1 + m_2\mathbf{b}_2 \tag{8.2}$$

and m_1, m_2 are integers. Thus $\mathbf{b}_1, \mathbf{b}_2, \mathbf{b}_3'$ are also basic vectors. We will use this freedom to vary $\mathbf{b}_3$ in the computations to follow.

Since $\mathbf{Cu} = \mathbf{u}$ it follows that

$$\mathbf{Cb}_3 - \mathbf{b}_3 = \mathbf{Cv} - \mathbf{v} = n_1\mathbf{b}_1 + n_2\mathbf{b}_2 \in L \cap P. \tag{8.3}$$

Relation (8.3) will enable us to compute $\mathbf{v}$, hence to enumerate the possibilities for $\mathbf{b}_3$. This enumeration depends on the value of n. If $n = 2$ then $\mathbf{Cv} = -\mathbf{v}$.

Solving for $\mathbf{v}$ in (8.3) and substituting into (8.1), we get

(8.4) $$\mathbf{b}_3 = \mathbf{u} + \tfrac{1}{2}n_1\mathbf{b}_1 + \tfrac{1}{2}n_2\mathbf{b}_2.$$

Using the freedom of (8.2)in selecting $\mathbf{b}_3$, we can add arbitrary integer multiples of $\mathbf{b}_1$ and $\mathbf{b}_2$ to (8.4). Thus, we can choose $\mathbf{b}_3$ such that $n_1, n_2 = 0, 1$. There are four possibilities:

(8.5) $$\begin{aligned} &(1)\quad \mathbf{b}_3 = \mathbf{u}, &&(2)\quad \mathbf{b}_3 = \mathbf{u} + \tfrac{1}{2}\mathbf{b}_1,\\ &(3)\quad \mathbf{b}_3 = \mathbf{u} + \tfrac{1}{2}\mathbf{b}_2, &&(4)\quad \mathbf{b}_3 = \mathbf{u} + \tfrac{1}{2}\mathbf{b}_1 + \tfrac{1}{2}\mathbf{b}_2. \end{aligned}$$

Only in case (1) is $\mathbf{b}_3$ perpendicular to the plane P. Since $\mathbf{b}_1$ and $\mathbf{b}_2$ have not been uniquely specified, these four cases are not all distinct. Under an interchange of $\mathbf{b}_1$ and $\mathbf{b}_2$, (2) and (3) coincide. Furthermore, if $\mathbf{b}_1$ and $\mathbf{b}_2$ are replaced by the new basic vectors $\mathbf{b}_1$ and $\mathbf{b}_1 + \mathbf{b}_2$ then (3) (4) coincide. However, the same lattice cannot have a primitive cell of the form (1) and a primitive cell of the form (2), (3), or (4).

The cases $n = 4, 6$ follow from the proof of Theorem 2.10. It was shown that we can choose $\mathbf{b}_1$ as the shortest nonzero vector in $P \cap L$ and $\mathbf{b}_2 = \mathbf{C}\mathbf{b}_1$. Then the expression for $\mathbf{v}$ is given by (7.6). If $n = 4$ then

$$\mathbf{b}_3 = \mathbf{u} + \tfrac{1}{2}(n_2 - n_1)\mathbf{b}_1 - \tfrac{1}{2}(n_2 + n_1)\mathbf{b}_2.$$

Now $n_2 \pm n_1$ are simultaneously odd or even integers. Thus, addition of integer multiples of $\mathbf{b}_1$ and $\mathbf{b}_2$ reduces $\mathbf{b}_3$ to two normal forms:

(8.6) $$(1)\quad \mathbf{b}_3 = \mathbf{u}, \qquad (2)\quad \mathbf{b}_3 = \mathbf{u} + \tfrac{1}{2}\mathbf{b}_1 + \tfrac{1}{2}\mathbf{b}_2.$$

If $n = 6$, (7.6) yields

$$\mathbf{b}_3 = \mathbf{u} + n_2\mathbf{b}_1 - (n_1 + n_2)\mathbf{b}_2.$$

Addition of $(n_1 + n_2)\mathbf{b}_2 - n_2\mathbf{b}_1$ reduces this to

(8.7) $$\mathbf{b}_3 = \mathbf{u}.$$

Thus, $\mathbf{b}_3$ can always be chosen perpendicular to $\mathbf{b}_1$ and $\mathbf{b}_2 = \mathbf{C}\mathbf{b}_1$.

Armed with this information, we examine the holohedries one at a time and determine the lattice types which correspond to them.

The cubic holohedry O_h. Let l be one of the fourfold axes of the group O_h and consider a lattice L with primitive cell corresponding to the choice (1) of (8.6). Thus, the basic vectors are chosen so that (1) they are mutually perpendicular, (2) $\mathbf{b}_3$ lies on l, and (3) $\mathbf{b}_1$ and $\mathbf{b}_2$ have the same (minimal) length. Now O_h has four rotation axes through $\boldsymbol{\theta}$ evenly spaced in the plane P spanned by $\mathbf{b}_1$ and $\mathbf{b}_2$, so at least one of these axes l_1 lies between $\mathbf{b}_1$ and $\mathbf{b}_2$ (Fig. 2.9). The axis l_1 is at least twofold. Let $\mathbf{R}$ be the rotation through π about l_1. Then $\mathbf{R}\mathbf{b}_1$ and $\mathbf{R}\mathbf{b}_2$ must lie in $L \cap P$. However, $L \cap P$ is a square grid, so l_1 must be at an angle of $\pi/4$ with both $\mathbf{b}_1$ and $\mathbf{b}_2$. Now l_1 cannot

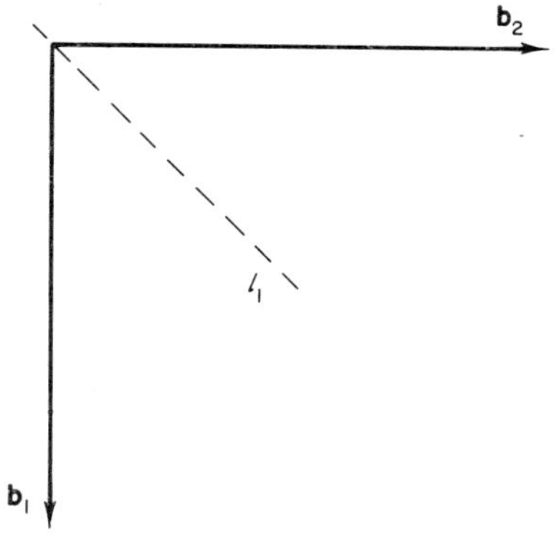

FIGURE 2.9

be a fourfold axis since a rotation of $\pi/2$ about l_1 maps $\mathbf{b}_1$ and $\mathbf{b}_2$ into points not in L. Thus, l_1 is a twofold axis and the two fourfold axes in P (which form an angle of $\pi/4$ with l_1) must lie along $\mathbf{b}_1$ and $\mathbf{b}_2$, respectively. A rotation of $\pi/2$ about the axis through $\mathbf{b}_2$ necessarily maps $\mathbf{b}_3$ onto $\mathbf{b}_1$ or $-\mathbf{b}_1$. Therefore, $\|\mathbf{b}_3\| = \|\mathbf{b}_1\|$ and the three basic vectors have the same length. (Fig. 2.10). The primitive cell is a cube Γ_c. It is now clear from the definition of O_h that a lattice of type Γ_c admits O_h as its holohedry. All such lattices can be designated by a single parameter, the length of one side of the primitive cell.

A second possibility is that the primitive cell of the lattice L about the fourfold axis l takes the form (2) in expression (8.6). Thus, (1) $\mathbf{b}_1 \perp \mathbf{b}_2$, (2) the basic vectors $\mathbf{b}_1$, $\mathbf{b}_2$ are perpendicular to l and have minimal length in the set of such lattice vectors, (3) $\|\mathbf{b}_1\| = \|\mathbf{b}_2\|$, and (4) $\mathbf{b}_3 = \mathbf{u} + \frac{1}{2}\mathbf{b}_1 + \frac{1}{2}\mathbf{b}_2$,

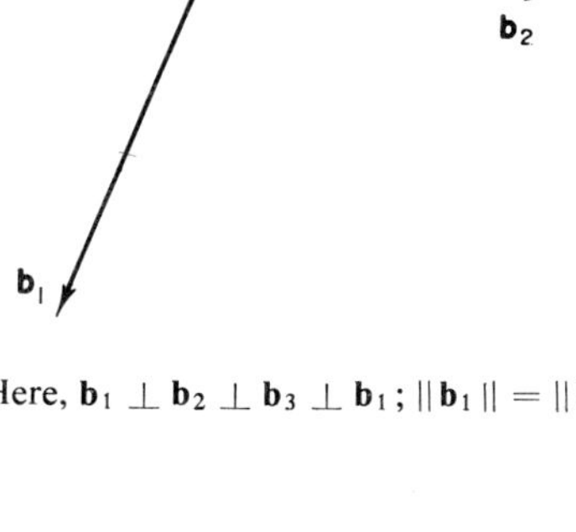

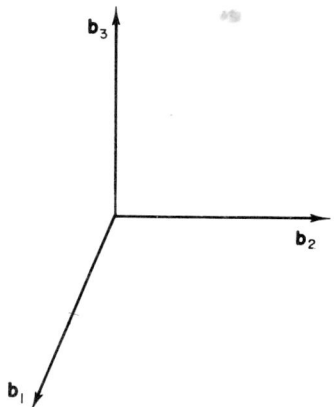

FIGURE 2.10. Here, $\mathbf{b}_1 \perp \mathbf{b}_2 \perp \mathbf{b}_3 \perp \mathbf{b}_1$; $\|\mathbf{b}_1\| = \|\mathbf{b}_2\| = \|\mathbf{b}_3\|$.

where $\mathbf{u}$ lies along l. It follows that the lattice vector $\mathbf{b}_4 = 2\mathbf{b}_3 - \mathbf{b}_1 - \mathbf{b}_2$ lies on l. Thus, $\mathbf{b}_1 \perp \mathbf{b}_2 \perp \mathbf{b}_4 \perp \mathbf{b}_1$. Just as above, we see that the twofold or fourfold axis l_1 perpendicular to l makes an angle $\pi/4$ with $\mathbf{b}_1$ and $\mathbf{b}_2$.

Suppose l_1 is a fourfold axis. From (4), $\mathbf{b}_3$ lies in the plane through l and l_1, so a rotation through the angle $\pi/2$ about l_1 must map $\mathbf{b}_3$ into either $\mathbf{b}_1$ or $\mathbf{b}_2$. We conclude that all of the basic vectors have the same length and $\|\mathbf{b}_4\| = \sqrt{2}\,\|\mathbf{b}_1\|$. The positions of the basic vectors $\mathbf{b}_1, \mathbf{b}_2, \mathbf{b}_3$ and the orientation of the axes of O_h have now been completely determined. In particular, the vectors $\mathbf{b}_1$ and $\mathbf{b}_2$ lie on twofold axes. We need only verify that the lattice L with these basic vectors actually admits O_h as a symmetry group. However, the symmetry properties of this primitive cell are not easy to visualize. It would be helpful if we could construct a cell all of whose vertices were on fourfold axes. This can be achieved by using the lattice vectors $\mathbf{b}_1 + \mathbf{b}_2$, $\mathbf{b}_1 - \mathbf{b}_2$, and $\mathbf{b}_4$ to generate a cube $\mathbf{\Gamma}_c^f$. (Note that $\mathbf{b}_1 + \mathbf{b}_2$ lies on l_1 and $\mathbf{b}_1 - \mathbf{b}_2$ lies on another fourfold axis.) The lattice points $\mathbf{b}_3$, $\mathbf{b}_3 - \mathbf{b}_2$, $\mathbf{b}_3 + \mathbf{b}_1$, and $\mathbf{b}_3 + \mathbf{b}_1 - \mathbf{b}_2$ are the midpoints of the four vertical faces of $\mathbf{\Gamma}_c^f$, $\mathbf{b}_1$ is the midpoint of the bottom face, and $\mathbf{b}_4 + \mathbf{b}_1$ is the midpoint of the top face. All of the remaining lattice points in $\mathbf{\Gamma}_c^f$ lie at the vertices. It is now easy to check that a lattice built from cells of type $\mathbf{\Gamma}_c^f$ admits O_h as a symmetry group. Lattices with type $\mathbf{\Gamma}_c^f$ symmetry are called **face-centered cubic**. The cell $\mathbf{\Gamma}_c^f$ is not primitive since its volume is four times that of a primitive cell. However, in practice the face-centered cubic cell is often preferable to a primitive cell since it exhibits O_h symmetry in a very explicit form. All lattices of type $\mathbf{\Gamma}_c^f$ can be described by a single parameter, the length of one side of the face-centered cube.

The only remaining possibility is that l_1 is a twofold axis. In this case fourfold axes lie along $\mathbf{b}_1$ and $\mathbf{b}_2$ and the lattice point $\mathbf{b}_4 = 2\mathbf{b}_3 - \mathbf{b}_1 - \mathbf{b}_2$ lies on l. It is easy to show that the lattice points on l consist of all integer multiples of a single lattice point $\mathbf{a}$. Here, $\mathbf{a} = m_1\mathbf{b}_1 + m_2\mathbf{b}_2 + m_3\mathbf{b}_3$ is characterized by the fact that $\pm\mathbf{a}$ are the lattice points on l closest to $\boldsymbol{\theta}$. The proof uses the Euclidean algorithm and is almost a copy of the proof of Lemma 2.2. Since the coefficients of $\mathbf{b}_4$ have no common integer divisor other than ± 1 it follows that $\mathbf{b}_4 = \pm\mathbf{a}$. Therefore, there is no lattice point on l between $\boldsymbol{\theta}$ and $\mathbf{b}_4$. As a consequence, a rotation through $\pi/2$ radians about the fourfold axis $\mathbf{b}_1$ must map $\mathbf{b}_4$ into either $\mathbf{b}_2$ or $-\mathbf{b}_2$. Hence, $\|\mathbf{b}_1\| = \|\mathbf{b}_2\| = \|\mathbf{b}_4\|$ and the positions of the basic vectors $\mathbf{b}_1, \mathbf{b}_2, \mathbf{b}_3$ are uniquely determined. Just as in the previous case, the symmetry properties of the primitive cell are not easy to visualize. To remedy this we consider the cube $\mathbf{\Gamma}_c^v$ generated by $\mathbf{b}_1, \mathbf{b}_2, \mathbf{b}_4$. The lattice point $\mathbf{b}_3$ lies at the center of $\mathbf{\Gamma}_c^v$ and the remaining lattice points in $\mathbf{\Gamma}_c^v$ lie at the vertices. It is now easy to check that a lattice constructed from type $-\mathbf{\Gamma}_c^v$ cells actually admits O_h as a symmetry group. Such lattices are called **body-centered cubic**. The volume of a $\mathbf{\Gamma}_c^v$ cell is twice that of a primitive cell. All lattices of type Γ_c^v can be described uniquely in terms of the length of one side of a body-centered cell.

We have shown that the crystal system with cubic holohedry O_h divides into three lattice types: primitive Γ_c, face-centered Γ_c^f, and body-centered Γ_c^v.

The hexagonal holohedry D_{6h}. Let l be the sixfold axis of D_{6h} and consider a lattice L admitting D_{6h} as a symmetry group. Since D_{6h} is not a proper subgroup of any possible holohedry, it must be the holohedry of L. According to (8.7) we can find vectors $\mathbf{b}_1, \mathbf{b}_2, \mathbf{b}_3$ for L such that (1) the angle between $\mathbf{b}_1$ and $\mathbf{b}_2$ is $\pi/3$, (2) $\mathbf{b}_1 \perp \mathbf{b}_3 \perp \mathbf{b}_2$, and (3) $\mathbf{b}_1$ and $\mathbf{b}_2$ have the same (minimal) length and lie in the plane P through $\boldsymbol{\theta}$ and perpendicular to l. Conversely, it is straightforward to show that a lattice L with basic vectors satisfying (1)–(3) actually admits D_{6h} as a symmetry group. The primitive cell just constructed is denoted Γ_h. Lattices of type Γ_h are uniquely determined by two parameters: $\|\mathbf{b}_1\|$ and $\|\mathbf{b}_3\|$.

The tetragonal holohedry D_{4h}. Let l be the fourfold axis of D_{4h} and L a lattice with primitive cell corresponding to choice (1) of (8.6). Then (1) $\mathbf{b}_1 \perp \mathbf{b}_2 \perp \mathbf{b}_3 \perp \mathbf{b}_1$ and (2) $\mathbf{b}_1$ and $\mathbf{b}_2$ have the same (minimal) length both lying in the plane P through $\boldsymbol{\theta}$ perpendicular to l. Conversely, it is easy to show that any lattice with primitive cell satisfying (1) and (2) admits D_{4h} as a symmetry group. In order that D_{4h} qualify as the holohedry of L it is necessary to require $\|\mathbf{b}_1\| \neq \|\mathbf{b}_3\|$. Otherwise L would have O_h as holohedry. The primitive cell just constructed is denoted Γ_q. Lattices of type Γ_q are determined by the two parameters $\|\mathbf{b}_1\|$ and $\|\mathbf{b}_3\|$.

Now suppose the primitive cell of L about the fourfold axis l takes the form (2) of expression (8.6). Then (1) $\mathbf{b}_1$ and $\mathbf{b}_2$ are vectors of minimal length in the plane P perpendicular to l and passing through $\boldsymbol{\theta}$, (2) $\mathbf{b}_1 \perp \mathbf{b}_2$, (3) $\|\mathbf{b}_1\| = \|\mathbf{b}_2\|$, and (4) $\mathbf{b}_3 = \mathbf{u} + \frac{1}{2}\mathbf{b}_1 + \frac{1}{2}\mathbf{b}_2$, where $\mathbf{u}$ lies on l. Clearly, the lattice vector $\mathbf{b}_4 = 2\mathbf{u} = 2\mathbf{b}_3 - \mathbf{b}_1 - \mathbf{b}_2$ also lies on l, and there is no lattice vector on l between $\mathbf{b}_4$ and $\boldsymbol{\theta}$. The vectors $\mathbf{b}_1, \mathbf{b}_2, \mathbf{b}_4$ are mutually orthogonal. The group D_{4h} has four twofold axes lying in P and it is obvious that one axis lies along $\mathbf{b}_1$, one axis lies along $\mathbf{b}_2$, and a third axis makes an angle of $\pi/4$ with $\mathbf{b}_1$ and $\mathbf{b}_2$. Let Γ_q^v be the rectangular parallelepiped (box) generated by $\mathbf{b}_1, \mathbf{b}_2, \mathbf{b}_4$. The lattice point $\mathbf{b}_3$ lies at the center of Γ_q^v but all other lattice points are located at the vertices. Clearly, a lattice constructed from Γ_q^v cells admits D_{4h} as a symmetry group. However, D_{4h} is the holohedry only if $\|\mathbf{b}_4\| \neq \|\mathbf{b}_1\|$, since otherwise the holohedry would be O_h. Note that Γ_q^v has twice the volume of a primitive cell. A lattice of type Γ_q^v is called **body-centered.** All lattices of type Γ_q^v can be described by the two parameters $\|\mathbf{b}_1\|$ and $\|\mathbf{b}_4\|$.

We have shown that the crystal system with tetragonal holohedry contains two lattice types: primitive Γ_q and body-centered Γ_q^v.

The rhombohedral holohedry D_{3d}. Let l be the threefold axis of D_{3d} and let P be the plane through $\boldsymbol{\theta}$ perpendicular to l. Suppose $\mathbf{C}$ is a rotation through the angle $2\pi/3$ about l. Then we can choose basic vectors $\mathbf{b}_1, \mathbf{b}_2, \mathbf{b}_3$ (Fig. 2.11) for L such that $\mathbf{b}_1$ has minimal nonzero length in $L \cap P$ and $\mathbf{b}_2 = \mathbf{C}\mathbf{b}_1$.

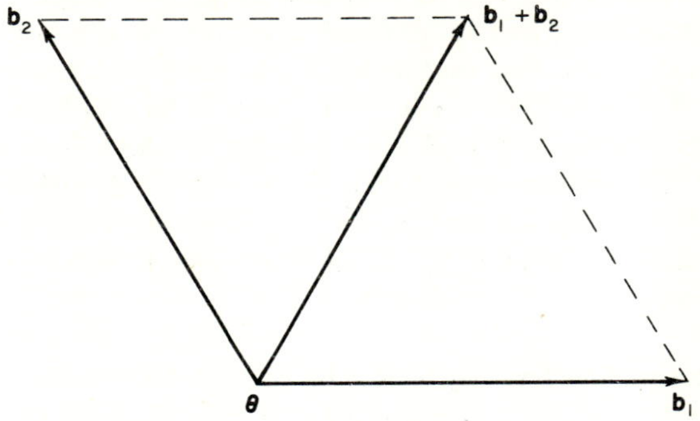

FIGURE 2.11

Write $\mathbf{b}_3 = \mathbf{u} + \mathbf{v}$, where $\mathbf{u}$ lies along l and $\mathbf{v}$ lies in P. Let $\mathbf{S} \in D_{3d}$ be the rotation through the angle $\pi/3$ about l followed by a reflection in P. Clearly, $\mathbf{S}\mathbf{u} = -\mathbf{u}$, $\mathbf{S}\mathbf{b}_1 = \mathbf{b}_1 + \mathbf{b}_2$, and $\mathbf{S}\mathbf{b}_2 = -\mathbf{b}_1$, In particular,

$$\mathbf{S}\mathbf{b}_3 + \mathbf{b}_3 = \mathbf{S}\mathbf{v} + \mathbf{v} = n_1\mathbf{b}_1 + n_2\mathbf{b}_2 \in L \cap P, \tag{8.8}$$

where n_1, n_2 are integers. Multiplying both sides of (8.8) by $\mathbf{S}^{-1}$, we obtain

$$\mathbf{v} + \mathbf{S}^{-1}\mathbf{v} = (n_2 - n_1)\mathbf{b}_1 + n_2\mathbf{b}_2. \tag{8.9}$$

Addition of (8.8) and (8.9) yields

$$\mathbf{S}\mathbf{v} + \mathbf{S}^{-1}\mathbf{v} + 2\mathbf{v} = (n_1 + n_2)\mathbf{b}_1 + (2n_2 - n_1)\mathbf{b}_2.$$

A computation exactly like that following (7.5) shows that the left-hand side is $3\mathbf{v}$. Thus,

$$\mathbf{b}_3 = \mathbf{u} + \tfrac{1}{3}(n_1 + n_2)\mathbf{b}_1 + \tfrac{1}{3}(2n_2 - n_1)\mathbf{b}_2. \tag{8.10}$$

As usual, we can subtract arbitrary integer multiples of $\mathbf{b}_1$ and $\mathbf{b}_2$ from $\mathbf{b}_3$ and still maintain a primitive cell. There are only three distinct possibilities:

(8.11)

(1) $\mathbf{b}_3 = \mathbf{u}$, (2) $\mathbf{b}_3 = \mathbf{u} + \tfrac{1}{3}(\mathbf{b}_1 - \mathbf{b}_2)$, (3) $\mathbf{b}_3 = \mathbf{u} - \tfrac{1}{3}(\mathbf{b}_1 - \mathbf{b}_2)$.

Case (1) is ruled out. For, if $\mathbf{b}_3 = \mathbf{u}$ it is easy to see that the primitive cell generated by $\mathbf{b}_1, \mathbf{b}_1 + \mathbf{b}_2, \mathbf{b}_3$ is just Γ_h, so L has holohedry D_{6h}. Furthermore, cases (2) and (3) are really the same since the lattice does not furnish us with an orientation with which to tell $\mathbf{b}_1$ and $\mathbf{b}_2$ apart. We choose (2) as our normal form. Note that the lattice point $3\mathbf{u} = 3\mathbf{b}_3 - \mathbf{b}_1 + \mathbf{b}_2$ lies on l and there are no lattice points on l between $3\mathbf{u}$ and $\boldsymbol{\theta}$. It is left to the reader to verify that the three twofold axes of D_{3d} must lie along $\mathbf{b}_1$, $\mathbf{b}_1 + \mathbf{b}_2$, and $\mathbf{b}_2$, respectively, and the three vertical reflection planes bisect the angles between adjacent twofold axes. The primitive cell generated by $\mathbf{b}_1, \mathbf{b}_2, \mathbf{b}_3$ is denoted $\boldsymbol{\Gamma}_{rh}$. It is now straightforward to check that a lattice of type $\boldsymbol{\Gamma}_{rh}$ does indeed have D_{3d} as its holohedry. All lattices of type $\boldsymbol{\Gamma}_{rh}$ are uniquely described by the two parameters $\|\mathbf{b}_1\|$ and $\|\mathbf{u}\|$.

The orthorhombic holohedry D_{2h}. Let l be a twofold axis of D_{2h}. Using Corollary 2.2, we choose a primitive cell for L such that $\mathbf{b}_1, \mathbf{b}_2$ lie in the plane P through $\boldsymbol{\theta}$, perpendicular to l. Assume the lattice corresponds to choice (1) of (8.5) and recall that D_{2h} has two perpendicular twofold axes l_1, l_2 in the plane P and two vertical reflection planes each containing l with one of the l_i. It is an elementary computation to show that there are only two possibilities: either (1) the l_i lie along $\mathbf{b}_1$ and $\mathbf{b}_2$, or (2) $\|\mathbf{b}_1\| = \|\mathbf{b}_2\|$ and the l_i are the two perpendicular bisectors of the angles between the vectors $\mathbf{b}_1, \mathbf{b}_2$ in P. In the first case the $\mathbf{b}_i$ are mutually orthogonal and generate a primitive right parallelepiped $\boldsymbol{\Gamma}_o$. In order that D_{2h} be the holohedry of a lattice constructed from type $\boldsymbol{\Gamma}_o$ cells it is necessary and sufficient that the lengths of no two sides of the primitive cell be equal. The type-$\boldsymbol{\Gamma}_o$ lattices are determined by three parameters.

Since $\|\mathbf{b}_1\| = \|\mathbf{b}_2\|$ in case (2), it is clear that the lattice vectors $\mathbf{b}_1 \pm \mathbf{b}_2$ and $\mathbf{b}_3$ generate a right parallelepiped $\boldsymbol{\Gamma}_o^b$. The lattice point $\mathbf{b}_1$ is the midpoint of the base and $\mathbf{b}_3 + \mathbf{b}_1$ is the midpoint of the top of $\boldsymbol{\Gamma}_o^b$. The only remaining lattice points in this cell are the vertices. For this reason $\boldsymbol{\Gamma}_o^b$ is called **base-centered.** The volume of $\boldsymbol{\Gamma}_o^b$ is twice the volume of a primitive cell. Clearly, a type-$\boldsymbol{\Gamma}_o^b$ lattice admits D_{2h} as a symmetry group, but D_{2h} is the holohedry of the lattice only if the angle between $\mathbf{b}_1$ and $\mathbf{b}_2$ is not $\pi/2$ or $\pi/3$. The type-$\boldsymbol{\Gamma}_o^b$ lattices are determined by three parameters: $\|\mathbf{b}_1\|$, $\|\mathbf{b}_3\|$, and the angle between $\mathbf{b}_1$ and $\mathbf{b}_2$.

As remarked in the discussion following expressions (8.5), the choices (2)–(4) for $\mathbf{b}_3$ are not distinct. For normalization purposes we choose $\mathbf{b}_3 = \mathbf{u} + \frac{1}{2}\mathbf{b}_1 + \frac{1}{2}\mathbf{b}_2$. Again there are two possibilities: either (1) the mutually orthogonal twofold axes l_1, l_2 in P lie along $\mathbf{b}_1$ and $\mathbf{b}_2$, or (2) the l_i are bisectors of the angles between $\mathbf{b}_1$, $\mathbf{b}_2$, and $\|\mathbf{b}_1\| = \|\mathbf{b}_2\|$. In case (1) we see that $\mathbf{b}_1$, $\mathbf{b}_2$, and $2\mathbf{u} = 2\mathbf{b}_3 - \mathbf{b}_1 - \mathbf{b}_2$ are mutually orthogonal lattice vectors which generate a right parallelepiped $\boldsymbol{\Gamma}_o^v$. The only lattice point in $\boldsymbol{\Gamma}_o^v$ other than the vertices is $\mathbf{b}_3$, the midpoint of $\boldsymbol{\Gamma}_o^v$. The cell $\boldsymbol{\Gamma}_o^v$ is called **body-centered.** It is now easy to check that a lattice constructed from type-$\boldsymbol{\Gamma}_o^v$ cells admits D_{2h} as a symmetry group. However, D_{2h} is a holohedry of such a lattice only if $\|\mathbf{b}_1\| \neq \|\mathbf{b}_2\|$, since otherwise D_{4h} would be a symmetry of the lattice. The volume of Γ_o^v is twice the volume of a primitive cell. The type-Γ_o^v lattices can be described by three parameters, the dimensions of the body-centered cell.

Corresponding to possibility (2) above we see that $\mathbf{b}_1 + \mathbf{b}_2, \mathbf{b}_1 - \mathbf{b}_2$, and $2\mathbf{u} = 2\mathbf{b}_3 - \mathbf{b}_1 - \mathbf{b}_2$ are mutually orthogonal and generate a right parallelepiped Γ_o^f. This cell clearly contains lattice points only at its vertices and at the midpoints of each of its six faces. For obvious reasons Γ_o^f is called **face-centered.** Type Γ_o^f lattices clearly admit D_{2h} as a symmetry group. However, D_{2h} is the holohedry of such lattices only if the base of Γ_o^f is not square. The volume of the face-centered cell is four times the volume of a primitive cell. The possible type Γ_o^f lattices are determined by three parameters, the dimensions of Γ_o^f.

The crystal system with holohedry D_{2h} thus contains four lattice types: primitive Γ_o, base-centered Γ_o^b, body-centered Γ_o^v, and face-centered Γ_o^f.

The monoclinic holohedry C_{2h}. Let l be the twofold axis of C_{2h}. We can choose basic vectors for the lattice L such that $\mathbf{b}_1, \mathbf{b}_2$ are vectors of minimal length in the plane P perpendicular to L and, by (8.5), we can assume that either $\mathbf{b}_3 = \mathbf{u}$ or $\mathbf{b}_3 = \mathbf{u} + \frac{1}{2}\mathbf{b}_2$. If $\mathbf{b}_3 = \mathbf{u}$ then $\mathbf{b}_3$ is perpendicular to $\mathbf{b}_1, \mathbf{b}_2$ and the basic vectors generate a primitive cell Γ_m. It is obvious that type Γ_m lattices admit C_{2h} as a symmetry group. However, C_{2h} is a holohedry only if the Γ_m cell does not coincide with $\Gamma_q, \Gamma_{rh}, \Gamma_h$, or Γ_c. The type-Γ_m lattices are determined by four parameters: $\|\mathbf{b}_1\|, \|\mathbf{b}_2\|, \|\mathbf{b}_3\|$, and the angle between $\mathbf{b}_1$ and $\mathbf{b}_2$.

If $\mathbf{b}_3 = \mathbf{u} + \frac{1}{2}\mathbf{b}_2$ then $\mathbf{b}_4 = 2\mathbf{b}_3 - \mathbf{b}_2$ lies on l. The lattice vectors $\mathbf{b}_1, \mathbf{b}_2, \mathbf{b}_4$ generate the **base-centered** cell Γ_m^b. This cell contains only the lattice point $\mathbf{b}_3$ as the center of one face and $\mathbf{b}_3 + \mathbf{b}_1$ as the center of the opposite face, in addition to the vertices. A type Γ_m^b lattice admits C_2^h as a symmetry group, but C_{2h} is the holohedry of such a lattice only if $\mathbf{b}_1$ is not perpendicular to $\mathbf{b}_2$ or $\mathbf{b}_1 + \mathbf{b}_2$. The type Γ_m^b lattices can be determined by four parameters: $\|\mathbf{b}_1\|, \|\mathbf{b}_2\|, \|\mathbf{b}_4\|$, and the angle between $\mathbf{b}_1$ and $\mathbf{b}_2$. (If we had chosen $\mathbf{b}_3$ in the form $\mathbf{b}_3 = \mathbf{u} + \frac{1}{2}\mathbf{b}_1 + \frac{1}{2}\mathbf{b}_2$ we would have been led naturally to a body-centered cell. The choice of Γ_m^b rather than a body-centered cell to designate this lattice type is a matter of custom rather than logical necessity.)

We have shown that the monoclinic crystal system contains two lattice types: primitive Γ_m and base-centered Γ_m^b.

The triclinic holohedry S_2. Since every lattice admits S_2 as a symmetry group, S_2 is the holohedry for all lattices which do not fall into one of the thirteen lattice types classified above. We can uniquely define basic vectors for such a lattice by requiring $\mathbf{b}_1, \mathbf{b}_2, \mathbf{b}_3$, not all lying in the same plane, to have minimal nonzero distance from $\boldsymbol{\theta}$. (A precise definition of the $\mathbf{b}_i$ to obtain uniqueness is a matter of taste.) The basic vectors generate a primitive lattice Γ_t. The lattices of type Γ_t can be designated by six parameters: the lengths of the three basic vectors and the three angles between pairs of basic vectors.

In conclusion, we have verified our list of 7 holohedries or crystal systems and have shown that there exist 14 lattice types (Bravais lattices).

The crystallographic point groups are just the possible subgroups of the 7 holohedries. It was shown in Section 2.7 that there are at most 32 such groups. Furthermore, it is easy to check that each of the 32 groups is a subgroup of at least one holohedry. Therefore, there are exactly 32 crystallographic point groups. Crystallographers say that there are 32 **crystal classes.**

TABLE 2.2 THE BRAVAIS LATTICES

Bravais lattice	Crystal classes	Basic vectors $\mathbf{b}_1, \mathbf{b}_2, \mathbf{b}_3$
1. Triclinic S_2	$S_2, C_1 = \{\mathbf{E}\}$	
Γ_t primitive		Arbitrary
2. Monoclinic C_{2h}	C_{2h}, C_2, C_{1h}	
Γ_m primitive		$\mathbf{b}_2 \perp \mathbf{b}_3 \perp \mathbf{b}_1$
$\Gamma_m{}^b$ base-centered		$\mathbf{b}_1 \perp \mathbf{b}_2 \perp (2\mathbf{b}_3 - \mathbf{b}_2) \perp \mathbf{b}_1$
3. Orthorhombic D_{2h}	D_{2h}, D_2, C_{2v}	
Γ_o primitive		$\mathbf{b}_1 \perp \mathbf{b}_2 \perp \mathbf{b}_3 \perp \mathbf{b}_1$
$\Gamma_o{}^b$ base-centered		$\mathbf{b}_3 \perp (\mathbf{b}_1 + \mathbf{b}_2) \perp (\mathbf{b}_1 - \mathbf{b}_3) \perp \mathbf{b}_3$, $\lVert\mathbf{b}_1\rVert = \lVert\mathbf{b}_2\rVert$
$\Gamma_o{}^v$ body-centered		$\mathbf{b}_1 \perp \mathbf{b}_2 \perp (2\mathbf{b}_3 - \mathbf{b}_1 - \mathbf{b}_2) \perp \mathbf{b}_1$, $\lVert\mathbf{b}_1\rVert = \lVert\mathbf{b}_2\rVert$
$\Gamma_o{}^f$ face-centered		$(\mathbf{b}_1 + \mathbf{b}_2) \perp (\mathbf{b}_1 - \mathbf{b}_2)$ $\perp (2\mathbf{b}_3 - \mathbf{b}_1 - \mathbf{b}_2) \perp (\mathbf{b}_1 + \mathbf{b}_2)$
4. Tetragonal D_{4h}	$D_{4h}, D_4, C_{4v}, C_{4h},$	
Γ_q primitive	C_4, D_{2d}, S_4	$\mathbf{b}_1 \perp \mathbf{b}_2 \perp \mathbf{b}_3 \perp \mathbf{b}_1, \lVert\mathbf{b}_1\rVert = \lVert\mathbf{b}_2\rVert$
$\Gamma_q{}^v$ body-centered		$\mathbf{b}_1 \perp (2\mathbf{b}_3 - \mathbf{b}_1 - \mathbf{b}_2) \perp \mathbf{b}_2 \perp \mathbf{b}_1$, $\lVert\mathbf{b}_1\rVert = \lVert\mathbf{b}_2\rVert$
5. Rhombohedral D_{3d}	$D_{3d}, D_3, C_{3v},$	
Γ_{rh} primitive	S_6, C_3	$\mathbf{b}_1 \perp (2\mathbf{b}_3 - \mathbf{b}_1 + \mathbf{b}_2) \perp \mathbf{b}_2$, $\lVert\mathbf{b}_1\rVert = \lVert\mathbf{b}_2\rVert, \angle\mathbf{b}_1\mathbf{b}_2 = 2\pi/3$
6. Hexagonal D_{6h}	$D_{6h}, C_{6h}, C_{6v},$	
Γ_h primitive	D_{3h}, C_{3h}, C_6, D_6	$\mathbf{b}_1 \perp \mathbf{b}_3 \perp \mathbf{b}_2, \lVert\mathbf{b}_1\rVert = \lVert\mathbf{b}_2\rVert$, $\angle\mathbf{b}_1\mathbf{b}_2 = \pi/3$
7. Cubic O_h	O_h, O, T_d, T_h, T	
Γ_c primitive		$\mathbf{b}_1 \perp \mathbf{b}_2 \perp \mathbf{b}_3 \perp \mathbf{b}_1$, $\lVert\mathbf{b}_1\rVert = \lVert\mathbf{b}_2\rVert = \lVert\mathbf{b}_3\rVert$
$\Gamma_c{}^v$ body-centered		$\mathbf{b}_1 \perp \mathbf{b}_2 \perp \mathbf{b}_4 \perp \mathbf{b}_1$, $\lVert\mathbf{b}_1\rVert = \lVert\mathbf{b}_2\rVert = \lVert\mathbf{b}_4\rVert$, $\mathbf{b}_4 = 2\mathbf{b}_3 - \mathbf{b}_1 - \mathbf{b}_2$
$\Gamma_c{}^f$ face-centered		$\mathbf{b}_1 \perp \mathbf{b}_2 \perp \mathbf{b}_4 \perp \mathbf{b}_1$, $\lVert\mathbf{b}_1\rVert = \lVert\mathbf{b}_2\rVert = (1/\sqrt{2})\lVert\mathbf{b}_4\rVert$, $\mathbf{b}_4 = 2\mathbf{b}_3 - \mathbf{b}_1$

A crystal class or point group K is said to belong to a crystal system with holohedry F if F is the smallest holohedry containing K. The distribution of crystal classes among the crystal systems is indicated in Table 2.2.

2.9 Crystal Structure

In this section the use of the term "crystal," introduced earlier, will be given a physical justification.

First we justify our restriction to discrete (even finite) point groups to describe the symmetries of a body S of finite extent. If we believe the atomic theory we can assume S is made up of n atoms. (For the purposes of this discussion think of an atom as a tiny billiard ball.) Every Euclidean symmetry of S must then induce a permutation of these n atoms. If the atoms do not all lie in a single plane then the symmetry $\mathbf{g}$ of S is uniquely determined by the permutation it induces. Since there are at most $n!$ permutations of the n atoms, the complete symmetry group G of S must be finite with order dividing $n!$.

If the atoms of S lie in a plane P but not along a single line then a symmetry $\mathbf{g}$ is determined by the permutation it induces, up to a possible reflection in P. The order of the complete symmetry group G of S is a divisor of $2n!$. The order of G is twice the order of the plane symmetry group obtained by considering S as a subset of R_2.

If the atoms of S lie on a single axis l then the symmetry group will no longer be discrete. Indeed, if S consists of a single atom the complete symmetry group is the orthogonal group $O(3)$. If S contains more than one atom strung along l then S admits the nondiscrete symmetry group $C_{\infty v}$ consisting of all rotations about l and reflections in all planes passing through l. If in addition S admits a reflection $\boldsymbol{\sigma}$ in a plane P perpendicular to l then $D_{\infty h} = C_{\infty v} \times \{\mathbf{E}, \boldsymbol{\sigma}\}$ is the complete symmetry group of S. If S does not admit a reflection $\boldsymbol{\sigma}$ then $C_{\infty v}$ is the complete symmetry group. This concludes our catalog of possible point groups.

Next we discuss the relationship between physical crystals and the lattice groups and crystallographic point groups introduced in the preceding sections. We first assume that the crystal occupies all space. (In this way we can avoid the consideration of the crystal boundary.) Roughly speaking, we can think of a crystal as formed by stacking together identical copies of a unit cell C so as to fill all space. The unit cell contains some given distribution of atoms, all atoms vibrating about equilibrium points in the crystal.

Two points $\mathbf{x}$ and $\mathbf{y}$ in a crystal are said to be **equivalent** if all of the (time-averaged) physical properties of the crystal are identical at $\mathbf{x}$ and $\mathbf{y}$. In other words $\mathbf{x}$ and $\mathbf{y}$ are equivalent points if there is no operational way of distinguishing between them. The crystal as viewed by an observer at $\mathbf{x}$ appears identical to that as viewed by an observer at $\mathbf{y}$. Only time-averaged properties are considered so as to avoid asymmetries due to the fluctuations of the atoms alone. (We can think of the atoms as frozen at their points of equilibrium.) It is clear that the point $\mathbf{x}$ is equivalent to at least all points of the form $\mathbf{x} + n_1\mathbf{b}_1 + n_2\mathbf{b}_2 + n_3\mathbf{b}_3$, where n_1, n_2, n_3 are arbitrary integers and $\mathbf{b}_1, \mathbf{b}_2, \mathbf{b}_3$ are the vectors generating C.

We define the complete symmetry group of a crystal as the subgroup of $E(3)$ consisting of those Euclidean motions that map each point $\mathbf{x}$ in the crystal onto a point $\mathbf{y}$ equivalent to $\mathbf{x}$. Clearly, in our model the lattice group

$$H = \{\mathbf{T_a} : \mathbf{a} = n_1\mathbf{b}_1 + n_2\mathbf{b}_2 + n_3\mathbf{b}_3\} \tag{9.1}$$

is a subgroup of the complete symmetry group. Now we forget about our hypothetical model and simply define an ideal crystal as a solid (filling all space) which satisfies the **lattice postulate:** There exist noncoplanar vectors $\mathbf{b}_1, \mathbf{b}_2, \mathbf{b}_3$ such that any point $\mathbf{x}$ is equivalent to all points

$$\mathbf{x} + n_1\mathbf{b}_1 + n_2\mathbf{b}_2 + n_3\mathbf{b}_3, \qquad n_1, n_2, n_3 \quad \text{integers.} \tag{9.2}$$

The lattice postulate amounts to the requirement that the lattice group H given by (9.1) is a symmetry group of the crystal. In addition we require that H is the maximal lattice symmetry group of the crystal, i.e., if $\mathbf{T}$ is a translational symmetry then $\mathbf{T} \in H$.

Let G be the complete symmetry group of a crystal. Clearly, H is a subgroup of G; in fact we shall see that it is a normal subgroup. G is called a **crystal group** or **space group.** A list of the possible isomorphism classes of space groups serves as a list of possible symmetry types of ideal crystals. Such a list is important because it serves as a classification scheme (something like the periodic table of the elements) into which all crystals can be fit. Furthermore, to a considerable extent the symmetry of a crystal serves to determine the physical propeties of the crystal.

Before proceeding to an analysis of the possible space groups G it is useful to clarify the meaning of some of the statements made above. First of all we shall be dealing exclusively with **ideal** crystals, i.e., crystals which satisfy the lattice postulate. The extent to which **real** crystals satisfy the lattice postulate is a matter for the physicist or crystallographer to determine. Although there is a great deal of evidence to support such a postulate, in the final analysis the validity of this postulate depends on experimental verification (and on the physicists definition of a crystal). We do not wish to claim that the possible crystal structures are determined *a priori* by group theory.

The H-orbit of every point $\mathbf{x}$ in a crystal is a Bravais lattice L, (9.2). However, it is not true that the crystal can be identified with L. In particular, the elements of the holohedry F of L may not be elements of the space group

G. To understand this, think of a model of a crystal formed by placing an identical pattern of atoms about each vertex in a lattice L. The translational symmetry is completely determined by the lattice. However, the total symmetry group G is also dependent on the pattern of atoms about each lattice point. [Some of the crystal symmetries may be of the form $\mathbf{g} = \{\mathbf{a}, \mathbf{O}\}$, where $\mathbf{O} \in \mathbf{O}(3)$ is not the identity and $\mathbf{a} \notin H$. Such symmetries are screw displacements or glide reflections.]

We examine the space group G of an ideal crystal in more detail. Every $\mathbf{g} \in G$ can be written uniquely in the form

$$\mathbf{g} = \{\mathbf{a}, \mathbf{O}\} = \mathbf{T_a O}, \tag{9.3}$$

where $\mathbf{O} \in O(3)$, the orthogonal group about $\boldsymbol{\theta}$. If $\mathbf{O} = \mathbf{E}$ then $\{\mathbf{a}, \mathbf{E}\} = \mathbf{T_a} \in H$, i.e., $\mathbf{a}$ can be written in the form (9.1). Since $T(3)$ is a normal subgroup of $E(3)$,

$$\mathbf{g T_a g}^{-1} \in T(3) \cap G = H, \tag{9.4}$$

so H is normal in G. Let K be the set of all $\mathbf{O} \in O(3)$ that occur in expression (9.3) as $\mathbf{g}$ runs over the elements of G. The relation

$$\mathbf{g}_1\mathbf{g}_2 = \{\mathbf{a}_1, \mathbf{O}_1\}\{\mathbf{a}_2, \mathbf{O}_2\} = \{\mathbf{a}_1 + \mathbf{O}_1\mathbf{a}_2, \mathbf{O}_1\mathbf{O}_2\} \tag{9.5}$$

for $\mathbf{g}_1, \mathbf{g}_2 \in G$ proves that K is a group. (However, K may not be a subgroup of G since $\mathbf{a}_1$ and $\mathbf{a}_2$ are not necessarily in the lattice group H.) Furthermore, the identity

$$\mathbf{g T_b g}^{-1} = \{\mathbf{Ob}, \mathbf{E}\} = \mathbf{T_{Ob}} \in H \tag{9.6}$$

valid for $\mathbf{b}$ in H and $\mathbf{g} \in G$ given by (9.3), shows that $\mathbf{O} \in K$ maps any lattice vector $\mathbf{b}$ in H into another lattice vector. This proves that K is one of the 32 crystallographic point groups. Thus, to each ideal crystal we can uniquely assign a crystallographic point group K.

The group K is isomorphic to the factor group G/H. Indeed, each (left or right) H-coset of G consists of all elements of the form $\{\mathbf{a}, \mathbf{O}\}$ for some fixed $\mathbf{O} \in K$. Furthermore, if $\mathbf{g}_1 = \{\mathbf{a}_1, \mathbf{O}\}$ and $\mathbf{g}_2 = \{\mathbf{a}_2, \mathbf{O}\}$ are in G then

$$\mathbf{g}_1\mathbf{g}_2^{-1} = \{\mathbf{a}_1 - \mathbf{a}_2, \mathbf{E}\} \in H, \tag{9.7}$$

so $\mathbf{g}_1$ and $\mathbf{g}_2$ are contained in the same coset. As a unique representative in each coset we can choose the element

$$\{\mathbf{a}', \mathbf{O}\}, \qquad \mathbf{a}' = \alpha_1\mathbf{b}_1 + \alpha_2\mathbf{b}_2 + \alpha_3\mathbf{b}_3, \tag{9.8}$$

where the α_i are real numbers such that $0 \leq \alpha_i < 1$. If $\alpha_1 = \alpha_2 = \alpha_3 = 0$ then $\mathbf{O} \in G \cap K$; otherwise $\mathbf{O} \notin G$. Note that the representative of the coset H is $\mathbf{E} = \{\boldsymbol{\theta}, \mathbf{E}\}$.

Remark. Since $K \cong G/H$ is a finite group and H is discrete it follows that G is discrete. Therefore, every space group is discrete.

In a real crystal the lengths of the basic vectors $\mathbf{b}_i$ are of the order of the distance between neighboring atoms of the crystal since unit cells contain comparatively few atoms. These lengths are so small that they cannot be detected by macroscopic observations. To the macroscopic observer the translational symmetry group of the crystal appears to be $T(3)$. That is, the physical properties of the crystal appear to be invariant under any translation. (Recall that we are neglecting boundary effects by assuming that the crystal occupies all space.) Similarly the vectors $\mathbf{a}'$ in (9.8) appear to the macroscopic observer to be $\boldsymbol{\theta}$, so the elements of $K \cong G/H$ appear to form the point symmetry group of the crystal. The complete symmetry group of the crystal appears to be the semidirect product of K and $T(3)$. Thus, as far as macroscopic observations are concerned, every crystal falls into exactly one of 32 possible crystal classes determined by the crystallographic point group K.

It is only when we carry out microscopic observations which can detect the existence of a primitive cell that the space group of the crystal becomes important. In particular, microscopic observations may show that K is *not* a symmetry group of the crystal.

Two crystals belonging to the same (macroscopic) crystal class may not have isomorphic space groups. In fact, it can be shown that the 32 crystal classes break up into 219 isomorphism classes of space groups. In the next section we shall indicate how these results are obtained.

The reader may be wondering why we classified point groups in conjugacy classes and then switched to the (cruder) cataloging of space groups in isomorphism classes. The reason is a practical one. Consider those space groups which are just lattice groups, i.e., $K = \{\mathbf{E}\}$. There are already a continuum number of conjugacy classes of lattice groups, so a listing of conjugacy classes of space groups is out of the question. On the other hand, all three-dimensional lattice groups form a single isomorphism class.

2.10 Space Groups

Definition. A **crystallographic space group (space group)** G is a discrete subgroup of $E(3)$ such that $H = G \cap T(3)$ is a three-dimensional lattice group.

According to Eq. (9.3)–(9.6) the space group G is a symmetry group of some ideal crystal. That is, the definition of space group given above is equivalent to the more intuitive definition of a symmetry group of an ideal crystal.

It is a tedious exercise to determine all isomorphism classes of space groups. Here we discuss only the basic ideas involved in such a classification.

Every space group G belongs to one of 14 lattice types and one of 32 crystal classes determined by the lattice group H and the crystallographic point group $K \cong G/H$. Recall from Table 2.2 that the lattice type severely restricts the possible crystal classes. In particular, a crystal class K is assigned to the lattice type with *smallest* holohedry F containing K. We shall examine the significance of this assignment shortly.

Given a lattice type H and crystal class K which leaves H invariant, we look for all space groups G with lattice subgroup H such that $G/H \cong K$. Clearly one such group is the semidirect product of K and H:

$$G = \{\mathbf{hk}: \mathbf{h} \in H, \quad \mathbf{k} \in K\} \tag{10.1}$$

$$(\mathbf{h}_1\mathbf{k}_1)(\mathbf{h}_2\mathbf{k}_2) = \mathbf{h}_1(\mathbf{k}_1\mathbf{h}_2\mathbf{k}_1^{-1})(\mathbf{k}_1\mathbf{k}_2) \in G. \tag{10.2}$$

In this case the point group K is actually a subgroup of G. Space groups of the form (10.1) are called **symmorphic groups.** There are 73 isomorphism classes of symmorphic groups.

Indeed, referring to Table 2.2, we see that the triclinic crystal system has one lattice type and two crystal classes, which yields a total of two symmorphic groups. The monoclinic system has three classes and two types, which yields six symmorphic groups, etc. Proceeding in this way, we get a total of 61 symmorphic groups. To find the remaining 12 groups, we must examine our procedure more carefully.

First we consider the reasoning behind the assignment of a crystal class K to the type with the smallest holohedry containing K. Suppose F and F' are holohedries such that $F \supset F' \supseteq K$. The reader can verify that (with one exception) a lattice type belonging to the holohedry F can always be changed to a lattice type belonging to F' by an arbitrarily small deformation of the basic lattice vectors. Thus, the semidirect product of F' and K must be isomorphic to the semidirect product of F and K, and we get no new isomorphism classes of symmorphic space groups by associating K with lattice types belonging to F. The single exception occurs in the case $F = D_{6h}$, $F' = D_{3d}$, where the lattice type $\boldsymbol{\Gamma}_h$ cannot be changed to type $\boldsymbol{\Gamma}_{rh}$ by an arbitrarily small deformation. In this case the five crystal classes belonging to the rhombohedral system can be combined with the hexagonal lattice group to yield five new isomorphism classes of symmorphic groups.

In certain cases the group K may act as a symmetry group of a lattice H in two physically distinct ways. This can occur if the smallest holohedry F containing K also contains another subgroup K' which is isomorphic but not conjugate to K. Then the semidirect product of K and H may not be isomorphic to the semidirect product of K' and H. For example the crystal class C_{2v} in the orthorhombic system can act on the base-centered cell $\boldsymbol{\Gamma}_o^b$ in two distinct ways depending on whether the twofold axis of C_{2v} is parallel or perpendicular to the lattice vector $\mathbf{b}_3$. Thus, there are two symmorphic space groups G of type $\boldsymbol{\Gamma}_o^b$ and crystal class C_{2v}. In the tetragonal system D_{2d} is the only crystal class without a fourfold axis. It is not difficult to show that D_{2d} can operate on each of the lattice types $\boldsymbol{\Gamma}_q$ and $\boldsymbol{\Gamma}_q^v$ in two distinct ways, yielding two more symmorphic groups. Finally each of the crystal classes C_{3v}, D_3, D_{3d}, and D_{3h} acts on the primitive hexagonal lattice $\mathbf{\Gamma}_h$ in two distinct ways. This yields four more groups, for a total of $61 + 5 + 1 + 2 + 4 = 73$ symmorphic groups.

The nonsymmorphic space groups G have the property that they contain elements of the form (9.8) with $\mathbf{a}' \neq \mathbf{\theta}$. In particular the crystal class K of G is *not* a subgroup of G. Let $\mathbf{O}_1, \ldots, \mathbf{O}_n$ be an enumeration of the elements of K. To find all space groups G with crystal class K and lattice group H it is enough to find all sets of vectors $\mathbf{a}_1', \ldots, \mathbf{a}_n'$ such that the product of any two group elements

$$\{\mathbf{a}_1', \mathbf{O}_1\}, \ldots, \quad \{\mathbf{a}_n', \mathbf{O}_n\} \tag{10.3}$$

can be written as the product of an element of H and some one of these group elements. [The $\mathbf{a}_i'$ are subject to the restriction (9.8). If all the $\mathbf{a}_i'$ are zero vectors then G is symmorphic. The nonsymmorphic space groups contain nontrivial screw displacements and glide reflections.] Once all such groups are determined it is necessary to sort them into isomorphism classes. It turns out that there are 146 isomorphism classes of nonsymmorphic space groups. This gives us a total of 219 space groups. (In particular there are only a finite number of space groups. For a proof see Burckhardt [1].)

To illustrate the methods used in this classification we will compute all space groups belonging to the crystal class C_4. Such groups must necessarily belong to the tetragonal crystal system, so the possible lattice types are primitive ($\mathbf{\Gamma}_q$) or body-centered ($\mathbf{\Gamma}_q^v$).

First we compute the groups of type $\mathbf{\Gamma}_q$. Recall that the basic vectors $\mathbf{b}_1$, $\mathbf{b}_2$, $\mathbf{b}_3$ of $\mathbf{\Gamma}_q$ are mutually orthogonal, $\|\mathbf{b}_1\| = \|\mathbf{b}_2\|$, and the fourfold axis l of C_4 lies along $\mathbf{b}_3$. Let $\mathbf{C}$ be a rotation through the angle $+\pi/2$ about l. According to the remarks following (10.3) we must find all possible triples of vectors $\mathbf{a}_1$, $\mathbf{a}_2$, $\mathbf{a}_3$ with

$$\mathbf{a}_i = \sum_{j=1}^{3} \alpha_{ji}\mathbf{b}_j, \qquad 0 \leq \alpha_{ji} < 1, \quad i = 1, 2, 3 \tag{10.4}$$

such that the product of any two of

$$\{\mathbf{\theta}, \mathbf{E}\}, \quad \{\mathbf{a}_1, \mathbf{C}\}, \quad \{\mathbf{a}_2, \mathbf{C}^2\}, \quad \{\mathbf{a}_3, \mathbf{C}^3\} \tag{10.5}$$

is equal to the product of an element of the lattice group H and some one of the elements (10.5). (In fact, $\{\mathbf{a}_1, \mathbf{C}\}$ and H generate the entire space group G.) Once the possible space groups have been determined they must be split into isomorphism classes.

The equation to be solved for the $\mathbf{a}_i$ reads

$$\{\mathbf{a}_i, \mathbf{C}^i\}\{\mathbf{a}_j, \mathbf{C}^j\} = \{\mathbf{h}_{ij}, \mathbf{E}\}\{\mathbf{a}_k, \mathbf{C}^{i+j}\}, \tag{10.6}$$

where $\mathbf{h}_{ij} \in H$, $k = i + j (\text{mod } 4)$, and $\mathbf{a}_0 = \mathbf{\theta}$. Carrying out the indicated multiplications in (10.6), we find

(10.7) $$\mathbf{a}_i + \mathbf{C}^i\mathbf{a}_j - \mathbf{a}_k \in H, \qquad 0 \leq i, j \leq 3.$$

Set $\mathbf{a}_i = \mathbf{u}_i + \mathbf{v}_i$, where $\mathbf{u}_i$ lies along l and $\mathbf{v}_i$ is perpendicular to l. Since $\mathbf{Cu}_i = \mathbf{u}_i$, Eq. (10.7) split into

(10.8a) $$\mathbf{u}_i + \mathbf{u}_j - \mathbf{u}_k \in H$$

(10.8b) $$\mathbf{v}_i + \mathbf{C}^i\mathbf{v}_j - \mathbf{v}_k \in H.$$

In particular expression (10.8a) is an integer multiple of $\mathbf{b}_3$ and (10.8b) is an integral linear combination of $\mathbf{b}_1$ and $\mathbf{b}_2$. Since $\{\mathbf{a}_1, \mathbf{C}\}^4 \in H$ we have $4\mathbf{u}_1 = n\mathbf{b}_3 \in H$. Therefore, there are four possibilities,

(10.9) $$\mathbf{u}_1 = \tfrac{1}{4}n\mathbf{b}_3, \qquad n = 0, 1, 2, 3,$$

and the rest of the $\mathbf{u}_i$ can be determined uniquely from $\mathbf{u}_1$ and (10.8a). Expressions (10.8b) are not so easy to solve. However, by making use of the fact that $\mathbf{C}^2\mathbf{v}_j = -\mathbf{v}_j$ we can check that these expressions are equivalent to

(10.10) $$\mathbf{v}_2 - \mathbf{v}_1 - \mathbf{Cv}_1 \in H, \qquad \mathbf{v}_3 - \mathbf{Cv}_1 \in H.$$

Expressions (10.8b) have an infinite number of solutions, since $\mathbf{v}_i$ can be chosen arbitrarily, in which case $\mathbf{v}_2$ and $\mathbf{v}_3$ are determined by (10.4) and (10.10).

Let G be the space group corresponding to a particular choice of $\mathbf{v}_1$. Note that for any $\mathbf{T}_\mathbf{a} \in T(3)$ the group $G' = \mathbf{T}_\mathbf{a}G\mathbf{T}_\mathbf{a}^{-1}$ is conjugate, hence isomorphic to G. Since $\mathbf{T}_\mathbf{a}\mathbf{h}\mathbf{T}_\mathbf{a}^{-1} = \mathbf{h}$ for $\mathbf{h} \in H$, G and G' have the same lattice groups. Moreover,

(10.11) $$\mathbf{T}_\mathbf{a}\{\mathbf{a}_1, \mathbf{C}\}\mathbf{T}_\mathbf{a}^{-1} = \{\mathbf{a}_1 - \mathbf{Ca} + \mathbf{a}, \mathbf{C}\} = \{\mathbf{a}_1', \mathbf{C}\}.$$

Writing $\mathbf{a}_1' = \mathbf{u}_1' + \mathbf{v}_1'$, we find

(10.12) $$\mathbf{u}_1' = \mathbf{u}_1, \qquad \mathbf{v}_1' = \mathbf{v}_1 - \mathbf{Ca} + \mathbf{a}.$$

An appropriate choice of $\mathbf{a}$ yields $\mathbf{v}_1' = \boldsymbol{\theta}$ [set $\mathbf{a} = -\tfrac{1}{2}(\mathbf{v}_1 + \mathbf{Cv}_1]$. Thus, G is conjugate to a space group with $\mathbf{v}_1 = \boldsymbol{\theta}$. Furthermore, (10.10) implies $\mathbf{v}_2 = \mathbf{v}_3 = \boldsymbol{\theta}$. We conclude that there are at most four isomorphism classes of space groups of type $\boldsymbol{\Gamma}_q$ and crystal class C_4. For each of these groups $C_4{}^j, j = 1, \ldots, 4$, we list the element $\{\mathbf{a}_1, \mathbf{C}\}$ which, together with H, generates the group:

(10.13)

$$\mathbf{C}_4{}^1\colon\ \{\boldsymbol{\theta}, \mathbf{C}\}, \qquad \mathbf{C}_4{}^2\colon\ \{\tfrac{1}{4}\mathbf{b}_3, \mathbf{C}\}, \qquad \mathbf{C}_4{}^3\colon\ \{\tfrac{1}{2}\mathbf{b}_3, \mathbf{C}\}, \qquad \mathbf{C}_4{}^4\colon\ \{\tfrac{3}{4}\mathbf{b}_3, \mathbf{C}\}.$$

The group $C_4{}^1$ is symmorphic, while the other groups contain nontrivial screw displacements. Instead of the generator we chose for $C_4{}^4$ we could have chosen $\{\tfrac{1}{4}\mathbf{b}_3, \mathbf{C}^{-1}\}$. This is called a left-handed screw displacement, just as the generator $\{\tfrac{1}{4}\mathbf{b}_3, \mathbf{C}\}$ of $C_4{}^2$ is called a right-handed screw displacement. It is clear that $C_4{}^2$ and $C_4{}^4$ differ only the winding-sense along their respective screw axes. In particular C_4^2 and C_4^4 are isomorphic groups. The isomorphism acts like the identity on H and takes $\{\frac{1}{4}\mathbf{b}_3, \mathbf{C}^{-1}\}$ into $\{\frac{1}{4}\mathbf{b}_3, \mathbf{C}\}$. Isomorphic pairs of space groups differing only in their winding-sense are called **enantiomorphic.** There are 11 pairs of enantiomorphic space groups. These pairs are distinguished by crystallographers because enantiomorphic crystals turn out to have physically very different properties. Thus, crystallographers recognize 230 space groups even through there are only 219 isomorphism classes (Burckhardt [1]).

We now pass to the determination of class C_4 space groups of type $\mathbf{\Gamma}_q^v$. Recall that the basic vectors $\mathbf{b}_1, \mathbf{b}_2, \mathbf{b}_3$ for a lattice of type $\mathbf{\Gamma}_q^v$ are chosen such that $\mathbf{b}_1$ and $\mathbf{b}_2$ are perpendicular to the fourfold axis l of C_4, $\mathbf{b}_1 \perp \mathbf{b}_2$, $||\mathbf{b}_1|| = ||\mathbf{b}_2||$, and $2\mathbf{b}_3 - \mathbf{b}_1 - \mathbf{b}_2$ lies along l. We consider the elements (10.4) and (10.5) now adapted to the lattice type $\mathbf{\Gamma}_q^v$. Then $\{\mathbf{a}_1, \mathbf{C}\}^4 = \{(\mathbf{E} + \mathbf{C} + \mathbf{C}^2 + \mathbf{C}^3)\mathbf{a}_1, \mathbf{E}\} \in H$. Furthermore, $\mathbf{Cb}_1 = \mathbf{b}_2$, $\mathbf{Cb}_2 = -\mathbf{b}_1$, $\mathbf{Cb}_3 = \mathbf{b}_3 - \mathbf{b}_1$. Thus, if $\mathbf{a}_1 = \alpha_1\mathbf{b}_1 + \alpha_2\mathbf{b}_2 + \alpha_3\mathbf{b}_3$ we find

$$(\mathbf{E} + \mathbf{C} + \mathbf{C}^2 + \mathbf{C}^3)\mathbf{a}_1 = 2\alpha_3(2\mathbf{b}_3 - \mathbf{b}_1 - \mathbf{b}_2) \in H.$$

The only possibilities are $\alpha_3 = 0$ or $\alpha_3 = \frac{1}{2}$. We can map G onto a conjugate space group $G' = \mathbf{T}_{\mathbf{a}}G\mathbf{T}_{\mathbf{a}}^{-1}$ such that the image $\{\mathbf{a}_1', \mathbf{C}\}$ of $\{\mathbf{a}_1, \mathbf{C}\}$ is given by

$$\mathbf{a}_1' = \mathbf{a}_1 - \mathbf{Ca} + \mathbf{a} = \alpha_1'\mathbf{b}_1 + \alpha_2'\mathbf{b}_2 + \alpha_3\mathbf{b}_3. \tag{10.14}$$

By choosing $\mathbf{a} \in R_3$ appropriately we can require that α_1' and α_2' take any desired values. [Note that $(\mathbf{a} - \mathbf{Ca}) \perp l$.] In the case $\alpha_3 = 0$ we choose $\alpha_1' = \alpha_2' = 0$ and obtain the symmorphic group

$$C_4^5\colon\quad \{\boldsymbol{\theta}, \mathbf{C}\}. \tag{10.15}$$

In the case $\alpha_3 = \frac{1}{2}$ it is most convenient to choose $\alpha_1' = \alpha_2' = -\frac{1}{4}$ so $\mathbf{a}_1' = \frac{1}{4}\mathbf{b}_4$, where $\mathbf{b}_4 = 2\mathbf{b}_3 - \mathbf{b}_1 - \mathbf{b}_2$. It is trivial to verify that $\{\frac{1}{4}\mathbf{b}_4, \mathbf{C}\}$ and H generate the space group

$$C_4^6\colon\quad \{\tfrac{1}{4}\mathbf{b}_4, \mathbf{C}\}. \tag{10.16}$$

We have shown that there are six space groups of crystal class C_4 (five isomorphism classes). The techniques used to derive these results are typical of those used to obtain the complete list of space groups.

Problems

2.1 Using only the definition (2.1), show that an isometry $\mathbf{T}$ maps a line segment with endpoints $\mathbf{x}_1, \mathbf{x}_2$ onto a line segment with endpoints $\mathbf{Tx}_1, \mathbf{Tx}_2$. Then show that $\mathbf{T}$ maps planes onto planes.

2.2 Prove: An isometry which fixes four noncoplanar points is the identity transformation.

2.3 Compute all point groups in two-dimensional space. (Answer: $C_n, D_n, n = 1, 2, \ldots.$)

2.4 Show that there are exactly two space groups belonging to crystal class C_{6h}.

2.5 Show that there are exactly four space groups belonging to crystal class D_{3h}.

The next three problems concern two-dimensional lattices and space groups. They are much simpler than the corresponding problems in three-space.

2.6 Prove that there are ten crystallographic point groups in the plane.

2.7 Verify that there are four holohedries C_2, D_2, D_4, and D_6 and six two-dimensional lattice types.

2.8 Verify the existence of exactly 13 isomorphism classes of symmorphic two-dimensional space groups and 4 nonsymmorphic groups.

2.9 Let $O'(3) = O(3) \times \{\mathbf{E}, \mathbf{R}\}$, where $\mathbf{R}^2 = \mathbf{E}$ and $\mathbf{E}$ is the identity operator. A **magnetic symmetry group (color group)** is a finite subgroup G of $O'(3)$ such that $\mathbf{O} \in O(3)$ satisfies the crystallographic restriction for each $g = \mathbf{O} \times \mathbf{E}$ or $g = \mathbf{O} \times \mathbf{R}$ in G. Using the proof of Theorem 2.5, show that the magnetic symmetry groups G fall into three classes: (1) the 32 crystallographic point groups, i.e., $G \subset O(3)$; (2) the 32 groups $K \times \{\mathbf{E}, \mathbf{R}\}$, where K is a crystallographic point group, i.e., $\mathbf{R} \in G$; (3) groups G such that $\mathbf{R} \notin G$ but $G \not\subset O(3)$. Show how to determine (in principle) all 58 groups in class (3). What is the physical significance of these groups? (See Hamermesh [1].)

2.10 Explain the rationale behind the Schöenflies notation for point groups, especially the meaning of the subscripts d, h, and n.

2.11 Prove Corollaries 2.3 and 2.4.

Chapter 3

Group Representation Theory

3.1 A Group Representation

Let V be a vector space, real or complex, and denote by $GL(V)$ the group of all nonsingular linear transformations of V onto itself.

Definition. A **representation (rep)** of a group G with **representation space** V is a homomorphism $\mathbf{T}$: $g \longrightarrow \mathbf{T}(g)$ of G into $GL(V)$. The **dimension** of the representation is the dimension of V.

As a consequence of this definition we have the following:

(1.1)

$$\mathbf{T}(g_1)\mathbf{T}(g_2) = \mathbf{T}(g_1 g_2), \quad \mathbf{T}(g)^{-1} = \mathbf{T}(g^{-1}), \quad \mathbf{T}(e) = \mathbf{E}, \quad g_1, g_2, g \in G,$$

where $\mathbf{E}$ is the identity operator on V. Unless otherwise specified, only finite-dimensional reps of finite groups will be studied in the present chapter. This finiteness restriction will be lifted later. It will also be assumed unless otherwise mentioned that V is defined over the complex field $\mathfrak{C}$.

Definition. An ***n*-dimensional matrix rep** of G is a homomorphism T: $g \longrightarrow T(g)$ of G into $GL(n, \mathfrak{C})$ [or $GL(n, R)$].

The $n \times n$ matrices $T(g)$, $g \in G$, satisfy multiplication properties analogous to (1.1). Any group rep $\mathbf{T}$ of G with rep space V defines many matrix reps. For, if $\{\mathbf{v}_1, \ldots, \mathbf{v}_n\}$ is a basis of V, the matrices $T(g) = (T(g)_{kj})$ defined by

$$\mathbf{T}(g)\mathbf{v}_k = \sum_{j=1}^{n} T(g)_{jk}\mathbf{v}_j, \qquad 1 \le k \le n. \tag{1.2}$$

form an n-dimensional matrix rep of G. Every choice of a basis for V yields a new matrix rep of G defined by **T**. However, any two such matrix reps T, T' are equivalent in the sense that there exists a matrix $S \in GL(n, \mathfrak{C})$ such that

$$T'(g) = ST(g)S^{-1} \tag{1.3}$$

for all $g \in G$. In fact if T, T' correspond to the bases $\{\mathbf{v}_i\}$, $\{\mathbf{v}_i'\}$ respectively, then for S we can take the matrix (S_{ji}) defined by

$$\mathbf{v}_i = \sum_{j=1}^{n} S_{ji}\mathbf{v}_j', \qquad i = 1, \ldots, n. \tag{1.4}$$

Definition. Two complex n-dimensional matrix reps T and T' are **equivalent** ($T \cong T'$) if there exists an $S \in GL(n, \mathfrak{C})$ such that (1.3) holds.

Equivalent matrix reps can be viewed as arising from the same operator rep.

Conversely, given an n-dimensional matrix rep $T(g)$ we can define many n-dimensional operator reps of G. If V is an n-dimensional vector space with basis $\{\mathbf{v}_i\}$ we can define the group rep **T** by expression (1.2), i.e., we define the operator $\mathbf{T}(g)$ by the right-hand side of (1.2). Every choice of a vector space V and a basis $\{\mathbf{v}_i\}$ for V yields a new operator rep defined by T. However, if V, V' are two such n-dimensional vector spaces with bases $\{\mathbf{v}_i\}$, $\{\mathbf{v}_i'\}$ respectively, then the reps **T** and **T**$'$ are related by

$$\mathbf{T}'(g) = \mathbf{S}\mathbf{T}(g)\mathbf{S}^{-1}, \tag{1.5}$$

where **S** is an invertible operator from V onto V' defined by

$$\mathbf{S}\mathbf{v}_i = \mathbf{v}_i', \qquad 1 \le i \le n.$$

Definition. Two n-dimensional group reps **T**, **T**$'$ of G on the spaces V, V' are **equivalent** ($\mathbf{T} \cong \mathbf{T}'$) if there exists an invertible linear transformation **S** of V onto V' such that expression (1.5) holds.

The reader can easily check that equivalent operator reps correspond to equivalent matrix reps, i.e., there is a 1–1 correspondence between classes of equivalent operator reps and classes of equivalent matrix reps. (Note: The above definitions can be modified in an obvious manner to yield definitions of equivalence classes of **real** operator and matrix reps and to establish their 1–1 correspondence.)

In order to determine all possible reps of a group G it is enough to find one rep **T** in each equivalence class. The remaining reps **T**′ in each class are given by (1.5), where **S** runs over all invertible operators from V to V' and V' runs over all n-dimensional vector spaces. It is a matter of choice whether we study operator reps or matrix reps. For theoretical purposes the operator reps are usually more convenient, while matrix reps are more useful for computations.

Most applications of groups to the physical sciences occur via representation theory. Group reps appear naturally in the study of physical problems with inherent symmetry and analysis of the reps aids the solution of these problems. We present some examples of group reps.

Example 1. The matrix groups $GL(n, \mathfrak{C})$, $SL(n, \mathfrak{C})$, $O(n)$, etc. are n-dimensional matrix reps of themselves.

Example 2. Any group of operators on a vector space is a rep of itself. In particular, the point groups considered as linear operators on the vector space R_3 define three-dimensional reps of themselves.

Example 3. Let G be a group of order n. We formally define an n-dimentional vector space R_G consisting of all elements of the form

$$\sum_{g \in G} x(g) \cdot g, \qquad x(g) \in \mathfrak{C}. \tag{1.6}$$

Two vectors $\sum x(g) \cdot g$ and $\sum y(g) \cdot g$ are equal if and only if $x(g) = y(g)$ for all $g \in G$. The sum of two vectors and the scalar multiple of a vector are defined by

$$\begin{aligned} \sum x(g) \cdot g + \sum y(g) \cdot g &= \sum [x(g) + y(g)] \cdot g \\ \alpha \sum x(g) \cdot g &= \sum \alpha x(g) \cdot g \end{aligned} \tag{1.7}$$

The zero vector of R_G is $\theta = \sum 0 \cdot g$. Furthermore, the vectors $1 \cdot g$, $g \in G$, form a natural basis for R_G. (From now on we write $1 \cdot g = g \in R_G$.) We define the **product** of two elements $x = \sum x(g) \cdot g$, $y = \sum y(h) \cdot h$ in a natural manner:

$$\begin{aligned} xy &= \left(\sum x(g) \cdot g\right) \sum y(h) \cdot h = \sum_{g, h \in G} x(g) y(h) \cdot gh \\ &= \sum_{k \in G} xy(k) \cdot k, \end{aligned} \tag{1.8}$$

where

$$xy(g) = \sum_{h \in G} x(h) y(h^{-1} g).$$

It is easy to verify the following relations:

(1.9)

$$\begin{aligned} &(x + y)z = xz + yz, \qquad x(y + z) = xy + xz, \qquad x, y, z \in R_G, \\ &(xy)z = x(yz), \qquad \alpha(xy) = (\alpha x)y = x(\alpha y), \qquad ex = xe = x, \qquad \alpha \in \mathfrak{C}, \end{aligned}$$

where e is the identity element of G. Thus, R_G is an algebra, called the **group algebra** or **group ring** of G. The mapping $\mathbf{L}$ of G into $GL(R_G)$ given by

$$\mathbf{L}(g)x = gx, \qquad x \in R_G, \tag{1.10}$$

defines an n-dimensional rep of G, the **(left) regular** rep. In fact,

$$\mathbf{L}(g_1 g_2)x = g_1 g_2 x = \mathbf{L}(g_1)g_2 x = \mathbf{L}(g_1)\mathbf{L}(g_2)x$$

and the $\mathbf{L}(g)$ are linear operators.

This example provides us with a rep of any finite group, and is of great importance for theoretical purposes. Another natural rep of G on R_G is the **(right) regular** rep defined by

$$\mathbf{R}(g)x = xg^{-1}, \qquad x \in R_G, \qquad g \in G. \tag{1.11}$$

[Check that g^{-1} is needed on the right-hand side of (1.11) to make $\mathbf{R}$ a rep.]

Example 4. Consider the Helmholtz equation

$$\Delta u(\mathbf{x}) + k^2 u(\mathbf{x}) = 0, \tag{1.12}$$

where $\mathbf{x} = (x_1, x_2, x_3) \in R_3$, $k \geq 0$, and

$$\Delta = \sum_{i=1}^{3} \frac{\partial^2}{\partial x_i{}^2}.$$

The set of all solutions $u(\mathbf{x})$ of (1.12) (defined for all $\mathbf{x} \in R_3$) forms an infinite-dimensional vector space V_k. In particular, any finite linear combination of solutions of (1.12) is a solution. We show that the operators $\mathbf{T}(g)$, $g \in E(3)$, given by

$$[\mathbf{T}(g)u](\mathbf{x}) = u(\mathbf{g}^{-1}\mathbf{x}), \tag{1.13}$$

where

$$\mathbf{gx} = \{\mathbf{a}, \mathbf{O}\}\mathbf{x} = \mathbf{Ox} + \mathbf{a}, \qquad \mathbf{O} \in O(3), \qquad \mathbf{a} \in T(3),$$

define an (infinite-dimensional) rep of $E(3)$ on V_k. The homomorphism property follows from

$$[\mathbf{T}(g_1 g_2)u](\mathbf{x}) = u(\mathbf{g}_2^{-1}\mathbf{g}_1^{-1}\mathbf{x}) = [\mathbf{T}(g_2)u](\mathbf{g}_1^{-1}\mathbf{x}) = [\mathbf{T}(g_1)\mathbf{T}(g_2)u](\mathbf{x}).$$

Note that $\mathbf{T}(g)\mathbf{u}$ is a function whose value at $\mathbf{x}$ is the value of u at $\mathbf{g}^{-1}\mathbf{x}$. The reader should check that use of $\mathbf{gx}$ on the right-hand side of (1.13) would *not* lead to a homomorphism.

In order to prove our assertion we must show that V_k is invariant under the operators $\mathbf{T}(g)$. Write $\mathbf{x}' = \mathbf{g}^{-1}\mathbf{x}$. Since

$$\mathbf{g}^{-1} = \{-\mathbf{O}^{-1}\mathbf{a}, \mathbf{O}^{-1}\}, \qquad \mathbf{O} \in O(3)$$

a simple computation gives

$$\frac{\partial}{\partial x_l} = \sum_{i=1}^{3} O_{li} \frac{\partial}{\partial x_i{}'}.$$

Therefore,

$$\Delta = \sum_{l,i,j=1}^{3} O_{li}O_{lj} \frac{\partial^2}{\partial x_i' \, \partial x_j'} = \sum_{i=1}^{3} \frac{\partial^2}{\partial (x_i')^2} = \Delta' \tag{1.14}$$

since O is an orthogonal matrix. If $u \in V_k$ then

$$\Delta[\mathbf{T}(g)u](\mathbf{x}) = \Delta[u(\mathbf{x}')] = \Delta' u(\mathbf{x}') = -k^2 u(x') = -k^2[\mathbf{T}(g)u](\mathbf{x}),$$

so $\mathbf{T}(g)u \in V_k$. The existence of this group rep has important consequences in the study of the solutions of the Helmholtz equation. These consequences will be explored in Chapter 8.

Example 5. The square integrable solutions $\Psi(\mathbf{x})$ of the Schrödinger equation

$$-\frac{\hbar^2}{2m} \Delta\Psi(\mathbf{x}) + V(\mathbf{x})\Psi(\mathbf{x}) = E\Psi(\mathbf{x}) \tag{1.15}$$

describing a particle of mass m and energy E subject to the potential field $V(\mathbf{x})$, form a vector space W_E. Suppose $V(\mathbf{x})$ is invariant under the action of some subgroup G of $O(3)$:

$$V(\mathbf{gx}) = V(\mathbf{x}), \qquad g \in G.$$

[For example, if $V(\mathbf{x})$ has rotational symmetry, G may be $O(3)$. Another possibility is a point group.] Then the operators

$$[\mathbf{T}(g)\Psi](\mathbf{x}) = \Psi(\mathbf{g}^{-1}\mathbf{x}), \qquad g \in G,$$

satisfy the homomorphism property and map solutions of (1.15) into other solutions. Furthermore, for $\Psi \in W_E$

$$\int_{R_3} |\Psi(\mathbf{g}^{-1}\mathbf{x})|^2 \, d^3\mathbf{x} = \int_{R_3} |\Psi(\mathbf{x})|^2 \, d^3\mathbf{x} < \infty, \qquad d^3\mathbf{x} = dx_1 \, dx_2 \, dx_3, \tag{1.16}$$

since the Jacobian of the coordinate transformation is $+1$. Therefore, $\mathbf{T}(g)\Psi \in W_E$ and the operators $\mathbf{T}(g)$ define a length-preserving rep of G on W_E, where the inner product $\langle -, - \rangle$ is given by

$$\langle \Psi_1, \Psi_2 \rangle = \int_{R_3} \Psi_1(\mathbf{x})\overline{\Psi_2(\mathbf{x})} \, d^3\mathbf{x}.$$

It is easy to verify that

$$\langle \mathbf{T}(g)\Psi_1, \mathbf{T}(g)\Psi_2 \rangle = \langle \Psi_1, \Psi_2 \rangle, \qquad g \in G.$$

Thus, the operators $\mathbf{T}(g)$ are **unitary** with respect to $\langle -, - \rangle$ and they define a **unitary rep** of G on W_E. In most quantum mechanical problems the eigenspaces W_E are zero-dimensional except for a countable number of values E_n (the bound-state energy levels) where they have finite nonzero dimension. We shall show later that a knowledge of the symmetry group of Schrödinger's equation furnishes us with important information about the eigenspaces W_{E_n} even in cases where (1.15) cannot be explicitly solved.

Let **T** be a rep of the finite group G on a finite-dimensional inner product space V. The rep **T** is said to be **unitary** if for all $g \in G$

$$\langle \mathbf{T}(g)\mathbf{v}, \mathbf{T}(g)\mathbf{w}\rangle = \langle \mathbf{v}, \mathbf{w}\rangle, \qquad \mathbf{v}, \mathbf{w} \in V, \tag{1.17}$$

i.e., if the operators $\mathbf{T}(g)$ are unitary. Recall that an **orthonormal (ON) basis** for the n-dimensional space V is a basis $\{\mathbf{v}_1, \ldots, \mathbf{v}_n\}$ such that $\langle \mathbf{v}_i, \mathbf{v}_j\rangle = \delta_{ij}$, where $\langle -, -\rangle$ is the inner product on V. The matrices $T(g)$ of the operators $\mathbf{T}(g)$ with respect to an ON basis $\{\mathbf{v}_i\}$ are unitary matrices

$$\overline{T(g)}_{ji} = T(g^{-1})_{ij} = [T(g)^{-1}]_{ij}.$$

Hence, they form a unitary matrix rep of G. Unitary operator and matrix reps have useful properties which make them desirable in both theoretical and computational problems. The following theorem shows that for finite groups at least, we can always restrict ourselves to unitary reps.

Theorem 3.1. Let **T** be a rep of G on the inner product space V. Then **T** is equivalent to a unitary rep on V.

Proof. First we define a new inner product $(-, -)$ on V with respect to which **T** is unitary. For $\mathbf{u}, \mathbf{v} \in V$ let

$$(\mathbf{u}, \mathbf{v}) = \frac{1}{n(G)} \sum_{g \in G} \langle \mathbf{T}(g)\mathbf{u}, \mathbf{T}(g)\mathbf{v}\rangle. \tag{1.18}$$

[Note that $(\mathbf{u}, \mathbf{v})$ is an average of the numbers $\langle \mathbf{T}(g)\mathbf{u}, \mathbf{T}(g)\mathbf{v}\rangle$ taken over the group.] It is easy to check that $(-, -)$ is an inner product on V. Furthermore,

$$\begin{aligned}(\mathbf{T}(h)\mathbf{u}, \mathbf{T}(h)\mathbf{v}) &= \frac{1}{n(G)} \sum_{g \in G} \langle \mathbf{T}(gh)\mathbf{u}, \mathbf{T}(gh)\mathbf{v}\rangle \\ &= \frac{1}{n(G)} \sum_{g' \in G} \langle \mathbf{T}(g')\mathbf{u}, \mathbf{T}(g')\mathbf{v}\rangle = (\mathbf{u}, \mathbf{v}),\end{aligned}$$

where the next to last equality follows from the fact that if g runs through the elements of G exactly once, then so does gh. Now **T** is unitary with respect to the new inner product, but not the old one. Let $\{\mathbf{u}_i\}$ be an ON basis of V with respect to $(-, -)$ and let $\{\mathbf{v}_i\}$ be an ON basis with respect to $\langle -, -\rangle$. Define the nonsingular linear operator $\mathbf{S}: V \longrightarrow V$ by $\mathbf{S}\mathbf{u}_i = \mathbf{v}_i$. Then for $\mathbf{w} = \sum \alpha_i \mathbf{u}_i$ and $\mathbf{x} = \sum \beta_j \mathbf{u}_j$ we find

$$\langle \mathbf{S}\mathbf{w}, \mathbf{S}\mathbf{x}\rangle = \sum_{i,j} \alpha_i \bar{\beta}_i \langle \mathbf{S}\mathbf{u}_i, \mathbf{S}\mathbf{u}_j\rangle = \sum_i \alpha_i \bar{\beta}_i = (\mathbf{w}, \mathbf{x}),$$

so

$$\begin{aligned}\langle \mathbf{S}\mathbf{T}(g)\mathbf{S}^{-1}\mathbf{w}, \mathbf{S}\mathbf{T}(g)\mathbf{S}^{-1}\mathbf{x}\rangle &= (\mathbf{T}(g)\mathbf{S}^{-1}\mathbf{w}, \mathbf{T}(g)\mathbf{S}^{-1}\mathbf{x}) \\ &= (\mathbf{S}^{-1}\mathbf{w}, \mathbf{S}^{-1}\mathbf{x}) = \langle \mathbf{w}, \mathbf{x}\rangle.\end{aligned}$$

Thus, the rep $\mathbf{T}'(g) = \mathbf{S}\mathbf{T}(g)\mathbf{S}^{-1}$ is unitary on V. Q.E.D.

We can always assume that a rep **T** on V is unitary. Indeed, we can always define an inner product on V with respect to which **T** is unitary. Moreover, if V is already equipped with a given inner product $\langle -, - \rangle$ then we can find a unitary rep (with respect to $\langle -, - \rangle$) which is equivalent to **T**.

Theorem 3.1 and its proof are also valid for reps on a real vector space V. In this case we can find an inner product on V with respect to which the operators $\mathbf{T}(g)$ are **orthogonal**.

3.2 Reducible Representations

In this section **T** will be a finite-dimensional rep of a finite group G acting on the (real or complex) vector space V.

Definition. A subspace W of V is **invariant** under **T** if $\mathbf{T}(g)\mathbf{w} \in W$ for every $g \in G$, $\mathbf{w} \in W$.

If W is invariant under **T** we can define a rep $\mathbf{T}' = \mathbf{T} \mid W$ of G on W by

$$\mathbf{T}'(g)\mathbf{w} = \mathbf{T}(g)\mathbf{w}, \qquad \mathbf{w} \in W. \tag{2.1}$$

This rep is called the **restriction** of **T** to W. If **T** is unitary so is $\mathbf{T}'$.

Definition. The rep **T** is **reducible** if there is a proper subspace W of V which is invariant under **T**. Otherwise, **T** is **irreducible (irred)**.

A rep is irred if the only invariant subspaces of V are $\{\boldsymbol{\theta}\}$ and V itself. One-dimensional and zero-dimensional reps are necessarily irred. However, the trivial zero-dimensional rep will be ignored in all the material to follow.

We now give a matrix interpretation of reducibility. Suppose **T** is reducible and W is a proper invariant subspace of V. If $\dim W = k$ and $\dim V = n$ we can find a basis $\mathbf{v}_1, \ldots, \mathbf{v}_n$ for V such that $\mathbf{v}_1, \ldots, \mathbf{v}_k$, $1 \leq k \leq n$, form a basis for W. Then the matrices of the operators $\mathbf{T}(g)$ with respect to this basis take the form

$$\begin{array}{cc} & \begin{array}{cc} k & n-k \end{array} \\ \begin{array}{c} k \\ n-k \end{array} & \begin{pmatrix} T'(g) & *** \\ Z & T''(g) \end{pmatrix} \end{array}$$

The $k \times k$ matrices $T'(g)$ and the $(n - k) \times (n - k)$ matrices $T''(g)$ separately define matrix reps of G. In particular $T'(g)$ is the matrix of the rep $\mathbf{T}'(g)$, (2.1), with respect to the basis $\mathbf{v}_1, \ldots, \mathbf{v}_k$ of W. Here Z is the zero matrix.

Every reducible rep can be decomposed into irred reps in an almost unique manner. Thus the problem of constructing all reps of G simplifies to the problem of constructing all irred reps. The irred reps emerge as fundamental building blocks for the theory of reps of finite groups. To prove these statements in the simplest fashion we assume, as we can, that **T** is unitary.

If W is a proper subspace of the inner product space V and

$$W^{\perp} = \{\mathbf{v} \in V : \langle \mathbf{v}, \mathbf{w} \rangle = 0, \quad \text{all} \quad \mathbf{w} \in W\} \tag{2.2}$$

is the subspace of all vectors perpendicular to W, it is an easy exercise in linear algebra to prove that $V = W \oplus W^{\perp}$ (V is the **direct sum** of W and $W^{\perp}$). That is, every $\mathbf{v} \in V$ can be written uniquely in the form

$$\mathbf{v} = \mathbf{w} + \mathbf{w}', \qquad \mathbf{w} \in W, \qquad \mathbf{w}' \in W^{\perp}.$$

Theorem 3.2. If **T** is a reducible unitary rep of G on V and W is a proper invariant subspace of V, then $W^{\perp}$ is also a proper invariant subspace of V. In this case we write $\mathbf{T} = \mathbf{T}' \oplus \mathbf{T}''$ and say that **T** is the **direct sum** of $\mathbf{T}'$ and $\mathbf{T}''$, where $\mathbf{T}', \mathbf{T}''$ are the (unitary) restrictions of **T** to $W, W^{\perp}$, respectively.

Proof. We must show $\mathbf{T}(g)\mathbf{u} \in W^{\perp}$ for every $g \in G$, $\mathbf{u} \in W^{\perp}$. Now for every $\mathbf{w} \in W$,

$$\langle \mathbf{T}(g)\mathbf{u}, \mathbf{w} \rangle = \langle \mathbf{u}, \mathbf{T}(g^{-1})\mathbf{w} \rangle = 0$$

since $\mathbf{T}(g^{-1})\mathbf{w} \in W$. The first equality follows from (1.17) and unitarity. Thus, $\mathbf{T}(g)\mathbf{u} \in W^{\perp}$. Q.E.D.

Suppose **T** is reducible and V_1 is a proper invariant subspace of V of smallest dimension. Then, necessarily, the restriction $\mathbf{T}_1$ of **T** to V_1 is irred and we have the direct sum decomposition $V = V_1 \oplus V_1^{\perp}$, where $V_1^{\perp}$ is invariant under **T**. If $V_1^{\perp}$ is not irred we can find a proper irred subspace V_2 of smallest dimension such that $V_1^{\perp} = V_2 \oplus V_2^{\perp}$ by repeating the above argument. We continue in this fashion until eventually we obtain the direct sum decomposition

$$V = V_1 \oplus V_2 \oplus \cdots \oplus V_l \quad \text{or} \quad \mathbf{T} = \mathbf{T}_1 \oplus \mathbf{T}_2 \oplus \cdots \oplus \mathbf{T}_l \tag{2.3}$$

where the V_i are mutually orthogonal proper invariant subspaces of V which transform irreducibly under the restrictions $\mathbf{T}_i$ of **T** to V_i. The decomposition process comes to an end after a finite number of steps because V is finite-dimensional. Some of the $\mathbf{T}_i$ may be equivalent. If a_1 of the reps $\mathbf{T}_i$ are equivalent to $\mathbf{T}_1$, a_2 to $\mathbf{T}_2, \ldots, a_k$ to $\mathbf{T}_k$ and $\mathbf{T}_1, \ldots, \mathbf{T}_k$ are pairwise nonequivalent, we write

$$\mathbf{T} = \sum_{j=1}^{k} \oplus\, a_j \mathbf{T}_j. \tag{2.4}$$

With this notation we are identifying equivalent reps. It is a straightforward exercise to show that, given a_j copies of $\mathbf{T}_j$, $1 \leq j \leq k$, one can construct a rep of G equivalent to **T**.

Theorem 3.3. Every finite-dimensional unitary rep of a finite group can be decomposed into a direct sum of irred unitary reps.

The above decomposition is not unique since the irred subspaces $V_1, \ldots, V_l$ are not uniquely determined. However, it will be shown in the next section that the integers a_j in (2.4) are uniquely determined. Thus, up to equivalence we can determine uniquely how many times a particular irred rep of G occurs in the decomposition of **T**. The integer a_j is called the **multiplicity** of $\mathbf{T}_j$ in **T**.

It follows from Theorem 3.1 that the statement of Theorem 3.3 still holds when the word "unitary" is deleted. To get a matrix interpretation of Theorem 3.3 choose a basis for V by combining bases $\{\mathbf{v}_i^{(j)}\}$, $j = 1, \ldots, l$, for $V_1, V_2, \ldots$, and V_l. In terms of this basis the matrix $T(g)$ of $\mathbf{T}(g)$ is given by

$$\begin{pmatrix} T_1(g) & & & & Z \\ & T_2(g) & & & \\ & & \cdot & & \\ & & & \cdot & \\ Z & & & & T_l(g) \end{pmatrix} \tag{2.5}$$

where $n_j = \dim V_j$ and $T_j(g)$ is the matrix of $\mathbf{T}_j(g)$ with respect to the basis $\{\mathbf{v}_i^{(j)}\}$, $1 \leq i \leq n_j$.

3.3 Irreducible Representations

The fundamental problem in the representation theory of a finite group is the construction of a complete set of nonequivalent irred reps. A secondary problem is the determination of a practical method for decomposing a reducible rep into irred reps. The following two theorems (Shur's lemmas) are crucial.

Theorem 3.4. Let **T**, **T**′ be irred reps of the group G on the finite-dimensional vector spaces V, V', respectively and let **A** be a nonzero linear transformation mapping V into V' such that

$$\mathbf{T}'(g)\mathbf{A} = \mathbf{A}\mathbf{T}(g) \tag{3.1}$$

for all $g \in G$. Then **A** is a nonsingular linear transformation of V onto V', so **T** and **T**′ are equivalent.

Proof. Let $N_\mathbf{A}$ be the **null space** and $R_\mathbf{A}$ the **range** of **A**:

$$N_\mathbf{A} = \{\mathbf{v} \in V: \mathbf{A}\mathbf{v} = \boldsymbol{\theta}\} \qquad R_\mathbf{A} = \{\mathbf{v}' \in V': \mathbf{v}' = \mathbf{A}\mathbf{v} \quad \text{for some} \quad \mathbf{v} \in V\}.$$

The subspace $N_\mathbf{A}$ of V is invariant under **T** since $\mathbf{A}\mathbf{T}(g)\mathbf{v} = \mathbf{T}'(g)\mathbf{A}\mathbf{v} = \boldsymbol{\theta}$ for all $g \in G$, $\mathbf{v} \in V$. Since **T** is irred, $N_\mathbf{A}$ is either V or $\{\boldsymbol{\theta}\}$. The first possibility implies $\mathbf{A} = \mathbf{Z}$, the zero operator, which is impossible. Therefore, $N_\mathbf{A} = \{\boldsymbol{\theta}\}$. The subspace $R_\mathbf{A}$ of V' is invariant under **T**′ because $\mathbf{T}'(g)\mathbf{A}\mathbf{v} =$

$\mathbf{AT}(g)\mathbf{v} \in R_{\mathbf{A}}$ for all $\mathbf{v} \in V$. But $\mathbf{T}'$ is irred so $R_{\mathbf{A}}$ is either V' or $\{\boldsymbol{\theta}\}$. If $R_{\mathbf{A}} = \{\boldsymbol{\theta}\}$ then $\mathbf{A} = \mathbf{Z}$, which is impossible. Therefore $R_{\mathbf{A}} = V'$ which implies that $\mathbf{T}$ and $\mathbf{T}'$ are equivalent. Q.E.D.

Corollary 3.1. Let $\mathbf{T}, \mathbf{T}'$ be nonequivalent finite-dimensional irred reps of G. If $\mathbf{A}$ is a linear transformation from V to V' which satisfies (3.1) for all $g \in G$ then $\mathbf{A} = \mathbf{Z}$.

The results in the remainder of this section apply only to complex reps.

Theorem 3.5. Let $\mathbf{T}$ be a rep of the group G on the finite-dimensional complex vector space V. Then $\mathbf{T}$ is irred if and only if the only transformations $\mathbf{A}: V \longrightarrow V$ such that

$$\mathbf{T}(g)\mathbf{A} = \mathbf{AT}(g) \tag{3.2}$$

for all $g \in G$ are $\mathbf{A} = \lambda\mathbf{E}$, where $\lambda \in \mathfrak{C}$ and $\mathbf{E}$ is the identity operator on V.

Proof. It is well known that a linear operator on a finite-dimensional complex vector space always has at least one eigenvalue. (This statement is false for a real vector space.) Let λ be an eigenvalue of an operator $\mathbf{A}$ which satisfies (3.2) and define the eigenspace C_λ by

$$C_\lambda = \{\mathbf{v} \in V: \mathbf{Av} = \lambda\mathbf{v}\}.$$

Clearly C_λ is a subspace of V and $\dim C_\lambda > 0$. Furthermore, C_λ is invariant under $\mathbf{T}$ because

$$\mathbf{AT}(g)\mathbf{v} = \mathbf{T}(g)\mathbf{Av} = \lambda\mathbf{T}(g)\mathbf{v}$$

for $\mathbf{v} \in C_\lambda$, $g \in G$, so $\mathbf{T}(g)\mathbf{v} \in C_\lambda$. If $\mathbf{T}$ is irred then $C_\lambda = V$ and $\mathbf{Av} = \lambda\mathbf{v}$ for all $\mathbf{v} \in V$.

Conversely, suppose $\mathbf{T}$ is reducible. Then there exists a proper invariant subspace V_1 of V and by Theorem 3.2, a proper invariant subspace V_2 such that $V = V_1 \oplus V_2$. Any $\mathbf{v} \in V$ can be written uniquely as $\mathbf{v} = \mathbf{v}_1 + \mathbf{v}_2$ with $\mathbf{v}_j \in V_j$. We define the projection operator $\mathbf{P}$ on V by $\mathbf{Pv} = \mathbf{v}_1 \in V_1$. Then $\mathbf{PT}(g)\mathbf{v} = \mathbf{T}(g)\mathbf{Pv} = \mathbf{T}(g)\mathbf{v}_1$ (verify this), and $\mathbf{P}$ is clearly not a multiple of $\mathbf{E}$. Q.E.D.

Choosing a basis for V and a basis for V' we can immediately translate Shur's lemmas into statements about irred matrix reps.

Corollary 3.2. Let T and T' be $n \times n$ and $m \times m$ complex irred matrix reps of the group G, and let A be an $m \times n$ matrix such that

$$T'(g)A = AT(g) \tag{3.3}$$

for all $g \in G$. If T and T' are nonequivalent then $A = Z$, the zero matrix.

(In particular, this is true if $n \neq m$.) If $\mathbf{T} = \mathbf{T}'$ then $A = \lambda E_n$, where $\lambda \in \mathfrak{C}$ and E_n is the $n \times n$ identity matrix.

Note that the proofs of Shur's lemmas use only the concept of irreducibility and the fact that the rep spaces are finite-dimensional. The homomorphism property of reps and the fact that G is finite are not needed.

Theorem 3.5 is extremely useful because it yields a practical method for determining if a group rep is irred. The original definition of irreducibility, while useful for theoretical purposes, is too complicated to verify directly in most practical problems. Theorem 3.5 can easily be translated to a theorem about matrix reps, whose obvious statement and proof are left to the reader.

Let G be a finite group and select one irred rep $\mathbf{T}^{(\mu)}$ of G in each equivalence class of irred reps. Then every irred rep is equivalent to some $\mathbf{T}^{(\mu)}$ and the reps $\mathbf{T}^{(\mu_1)}$, $\mathbf{T}^{(\mu_2)}$ are nonequivalent if $\mu_1 \neq \mu_2$. The parameter μ indexes the equivalence classes of irred reps. (We will soon show that there are only a finite number of these classes.) Introduction of a basis in each rep space $V^{(\mu)}$ leads to a matrix rep $T^{(\mu)}$. The $T^{(\mu)}$ form a complete set of irred $n_\mu \times n_\mu$ matrix reps of G, one from each equivalence class. Here $n_\mu = \dim V^{(\mu)}$. If we wish, we can choose the $T^{(\mu)}$ to be unitary.

The following trick leads to an extremely useful set of relations in rep theory, the orthogonality relations. Given two irred matrix reps $T^{(\mu)}$, $T^{(\nu)}$ of G, choose an arbitrary $n_\mu \times n_\nu$ matrix B and form the $n_\mu \times n_\nu$ matrix

$$A = N^{-1} \sum_{g \in G} T^{(\mu)}(g) B T^{(\nu)}(g^{-1}) \tag{3.4}$$

where $N = n(G)$. Here, A is just the average of the matrices $T^{(\mu)}(g)BT^{(\nu)}(g^{-1})$ over the group G. We will show that A satisfies

$$T^{(\mu)}(h)A = AT^{(\nu)}(h) \tag{3.5}$$

for all $h \in G$. This result and Corollary 3.2 imply that if $\mu \neq \nu$ then $A = Z$, whereas if $\mu = \nu$ then $A = \lambda E_{n_\mu}$ for some $\lambda \in \mathfrak{C}$. The verification of (3.5) follows from

$$\begin{aligned} T^{(\mu)}(h)A &= N^{-1} \sum_{g \in G} T^{(\mu)}(h)T^{(\mu)}(g)BT^{(\nu)}(g^{-1}) \\ &= N^{-1} \sum_{g \in G} T^{(\mu)}(hg)BT^{(\nu)}((hg)^{-1})T^{(\nu)}(h) = AT^{(\nu)}(h). \end{aligned}$$

We have used the fact that as g runs over each of the elements of G exactly once, so does $g' = hg$. Applying Corollary 3.2, we obtain the result $A = \lambda(\mu, B)\delta_{\mu\nu}E_{n_\mu}$ where $\delta_{\mu\nu}$ is the Kronecker delta, and the constant $\lambda \in \mathfrak{C}$ depends on μ and B. To derive all possible consequences of this identity it is enough to let B run through the $n_\mu \times n_\nu$ matrices $B^{(l,m)} = (B^{(l,m)}_{jk})$, where

$$B^{(l,m)}_{jk} = \begin{cases} 1 & \text{if} \quad j = l, \quad k = m, \qquad 1 \le j \le n_\mu, \quad 1 \le k \le n_\nu, \\ 0 & \text{otherwise.} \end{cases} \tag{3.6}$$

Making these substitutions, we obtain

$$(3.7)\qquad \sum_{g\in G} T^{(\mu)}_{il}(g)T^{(\nu)}_{ms}(g^{-1}) = N\lambda\delta_{\mu\nu}\delta_{is}, \qquad 1\le i,l\le n_\mu, \quad 1\le m,s\le n_\nu.$$

Here, λ may depend on μ, l, and m, but not on i or s. To evaluate λ, set $\nu=\mu$, $s=i$, and sum on i to obtain

$$n_\mu N\lambda = \sum_{g\in G}\sum_{i=1}^{n_\mu} T^{(\mu)}_{mi}(g^{-1})T^{(\mu)}_{il}(g) = \sum_{g\in G} T^{(\mu)}_{ml}(e) = N\delta_{ml}$$

since $N = n(G)$. Therefore, $\lambda = \delta_{ml}n_\mu^{-1}$. We can simplify (3.7) slightly if we assume (as we can) that all of the matrix reps $T^{(\nu)}(g)$ are unitary. Then

$$T^{(\nu)}_{ms}(g^{-1}) = \overline{T^{(\nu)}_{sm}(g)}$$

and (3.7) reduces to

$$(3.8)\qquad \sum_{g\in G} T^{(\mu)}_{il}(g)\overline{T^{(\nu)}_{sm}(g)} = (N/n_\mu)\delta_{is}\,\delta_{lm}\,\delta_{\mu\nu}.$$

Equations (3.7) and (3.8) are called the **orthogonality relations** for the matrix elements of irred reps of G. We have derived these remarkable relations without any detailed knowledge of the structure of G.

To better understand the orthogonality relations it is convenient to consider the elements x of the group ring R_G as complex-valued functions $x(g)$ on the group G. The relation between this approach and the definition of R_G as given in Example 3, Section 3.1, is provided by the correspondence

$$(3.9)\qquad x = \sum_{g\in G} x(g)\cdot g \longleftrightarrow x(g).$$

The elements of the N-tuple $(x(g_1), \ldots, x(g_N))$, where g_i ranges over G, can be regarded as the components of $x \in R_G$ in the natural basis provided by the elements of G. Furthermore the 1–1 mapping (3.9) leads to the relations

$$(3.10)\qquad \begin{aligned} &x+y \longleftrightarrow x(g)+y(g), \qquad \alpha x \longleftrightarrow \alpha x(g),\\ &xy \longleftrightarrow xy(g) = \sum_{h\in G} x(h)y(h^{-1}g) \end{aligned}$$

where the expression defining $xy(g)$ is called the **convolution product** of $x(g)$ and $y(g)$. Thus, we can consider R_G as the ring of all complex-valued functions $x(g)$ on G where addition, scalar multiplication, and convolution product are defined by (3.10). Indeed, the ring of functions just constructed is algebraically isomorphic to R_G with the isomorphism given by (3.9). Under this isomorphism the element $h = 1\cdot h \in R_G$ is mapped into the function

$$h(g) = \begin{cases} 1 & \text{if } g = h \\ 0 & \text{otherwise.} \end{cases}$$

Now consider the right regular rep on R_G. Writing

$$(3.11)\qquad \mathbf{R}(h)x = \sum_{g\in G} [\mathbf{R}(h)x](g)\cdot g = xh^{-1} = \sum_{g} x(gh)\cdot g \qquad x\in R_G,$$

we obtain

$$[\mathbf{R}(h)x](g) = x(gh), \qquad h \in G, \tag{3.12}$$

as the action of $\mathbf{R}(h)$ on our new model of R_G. From Theorem 3.1, there exists an inner product on the N-dimensional vector space R_G with respect to which the right regular rep $\mathbf{R}$ is unitary. In particular, the following inner product works:

$$\langle x, y\rangle = N^{-1} \sum_{g \in G} x(g)\overline{y(g)}, \qquad x, y \in R_G. \tag{3.13}$$

The reader can easily verify that this is an inner product and that the $\mathbf{R}(h)$ operators are unitary with respect to it. Now note that for fixed μ, i, j with $1 \le i, j \le n_\mu$ the matrix element $T_{ij}^{(\mu)}(g)$ defines a function on G, hence an element of R_G. Furthermore, comparing (3.13) with (3.8), we see that the functions

$$\varphi_{ij}^{(\mu)}(g) = n_\mu^{1/2} T_{ij}^{(\mu)}(g), \qquad 1 \le i, j \le n_\mu, \tag{3.14}$$

where μ ranges over all equivalence classes of irred reps of G, form an ON set in R_G. Since R_G is N-dimensional the ON set can contain at most N elements. Thus there are only a finite number, say α, of nonequivalent irred reps of G. Each irred matrix rep μ yields $n_\mu{}^2$ vectors of the form (3.14). The full ON set $\{\varphi_{ij}^{(\mu)}\}$ spans a subspace of R_G of dimension

$$n_1{}^2 + n_2{}^2 + \cdots + n_\alpha{}^2 \le N. \tag{3.15}$$

The inequality (3.15) is a strong restriction on the possible number and dimensions of irred reps of G. This result can be strengthened even more by showing that the ON set $\{\varphi_{ij}^{(\mu)}\}$ is actually a **basis** for R_G. Since the dimension N of R_G is equal to the number of basis vectors, we obtain the equality

$$n_1{}^2 + n_2{}^2 + \cdots + n_\alpha{}^2 = N. \tag{3.16}$$

To prove this result, let V be the subspace of R_G spanned by the ON set $\{\varphi_{ij}^{(\mu)}\}$. From (3.14) and the homomorphism property of the matrices $T^{(\mu)}(g)$ there follows

$$[\mathbf{R}(h)\varphi_{ij}^{(\mu)}](g) = \varphi_{ij}^{(\mu)}(gh) = \sum_{k=1}^{n_\mu} T_{kj}^{(\mu)}(h)\varphi_{ik}^{(\mu)}(g) \in V. \tag{3.17}$$

Thus, V is invariant under $\mathbf{R}$. According to Theorem 3.2, $V^\perp$ is also invariant under $\mathbf{R}$ and $R_G = V \oplus V^\perp$. Here, $V^\perp$ is defined with respect to the inner product (3.13). If $V^\perp \ne \{\boldsymbol{\theta}\}$ then it contains a subspace W transforming under some irred rep $\mathbf{T}^{(\nu)}$ of G. Thus, there exists an ON basis $x_1, \ldots, x_{n_\nu}$ for W such that

$$[\mathbf{R}(g)x_i](h) = x_i(hg) = \sum_{j=1}^{n_\mu} T_{ji}^{(\nu)}(g)x_j(h), \qquad 1 \le i \le n_\nu. \tag{3.18}$$

Setting $h = e$ in (3.18) and letting g run over G, we find

$$x_i(g) = \sum_j x_j(e)T^{(\nu)}_{ji}(g) = \sum_j x_j(e)\varphi^{(\nu)}_{ji}(g)/n_\nu^{1/2},$$

so $x_i \in V$. Thus, $W \subseteq V \cap V^\perp$. This is possible only if $W = \{\boldsymbol{\theta}\}$. Therefore, $V^\perp = \{\boldsymbol{\theta}\}$ and $V = R_G$.

Theorem 3.6. The functions

$$\{\varphi^{(\mu)}_{ij}(g)\}, \qquad \mu = 1, \ldots, \alpha, \quad 1 \le i, j \le n_\mu,$$

form an ON basis for R_G. Every function $x \in R_G$ can be written uniquely in the form

$$x(g) = \sum_{i,j,\mu} a^\mu_{ij}\varphi^{(\mu)}_{ij}(g), \qquad a^\mu_{ij} = \langle x, \varphi^{(\mu)}_{ij}\rangle.$$

Relation (3.17) also yields the interesting fact that for fixed μ and i, $1 \le i \le n_\mu$, the n_μ vectors $\{\varphi^{(\mu)}_{ij} : 1 \le j \le n_\mu\}$ form an ON basis for a subspace $V^{(\mu)}_i$ of R_G which transforms under the irred rep $\mathbf{T}^{(\mu)}$ of G. Thus,

$$R_G = \sum_{\mu, i} \oplus V^{(\mu)}_i, \qquad 1 \le \mu \le \alpha, \quad 1 \le i \le n_\mu,$$

and the rep $\mathbf{T}^{(\mu)}$ occurs with multiplicity n_μ in the right regular rep $\mathbf{R}$.

(3.19) $$\mathbf{R} = \sum_{\mu=1}^{\alpha} \oplus\, n_\mu \mathbf{T}^{(\mu)}.$$

3.4 Group Characters

The orthogonality relations and decomposition theorems of the preceding section suffer from the defect that they are basis-dependent. To determine in what sense our results are unique we free them from a dependence on the choice of basis vectors for V.

Let $\mathbf{T}$ be a rep of the finite group G on the n-dimensional vector space V. With respect to some fixed basis in V the operators $\mathbf{T}(g)$ define a matrix rep in terms of $n \times n$ matrices $T(g)$. We define the **character** of $\mathbf{T}$ as the function

(4.1) $$\chi(g) = \operatorname{tr} T(g), \qquad g \in G.$$

Since the trace satisfies

$$\operatorname{tr}(AB) = \operatorname{tr}(BA)$$

for any two $n \times n$ matrices, we find

(4.2) $$\operatorname{tr}(ST(g)S^{-1}) = \operatorname{tr}(T(g)S^{-1}S) = \operatorname{tr}(T(g))$$

for all nonsingular $n \times n$ matrices S. Thus, equivalent matrix reps have the same character and $\chi(g)$ is independent of basis. Furthermore, we will soon show that two reps with equal characters are equivalent. Thus, there is a 1–1 relationship between equivalence classes of reps of G and group characters on G. If χ is the character of an irred rep it is called **simple**; if the rep is reducible, χ is a **compound** character.

The orthogonality relations for matrix elements immediately lead to orthogonality relations for characters. Let $\chi^{(\mu)}$ be the character of the irred rep $T^{(\mu)}$, $\mu = 1, \ldots, \alpha$. Setting $i = l$ and $m = s$ in (3.7) and summing i from 1 to n_μ, and m from 1 to n_ν, we obtain

$$\sum_{g \in G} \chi^{(\mu)}(g)\chi^{(\nu)}(g^{-1}) = N\delta_{\mu\nu}. \tag{4.3}$$

If we assume, as we can, that the matrix rep $T(g)$ is unitary then

$$\chi(g^{-1}) = \operatorname{tr} T(g^{-1}) = \operatorname{tr} \overline{T(g)}^t = \overline{\operatorname{tr} T(g)} = \overline{\chi(g)}, \tag{4.4}$$

a result which is now seen to be valid independent of basis. Substituting this result into (4.3), we obtain the orthogonality relations

$$\langle \chi^{(\mu)}, \chi^{(\nu)} \rangle = \delta_{\mu\nu}, \qquad 1 \leq \mu, \nu \leq \alpha, \tag{4.5}$$

where the inner product is defined by (3.17). Thus, simple characters of G form an ON set in R_G.

Now let $\mathbf{T}(g)$ be an arbitrary rep with character $\chi(g)$. It follows from (2.4) and (2.5) that with respect to one basis at least, we can write

$$\chi(g) = \sum_{\mu=1}^{\alpha} a_\mu \chi^{(\mu)}(g) \tag{4.6}$$

where a^μ is the multiplicity of $\mathbf{T}^{(\mu)}$ in $\mathbf{T}$. However, the orthogonality relations (4.5) imply

$$\langle \chi, \chi^{(\mu)} \rangle = a_\mu, \qquad 1 \leq \mu \leq \alpha. \tag{4.7}$$

Since the left-hand side of (4.7) is basis-independent, so is the right-hand side.

Theorem 3.7. The multiplicity a_μ of the irred rep $\mathbf{T}^{(\mu)}$ in $\mathbf{T}$ is given by (4.7). Since reps with the same multiplicities are equivalent, reps with equal characters are equivalent.

Thus, the multiplicities a_μ are unique even though the exact decomposition of the rep space into irred subspaces may be nonunique.

Corollary 3.3. Let $\chi(g)$ be a group character of G. Then $\langle \chi, \chi \rangle$ is a nonnegative integer and $\chi(g)$ corresponds to an irred rep if and only if $\langle \chi, \chi \rangle = 1$.

Proof. We can write χ as a unique sum of simple characters:

$$\chi(g) = \sum_{\mu=1}^{\alpha} a_\mu \chi^{(\mu)}(g).$$

Since the $\chi^{(\mu)}$ form an ON set there follows

$$\langle \chi, \chi \rangle = \sum_{\mu=1}^{\alpha} a_\mu{}^2. \tag{4.8}$$

The right-hand side equals one if and only if one of the a_μ is one and the rest are zero. Q.E.D.

We list a few additional properties of characters. If $\mathbf{T}$ is an n-dimensional rep with character χ then

$$\chi(e) = \operatorname{tr} E_n = n. \tag{4.9}$$

Thus $\chi(e)$ is always equal to the dimension of the rep. Furthermore,

$$\chi(hgh^{-1}) = \operatorname{tr}[T(h)T(g)T(h)^{-1}] = \operatorname{tr} T(g) = \chi(g), \qquad g, h \in G, \tag{4.10}$$

so χ is constant on each conjugacy class of G. Suppose G has k conjugacy classes containing $m_1, \ldots, m_k$ elements, respectively, with $m_1 + \cdots + m_k = N$. Then the orthogonality relations (4.5) read

$$N^{-1} \sum_{i=1}^{k} \chi_i^{(\mu)} \overline{\chi_i^{(\nu)}} m_i = \delta_{\mu\nu} \tag{4.11}$$

where $\chi_i^{(\mu)}$ is the value of $\chi^{(\mu)}(g)$ with g in the ith conjugacy class. Relations of the form (4.11) are not as esthetically pleasing as (4.5) but they are useful for practical computations.

Let us examine the relationship between group characters and the subspace F of R_G consisting of all functions $\psi(g)$ such that

$$\psi(hgh^{-1}) = \psi(g), \qquad g, h \in G,$$

i.e., all functions which are constant on conjugacy classes. Each $\psi \in F$ is uniquely determined by k complex numbers, the value assumed by ψ on the k conjugacy classes of G. Thus F is k-dimensional. Clearly, the α simple characters of G form an ON set in F with respect to the inner product $\langle -, - \rangle$. In fact, $\alpha = k$ and these characters form an ON basis for F.

Theorem 3.8. The number α of nonequivalent irred reps of G is equal to the number of conjugacy classes in G.

Proof. Let $\psi \in F$. Since $F \subseteq R_G$ we can expand ψ in the form

$$\psi(g) = \sum_{\mu, i, j} a^{\mu}_{ij} T^{(\mu)}_{ij}(g)$$

where the $T^{(\mu)}_{ij}(g)$ are the matrix elements of a complete set of nonequivalent unitary irred reps of G. Since (summing over repeated indices)

$$\begin{aligned}\psi(g) = N^{-1} \sum_{h \in G} \psi(hgh^{-1}) &= N^{-1} \sum_{h} a^{\mu}_{ij} T^{(\mu)}_{il}(h) T^{(\mu)}_{lm}(g) T^{(\mu)}_{mj}(h^{-1}) \\ &= a^{\mu}_{ij} T^{(\mu)}_{lm}(g) \langle T^{(\mu)}_{il}, T^{(\mu)}_{jm} \rangle = \sum_{i, \mu} (a^{\mu}_{ii}/n_\mu) \chi^{(\mu)}(g),\end{aligned}$$

$\psi(g)$ is a linear combination of simple characters. Therefore, the simple characters form an ON basis for F. Q.E.D.

In terms of the $\alpha \times \alpha$ matrix A with elements

$$A_{\mu i} = (m_i/N)^{1/2}\chi_i^{(\mu)}, \qquad 1 \leq i, \mu \leq \alpha,$$

the first orthogonality relation (4.11) reads $A\bar{A}^t = E_\alpha$. Thus, $\bar{A}^t = A^{-1}$ and $\bar{A}^t A = E_\alpha$, or

$$\sum_{\mu=1}^{\alpha} \bar{\chi}_i^{(\mu)}\chi_j^{(\mu)} = (N/m_i)\delta_{ij}, \qquad 1 \leq i, j \leq \alpha. \tag{4.12}$$

This is known as the **second orthogonality relation** for characters.

As an example of character methods we verify expression (3.19) for the decomposition of the right regular rep $\mathbf{R}$ into irred reps. We begin by computing the character χ of $\mathbf{R}$ in the natural basis for R_G provided by the group elements. Now,

$$\mathbf{R}(h)g = gh^{-1}, \qquad h, g \in G,$$

so $\mathbf{R}(h)$ acts on the natural basis by permuting the basis vectors. If $h \neq e$ then no basis vector is left fixed under $\mathbf{R}(h)$. Thus, the matrix of $\mathbf{R}(h)$ in the natural basis has matrix elements which are zeros and ones, and if $h \neq e$ the diagonal matrix elements are all zero:

$$\chi(h) = \begin{cases} N & \text{if} \quad h = e \\ 0 & \text{if} \quad h \neq e. \end{cases}$$

Writing

$$\chi = \sum_{\mu=1}^{\alpha} a_\mu \chi^{(\mu)}$$

we obtain

$$a_\mu = \langle \chi, \chi^{(\mu)} \rangle = \overline{\chi^{(\mu)}(e)} = n_\mu.$$

Therefore, the multiplicity of $\mathbf{T}^{(\mu)}$ in $\mathbf{R}$ equals the dimension of $\mathbf{T}^{(\mu)}$. (The results for the decomposition of the left regular rep $\mathbf{L}$ are the same, so $\mathbf{R}$ and $\mathbf{L}$ are equivalent reps.)

Later we shall present a detailed derivation of the simple characters for the crystallographic point groups and the symmetric groups. Here we consider only the simple case where G is an abelian group of order N. Then G contains N conjugacy classes with one element each. Thus $\alpha = N$ and the relation

$$n_1^2 + n_2^2 + \cdots + n_N^2 = N$$

implies $n_1 = \cdots = n_N = 1$. The N nonequivalent irred reps of G are one-dimensional. In this special case the simple characters $\chi^{(\mu)}(g)$ coincide with the irreducible 1×1 matrix reps. Thus, the characters satisfy the homomorphism property

$$\chi^{(\mu)}(g_1)\chi^{(\mu)}(g_2) = \chi^{(\mu)}(g_1 g_2). \tag{4.13}$$

Since $g^N = e$ for every $g \in G$ there follows

$$[\chi^{(\mu)}(g)]^N = \chi^{(\mu)}(g^N) = \chi^{(\mu)}(e) = 1,$$

so $\chi^{(\mu)}(g)$ is an Nth root of unity. In order to explicitly list the simple characters for any abelian group it would be necessary to study the structure theory of such groups. However, if G is cyclic it is easy to give complete results. Let g_0 be an element of order N which generates G. Then

$$[\chi^{(\mu)}(g_0)]^N = 1 \tag{4.14}$$

for each of the N simple characters of G. Furthermore, the numbers $\chi^{(\mu)}(g_0)$ uniquely determine $\chi^{(\mu)}$, so these N numbers must be distinct. The equation $\omega^N = 1$ has exactly N solutions,

$$\omega_\mu = \exp(2\pi i\mu/N), \qquad \mu = 0, 1, \ldots, N-1.$$

Thus, the simple characters can be uniquely defined by

$$\chi^{(\mu)}(g_0{}^n) = \exp(2\pi i n\mu/N), \qquad \mu, n = 0, 1, \ldots, N-1. \tag{4.15}$$

The reader should understand that the above discussion applies only to complex reps of a group G. Character arguments can be applied to real reps only with special care. To understand the difficulties involved here, consider a real irred matrix rep T of G. We can also consider T as a complex matrix rep T^c of G. However, T^c may not be irred. For example, in an appropriate basis the generator $\mathbf{C}(\pi/2)$ of the cyclic group C_4, considered as a transformation group in the plane, corresponds to the matrix

$$\begin{pmatrix} 0 & -1 \\ 1 & 0 \end{pmatrix}.$$

The two dimensional real rep generated by this matrix is irred since the matrix has complex eigenvalues $\pm i$ and cannot be diagonalized by a real similarity transformation. However, considered as a complex matrix rep it is reducible.

3.5 New Representations from Old Ones

Let G be a group of order N. We discuss some methods for using known reps of G to construct new reps. The right and left regular reps and the identity rep are already familiar. (The **identity representation** of G is the irred one-dimensional rep defined by mapping each $g \in G$ into 1.) Furthermore, if G is defined as a matrix group, this matrix realization automatically yields a rep. To construct more reps we will probably have to rely on one of the methods presented below. Ultimately, we want to explicitly construct all irred reps of G (or at least their characters) and to explicitly decompose an arbitrary rep of G into irred reps.

First we review two techniques which have been studied earlier. If $\mathbf{T}_1, \mathbf{T}_2$ are reps of G on the vector spaces V_1, V_2, respectively, the **direct sum** $\mathbf{T}_1 \oplus \mathbf{T}_2$ is a rep acting on $V_1 \oplus V_2$ (vector space direct sum) and defined by

$$[\mathbf{T}_1 \oplus \mathbf{T}_2(g)]\mathbf{v}_1 \oplus \mathbf{v}_2 = \mathbf{T}_1(g)\mathbf{v}_1 \oplus \mathbf{T}_2(g)\mathbf{v}_2, \qquad \mathbf{v}_i \in V_i. \tag{5.1}$$

It is easy to show that the character χ of this rep is $\chi(g) = \chi_1(g) + \chi_2(g)$, where χ_1, χ_2 are the characters of $\mathbf{T}_1, \mathbf{T}_2$, respectively. This procedure can easily be extended to define the direct sum of any finite number of reps. We know already that every rep $\mathbf{T}$ is equivalent to a rep

$$\sum_{\mu=1}^{\alpha} \oplus a_\mu \mathbf{T}^{(\mu)}$$

where the $\mathbf{T}^{(\mu)}$ are a complete set of nonequivalent irred reps of G and the multiplicities a_μ are uniquely determined.

If $\mathbf{T}$ has rep space V and W is a proper invariant subspace of V then the rep $\mathbf{T}' = \mathbf{T} \mid W$ on W defined by

$$\mathbf{T}'(g)\mathbf{w} = \mathbf{T}(g)\mathbf{w}, \qquad g \in G, \quad \mathbf{w} \in W \tag{5.2}$$

is called the **restriction** of $\mathbf{T}$ to W. We have seen that every reducible rep of G can be written as a direct sum of certain of its irred restrictions.

Let V and V' be vector spaces of dimensions n, n', respectively and let $\{\mathbf{v}_i\}, \{\mathbf{v}_j'\}$ be bases for these spaces. We define $V \otimes V'$, the **tensor product** of V and V', as the nn'-dimensional space with basis $\{\mathbf{v}_i \otimes \mathbf{v}_j'\}$, $1 \leq i \leq n$, $1 \leq j \leq n'$. Thus, any $\mathbf{w} \in V \otimes V'$ can be written uniquely in the form

$$\mathbf{w} = \sum_{ij} \alpha_{ij} \mathbf{v}_i \otimes \mathbf{v}_j'. \tag{5.3}$$

If $\mathbf{v} = \sum \alpha_i \mathbf{v}_i$ and $\mathbf{v}' = \sum \beta_j \mathbf{v}_j'$ we define the vector $\mathbf{v} \otimes \mathbf{v}' \in V \otimes V'$ by

$$\mathbf{v} \otimes \mathbf{v}' = \sum_{i=1}^{n} \sum_{j=1}^{n'} \alpha_i \beta_j \mathbf{v}_i \otimes \mathbf{v}_j'. \tag{5.4}$$

If $\mathbf{w} \in V \otimes V'$ can be written in the form $\mathbf{w} = \mathbf{v} \otimes \mathbf{v}'$ then $\mathbf{w}$ is said to be **indecomposable.** The example $\mathbf{w} = \mathbf{v}_1 \otimes \mathbf{v}_2' + \mathbf{v}_2 \otimes \mathbf{v}_1'$ shows that if n, $n' \geq 2$ not every $\mathbf{w}$ is indecomposable. As a consequence of definition (5.4) it is easy to verify the following properties:

$$\alpha(\mathbf{v} \otimes \mathbf{v}') = (\alpha\mathbf{v}) \otimes \mathbf{v}' = \mathbf{v} \otimes (\alpha\mathbf{v}'), \qquad \alpha \in \mathfrak{C}, \tag{5.5a}$$

$$(\mathbf{u} + \mathbf{v}) \otimes \mathbf{v}' = \mathbf{u} \otimes \mathbf{v}' + \mathbf{v} \otimes \mathbf{v}', \qquad \mathbf{u}, \mathbf{v} \in V, \tag{5.5b}$$

$$\mathbf{v} \otimes (\mathbf{u}' + \mathbf{v}') = \mathbf{v} \otimes \mathbf{u}' + \mathbf{v} \otimes \mathbf{v}', \qquad \mathbf{u}', \mathbf{v}' \in V'. \tag{5.5c}$$

Although our definition of tensor product appears to depend on the choice of bases $\{\mathbf{v}_i\}$ and $\{\mathbf{v}_j'\}$ it is actually independent of this choice. For, let $\{\mathbf{u}_i\}$,

$\{\mathbf{u}_j'\}$ be new bases related to the old bases by

$$\mathbf{v}_l = \sum_{i=1}^{n} A_{il}\mathbf{u}_i, \qquad \mathbf{v}_k' = \sum_{j=1} A'_{jk}\mathbf{u}_j'$$

where the matrices A and A' are nonsingular. From relations (5.5a)–(5.5c) the basis vectors $\mathbf{v}_l \otimes \mathbf{v}_k'$ can all be expressed as linear combinations of the nn' vectors $\mathbf{u}_i \otimes \mathbf{u}_j'$. Since $V \otimes V'$ is nn'-dimensional it follows that the set $\{\mathbf{u}_i \otimes \mathbf{u}_j'\}$ is also a basis for $V \otimes V'$. This shows that the definition of $V \otimes V'$ is independent of basis. The definition of an indecomposable vector is also independent of basis.

Suppose $\mathbf{T}, \mathbf{T}'$ are reps of G on the spaces V, V', respectively. The **tensor product** $\mathbf{T} \otimes \mathbf{T}'$ is the rep of G on $V \otimes V'$ defined by

$$[\mathbf{T} \otimes \mathbf{T}'(g)]\mathbf{v} \otimes \mathbf{v}' = \mathbf{T}(g)\mathbf{v} \otimes \mathbf{T}'(g)\mathbf{v}', \qquad g \in G, \tag{5.6}$$

and linearity of the operator $\mathbf{T} \otimes \mathbf{T}'(g)$. It is straightforward to verify the rep property of these operators. Let $\{\mathbf{v}_i\}$, $\{\mathbf{v}_j'\}$ be bases of V, V', and let $T(g)$, $T'(g)$ be the corresponding matrix reps of $\mathbf{T}, \mathbf{T}'$. Then the matrix rep of $\mathbf{T} \otimes \mathbf{T}'$ with respect to $\{\mathbf{v}_i \otimes \mathbf{v}_j'\}$ is defined by

$$[\mathbf{T} \otimes \mathbf{T}'(g)]\mathbf{v}_i \otimes \mathbf{v}_j' = \sum_{l=1}^{n} \sum_{k=1}^{n'} T_{li}(g)T'_{kj}(g)\mathbf{v}_l \otimes \mathbf{v}_k' \tag{5.7}$$

or

$$[T \otimes T'(g)]_{lk,ij} = T_{li}(g)T'_{kj}(g).$$

(Note the double-suffix notation.) The character $\chi \otimes \chi'(g)$ is

$$\chi \otimes \chi'(g) = \sum_{l=1}^{n} \sum_{k=1}^{n'} T_{ll}(g)T'_{kk}(g) = \chi(g)\chi'(g). \tag{5.8}$$

Thus, the character of the tensor product is the product of the characters of the factors. As an immediate consequence of this result we see that the reps $\mathbf{T} \otimes \mathbf{T}'$ and $\mathbf{T}' \otimes \mathbf{T}$ are equivalent.

The above definitions have obvious generalizations to define n-fold tensor products $V^{(1)} \otimes \cdots \otimes V^{(n)}$ and tensor product reps $\mathbf{T}^{(1)} \otimes \cdots \otimes \mathbf{T}^{(n)}$ of the reps $\mathbf{T}^{(j)}$ on $V^{(j)}$. The dimension of the tensor product space is the product of the dimensions of the factor spaces $V^{(j)}$.

Let $\{\mathbf{T}^{(\mu)}\}$, $1 \le \mu \le \alpha$, be a complete set of nonequivalent irred reps of G. We can form tensor product reps $\mathbf{T}^{(\mu)} \otimes \mathbf{T}^{(\nu)}$, $1 \le \mu, \nu \le \alpha$, with characters $\chi^{(\mu)} \otimes \chi^{(\nu)}(g) = \chi^{(\mu)}(g)\chi^{(\nu)}(g)$. These reps can then be decomposed into irred reps

$$\mathbf{T}^{(\mu)} \otimes \mathbf{T}^{(\nu)} \cong \sum_{\xi=1}^{\alpha} \oplus\, a_\xi \mathbf{T}^{(\xi)} \tag{5.9}$$

where

$$a_\xi = \langle \chi^{(\mu)}\chi^{(\nu)}, \chi^{(\xi)} \rangle.$$

The expansion (5.9) is called a **Clebsch–Gordan series**. Many important problems in mathematical physics reduce to the computation of the multiplicities a_ξ. Suppose we have agreed on a complete set of nonequivalent irred matrix reps $T^{(\xi)}(g)$ of G. Let $\{\mathbf{v}_i^{(\mu)}\}$, $\{\mathbf{v}_j^{(\nu)}\}$ be bases of $V^{(\mu)}$, $V^{(\nu)}$ whose associated matrix reps of $\mathbf{T}^{(\mu)}, \mathbf{T}^{(\nu)}$ are $T^{(\mu)}$, $T^{(\nu)}$, respectively. The set $\{\mathbf{v}_i^{(\mu)} \otimes \mathbf{v}_j^{(\nu)}\}$ clearly defines a basis of $V^{(\mu)} \otimes V^{(\nu)}$. On the other hand, by (5.9) there exists another basis $\{\mathbf{w}_l^{\xi,s}\}$, $1 \le \xi \le \alpha$, $1 \le s \le a_\xi$, for $V^{(\mu)} \otimes V^{(\nu)}$ such that for fixed ξ and s, the vectors $\{\mathbf{w}_l^{\xi,s}, 1 \le l \le n_\xi\}$ form a basis for an irred subspace transforming under $\mathbf{T}^{(\xi)}$ and inducing the matrix rep $T^{(\xi)}$. (We have by no means uniquely defined the basis $\{\mathbf{w}_l^{\xi,s}\}$. For practical computations it is necessary to be explicit as to how each basis vector is chosen. This matter will be taken up later.) In terms of the "natural basis" $\{\mathbf{v}_i^{(\mu)} \otimes \mathbf{v}_j^{(\nu)}\}$ the tensor product induces the $n_\mu n_\nu$-dimensional matrix rep

(5.10) $$[T^{(\mu)} \otimes T^{(\nu)}(g)]_{lk,ij} = T_{li}^{(\mu)}(g) T_{kj}^{(\nu)}(g)$$

while in terms of the $\{\mathbf{w}_l^{\xi,s}\}$ basis the matrix rep is

(5.11) $$\begin{pmatrix} \overbrace{T^{(1)}(g) \; \ddots \; T^{(1)}(g)}^{a_1} & & & Z \\ & \overbrace{T^{(2)}(g) \; \ddots}^{a_2} & & \\ & & \ddots & \\ Z & & & \overbrace{T^{(\alpha)}(g) \; \ddots \; T^{(\alpha)}(g)}^{a_\alpha} \end{pmatrix}$$

These two bases are related by expressions of the form

(5.12) $$\mathbf{w}_l^{\xi,s} = \sum_{ij} (\mu i, \nu j \mid \xi s l) \mathbf{v}_i^{(\mu)} \otimes \mathbf{v}_j^{(\nu)}.$$

The expansion coefficients $(\mu i, \nu j \mid \xi s l)$ are called **Clebsch–Gordan (CG) coefficients**. These coefficients form an $n_\mu n_\nu \times n_\mu n_\nu$ matrix. This matrix is clearly invertible with inverse matrix elements defined by

(5.13) $$\mathbf{v}_i^{(\mu)} \otimes \mathbf{v}_j^{(\nu)} = \sum_{\xi k l} (\xi s l \mid \mu i, \nu j) \mathbf{w}_l^{\xi,s}, \qquad 1 \le i \le n_\mu, \quad 1 \le j \le n_\nu.$$

As an immediate consequence we have the relations

(5.14) $$\begin{aligned} &\sum_{\xi s l} (\mu i, \nu j \mid \xi s l)(\xi s l \mid \mu i', \nu j') = \delta_{ii'} \delta_{jj'} \\ &\sum_{ij} (\xi s l \mid \mu i, \nu j)(\mu i, \nu j \mid \xi' s' l') = \delta_{\xi\xi'} \delta_{ss'} \delta_{ll'}. \end{aligned}$$

Furthermore, if we assume, as we can, that the $T^{(\xi)}(g)$ are unitary matrices and the above bases are ON with respect to an inner product $\langle -, - \rangle$ on $V^{(\mu)} \otimes V^{(\nu)}$ then the matrix formed by the CG coefficients is unitary, i.e.,

(5.15) $$(\xi s l \mid \mu i, \nu j) = \overline{(\mu i, \nu j \mid \xi s l)}.$$

Although the "natural basis" is the easiest to compute it is the **w**-basis which is the most useful in applications, since this basis explicitly exhibits the decomposition of $\mathbf{T}^{(\mu)} \otimes \mathbf{T}^{(\nu)}$ into irred reps. By (5.12), to obtain this new basis from $\{\mathbf{v}_i^{(\mu)} \otimes \mathbf{v}_j^{(\nu)}\}$ it is sufficient to know the CG coefficients. For this reason much effort has been expended in the compilation of CG coefficients. We will return to this problem later.

If $\mathbf{T}_1$ and $\mathbf{T}_2$ are reps of the groups G_1 and G_2, respectively, we can define a rep $\mathbf{T}$ of the direct product group $G_1 \times G_2$ on $V_1 \otimes V_2$ by

$$(5.16) \qquad \mathbf{T}(g_1 g_2)\mathbf{v}_1 \otimes \mathbf{v}_2 = \mathbf{T}_1(g_1)\mathbf{v}_1 \otimes \mathbf{T}_2(g_2)\mathbf{v}_2 \qquad g_i \in G_i, \quad \mathbf{v}_i \in V_i.$$

If $\mathbf{T}_1$ is n_1-dimensional and $\mathbf{T}_2$ is n_2-dimensional then $\mathbf{T}$ is $n_1 n_2$-dimensional. Furthermore, an elementary computation similar to (5.8) shows that the character χ of $\mathbf{T}$ is

$$(5.17) \qquad \chi(g_1 g_2) = \chi_1(g_1)\chi_2(g_2)$$

where χ_i is the character of $\mathbf{T}_i$. If $\mathbf{T}_1$ and $\mathbf{T}_2$ are irred then

$$\langle \chi, \chi \rangle_{G_1 \times G_2} = \langle \chi_1, \chi_1 \rangle_{G_1} \langle \chi_2, \chi_2 \rangle_{G_2} = 1$$

so χ is irred. Let $\chi_1^{(\mu)}$, $1 \le \mu \le \alpha_1$, be the simple characters of G_1 corresponding to reps of dimension $n_\mu^{(1)}$. Let $\chi_2^{(\nu)}$, $1 \le \nu \le \alpha_2$, and $n_\nu^{(2)}$ be similar quantities for G_2. Then the characters $\chi^{(\mu,\nu)}(g_1 g_2) = \chi_1^{(\mu)}(g_1)\chi_2^{(\nu)}(g_2)$ belong to $\alpha_1 \alpha_2$ nonequivalent irred reps $\mathbf{T}^{(\mu,\nu)}$ of $G_1 \times G_2$, since

$$\langle \chi^{(\mu,\nu)}, \chi^{(\mu',\nu')} \rangle_{G_1 \times G_2} = \delta_{\mu\mu'} \delta_{\nu\nu'}.$$

Now G_1 has α_1 conjugacy classes and G_2 has α_2 conjugacy classes, so $G_1 \times G_2$ must have exactly $\alpha_1 \alpha_2$ conjugacy classes. Thus, every irred rep of $G_1 \times G_2$ is equivalent to exactly one of the irred reps $\mathbf{T}^{(\mu,\nu)}$. We have shown that a knowledge of the irred characters and reps of the factors G_1, G_2 immediately yields the irred characters and reps of $G_1 \times G_2$.

If $\mathbf{T}$ is a rep of G on V we can obtain a rep $\mathbf{T}_H$ of any subgroup H of G by restricting $\mathbf{T}$ to H,

$$(5.18) \qquad \mathbf{T}_H(h) = \mathbf{T}(h), \qquad h \in H.$$

We sometimes write $\mathbf{T}_H = \mathbf{T} \mid H$. The character $\chi_H = \chi \mid H$ of this rep is given by $\chi_H(h) = \chi(h)$.

On the other hand, there is a method due to Frobenius for constructing a rep of G from a rep of the subgroup H. Let $\mathbf{T}$ be a rep of H on the space V. Denote by $\mathcal{V}^G$ the vector space of all functions $\mathbf{f}(g)$ with domain G and range contained in V where addition and scalar multiplication of functions are the vector operations. Here, for a fixed $g \in G$, $\mathbf{f}(g)$ is a vector in V. Let V^G be the subspace of $\mathcal{V}^G$ defined by

$$(5.19) \qquad V^G = \{\mathbf{f} \in \mathcal{V}^G : \mathbf{f}(hg) = \mathbf{T}(h)\mathbf{f}(g) \quad \text{for all} \quad h \in H, \quad g \in G\}$$

We define a rep $\mathbf{T}^G$ of G on V^G by

$$[\mathbf{T}^G(g)\mathbf{f}](g') = \mathbf{f}(g'g), \qquad g, g' \in G, \quad \mathbf{f} \in V^G. \tag{5.20}$$

It is clear that V^G is invariant under G and the operators $\mathbf{T}^G(g)$ satisfy the homomorphism property. Here, $\mathbf{T}^G$ is called an **induced representation**. Let

$$Hg_1, \ldots, Hg_m, \qquad n(G) = m \cdot n(H)$$

be the distinct right cosets of H, where $g_1 = e$. Any $\mathbf{f} \in V^G$ is uniquely determined by the m vectors $\mathbf{f}(g_1), \ldots, \mathbf{f}(g_m)$ since for $g = hg_i$ in the right coset Hg_i we have

$$\mathbf{f}(g) = \mathbf{f}(hg_i) = \mathbf{T}(h)\mathbf{f}(g_i).$$

Let $\{\mathbf{v}_j\}$, $1 \le j \le d$, be a basis for V and define elements $\mathbf{e}_j{}^k(g)$ of V^G by

$$\mathbf{e}_j{}^k(g_i) = \delta_{ik}\mathbf{v}_j, \qquad 1 \le i, k \le m, \quad 1 \le j \le d. \tag{5.21}$$

The functions $\{\mathbf{e}_j{}^k\}$ form a basis for V^G, so the induced rep is md-dimensional. Let $T(h)$ be the matrix of $\mathbf{T}(h)$ relative to the $\{\mathbf{v}_j\}$ basis. We will use $T(h)$ to compute the matrix rep of G defined by $\mathbf{T}^G$ relative to the $\{\mathbf{e}_j{}^k\}$ basis:

$$\begin{aligned}[\mathbf{T}^G(g)\mathbf{e}_j{}^k](g_s) &= \mathbf{e}_j{}^k(g_s g) = \mathbf{e}_j{}^k(hg_r) = \mathbf{T}(h)\mathbf{e}_j{}^k(g_r) \\ &= \sum_{i=1}^{d} T_{ij}(h)\mathbf{e}_i{}^r(g_k) = \sum_{i=1}^{d} T_{ij}(h)\mathbf{e}_i{}^l(g_s)\end{aligned}$$

where g_r and g_l are the representatives of the right cosets containing $g_s g$ and $g_k g^{-1}$, respectively. From (5.21) we have $h = g_l g g_k^{-1}$ for $s = l$, i.e., $r = k$. We conclude that

$$\mathbf{T}^G(g)\mathbf{e}_j{}^k = \sum_{i=1}^{d} T_{ij}(h)\mathbf{e}_i{}^l = \sum_{i,l} \dot{T}_{ij}(g_l g g_k^{-1})\mathbf{e}_i{}^l \tag{5.22}$$

where

$$\dot{T}_{ij}(g) = \begin{cases} T_{ij}(g) & \text{if} \quad g \in H \\ 0 & \text{if} \quad g \notin H. \end{cases} \tag{5.23}$$

If we order the basis $\{\mathbf{e}_j{}^k\}$ in the sequence

$$\mathbf{e}_1{}^1, \ldots, \mathbf{e}_d{}^1, \mathbf{e}_1{}^2, \ldots, \mathbf{e}_d{}^2, \ldots, \mathbf{e}_1{}^m, \ldots, \mathbf{e}_d{}^m$$

then the matrix of $\mathbf{T}^G(g)$ with respect to this basis is

$$T^G(g) = \begin{pmatrix} \dot{T}(g_1 g g_1^{-1}) & \cdots & \dot{T}(g_1 g g_m^{-1}) \\ \cdot & & \cdot \\ \cdot & & \cdot \\ \cdot & & \cdot \\ \dot{T}(g_m g g_1^{-1}) & \cdots & \dot{T}(g_m g g_m^{-1}) \end{pmatrix}. \tag{5.24}$$

That is, the $md \times md$ matrix $T^G(g)$ is partitioned into an $m \times m$ array of $d \times d$ matrix blocks. The block in the jth row and kth column of the array is $\dot{T}(g_j g g_k^{-1})$. The character is clearly

$$\chi^G(g) = \sum_{k=1}^{m} \sum_{i=1}^{d} \dot{T}_{ii}(g_k g g_k^{-1}) = \sum_{k=1}^{m} \dot{\chi}(g_k g g_k^{-1}) \tag{5.25}$$

where $\chi(h)$ is the character of $\mathbf{T}$ and

$$\dot{\chi}(g) = \begin{cases} \chi(g) & \text{if} \quad g \in H \\ 0 & \text{if} \quad g \notin H. \end{cases} \tag{5.26}$$

We can write (5.25) in a more convenient form by noting that

$$\dot{\chi}(hg_k g(hg_k)^{-1}) = \dot{\chi}(g_k g g_k^{-1}), \qquad h \in H.$$

Therefore,

$$\chi^G(g) = [n(H)]^{-1} \sum_{t \in G} \dot{\chi}(tgt^{-1}). \tag{5.27}$$

To recapitulate, given the character χ corresponding to a rep $\mathbf{T}$ of H, we can define the character χ^G of the induced rep $\mathbf{T}^G$ by expression (5.27). One of the most useful induced reps is that obtained from the one-dimensional identity rep of H. Then $\chi(h) = 1$ for all $h \in H$ and

$$\chi^G(g) = \frac{n(G)}{n(H)} \frac{m_g}{n_g} \tag{5.28}$$

where n_g is the number of elements in G conjugate to g and m_g is the number of elements in $H \cap G$ conjugate to g. (Prove it!)

An important result on induced reps is the **Frobenius reciprocity theorem.** Let H be a subgroup of G and let $\mathbf{T}$, $\mathbf{Q}$ be irred reps of H and G with characters χ, ψ, respectively.

Theorem 3.9. The multiplicity of the irred rep $\mathbf{Q}$ in $\mathbf{T}^G$ is equal to the multiplicity of the irred rep $\mathbf{T}$ in $\mathbf{Q} \mid H = \mathbf{Q}_H$.

Proof. It is enough to show that

$$\langle \chi^G, \psi \rangle_G = \langle \chi, \psi_H \rangle_H \tag{5.29}$$

since the left-hand side is the multiplicity of $\mathbf{Q}$ in $\mathbf{T}^G$ and the right-hand side is the multiplicity of $\mathbf{T}$ in $\mathbf{Q}_H$. Using (5.26) and (5.27) we have

$$\begin{aligned} \langle \chi^G, \psi \rangle_G &= [n(G)]^{-1} \sum_{g \in G} \chi^G(g)\bar{\psi}(g) \\ &= [n(G)n(H)]^{-1} \sum_{g, s \in G} \dot{\chi}(sgs^{-1})\bar{\psi}(g). \end{aligned}$$

Since $\psi(sgs^{-1}) = \psi(g)$ and sgs^{-1} ranges over G as g does for fixed $s \in G$, there follows

$$\begin{aligned} \langle \chi^G, \psi \rangle_G &= [n(H)]^{-1} \sum_{t \in G} \dot{\chi}(t)\bar{\psi}(t) \\ &= [n(H)]^{-1} \sum_{t \in H} \chi(t)\bar{\psi}(t) = \langle \chi, \psi_H \rangle_H. \quad \text{Q.E.D.} \end{aligned}$$

The Frobenius reciprocity theorem is important because it enables one to decompose any induced rep into a direct sum of irred reps.

3.6 Character Tables

We now apply the results of the preceding sections to compute the simple characters and reps of the crystallographic point groups. For many applications only the simple characters are needed, not the rep matrices themselves. Furthermore, a simple character yields much information about its corresponding group rep and it is often possible to construct the group rep rather easily once the character is known. Although we study primarily the crystallographic point groups, the techniques used in the construction are applicable to any finite group.

Let G be a finite group of order N and $\mathbf{T}^{(1)}, \ldots, \mathbf{T}^{(\alpha)}$ a complete set of nonequivalent irred reps with dimensions $n_1, \ldots, n_\alpha$. The group G has α conjugacy classes and

(6.1) $$n_1^2 + n_2^2 + \cdots + n_\alpha^2 = N.$$

Furthermore, the characters $\chi^{(\mu)}$ obey the orthogonality relations

(6.2) $$\langle \chi^{(\mu)}, \chi^{(\nu)} \rangle = N^{-1} \sum_{i=1}^{\alpha} m_i \chi_i^{(\mu)} \overline{\chi_i^{(\nu)}} = \delta_{\mu\nu}, \qquad 1 \leq \mu, \nu \leq \alpha,$$

and

(6.3) $$\sum_{\mu=1}^{\alpha} \chi_i^{(\mu)} \overline{\chi_j^{(\nu)}} = \delta_{ij} N/m_i, \qquad 1 \leq i, j \leq \alpha,$$

where $\chi_i^{(\mu)}$ is the value of $\chi^{(\mu)}(g)$ for g an element of the ith conjugacy class $\mathcal{K}_i$ and m_i is the number of elements in $\mathcal{K}_i$. We assume $\mathcal{K}_1 = \{e\}$ and $m_1 = 1$. Thus, $\chi_1^{(\mu)} = \chi^{(\mu)}(e) = n_\mu$. A **character table** for G is a table of the form

(6.4)

	$\mathcal{K}_1$	$m_2\mathcal{K}_2$	$\cdots$	$m_\alpha\mathcal{K}_\alpha$
$\chi^{(1)}$	$\chi_1^{(1)}$	$\chi_2^{(1)}$	$\cdots$	$\chi_\alpha^{(1)}$
$\chi^{(2)}$	$\chi_1^{(2)}$	$\chi_2^{(2)}$	$\cdots$	$\chi_\alpha^{(2)}$
$\vdots$				
$\chi^{(\alpha)}$	$\chi_1^{(\alpha)}$	$\chi_2^{(\alpha)}$	$\cdots$	$\chi_\alpha^{(\alpha)}$

listing all simple characters of G. We already know $\chi^{(1)}(g) = 1$ for all $g \in G$, the character of the one-dimensional rep in which $\mathbf{T}(g) = \mathbf{E}$. To obtain the rest of the table we use the orthogonality relations and various devices for constructing reps which were discussed in the previous section. We always assume the characters are ordered so that $1 = n_1 \leq n_2 \leq \cdots \leq n_\alpha$.

It is worth noting that isomorphic groups have the same reps. The rep theory of a group is determined by its abstract group structure alone. Thus, the isomorphic groups D_n and C_{nv} have the same character tables even though these groups are not conjugate subgroups of $E(3)$. Similarly, the following pairs of isomorphic groups have the same character tables:

$$S_{2n} \cong C_{2n}, \qquad C_{2h} \cong D_2, \qquad T_d \cong O, \qquad D_{2nd} \cong D_{4n}. \tag{6.5}$$

A number of the crystallographic point groups can be expressed as direct products of groups of lower order:

$$\begin{gathered} C_6 \cong C_3 \times C_2, \qquad C_{nh} \cong C_n \times C_2, \qquad D_{nh} \cong D_n \times C_2 \\ D_{2n+1,d} \cong D_{2n+1} \times C_2, \qquad T_h \cong T \times C_2, \qquad O_h \cong O \times C_2 \\ D_2 \cong C_2 \times C_2, \qquad D_6 \cong D_3 \times C_2. \end{gathered} \tag{6.6}$$

According to the discussion following expression (5.16) the simple characters of each of the direct product groups can be obtained by forming all possible products of simple characters belonging to the factors. Thus, to derive character tables for each of the 32 crystallographic point groups it is enough to study the groups $C_2, C_3, C_4, D_3, D_4, T, O$.

The character tables of the cyclic groups C_n follow from Eq. (4.15). Let g be a generator of C_2, $g^2 = e$. Then the conjugacy classes of C_2 are

$$\mathcal{E} = \{e\}, \qquad \mathcal{C}_2 = \{g\} \tag{6.7}$$

and the character table reads

(6.8)

C_2	$\mathcal{E}$	$\mathcal{C}_2$
$\chi^{(1)}$	1	1
$\chi^{(2)}$	1	-1

Let g be a generator of C_3, $g^3 = e$. The conjugacy classes are

$$\mathcal{E} = \{e\}, \qquad \mathcal{C}_3 = \{g\}, \qquad \mathcal{C}_3^{\,2} = \{g^2\} \tag{6.9}$$

and we obtain

(6.10)

C_3	$\mathcal{E}$	$\mathcal{C}_3$	$\mathcal{C}_3^{\,2}$
$\chi^{(1)}$	1	1	1
$\chi^{(2)}$	1	ε	ε^2
$\chi^{(3)}$	1	ε^2	ε

$\varepsilon = \exp(2\pi i/3)$.

Finally, let g be a generator of C_4, $g^4 = e$. The conjugacy classes are

$$\mathcal{E} = \{e\}, \qquad \mathcal{C}_4 = \{g\}, \qquad \mathcal{C}_4^{\,2} = \{g^2\}, \qquad \mathcal{C}_4^{\,3} = \{g^3\} \tag{6.11}$$

and the character table is

(6.12)

C_4	$\mathcal{E}$	$\mathcal{C}_4$	$\mathcal{C}_4{}^2$	$\mathcal{C}_4{}^3$
$\chi^{(1)}$	1	1	1	1
$\chi^{(2)}$	1	i	-1	$-i$
$\chi^{(3)}$	1	-1	1	-1
$\chi^{(4)}$	1	$-i$	-1	i

The group D_3 (of order six) is not quite so easy to handle. The elements g, h with $g^3 = h^2 = e$ and $hgh = g^{-1}$ generate D_3. The conjugacy classes are

$$\mathcal{E} = \{e\}, \qquad \mathcal{C}_3 = \{g, g^2\}, \qquad \mathcal{C}_2 = \{h, gh, g^2h\}. \tag{6.13}$$

Thus, there are three irred reps of dimensions n_1, n_2, n_3 with $n_1 = 1$ (the identity rep) and

$$n_1{}^2 + n_2{}^2 + n_3{}^2 = 6.$$

The only possible solution is $n_1 = n_2 = 1$, $n_3 = 2$. There is another one-dimensional rep in addition to the identity rep. This can easily be found by inspection: $\chi^{(2)}(g) = 1$, $\chi^{(2)}(h) = -1$.

To obtain the third character we use the orthogonality relations

$$0 = 6\langle\chi^{(3)}, \chi^{(1)}\rangle = 2 + 2\chi_2^{(3)} + 3\chi_3^{(3)}$$
$$0 = 6\langle\chi^{(3)}, \chi^{(2)}\rangle = 2 + 2\chi_2^{(3)} - 3\chi_3^{(3)}.$$

Solving these equations simultaneously, we find $\chi_2^{(3)} = -1$, $\chi_3^{(3)} = 0$. The complete table is

(6.14)

D_3	$\mathcal{E}$	$2\mathcal{C}_3$	$3\mathcal{C}_2$
$\chi^{(1)}$	1	1	1
$\chi^{(2)}$	1	1	-1
$\chi^{(3)}$	2	-1	0

The two-dimensional rep $\mathbf{T}^{(3)}$ is equivalent to the 2×2 matrix rep one obtains by considering D_3 as a transformation group in the plane, i.e., as the symmetry group of an equilateral triangle. Indeed, with respect to the basis pictured in Fig. 3.1, we can associate the matrices

$$g \sim \begin{pmatrix} -\frac{1}{2} & \frac{1}{2}\sqrt{3} \\ -\frac{1}{2}\sqrt{3} & -\frac{1}{2} \end{pmatrix}, \qquad h \sim \begin{pmatrix} -1 & 0 \\ 0 & 1 \end{pmatrix}. \tag{6.15}$$

Thus, the character χ satisfies $\chi(e) = \chi_1 = 2$, $\chi(g) = \chi_2 = -1$, $\chi(h) = \chi_3 = 0$, so $\chi = \chi^{(3)}$.

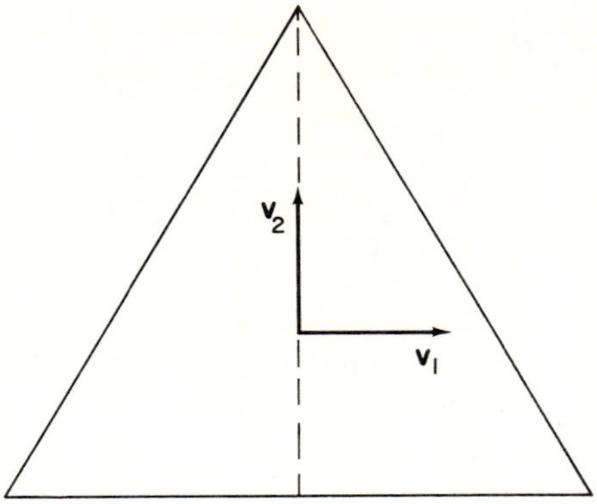

FIGURE 3.1

The group D_4 of order eight is generated by elements g, h such that $g^4 = h^2 = e$ and $(gh)^2 = e$. This group has five conjugacy classes:

(6.16)
$$\mathcal{E} = \{e\}, \quad \mathcal{C}_4{}^2 = \{g^2\}, \quad \mathcal{C}_4 = \{g, g^3\},$$
$$\mathcal{C}_2 = \{h, g^2h\}, \quad \mathcal{C}_2{}' = \{gh, g^3h\}.$$

Thus, there are five irred reps of dimensions $n_1 = 1 \leq n_2 \leq \cdots \leq n_5$ such that

$$n_1{}^2 + n_2{}^2 + n_3{}^2 + n_4{}^2 + n_5{}^2 = 8.$$

The only possibility is $n_1 = n_2 = n_3 = n_4 = 1, n_5 = 2$. The one-dimensional reps can be determined by inspection. Thus,

(6.17)

D_4	$\mathcal{E}$	$\mathcal{C}_4{}^2$	$2\mathcal{C}_4$	$2\mathcal{C}_2$	$2\mathcal{C}_2{}'$
$\chi^{(1)}$	1	1	1	1	1
$\chi^{(2)}$	1	1	1	-1	-1
$\chi^{(3)}$	1	1	-1	1	-1
$\chi^{(4)}$	1	1	-1	-1	1
$\chi^{(5)}$	2	$\chi_2^{(5)}$	$\chi_3^{(5)}$	$\chi_4^{(5)}$	$\chi_5^{(5)}$

The orthogonality relations $\langle \chi^{(5)}, \chi^{(j)} \rangle = 0$, $j = 1, 2, 3, 4$, imply $\chi_2^{(5)} = -2$, $\chi_3^{(5)} = \chi_4^{(5)} = \chi_5^{(5)} = 0$. The reader can explicitly construct a 2×2 irred matrix rep $T^{(5)}$ in a manner similar to (6.15). One realization of $T^{(5)}$ is generated by

(6.18)
$$T^{(5)}(g) = \begin{pmatrix} 0 & 1 \\ -1 & 0 \end{pmatrix}, \qquad T^{(5)}(h) = \begin{pmatrix} 1 & 0 \\ 0 & -1 \end{pmatrix}.$$

The tetrahedral group T contains 12 elements. Realizing the elements of T as permutations of the four vertices of a tetrahedron, we obtain the four conjugacy classes

(6.19)
$$\mathcal{K}_1 = \{1\}, \qquad \mathcal{K}_2 = \{(12)(34), (13)(24), (14)(23)\}$$
$$\mathcal{K}_3 = \{(123), (142), (134), (243)\}, \qquad \mathcal{K}_4 = \{(132), (124), (143), (234)\}.$$

There are four irred reps of dimensions $n_1 = 1 \leq n_2 \leq \cdots \leq n_4$ such that

$$n_1{}^2 + n_2{}^2 + n_3{}^2 + n_4{}^2 = 12.$$

The only possible solution is $n_1 = n_2 = n_3 = 1$, $n_4 = 3$. Note that the subgroup

$$D = \{1, (12)(34), (13)(24), (14)(23)\}$$

(the identity element and the three rotations of 180°) is normal in T. Therefore, the factor group T/D is cyclic of order three and has three one-dimensional nonequivalent irred reps given by (6.10). These reps $\mathbf{T}^{(i)\prime}$, $i = 1, 2, 3$, are defined by mapping a generator g of C_3 into 1, ε, or ε^2, respectively, where $\varepsilon = \exp(2\pi i/3)$. The combined homomorphisms

$$T \longrightarrow T/D \xrightarrow{\mathbf{T}^{(i)\prime}} \mathfrak{C}$$

yield three one-dimensional nonequivalent reps of T such that D is mapped into the identity operator. These reps are necessarily irred and they exhaust the possible one-dimensional reps of T. A simple computation gives the character table

(6.20)

T	$\mathcal{K}_1$	$3\mathcal{K}_2$	$4\mathcal{K}_3$	$4\mathcal{K}_4$
$\chi^{(1)}$	1	1	1	1
$\chi^{(2)}$	1	1	ε	ε^2
$\chi^{(3)}$	1	1	ε^2	ε
$\chi^{(4)}$	3	$\chi_2^{(4)}$	$\chi_3^{(4)}$	$\chi_4^{(4)}$

The fourth line of the table can be obtained from the orthogonality relations $\langle\chi^{(4)}, \chi^{(j)}\rangle = 0$, $j = 1, 2, 3$, which have the solution $\chi_2^{(4)} = -1$, $\chi_3^{(4)} = \chi_4^{(4)} = 0$. Note that a rep of the form

$$\mathbf{T} \cong a_1\mathbf{T}^{(1)} + a_2\mathbf{T}^{(2)} + a_3\mathbf{T}^{(3)}$$

has the property that $\mathbf{T}(g_1)$ and $\mathbf{T}(g_2)$ commute for all g_1, g_2 in the tetrahedral group. Indeed, the reps $\mathbf{T}^{(i)}$, $i = 1, 2, 3$, are one-dimensional so that with respect to a suitable basis the matrices of the operators can be simultaneously diagonalized. The natural three-dimensional rep of the tetrahedral group as a transformation group on R_3 is not commutative. Therefore, this natural rep must be $\mathbf{T}^{(4)}$.

The octahedral group O, the direct symmetry group of the cube, contains 24 elements in five conjugacy classes. They are

$$\begin{aligned}
\mathcal{E} = \{e\}, \qquad \mathcal{C}_4{}^2 &= \{\text{three rotations of } 180^\circ \text{ about fourfold axes}\} \\
\mathcal{C}_2 &= \{\text{six rotations of } 180^\circ \text{ about twofold axes}\} \\
\mathcal{C}_4 &= \{\text{three rotations of } 90^\circ \text{ and three rotations of } 270^\circ \text{ about fourfold axes}\} \\
\mathcal{C}_3 &= \{\text{four rotations of } 120^\circ \text{ and four rotations of } 240^\circ \text{ about threefold axes}\}.
\end{aligned} \tag{6.21}$$

There are five irred reps of dimensions $1 \leq n_1 \leq n_2 \leq \cdots \leq n_5$ such that

$$n_1{}^2 + \cdots + n_5{}^2 = 24.$$

The only possibility is $n_1 = n_2 = 1$, $n_3 = 2$, $n_4 = n_5 = 3$. It is clear from the drawing of a cube in Fig. 3.2 that points $ABCD$ are the vertices of a tetra-

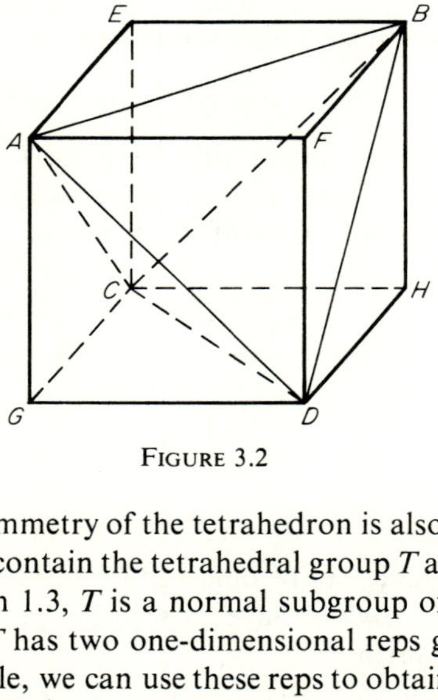

FIGURE 3.2

hedron. Every direct symmetry of the tetrahedron is also a direct symmetry of the cube. Thus O must contain the tetrahedral group T as a subgroup of index $24/12 = 2$. By Theorem 1.3, T is a normal subgroup of O and O/T is cyclic of order two. Now O/T has two one-dimensional reps given by (6.8). Just as in the preceding example, we can use these reps to obtain two one-dimensional irred reps of O such that T is mapped into the identity operator:

O	$\mathcal{E}$	$3\mathcal{C}_4{}^2$	$6\mathcal{C}_2$	$6\mathcal{C}_4$	$8\mathcal{C}_3$
$\chi^{(1)}$	1	1	1	1	1
$\chi^{(2)}$	1	1	-1	-1	1

Next we construct the two-dimensional irred rep $\mathbf{T}^{(3)}$. Consider the simple character $\chi^{(2)}$ of T, table (6.20). The corresponding induced character χ^O is associated with a two-dimensional rep of O and is easily shown to be given by

O	$\mathcal{E}$	$3\mathcal{C}_4{}^2$	$6\mathcal{C}_2$	$6\mathcal{C}_4$	$8\mathcal{C}_3$
χ^O	2	2	0	0	-1

[See expression (5.27) defining an induced character.] Now

$$\langle \chi^O, \chi^O \rangle = (1/24)(2\cdot 2 + 3\cdot 2\cdot 2 + 6\cdot 0 + 6\cdot 0 + 8\cdot(-1)\cdot(-1)) = 1$$

so χ^O is simple. Since there is only one irred rep of dimension two we must have $\chi^O = \chi^{(3)}$. By means of (5.24) we could explicitly construct the irred rep $\mathbf{T}^{(3)}$.

The natural rep of O as a transformation group on R_3 must be irred since its restriction to T is the irred rep whose character is given by the bottom row of table (6.20). It follows that the character $\chi^{(4)}$ of this rep is

O	$\mathcal{E}$	$3\mathcal{C}_4{}^2$	$6\mathcal{C}_2$	$6\mathcal{C}_4$	$8\mathcal{C}_3$
$\chi^{(4)}$	3	-1	$\chi_3^{(4)}$	$\chi_4^{(4)}$	0

The values $\chi_3^{(4)}$, $\chi_4^{(4)}$ are not immediately determined since no elements of T lie in $\mathcal{C}_2$ or $\mathcal{C}_4$. To obtain these elements note that in suitable coordinate systems, a rotation of 180° about a twofold axis and a rotation of 90° about a fourfold axis can be represented by the matrices

$$\begin{pmatrix} 1 & 0 & 0 \\ 0 & -1 & 0 \\ 0 & 0 & -1 \end{pmatrix}, \qquad \begin{pmatrix} 0 & 1 & 0 \\ -1 & 0 & 0 \\ 0 & 0 & 1 \end{pmatrix},$$

respectively. Taking traces we obtain $\chi_3^{(4)} = -1$, $\chi_4^{(4)} = 1$.

There remains only a single irred rep of dimension three. We could obtain the character of this rep by using the orthogonality relations. However, it is more informative to consider the tensor product $\mathbf{T}^{(5)} \cong \mathbf{T}^{(2)} \otimes \mathbf{T}^{(4)}$. This is a three-dimensional rep with character $\chi^{(5)}(g) = \chi^{(2)}(g)\chi^{(4)}(g)$. Thus,

O	$\mathcal{E}$	$3\mathcal{C}_4{}^2$	$6\mathcal{C}_2$	$6\mathcal{C}_4$	$8\mathcal{C}_3$
$\chi^{(5)}$	3	-1	1	-1	0

Since

$$\langle \chi^{(5)}, \chi^{(5)} \rangle = (1/24)(9 + 3 + 6 + 6 + 0) = 1$$

it follows that $\chi^{(5)}$ is a simple character distinct from $\chi^{(4)}$. The complete character table is therefore

(6.22)

O	$\mathcal{E}$	$3\mathcal{C}_4^{\,2}$	$6\mathcal{C}_2$	$6\mathcal{C}_4$	$8\mathcal{C}_3$
$\chi^{(1)}$	1	1	1	1	1
$\chi^{(2)}$	1	1	−1	−1	1
$\chi^{(3)}$	2	2	0	0	−1
$\chi^{(4)}$	3	−1	−1	1	0
$\chi^{(5)}$	3	−1	1	−1	0

The above results enable us to compute the simple characters and reps for each of the crystallographic point groups. Similar techniques yield the irred reps of any finite group, although in practice the required computations may be extremely difficult.

3.7 The Method of Projection Operators

Let G be a finite group and $\mathbf{T}$ a reducible unitary rep of G on the vector space V. Suppose $\mathbf{T}^{(1)}, \ldots, \mathbf{T}^{(\alpha)}$ are a complete set of nonequivalent irred unitary reps of G. Then V can be decomposed into a direct sum of invariant subspaces

(7.1)
$$V = \sum_{\mu=1}^{\alpha} \sum_{i=1}^{a_\mu} \oplus V_i^{(\mu)}$$

where the restriction of $\mathbf{T}$ to $V_i^{(\mu)}$ is equivalent to the irred rep $\mathbf{T}^{(\mu)}$. Here a_μ is the multiplicity of $\mathbf{T}^{(\mu)}$ in $\mathbf{T}$. In this section we study a method which allows us to explicitly perform the decomposition (7.1). Furthermore, we examine the subspaces $V_i^{(\mu)}$ and determine to what extent they are unique.

Let $\mathbf{A}$ be a linear operator on the finite-dimensional inner product space V. Recall that $\mathbf{A}^*$, the **adjoint** of A, is the linear operator on V uniquely defined by

(7.2)
$$\langle \mathbf{Au}, \mathbf{v} \rangle = \langle \mathbf{u}, \mathbf{A}^*\mathbf{v} \rangle$$

for all $\mathbf{u}, \mathbf{v} \in V$, where $\langle -, - \rangle$ is the inner product. With respect to an ON basis for V the matrix for $\mathbf{A}^*$ is the conjugate transpose of the matrix for $\mathbf{A}$. The operator $\mathbf{A}$ is **self-adjoint** if $\mathbf{A} = \mathbf{A}^*$. If $\mathbf{A}^2 = \mathbf{A}$ then $\mathbf{A}$ is a **projection operator.** The **range** $R_\mathbf{A}$ and the **null space** $N_\mathbf{A}$ of a linear transformation are the V subspaces

(7.3)
$$R_\mathbf{A} = \{\mathbf{w} \in V : \mathbf{w} = \mathbf{Av} \quad \text{for some} \quad \mathbf{v} \in V\}, \qquad N_\mathbf{A} = \{\mathbf{v} \in V : \mathbf{Av} = \boldsymbol{\theta}\}.$$

Let $\mathbf{P}$ be a projection operator on V and let $W = R_\mathbf{P}$. Any $\mathbf{v} \in V$ can be written uniquely in the form

$$\mathbf{v} = \mathbf{Pv} + (\mathbf{E} - \mathbf{P})\mathbf{v} = \mathbf{w} + \mathbf{w}'$$

where $\mathbf{w} \in W$, $\mathbf{w}' \in W' = R_{(\mathbf{E}-\mathbf{P})}$. In particular, $\mathbf{Pw} = \mathbf{P}^2\mathbf{v} = \mathbf{Pv} = \mathbf{w}$ for $\mathbf{w} \in W$ and $\mathbf{Pw}' = \mathbf{P}(\mathbf{E} - \mathbf{P})\mathbf{v} = \boldsymbol{\theta}$ for $\mathbf{w}' \in W'$. If $\mathbf{u} \in W \cap W'$, then $\mathbf{Pu} = \mathbf{u}$ since $\mathbf{u} \in W$ and $\mathbf{Pu} = \boldsymbol{\theta}$ since $\mathbf{u} \in W'$. Thus, $W \cap W' = \{\boldsymbol{\theta}\}$ and

$$V = W \oplus W'. \tag{7.4}$$

We have shown that the projection $\mathbf{P}$ induces a direct sum decomposition of V. Conversely, if W and W' are subspaces of V such that (7.4) is valid, then the assignment $\mathbf{Pv} = \mathbf{w}$ defined by the decomposition

$$\mathbf{v} = \mathbf{w} + \mathbf{w}', \qquad \mathbf{v} \in V, \qquad \mathbf{w} \in W, \qquad \mathbf{w}' \in W'$$

determines a projection operator on V. Indeed, $\mathbf{Pw} = \mathbf{w}$ so $\mathbf{P}^2\mathbf{v} = \mathbf{Pw} = \mathbf{w} = \mathbf{Pv}$ and $\mathbf{P}^2 = \mathbf{P}$. Different choices of the supplementary space W' lead to different projection operators $\mathbf{P}$. If $W' = W^{\perp}$, then the corresponding projection operator is self-adjoint.

Theorem 3.10. There is a 1–1 relationship between subspaces W of V and self-adjoint projection operators $\mathbf{P}$ on V, given by $W = R_{\mathbf{P}}$, $W^{\perp} = N_{\mathbf{P}}$.

Proof. The decomposition $V = W \oplus W^{\perp}$ defines a projection $\mathbf{P}$ with $W = R_{\mathbf{P}}$, $W^{\perp} = N_{\mathbf{P}}$. If $\mathbf{v}_1, \mathbf{v}_2 \in V$ with

$$\mathbf{v}_i = \mathbf{w}_i + \mathbf{w}_i', \qquad \mathbf{w}_i \in W, \qquad \mathbf{w}_i' \in W^{\perp}, \qquad i = 1, 2,$$

then

$$\langle \mathbf{Pv}_1, \mathbf{v}_2 \rangle = \langle \mathbf{w}_1, \mathbf{w}_2 + \mathbf{w}_2' \rangle = \langle \mathbf{w}_1 + \mathbf{w}_1', \mathbf{w}_2 \rangle = \langle \mathbf{v}_1, \mathbf{Pv}_2 \rangle$$

so $\mathbf{P}^* = \mathbf{P}$.

Conversely, suppose $\mathbf{P}$ is a self-adjoint projection operator on V and set $W = R_{\mathbf{P}}$. Since $\mathbf{E} = \mathbf{P} + (\mathbf{E} - \mathbf{P})$ we can write

$$\mathbf{v} = \mathbf{Pv} + (\mathbf{E} - \mathbf{P})\mathbf{v} = \mathbf{w} + \mathbf{w}', \qquad \mathbf{v} \in V. \tag{7.5}$$

By definition $\mathbf{Pv} \in W$, while for any vector $\mathbf{Pu}, \mathbf{u} \in V$, in W we obtain

$$\langle \mathbf{Pu}, (\mathbf{E} - \mathbf{P})\mathbf{v} \rangle = \langle \mathbf{u}, \mathbf{P}(\mathbf{E} - \mathbf{P})\mathbf{v} \rangle = \langle \mathbf{u}, (\mathbf{P} - \mathbf{P})\mathbf{v} \rangle = 0 \tag{7.6}$$

so $(\mathbf{E} - \mathbf{P})\mathbf{v} \in W^{\perp}$. Thus $N_{\mathbf{P}} = R_{(\mathbf{E}-\mathbf{P})} = W^{\perp}$. Q.E.D.

Theorem 3.11. Let $\mathbf{T}$ be a finite-dimensional rep of G on the inner product space V. If V can be decomposed in the form $V = W_1 \oplus W_2$, where W_1 and W_2 are invariant under $\mathbf{T}$ then the projection operator $\mathbf{P}$ on V defined by

$$\mathbf{Pv} = \mathbf{w}_1 \quad \text{for} \quad \mathbf{v} = \mathbf{w}_1 + \mathbf{w}_2, \quad \mathbf{w}_i \in W_i \tag{7.7}$$

satisfies

$$\mathbf{T}(g)\mathbf{P} = \mathbf{PT}(g) \qquad \text{for all} \quad g \in G. \tag{7.8}$$

Conversely, if $\mathbf{P}$ is a projection operator on V satisfying (7.8) then $V = W_1 \oplus W_2$, where $W_1 = R_{\mathbf{P}}$ and $W_2 = N_{\mathbf{P}}$ are invariant under $\mathbf{T}$.

Proof. Suppose W_1, W_2 are invariant under $\mathbf{T}$ and $V = W_1 \oplus W_2$. Then any $\mathbf{v} \in V$ can be written uniquely in the form $\mathbf{v} = \mathbf{w}_1 + \mathbf{w}_2$, $\mathbf{w}_i \in W_i$, and

$$\mathbf{PT}(g)\mathbf{v} = \mathbf{P}(\mathbf{T}(g)\mathbf{w}_1 + \mathbf{T}(g)\mathbf{w}_2) = \mathbf{T}(g)\mathbf{w}_1 = \mathbf{T}(g)\mathbf{Pv},$$

where $\mathbf{P}$ is defined by (7.7). Conversely, if $\mathbf{P}$ is a projection operator satisfying (7.8) and $\mathbf{w}_1 \in W_1 = R_{\mathbf{P}}$, then

$$\mathbf{T}(g)\mathbf{w}_1 = \mathbf{T}(g)\mathbf{Pw}_1 = \mathbf{PT}(g)\mathbf{w}_1 \in R_{\mathbf{P}}$$

so W_1 is invariant under $\mathbf{T}$. Similarly, if $\mathbf{w}_2 \in W_2 = N_{\mathbf{P}}$, then

$$\mathbf{PT}(g)\mathbf{w}_2 = \mathbf{T}(g)\mathbf{Pw}_2 = \boldsymbol{\theta}$$

so $N_{\mathbf{P}}$ is invariant under $\mathbf{T}(g)$. Q.E.D.

This theorem establishes a 1–1 relationship between decompositions of V into a direct sum of two invariant subspaces and projection operators on V which commute with the operators $\mathbf{T}(g)$. We now determine which operators correspond to subspaces which transform irreducibly under $\mathbf{T}$. Let $P(\mathbf{T})$ be the set of all projection operators on V which commute with $\mathbf{T}(g)$ for all $g \in G$ and let $IP(\mathbf{T})$ be the set of all $\mathbf{P} \in P(\mathbf{T})$ which **cannot** be written in the form

(7.9)
$$\mathbf{P} = \mathbf{P}_1 + \mathbf{P}_2, \qquad \mathbf{P}_i \in P(\mathbf{T}), \qquad \mathbf{P}_1\mathbf{P}_2 = \mathbf{P}_2\mathbf{P}_1 = \mathbf{Z}, \qquad \mathbf{P}_1, \mathbf{P}_2 \neq \mathbf{Z},$$

where $\mathbf{Z}$ is the zero operator on V.

Theorem 3.12. Let W be a proper invariant subspace of V and $\mathbf{P} \in P(\mathbf{T})$ a projection operator on W. Then W is irred under $\mathbf{T}$ if and only if $\mathbf{P} \in IP(\mathbf{T})$.

Proof. Suppose W is reducible under $\mathbf{T}$. Then W contains proper invariant subspaces W_1 and W_2 such that $W = W_1 \oplus W_2$. Also, the invariant subspace $W' = N_{\mathbf{P}}$ satisfies

$$V = W \oplus W' = W_1 \oplus W_2 \oplus W'.$$

Defining the projection operators $\mathbf{P}_1, \mathbf{P}_2 \in P(\mathbf{T})$ by

$$\mathbf{P}_1\mathbf{v} = \mathbf{w}_1, \qquad \mathbf{P}_2\mathbf{v} = \mathbf{w}_2,$$

where $\mathbf{v} = \mathbf{w}_1 + \mathbf{w}_2 + \mathbf{w}'$, with $\mathbf{w}_i \in W_i$, $\mathbf{w}' \in W'$, we obtain $\mathbf{P} = \mathbf{P}_1 + \mathbf{P}_2$. Furthermore, $\mathbf{P}_1\mathbf{P}_2 = \mathbf{P}_2\mathbf{P}_1 = \mathbf{Z}$, since $W_1 \cap W_2 = \{\boldsymbol{\theta}\}$.

Conversely, suppose $\mathbf{P} = \mathbf{P}_1 + \mathbf{P}_2$, with $\mathbf{P}_1, \mathbf{P}_2 \in P(\mathbf{T})$, $\mathbf{P}_1\mathbf{P}_2 = \mathbf{P}_2\mathbf{P}_1 = \mathbf{Z}$, and $\mathbf{P}, \mathbf{P}_1, \mathbf{P}_2 \neq \mathbf{Z}$. Then

$$\mathbf{P}^2 = (\mathbf{P}_1 + \mathbf{P}_2)^2 = \mathbf{P}_1{}^2 + \mathbf{P}_1\mathbf{P}_2 + \mathbf{P}_2\mathbf{P}_1 + \mathbf{P}_2{}^2 = \mathbf{P}_1 + \mathbf{P}_2 = \mathbf{P}$$

so $\mathbf{P} \in P(\mathbf{T})$. Set $W = R_{\mathbf{P}}$, $W_1 = R_{\mathbf{P}_1}$, $W_2 = R_{\mathbf{P}_2}$. For any $\mathbf{w} \in W$,

$$\mathbf{w} = \mathbf{P}\mathbf{w} = \mathbf{P}_1\mathbf{w} + \mathbf{P}_2\mathbf{w} = \mathbf{w}_1 + \mathbf{w}_2$$

with $\mathbf{w}_i = \mathbf{P}_i\mathbf{w} \in W_i$. If $\mathbf{w} \in W_1 \cap W_2$ then $\mathbf{w} = \mathbf{P}_1\mathbf{P}_2\mathbf{w} = \boldsymbol{\theta}$ since $\mathbf{P}_1\mathbf{P}_2 = \mathbf{Z}$. Thus $W = W_1 \oplus W_2$, where W, W_1, W_2 are nonzero invariant subspaces of V. Q.E.D.

We now return to the decomposition (7.1) of V into irred subspaces under the action of the operators $\mathbf{T}(g)$. Let $T^{(\mu)}(g)$, $1 \leq \mu \leq \alpha$, be a complete set of nonequivalent unitary matrix reps of G corresponding to the operator reps $\mathbf{T}^{(\mu)}$. We can find an ON basis $\{\mathbf{v}_{ij}^{(\mu)}, 1 \leq j \leq n_\mu\}$ for each irred subspace $V_i^{(\mu)}$ such that

$$\mathbf{T}(g)\mathbf{v}_{ij}^{(\mu)} = \sum_{k=1}^{n_\mu} T_{kj}^{(\mu)}(g)\mathbf{v}_{ik}^{(\mu)}, \qquad 1 \leq j \leq n_\mu. \tag{7.10}$$

Corresponding to each simple character $\chi^{(\mu)}$ of G we define the linear operator $\mathbf{P}_\mu$ on V by

$$\mathbf{P}_\mu = \frac{n_\mu}{n(G)} \sum_{g \in G} \overline{\chi^{(\mu)}(g)}\mathbf{T}(g). \tag{7.11}$$

Since $hg = h(gh)h^{-1}$, for $g, h \in G$, i.e., the elements gh and hg lie in the same conjugacy class of G, it follows easily that $\mathbf{P}_\mu$ commutes with the operators $\mathbf{T}(h)$. Now

$$\mathbf{P}_\mu\mathbf{v}_{ij}^{(\nu)} = \frac{n_\mu}{n(G)} \sum_g \sum_{l=1}^{n_\mu} \sum_{k=1}^{n_\nu} T_{kj}^{(\nu)}(g)\overline{T_{ll}^{(\mu)}(g)}\mathbf{v}_{ik}^{(\nu)} = \delta_{\mu\nu}\mathbf{v}_{ij}^{(\nu)}, \tag{7.12}$$

where we have used (7.10), (7.11), and the orthogonality relations for matrix elements. It follows from this result and Theorems 3.10 and 3.11, that $\mathbf{P}_\mu \in P(\mathbf{T})$ is the self-adjoint projection operator on the invariant subspace

$$V^{(\mu)} = \sum_{i=1}^{a_\mu} \oplus V_i^{(\mu)}. \tag{7.13}$$

In general the spaces $V_i^{(\mu)}$ occurring in the decomposition (7.1) are not uniquely determined. However, since the definition of the projection operators $\mathbf{P}_\mu$ is basis-independent, the spaces $V^{(\mu)}$ *are* uniquely determined. The ambiguity occurs in the decomposition of each $V^{(\mu)}$ into irreducible subspaces. If $a_\mu > 1$ there is no unique way to perform the decomposition (7.13).

To carry out this nonunique decomposition we define operators

$$\mathbf{P}_\mu^{lk} = \frac{n_\mu}{n(G)} \sum_{g \in G} \overline{T_{lk}^{(\mu)}(g)}\mathbf{T}(g), \qquad l, k = 1, \ldots, n_\mu. \tag{7.14}$$

Since

$$\mathbf{P}_\mu^{lk}\mathbf{v}_{ij}^{(\nu)} = \frac{n_\mu}{n(G)} \sum_{g \in G} \sum_{m=1}^{n_\mu} T_{mj}^{(\nu)}(g)\overline{T_{lk}^{(\mu)}(g)}\mathbf{v}_{im}^{(\nu)} = \delta_{\mu\nu}\delta_{jk}\mathbf{v}_{il}^{(\nu)}, \tag{7.15}$$

it follows that $\mathbf{P}_\mu^{lk}$ is the self-adjoint projection operator on the a_μ-dimensional space $W_k^{(\mu)}$ spanned by the ON basis vectors $\{\mathbf{v}_{ik}^{(\mu)}, i = 1, \ldots, a_\mu\}$. Furthermore, the relations

(7.16) $$\mathbf{P}_\mu^{lk}\mathbf{P}_{\mu'}^{l'k'} = \delta_{\mu\mu'}\delta_{kl'}\mathbf{P}_\mu^{lk'}, \quad (\mathbf{P}_\mu^{lk})^* = \mathbf{P}_\mu^{kl},$$

(7.17) $$\mathbf{P}_\mu = \sum_{k=1}^{n_\mu} \mathbf{P}_\mu^{kk}$$

follow easily from (7.15).

If $\mathbf{v} \in V$ such that for some l, k, μ we have $\mathbf{P}_\mu^{lk}\mathbf{v} = \mathbf{w}_l \neq \boldsymbol{\theta}$, then the n_μ vectors $\mathbf{w}_j = \mathbf{P}_\mu^{jk}\mathbf{v} = \mathbf{P}_\mu^{jl}\mathbf{w}_l$, $1 \leq j \leq n_\mu$, are all nonzero and span an invariant subspace of V which transforms irreducibly according to the rep $\mathbf{T}^{(\mu)}$. Indeed, if

$$\mathbf{v} = \sum_{\nu lm} \alpha_{lm}^{(\nu)}\mathbf{v}_{lm}^{(\nu)}$$

then

$$\mathbf{w}_j = \mathbf{P}_\mu^{jk}\mathbf{v} = \sum_{l=1}^{a_\mu} \alpha_{lk}^{(\mu)}\mathbf{v}_{lj}^{(\mu)}$$

and

$$\mathbf{T}(g)\mathbf{w}_j = \sum_{l=1}^{a_\mu} \alpha_{lk}^{(\mu)}\mathbf{T}(g)\mathbf{v}_{lj}^{(\mu)} = \sum_{l=1}^{n_\mu} T_{ij}^{(\mu)}(g)\mathbf{w}_i.$$

Furthermore, if $\mathbf{v}, \mathbf{v}' \in V$ are such that the vectors $\mathbf{w}_l = \mathbf{P}_\mu^{lk}\mathbf{v}$ and $\mathbf{w}_l' = \mathbf{P}_\mu^{lk}\mathbf{v}'$ are orthogonal for some fixed l, k, μ then

(7.18) $$\begin{aligned}\langle \mathbf{P}_\mu^{jk}\mathbf{v}, \mathbf{P}_\mu^{ik}\mathbf{v}'\rangle &= \langle \mathbf{P}_\mu^{jl}\mathbf{P}_\mu^{lk}\mathbf{v}, \mathbf{P}_\mu^{il}\mathbf{P}_\mu^{lk}\mathbf{v}'\rangle \\ &= \langle \mathbf{P}_\mu^{jl}\mathbf{w}_l, \mathbf{P}_\mu^{il}\mathbf{w}_l'\rangle = \delta_{ij}\langle \mathbf{w}_l, \mathbf{w}_l'\rangle = 0\end{aligned}$$

so $\mathbf{w}_j \perp \mathbf{w}_i'$ for $1 \leq i, j \leq n_\mu$.

From the above considerations we can decompose V into irred subspaces as follows: For each $\mu = 1, \ldots, \alpha$ apply the projection operator $\mathbf{P}_\mu^{11}$ to V and let $W_1^{(\mu)}$ be the range of this operator. Choose an ON basis $\{\mathbf{w}_{l1}^{(\mu)}, 1 \leq l \leq a_\mu\}$ for the a_μ-dimensional space $W_1^{(\mu)}$. Then the vectors $\{\mathbf{w}_{lj}^{(\mu)}, 1 \leq j \leq n_\mu\}$, where $\mathbf{w}_{lj}^{(\mu)} = \mathbf{P}_\mu^{j1}\mathbf{w}_{l1}^{(\mu)}$, form an ON basis for an invariant subspace $V_l^{(\mu)}$ of V such that the restriction of $\mathbf{T}$ to $V_l^{(\mu)}$ is equivalent to $\mathbf{T}^{(\mu)}$. In fact,

$$\mathbf{T}(g)\mathbf{w}_{lj}^{(\mu)} = \sum_k T_{kj}^{(\mu)}(g)\mathbf{w}_{lk}^{(\mu)}, \qquad 1 \leq j \leq n_\mu.$$

Furthermore,

(7.19) $$V = \sum_{\mu=1}^{\alpha}\sum_{l=1}^{a_\mu} \oplus V_l^{(\mu)}$$

and the $V_l^{(\mu)}$ are mutually orthogonal. The totality of ON vectors $\{\mathbf{w}_{lj}^{(\mu)}\}$ form a basis for V since the number of elements in this set is equal to the dimension of V. Since this decomposition depends on the choice of basis vectors $\{\mathbf{w}_{l1}^{(\mu)}\}$ for $W_1^{(\mu)}$ and matrix reps $T^{(\mu)}(g)$, it is not unique.

An interesting special case occurs when **T** is the left regular rep **L** of G on the group ring R_G. We can use the convolution structure on R_G to derive additional information about the projection operators and irred subspaces. Recall that the action of **L** on R_G is given by

$$\mathbf{L}(g)x = gx, \qquad x \in R_G, \quad g \in G.$$

We can consider **L** as a unitary rep on R_G with inner product defined by (3.13). The multiplicity of the irred rep $\mathbf{T}^{(\mu)}$ in **L** is n_μ. All of our results concerning projection operators can immediately be specialized to R_G. For example the projection operators $\mathbf{P}_\mu$, (7.11), become

$$\mathbf{P}_\mu x = \frac{n_\mu}{n(G)} \sum_{g \in G} \overline{\chi^{(\mu)}(g)} \mathbf{L}(g)x = p_\mu x, \qquad x \in R_G, \tag{7.20}$$

where

$$p_\mu = \frac{n_\mu}{n(G)} \sum_{g \in G} \overline{\chi^{(\mu)}(g)} \cdot g \in R_G.$$

However, we can analyze such operators in another manner.

Let W be an invariant subspace of R_G and let **P** be a projection operator determined by the decomposition $R_G = W \oplus W'$ with W' also invariant under **L**, i.e., $W = R_\mathbf{P}$, $W' = N_\mathbf{P}$. Then **P** commutes with $\mathbf{L}(g)$ and

$$g(\mathbf{P}x) = \mathbf{P}(gx), \qquad x \in R_G, \quad g \in G.$$

Let $e' = \mathbf{P}e$, where e is the identity element of R_G. Then for any $x = \sum x(g)\cdot g$ we have

$$\mathbf{P}x = \sum x(g)\cdot \mathbf{P}g = \sum x(g)\cdot g\mathbf{P}e = \sum x(g)\cdot ge' = xe'. \tag{7.21}$$

Furthermore, $\mathbf{P}^2 x = \mathbf{P}(xe') = x(e')^2 = \mathbf{P}x = xe'$ and setting $x = e$ we obtain

$$(e')^2 = e' \tag{7.22}$$

so e' is an idempotent. (An element y of R_G is called an **idempotent** if $y^2 = yy = y$.) Note that $W = \{xe' : x \in R_G\}$. Conversely, if $e' \in R_G$ satisfies (7.22) it is easy to verify that **P** defined by

$$\mathbf{P}x = xe', \qquad x \in R_G$$

is a projection operator which commutes with left multiplication by elements of G. It follows that there is a 1–1 correspondence between idempotents e' and projections **P** commuting with **L**. The idempotent e is associated with the operator **E**. If the idempotents e_1, e_2 are associated with projection operators $\mathbf{P}_1$ and $\mathbf{P}_2 = \mathbf{E} - \mathbf{P}_1$, respectively, then the relation $\mathbf{E} = \mathbf{P}_1 + \mathbf{P}_2$ implies $e = e_1 + e_2$. Furthermore, the relation $\mathbf{P}_1\mathbf{P}_2 = \mathbf{Z}$ implies $e_2 e_1 = 0$ since

$$0 = \mathbf{P}_1\mathbf{P}_2 e = (\mathbf{P}_2 e)e_1 = ee_2e_1 = e_2e_1.$$

Proceeding in this manner we see that properties of projection operators **P** on R_G which commute with left multiplication can be translated into properties of the corresponding generating idempotents e'. For example, Theorem 3.12 yields the following theorem.

Theorem 3.13. Let W be a subspace of R_G invariant under **L** and let **P** be a projection operator on R_G such that $W = R_{\mathbf{P}}$, and $\mathbf{P}x = xe'$, all $x \in R_G$. Then W is irred if and only if there do not exist elements e_1, e_2 of R_G such that

(7.23)

$$e' = e_1 + e_2, \qquad e_1{}^2 = e_1 \neq 0, \qquad e_2{}^2 = e_2 \neq 0, \qquad e_1 e_2 = e_2 e_1 = 0.$$

An idempotent e' is called **primitive** if there exists no decomposition of the form (7.23). Thus, if e' is primitive then the set

$$W = \{xe' : x \in R_G\}$$

is an irred subspace of R_G under **L**. Conversely, if W is irred then every idempotent that generates W is primitive.

We shall now give another criterion for a primitive idempotent which is frequently simpler to verify than (7.23). If e' is an idempotent and $\mathbf{P}x = xe'$ is the corresponding projection operator then we can write $R_G = W_1 \oplus W_2$, where the invariant subspaces W_1 and W_2 can be characterized by

$$W_1 = R_{\mathbf{P}} = \{xe' : x \in R_G\}, \qquad W_2 = N_{\mathbf{P}} = \{x(e - e') : x \in R_G\}.$$

Furthermore $xe' = x$ for all $x \in W_1$, and $ye' = 0$ for all $y \in W_2$.

Theorem 3.14. If e' is a primitive idempotent then $e'xe' = \lambda_x e'$, $\lambda_x \in \mathfrak{C}$, for each $x \in R_G$. Conversely, if e' is idempotent and $e'xe' = \lambda_x e'$ for each $x \in R_G$ then e' is primitive.

Proof. Suppose e' is a primitive idempotent. Then for any $x \in R_G$ the operator **A** defined by

$$\mathbf{A}y = ye'xe', \qquad y \in R_G$$

commutes with the $\mathbf{L}(g)$. Furthermore, $\mathbf{A}y \in W_1$ for $y \in W_1$ and $\mathbf{A}y = 0$ for $y \in W_2$. Thus $\mathbf{A}_1$, the restriction of **A** to W_1, commutes with the $\mathbf{L}(g)$ and maps the irred space W_1 into itself. Theorem 3.5 implies $\mathbf{A}_1 = \lambda_x \mathbf{E}_{W_1}$ for some $\lambda_x \in \mathfrak{C}$. Thus $\mathbf{A} = \lambda_x \mathbf{P}$ or $e'xe' = \lambda_x e'$.

Conversely, suppose e' is idempotent and $e'xe' = \lambda_x e'$ for each $x \in R_G$. Let $e' = e_1 + e_2$, with $e_1{}^2 = e_1$, $e_2{}^2 = e_2$, $e_1 e_2 = e_2 e_1 = 0$. Then $e'e_1e' = (e_1 + e_2)e_1(e_1 + e_2) = e_1$, so $e_1 = \lambda e'$. Since e_1 and e' are idempotent it follows that $\lambda^2 = \lambda$ or $\lambda = 0, 1$. Thus, there exists no decomposition of e' for which e_1, e_2 are both nonzero and e' is primitive. Q.E.D.

Theorem 3.14 will prove very useful in Chapter 4 when we discuss the rep theory of the symmetric groups.

Suppose W_1 and W_2 are invariant subspaces of R_G which define equivalent reps of G. Then there exists an invertible transformation $\mathbf{S}$ from W_1 onto W_2 such that $g\mathbf{S}w = \mathbf{S}gw$ for all $g \in G$, $w \in W_1$. Furthermore $x\mathbf{S}w = \mathbf{S}(xw)$ for all $x \in R_G$. Let e_1, e_2 be generating idempotents for W_1 and W_2, respectively, and set $c = \mathbf{S}e_1 \in W_2$. Then for any $w \in W_1$ we have $\mathbf{S}w = \mathbf{S}we_1 = w\mathbf{S}e_1 = wc$. Therefore, any equivalence mapping $\mathbf{S}$ from W_1 to W_2 is given in terms of right multiplication by a ring element c. Since $c \in W_2$ we have $c = ce_2$. Furthermore $c = \mathbf{S}e_1 = \mathbf{S}e_1e_1 = e_1\mathbf{S}e_1 = e_1c$, so $c = e_1ce_2$. Thus we can assume that the equivalence is given by a nonzero element of the form e_1xe_2, $x \in R_G$. If W_1 and W_2 are irred we can say more.

Theorem 3.15. Two irred subspaces W_1, W_2 with primitive idempotents e_1, e_2 define equivalent reps if and only if there exist nonzero elements e_1xe_2, $x \in R_G$. Each such element defines an equivalence mapping $\mathbf{S}$ from W_1 to W_2.

Proof. If W_1 and W_2 define equivalent reps, then by the argument in the preceding paragraph, $e_1ce_2 \neq 0$, where $c = \mathbf{S}e_1$. Conversely, if $e_1xe_2 \neq 0$ then $\mathbf{S}w = we_1xe_2$, $w \in W_1$, is a nonzero mapping from W_1 into W_2 which commutes with the operators $\mathbf{L}(g)$. By Theorem 3.4 and the hypothesis of irreducibility, W_1 and W_2 define equivalent reps. Q.E.D.

The subspaces W of R_G which are invariant under the left regular rep are called left ideals.

Definition. A **left ideal** W is a subspace of R_G such that $xw \in W$ for all $x \in R_G$, $w \in W$.

If $c \in R_G$ the set

$$R_Gc = \{xc : x \in R_G\} \tag{7.24}$$

is clearly a left ideal. Moreover, we have shown that every left ideal can be obtained in this form. A left ideal W is said to be **minimal** if it contains no proper left ideal, i.e., if W is irred under the left regular rep. There is a similar definition of **right ideals**, which are just the subspaces of R_G invariant under the right regular rep.

A two-sided ideal is a subspace of R_G which is invariant under both the left and the right regular reps.

Definition. A **two-sided ideal** U is a subspace of R_G such that $xuy \in U$ for all $x, y \in R_G$, $u \in U$.

Let n be the multiplicity of the irred rep $\mathbf{T}$ in $\mathbf{L}$. Then there exist n linearly independent irred subspaces W_j, $1 \le j \le n$, each transforming under $\mathbf{T}$ so that the space

$$U = W_1 \oplus \cdots \oplus W_n$$

contains all irred subspaces W of R_G such that $\mathbf{L} \mid W$ is equivalent to $\mathbf{T}$. We have shown above that U is independent of the choice of the W_i. We will now show that U is a two-sided ideal. Let $u \in U$. Then

$$u = u_1 + \cdots + u_n, \qquad u_i \in W_i.$$

If $y \in R_G$ it follows from Theorem 3.15 that $u_i y$ is either zero or an element of a left ideal equivalent to W_i. In either case, $u_i y \in U$ for $1 \le i \le n$. Thus $uy \in U$ for all $u \in U$, $y \in R_G$. This proves that U is a right ideal. On the other hand, U is a left ideal since each of the W_i is a left ideal.

Finally, we will show that U is a **minimal** two-sided ideal. That is, U contains no proper two-sided ideal U'. For, if $U' \subseteq U$ and $U' \neq \{0\}$ then U' contains a minimal left ideal W. By Theorem 3.15 there exist ring elements $c_1, \ldots, c_n$ such that $W_i = Wc_i$, $1 \le i \le n$. Since U' is a right ideal, $W_i \subseteq U'$. Therefore, $U = U'$.

Let U_μ be the minimal two-sided ideal corresponding to the irred rep $\mathbf{T}^{(\mu)}$. [Note that $U_\mu = V^{(\mu)}$, (7.13), in the case $V = R_G$.] Then

$$R_G = U_1 \oplus \cdots \oplus U_\alpha \tag{7.25}$$

and $U_\mu U_\nu = \{0\}$ for $\mu \neq \nu$. Indeed, the first expression follows from (7.13) and (7.19). To prove the second formula note that $U_\mu U_\nu \subseteq U_\mu \cap U_\nu = \{0\}$ since U_μ and U_ν are disjoint two-sided ideals. A proof of the relation $U_\mu U_\mu = U_\mu$ is left to the reader.

3.8 Applications

We now study several applications of the rep theory of finite groups to problems in theoretical physics. These examples have been selected so that they can be understood without an extensive knowledge of physics. Some of the most important applications which require a knowledge of the rep theory of certain Lie groups, particularly of the group $SO(3)$, will be discussed in later chapters.

Our first application concerns the use of symmetry groups to determine the structure of tensors occurring in physical theories. We start by defining a tensor.

Let V be an m-dimensional vector space, real or complex, and consider the n-fold tensor product

$$V^{\otimes n} = V \otimes V \otimes \cdots \otimes V \quad (n\text{-times}). \tag{8.1}$$

If $\{\mathbf{v}_j, 1 \leq j \leq m\}$ is a basis for V then the mn vectors $\{\mathbf{v}_{j_1} \otimes \cdots \otimes \mathbf{v}_{j_n}, 1 \leq j_1, \ldots, j_n \leq m\}$ form a basis for $V^{\otimes n}$. The elements of $V^{\otimes n}$ are called **(contravariant) tensors** of **rank** n. Every tensor $\mathbf{a}$ can be written uniquely in the form

$$\mathbf{a} = \sum_{j_1 \cdots j_n} a_{j_1 \cdots j_n} \mathbf{v}_{j_1} \otimes \cdots \otimes \mathbf{v}_{j_n}. \tag{8.2}$$

In terms of a new basis $\{\mathbf{v}_k'\}$ for V related to $\{\mathbf{v}_j\}$ by

$$\mathbf{v}_j = \sum_{j=1}^{m} g_{kj} \mathbf{v}_k', \qquad 1 \leq j \leq m, \tag{8.3}$$

we find

$$\mathbf{a} = \sum_{k_1 \cdots k_n} a'_{k_1 \cdots k_n} \mathbf{v}'_{k_1} \otimes \cdots \otimes \mathbf{v}'_{k_n} \tag{8.4}$$

where the tensor components a and a' are related by

$$a'_{k_1 \cdots k_n} = \sum_{j_1 \cdots j_n} a_{j_1 \cdots j_n} g_{k_1 j_1} \cdots g_{k_n j_n}. \tag{8.5}$$

The matrices $g = (g_{kj})$ are nonsingular and any nonsingular matrix defines a change of basis. One should carefully distinguish between the tensor $\mathbf{a}$ and the tensor components $a_{j_1 \cdots j_n}$. The components of a fixed tensor depend on the basis chosen in $V^{\otimes n}$.

Let G be a group of linear transformations $\mathbf{g}$ on V. (We do not require that G be finite.) Then, as discussed in Section 3.5, we can define a rep $\mathbf{T}^{\otimes n}$ of G on $V^{\otimes n}$ by

$$\mathbf{T}^{\otimes n}(g)\mathbf{w}_1 \otimes \mathbf{w}_2 \otimes \cdots \otimes \mathbf{w}_n = \mathbf{g}\mathbf{w}_1 \otimes \mathbf{g}\mathbf{w}_2 \otimes \cdots \otimes \mathbf{g}\mathbf{w}_n, \qquad g \in G, \tag{8.6}$$

for all $\mathbf{w}_1, \ldots, \mathbf{w}_n \in V$. As usual, we choose a basis $\{\mathbf{v}_j\}$ for V and define the matrix (g_{kj}) corresponding to each $g \in G$ by

$$\mathbf{g}\mathbf{v}_j = \sum_{k=1}^{m} g_{kj} \mathbf{v}_k, \qquad 1 \leq j \leq m. \tag{8.7}$$

Then the tensor $\mathbf{a}$, Eq. (8.2) is transformed into the tensor $\mathbf{T}^{\otimes n}(g)\mathbf{a}$, where

$$\mathbf{T}^{\otimes n}(g)\mathbf{a} = \sum_{k_1 \cdots k_n} a'_{k_1 \cdots k_n} \mathbf{v}_{k_1} \otimes \cdots \otimes \mathbf{v}_{k_n}, \tag{8.8}$$

$$a'_{k_1 \cdots k_n} = \sum_{j_1 \cdots j_n} a_{j_1 \cdots j_n} g_{k_1 j_1} \cdots g_{k_n j_n}. \tag{8.9}$$

Expressions (8.5) and (8.9) are identical, but their interpretations are different. In the first case (**passive**) the tensor is fixed and the basis is changed. In the second case (**active**) the basis remains fixed while the tensor $\mathbf{a}$ is mapped into a new tensor by the operator $\mathbf{T}^{\otimes n}(g)$. For the present we consider only the active case and fix the basis $\{\mathbf{v}_j\}$. Then every tensor $\mathbf{a}$ is uniquely determined by its components $a_{j_1 \cdots j_n}$ with respect to the basis and we can consider the rep $\mathbf{T}^{\otimes n}$ to be defined by (8.9). Another useful rep of G is obtained by forming the tensor product $\mathbf{Q} \otimes \mathbf{T}^{\otimes n}$, where $\mathbf{Q}$ is the one-dimensional rep $\mathbf{Q}(g) = \det g$.

(Recall that the value of the determinant is independent of basis in V.) The basis space for this rep is $V^{\otimes n}$ again and $g \in G$ acts on the tensor **a** with components $a_{j_1 \cdots j_n}$ to transform it into a tensor with components

(8.10) $$a'_{k_1 \cdots k_n} = \det(g) \sum_{j_1 \cdots j_n} a_{j_1 \cdots j_n} g_{k_1 j_1} \cdots g_{k_n j_n}.$$

Frequently the above reps occur in physical theories where V is a real three-dimensional inner product space and $G = O(3)$, the group of all length-preserving linear transformations on V. Then with respect to an ON basis $\mathbf{v}_1, \mathbf{v}_2, \mathbf{v}_3$ for V, the matrix O of each $\mathbf{O} \in O(3)$ is orthogonal: $O^t = O^{-1}$. The tensors $\mathbf{a} \in V^{\otimes n}$ which transform according to the rep $\mathbf{T}^{\otimes n}$,

(8.11) $$a'_{k_1 \cdots k_n} = \sum_{j_1 \cdots j_n = 1}^{3} a_{j_1 \cdots j_n} O_{k_1 j_1} \cdots O_{k_n j_n}$$

are called **polar tensors** of rank n. Those which transform according to

(8.12) $$a'_{k_1 \cdots k_n} = \det O \sum_{j_1 \cdots j_n = 1}^{3} a_{j_1 \cdots j_n} O_{k_1 j_1} \cdots O_{k_n j_n}$$

are called **axial tensors** of rank n. Note that $\det O = \pm 1$.

We give some familiar examples of polar and axial tensors. The action of $O(3)$ as a transformation group on R_3, considered in Chapter 2, defines a rep of $O(3)$ in which each $\mathbf{v} \in R_3$ transforms as a polar vector (polar tensor of rank 1). The well-known vector product or cross product $\mathbf{u} \times \mathbf{v}$ of two polar vectors transforms as an axial vector. In particular, under the inversion operator $\mathbf{I} \in O(3)$, $\mathbf{u} \longrightarrow -\mathbf{u}$, $\mathbf{v} \longrightarrow -\mathbf{v}$, and $\mathbf{u} \times \mathbf{v} \longrightarrow (-\mathbf{u}) \times (-\mathbf{v}) = \mathbf{u} \times \mathbf{v}$. The scalar product of two polar vectors transforms as a **scalar** (polar tensor of rank zero), while the scalar product of a polar vector and an axial vector transforms as a **pseudoscalar** (axial tensor of rank zero).

Let S be some physical system (molecule, crystal, garbage truck, etc.) in three-dimensional space R_3. We choose an arbitrary point, say $\boldsymbol{\theta}$, as the origin in R_3 and construct an orthogonal coordinate system at $\boldsymbol{\theta}$ with ON basis vectors $\mathbf{v}_1, \mathbf{v}_2, \mathbf{v}_3$ pointing along the coordinate axes. We position ourselves at the point $\boldsymbol{\theta}$ and measure various physical properties of the system S. These measurements are performed very rapidly; so fast that they all take place in a single instant of time t_0. (We will be concerned with determining the properties of the system at a given instant of time and not with the time evolution of the system.) Once we have measured some physical property ρ of S with respect to the ON basis $\{\mathbf{v}_i\}$ we can perform an orthogonal transformation $\mathbf{O}$ on R_3 and then measure the same physical property of the system $\mathbf{O}S$ with respect to $\{\mathbf{v}_i\}$. Hopefully, there will be some functional relationship $\mathbf{O}\rho = f(\mathbf{O}, \rho)$ between the new measurement $\mathbf{O}\rho$ and the old one $\mathbf{x}$. Indeed, the measurable quantities ρ frequently transform as components of an axial or polar tensor. We mention some examples, at least a few of which should be familiar to the reader. The temperature at $\boldsymbol{\theta}$ is a scalar, while the rotary power of an optically active crystal is a pseudoscalar. The electric field and the current density are polar vectors, while the magnetic field is an axial vector. The resistivity and moment of inertia tensors are polar tensors of rank two, as are the stress and strain tensors. The optical gyration tensor is axial of rank two.

Let G be the point symmetry group of the physical system S. (This means that S and $\mathbf{g}S$ are physically indistinguishable for all $\mathbf{g}$ in the point group G.) if $\mathbf{a} \in V^{\otimes n}$ is any tensor describing a physical property of S it necessarily follows that $\mathbf{T}^{\otimes n}(g)\mathbf{a} = \mathbf{a}$. In terms of tensor components this relation becomes

$$(8.13) \qquad a_{k_1 \cdots k_n} = \sum_{j_1 \cdots j_n} a_{j_1 \cdots j_n} g_{k_1 j_1} \cdots g_{k_n j_n}, \qquad k_i = 1, 2, 3,$$

valid for all $g \in G$, where g_{kj} are the matrix elements of $\mathbf{g}$ with respect to the $\mathbf{v}_i$ basis of R_3. [Equation (8.13) is valid for polar tensors of rank n. The results are modified in an obvious manner if $\mathbf{a}$ is an axial tensor.] This relation places a restriction on the tensor components of $\mathbf{a}$. If $G = \{e\}$ and the tensor components are subject to no symmetry requirements then the nth-rank tensors have 3^n independent components. However, if G is a nontrivial symmetry group then by (8.13) the 3^n components are not all independent. We can use the symmetry group to determine the maximal number of linearly independent components of an nth-rank tensor, hence the number of parameters needed to uniquely determine a physical property of S associated with the tensor.

As an example of the restrictions provided by the symmetries of S, suppose G contains the inversion I. Then (8.13) with $g = I$ yields

$$a_{k_1 \cdots k_n} = (-1)^n a_{k_1 \cdots k_n}$$

which shows that all polar tensors of odd rank are identically zero. Thus if some physical property of S is described by a polar tensor of odd rank, it follows from symmetry conditions alone that this tensor is identically zero. Similarly, if $I \in G$ then all axial tensors of even rank are zero.

The most common method used for computing the possible tensors invariant under G is the brute force method. One chooses a set $g_1, \ldots, g_l$ whose elements generate G and then substitutes each of these elements into (8.13) to obtain a system of identities relating the tensor components of $\mathbf{a}$. These identities must then be solved to determine the number of linearly independent components of $\mathbf{a}$ and the dependence of all components on a suitably chosen set of independent ones. If the components of $\mathbf{a}$ satisfy Eq. (8.13) for the generators of G then the components will automatically satisfy these equations for any $g \in G$.

We can develop more sophisticated methods to solve this problem by noting that the solutions $\mathbf{a} \in V^{\otimes n}$ of the equation

$$\mathbf{T}^{\otimes n}(g)\mathbf{a} = \mathbf{a}, \qquad \text{all} \quad g \in G,$$

form a subspace $V^{(1)}$ which is invariant under $\mathbf{T}^{\otimes n}$. Let dim $V^{(1)} = q$ and let $\mathbf{a}_1, \ldots, \mathbf{a}_q$ be a basis for $V^{(1)}$. Then each of the $\mathbf{a}_i$ generates a one-dimensional invariant subspace W_i of $V^{(1)}$ such that the action of $\mathbf{T}^{\otimes n}$ on W_i is equivalent to the irred identity rep $\mathbf{T}^{(1)}$:

$$V^{(1)} = W_1 \oplus \cdots \oplus W_q.$$

Thus, q is the multiplicity of the identity rep of G in $\mathbf{T}^{\otimes n}$ and the number of linearly independent tensor components for solutions $\mathbf{a}$ of (8.13). To find $V^{(1)}$ we can make use of the projection operator $\mathbf{P}_1$ of Section 3.7:

$$\mathbf{P}_1 = \frac{1}{n(G)} \sum_{g \in G} \overline{\chi^{(1)}(g)} \mathbf{T}^{\otimes n}(g) = \frac{1}{n(G)} \sum_{g \in G} \mathbf{T}^{\otimes n}(g). \tag{8.14}$$

Then

$$V^{(1)} = \{\mathbf{P}_1 \mathbf{b} : \mathbf{b} \in V^{\otimes n}\} \tag{8.15}$$

or $V^{(1)}$ is the space of all solutions $\mathbf{a}$ of the equation

$$\mathbf{P}_1 \mathbf{a} = \mathbf{a}. \tag{8.16}$$

We can use the orthogonality relations for characters to obtain a simple expression for q. Let χ be the character of the natural three-dimensional rep $\mathbf{T}$ of G as a transformation group on R_3,

$$\chi(g) = \operatorname{tr} \mathbf{g}, \qquad g \in G.$$

It follows from (1.13), Section 2.1, that

$$\chi(g) = 1 + 2\cos\varphi \tag{8.17}$$

for $g = C_k(\varphi) \in G$ and

$$\chi(g) = -1 + 2\cos\varphi \tag{8.18}$$

for $g = S_k(\varphi) \in G$, so the character χ is immediately determined from a description of the action of each g. Since $\mathbf{T}^{\otimes n}$ is the tensor product of n copies of $\mathbf{T}$, the character χ^n of this rep is

$$\chi^n(g) = [\chi(g)]^n. \tag{8.19}$$

[This result is correct for polar tensors of rank n. For axial tensors of rank n the character is

$$\chi'^n(g) = \varepsilon_g \chi^n(g) \tag{8.20}$$

where $\varepsilon_g = 1$ if $g \in G \cap SO(3)$ and $\varepsilon_g = -1$ if g is an improper rotation.] Then we have

$$q = \langle \chi^n, \chi^{(1)} \rangle = \frac{1}{n(G)} \sum_{g \in G} [\chi(g)]^n. \tag{8.21}$$

The reader may be wondering why we have applied character theory to the **real** rep $\mathbf{T}^{\otimes n}$ since we emphasized earlier that this theory applies only to

complex reps. To verify that Eq. (8.21) is valid we formally define the complex m-dimensional vector space V_c by

$$V_c = \{\mathbf{v}_1 + i\mathbf{v}_2 : \mathbf{v}_1, \mathbf{v}_2 \in V\}.$$

Then every element $\mathbf{a}$ of the 3^n-dimensional tensor product space $V_c^{\otimes n}$ can be written uniquely in the form

$$\mathbf{a} = \mathbf{b}_1 + i\mathbf{b}_2, \qquad \mathbf{b}_1, \mathbf{b}_2 \in V^{\otimes n}. \tag{8.22}$$

We define the complex rep $\mathbf{T}_c^{\otimes n}$ of G on $V_c^{\otimes n}$ by

$$\mathbf{T}_c^{\otimes n}(g)\mathbf{a} = \mathbf{T}^{\otimes n}(g)\mathbf{b}_1 + i\mathbf{T}^{\otimes n}(g)\mathbf{b}_2. \tag{8.23}$$

If $\{\mathbf{v}_j\}$ is an ON basis for V then it is also an ON basis for V_c and the set $\{\mathbf{v}_{j_1} \otimes \cdots \otimes \mathbf{v}_{j_n}\}$ is an ON basis for $V_c^{\otimes n}$. In terms of tensor components the action of $\mathbf{T}_c^{\otimes n}(g)$ is given by relations (8.7)–(8.9) where now the components $a_{j_1 \cdots j_n}$ take complex values. The multiplicity of the identity rep $\mathbf{T}^{(1)}$ in $\mathbf{T}^{\otimes n}$ is now given by the right-hand side of (8.21). Let this multiplicity be r. Then we can find a basis $\mathbf{a}_1, \ldots, \mathbf{a}_r$ for the subspace $V_c^{(1)}$ consisting of all elements of $V_c^{\otimes n}$ which are fixed under $\mathbf{T}_c^{\otimes n}$. By definition

$$\overline{\mathbf{T}_c^{\otimes n}(g)\mathbf{a}} = \mathbf{T}_c^{\otimes n}(g)\bar{\mathbf{a}} \tag{8.24}$$

where $\bar{\mathbf{a}} = \mathbf{b}_1 - i\mathbf{b}_2$ and $\mathbf{a} = \mathbf{b}_1 + i\mathbf{b}_2$. Thus, if $\mathbf{a} \in V_c^{(1)}$ then $\bar{\mathbf{a}} \in V_c^{(1)}$. Clearly, the $2r$ real tensors

$$\tfrac{1}{2}(\mathbf{a}_j + \bar{\mathbf{a}}_j), \qquad \tfrac{1}{2}i(\mathbf{a}_j - \bar{\mathbf{a}}_j), \qquad 1 \leq j \leq r, \tag{8.25}$$

span $V_c^{(1)}$ and also lie in $V^{(1)}$. Among these $2r$ real tensors there must be r which form a basis for $V_c^{(1)}$. Thus, $r \leq q$. However $V^{(1)} \subseteq V_c^{(1)}$ so $r = q$. This justifies our use of characters to compute q.

As an example we compute the dimension q of the space $V^{(1)}$ of polar tensors of rank two, invariant under C_{4v}. Since $C_{4v} \cong D_4$ we can use the characters and conjugacy classes (6.16) of D_4 to perform the computation. From (8.17) and (8.18) the character χ of the natural rep of C_{4v} as a transformation group on R_3 is

(8.26)

	$\mathcal{E}$	$\mathcal{C}_4^{\,2}$	$2\mathcal{C}_4$	$2\mathcal{C}_2$	$2\mathcal{C}_{2'}$
χ	3	−1	1	1	1

Thus

$$q = \langle \chi^2, \chi^{(1)} \rangle = \tfrac{1}{8} \sum_{g \in D_4} \chi^2(g) = 2.$$

Tensors describing physical phenomena frequently possess internal symmetry properties which are independent of their external point symmetry properties. For example, the moment of inertia and stress and strain tensors are all **symmetric** polar tensors of rank two, i.e., $a_{jk} = a_{kj}$ for $1 \leq j, k \leq 3$. Other physically interesting tensors of rank two are **skew-symmetric**, $a_{jk} =$

$-a_{kj}$ for $1 \le j, k \le 3$. Here we consider only polar tensors of rank two. The procedure needed to extend our results to tensors of higher order should be clear after an examination of this simple case.

It follows easily from the transformation law (8.11) that every symmetric tensor $\mathbf{a} \in V^{\otimes 2}$ is mapped into a symmetric tensor $\mathbf{T}^{\otimes 2}(O)\mathbf{a}$ by any $O \in O(3)$. Similarly, a skew-symmetric tensor is mapped into a skew-symmetric tensor. If the tensor

$$\mathbf{a} = \sum_{j,k=1}^{3} a_{jk}\mathbf{v}_j \otimes \mathbf{v}_k$$

is both symmetric and skew symmetric then

$$a_{jk} = -a_{kj} = a_{kj} = 0$$

so $\mathbf{a} = \boldsymbol{\theta}$. Given any $\mathbf{a} \in V^{\otimes 2}$ we can construct a symmetric tensor $\mathbf{a}^S$ with components

$$a^S_{jk} = \tfrac{1}{2}(a_{jk} + a_{kj})$$

and a skew-symmetric tensor $\mathbf{a}^A$ with components

$$a^A_{jk} = \tfrac{1}{2}(a_{jk} - a_{kj}),$$

so $\mathbf{a} = \mathbf{a}^S + \mathbf{a}^A$. This shows that

$$V^{\otimes 2} = W^S \oplus W^A$$

where W^S and W^A are the invariant subspaces of all symmetric and skew-symmetric tensors, respectively. Here $\dim W^S = 6$, $\dim W^A = 3$. (Prove it!) The character χ^2 of the rep $\mathbf{T}^{\otimes 2}$ acting on $V^{\otimes 2}$ can be written

$$\chi^2(g) = \chi^S(g) + \chi^A(g), \tag{8.27}$$

where χ^S is the character of $\mathbf{T}^{\otimes 2} \mid W^S$ and χ^A is the character of $\mathbf{T}^{\otimes 2} \mid W^A$. The symmetric tensors which are fixed under G form a subspace $V^{(1)}_S$ of W^S, while the skew-symmetric tensors fixed by G form a subspace $V^{(1)}_A$ of W^A. If $q_S = \dim V^{(1)}_S$, $q_A = \dim V^{(1)}_A$, then $q_S + q_A = q$ and

$$q_S = \langle \chi^S, \chi^{(1)} \rangle, \qquad q_A = \langle \chi^A, \chi^{(1)} \rangle.$$

We will compute the character χ^S directly. It is easy to show that the set of six symmetric tensors

$$\{\mathbf{v}_j \otimes \mathbf{v}_k + \mathbf{v}_k \otimes \mathbf{v}_j, \quad 1 \le j \le k \le 3\}$$

is a basis for W^S. Then

$$\mathbf{T}^{\otimes 2}(g)(\mathbf{v}_j \otimes \mathbf{v}_k + \mathbf{v}_k \otimes \mathbf{v}_j) = \tfrac{1}{2} \sum_{lh} (T_{lj}(g)T_{hk}(g) + T_{lk}(g)T_{hj}(g)) \times (\mathbf{v}_l \otimes \mathbf{v}_h + \mathbf{v}_h \otimes \mathbf{v}_l).$$

Taking the trace of this transformation, we find

$$\chi^S(g) = \tfrac{1}{2} \sum_{jk} (T_{jj}(g)T_{kk}(g) + T_{jk}(g)T_{kj}(g)) = \tfrac{1}{2}(\chi^2(g) + \chi(g^2)). \tag{8.28}$$

Furthermore,

(8.29) $$\chi^A(g) = \chi^2(g) - \chi^S(g) = \tfrac{1}{2}(\chi^2(g) - \chi(g^2)).$$

We apply these results to compute the dimension q_S of the space $V_S^{(1)}$ of symmetric polar tensors of rank two which are invariant under C_{4v}. From (8.26) and (6.16) we find

(8.30)

	$\mathcal{E}$	$\mathcal{C}_4{}^2$	$2\mathcal{C}_4$	$2\mathcal{C}_2$	$2\mathcal{C}_{2'}$
χ^2	9	1	1	1	1
χ^S	6	2	0	2	2
χ^A	3	-1	1	-1	-1

Thus,

$$q_S = \langle \chi^S, \chi^{(1)} \rangle = 2, \qquad q_A = 0.$$

Since $q_S = q = 2$ it follows that all second-rank polar tensors fixed under C_{4v} are symmetric. A skew-symmetric tensor of this type describing a physical property of a system with C_{4v} symmetry is zero. One physical consequence of this computation is that all solids with C_{4v} symmetry have moment of inertia tensors which are determined by two parameters. The homogeneous four-pyramid is such a solid. (In this special case we found $q = q_S$. However, this equality is the exception rather than the rule.)

Our next application of group rep theory pertains to perturbation theory in quantum mechanics. So as not to interrupt the continuity of our presentation we assume that the reader understands a few basic facts about Hilbert space, Lebesgue integration, and the Hamiltonian operator in quantum mechanics. The relevant definitions are presented in the appendix. We concentrate on algebraic and group-theoretic questions and ignore certain analytic difficulties pertaining to unbounded operators in Hilbert space.

Consider a nonrelativistic quantum mechanical system consisting of k particles with masses $m_1, \ldots, m_k$, respectively. We suppose that the interaction between the particles is described by a real-valued potential function $V(\mathbf{x}_1, \ldots, \mathbf{x}_k)$, where $\mathbf{x}_j \in R_3$ refers to the coordinates of the jth particle. The possible (pure) states of this sytem are elements of the Hilbert space $\mathcal{H}$ consisting of all complex valued functions

$$\Psi(\mathbf{x}_1, \ldots, \mathbf{x}_k),$$

such that

$$\int_{(R_3)^k} |\Psi(\mathbf{x}_1, \ldots, \mathbf{x}_k)|^2 \, d^3\mathbf{x}_1 \cdots d^3\mathbf{x}_k < \infty.$$

The inner product $(-, -)$ on $\mathcal{H}$ is defined by

(8.31) $$(\Psi, \Phi) = \int_{(R_3)^k} \Psi(\mathbf{x}_1, \ldots, \mathbf{x}_k)\overline{\Phi(\mathbf{x}_1, \ldots, \mathbf{x}_k)} \, d^3\mathbf{x}_1 \cdots d^3\mathbf{x}_k.$$

The Hamiltonian operator **H** of this system is defined by

$$\mathbf{H}\Psi(\mathbf{x}_1, \ldots, \mathbf{x}_k) = -\sum_{j=1}^{k} \frac{\hbar^2}{2m_j} \Delta_j \Psi(\mathbf{x}_1, \ldots, \mathbf{x}_k) + V(\mathbf{x}_1, \ldots, \mathbf{x}_k)\Psi(\mathbf{x}_1, \ldots \mathbf{x}_k) \tag{8.32}$$

where

$$\Delta_j = \sum_{l=1}^{3} \frac{\partial}{\partial x_{lj}^2}, \qquad \mathbf{x}_j = (x_{1j}, x_{2j}, x_{3j}),$$

and $h = 2\pi\hbar$ is Planck's constant. In the remainder of this book we choose units such that $\hbar = 1$.

Note. The Hamiltonian operator is not defined for all $\Psi \in \mathcal{H}$. Expression (8.32) defines an element of $\mathcal{H}$ only if the function $\mathbf{H}\Psi$ is square integrable. A precise definition of the domain of **H** is difficult and we refer the interested reader to Helwig [1]. For most potential functions $V(\mathbf{x}_1, \ldots, \mathbf{x}_k)$ which occur in quantum mechanics it is possible to define the Hamiltonian in a satisfactory manner such that the domain of **H** is dense in $\mathcal{H}$. Furthermore, it can be shown that **H** is a **symmetric operator**, i.e.,

$$(\Psi, \mathbf{H}\Phi) = (\mathbf{H}\Psi, \Phi) \tag{8.33}$$

for all Ψ, Φ in the domain of **H**. Equation (8.33) is easy to obtain formally but difficult to prove by a rigorous computation. In some of the following arguments we shall also proceed formally, as do almost all textbooks on applications of group theory to quantum mechanics. The needed rigor can be supplied by Helwig [1] and Kato [1].

In analogy with Example 5 of Section 3.1 we define a unitary rep of $E(3)$ on $\mathcal{H}$ by

(8.34)

$$\mathbf{T}(g)\Psi(\mathbf{x}_1, \ldots, \mathbf{x}_k) = \Psi(g^{-1}\mathbf{x}_1, \ldots, g^{-1}\mathbf{x}_k), \qquad g = \{\mathbf{a}, \mathbf{O}\} \in E(3).$$

The relation

$$(\mathbf{T}(g)\Psi, \mathbf{T}(g)\Phi) = (\Psi, \Phi), \qquad \Psi, \Phi \in \mathcal{H} \tag{8.35}$$

follows from (8.31) and a simple change of variable. Let G be any subgroup of $E(3)$ consisting of transformations **g** such that

$$V(\mathbf{g}\mathbf{x}_1, \ldots, \mathbf{g}\mathbf{x}_k) = V(\mathbf{x}_1, \ldots, \mathbf{x}_k), \qquad \text{all} \quad \mathbf{x}_j \in R_3.$$

Then by an elementary computation similar to that carried out in Examples 4 and 5 of Section 3.1 we can show

$$\mathbf{T}(g)\mathbf{H}\Psi = \mathbf{H}\mathbf{T}(g)\Psi, \qquad g \in G,$$

for all vectors Ψ in the domain of **H**. Thus, the operators $\mathbf{T}(g)$ define a unitary rep of G on $\mathcal{H}$ and these operators commute with **H**. The group G is called a symmetry group of the Hamiltonian.

The fundamental problem for this quantum mechanical system is the determination of the eigenvalues and eigenvectors of **H**, i.e., the solutions of the eigenvalue problem

$$\mathbf{H}\Psi = \lambda\Psi, \qquad \Psi \in \mathcal{H}. \tag{8.36}$$

(We study only the point spectrum of **H**. Group-theoretic methods also apply to the continuous spectrum but such a treatment is beyond the scope of this book.) Equation (8.36) is called the (time-independent) **Schrödinger equation.** Since **H** is symmetric the eigenvalues are real. Indeed, suppose λ is an eigenvalue of **H** with eigenvector Ψ. We can normalize Ψ so that $(\Psi, \Psi) = 1$. Then

$$\lambda = (\mathbf{H}\Psi, \Psi) = (\Psi, \mathbf{H}\Psi) = \bar{\lambda}$$

so λ is real. Furthermore, if λ, μ are eigenvalues of **H** with corresponding eigenvectors Ψ, Φ then

$$\lambda(\Psi, \Phi) = (\mathbf{H}\Psi, \Phi) = (\Psi, \mathbf{H}\Phi) = \mu(\Psi, \Phi)$$

so $(\Psi, \Phi) = 0$, if $\lambda \neq \mu$.

Let λ be an eigenvalue of **H** and define the eigenspace $W_\lambda \subset \mathcal{H}$ by

$$W_\lambda = \{\Psi \in \mathcal{H} : \mathbf{H}\Psi = \lambda\Psi\}.$$

If $g \in G$ we have

$$\mathbf{H}\mathbf{T}(g)\Psi = \mathbf{T}(g)\mathbf{H}\Psi = \lambda\mathbf{T}(g)\Psi \tag{8.37}$$

for all $\Psi \in W_\lambda$. Therefore $\mathbf{T}(g)\Psi \in W_\lambda$ and W_λ is invariant under the unitary rep **T** of the symmetry group G. Suppose W_λ is finite-dimensional and G is a point group. Then we can decompose W_λ into a direct sum of subspaces which transform irreducibly under G:

$$W_\lambda = \sum_{\mu=1}^{\alpha} \sum_{i=1}^{a_\mu} \oplus W_i^{(\mu)}$$

Here, the restriction of **T** to $W_i^{(\mu)}$ is equivalent to the irred rep $\mathbf{T}^{(\mu)}$ of G and a_μ is the multiplicity of $\mathbf{T}^{(\mu)}$ in **T**. The reps $\mathbf{T}^{(1)}, \ldots, \mathbf{T}^{(\alpha)}$ form a complete set of nonequivalent irred unitary reps of G. If $T^{(1)}, \ldots, T^{(\alpha)}$ are a corresponding complete set of unitary matrix reps we can find an ON basis $\{\mathbf{w}_{ij}^{(\mu)}, 1 \leq j \leq n_\mu\}$ for each space $W_i^{(\mu)}$ such that

$$\mathbf{T}(g)\mathbf{w}_{ij}^{(\mu)} = \sum_{l=1}^{n_\mu} T_{lj}^{(\mu)}(g)\mathbf{w}_{il}^{(\mu)}.$$

Then the complete set of symmetry-adapted basis vectors $\{\mathbf{w}_{ij}^{(\mu)}\}$ forms an ON basis for W_λ. In this way we use the irred reps of G to label the eigenvectors of **H**.

The **complete symmetry group** of the Hamiltonian **H** is the group K of all unitary operators **U** on $\mathcal{H}$ such that

$$\mathbf{U}\mathbf{H} = \mathbf{H}\mathbf{U}. \tag{8.38}$$

Since the elements of K are unitary operators, K defines a unitary rep of itself. Just as in (8.37), we can show that W_λ is invariant (even irred) under K, so the reps of K can be used to label the elements of W_λ. However, it may be very difficult to determine all elements of K. Thus we usually restrict ourselves to consideration of the subgroup G' consisting of all symmetries of $\mathbf{H}$ taking the form (8.34). The symmetry group G' can be determined by inspection. (Later we shall include in G' spin transformations and permutations of indistinguishable particles when this is appropriate.)

The eigenspace W_λ is usually irred under G'. The degeneracy of the eigenvalue λ, i.e., the dimension of W_λ, is then equal to the dimension of some irred rep of G'. For certain specially chosen potential functions $V(\mathbf{x}_1, \ldots, \mathbf{x}_k)$ it is possible to find eigenvalues λ of $\mathbf{H}$ for which W_λ is not irred, but this is rare. In such a case the eigenvalue λ has an **accidental degeneracy**, i.e., a degeneracy which does not follow from the symmetry of the Hamiltonian. (See the discussion of the hydrogen atom, Section 9.7.) Accidental degeneracy can be removed by a slight alteration of the potential function which does not change the symmetry group of the Hamiltonian.

If a point symmetry group G is a proper subgroup of G' then W_λ need not transform irreducibly under G and in general W_λ will break up into a direct sum of irred reps of G. In practice, if a physicist finds that W_λ is not irred under the action of G he has strong reason for suspecting the existence of a larger symmetry group. Thus, he is likely to search for additional symmetries of $\mathbf{H}$.

The eigenvalue equation

$$\mathbf{H}\Psi = \lambda\Psi \tag{8.39}$$

has been solved exactly for only a few simple Hamiltonians. The Hamiltonians $\mathbf{H}$ for which (8.39) can be solved usually correspond to physical systems which exhibit a high degree of symmetry. The two most important examples are the hydrogen atom and the harmonic oscillator, which will be discussed in later chapters. For systems with lower symmetry, Eq. (8.39) usually cannot be solved explicitly and some sort of approximation has to be employed. Group-theoretic methods are of the utmost importance here because they yield information about the multiplicities of the eigenvalues even in those cases where (8.39) cannot be solved exactly.

As an example, suppose $\mathbf{H}$ admits the point group G as a symmetry group and suppose we can find an eigenvector Ψ of $\mathbf{H}$ with eigenvalue λ. Furthermore, suppose Ψ is an element of a subspace $V^{(\mu)}$ of $\mathcal{H}$ such that the action of G on $V^{(\mu)}$ is equivalent to the irred rep $T^{(\mu)}$. Then the nonzero subspace $V^{(\mu)} \cap W_\lambda$ is invariant under G. Since $V^{(\mu)}$ is irred it follows that $V^{(\mu)} = V^{(\mu)} \cap W_\lambda$, so $V^{(\mu)} \subseteq W_\lambda$ and $\dim W_\lambda \geq n_\mu = \dim T^{(\mu)}$. Thus, the eigenvalue λ has multiplicity at least n_μ. The reader should be able to construct an ON set of n_μ eigenvectors by a judicious application of the projection operators $\mathbf{P}_\mu^{jk}$, (7.14), to Ψ.

Consider a physical system with Hamiltonian

$$\mathbf{H} = \mathbf{H}_1 + \mathbf{H}_2 \tag{8.40}$$

where

$$\mathbf{H}_1 = -\sum_{j=1}^{k} \frac{1}{2m_j} \Delta_j + V_1(\mathbf{x}_1, \ldots, \mathbf{x}_k), \qquad \mathbf{H}_2 = V_2(\mathbf{x}_1, \ldots, \mathbf{x}_k) \tag{8.41}$$

and suppose the eigenvalue equation

$$\mathbf{H}_1\Psi = \lambda\Psi \tag{8.42}$$

can be solved explicitly. We think of the physical system with Hamiltonian $\mathbf{H}$ as obtained from the system with Hamiltonian $\mathbf{H}_1$ by the addition of a "small" perturbing potential V_2. If the perturbation is not too large we would expect the eigenvalues and eigenfunctions of the Hamiltonian $\mathbf{H}$ to be "close" to those of $\mathbf{H}_1$. Proceeding formally, let us consider a family of Hamiltonians

$$\mathbf{H}(t) = \mathbf{H}_1 + t\mathbf{H}_2 \tag{8.43}$$

where the real parameter t runs from 0 to 1. Then $\mathbf{H}(0) = \mathbf{H}_1$, $\mathbf{H}(1) = \mathbf{H}$. If the perturbing potential is not too big it is reasonable to suppose that the eigenvalues and eigenfunctions of $\mathbf{H}(t)$ will be continuous functions of t. To be more precise, let λ_0 be any isolated eigenvalue of $\mathbf{H}_1$ with finite multiplicity m. Then we suppose there exist m continuous functions $\lambda_1(t), \ldots, \lambda_m(t)$ and m eigenvectors $\Psi_{1t}, \ldots, \Psi_{mt}$ in $\mathcal{H}$ which are continuous functions of t (in the norm $\|\cdot\|$) and satisfy

$$\mathbf{H}(t)\Psi_{lt} = \lambda_l(t)\Psi_{lt}, \qquad 0 \leq t \leq 1, \quad 1 \leq l \leq m. \tag{8.44}$$

It is assumed that the set $\{\Psi_{lt}\}$ is ON for each t and $\lambda_l(0) = \lambda_0$. It can be shown (Kato [1]) that for a wide variety of perturbing potentials V_2 the above suppositions are correct and in fact, the $\lambda_l(t)$ can be expanded in power series in t. Physicists commonly employ a perturbation theory to compute the first few terms in the power series and get an approximation for the desired eigenvalues $\lambda_l(1)$. One of the most important problems is to determine the multiplicities of the eigenvalues $\lambda_l(1)$ of $\mathbf{H}$, that is, to determine how the m-fold degenerate eigenvalue λ_0 of $\mathbf{H}_1$ splits into eigenvalues $\lambda_l(1)$ as the perturbing potential V_2 is turned on. Group theory yields exact information about this splitting.

Suppose the point group G is a symmetry group of both Hamiltonians $\mathbf{H}_1$ and $\mathbf{H}$. Then the operators $\mathbf{T}(g)$ will commute with both $\mathbf{H}_1$ and $\mathbf{H}_2$, so G will be a symmetry of $\mathbf{H}(t)$ for all t. Let W_0 be the eigenspace of $\mathbf{H}_1$ corresponding to eigenvalue λ_0. Then the restriction of $\mathbf{T}$ to W_0 can be decomposed into irred reps,

$$\mathbf{T}\,|\,W_0 \cong \sum_{\mu=1}^{\alpha} \oplus\, a_\mu{}^0\mathbf{T}^{(\mu)}, \tag{8.45}$$

where $a_\mu{}^0$ is the multiplicity of $\mathbf{T}^{(\mu)}$ in $\mathbf{T}\,|\,W_0$. For each t between 0 and 1 the eigenvectors $\{\Psi_{lt}\}$ form an ON basis for the direct sum W_t of the eigenspaces of $\mathbf{H}(t)$ corresponding to the eigenvalues $\lambda_1(t), \ldots, \lambda_m(t)$. Thus W_t is invariant under $\mathbf{T}$ and we have the decomposition

$$\mathbf{T}\,|\,W_t \cong \sum_{\mu=1}^{\alpha} \oplus\, a_\mu{}^t\mathbf{T}^{(\mu)}. \tag{8.46}$$

The integers $a_\mu{}^t$ must remain fixed as t varies from 0 to 1. To see this we compute the character $\chi_t(g)$ of $\mathbf{T}\,|\,W_t$. Since the ON basis vectors $\{\Psi_{lt}\}$ are continuous in t the matrix elements

$$\langle \mathbf{T}(g)\Psi_{lt}, \Psi_{jt}\rangle = T^t_{jl}(g)$$

are continuous functions of t for fixed g. Thus the character $\chi_t(g)$ is continuous in t, as is $a_\mu{}^t = \langle \chi_t, \chi^{(\mu)}\rangle$. Since $a_\mu{}^t$ is an integer it must remain constant: $a_\mu{}^t = a_\mu{}^0 = a_\mu$.

Therefore, the reps $\mathbf{T}\,|\,W_0$ and $\mathbf{T}\,|\,W_1$ are equivalent. This result shows that the perturbation V_2 splits the m-fold degenerate eigenspace W_0 of H_1 into $a_1 + \cdots + a_\alpha$ eigenspaces of $\mathbf{H}$. There are a_1 eigenvalues of $\mathbf{H}$, each with multiplicity $n_1, \ldots$, and a_α eigenvalues of $\mathbf{H}$, each with multiplicity n_α. At most $a_1 + \cdots + a_\alpha$ of these eigenvalues are distinct, i.e., some of them may be equal. If the original rep $\mathbf{T}\,|\,W_0$ is irred then $\mathbf{T}\,|\,W_1$ is also irred and the m-fold eigenvalue λ_0 of $\mathbf{H}_1$ is perturbed to an m-fold eigenvalue λ_1 of $\mathbf{H}$.

Now suppose G_1 is the largest point symmetry group of $\mathbf{H}_1$. Let λ_0 be an eigenvalue of $\mathbf{H}_1$ and suppose the corresponding m-dimensional eigenspace transforms according to the irred rep $\mathbf{Q}$ of G_1. Furthermore, suppose $\mathbf{H}_2$ does not admit G_1 as a symmetry group but only a proper subgroup G of G_1. Then G is the maximal point symmetry group of $\mathbf{H}(t) = \mathbf{H}_1 + t\mathbf{H}_2$. The restriction of $\mathbf{Q}$ to the subgroup G splits into a direct sum of irred reps $\mathbf{T}^{(\mu)}$ of G:

$$\mathbf{Q}\,|\,G \cong \sum_{\mu=1}^{\alpha} \oplus\, a_\mu\mathbf{T}^{(\mu)}. \tag{8.47}$$

It follows from the analysis of expression (8.45) that the m-fold eigenvalue λ_0 of $\mathbf{H}_1$ splits into a_1 eigenvalues of $\mathbf{H}$, each of multiplicity $n_1, \ldots$, and a_α eigenvalues of $\mathbf{H}$, each of multiplicity n_α. Unless there is accidental degeneracy, there will be $a_1 + \cdots + a_\alpha$ distinct eigenvalues.

As an example, consider the case where $\mathbf{H}_1$ has octahedral symmetry O while the perturbing potential $\mathbf{H}_2$ has only tetragonal symmetry D_4, a subgroup of O. (There are several subgroups of O which are isomorphic to

D_4, but any two of these subgroups are conjugate, so it makes no difference which one we choose.) The character tables of D_4 and O are given in (6.17) and (6.22). We denote the irred reps of D_4 by $\mathbf{T}^{(1)}, \ldots, \mathbf{T}^{(5)}$ and those of O by $\mathbf{Q}^{(1)}, \ldots, \mathbf{Q}^{(5)}$. Since O has irred reps of dimensions one, two, and three, these are the possible multiplicities for eigenvalues of $\mathbf{H}_1$. To determine the manner in which each eigenspace of $\mathbf{H}_1$ splits into eigenspaces of $\mathbf{H}$ we must determine the multiplicities a_μ of $\mathbf{T}^{(\mu)}$ in $\mathbf{Q}^{(\nu)} \mid D_4$. These multiplicities can easily be computed from the character tables. We have

$$
\begin{aligned}
&\mathbf{Q}^{(1)} \mid D_4 \cong \mathbf{T}^{(1)}, \quad n_1 = 1; \qquad \mathbf{Q}^{(2)} \mid D_4 \cong \mathbf{T}^{(3)}, \quad n_3 = 1;\\
&\mathbf{Q}^{(3)} \mid D_4 \cong \mathbf{T}^{(1)} \oplus \mathbf{T}^{(3)}, \quad n_1 = n_3 = 1;\\
&\mathbf{Q}^{(4)} \mid D_4 \cong \mathbf{T}^{(2)} \oplus \mathbf{T}^{(5)}, \quad n_2 = 1, \quad n_5 = 2;\\
&\mathbf{Q}^{(5)} \mid D_4 \cong \mathbf{T}^{(4)} \oplus \mathbf{T}^{(5)}, \quad n_4 = 1, \quad n_5 = 2.
\end{aligned} \tag{8.48}
$$

For example, the character χ of $\mathbf{Q}^{(5)}$ restricted to D_4 is

	$\mathcal{E}$	$\mathcal{C}_4^2$	$2\mathcal{C}_4$	$2\mathcal{C}_2$	$2\mathcal{C}_{2'}$
χ	3	-1	-1	-1	1

From (6.17) we obtain $\chi(g) = \chi^{(4)}(g) + \chi^{(5)}(g)$, which yields the last line in (8.48). These results show that under the perturbing potential an eigenvalue of multiplicity two splits into two simple eigenvalues, while an eigenvalue of multiplicity three splits into an eigenvalue of multiplicity two and a simple eigenvalue. We can reduce these degeneracies still further by adding to the Hamiltonian a perturbation $\mathbf{H}_3$ with lower symmetry D_2. Then the restrictions of the reps $\mathbf{T}^{(\mu)}$ to D_2 will split into direct sums of irred reps $\mathbf{S}^{(1)}, \mathbf{S}^{(2)}, \mathbf{S}^{(3)}, \mathbf{S}^{(4)}$ of D_2. In fact,

$$
\begin{aligned}
&\mathbf{T}^{(1)} \mid D_2 \cong \mathbf{S}^{(1)}, \qquad \mathbf{T}^{(2)} \mid D_2 \cong \mathbf{S}^{(3)}, \qquad \mathbf{T}^{(3)} \mid D_2 \cong \mathbf{S}^{(1)}\\
&\mathbf{T}^{(4)} \mid D_2 \cong \mathbf{S}^{(3)}, \qquad \mathbf{T}^{(5)} \mid D_2 \cong \mathbf{S}^{(2)} \oplus \mathbf{S}^{(4)}
\end{aligned} \tag{8.49}
$$

where the character of $\mathbf{S}^{(j)}$ is $\chi^{(j)}$. The $\mathbf{S}^{(j)}$ is one-dimensional, so each of the multiply degenerate eigenvalues of $\mathbf{H}_1$ is split into simple eigenvalues of $\mathbf{H}' = \mathbf{H}_1 + \mathbf{H}_2 + \mathbf{H}_3$. The only possible degeneracies of eigenvalues of $\mathbf{H}'$ are accidental. The introduction of symmetry group lattices such as $O \supset D_4 \supset D_2$ is very useful in quantum mechanics for predicting the distribution of eigenvalues. For instance if the perturbation $\mathbf{H}_3$ is small with respect to $\mathbf{H}_2$ then one can predict that a triply degenerate eigenvalue of $\mathbf{H}_1$ will split into three simple eigenvalues, but that two of these eigenvalues will lie very close together in relation to the third.

The above ideas relating symmetry to perturbation theory are applicable to any symmetry group of the Hamiltonian, not only to the point groups. Indeed, the results which are of most importance in quantum theory relate to the rotation group $SO(3)$ and the symmetric group S_N whose reps will be studied later. In Chapter 7 we will return to the study of symmetry in perturbation theory.

The routine proofs of the following statements are omitted. Let W_λ be an eigenspace of the Hamiltonian (8.32). Since the potential V is real, the complex conjugate function $\bar{\Psi}(\mathbf{x}_1, \ldots, \mathbf{x}_k) \in W_\lambda$ for all $\Psi(\mathbf{x}_1, \ldots, \mathbf{x}_k) \in W_\lambda$. If $V^{(\mu)}$ is a subspace of W_λ transforming under the irred rep $\mathbf{T}^{(\mu)}$ of the symmetry group G then the complex conjugate space $\bar{V}^{(\mu)} \subseteq W_\lambda$ transforms under the irred rep $\bar{\mathbf{T}}^{(\mu)}$. If the simple character $\chi^{(\mu)}$ is real-valued then $\mathbf{T}^{(\mu)}$ is equivalent to $\bar{\mathbf{T}}^{(\mu)}$. However, if $\chi^{(\mu)} \neq \bar{\chi}^{(\mu)}$ then $\mathbf{T}^{(\mu)}$ and $\bar{\mathbf{T}}^{(\mu)}$ are nonequivalent irred reps and the eigenspace W_λ is not irred. This degeneracy is due to the fact that the (nonlinear) complex-conjugation operator commutes with $\mathbf{H}$, and is not considered accidental. Reps of G with complex characters always occur in complex conjugate pairs. (However, a rep with a real character need not be real; see Hamermesh [1, p. 138].)

Problems

3.1 Let T be an irred matrix rep of the finite group G and let C be a conjugacy class in G. Show that $\sum_{g \in C} T(g)$ is a multiple of the identity matrix.

3.2 Let G be a finite group with commutator subgroup G_C. (See Problem 1.8.) Show that the number of one-dimensional reps of G is equal to the index of G_C in G.

3.3 Let $\mathbf{T}_j, \mathbf{T}_j'$, $(j = 1, 2)$ be reps of the groups G such that $\mathbf{T}_j \cong \mathbf{T}_j'$. Show that $\mathbf{T}_1 \oplus \mathbf{T}_2 \cong \mathbf{T}_1' \oplus \mathbf{T}_2'$.

3.4 Let $\mathbf{T}, \mathbf{T}'$ be unitary reps of G on the inner product spaces V, V', respectively. If $\langle -, - \rangle, \langle -, - \rangle'$ are the inner products on V, V' show that $(\mathbf{u} \otimes \mathbf{u}', \mathbf{v} \otimes \mathbf{v}') = \langle \mathbf{u}, \mathbf{v} \rangle \langle \mathbf{u}', \mathbf{v}' \rangle'$ defines an inner product on $V \otimes V'$ with respect to which $\mathbf{T} \otimes \mathbf{T}'$ is unitary.

3.5 Prove: If $\mathbf{T}$ is an irred rep and $\mathbf{Q}$ a one-dimensional rep of G then $\mathbf{T} \otimes \mathbf{Q}$ is irred.

3.6 Let $\mathbf{T}_1, \mathbf{T}_2$ be irred reps of the finite group G with dimensions $d_1 > d_2$. Show that $\mathbf{T}_1 \otimes \mathbf{T}_2$ contains no irred rep $\mathbf{T}_3$ with $d_3 < d_1/d_2$.

3.7 Compute the character table of the icosahedral group Y.

3.8 Prove: The dimensions n_i of the irred reps of the finite group G are divisors of $n(G)$. (This is a difficult theorem. See Hall [1, Section 16.8].)

3.9 Determine the dimensions of the following subspaces of second-rank tensors which are fixed under Y: (a) polar, (b) symmetric polar, (c) axial, (d) symmetric axial. Repeat for the group C_{2v}.

3.10 Consider a quantum mechanical system with octahedral symmetry O. Suppose a perturbation is applied which reduces the symmetry to (a) T, (b) D_3, (c) C_4. In each case determine how the possible energy levels of the original system are split by the perturbation.

3.11 Let K be a subgroup of H and H a subgroup of the finite group G. Prove the following properties of induced reps: (a) If $\mathbf{T}$ is a rep of K then $(\mathbf{T}^H)^G \cong \mathbf{T}^G$. (b) If $\mathbf{R}$ is a rep of H and $\mathbf{S}$ a rep of G then $\mathbf{R}^G \otimes \mathbf{S} \cong (\mathbf{R} \otimes (\mathbf{S} \mid H))^G$.

3.12 Let G be a group of order N and $\chi(g)$ a character of G. Prove that $N^{-1} \sum_{g \in G} [\chi(g)]^n$ is a nonnegative integer for each $n = 1, 2, \ldots$.

3.13 Let $T_1(g)$ and $T_2(g)$ be $n \times n$ matrix reps of G with real matrix elements. These reps are **real equivalent** if there is a real nonsingular matrix S such that $T_1(g)S = ST_2(g)$ for all $g \in G$. Show that T_1 and T_2 are complex equivalent if and only if they are real equivalent. (Hint: Write $S = A + iB$, where A and B are real, and show that $A + tB$ is invertible for some real number t.)

3.14 Show that the matrix elements of two real irred reps of a group **G** which are not real equivalent satisfy an orthogonality relation. Show that every real irred rep is real equivalent to a rep by real orthogonal matrices.

Chapter 4

Representations of the Symmetric Groups

4.1 Conjugacy Classes in S_n

The symmetric groups occur as symmetry groups of quantum mechanical systems which contain n identical particles. Furthermore, the irred reps of S_n are intimately bound up with the irred reps of certain Lie groups, most notably $GL(n, \mathfrak{C})$ and $O(n, \mathfrak{C})$. (See Section 4.3, where we discuss the relation between S_n and symmetry classes of tensors.) For these reasons a knowledge of the rep theory of S_n is indispensable for an understanding of the role of groups in modern physical theories.

For low values of n, say $n \leq 5$, we could use the methods of Section 3.6 to compute the character tables and rep matrices of S_n. For example S_4 is isomorphic to the octahedral group O and has the character table (6.22). (Every $g \in O$ is uniquely determined by a permutation of the four threefold axes.) However, the construction of character tables becomes rapidly more difficult as n increases.

To obtain the irred reps of S_n for all n simultaneously, we develop new tools which exploit the structure of these groups. In distinction to the general methods of Chapter 3, the methods introduced in this chapter apply to the symmetric groups alone. Furthermore, the proofs of the basic facts about the rep theory of S_n are somewhat complicated, although the final results are not difficult to state. We will not give a complete coverage of the symmetric groups, but merely determine the primitive idempotents in the group ring of S_n, compute the simple characters, and study the relation between S_n and symmetry classes of tensors. The principal omissions, which the reader can fill in by consulting Boerner [1], Hamermesh [1], Robinson [1], or Rutherford [1], are a construction of the matrix elements of irred reps and a detailed study of the computational problems involved in construction and decomposition of reps. The theory developed here is sufficient for all subsequent applications of S_n which occur in this book and for the majority of applications to modern physical theories.

To begin we investigate the structure of S_n in more detail. We will use the notation for permutations introduced in Example 5, Section 1.1. Let

$$s = \begin{pmatrix} 1 & 2 & \cdots & n \\ s(1) & s(2) & \cdots & s(n) \end{pmatrix} \tag{1.1}$$

be an element of S_n and

$$h(\mathbf{x}) = \prod_{1 \le \mu < \nu \le n} (x_\nu - x_\mu), \qquad x = (x_1, \ldots, x_n), \tag{1.2}$$

where the x_j are arbitrary variables. If f is any function of $\mathbf{x}$ we define the new function $T_s f$ by

$$T_s f(\mathbf{x}) = f(\mathbf{x}_s), \qquad \mathbf{x}_s = (\mathbf{x}_{s(1)}, \ldots, \mathbf{x}_{s(n)}). \tag{1.3}$$

This mapping satisfies the homomorphism property since

$$[T_{st} f](\mathbf{x}) = f(\mathbf{x}_{st}) = \{T_s[T_t f]\}(\mathbf{x}) \tag{1.4}$$

for $s, t \in S_n$. Indeed $s[t(i)] = st(i)$ and $[T_t f](\mathbf{x}_s) = [T_t f](\mathbf{y}) = f(\mathbf{y}_t) = f(\mathbf{x}_{st})$ since $y_i = x_{s(i)}$ and $y_{t(i)} = x_{st(i)}$. Now $[T_s h](\mathbf{x}) = \pm h(\mathbf{x})$, for every $s \in S_n$, where h is the function (1.2). That is, an arbitrary permutation of the indices of $\mathbf{x}$ either leaves h fixed or changes its sign. The restriction of the operators T_s to the one-dimensional vector space generated by h yields an irred rep of S_n called the **alternating** rep. We can regard this rep as a homomorphism μ of S_n into the cyclic group of order two containing the elements $\{\pm 1\}$. The permutation s is **even** if μ maps s into $+1$ and **odd** if s is mapped into -1. The reader can verify that the permutation $s = (12)$ in cycle notation is odd. This proves that μ is onto for $n \ge 2$. Let A_n be the kernel of μ, i.e., the set of even permutations. By Theorem 1.3, A_n is a normal subgroup of index two in S_n. Thus $\{A_n, (12)A_n\}$ is a coset decomposition of S_n.

A two-cycle $(i_1\ i_2) \in S_n$ is called a **transposition.** If s is the permutation (1.1) then $s(12)s^{-1} = (s(1), s(2))$. Thus, any two transpositions in S_n are conjugate. Since $(12) \notin A_n$ all transpositions are odd. It follows that a product of an odd number of transpositions is odd while a product of an even number of transpositions is even. Every permutation s is a product of transpositions. Indeed s is a product of cycles and any cycle

$$(i_1 i_2 \cdots i_j) = (i_1 i_2)(i_2 i_3) \cdots (i_{j-1} i_j) \tag{1.5}$$

is a product of transpositions. A permutation can be written as a product of transpositions in many ways but the number of factors is always even or odd depending on the parity of the permutation.

The conjugacy classes of S_n are easily described. If $s, t \in S_n$ then

$$[sts^{-1}](s(j)) = sts^{-1}s(j) = s[t(j)], \qquad 1 \le j \le n,$$

so

$$sts^{-1} = \begin{pmatrix} s(1) & s(2) & \cdots & s(n) \\ st(1) & st(2) & \cdots & st(n) \end{pmatrix}. \tag{1.6}$$

Thus sts^{-1} is obtained from t by applying s to the numbers in the two rows of

$$t = \begin{pmatrix} 1 & 2 & \cdots & n \\ t(1) & t(2) & \cdots & t(n) \end{pmatrix}.$$

In terms of cycle notation the results are even more transparent. For example, if $t = (13642)(57)(8) \in S_8$ and s is given by (1.1) with $n = 8$, then

$$sts^{-1} = (s_1 s_3 s_6 s_4 s_2)(s_5 s_7)(s_8), \qquad s(j) = s_j. \tag{1.7}$$

Two elements of S_n are conjugate if and only if they have the same cycle structure. Furthermore, the elements of a conjugacy class are either all even or all odd.

As an illustration we list the five conjugacy classes of S_4:

$$\begin{aligned} &\{e\}, \qquad \{(12), (13), (14), (23), (24), (34)\}, \\ &\{(12)(34), (13)(24), (14)(23)\}, \\ &\{(123), (124), (132), (134), (142), (143), (234), (243)\} \\ &\{(1234), (1243), (1324), (1342), (1423), (1432)\}. \end{aligned}$$

To each set of nonnegative integers $(\nu_1, \nu_2, \ldots, \nu_n)$ such that

$$n = \nu_1 + 2\nu_2 + \cdots + n\nu_n \tag{1.8}$$

there corresponds a conjugacy class in S_n. This class consists of those elements with ν_1 one-cycles, ν_2 two-cycles, . . . , and ν_n n-cycles. According to Section 1.2, the number of elements in the conjugacy class (ν_i) is $m_\nu = n!/n_\nu$, where n_ν is the order of the group

$$H^s = \{t \in S_n : tst^{-1} = s\}$$

and s is an element in the conjugacy class (ν_i). We compute the number of possible permutations t. Any cycle of length i in s remains invariant under any one of the i cyclic permutations of its digits. Each of the ν_i i-cycles can be acted on independently in this fashion and the i-cycles can also be permuted among themselves. Thus, there are a total of $i^{\nu_i}\nu_i!$ permutations which preserve the cycles of length i in s. Since cycles of different length can be considered independently, we find

$$n_\nu = 1^{\nu_1}\nu_1!\, 2^{\nu_2}\nu_2! \cdots n^{\nu_n}\nu_n!, \qquad m_\nu = n!/n_\nu. \tag{1.9}$$

The number of conjugacy classes in S_n, hence the number of nonequivalent irred reps is just the number of sets of nonnegative integers (ν_i) satisfying (1.8). The structure of such solutions (ν_i) is more easily comprehended in terms of the nonnegative integers λ_i,

$$\begin{aligned} \lambda_1 &= \nu_1 + \nu_2 + \cdots + \nu_n \\ \lambda_2 &= \nu_2 + \nu_3 + \cdots + \nu_n \\ \lambda_3 &= \nu_3 + \nu_4 + \cdots + \nu_n \\ &\vdots \\ \lambda_n &= \nu_n. \end{aligned} \tag{1.10}$$

Clearly,

$$\lambda_1 + \lambda_2 + \cdots + \lambda_n = n, \qquad \lambda_1 \geq \lambda_2 \geq \cdots \geq \lambda_n \geq 0. \tag{1.11}$$

The integers $\{\lambda_i\}$ satisfying (1.11) are said to form a **partition** of n. We have shown that each conjugacy class (ν_i) corresponds to a partition of n. Conversely if $\{\lambda_1, \ldots, \lambda_n\}$ form a partition of n then the integers

$$\nu_i = \lambda_i - \lambda_{i+1}, \qquad 1 \leq i \leq n-1, \quad \nu_n = \lambda_n, \tag{1.12}$$

determine a conjugacy class (ν_i) in S_n. Thus the number of conjugacy classes of S_n is equal to the number of partitions (1.11) of n.

Ordinarily a partition $\{\lambda_1, \ldots, \lambda_r, 0, \ldots, 0\}$ of n is written $\{\lambda_1, \ldots, \lambda_r\}$, i.e., we leave out the λ_i that are 0. Also, if several of the λ_i are equal we use exponents to shorten the notation. Thus, the partitions {22100}, {21110}, {31100} of 5 are usually written in the abbreviated forms

$$\{2^2 1\}, \quad \{21^3\}, \quad \{31^2\},$$

respectively. The cycle structure of the conjugacy classes corresponding to $\{\lambda_i\}$ can be recovered from (1.12). As an example we list the five partitions of 4, or what is the same thing, the five conjugacy classes of S_4:

$$\{4\}, \quad \{31\}, \quad \{2^2\}, \quad \{21^2\}, \quad \{1^4\}.$$

In this example as in the rest of this chapter we adopt a dictionary ordering of partitions. That is, the partition $\{\lambda_1, \ldots, \lambda_n\}$ precedes (or is greater than) the partition $\{\lambda_1', \ldots, \lambda_n'\}$ if the first nonzero difference $\lambda_i - \lambda_i'$, $i = 1, \ldots, n$, is positive.

4.2 Young Tableaux

We now proceed to determine the irred reps of S_n by the method of Young as simplified by Von Neumann (Boerner [1], Weyl [3]). In this approach one computes the primitive idempotents in the group ring. As we have shown in

Section 3.7 each such idempotent generates an irred rep of S_n. We will obtain the simple characters indirectly via an examination of symmetry classes of tensors.

There is another approach to this theory, due to Frobenius, in which the method of induced reps is employed to compute the simple characters directly. The primitive idempotents and matrix elements of irred reps are then derived from the characters. See Hamermesh [1] and Littlewood [1] for an exposition of this method.

Let $R_n = R_{S_n}$ be the group ring of S_n. Every $x \in R_n$ can be written uniquely in the form $x = \sum x(s) \cdot s$, where s runs over the $n!$ elements of S_n. According to the results of Section 3.7 any primitive idempotent in R_n generates an irred rep of S_n and every irred rep can be so generated. We already know two irred reps: the one-dimensional identity and alternating reps. Each of these reps is contained exactly once in the decomposition of the left regular rep $\mathbf{L}$ of S_n on R_n. The corresponding idempotents are easily constructed. Consider the element $c = (n!)^{-1} \sum s$, where the sum extends over S_n. Clearly, $sc = cs = c$ for all $s \in S_n$. It follows that $c^2 = c$, so c is idempotent. Furthermore, c is primitive idempotent because the invariant subspace it generates consists of the elements λc, $\lambda \in \mathfrak{C}$. Since $\mathbf{L}(s)c = sc = c$, the restriction of $\mathbf{L}$ to $\{\lambda c\}$ is equivalent to the identity rep of S_n.

Similarly the element $c' = (n!)^{-1} \sum \delta_s s$, where $\delta_s = +1$ if s is even and $\delta_s = -1$ if s is odd, satisfies $sc' = c's = \delta_s c'$ for all $s \in S_n$. Thus, $(c')^2 = c'$ and c' is idempotent. The reader can check that c' generates an invariant subspace under $\mathbf{L}$ which transforms according to the alternating rep of S_n.

Unfortunately the remaining idempotents are not so easy to find. To simplify our discussion slightly we introduce the concept of essential idempotence. An element c is **essentially idempotent** if there exists a nonzero constant λ such that $c^2 = \lambda c$. If c is essentially idempotent then $c' = \lambda^{-1} c$ is idempotent since $(c')^2 = \lambda^{-2} c^2 = \lambda^{-1} c = c'$. We shall find it convenient to work with essential idempotents c and there is no loss of generality in doing so since c can be normalized to an idempotent element.

There are exactly as many irred reps of S_n as there are partitions $\{\lambda_j\}$ of n, $\lambda_1 + \lambda_2 + \cdots + \lambda_n = n$, $\lambda_1 \geq \lambda_2 \geq \cdots \geq \lambda_n \geq 0$, so it seems reasonable that each partition should be related to an irred rep. We shall describe this relationship first and then verify its validity.

Consider the partition $\{\lambda_j\}$ of n with $\lambda_1 \geq \lambda_2 \geq \cdots \geq \lambda_r > 0$, $\lambda_{r+1} = \lambda_{r+2} = \cdots = \lambda_n = 0$. To this partition we associate a **frame** consisting of n squares arranged in r rows. The first row consists of λ_1 squares, the second of λ_2 squares, . . . , and the rth of λ_r squares. For example the partition $\{3, 2^2, 1\} = \{3, 2, 2, 1, 0, 0, 0, 0\}$ of $n = 8$ is associated with the frame shown in Fig. 4.1. A **Young tableau** is obtained by filling in the n squares of the frame

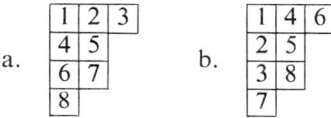

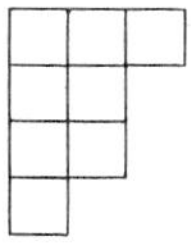

FIGURE 4.1

with the digits 1, 2, . . . , n taken in any order. Each digit is used exactly once. As an example Fig. 4.2 shows two tableaux each of whose frame is that of Fig. 4.1. Clearly, there are $n!$ tableaux associated with the frame $\{\lambda_j\}$.

a.

1	2	3
4	5	
6	7	
8		

b.

1	4	6
2	5	
3	8	
7		

FIGURE 4.2

Given a tableau T we define two sets of permutations $R(T)$ and $C(T)$. Here $R(T)$ consists of all $p \in S_n$ that permute the digits in each row of T among themselves without altering the row in which a digit lies. The elements p are called **row permutations**. The set $C(T)$ consists of all $q \in S_n$ that permute the digits in each column of T. The q are called **column permutations**. It is easy to verify that $R(T)$ and $C(T)$ are subgroups of S_n.

For example, the tableau shown in Fig. 4.3 is associated with

$$R(T) = \{(146), (164), (14), (16), (46), (1), (146)(25), (164)(25), (14)(25), (16)(25), (46)(25), (25)\}$$

$$C(T) = \{(123), (132), (12), (13), (23), (1), (123)(45), (132)(45), (12)(45), (13)(45), (23)(45), (45)\}.$$

1	4	6
2	5	
3		

FIGURE 4.3

Corresponding to the tableau T we construct the elements

$$P = \sum_{p \in R(T)} p, \qquad Q = \sum_{q \in C(T)} \delta_q q \tag{2.1}$$

in the group ring R_n, where $\delta_q = +1$ if q is even and -1 if q is odd.

Theorem 4.1. The ring element $c = PQ$ is essentially idempotent and the invariant subspace $R_n c$ determines an irred rep of S_n. Reps determined by different tableaux with the same frame are equivalent, while those determined by tableaux with different frames are nonequivalent.

According to this theorem there is a 1–1 correspondence between irred reps of S_n and frames $\{\lambda_j\}$. The proof is complicated and relies heavily on the methods of Section 3.7. We verify the theorem through a series of lemmas.

Note that $R(T) \cap C(T) = \{e\}$ since only the identity element is simultaneously a row and a column permutation. If $pq = p'q'$, where $p, p' \in R(T)$ and $q, q' \in C(T)$, then $q(q')^{-1} = p^{-1}p' = e$, so $q = q'$, $p = p'$. Thus, each of the terms pq in

$$c = \sum_{pq} \delta_q pq \tag{2.2}$$

is a distinct element of S_n, so $c \neq 0$.

If T is a tableau and $s \in S_n$, let $T' = sT$ be the tableau obtained by applying s to the digits of T. Thus, if T is the tableau pictured in Fig. 4.2(b) and $s = (257)(34)(16)$, then we have the situation shown in Fig. 4.4.

$$sT = \begin{array}{|c|c|c|}\hline 6 & 3 & 1 \\ \hline 5 & 7 \\ \cline{1-2} 4 & 8 \\ \cline{1-2} 2 \\ \cline{1-1} \end{array}$$

FIGURE 4.4

We say that a digit m is at position (i, j) in T if m lies in the ith row and jth column of T. In Figure 4.3, 4 is in position (1, 2) and 5 is in position (2, 2).

Lemma 4.1. Let $r, s \in S_n$ and $T' = sT$. If the digit m at position (i, j) in T lies at position (i_1, j_1) in rT then the digit $s(m)$ at position (i, j) in T' lies at position (i_1, j_1) in $r'T'$, where $r' = srs^{-1}$. We say that r' is the permutation **corresponding** to r for T'.

Proof. If the digit m lies at position (i, j) in T and (i_1, j_1) in rT then the digit k at (i_1, j_1) in T satisfies $r(k) = m$. Therefore, the digit at position (i_1, j_1) of $r'T' = (srs^{-1})sT$ is $srs^{-1} \cdot s(k) = s[r(k)] = s(m)$, which is the digit at position (i, j) of T'. Q.E.D.

Example. Let $s = (257)(34)(16)$, $r = (135)$ and let T be the tableau in Fig. 4.2(b). Then $r' = srs^{-1} = (647)$.

$$T = \begin{array}{|c|c|c|}\hline 1 & 4 & 6 \\ \hline 2 & 5 \\ \cline{1-2} 3 & 8 \\ \cline{1-2} 7 \\ \cline{1-1} \end{array}, \qquad rT = \begin{array}{|c|c|c|}\hline 3 & 4 & 6 \\ \hline 2 & 1 \\ \cline{1-2} 5 & 8 \\ \cline{1-2} 7 \\ \cline{1-1} \end{array},$$

$$T' = sT = \begin{array}{|c|c|c|}\hline 6 & 3 & 1 \\ \hline 5 & 7 \\ \cline{1-2} 4 & 8 \\ \cline{1-2} 2 \\ \cline{1-1}\end{array}\,, \qquad r'T' = srT = \begin{array}{|c|c|c|}\hline 4 & 3 & 1 \\ \hline 5 & 6 \\ \cline{1-2} 7 & 8 \\ \cline{1-2} 2 \\ \cline{1-1}\end{array}\,.$$

Now the digit 3 at position (3, 1) of T goes to position (1, 1) of rT, while the digit $4 = s(3)$ at position (3, 1) of T' goes to position (1, 1) of $r'T'$. Similarly, the reader can check the validity of the lemma for all entries of T.

Corollary 4.1. If $T' = sT$ then $R(T') = sR(T)s^{-1}$, $C(T') = sC(T)s^{-1}$, $P' = sPs^{-1}$, $Q' = sQs^{-1}$, $c' = scs^{-1}$.

Proof. By the lemma, if p is a row permutation of T then $p' = sps^{-1}$ is a row permutation of T'. Similarly, if $q \in C(T)$ then $q' = sqs^{-1} \in C(T')$. The corollary follows easily from these remarks and the elementary fact that $\delta_{q'} = \delta_q$. Q.E.D.

Lemma 4.2. An element s of S_n can be written $s = pq$, where p, q are row and column permutations of the tableau T, if and only if no two digits in the same row of T lie in the same column of $T' = sT$.

Proof. Suppose $s = pq$ and m_1, m_2 are distinct digits in the ith row of T. Then $sT = (pqp^{-1})pT = q'pT$. According to Lemma 4.1 we can obtain the tableau sT from T by performing a row permutation p on T followed by a column permutation $q' = pqp^{-1}$ on the resultant tableau pT. Clearly, m_1 and m_2 are still in the ith row of pT, so they must lie in different columns of $q'pT$.

Note. The tableau pqT may *not* be obtained by applying a row permutation to qT. The element p is a row permutation of T but not necessarily an element of $R(qT)$.

Conversely, suppose no two digits in the same row of T lie in the same column of T'. Then the digits in the first column of T' lie in different rows of T. Applying a suitable row permutation to T we can move these digits into the first column. Leaving the first column fixed, we can apply the same procedure to the second column, and so on. Thus there is a $p \in R(T)$ such that the digits in each column of T' are the same as the digits in the corresponding column of pT, though not necessarily in the same order. Finally we can apply a column permutation q' to pT such that $T' = q'pT$. Writing $q = p^{-1}q'p$ we note from Corollary 4.1 that $q \in C(T)$. Therefore, $T' = pqT$. Q.E.D.

Example. The permutation $s = (257)(34)(16)$ is not a pq for the tableau of Fig. 4.2(b) since the digits 2, 5 which lie in row two of T also lie in column one of sT (Fig. 4.4).

Lemma 4.3. If T belongs to the frame $\{\lambda_j\}$ and T' belongs to the frame $\{\lambda_j'\}$ with $\{\lambda_j\} > \{\lambda_j'\}$, then there exist two digits which lie in the same row of T and the same column of T'.

Proof. If such a pair of digits did not exist then the λ_1 digits in the first row of T would lie in different columns of T'. Thus, $\lambda_1' \geq \lambda_1$ but since $\{\lambda_j\} > \{\lambda_j'\}$ it follows that $\lambda_1' = \lambda_1$. By means of a column permutation of T' we can transform T' to a tableau T'' with the same first row as T. Since this permutation does not change the distribution of digits among the columns, we can repeat our argument on the second row of T'' to obtain $\lambda_2' = \lambda_2$. Similarly $\lambda_j' = \lambda_j$ for all j and $\{\lambda_j'\} = \{\lambda_j\}$, which is impossible. Q.E.D.

Let T and T' be two tableaux associated with S_n, not necessarily with the same frame. Let P, Q, c, respectively P', Q', c', be the ring elements defined by (2.1) and (2.2).

Lemma 4.4. If there exist two digits which lie in one row of T and one column of T' then $c'c = 0$.

Proof. First we remark that

(2.3) $$pP = Pp = P, \qquad qQ = Qq = \delta_q Q$$

(2.4) $$pcq = pPQq = \delta_q c$$

for all $p \in R(T)$ and $q \in C(T)$. The occurrence of the parity δ_q follows from $\delta_{s_1 s_2} = \delta_{s_1}\delta_{s_2}$.

By hypothesis there exist two digits m, k such that the transposition $t = (mk) \in R(T) \cap C(T')$. Thus, $Q'P = Q't \cdot tP = -Q'P$, so $Q'P = 0$, since $t^2 = e$ and $\delta_t = -1$. Therefore $c'c = P'Q'PQ = 0$. Q.E.D.

We shall now show that property (2.4) characterizes c up to a scalar multiple.

Lemma 4.5. If $x \in R_n$ such that $pxq = \delta_q x$ for all $p \in R(T)$ and $q \in C(T)$ then $x = \lambda c$ for some $\lambda \in \mathfrak{C}$.

Proof. Let $x = \sum x(s) \cdot s$. Then

$$x = \delta_q p^{-1} x q^{-1} = \delta_q \sum_s x(s) \cdot p^{-1} s q^{-1} = \delta_q \sum_s x(psq) \cdot s$$

so

(2.5) $$x(s) = \delta_q x(psq)$$

for every $s \in S_n$, $p \in R(T)$, $q \in C(T)$. Setting $s = e$ we see that $x(pq) = \delta_q \lambda$, where $\lambda = x(e)$. A comparison of this result with (2.2) shows that the lemma is true provided we can show $x(s) = 0$ whenever s is not a pq.

If s is not a pq there exist two digits in the same row of T and the same column of sT. The transposition p of these digits belongs to $R(T) \cap C(sT)$. Similarly the transposition $q^{-1} = s^{-1}ps \in C(T)$ by Corollary 4.1. Thus $s = psq$ and $x(s) = x(psq) = -x(s)$ by (2.5), so $x(s) = 0$. Q.E.D.

Lemma 4.6. The ring element $c = PQ$ corresponding to the tableau T is essentially idempotent, and the invariant subspace $R_n c$ yields an irred rep of S_n whose degree divides $n!$.

Proof. Since $pc^2q = pccq = \delta_q c^2$ for all $p \in R(T)$, $q \in C(T)$, it follows from Lemma 4.5 that $c^2 = \lambda c$ for some $\lambda \in \mathfrak{C}$. Thus, c is essentially idempotent if $\lambda \neq 0$. [Since the coefficients $c(pq)$ of c are ± 1 it follows that λ is an integer.]

Consider the linear transformation $\mathbf{A}$ on R_n defined by $\mathbf{A}x = xc$, $x \in R_n$. We will compute the trace of $\mathbf{A}$ in the natural basis $\{s_j\}$ of elements of S_n. Writing $c = \sum_{j=1}^{n!} c(s_j) \cdot s_j = \sum \delta_q \cdot pq$, we find

$$[\mathbf{A}s_j](s_j) = [s_j c](s_j) = c(e) = 1, \qquad 1 \leq j \leq n!.$$

Therefore the trace of the matrix describing $\mathbf{A}$ is $n!$.

Next we compute the trace of $\mathbf{A}$ with respect to a basis $v_1, \ldots, v_{n!}$, where $v_1, \ldots, v_f$ form a basis for the f-dimensional space $R_n c$. If $x = yc \in R_n c$ then $\mathbf{A}x = xc = yc^2 = \lambda yc = \lambda x$, so $\mathbf{A}v_j = \lambda v_j$, $1 \leq j \leq f$. Furthermore $v_j \notin R_n c$ for $f + 1 \leq j \leq n!$ and $\mathbf{A}v_j \in R_n c$ for all j. Thus the trace of $\mathbf{A}$ in the v-basis is λf. We conclude that $\lambda = n!/f > 0$. Since λ is an integer, f divides $n!$.

By Theorem 3.14, to show that $\lambda^{-1}c$ is a primitive idempotent, it is enough to verify that $\tilde{x} = cxc$ is a multiple of c for every $x \in R_n$. For any $p \in R(T)$ and $q \in C(T)$ we have

$$p\tilde{x}q = pcxcq = \delta_q cxc = \delta_q \tilde{x}.$$

Therefore, by Lemma 4.5, $\tilde{x}$ is a multiple of c. Q.E.D.

Lemma 4.7. Tableaux corresponding to different frames yield nonequivalent reps of S_n, while those corresponding to the same frame yield equivalent reps.

Proof. Suppose T is a tableau with frame $\{\lambda_j\}$ and T' is a tableau with different frame $\{\lambda_j'\}$. Without loss of generality we can assume $\{\lambda_j\} > \{\lambda_j'\}$. By Theorem 3.15, to prove that the reps determined by T and T' are nonequivalent it is enough to show $c'xc = 0$ for all $x \in R_n$. From Lemmas 4.3 and 4.4 there follows $c'c = 0$. Furthermore sT has the same frame as T and essential idempotent scs^{-1} for each $s \in S_n$. Thus, $c'scs^{-1} = 0$ or $c'sc = 0$ for all $s \in S_n$. Therefore $c'xc = \sum x(s) \cdot c'sc = 0$ for all $x \in R_n$.

If T and T' have the same frame then $T' = sT$ for some $s \in S_n$ and

$c' = scs^{-1}$. Thus $c'sc = (scs^{-1})sc = sc^2 = \lambda sc \neq 0$ since $\lambda c \neq 0$. By Theorem 3.15, c and c' generate equivalent reps of S_n. Q.E.D.

Lemmas 4.1–4.7 constitute a proof of Theorem 4.1. Note that the identity rep of S_n corresponds to the frame $\{n\}$, i.e., the frame with one row of n squares. The alternating rep corresponds to the frame $\{1^n\}$, i.e., the frame with one column of n squares.

Since there are $n!$ tableaux with the same frame, Theorem 4.1 enables us to construct $n!$ subspaces in the $n!$-dimensional space R_n which transform under a given f-dimensional irred rep of S_n. The multiplicity of this irred rep in R_n is only f, so these subspaces are not all independent of one another. We shall show that it is possible to select f generating idempotents $c_1, \ldots, c_f$ corresponding to the given frame such that $R_nc_1, \ldots, R_nc_f$ are linearly independent. Then, according to the general theory of Section 3.7, every irred subspace R_nc corresponding to this frame is a subspace of

$$R_nc_1 \oplus \cdots \oplus R_nc_f.$$

As an example, we already know that the one-dimensional identity rep has multiplicity one in R_n. Therefore, the $n!$ generating idempotents c_j corresponding to the frame $\{n\}$ must all generate the same one-dimensional subspace of R_n. The reader can easily show that $c_1 = c_2 = \cdots = c_{n!}$.

Let $\{\lambda_1, \ldots, \lambda_n\}$ be a frame with corresponding (essential) generating idempotents $c_1, \ldots, c_{n!}$. We shall first show that the $n!$ left ideals R_nc_j span the minimal two sided ideal U which contains all left ideals transforming under the irred rep $\{\lambda_j\}$.

Lemma 4.8. Let R_nc be an irred subspace of R_n transforming according to the irred rep $\{\lambda_j\}$ under **L**. Then

$$R_nc \subseteq R_nc_1 + R_nc_2 + \cdots + R_nc_{n!},$$

i.e., each $x \in R_nc$ is a linear combination of elements in the R_nc_j.

Proof. Since R_nc and R_nc_1 correspond to equivalent reps of S_n under **L** it follows from Theorem 3.15 that there exists $y \in R_n$ such that $R_nc = R_nc_1y$ (as vector spaces). Now $y = \sum y(s)\cdot s$, so the lemma will be proved if we can show that for each s, $R_nc_1s = R_nc_j$, where c_j is one of the generating idempotents. If T_1 is the tableau with idempotent c_1 then $s^{-1}T_1 = T_j$ is a tableau with the same frame corresponding to the idempotent (say) c_j. Thus $c_j = s^{-1}c_1s$, so $xc_1s = xsc_j \subseteq R_nc_j$ for all $x \in R_n$. Since both R_nc_1s and R_nc_j are minimal left ideals it follows that $R_nc_1s = R_nc_j$. Q.E.D.

The left ideals R_nc_j span the two-sided ideal U but they are not linearly independent. We can obtain a linearly independent set of left ideals which span U by considering the standard tableaux. A tableau T is called a **standard tableau** if the digits in each row of T increase from left to right and the digits in each column increase from top to bottom. For example, the tableaux in Fig. 4.2 and 4.3 are standard, while the tableau in Fig. 4.4 is not.

We have already defined a dictionary ordering for frames. Similarly we can define a dictionary ordering for the standard tableaux belonging to a given frame. Given two such tableaux T, T' we compare their corresponding digits, starting at the left end of the first row and going from left to right. If the first nonzero difference $m - m'$ is positive for corresponding digits m in T and m' in T' we say $T > T'$. If all corresponding digits in the first row are equal we compare digits in the second row, etc. As an example we list the standard tableaux in increasing order of the frame $\{3, 1^2\}$:

1 2 3	1 2 4	1 2 5	1 3 4	1 3 5	1 4 5
4	3	3	2	2	2
5	5	4	5	4	3

Theorem 4.2. The dimension f of the irred rep corresponding to the frame $\{\lambda_j\}$ is equal to the number of standard tableaux $T_1, \ldots, T_f$ belonging to this frame.

A proof of this theorem will be given in Section 4.4. The theorem implies, for example, that the dimension of the rep $\{3, 1^2\}$ is six. More important, it says that the multiplicity of the rep $\{\lambda_j\}$ is equal to the number f of standard tableaux belonging to the frame $\{\lambda_j\}$. We shall show that the generating idempotents $c_1, \ldots, c_f$ of the standard tableaux generate linearly independent left ideals $R_n c_1, \ldots, R_n c_f$.

Lemma 4.9. If $T_i < T_l$ then $c_l c_i = 0$.

Proof. By Lemma 4.4 it is enough to show that there exist two digits in the same row of T_l and the same column of T_i. Consider the first space (j, k) (row j, column k) which is occupied by different digits, m in T_i and m' in T_l. Clearly, $m' > m$ since $T_l > T_i$. If $k = 1$ then the entry at $(j, 1)$ is the smallest integer not in the first $j - 1$ rows. Thus the entries m, m' at $(j, 1)$ would be the same for the two tableaux, which is impossible. Thus, $k > 1$. The digit m lies at (j, k) in T_i. We will determine the position (a, b) of m in T_l. Since $m < m'$ we cannot have $a \geq j$ and $b \geq k$, i.e., m cannot lie below and to the right of (j, k). Also, by definition of (j, k), m cannot lie at a position which comes before (j, k) in the dictionary ordering. Thus the only possibility is $a < j$, $b < k$; so m lies below and to the left of (j, k) (Fig. 4.5). The position (j, b) comes before (j, k) in the dictionary ordering, so the digit q at this

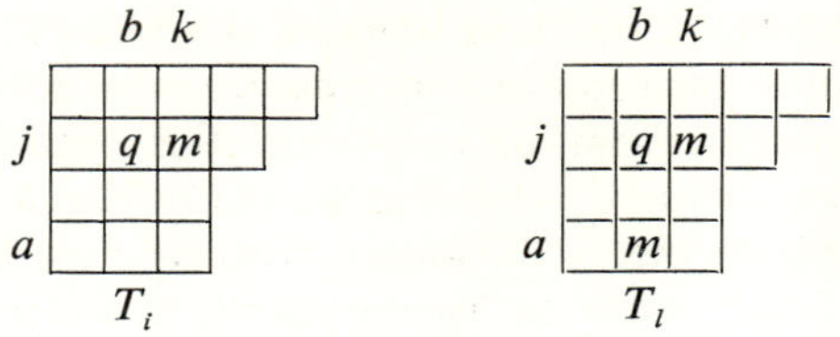

FIGURE 4.5

position is the same for both T_i and T_l. Thus, the digits q, m lie in the jth row of T_i and the bth column of T_l. Q.E.D.

Theorem 4.3. The left ideals $R_nc_1, \ldots, R_nc_f$ corresponding to the standard tableaux are linearly independent and

$$U = R_nc_1 \oplus \cdots \oplus R_nc_f.$$

Proof. By Theorem 4.2 it is enough to show that if

$$(2.6) \qquad x_1c_1 + \cdots + x_fc_f = 0, \qquad x_j \in R_n,$$

then each term $x_jc_j = 0$. If we multiply on the right by c_1, then Lemma 4.9 implies that all terms c_jc_1 are zero unless $j = 1$. Thus $x_1c_1{}^2 = 0$ or $\lambda x_1c_1 = 0$, with $\lambda \neq 0$. Similarly, multiplication of (2.6) on the right by c_2 yields $x_2c_2 = 0$. Continuing in this way, we can show that $x_jc_j = 0$, $1 \leq j \leq f$. Q.E.D.

4.3 Symmetry Classes of Tensors

An important application of S_n rep theory to physics is in the construction of symmetry classes of tensors. As we shall see, this subject is closely related to the rep theory of the general linear groups. A formula for the simple characters of S_n will arise as a by-product of our analysis.

Let V be a complex m-dimensional vector space and consider the rep defined on V by the group $GL(m, \mathfrak{C})$ of all invertible linear operators, $\mathbf{g} : V \rightarrow V$. In terms of a basis $\{\mathbf{v}_i\}$ for V the rep matrices $g = (g_i{}^j)$ are defined by

$$(3.1) \qquad \mathbf{g}\mathbf{v}_i = \sum_{j=1}^{m} g_i{}^j \mathbf{v}_j, \qquad 1 \leq i \leq m.$$

(It will prove convenient to adopt this superscript–subscript notation for indices.) As $\mathbf{g}$ runs over the group of invertible operators on V, g runs over the group of complex $m \times m$ nonsingular matrices. We will refer to either of these isomorphic groups as $GL(m, \mathfrak{C})$.

Consider the tensor product rep $\mathbf{T}$ of $GL(m, \mathfrak{C})$ on $V^{\otimes\alpha}$ defined by

$$(3.2) \qquad \mathbf{T}(g)[\mathbf{w}_1 \otimes \cdots \otimes \mathbf{w}_\alpha] = \mathbf{g}\mathbf{w}_1 \otimes \cdots \otimes \mathbf{g}\mathbf{w}_\alpha$$

for any $\mathbf{w}_j \in \mathbf{V}$ (see Section 3.5). In terms of the basis $\{\mathbf{v}_i\}$ for V,

$$\mathbf{w} = \sum_{j_1 \cdots j_\alpha = 1}^{m} a^{j_1 \cdots j_\alpha} \mathbf{v}_{j_1} \otimes \cdots \otimes \mathbf{v}_{j_\alpha} \in V^{\otimes \alpha}, \tag{3.3}$$

$$\mathbf{T}(g)\mathbf{w} = \sum_{i_1 \cdots i_\alpha = 1}^{m} [T(g)a]^{i_1 \cdots i_\alpha} \mathbf{v}_{i_1} \otimes \cdots \otimes \mathbf{v}_{i_\alpha}, \tag{3.4}$$

$$[T(g)a]^{i_1 \cdots i_\alpha} = \sum_{j_1 \cdots j_\alpha = 1}^{m} a^{j_1 \cdots j_\alpha} g^{i_1}_{j_1} \cdots g^{i_\alpha}_{j_\alpha}. \tag{3.5}$$

We can view this rep either as acting on tensors of rank α [Eq. (3.2)], or on the tensor components [Eq. (3.5)]. It will be convenient to shift back and forth between these equivalent interpretations.

The definitions of reducible and irred reps given in Chapter 3 hold for the infinite groups $GL(m, \mathfrak{C})$ as well as for finite groups. Also, the Schur lemmas are immediately applicable to infinite groups. However, those results which explicitly use the finiteness property of a group, such as the character theory, cannot be directly applied to $GL(m, \mathfrak{C})$. Indeed, Theorem 3.3, which states that any rep of a finite group can be decomposed into a direct sum of irred reps, is *not* true for $GL(m, \mathfrak{C})$. Fortunately, we can show that the tensor product rep $\mathbf{T}$ of $GL(m, \mathfrak{C}) = G_m$ on $V^{\otimes \alpha}$ is decomposable into a direct sum of irred reps. The symmetric group S_α figures strongly in this decomposition.

To clarify the relationship between S_α and $G_m = GL(m, \mathfrak{C})$ we define a rep of S_α on $V^{\otimes \alpha}$. For any $s \in S_\alpha$,

$$s = \begin{pmatrix} 1 & \cdots & \alpha \\ s(1) & \cdots & s(\alpha) \end{pmatrix}, \tag{3.6}$$

let $\mathbf{s}$ be the linear operator on $V^{\otimes \alpha}$ defined by

$$\mathbf{s}\mathbf{w} = \mathbf{w}_{s^{-1}(1)} \otimes \cdots \otimes \mathbf{w}_{s^{-1}(\alpha)}$$

for $\mathbf{w} = \mathbf{w}_1 \otimes \cdots \otimes \mathbf{w}_\alpha$ any indecomposable element of $V^{\otimes \alpha}$. These operators are well-defined and yield a rep of S_α. If an arbitrary tensor $\mathbf{w}$ is given by (3.3) then the action of s on the tensor components $a^{j_1 \cdots j_\alpha}$ of $\mathbf{w}$ is

$$(sa)^{j_1 \cdots j_\alpha} = a^{j_{s(1)} \cdots j_{s(\alpha)}}. \tag{3.7}$$

(Verify this.) For example, if $\alpha = 4$, $s = (12)(34)$, then $(sa)^{2331} = a^{3213}$, $(sa)^{1111} = a^{1111}$. If $s = (123)$ then $(sa)^{2331} = a^{3321}$. The reader should check these examples carefully to make sure he understands Eq. (3.7).

The **symmetric tensors** are those $\mathbf{w} \in V^{\otimes \alpha}$ such that $\mathbf{s}\mathbf{w} = \mathbf{w}$ for all $s \in S_\alpha$. Clearly, these tensors form a subspace $\mathfrak{S}$ of $V^{\otimes \alpha}$. It follows from (3.7) that with respect to the fixed basis $\{\mathbf{v}_{j_1} \otimes \cdots \otimes \mathbf{v}_{j_\alpha}\}$, the elements of $\mathfrak{S}$ are those tensors $\mathbf{w}$ whose components differing only in the order of the indices are equal. Thus $\mathbf{w} \in \mathfrak{S}$ is uniquely determined by the independent components $a^{j_1 \cdots j_\alpha}$ with $j_1 \leq j_2 \leq \cdots \leq j_\alpha$. To compute the dimension of $\mathfrak{S}$

note that the integers $j_1, j_2 + 1, j_3 + 2, \ldots, j_\alpha + \alpha - 1$ are α distinct numbers chosen from $1, 2, \ldots, m + \alpha - 1$, and every such choice labels a component. Therefore $\dim \mathcal{S} = (m + \alpha - 1)!/[\alpha!(m-1)!]$, the number of combinations of $m + \alpha - 1$ objects taken α at a time.

The tensors $\mathbf{v} \otimes \mathbf{v} \otimes \cdots \otimes \mathbf{v}$, $\mathbf{v} \in V$, obviously lie in $\mathcal{S}$. We shall show that every $\mathbf{w} \in \mathcal{S}$ is a linear combination of such tensors.

Lemma 4.10. The set of all tensors $\mathbf{v} \otimes \mathbf{v} \otimes \cdots \otimes \mathbf{v}$ spans $\mathcal{S}$.

Proof. If $\mathbf{v} = \sum a^j \mathbf{v}_j$ then $\mathbf{v} \otimes \cdots \otimes \mathbf{v}$ has tensor components $a^{j_1} a^{j_2} \cdots a^{j_\alpha}$. By symmetry we can restrict ourselves to the components for which $1 \leq j_1 \leq j_2 \leq \cdots \leq j_\alpha \leq m$. If the set of all such tensors does not span $\mathcal{S}$ then there must exist constants $C_{j_1 \cdots j_\alpha}$, not all zero, such that

$$\sum_{j_1 \leq j_2 \leq \cdots \leq j_\alpha} C_{j_1 \cdots j_\alpha} a^{j_1} \cdots a^{j_\alpha} \equiv 0 \tag{3.8}$$

for all numbers a^j, $1 \leq j \leq m$. (Prove it!) In each term $a^{j_1} \cdots a^{j_\alpha}$ let k_1 be the number of j_l equal to one, k_2 the number of j_l equal to two, etc., and write

$$a^{j_1} \cdots a^{j_\alpha} = (a^1)^{k_1} \cdots (a^m)^{k_m}, \qquad C_{j_1 \cdots j_\alpha} = C_{k_1 \cdots k_m}.$$

Then (3.8) becomes

$$\sum_{k_1 \cdots k_m} C_{k_1 \cdots k_m} (a^1)^{k_1} \cdots (a^m)^{k_m} \equiv 0, \qquad k_1 + \cdots + k_m = \alpha. \tag{3.9}$$

It is a well-known result from algebra (Van der Waerden [1]) that this homogeneous polynomial of degree α can be identically zero for all $a^1, \ldots, a^m$ only if all the coefficients $C_{k_1 \cdots k_m} = 0$. Q.E.D.

Expression (3.2) for the operators $\mathbf{T}(g)$ makes sense for all linear operators $\mathbf{g}$ on V, invertible or not. Furthermore, the homomorphism property $\mathbf{T}(g_1 g_2) = \mathbf{T}(g_1)\mathbf{T}(g_2)$ holds even if g_1, g_2 are not invertible. The set $\tilde{G}_m$ of all linear operators on V is said to form a **semigroup**, that is, $\tilde{G}_m$ satisfies the group axioms except that an element of $\tilde{G}_m$ need not have an inverse. Thus, we can define a rep of $\tilde{G}_m$ on V by means of the operators $\mathbf{T}(g)$.

Note that $\mathbf{sT}(g)\mathbf{w} = \mathbf{T}(g)\mathbf{sw}$ for all $s \in S_\alpha$, $g \in \tilde{G}_m$, $\mathbf{w} \in V^{\otimes \alpha}$, i.e., the operators $\mathbf{s}$ and $\mathbf{T}(g)$ always commute. The proof follows from

$$\begin{aligned} \mathbf{sT}(g)[\mathbf{v}_{j_1} \otimes \cdots \otimes \mathbf{v}_{j_\alpha}] &= \sum_{i_1 \cdots i_\alpha} g^{i_1}_{j_1} \cdots g^{i_\alpha}_{j_\alpha} \mathbf{v}_{i_{s^{-1}(1)}} \otimes \cdots \otimes \mathbf{v}_{i_{s^{-1}(\alpha)}} \\ &= \sum g^{i_{s(1)}}_{j_1} \cdots g^{i_{s(\alpha)}}_{j_\alpha} \mathbf{v}_{i_1} \otimes \cdots \otimes \mathbf{v}_{i_\alpha} \end{aligned} \tag{3.10}$$

and

$$\begin{aligned} \mathbf{T}(g)\mathbf{s}[\mathbf{v}_{j_1} \otimes \cdots \otimes \mathbf{v}_{j_\alpha}] &= \sum_{i_1 \cdots i_\alpha} g^{i_1}_{j_{s^{-1}(1)}} \cdots g^{i_\alpha}_{j_{s^{-1}(\alpha)}} \mathbf{v}_{i_1} \otimes \cdots \otimes \mathbf{v}_{i_\alpha} \\ &= \sum_{i_1 \cdots i_\alpha} g^{i_{s(1)}}_{j_1} \cdots g^{i_{s(\alpha)}}_{j_\alpha} \mathbf{v}_{i_1} \otimes \cdots \otimes \mathbf{v}_{i_\alpha}. \end{aligned} \tag{3.11}$$

We now determine the largest set A_α of linear operators on $V^{\otimes\alpha}$ which commute with all permutations **s**. Each such operator $\mathfrak{A}$ has matrix elements defined by

(3.12) $$\mathfrak{A}[\mathbf{v}_{j_1} \otimes \cdots \otimes \mathbf{v}_{j_\alpha}] = \sum_{i_1 \cdots i_\alpha} \mathfrak{A}^{i_1 \cdots i_\alpha}_{j_1 \cdots j_\alpha} \mathbf{v}_{i_1} \otimes \cdots \otimes \mathbf{v}_{i_\alpha}.$$

The requirement $\mathbf{s}\mathfrak{A} = \mathfrak{A}\mathbf{s}$ for all $s \in S_\alpha$ implies

(3.13) $$\mathfrak{A}^{i_{s(1)} \cdots i_{s(\alpha)}}_{j_{s(1)} \cdots j_{s(\alpha)}} = \mathfrak{A}^{i_1 \cdots i_\alpha}_{j_1 \cdots j_\alpha}$$

as the reader can verify. The elements of A_α are called **bisymmetric transformations**. Clearly, the operators $\mathbf{T}(g)$ are elements of A_α. It is evident that A_α is a vector space since linear combinations of bisymmetric transformations are bisymmetric. Moreover, A_α is an algebra, i.e., if $\mathfrak{A}, \mathfrak{B} \in A_\alpha$ then $\mathfrak{A}\mathfrak{B} \in A_\alpha$.

We now show that the relation between $V^{\otimes\alpha}$ and the symmetric tensors provided by Lemma 4.10 is analogous to the relation between the bisymmetric transformations and the operators $\mathbf{T}(g)$.

Theorem 4.4. The set of all operators $\mathbf{T}(g)$, $g \in \tilde{G}_m$, spans A_α.

Proof. Designate the matrix elements of $\mathfrak{A} \in A_\alpha$ by $\mathfrak{A}^{i_1 \cdots i_\alpha}_{j_1 \cdots j_\alpha} = \mathfrak{A}_{\mu_1 \cdots \mu_\alpha}$ where the pair of indices (i_k, j_k) is considered as a single index μ_k which takes m^2 values. According to (3.13) we can consider A_α as the subspace of all symmetric tensors in $W^{\otimes\alpha}$, where $\dim W = m^2$. The totality of all operators $\mathbf{T}(g)$ forms a subset of A_α with matrix elements $g^{i_1}_{j_1} \cdots g^{i_\alpha}_{j_\alpha} = g_{\mu_1} \cdots g_{\mu_\alpha}$, where the m^2 values g_μ range over all complex numbers. By Lemma 4.10 the tensors $g_{\mu_1} \cdots g_{\mu_\alpha}$ span the subspace of symmetric tensors in $W^{\otimes\alpha}$. Hence, they span A_α. Q.E.D.

We have shown that the algebra of bisymmetric transformations A_α is generated by the operators $\mathbf{T}(g)$, $g \in \tilde{G}_m$. Similarly, we can consider the algebra B_α of operators $\mathbf{x}$ generated by the permutations s:

(3.14) $$\mathbf{x}\mathbf{w} = \sum_{s \in S_\alpha} x(s) \cdot \mathbf{s}\mathbf{w}, \qquad x(s) \in \mathfrak{C}, \quad \mathbf{w} \in V^{\otimes\alpha}.$$

Clearly, B_α is a homomorphic image of the group ring R_α. The mapping

$$x = \sum x(s) \cdot s \longrightarrow \mathbf{x}$$

is not only a vector space homomorphism, but a ring homomorphism. That is, the product transformation $\mathbf{x}(\mathbf{y}\mathbf{w}) = (\mathbf{x}\mathbf{y})\mathbf{w}$ corresponds to the convolution product for $x, y \in R_\alpha$. The mapping may not be an isomorphism since a nonzero element x in R_α may be mapped into the zero operator on $V^{\otimes\alpha}$.

Any operator on $V^{\otimes\alpha}$ which commutes with all permutations **s** must commute with every element of B_α. This proves the following theorem.

Theorem 4.5. The linear operator $\mathbf{C}$ on $V^{\otimes\alpha}$ commutes with all elements of B_α if and only if $\mathbf{C} \in A_\alpha$.

Let A be an associative algebra with multiplicative identity e and let W be a complex vector space. A **representation T** of A on W is determined by a set of linear operators $\mathbf{T}(a)$ on W such that

(1) $\mathbf{T}(\gamma a + \mu b) = \gamma \mathbf{T}(a) + \mu \mathbf{T}(b), \quad a, b \in A, \quad \gamma, \mu \in \mathfrak{C};$
(2) $\mathbf{T}(ab) = \mathbf{T}(a)\mathbf{T}(b);$
(3) $\mathbf{T}(e) = \mathbf{E}.$

The notions of reducibility, irreducibility, and equivalence of reps of A are analogous to those for group reps.

The rep of S_α on $V^{\otimes\alpha}$ defined by the operators $\mathbf{s}$ induces a rep of the group ring R_α by the operators in B_α. We know that $V^{\otimes\alpha}$ can be decomposed into a direct sum of subspaces such that each subspace is irred under S_α.

The reader can verify the following facts: (1) Every rep $\mathbf{T}$ of S_α determines a rep $\mathbf{T}$ of R_α. [Set $\mathbf{T}(x) = \sum x(s)\mathbf{T}(s)$.] (2) Every rep of R_α determines a rep of S_α. [Restrict the operators $\mathbf{T}(x)$ to $x = 1 \cdot s$.] (3) Equivalent reps $\mathbf{T}, \mathbf{T}'$ of S_α correspond to equivalent reps of R_α. [$\mathbf{U}\mathbf{T}(x)\mathbf{U}^{-1} = \mathbf{T}'(x)$ for all $x = \sum x(s) \cdot s$ if and only if $\mathbf{U}\mathbf{T}(s)\mathbf{U}^{-1} = \mathbf{T}'(s)$ for all $s \in S_\alpha$.]

According to the above remarks, the rep of R_α provided by the operators in B_α can also be decomposed into a direct sum of irred reps. The irred subspaces of $V^{\otimes\alpha}$ under R_α are just the irred subspaces under S_α. We will show that this decomposition of $V^{\otimes\alpha}$ into a direct sum of irred subspaces induces a similar decomposition for the rep of the algebra A_α defined by (3.12).

Let $D^{(1)}, \ldots, D^{(\mu)}$ be a complete set of nonequivalent irred matrix reps of S_α. Then with respect to a suitable basis for $V^{\otimes\alpha}$ the $m^\alpha \times m^\alpha$ matrix corresponding to each permutation $\mathbf{s}$ is

$$
\begin{pmatrix}
\overbrace{D^{(1)}(s) \;\ddots\; D^{(1)}(s)}^{a_1} & & & Z \\
& \overbrace{D^{(2)}(s) \;\ddots\; D^{(2)}(s)}^{a_2} & & \\
& & \ddots & \\
Z & & & \overbrace{D^{(\mu)}(s) \;\ddots\; D^{(\mu)}(s)}^{a_\mu}
\end{pmatrix}
\tag{3.15}
$$

That is, the permutation rep decomposes into a direct sum of irred reps,

$$a_1 D^{(1)} \oplus a_2 D^{(2)} \oplus \cdots \oplus a_\mu D^{(\mu)}.$$

The matrices corresponding to A_α are just those which commute with the matrices (3.15) for all $s \in S_\alpha$. It is instructive to work out a simple example.

Suppose the matrices (3.15) take the form

$$s = \begin{pmatrix} D^{(1)}(s) & Z & Z \\ Z & D^{(2)}(s) & Z \\ Z & Z & D^{(2)}(s) \end{pmatrix} \begin{matrix} n_1 \\ n_2 \\ n_3 \end{matrix} \qquad \text{(columns } n_1,\ n_2,\ n_3\text{)} \tag{3.16}$$

with $n_1 = \dim D^{(1)}$, $n_2 = n_3 = \dim D^{(2)}$. The matrix of a bisymmetric transformation $\mathcal{A}$ can be written

$$\mathcal{A} = \begin{pmatrix} \mathcal{A}_{11} & \mathcal{A}_{12} & \mathcal{A}_{13} \\ \mathcal{A}_{21} & \mathcal{A}_{22} & \mathcal{A}_{23} \\ \mathcal{A}_{31} & \mathcal{A}_{32} & \mathcal{A}_{33} \end{pmatrix} \tag{3.17}$$

where $\mathcal{A}_{jk}$ is an $n_j \times n_k$ matrix. The condition $\mathcal{A} \in A_\alpha$ is just that the matrices $\mathcal{A}$ and s commute for all $s \in S_\alpha$. Now

$$\begin{aligned} \mathcal{A}s &= \begin{pmatrix} \mathcal{A}_{11}D^{(1)}(s) & \mathcal{A}_{12}D^{(2)}(s) & \mathcal{A}_{13}D^{(2)}(s) \\ \mathcal{A}_{21}D^{(1)}(s) & \mathcal{A}_{22}D^{(2)}(s) & \mathcal{A}_{23}D^{(2)}(s) \\ \mathcal{A}_{31}D^{(1)}(s) & \mathcal{A}_{32}D^{(2)}(s) & \mathcal{A}_{33}D^{(2)}(s) \end{pmatrix} \\ s\mathcal{A} &= \begin{pmatrix} D^{(1)}(s)\mathcal{A}_{11} & D^{(1)}(s)\mathcal{A}_{12} & D^{(1)}(s)\mathcal{A}_{13} \\ D^{(2)}(s)\mathcal{A}_{21} & D^{(2)}(s)\mathcal{A}_{22} & D^{(2)}(s)\mathcal{A}_{23} \\ D^{(2)}(s)\mathcal{A}_{31} & D^{(2)}(s)\mathcal{A}_{32} & D^{(2)}(s)\mathcal{A}_{33} \end{pmatrix} \end{aligned} \tag{3.18}$$

The requirement $\mathcal{A}s = s\mathcal{A}$ leads to a series of relations of the type

$$\mathcal{A}_{11}D^{(1)}(s) = D^{(1)}(s)\mathcal{A}_{11}, \qquad \mathcal{A}_{12}D^{(2)}(s) = D^{(1)}(s)\mathcal{A}_{12}.$$

Since $D^{(1)}$ and $D^{(2)}$ are nonequivalent irred reps of S_α, the Schur lemmas, Section 3.3, imply $\mathcal{A}_{12} = Z$ and $\mathcal{A}_{11} = \lambda_{11}E_{n_1}$, where λ_{11} is any complex number and E_{n_1} is the $n_1 \times n_1$ identity matrix. These considerations lead to the result

$$\mathcal{A} = \begin{pmatrix} \lambda_{11}E_{n_1} & Z & Z \\ Z & \lambda_{22}E_{n_2} & \lambda_{23}E_{n_2} \\ Z & \lambda_{32}E_{n_2} & \lambda_{33}E_{n_2} \end{pmatrix}, \qquad \lambda_{jk} \in \mathfrak{C}. \tag{3.19}$$

A simple rearrangement of rows and columns yields the matrix realization

$$\mathcal{A} = \begin{pmatrix} \overbrace{\lambda_{11} \; \ddots \; \lambda_{11}}^{n_1} & & & Z \\ & \begin{pmatrix} \lambda_{22} & \lambda_{23} \\ \lambda_{32} & \lambda_{33} \end{pmatrix} & & \\ & & \ddots \;{}^{n_2} & \\ Z & & & \begin{pmatrix} \lambda_{22} & \lambda_{23} \\ \lambda_{32} & \lambda_{33} \end{pmatrix} \end{pmatrix}. \tag{3.20}$$

We shall see that (3.20) is an explicit decomposition of the rep of A_α on $V^{\otimes\alpha}$ into irred reps. Note that the multiplicities n_1, n_2 of the matrix blocks in (3.20) are just the dimensions of $D^{(1)}$ and $D^{(2)}$, while the multiplicities 1, 2 of $D^{(1)}$ and $D^{(2)}$ are the dimensions of the matrix blocks in (3.20).

The general case is now clear. If the permutation rep of S_α decomposes in the form (3.15),

$$s \sim a_1 D^{(1)}(s) \oplus \cdots \oplus a_\mu D^{(\mu)}(s),$$

where a_j is the multiplicity of $D^{(j)}$ and $n_j = \dim D^{(j)}$, then the elements of A_α are all those of the form

$$\mathfrak{A} \sim n_1 C^{(1)}(\mathfrak{A}) \oplus \cdots \oplus n_\mu C^{(\mu)}(\mathfrak{A}) \tag{3.21}$$

where $C^{(k)}(\mathfrak{A})$ runs over all $a_k \times a_k$ matrices as $\mathfrak{A}$ runs over A_α. The matrix block $C^{(k)}(\mathfrak{A})$ occurs n_k times along the diagonal in the matrix expression for $\mathfrak{A}$ analogous to (3.20). Evidently, each of the matrix blocks $C^{(k)}$ is itself a matrix rep of A_α. Furthermore, this rep is irred. Indeed any $a_k \times a_k$ matrix B with the property $BC^{(k)}(\mathfrak{A}) = C^{(k)}(\mathfrak{A})B$ for all $\mathfrak{A} \in A_\alpha$ must be a multiple of E_{a_k}, since the $C^{(k)}(\mathfrak{A})$ run over *all* $a_k \times a_k$ matrices.

The irred reps $C^{(j)}$, $C^{(k)}$ for $j \neq k$ must be nonequivalent because $C^{(j)}(\mathfrak{A})$ and $C^{(k)}(\mathfrak{A})$ run over all $a_j \times a_j$ and $a_k \times a_k$ matrices completely independent of one another.

Theorem 4.6. The algebra A_α of bisymmetric transformations acting on $V^{\otimes\alpha}$ can be decomposed into the direct sum (3.21) of irred reps $C^{(k)}$.

According to Theorem 4.4 the irred matrix reps $C^{(k)}(\mathfrak{A})$ must remain irred when $\mathfrak{A}$ is restricted to elements of the form $\mathbf{T}(g)$, $g \in \tilde{G}_m$. This is because an arbitrary $\mathfrak{A} \in A_\alpha$ can be written in the form $\mathfrak{A} = \sum \beta_i \mathbf{T}(g_i)$, $g_i \in \tilde{G}_m$. If the restriction of $C^{(k)}$ to $\tilde{G}_m$ were reducible then the property $C^{(k)}(\mathfrak{A}) = \sum \beta_i C^{(k)}(g_i)$ would imply that $C^{(k)}$ was itself reducible. [For simplicity we write $C^{(k)}(g)$ for $C^{(k)}(T(g))$.]

If there were a nonsingular matrix B such that $C^{(k)}(g) = BC^{(j)}(g)B^{-1}$ for some $j \neq k$ and all $g \in \tilde{G}_m$, then by the argument in the preceding paragraph, $C^{(k)}$ and $C^{(j)}$ would be equivalent reps of A_α. Since this is false, the restrictions of the $C^{(k)}$ to $\tilde{G}_m$ remain nonequivalent. Finally, we can restrict the $C^{(k)}$ to $G_m = GL(m, \mathfrak{C})$, or more precisely to the elements $\mathbf{T}(g)$, $g \in G_m$. The matrix elements of $C^{(k)}(g)$ are homogeneous polynomials of order α in the $g_i{}^j$. Let B be an $a_k \times a_k$ matrix such that $BC^{(k)}(g) = C^{(k)}(g)B$ for all $g \in G_m$. This relation leads to a number of identities between the matrix elements of $C^{(k)}(g)$ which remain valid for singular matrices $g \in \tilde{G}_m$. However, the restriction of $C^{(k)}$ to $\tilde{G}_m$ is irred, so B is a multiple of the identity and the $C^{(k)}(g)$ define an irred rep of G_m. A similar argument shows that $C^{(1)}, \ldots, C^{(\mu)}$ yield nonequivalent irred reps of G_m.

Theorem 4.7. The rep **T** of $GL(m, \mathfrak{C})$ on $V^{\otimes\alpha}$ can be decomposed into a direct sum of irred reps

$$\mathbf{T} \cong n_1 C^{(1)} \oplus \cdots \oplus n_\mu C^{(\mu)} \tag{3.22}$$

analogous to the decomposition

$$a_1 D^{(1)} \oplus \cdots \oplus a_\mu D^{(\mu)}$$

of the permutation rep of S_α on $V^{\otimes\alpha}$. Here $a_k = \dim C^{(k)}$ and $n_k = \dim D^{(k)}$.

It will be shown in Chapter 9 that the decomposition (3.22) is essentially unique, i.e., the irred reps $C^{(k)}$ occurring in the decompositon and their multiplicities are uniquely determined. This does not follow from the results of Chapter 3, since $GL(m, \mathfrak{C})$ is not a finite group. We could prove the uniqueness directly at this point by making use of the rep theory of complete matrix algebras, but the proof will be deferred to save space.

A proof of the following theorem will also be deferred to Section 9.1. (See Boerner [1, p. 137] for a direct proof.)

Theorem 4.8. Let W be a subspace of $V^{\otimes\alpha}$ which is invariant under the rep **T** of G_m. Then there exists an invariant subspace W' such that $V^{\otimes\alpha} = W \oplus W'$.

It is clear that A_α consists of all linear transformations that commute with every $\mathbf{x} \in B_\alpha$. On the other hand, we have the following result.

Theorem 4.9. B_α consists of all linear transformations on $V^{\otimes\alpha}$ that commute with each $\mathfrak{A} \in A_\alpha$.

The major steps in the proof of this theorem are provided by the following lemmas, which are of independent interest.

Lemma 4.11. Let D be an irred $n \times n$ matrix rep of the finite group H. Then the matrices $D(h)$, $h \in H$, span the n^2-dimensional space of all $n \times n$ matrices, i.e., every matrix can be expressed as a linear combination of the matrices $D(h)$.

Proof. If the matrices $D(h)$ do not span an n^2-dimensional space there must exist a relation

$$\sum_{i,j=1}^{n} c_{ij} D_{ij}(h) = 0, \qquad \text{all} \quad h \in H, \tag{3.23}$$

among the matrix elements of $D(h)$ which is satisfied for constants c_{ij} not all zero. However, the orthogonality relations (3.7), Section 3.3, and (3.23) lead to

$$0 = n(H)^{-1} \sum_{h \in H} [\sum_{i,j} c_{ij} D_{ij}(h)]\, D_{lm}(h^{-1}) = c_{ml}/n$$

so $c_{ml} = 0$ for $m, l = 1, \ldots, n$. Q.E.D.

According to this result the matrix rep of the group ring R_H determined by D has the property that $D(x)$ runs over *all* $n \times n$ matrices as $x = \sum x(h) \cdot h$ runs over R_H. Indeed, $D(x) = \sum x(h) D(h)$ and the $x(h)$ range over all complex numbers as x runs over R_H.

Let H be a finite group and

$$R_H = U_1 \oplus U_2 \oplus \cdots \oplus U_\gamma$$

the decomposition of R_H into minimal two-sided ideals as described at the end of Section 3.7. The ideal U_ν corresponds to the irred rep $\mathbf{D}^{(\nu)}$ of H. Indeed, under the left regular rep, U_ν decomposes into a direct sum of n_ν left ideals, each left ideal transforming according to $\mathbf{D}^{(\nu)}$. Let $\mathbf{D}$ be a rep of R_H on the vector space W and let W_ν be an irred subspace of W such that $\mathbf{D} \mid W_\nu \cong \mathbf{D}^{(\nu)}$. (Recall that every irred rep of R_H remains irred when restricted to H.)

Lemma 4.12. If $\nu \neq \mu$, then $\mathbf{D}(y)\mathbf{w} = \boldsymbol{\theta}$ for all $\mathbf{w} \in W_\nu$ and all $y \in U_\mu$.

Proof. Let W_ν' be the subspace of W_ν spanned by all vectors of the form $\mathbf{D}(x)\mathbf{w}$, $x \in U_\nu$, $\mathbf{w} \in W_\nu$. Since W_ν is irred, either $W_\nu' = \{\boldsymbol{\theta}\}$ or $W_\nu' = W_\nu$. We shall show $W_\nu' \neq \{\boldsymbol{\theta}\}$.

From Section 3.7, $\mathbf{P}_\nu \mathbf{w} = \mathbf{w}$ for all $\mathbf{w} \in W_\nu$, where the projection operator is

(3.24) $$\mathbf{P}_\nu = \frac{n_\nu}{n(H)} \sum_{h \in H} \overline{\chi^{(\nu)}(h)}\, \mathbf{D}(h) = \mathbf{D}(p_\nu),$$

(3.25) $$p_\nu = \frac{n_\nu}{n(H)} \sum_{h \in H} \overline{\chi^{(\nu)}(h)} \cdot h \in R_H.$$

If we assume that the matrices $D^{(\nu)}(h)$ are unitary, the relation

$$R(g)\bar{D}^{(\nu)}_{jk}(h) = \bar{D}^{(\nu)}_{jk}(hg^{-1}) = \sum_l D^{(\nu)}_{kl}(g)\bar{D}^{(\nu)}_{jl}(h)$$

implies that the ring elements $\bar{D}^{(\nu)}_{jl} = \sum \bar{D}^{(\nu)}_{jl}(h) \cdot h$ for $1 \leq l \leq n_\nu$ and fixed j form a basis for the rep $\mathbf{D}^{(\nu)}$. Thus each of these ring elements lies in U_ν and it follows that $p_\nu \in U_\nu$. Therefore, $W_\nu' = W_\nu$.

If $\mathbf{y} \in U_\mu$, $\mu \neq \nu$, and $x \in U_\nu$ then $yx \in U_\mu \cap U_\nu = \{0\}$, so $\mathbf{D}(y)\mathbf{D}(x)\mathbf{w} = \mathbf{D}(yx)\mathbf{w} = \boldsymbol{\theta}$ for all $\mathbf{w} \in W_\nu$. But then $\mathbf{D}(y)\mathbf{w}' = \boldsymbol{\theta}$ for all $\mathbf{w}' \in W_\nu$ since $\mathbf{w}'$ can be expressed as a finite linear combination of elements of the form $\mathbf{D}(x)\mathbf{w}$. Q.E.D.

Now we turn to the proof of Theorem 4.9. To clarify the argument we consider the example given by expressions (3.17)–(3.20). Suppose the bisymmetric transformations have the matrix realization (3.19). To find all operators which commute with every element of A_α it is enough to determine those matrices B that commute with all matrices (3.19). The result is easily shown to be

$$B = \begin{pmatrix} B^{(1)} & Z & Z \\ Z & B^{(2)} & Z \\ Z & Z & B^{(2)} \end{pmatrix} \begin{matrix} n_1 \\ n_2 \\ n_2 \end{matrix} \qquad \text{(columns } n_1, n_2, n_2\text{)} \tag{3.26}$$

where $B^{(1)}$ and $B^{(2)}$ independently range over all $n_1 \times n_1$ matrices, $n_2 \times n_2$ matrices, respectively. Comparing (3.26) with (3.17) and using the lemmas we see that each matrix B corresponds to an element of B_α. Indeed as $x = \sum x(s)\cdot s$ ranges over R_α the matrices $D^{(i)}(x) = \sum x(s)D^{(i)}(s)$, $i = 1, 2$, range over all $n_i \times n_i$ matrices $B^{(i)}$. Moreover, according to Lemma 4.12 the matrices $B^{(1)}$ and $B^{(2)}$ are independent. That is, given any two matrices $B^{(1)}$ and $B^{(2)}$ there is an $x \in R_\alpha$ with $D^{(1)}(x) = B^{(1)}$ and $D^{(2)}(x) = B^{(2)}$. Thus the matrix B, (3.26), corresponds to an element of B_α. The argument for the general case proceeds exactly as in our example. Q.E.D.

We now resume the analysis of the rep $\mathbf{T}$ of $G_m = GL(m, \mathfrak{C})$ on $V^{\otimes\alpha}$. Our previous results have shown that $\mathbf{T}$ can be decomposed into a direct sum of irred reps and that this decomposition is closely related to the decomposition of $V^{\otimes\alpha}$ into subspaces irred under the permutation rep of S_α. However, these results are of a theoretical character and do not lend themselves to a practical method for decomposing tensor reps of G_m.

Theorem 4.9 provides us with the proper tool to obtain such a practical decomposition. Let W_1 be a subspace of $V^{\otimes\alpha}$ which is invariant under $\mathbf{T}$. According to Theorem 4.8 there exists a $\mathbf{T}$-invariant subspace W_2 of $V^{\otimes\alpha}$ such that $V^{\otimes\alpha} = W_1 \oplus W_2$. Let $\mathbf{P}$ be the projection operator on W_1 defined by $R_\mathbf{P} = W_1$, $N_\mathbf{P} = W_2$. From the results of Section 3.7, $\mathbf{T}(g)\mathbf{P} = \mathbf{P}\mathbf{T}(g)$ for all $g \in G_m$, hence $\mathbf{P}$ commutes with all elements of A_α. Theorem 4.9 yields the important conclusion: $\mathbf{P} \in B_\alpha$.

We can immediately apply Theorems 3.10–3.12 to the rep $\mathbf{T}$. Let $P(\mathbf{T})$ be the set of all projection operators on $V^{\otimes\alpha}$ that commute with the operators $\mathbf{T}(g)$.

Theorem 4.10. (1) There is a 1–1 relationship between projections $\mathbf{P}$ in $P(\mathbf{T})$ and decompositions $V^{\otimes\alpha} = W_1 \oplus W_2$ into $\mathbf{T}$-invariant subspaces, given by $R_\mathbf{P} = W_1$, $N_\mathbf{P} = W_2$.

(2) If $\mathbf{P} \in P(\mathbf{T})$ then $\mathbf{P} \in B_\alpha$. Conversely, if $\mathbf{Q} \in B_\alpha$ and $\mathbf{Q}^2 = \mathbf{Q} \neq \mathbf{Z}$ then $\mathbf{Q} \in P(\mathbf{T})$.

(3) Let W_1 be a **T**-invariant subspace of $V^{\otimes\alpha}$ and let $\mathbf{P} \in P(\mathbf{T})$ be a projection operator on W_1. Then W_1 is irred if and only if there do not exist nonzero operators $\mathbf{P}_1, \mathbf{P}_2 \in P(\mathbf{T})$ such that $\mathbf{P} = \mathbf{P}_1 + \mathbf{P}_2$ and $\mathbf{P}_1\mathbf{P}_2 = \mathbf{P}_2\mathbf{P}_1 = \mathbf{Z}$. [Recall that the projection operators **P** corresponding to irred subspaces W_1 are the elements of the set $IP(\mathbf{T})$.]

This theorem implies that any T-irred subspace W of $V^{\otimes\alpha}$ is given by

$$W = \{\mathbf{x}\mathbf{v} : \mathbf{v} \in V^{\otimes\alpha}\} \tag{3.27}$$

where **x** is a primitive idempotent in the algebra B_α. (Usually **x** is not uniquely determined by W.) Conversely, each primitive idempotent **x** in B_α uniquely determines a **T**-irred subspace W by (3.27).

Next we investigate the relationship between B_α and the group ring R_α. To clarify the discussion we introduce the notation $\mathbf{D}(x) = \mathbf{x}$, $x \in R_\alpha$, for the elements (3.14) of B_α. The $\mathbf{D}(x)$ define a rep of R_α which may not be faithful. That is, we may have $\mathbf{D}(x) = \mathbf{D}(y)$ for $x \neq y$. Let $\mathcal{O}_\alpha$ be the set of all $x \in R_\alpha$ such that $\mathbf{D}(x) = \mathbf{Z}$.

Lemma 4.13. $\mathcal{O}_\alpha$ is a two-sided ideal in R_α. There exists a two-sided ideal $\mathcal{R}_\alpha$ in R_α such that

$$R_\alpha = \mathcal{R}_\alpha \oplus \mathcal{O}_\alpha, \tag{3.28}$$

and the map $x \longrightarrow D(x)$ of $\mathcal{R}_\alpha$ into B_α is 1–1 and onto.

Proof. If $x \in \mathcal{O}_\alpha$, $y \in R_\alpha$, then

$$\mathbf{D}(yx) = \mathbf{D}(y)\mathbf{D}(x) = \mathbf{Z}, \qquad \mathbf{D}(xy) = \mathbf{D}(x)\mathbf{D}(y) = \mathbf{Z}$$

since $\mathbf{D}(x) = \mathbf{Z}$. Thus $yx, xy \in \mathcal{O}_\alpha$ and $\mathcal{O}_\alpha$ is a two-sided ideal.

According to the results at the end of Section 3.7 the group ring can be expressed as a direct sum of minimal two-sided ideals

$$R_\alpha = U_1 \oplus \cdots \oplus U_k \tag{3.29}$$

where k is the number of conjugacy classes in S_α and the number of partitions (frames) $\{\lambda_1, \ldots, \lambda_\alpha\}$ of α. Taking the frames $\{\lambda_j\}$ in dictionary order, we label the two-sided ideals such that U_i consists of those minimal left ideals (irred subspaces under **L**) that correspond to the ith frame. The ideal $\mathcal{O}_\alpha$ can be written as a direct sum of minimal left ideals. It follows from the last paragraph of Section 3.7 that if $\mathcal{O}_\alpha$ contains a minimal left ideal corresponding to the ith frame then $\mathcal{O}_\alpha$ contains all minimal left ideals corresponding to the ith frame, i.e., $U_i \subseteq \mathcal{O}_\alpha$. Thus,

$$\mathcal{O}_\alpha = U_{j_1} \oplus U_{j_2} \oplus \cdots \oplus U_{j_l} \tag{3.30}$$

where $1 \leq j_1 < j_2 < \cdots < j_l \leq k$. Choosing integers $j_{l+1} < \cdots < j_k$ such that $\{j_1, \ldots, j_k\}$ is a permutation of $\{1, \ldots, k\}$ we see

$$R_\alpha = (U_{j_1} \oplus \cdots \oplus U_{j_l}) \oplus (U_{j_{l+1}} \oplus \cdots \oplus U_{j_k}) = \mathcal{O}_\alpha \oplus \mathcal{R}_\alpha$$

where

$$\mathcal{R}_\alpha = U_{j_{l+1}} \oplus \cdots \oplus U_{j_k} \tag{3.31}$$

is a two-sided ideal.

Let $x_1, x_2 \in \mathcal{R}_\alpha$ such that $\mathbf{D}(x_1) = \mathbf{D}(x_2)$. Then $\mathbf{D}(x_1 - x_2) = \mathbf{Z}$ so $x_1 - x_2 \in \mathcal{O}_\alpha$. Thus, $x_1 - x_2 \in \mathcal{O}_\alpha \cap \mathcal{R}_\alpha = \{0\}$ or $x_1 = x_2$. This proves that the map $x \longrightarrow \mathbf{D}(x)$ is an isomorphism of $\mathcal{R}_\alpha$ and B_α. Q.E.D.

Because of the isomorphism between $\mathcal{R}_\alpha$ and B_α we can identify the operator $\mathbf{D}(x) \in B_\alpha$ with $x \in \mathcal{R}_\alpha$. Thus, there is a 1–1 correspondence between projection operators $\mathbf{P} \in IP(\mathbf{T})$ and primitive idempotents c in $\mathcal{R}_\alpha$, given by $\mathbf{P} = \mathbf{D}(c)$. In the previous section we have already discussed the determination of primitive idempotents in R_α. Our analysis of symmetry classes of tensors will be complete if we can develop a practical method for determining $\mathcal{R}_\alpha$ and for decomposing $V^{\otimes\alpha}$ into a direct sum of **T**-irred subspaces W_i,

$$V^{\otimes\alpha} = W_1 \oplus \cdots \oplus W_\gamma .$$

The basic problem remaining to be solved is this: How do we choose the primitive idempotents $c_i \in \mathcal{R}_\alpha$ such that $W_i = \{\mathbf{D}(c_i)\mathbf{w}\colon \mathbf{w} \in V^{\otimes\alpha}\}$? A clue to the solution of this problem is the observation that there is a 1–1 relationship between **T**-invariant subspaces W of $V^{\otimes\alpha}$ and *right* ideals $\mathcal{I}$ in $\mathcal{R}_\alpha$.

Indeed, let W be a nontrivial **T**-invariant subspace and let

$$\mathcal{I}_W = \{x \in \mathcal{R}_\alpha\colon \mathbf{x}\mathbf{v} \in W \quad \text{for all} \quad \mathbf{v} \in V^{\otimes\alpha}\} \tag{3.32}$$

[Recall that $\mathbf{D}(x)\mathbf{v} = \mathbf{x}\mathbf{v}$.] If $x \in \mathcal{I}_W$ and $y \in \mathcal{R}_\alpha$ then $(\mathbf{x}\mathbf{y})\mathbf{v} = \mathbf{x}(\mathbf{y}\mathbf{v}) \in W$ for all $\mathbf{v}$, so $xy \in \mathcal{I}_W$ and $\mathcal{I}_W$ is a right ideal. Moreover, if $\mathbf{P} = \mathbf{D}(z)$, $z \in \mathcal{R}_\alpha$, is a projection operator on W which commutes with the $\mathbf{T}(g)$ operators, then $z \in \mathcal{I}_W$, so $\mathcal{I}_W \neq \{0\}$.

On the other hand, given a nontrivial right ideal $\mathcal{I}$ in $\mathcal{R}_\alpha$ we can define a **T**-invariant subspace $W_\mathcal{I}$ of $V^{\otimes\alpha}$ generated by all elements of the form $\mathbf{x}\mathbf{v}$, $x \in \mathcal{I}$, $\mathbf{v} \in V^{\otimes\alpha}$.

Before proceeding with this analysis we remark that the basic facts concerning left ideals proved in Section 3.7 have an obvious modification for right ideals in a group ring R_G. A right ideal is just an invariant subspace of the group ring under the right regular rep. If $x \in R_G$ then the set xR_G is a right ideal. Conversely, every right ideal is of this form. If $\mathcal{I}$ is a right ideal then there is a **generating idempotent** $c \in R_G$ (not unique) such that

$\mathcal{I} = cR_G$. Also, $\mathcal{I}$ is a minimal right ideal if and only if c is a primitive idempotent.

Let $\mathcal{I}$ be a right ideal in $\mathcal{R}_\alpha$ and W a **T**-invariant subspace of $V^{\otimes\alpha}$.

Theorem 4.11. (1) $W_{(\mathcal{I}_W)} = W$.

(2) $\mathcal{I}_{(W_\mathcal{I})} = \mathcal{I}$.

That is, the relation between right ideals and **T**-invariant subspaces defined above is 1–1.

(3) Let $\mathcal{I} = \mathcal{I}_W$. Then $\mathcal{I}$ is minimal if and only if W is **T**-irred.

(4) Let $\mathcal{I} = \mathcal{I}_W$, $\mathcal{I}' = \mathcal{I}_{W'}$. Then $\mathcal{I}$ and $\mathcal{I}'$ are equivalent right ideals in $\mathcal{R}_\alpha$ if and only if the **T**-invariant subspaces W and W' define equivalent reps of G_m.

Proof. (1) Let c be the generating idempotent of $\mathcal{I}_W$. Then $W = \mathbf{c}V^{\otimes\alpha}$ since every $x \in \mathcal{I}_W$ can be written $x = cx$. Now $W_{(\mathcal{I}_W)}$ is the space generated by all elements of the form $\mathbf{xv} = \mathbf{cxv}$, $x \in \mathcal{I}_W$, $\mathbf{v} \in V^{\otimes\alpha}$, which is obviously $\mathbf{c}V^{\otimes\alpha}$.

(2) If c is a generating idempotent of $\mathcal{I}$, then $W_\mathcal{I} = \mathbf{c}V^{\otimes\alpha}$. If $x \in \mathcal{I}$, $\mathbf{v} \in V^{\otimes\alpha}$, then $\mathbf{xv} = (\mathbf{cx})\mathbf{v} = \mathbf{c}(\mathbf{xv}) \in W_\mathcal{I}$, so $\mathcal{I} \subseteq \mathcal{I}_{(W_\mathcal{I})}$. Conversely, suppose $y \in \mathcal{I}_{(W_\mathcal{I})}$. Then $\mathbf{yv} \in W_\mathcal{I}$, so $\mathbf{yv} = \mathbf{c}(\mathbf{yv}) = (\mathbf{cy})\mathbf{v}$ for all $\mathbf{v} \in V^{\otimes\alpha}$. This shows $y = cy \in \mathcal{I}$, so $\mathcal{I} = \mathcal{I}_{(W_\mathcal{I})}$.

(3) This follows immediately from Theorem 3.13 and the fact that $\mathbf{D}(c)$ is a projection on a **T**-irred subspace W if and only if c is a primitive idempotent.

(4) By property (3) it is sufficient to prove this assertion for primitive right ideals $\mathcal{I}, \mathcal{I}'$. Suppose $\mathcal{I}$ and $\mathcal{I}'$ are equivalent ideals with primitive generating idempotents c, c'. Then there is a nonzero $x = c'xc \in \mathcal{I}'$ such that $\mathcal{I}' = x\mathcal{I} = x\mathcal{R}_\alpha$. Thus, $W = \mathbf{c}V^{\otimes\alpha}$ and $W' = \mathbf{c}'V^{\otimes\alpha} = \mathbf{x}V^{\otimes\alpha} = \mathbf{xc}V^{\otimes\alpha} = \mathbf{D}(x)W$. Since W and W' are **T**-irred and the nonzero operator $\mathbf{D}(x)$ from W to W' commutes with the $\mathbf{T}(g)$, W and W' define equivalent reps. Conversely, if W is equivalent to W' there is a nonzero mapping $\mathbf{A}$ of W onto W' which commutes with the $\mathbf{T}(g)$. Let W'' be a **T**-invariant subspace such that $V^{\otimes\alpha} = W \oplus W''$. We can extend $\mathbf{A}$ to $V^{\otimes\alpha}$ by requiring $\mathbf{A}(\mathbf{w} + \mathbf{w}'') = \mathbf{Aw}$ for all $\mathbf{w} \in W$, $\mathbf{w}'' \in W''$. Clearly, the extended operator $\mathbf{A}$ commutes with the $\mathbf{T}(g)$, so $\mathbf{A} \in B_\alpha$ and $\mathbf{A} = \mathbf{D}(x)$, $x \in \mathcal{R}_\alpha$. This shows that $W' = \mathbf{A}W = \mathbf{x}W$. Since $\mathcal{I} = \{y \in \mathcal{R}_\alpha : \mathbf{y}V^{\otimes\alpha} \subseteq W\}$ with a similar definition for $\mathcal{I}'$, it follows that $\{0\} \subset x\mathcal{I} \subseteq \mathcal{I}'$. But $\mathcal{I}$ and $\mathcal{I}'$ are minimal, so $\mathcal{I}' = x\mathcal{I}$ and the ideals are equivalent. Q.E.D.

We can now explicitly decompose $V^{\otimes\alpha}$ into **T**-irred subspaces. Let $\mathcal{R}_\alpha = \mathcal{I}_1 \oplus \cdots \oplus \mathcal{I}_q$ be a decomposition of $\mathcal{R}_\alpha$ into minimal right ideals. Suppose $\mathcal{I}_j = c_j\mathcal{R}_\alpha$, $1 \leq j \leq q$, where the generators c_j are primitive idempotents.

Then

$$V^{\otimes\alpha} = W_1 \oplus \cdots \oplus W_q \tag{3.34}$$

where $W_j = \mathbf{c}_j V^{\otimes\alpha}$ is a **T**-irred subspace. Indeed it is evident from Theorem 4.10 that the sum $W_1 \oplus \cdots \oplus W_q$ is direct. The spaces $W_1, \ldots, W_q$ span $V^{\otimes\alpha}$ since the identity operator **E** belongs to $B_\alpha \cong \mathcal{R}_\alpha$. Thus, $\mathbf{E} = \mathbf{D}(x)$, $x \in \mathcal{R}_\alpha$, where x has the unique decomposition

$$x = x_1 + \cdots + x_q = c_1 x_1 + \cdots + c_q x_q,$$

$x_j \in \mathcal{I}_j$. If $\mathbf{v} \in V^{\otimes\alpha}$, then

$$\mathbf{v} = \mathbf{E}\mathbf{v} = \mathbf{x}\mathbf{v} = \mathbf{c}_1(\mathbf{x}_1\mathbf{v}) + \cdots + \mathbf{c}_q(\mathbf{x}_q\mathbf{v})$$

where $\mathbf{c}_j(\mathbf{x}_j\mathbf{v}) \in W_j$. The **T**-irred subspaces W_j are called **symmetry classes** of tensors.

To conclude our analysis we determine the relationship between Theorems 4.7 and 4.11. The first of these theorems decomposes **T** in terms of minimal **left** ideals of R_α, while the latter decomposition is in terms of minimal **right** ideals. In particular the **T**-irred subspaces W_j are not in general invariant under the operators $\mathbf{D}(s)$, $s \in S_\alpha$.

Let us fix our attention on one of the **T**-irred subspaces $W_j = \mathbf{c}_j V^{\otimes\alpha}$ in (3.34). According to Theorem 4.7 the restriction of **T** to W_j is equivalent to the irred rep $\mathbf{C}^{(\beta)}$ of G_m. Let $\mathbf{D}^{(\beta)}$ be the corresponding rep of S_α determined by this theorem and let U_β be the minimal two-sided ideal in R_α consisting of all minimal left ideals (**L**-irred subspaces) that transform according to $\mathbf{D}^{(\beta)}$. By Lemma 4.12, $\mathbf{D}(x)\mathbf{w} = \boldsymbol{\theta}$ for all $x \in U_\nu$, $\nu \neq \beta$, and all $\mathbf{w} \in W_j$. Let $\mathcal{I}$ be the minimal right ideal in $\mathcal{R}_\alpha$ associated with W_j:

$$\mathcal{I} = \{x \in \mathcal{R}_\alpha : \mathbf{x}\mathbf{v} \in W_j \quad \text{for all} \quad \mathbf{v} \in V^{\otimes\alpha}\}$$

Clearly, $\mathbf{x}W_j \neq \{\boldsymbol{\theta}\}$ for nonzero $x \in \mathcal{I}$, so the minimal two-sided ideal in $\mathcal{R}_\alpha$ containing $\mathcal{I}$ must be U_β. Thus, the minimal left and right ideals associated with W_j lie in the same two-sided ideal U_β.

Let $D^{(\beta)}$ be a unitary matrix rep corresponding to the operator rep $\mathbf{D}^{(\beta)}$. Then the matrix elements $\bar{D}^{(\beta)}_{jk}(s)$ considered as element of R_α satisfy the relations

$$L(t)\bar{D}^{(\beta)}_{jk}(s) = \bar{D}^{(\beta)}_{jk}(t^{-1}s) = \sum_{l=1}^{n_\beta} D^{(\beta)}_{lj}(t)\bar{D}^{(\beta)}_{lk}(s), \qquad 1 \leq j, k \leq n_\beta. \tag{3.35}$$

Thus, the ring elements $\bar{D}^{(\beta)}_{jk} = \sum \bar{D}^{(\beta)}_{jk}(s)\cdot s$ for fixed k generate a minimal left ideal in U_β, and the totality of these $n_\beta{}^2$ elements form a basis for U_β. The minimal right ideals in U_β all transform irreducibly under the right regular rep **R**, and the irred reps determined by these right ideals are equivalent. We will identify this irred rep. A simple computation yields

$$R(t)\bar{D}^{(\beta)}_{jk}(s) = \bar{D}^{(\beta)}_{jk}(st) = \sum_{l=1}^{n_\beta} \bar{D}^{(\beta)}_{lk}(t)\bar{D}^{(\beta)}_{jl}(s), \qquad 1 \leq j, k \leq n_\beta. \tag{3.36}$$

Thus, for fixed j the ring elements $\bar{D}_{jk}^{(\beta)}$ form a basis for a right ideal transforming according to $\bar{D}^{(\beta)}$, i.e., the matrices of this rep are the complex conjugates $\bar{D}^{(\beta)}(t)$. The character $\bar{\chi}^{(\beta)}$ of this rep is the complex conjugate of the simple character $\chi^{(\beta)}$. Since $\langle \bar{\chi}^{(\beta)}, \bar{\chi}^{(\beta)} \rangle = 1$ it follows that $\bar{\chi}^{(\beta)}$ is also a simple character.

The computations in the preceding paragraph are valid for any finite group H. In the following section we will explicitly compute the simple characters of S_α and show that they are all real. That is, $\bar{\chi}^{(\beta)}(s) = \chi^{(\beta)}(s)$ for $s \in S_\alpha$. Anticipating this result we see that the minimal left and right ideals in U_β all define irred reps of S_α equivalent to $\mathbf{D}^{(\beta)}$.

To clarify the relation between minimal left and right ideals in U_β consider the mapping $x \longrightarrow \hat{x}$ of R_α onto R_α defined by

$$\hat{x} = \sum_{s \in S_\alpha} x(s) \cdot s^{-1} = \sum_s x(s^{-1}) \cdot s \tag{3.37}$$

where $x = \sum x(s) \cdot s$. This transformation is a vector space isomorphism. (Prove it.) However, it is not a homomorphism of the group ring onto itself since $\widehat{xy} = \hat{y}\hat{x}$. In particular $\widehat{st} = (st)^{-1} = t^{-1}s^{-1} = \hat{t}\hat{s}$ for $s, t \in S_\alpha$. Such a map is sometimes called an **inverted isomorphism**, since it inverts the order of ring multiplication. It is clear that each left ideal is transformed into a right ideal by this mapping. Since $(\hat{x})\hat{} = x$ it also follows that each right ideal is the image of some left ideal. The left ideal $R_\alpha c$ with generating idempotent c is mapped onto the right ideal $\hat{c}\hat{R}_\alpha = \hat{c}R_\alpha$ with generating idempotent $\hat{c}$. [Note that $\hat{c}^2 = (c^2)\hat{} = \hat{c}$.] The idempotent c is primitive if and only if $\hat{c}$ is primitive.

Let $\mathfrak{L}$ be a minimal left ideal in U_β. There exists a basis $\{x_i\}$ for $\mathfrak{L}$ such that

$$sx_i = \sum_{j=1}^{n_\beta} D_{ji}^{(\beta)}(s)x_j, \qquad 1 \leq i \leq n_\beta. \tag{3.38}$$

Under the transformation $x \longrightarrow \hat{x}$, $\mathfrak{L}$ is mapped onto a minimal right ideal $\hat{\mathfrak{L}}$ with basis $\{\hat{x}_i\}$ such that

$$\hat{x}_i s^{-1} = \sum_{j=1}^{n_\beta} D_{ji}^{(\beta)}(s)\hat{x}_j, \tag{3.39}$$

[obtained by applying the inverted isomorphism to (3.38)]. Since the right ideal $\hat{\mathfrak{L}}$ yields the irred rep $\mathbf{D}^{(\beta)}$ it follows that $\hat{\mathfrak{L}} \subseteq U_\beta$. Thus, $x \longrightarrow \hat{x}$ maps the two-sided ideal U_β onto itself.

According to Section 4.2, the irred rep $\mathbf{D}^{(\beta)}$ of S_α corresponds to a partition $\{\lambda_1, \ldots, \lambda_\alpha\}$ of α. Recall that $\lambda_1 \geq \lambda_2 \geq \cdots \geq \lambda_\alpha \geq 0$ and $\lambda_1 + \cdots + \lambda_\alpha = \alpha$. We use the same notation $\{\lambda_1, \ldots, \lambda_\alpha\}$ to denote the frame corresponding to this partition. Let $T_1, \ldots, T_f$ be the f standard tableaux of the frame in dictionary order, and let $c_1, \ldots, c_f$ be the corresponding essential idempotents. According to Theorem 4.3,

$$U_\beta = R_\alpha c_1 \oplus \cdots \oplus R_\alpha c_f$$

is a decomposition of U_β as a direct sum of minimal left ideals. Applying the inverted isomorphism we immediately conclude that

$$U_\beta = \hat{c}_1 R_\alpha \oplus \cdots \oplus \hat{c}_f R_\alpha$$

is a decomposition of U_β into a direct sum of minimal right ideals. Thus, $W_j = \hat{c}_j V^{\otimes\alpha}$, $1 \le j \le f$, are **T**-irred subspaces of $V^{\otimes\alpha}$ all transforming according to equivalent reps of G_m. Applying this process to each minimal two-sided ideal in R_α, we obtain a decomposition of $V^{\otimes\alpha}$ into a direct sum of **T**-irred subspaces. (Note that an essential idempotent determines the same **T**-invariant subspace as does the corresponding idempotent.)

Let T be a standard tableau belonging to the frame $\{\lambda_1, \ldots, \lambda_\alpha\}$ and let c be the essential idempotent corresponding to T. From Theorem 4.1, $c = PQ$, where

$$P = \sum_{p \in R(T)} p, \qquad Q = \sum_{q \in C(T)} \delta_q q.$$

It follows that $\hat{c} = \hat{Q}\hat{P}$, where

$$\hat{P} = \sum_p \hat{p} = \sum_p p^{-1}, \qquad \hat{Q} = \sum_q \delta_q \hat{q} = \sum_q \delta_q q^{-1}.$$

However, p^{-1} ranges over $R(T)$ as p does, q^{-1} ranges over $C(T)$ as q does, and $\delta_{q^{-1}} = \delta_q$. Thus, $\hat{P} = P$, $\hat{Q} = Q$, and

$$\hat{c} = QP = \sum_{q,p} \delta_q qp. \tag{3.40}$$

The essential idempotent $\hat{c}$ is obtained from $c = PQ$ by interchanging the ring elements P and Q.

Finally, we note from Theorem 4.7 that there is a 1–1 relationship between frames $\{\lambda_1, \ldots, \lambda_\alpha\}$, $\lambda_1 + \cdots + \lambda_\alpha = \alpha$, and equivalence classes of irred reps of G_m. Therefore, we can use the frames to label the tensor irred reps of G_m.

Examples. Consider the case $\alpha = 2$, $m \ge 2$. There are two frames corresponding to S_2:

i.e., $\{2, 0\} = \{2\}$ and $\{1, 1\}$. The only standard tableaux are

$$T = \boxed{1}\boxed{2}, \qquad T' = \begin{array}{|c|}\hline 1 \\ \hline 2 \\ \hline\end{array},$$

so each frame defines a one-dimensional rep of S_2. The corresponding essential idempotents are

$$\hat{c} = QP = e[e + (12)] = e + (12), \qquad \hat{c}' = Q'P' = [e - (12)]e = e - (12).$$

The space of second-rank tensors is decomposed into two **T**-irred subspaces $W_1 = \hat{\mathbf{c}} V^{\otimes 2}$ and $W_2 = \hat{\mathbf{c}}' V^{\otimes 2}$. Introducing a basis $\{\mathbf{v}_j\}$ for V and $\{\mathbf{v}_j \otimes \mathbf{v}_k\}$

for $V^{\otimes 2}$ we see from (3.7) that the elements of W_1 are those tensors $\mathbf{b}$ whose components can be represented in the form

$$b^{i_1 i_2} = a^{i_1 i_2} + a^{i_2 i_1}$$

where the $a^{i_1 i_2}$ are arbitrary, i.e., the symmetric tensors. The elements of W_2 are tensors $\mathbf{b}$ whose components take the form

$$b^{\begin{smallmatrix} i_1 \\ i_2 \end{smallmatrix}} = a^{i_1 i_2} - a^{i_2 i_1},$$

the skew-symmetric tensors. (It is convenient to arrange the superscripts in the shape of the frame.) We have shown that $V^{\otimes 2} = W_1 \oplus W_2$, which we already knew, and that these two subspaces define irred reps of G_m.

Now consider the less trivial case $\alpha = 3$, $m \geq 2$. There are three frames corresponding to $\alpha = 3$: $\{3\}$, $\{2, 1\}$, $\{1^3\}$. The frames $\{3\}$, $\{1^3\}$ have one standard tableau each:

$$T = \begin{array}{|c|c|c|}\hline 1 & 2 & 3 \\ \hline\end{array}\,, \qquad T'' = \begin{array}{|c|}\hline 1 \\ \hline 2 \\ \hline 3 \\ \hline\end{array}\,.$$

The frame $\{2, 1\}$ has two standard tableaux:

$$T_1{}' = \begin{array}{|c|c|}\hline 1 & 2 \\ \hline 3 & \multicolumn{1}{c}{} \\ \cline{1-1}\end{array}\,, \qquad T_2 = \begin{array}{|c|c|}\hline 1 & 3 \\ \hline 2 & \multicolumn{1}{c}{} \\ \cline{1-1}\end{array}\,.$$

The essential idempotents are

$$\hat{c} = e + (12) + (13) + (23) + (123) + (132)$$

$$\hat{c}'' = e - (12) - (13) - (23) + (123) + (132)$$

$$\hat{c}_1{}' = e + (12) - (13) - (13)(12), \qquad \hat{c}_2{}' = e - (12) + (13) - (12)(13)$$

Thus,

$$V^{\otimes 3} = \hat{\mathbf{c}}V^{\otimes 3} \oplus \hat{\mathbf{c}}_1{}'V^{\otimes 3} \oplus \hat{\mathbf{c}}_2{}'V^{\otimes 3} \oplus \hat{\mathbf{c}}''V^{\otimes 3} \tag{3.41}$$

is a decomposition of $V^{\otimes 3}$ into $\mathbf{T}$-irred subspaces. The subspace corresponding to frame $\{3\}$ consists of those tensors whose tensor components

$$b^{i_1 i_2 i_3} \tag{3.42}$$

are completely symmetric with respect to the interchange of any two indices. Furthermore the subspace corresponding to $\{1^3\}$ consists of the completely skew-symmetric tensors

$$b^{\begin{smallmatrix} i_1 \\ i_2 \\ i_3 \end{smallmatrix}} \tag{3.43}$$

which change sign upon the transposition of any pair of indices. There are two subspaces corresponding to the frame $\{2, 1\}$. One space consists of all tensors of the form

$$b^{\begin{smallmatrix} i_1 i_2 \\ i_3 \end{smallmatrix}} = a^{i_1 i_2 i_3} + a^{i_2 i_1 i_3} - a^{i_3 i_2 i_1} - a^{i_2 i_3 i_1}, \qquad a^{ijk} \quad \text{arbitrary,} \tag{3.44}$$

and the other consists of all tensors

$$(b')^{i_1 i_3}_{i_2} = a^{i_1 i_2 i_3} - a^{i_2 i_1 i_3} + a^{i_3 i_2 i_1} - a^{i_3 i_1 i_2}. \tag{3.45}$$

Note that both of these tensor classes are skew-symmetric with respect to the transposition of indices in the same column. However, they are **not** symmetric with respect to the transposition of indices in the same row.

If $m \geq 3$ it is easy to see that each of the subspaces in the decomposition (3.41) is nonzero. Thus, $\mathcal{R}_3 = R_3$, $\mathcal{O}_3 = \{0\}$ and we have decomposed the tensor rep $\mathbf{T}$ of G_m on $V^{\otimes 3}$ into three irred reps, one with multiplicity two. The corresponding tensor subspaces are called symmetry classes of tensors.

However, if $m = 2$, the tensors (3.43) are identically zero. Indeed, any tensor component must have at least two equal indices. A transposition of these indices obviously leads to the same component. On the other hand such a transposition changes the sign of the component since the tensors are completely skew-symmetric. Thus, the tensors (3.43) are identically zero and $\hat{c}''R_3 \subseteq \mathcal{O}_3$. The reader can easily check that the other symmetry classes of tensors are nonzero, so $\hat{c}''R_3 = \mathcal{O}_3$ and

$$\mathcal{R}_3 = \hat{c}R_3 \oplus \hat{c}_1{}'R_3 \oplus \hat{c}_2{}'R_3$$

for the case $m = 2$.

How do we determine the two-sided ideals $\mathcal{O}_\alpha$ and $\mathcal{R}_\alpha$ in the general case? Let T be a Young tableau with frame $\{\lambda_1, \ldots, \lambda_\alpha\}$ and Young operator $\hat{c} = QP$,

(3.46) $T =$ 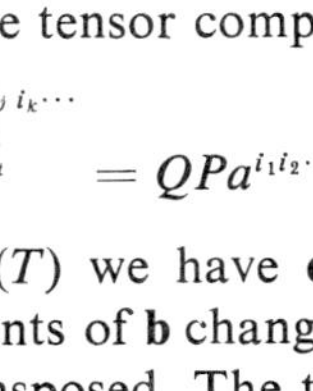

Then the elements $\mathbf{b}$ of $\hat{\mathbf{c}}V^{\otimes \alpha}$ have tensor components

$$b^J = b^{i_j i_k \cdots}_{\vdots\; i_l} = QPa^{i_1 i_2 \cdots i_\alpha} \tag{3.47}$$

Since $qQ = \delta_q Q$ for any $q \in C(T)$ we have $\mathbf{qb} = \mathbf{qQPa} = \delta_q\mathbf{QPa} = \delta_q\mathbf{b}$. In particular, the tensor components of $\mathbf{b}$ change sign whenever two indices in the same column of J are transposed. The tensors $\mathbf{b}$ are **skew-symmetric in the columns of** J. In general, the tensors $\mathbf{b}$ are not symmetric in the rows of J unless J has only one row.

These remarks enable us to determine $\mathcal{O}_\alpha$ and $\mathcal{R}_\alpha$ for the representation $\mathbf{T}$ of G_m on $V^{\otimes \alpha}$ with $\dim V = m$. Each minimal two-sided ideal U_β in the group ring R_α is uniquely associated with a frame $\{\lambda_1, \ldots, \lambda_\alpha\}$ and $R_\alpha = \mathcal{R}_\alpha \oplus \mathcal{O}_\alpha$.

Theorem 4.12. The two-sided ideals U_β corresponding to frames with r rows, $r > m$, lie in $\mathcal{O}_\alpha$. If $r \leq m$ then $U_\beta \subseteq \mathcal{R}_\alpha$. Thus, the decomposition of $V^{\otimes\alpha}$ into **T**-irred subspaces is determined by those frames of S_α with $r \leq m$ rows.

Proof. Let **b** be a tensor corresponding to a tableau T, (3.46), with $r > m$ rows. Then each tensor component b^J, (3.47), of **b** has $r > m$ indices in the first column of J and at least two of these indices must be equal. A transposition $q \in C(T)$ of two equal indices obviously leaves b^J fixed, $qb^J = b^J$. On the other hand $qb^J = -b^J$ since b^J is skew-symmetric in the columns of J. Therefore, $b^J = 0$.

Now suppose the tableau has $r \leq m$ rows and consider the tensor $\mathbf{a} \in V^{\otimes\alpha}$ with components $a^{J_0} = 1$,

$$J_0 = \begin{matrix} 1 & 1 & 1 & \cdots & 1 \\ 2 & 2 & \cdots & 2 & \\ \cdot & & & & \\ \cdot & & & & \\ \cdot & & & & \\ r & r & \cdots & r & \end{matrix}$$

and $a^J = 0$ for all indices $J \neq J_0$. (Here we arrange the indices in the shape of the tableau T.) We will show that $\hat{\mathbf{c}}\mathbf{a} = \mathbf{QPa} \neq \boldsymbol{\theta}$, where $\hat{\mathbf{c}}$ is the Young operator corresponding to T. This proves $U_\beta \subseteq R_\alpha$. Clearly, $(Pa)^{J_0} = n > 0$, where n is the order of $R(T)$ and $(Pa)^J = 0$, $J \neq J_0$. Thus, $\mathbf{Pa} = n\mathbf{a}$, so $\hat{\mathbf{c}}\mathbf{a} = n\mathbf{Qa} = n \sum \delta_q \mathbf{qa}$. The reader can easily check that each tensor **qa** has exactly one nonzero component and **qa**, **q′a** have the same nonzero component if and only if $q = q'$. This follows from the fact that $r \leq m$. Thus, the sum $\sum \delta_q \mathbf{qa}$ is nonzero and $\hat{\mathbf{c}}\mathbf{a} \neq \boldsymbol{\theta}$. Q.E.D.

In summary, we have the following result.

Theorem 4.13. Let T be a tableau of S_α with $r \leq m$ rows. Then the subspace $\hat{\mathbf{c}}_T V^{\otimes\alpha}$ transforms according to an irred rep of G_m, where $\hat{c}_T = QP$ and Q, P are obtained from T by (2.1). Tableaux T and T' determine equivalent reps of G_m if and only if they belong to the same frame $\{\lambda_j\}$. Furthermore,

$$V^{\otimes\alpha} = \sum_T \oplus \, \hat{\mathbf{c}}_T V^{\otimes\alpha}$$

where T runs over all standard tableaux of S_α with $r \leq m$ rows. The multiplicity of the irred rep $\{\lambda_j\}$ of G_m in $\mathbf{T} | V^{\otimes\alpha}$ is equal to the number of standard tableaux with frame $\{\lambda_j\}$.

We shall examine this construction from another point of view in Chapter 9, where we study the irred reps of G_m by Lie theory methods.

4.4 The Simple Characters of S_α

We now use the results of Section 4.3 to compute the simple characters of S_α. As usual we consider the reps $\mathbf{T}$ of G_m and $\mathbf{D}$ of S_α on $V^{\otimes\alpha}$ defined by (3.2)–(3.7). The irred reps $\mathbf{C}^{\{\lambda_j\}}$ of G_m and $\mathbf{D}^{\{\lambda_j\}}$ of S_α that occur with nonzero multiplicty in $\mathbf{T}$ and $\mathbf{D}$ are labeled by those frames $\{\lambda_1, \ldots, \lambda_\alpha\}$ of S_α that have $r \leq m$ rows. According to Theorem 4.7, the multiplicity of $\mathbf{C}^{\{\lambda_j\}}$ in $\mathbf{T}$ is equal to dim $\mathbf{D}^{\{\lambda_j\}}$ and the multiplicity of $\mathbf{D}^{\{\lambda_j\}}$ in $\mathbf{D}$ is equal to dim $\mathbf{C}^{\{\lambda_j\}}$.

According to (3.10) and (3.11), $\mathbf{sT}(g) = \mathbf{T}(g)\mathbf{s}$ for all $g \in G_m$, $s \in S_\alpha$. Thus we can define a rep $\mathbf{T}'$ of the direct product group $S_\alpha \times G_m$ on $V^{\otimes\alpha}$ by

$$\mathbf{T}'(sg) = \mathbf{sT}(g).$$

In terms of a basis $\{\mathbf{v}_j : 1 \leq j \leq m\}$ for V we have

$$\mathbf{sT}(g)[\mathbf{v}_{j_1} \otimes \cdots \otimes \mathbf{v}_{j_\alpha}] = \sum_{i_1 \cdots i_\alpha = 1}^{m} g_{j_1}^{i_{s(1)}} \cdots g_{j_\alpha}^{i_{s(\alpha)}} \mathbf{v}_{i_1} \otimes \cdots \otimes \mathbf{v}_{i_\alpha}, \tag{4.1}$$

[see (3.10)]. We will compute the character χ of $\mathbf{T}'$ in two different ways.

First it follows from the proof of Theorem 4.7 that if $d = \dim \mathbf{D}^{\{\lambda_j\}}$ and $c = \dim \mathbf{C}^{\{\lambda_j\}}$ then there exist cd linearly independent tensors

$$\{\mathbf{v}_l^k : 1 \leq k \leq d, \quad 1 \leq l \leq c\} \tag{4.2}$$

such that

$$\mathbf{T}(g)\mathbf{v}_l^k = \sum_{l'=1}^{c} C_{l'l}^{\{\lambda_j\}}(g)\mathbf{v}_{l'}^k, \qquad \mathbf{s}\mathbf{v}_l^k = \sum_{k'=1}^{d} D_{k'k}^{\{\lambda_j\}}(g)\mathbf{v}_l^{k'}, \tag{4.3}$$

where the matrix reps $C^{\{\lambda_j\}}$, $D^{\{\lambda_j\}}$ correspond to the operator reps $\mathbf{C}^{\{\lambda_j\}}$, $\mathbf{D}^{\{\lambda_j\}}$. Choosing vectors $\{\mathbf{v}_l^k\}$ for each frame $\{\lambda_1, \ldots, \lambda_\alpha\}$ of $r \leq m$ rows we get a total of m^α linearly independent vectors, which form a basis for $V^{\otimes\alpha}$.

We compute the trace of the operator $\mathbf{T}'(sg)$ restricted to the subspace $W^{\{\lambda_j\}}$ spanned by the basis vectors (4.2). Since

$$\mathbf{sT}(g)\mathbf{v}_l^k = \sum_{l'=1}^{c} \sum_{k'=1}^{d} C_{l'l}^{\{\lambda_j\}}(g) D_{k'k}^{\{\lambda_j\}}(s)\mathbf{v}_{l'}^{k'}$$

we find

$$\operatorname{tr}[\mathbf{sT}(g)]_{W^{\{\lambda_j\}}} = \sum_{l,k} C_{ll}^{\{\lambda_j\}}(g) D_{kk}^{\{\lambda_j\}}(s) = \varphi^{\{\lambda_j\}}(g)\chi^{\{\lambda_j\}}(s) \tag{4.4}$$

where $\varphi^{\{\lambda_j\}}$, $\chi^{\{\lambda_j\}}$ are the characters of $\mathbf{C}^{\{\lambda_j\}}$ and $\mathbf{D}^{\{\lambda_j\}}$, respectively. Thus,

$$\chi(sg) = \operatorname{tr}[\mathbf{sT}(g)] = \sum \varphi^{\{\lambda_j\}}(g)\chi^{\{\lambda_j\}}(s) \tag{4.5}$$

where the sum is taken over all partitions $\{\lambda_1, \ldots, \lambda_m\}$ with $\lambda_1 + \cdots + \lambda_m = \alpha$ and $\lambda_1 \geq \lambda_2 \geq \cdots \geq \lambda_m \geq 0$.

Now we compute $\chi(sg)$ another way. From (4.1) there follows

$$\chi(sg) = \sum_{j_1 \cdots j_\alpha = 1}^{m} g_{j_1}^{j_{s(1)}} \cdots g_{j_\alpha}^{j_{s(\alpha)}}. \tag{4.6}$$

We know that χ depends only on the conjugacy class in which s lies, not on s itself. Suppose s belongs to the conjugacy class $(\nu_1, \nu_2, \ldots, \nu_\alpha)$, i.e., ν_1 one-cycles, ν_2 two-cycles, etc. Clearly, to each μ-cycle $(klp \cdots z)$ in s there corresponds a closed sum

$$\sum_{j_k \cdots j_z = 1}^{m} g_{j_z}^{j_k} g_{j_k}^{j_l} g_{j_l}^{j_p} \cdots g_{j_v}^{j_z} = \operatorname{tr}(g^\mu) = \sigma_\mu. \tag{4.7}$$

Therefore,

$$\chi(sg) = (\sigma_1)^{\nu_1}(\sigma_2)^{\nu_2} \cdots (\sigma_\alpha)^{\nu_\alpha}$$

where σ_μ is the trace of the matrix g^μ. We have derived the formula

$$(\sigma_1)^{\nu_1}(\sigma_2)^{\nu_2} \cdots (\sigma_\alpha)^{\nu_\alpha} = \sum_{\{\lambda_j\}} \varphi^{\{\lambda_j\}}(g)\chi^{\{\lambda_j\}}(s). \tag{4.8}$$

Another identity is obtained by using the orthogonality relations for the characters $\chi^{\{\lambda_j\}}$. By (1.9) the conjugacy class $(\nu_1, \ldots, \nu_\alpha)$ contains

$$m_\nu = \alpha!/(1^{\nu_1}\nu_1! \cdots \alpha^{\nu_\alpha}\nu_\alpha!)$$

elements. Taking the inner product of (4.8) with $\chi^{\{\lambda_k\}}(s)$ we obtain

$$\varphi^{\{\lambda_j\}}(g) = \sum_{(\nu)} \frac{\chi_\nu^{\{\lambda_j\}}}{\nu_1! \cdots \nu_\alpha!} \left(\frac{\sigma_1}{1}\right)^{\nu_1} \cdots \left(\frac{\sigma_\alpha}{\alpha}\right)^{\nu_\alpha} \tag{4.9}$$

where the sum goes over all conjugacy classes $(\nu) = (\nu_1, \ldots, \nu_\alpha)$ in S_α and $\chi_\nu^{\{\lambda_j\}}$ is the value of the simple character $\chi^{\{\lambda_j\}}(s)$ for s in the conjugacy class (ν).

From (4.9) the simple character $\varphi^{\{\lambda_j\}}(g)$ is a generating function for the characters $\chi_\nu^{\{\lambda_j\}}$ of S_α. In Section 9.2 we shall compute $\varphi^{\{\lambda_j\}}(g)$ by a method entirely distinct from that used here. The result is as follows: Let $\varepsilon_1, \ldots, \varepsilon_m$ be the eigenvalues of $g \in G_m$. Then

$$\varphi^{\{\lambda_j\}}(\varepsilon_1, \ldots, \varepsilon_m) = \frac{|\varepsilon^{l_1}, \varepsilon^{l_2}, \ldots, \varepsilon^{l_{m-1}}, \varepsilon^{l_m}|}{|\varepsilon^{n-1}, \varepsilon^{n-2}, \ldots, \varepsilon, 1|}, \tag{4.10}$$

where $l_j = \lambda_j + m - j$, $j = 1, \ldots, m$, and

$$|\varepsilon^{l_1}, \ldots, \varepsilon^{l_m}| = \det \begin{pmatrix} \varepsilon_1^{l_1} & \varepsilon_1^{l_2} & \cdots & \varepsilon_1^{l_m} \\ \varepsilon_2^{l_1} & \varepsilon_2^{l_2} & \cdots & \varepsilon_2^{l_m} \\ \cdot & & & \cdot \\ \cdot & & & \cdot \\ \cdot & & & \cdot \\ \varepsilon_m^{l_1} & \varepsilon_m^{l_2} & \cdots & \varepsilon_m^{l_m} \end{pmatrix} \tag{4.11}$$

with a similar interpretation of the denominator in (4.10).

Thus (4.8) implies

$$\sigma_1^{\nu_1} \cdots \sigma_\alpha^{\nu_\alpha} |\varepsilon^{m-1}, \ldots, \varepsilon, 1| = \sum_{\{\lambda_j\}} \chi_\nu^{\{\lambda_j\}} |\varepsilon^{l_1}, \ldots, \varepsilon^{l_m}|, \tag{4.12}$$

$$\sigma_\mu = \operatorname{tr}(g^\mu) = \varepsilon_1{}^\mu + \varepsilon_2{}^\mu + \cdots + \varepsilon_m{}^\mu. \tag{4.13}$$

There is a similar formula for (4.9). These results are correct whether or not g can be diagonalized.

The expressions (4.12) can be used to compute the characters $\chi^{\{\lambda_j\}}$ directly. Indeed, as g ranges over G_m the $\varepsilon_1, \ldots, \varepsilon_m$ independently range over all nonzero complex numbers, so the coefficients $\chi_\nu^{\{\lambda_j\}}$ on the right-hand side of (4.12) are uniquely determined. It is evident from an examination of this expression that the characters of S_α are real. In fact the characters take on only integer values. Although $\chi_\nu^{\{\lambda_j\}}$ can be computed directly from (4.12) by expanding in powers of the ε_j, this process is difficult even for low values of m and α. In practice the characters are usually determined by graphical procedures or recursion relations which are derived from (4.12).

As an example we show how to compute the dimension of the irred rep $\{\lambda_j\}$ of S_α. This number is equal to $\chi^{\{\lambda_j\}}(s)$ for $s = e$ in the conjugacy class $(\alpha, 0, \ldots, 0)$, i.e., $\nu_1 = \alpha$, $\nu_2 = \cdots = \nu_\alpha = 0$. Therefore, the dimension $N^{\{\lambda_j\}}$ is determined by

(4.14) $$(\varepsilon_1 + \cdots + \varepsilon_\alpha)^\alpha \,|\varepsilon^{m-1}, \ldots, \varepsilon, 1| = \sum_{\{\lambda_j\}} N^{\{\lambda_j\}} \,|\varepsilon^{l_1}, \ldots, \varepsilon^{l_m}|.$$

[In order that the coefficient $N^{\{\lambda_j\}}$ appear in (4.14) it is necessary that all but the first m terms $\lambda_1, \ldots, \lambda_m$ of $\{\lambda_1, \ldots, \lambda_\alpha\}$ be zero. Otherwise, the value of m is immaterial. To be definite we can choose $m = \alpha$.] The determinant $|\varepsilon^{m-1}, \ldots, \varepsilon, 1|$ changes sign under the interchange of two rows. Thus, the left-hand side of (4.14) is a skew-symmetric function of the ε_j, i.e., it changes sign under the interchange of any two of these variables. If we expand $\sigma_1{}^\alpha \,|\varepsilon^{m-1}, \ldots, \varepsilon, 1|$ as a sum of monomials

(4.15) $$\varepsilon_1^{\beta_1}\varepsilon_2^{\beta_2} \cdots \varepsilon_m^{\beta_m}$$

no terms with $\beta_i = \beta_j$ for $i \neq j$ can occur with nonzero coefficient in the expansion. Indeed if the variables ε_i, ε_j are raised to equal powers then the term is invariant under an interchange of ε_i and ε_j while $\sigma_1{}^\alpha \,|\varepsilon^{m-1}, \ldots, 1|$ changes sign, so the offending term must have zero coefficient. Furthermore, by skew-symmetry the occurrence of the monomial (4.15) on the right-hand side with coefficient c implies the occurrence of each of the monomials

$$\delta_t \varepsilon_{t(1)}^{\beta_1}\varepsilon_{t(2)}^{\beta_2} \cdots \varepsilon_{t(m)}^{\beta_m}$$

with coefficient c, where t is any permutation of the integers $1, \ldots, m$ and δ_t is the parity of t.

The right-hand side of (4.14) reads

$$\sum_{\{\lambda_j\}} N^{\{\lambda_j\}} \sum_{t \in S_m} \delta_t \varepsilon_{t(1)}^{\lambda_1+m-1} \varepsilon_{t(2)}^{\lambda_2+m-2} \cdots \varepsilon_{t(m)}^{\lambda_m},$$

so $N^{\{\lambda_j\}}$ is the coefficient of the term

(4.16) $$\varepsilon_1^{\lambda_1+m-1}\varepsilon_2^{\lambda_2+m-2} \cdots \varepsilon_m^{\lambda_m}$$

in which the highest power comes first, the next highest power comes second, and so on. The coefficients of the terms which are not in the ordered form (4.16) are obtained from the skew-symmetry.

To obtain $N^{\{\lambda_j\}}$ we multiply

$$|\varepsilon^{m-1}, \ldots, \varepsilon, 1| = \sum_{t \in S_m} \delta_t \varepsilon_{t(1)}^{m-1} \varepsilon_{t(2)}^{m-2} \cdots \varepsilon_{t(m-1)} \cdot 1 \tag{4.17}$$

α times by $\sigma_1 = \varepsilon_1 + \cdots + \varepsilon_m$ and compute the coefficient of (4.16) in the resulting expression. The only term in (4.17) which contributes to the result is the ordered term

$$\varepsilon_1^{m-1} \varepsilon_2^{m-2} \cdots \varepsilon_{m-1} \cdot 1 \tag{4.18}$$

since at some stage of the multiplication process each of the other monomials would have two variables raised to the same power. Thus, $N^{\{\lambda_j\}}$ is equal to the total number of ways we can obtain (4.16) from (4.18) by means of α successive multiplications by the variables ε_i, making sure that at each step no two variables are raised to the same power. Clearly, the number of times we multiply by ε_1 is greater than or equal to the number of times we multiply by ε_2, etc.

An example should clarify the situation. Let us compute the dimension of the rep $\{2, 1^3\}$ of S_5. For $m = \alpha = 5$, (4.18) becomes $\varepsilon_1{}^4 \varepsilon_2{}^3 \varepsilon_3{}^2 \varepsilon_4$. We must end up with $\varepsilon_1{}^6 \varepsilon_2{}^4 \varepsilon_3{}^3 \varepsilon_4{}^2$, Eq. (4.16), by multiplying by one variable ε_i at a time, making sure that at each step the exponents of no two variables are equal. The possibilities are as follows:

$$\begin{array}{llll} (1) & \varepsilon_4 \varepsilon_3 \varepsilon_2 \varepsilon_1 \varepsilon_1 (\varepsilon_1{}^4 \varepsilon_2{}^3 \varepsilon_3{}^2 \varepsilon_4), & (2) & \varepsilon_4 \varepsilon_3 \varepsilon_1 \varepsilon_2 \varepsilon_1 (\varepsilon_1{}^4 \varepsilon_2{}^3 \varepsilon_3{}^2 \varepsilon_4) \\ (3) & \varepsilon_4 \varepsilon_1 \varepsilon_3 \varepsilon_2 \varepsilon_1 (\varepsilon_1{}^4 \varepsilon_2{}^3 \varepsilon_3{}^2 \varepsilon_4), & (4) & \varepsilon_1 \varepsilon_4 \varepsilon_3 \varepsilon_2 \varepsilon_1 (\varepsilon_1{}^4 \varepsilon_2{}^3 \varepsilon_3{}^2 \varepsilon_4). \end{array} \tag{4.19}$$

We conclude that the dimension of $\{2, 1^3\}$ is four.

Our method has a graphical interpretation. From our rule for obtaining $N^{\{\lambda_j\}}$, this integer is just the number of distinct ways we can fill in the frame of $\{\lambda_1, \ldots, \lambda_\alpha\}$ successively with α dots, making sure that at each step every row in the frame has at least as many dots as the rows below it. The dots are filled in from left to right in any given row. Multiplication by ε_i corresponds to the application of a dot in row i. As an example we again consider the frame $\{2, 1^3\}$. If we number the dots from 1 to 5 in the order of their application we obtain the results

$$(1)\ \begin{array}{|c|c|}\hline 1 & 2 \\ \hline 3 \\ \cline{1-1} 4 \\ \cline{1-1} 5 \\ \cline{1-1}\end{array} \qquad (2)\ \begin{array}{|c|c|}\hline 1 & 3 \\ \hline 2 \\ \cline{1-1} 4 \\ \cline{1-1} 5 \\ \cline{1-1}\end{array} \qquad (3)\ \begin{array}{|c|c|}\hline 1 & 4 \\ \hline 2 \\ \cline{1-1} 3 \\ \cline{1-1} 5 \\ \cline{1-1}\end{array} \qquad (4)\ \begin{array}{|c|c|}\hline 1 & 5 \\ \hline 2 \\ \cline{1-1} 3 \\ \cline{1-1} 4 \\ \cline{1-1}\end{array} \tag{4.20}$$

which correspond exactly to the expressions (1)–(4) of (4.19). Note that the tableaux (4.20) are just the standard tableaux associated with the frame $\{2, 1^3\}$.

The reader should now recognize that the number of ways we can fill in a frame $\{\lambda_j\}$ with dots satisfying the above rules is exactly the number of standard tableaux associated with $\{\lambda_j\}$. This proves Theorem 4.2.

Similar but more complicated graphical methods can be used to compute the characters $\chi_\nu^{\{\lambda_j\}}$ (see Hamermesh [1], Boerner [1], Murnaghan [1]).

Problems

4.1 Compute the conjugacy classes of S_6 and the number of elements in each class.

4.2 Apply the methods of Chapter 3 to deduce the character tables of S_3, S_4, and S_5. Hint: Use formula (5.27), Section 3.5, to derive characters of S_n from the simple characters of $S_{n-1} \subset S_n$.

4.3 List the possible frames of S_5 and the standard tableaux corresponding to each frame.

4.4 Let c be an essential idempotent corresponding to the frame $\{2, 1\}$. Determine the invariant subspace $R_3 c$ directly and use it to compute a matrix rep of S_3 equivalent to $\{2, 1\}$.

4.5 Construct the ring element $c = PQ$ corresponding to the tableau of Fig. 4.3 and verify directly that c is an essential idempotent. What is the dimension of the irred rep of S_6 determined by c?

4.6 Compute the dimension of the rep $\{4, 3, 1, 1\}$ of S_9.

4.7 Decompose the space $V^{\otimes 4}$ into subspaces irred under $GL(m, \mathfrak{C})$, dim $V = m \geq 2$.

4.8 Use formula (4.12) to obtain the character tables for S_3 and S_4.

4.9 Let G be a group, not necessarily finite, and let $T^{(1)}, \ldots, T^{(r)}$ be irred nonequivalent matrix reps of G. Show that the matrix elements $\{T_{ij}^{(k)}(g)\}$ $(1 \leq i, j \leq n_k;\ 1 \leq k \leq r)$ are linearly independent functions on G. (This is not easy. See Curtis and Reiner [1, Chapter IV].)

Chapter 5

Lie Groups and Lie Algebras

5.1 The Exponential of a Matrix

Let **A** be a linear operator on the n-dimensional inner product space V. The **norm** $||\mathbf{A}||$ of **A** is defined by the expression

$$(1.1)\qquad ||\mathbf{A}|| = \max_{\mathbf{v}\in V, ||\mathbf{v}||=1} \{||\mathbf{A}\mathbf{v}||\}.$$

[Recall that the norm of a vector is $||\mathbf{v}|| = (\mathbf{v}, \mathbf{v})^{1/2}$.] To see that this definition makes sense choose an ON basis $\{\mathbf{v}_j\}$ for V with respect to which **A** has matrix elements A_{ij} and **v** has vector components a_j. Then

$$(1.2)\qquad ||\mathbf{A}|| = \max_{|a_1|^2+\cdots+|a_n|^2=1} [\sum_{i=1}^{n} | \sum_{j=1}^{n} A_{ij}a_j |^2]^{1/2}$$

This maximum clearly exists, so $||\mathbf{A}||$ is well-defined. (We have assumed that V is complex in this computation. There is an analogous result for V real.)

If **v** is a nonzero element of V then $\mathbf{v}/||\mathbf{v}|| = \mathbf{w}$ has norm one. Thus $||\mathbf{A}\mathbf{w}|| \leq ||\mathbf{A}||$ from (1.1), so

$$(1.3)\qquad ||\mathbf{A}\mathbf{v}|| \leq ||\mathbf{A}|| \cdot ||\mathbf{v}||,$$

where we have used the fact $||\alpha\mathbf{v}|| = \alpha||\mathbf{v}||$ for $\alpha \geq 0$. Expression (1.3) is also valid for $\mathbf{v} = \boldsymbol{\theta}$.

Lemma 5.1. Let **A** and **B** be linear operators on V. Then $||\mathbf{AB}|| \leq ||\mathbf{A}|| \cdot ||\mathbf{B}||$.

Proof. Let $\mathbf{v} \in V$ with $\|\mathbf{v}\| = 1$. By (1.3), $\|(\mathbf{AB})\mathbf{v}\| = \|\mathbf{A}(\mathbf{Bv})\| \leq \|\mathbf{A}\| \cdot \|\mathbf{Bv}\| \leq \|\mathbf{A}\| \cdot \|\mathbf{B}\|$. Therefore $\|\mathbf{AB}\| \leq \|\mathbf{A}\| \cdot \|\mathbf{B}\|$. Q.E.D.

Corollary 5.1. $\|\mathbf{A}^m\| \leq \|\mathbf{A}\|^m$, $m = 1, 2, \ldots$.

Proof. Lemma 5.1 with $\mathbf{A} = \mathbf{B}$, and induction on m.

Lemma 5.2. Let $\mathbf{A}$ and $\mathbf{B}$ be linear operators on V. Then $\|\alpha\mathbf{A}\| = \alpha\|\mathbf{A}\|$ for all $\alpha \geq 0$ and $\|\mathbf{A} + \mathbf{B}\| \leq \|\mathbf{A}\| + \|\mathbf{B}\|$.

Proof. These results follow directly from the definition (1.1) and the relations $\|\alpha\mathbf{Av}\| = \alpha\|\mathbf{Av}\|$ and $\|\mathbf{Av} + \mathbf{Bv}\| \leq \|\mathbf{Av}\| + \|\mathbf{Bv}\|$.

We can define the matrix norm $\|A\|$ of an $n \times n$ matrix $A = (A_{ij})$ by $\|A\| = \|\mathbf{A}\|$, where $\mathbf{A}$ is the linear operator on V determined by A and the ON basis $\{\mathbf{v}_j\}$. In particular, $\|A\|$ is given explicitly by (1.2). Thus, the results proved above for operator norms hold as well for matrix norms.

Recall that a sequence of complex numbers $\{\alpha_j\}$ is a **Cauchy sequence** if for any $\varepsilon > 0$ there is an integer $N_\varepsilon > 0$ such that $|\alpha_i - \alpha_j| < \varepsilon$ for all $i, j > N_\varepsilon$. Every Cauchy sequence converges, i.e., if $\{\alpha_j\}$ is Cauchy then there is an $\alpha \in \mathfrak{C}$ such that $\alpha_j \longrightarrow \alpha$. A sequence $\{A_k\}$ of $n \times n$ matrices is said to be a **Cauchy sequence in the norm** $\|\cdot\|$ if for every $\varepsilon > 0$ there is an integer $N_\varepsilon > 0$ such that $\|A_i - A_j\| < \varepsilon$ for all $i, j > N_\varepsilon$. The sequence $\{A_k\}$ is Cauchy in the norm if and only if for each $i, j = 1, \ldots, n$ the matrix elements $\{A_{k,ij}, k = 1, 2, \ldots\}$ in the ith row and jth column form a Cauchy sequence of numbers. (Prove it.) Thus, if $\{A_k\}$ is a Cauchy sequence of matrices, the limits $A_{ij} = \lim_{k\to\infty} A_{k,ij}$ exist and $A_k \longrightarrow A$ with $A = (A_{ij})$. Then for any $\varepsilon > 0$ there is an integer $M_\varepsilon > 0$ such that $\|A - A_k\| < \varepsilon$ whenever $k > M_\varepsilon$. We have sketched a proof of the fact that every Cauchy sequence of matrices converges. Passing from matrices to linear operators, we immediately obtain the result that every Cauchy sequence of operators converges. Most of the results obtained in this section are valid for both operators and matrices, but the formulation of the corresponding operator results is left to the reader.

The exponential $\exp A$ of an $n \times n$ matrix A is defined by the sum

$$\exp A = \sum_{j=0}^{\infty} \frac{A^j}{j!}, \tag{1.4}$$

where $A^0 = E_n$ (the $n \times n$ identity matrix). To justify this definition consider the finite sums

$$S_m = \sum_{j=0}^{m} \frac{A^j}{j!}, \qquad m = 0, 1, \ldots,$$

which are well-defined $n \times n$ matrices. For $k > m$ we have

$$\|S_k - S_m\| = \left\| \sum_{j=m+1}^{k} \frac{A^j}{j!} \right\| \leq \sum_{j=m+1}^{k} \frac{\|A\|^j}{j!}.$$

Since

$$e^{\|A\|} = \sum_{j=0}^{\infty} \frac{\|A\|^j}{j!} < \infty,$$

then for any $\varepsilon > 0$ there is an integer $N_\varepsilon > 0$ with the property $\sum_{j=q}^{\infty}(\|A\|^j/j!) < \varepsilon$ for every $q > N_\varepsilon$. This proves that the sequence of matrices $\{S_m\}$ is Cauchy and $\exp A = \lim_{m\to\infty} S_m$ exists.

Theorem 5.1. If A and B are $n \times n$ matrices and B is nonsingular then $\exp(BAB^{-1}) = B(\exp A)B^{-1}$.

Proof.

$$\exp(BAB^{-1}) = \sum_{j=0}^{\infty} \frac{(BAB^{-1})^j}{j!} = \sum_{j=0}^{\infty} \frac{BA^jB^{-1}}{j!} = B(\exp A)B^{-1},$$

since

$$\|BS_mB^{-1} - B(\exp A)B^{-1}\| \leq \|B\|\cdot\|S_m - \exp A\|\cdot\|B^{-1}\| \longrightarrow 0$$

as $m \longrightarrow \infty$. Q.E.D.

We review a few basic facts about the eigenvalues of matrices. Let $\lambda_1, \ldots, \lambda_n$ be the eigenvalues of the $n \times n$ matrix A, each eigenvalue repeated a number of times equal to its multiplicity. The λ_j are the solutions of the characteristic equation

$$P_A(\lambda) = \det(A - \lambda E_n) = 0.$$

If the matrix C is similar to A, i.e., if there exists a nonsingular $n \times n$ matrix B such that $C = BAB^{-1}$, then

$$\begin{aligned} P_C(\lambda) &= \det(BAB^{-1} - \lambda BE_nB^{-1}) = (\det B)\det(A - \lambda E_n)\det B^{-1} \\ &= \det(A - \lambda E_n) = P_A(\lambda). \end{aligned}$$

Thus, similar matrices have the same eigenvalues. The eigenvalues of a diagonal matrix are just the diagonal elements. We say that a matrix A can be diagonalized by similarity transformations if A is similar to a diagonal matrix D. The diagonal elements of D are just the eigenvalues of A.

Not all $n \times n$ matrices can be diagonalized. For example the matrix

$$A = \begin{pmatrix} 0 & 1 \\ 0 & 0 \end{pmatrix}$$

has eigenvalues $\lambda_1 = \lambda_2 = 0$. Since A is obviously not similar to the zero matrix, it cannot be diagonalized. It *is* true that every complex matrix A is similar to a matrix C in upper triangular form. An $n \times n$ matrix C is **upper**

triangular if $C_{ij} = 0$ for all $1 \leq j < i \leq n$, i.e., if all matrix elements below the main diagonal are zero. The eigenvalues of a upper triangular matrix are just the diagonal elements:

$$C = \begin{pmatrix} \lambda_1 & & * & * & * \\ & \lambda_2 & & * & * \\ & & \ddots & & * \\ & Z & & & \\ & & & & \lambda_n \end{pmatrix}. \tag{1.5}$$

Theorem 5.2. Let A be a complex $n \times n$ matrix with eigenvalues $\lambda_1, \ldots, \lambda_n$. Then A is similar to an upper triangular matrix C with diagonal elements $\lambda_1, \ldots, \lambda_n$.

Proof. Induction on n. The theorem is obvious for $n = 1$. Assume it is true for all $(n-1) \times (n-1)$ matrices and let A be a matrix of degree n. Since λ_1 is an eigenvalue of A there is a nonzero column vector $v = (v_1, \ldots, v_n)$ such that $Av = \lambda_1 v$, i.e., $\sum_{j=1}^{n} A_{ij} v_j = \lambda_1 v_i$. Let e_1 be the column vector defined by the n-tuple $(1, 0, \ldots, 0)$. Clearly, there exists an $n \times n$ nonsingular matrix B_1 such that $B_1 v = e_1$. Then the matrix $B_1 A B_1^{-1}$ is similar to A and takes the form

$$B_1 A B_1^{-1} = \begin{pmatrix} \lambda_1 & * & \cdots & * \\ 0 & & & \\ \vdots & & A' & \\ 0 & & & \end{pmatrix},$$

where A' is an $(n-1) \times (n-1)$ matrix. Indeed, $B_1 A B_1^{-1} e_1 = B_1 A v = \lambda_1 B_1 v = \lambda_1 e_1$. Since $\det(A - \lambda E_n) = \det(B_1 A B_1^{-1} - \lambda E_n) = (\lambda_1 - \lambda) \times \det(A' - \lambda E_{n-1})$, the numbers $\lambda_2, \ldots, \lambda_n$ are the eigenvalues of A'. Furthermore, by the induction hypothesis there exists an $(n-1) \times (n-1)$ nonsingular matrix B_2' such that $B_2' A' (B_2')^{-1}$ is upper triangular. Defining the $n \times n$ matrix

$$B_2 = \begin{pmatrix} 1 & 0 & \cdots & 0 \\ 0 & & & \\ \vdots & & B_2' & \\ 0 & & & \end{pmatrix},$$

we find $B_2(B_1 A B_1^{-1}) B_2^{-1} = C$, where C is an $n \times n$ upper triangular matrix, Eq. (1.5). Thus $BAB^{-1} = C$, where $B = B_2 B_1$. Q.E.D.

Corollary 5.2. Let A be an $n \times n$ matrix with eigenvalues $\lambda_1, \ldots, \lambda_n$. Then $\exp A$ has eigenvalues $e^{\lambda_1}, \ldots, e^{\lambda_n}$.

Proof. The matrix A is similar to an upper triangular matrix C [Eq. (1.5)]. An elementary computation yields

$$C^m = \begin{pmatrix} \lambda_1{}^m & & * & * & * \\ & \lambda_2{}^m & & * & * \\ & & \ddots & & \\ & Z & & & \\ & & & & \lambda_n{}^m \end{pmatrix}, \qquad m = 1, 2, \ldots,$$

$$\exp C = \sum_{m=0}^{\infty} \frac{C^m}{m!} = \begin{pmatrix} e^{\lambda_1} & & * & * & * \\ & e^{\lambda_2} & & * & * \\ & & \ddots & & \\ & Z & & & \\ & & & & e^{\lambda_n} \end{pmatrix}.$$

Thus the eigenvalues of $\exp C$ are $e^{\lambda_1}, \ldots, e^{\lambda_n}$. Since A and C are similar there exists a nonsingular matrix B such that $BAB^{-1} = C$. By Theorem 5.1, $B(\exp A)B^{-1} = \exp(BAB^{-1}) = \exp C$, so $\exp A$ and $\exp C$ are similar. Similar matrices have the same eigenvalues. Q.E.D.

Corollary 5.3. $\det(\exp A) = \exp(\operatorname{tr} A) = e^{\operatorname{tr} A}$. In particular, $\exp A$ is a nonsingular matrix.

Proof. The determinant and trace of a matrix are the product and sum of its eigenvalues. By Theorem 5.2, $\det(\exp A) = \exp(\lambda_1 + \cdots + \lambda_n) = \exp(\operatorname{tr} A)$. Q.E.D.

We know from complex variable theory that corresponding to any power series

$$f(z) = \sum_{j=0}^{\infty} c_j z^j, \qquad c_j \in \mathfrak{C}, \tag{1.6}$$

there is associated a number $r \geq 0$, the **radius of convergence**, such that the power series converges absolutely if $|z| < r$ and diverges if $|z| > r$. In case the power series converges for all z, we set $r = +\infty$ (see Ahlfors [1].) By convergence of the power series for fixed z we mean that the partial sums

$$S_m(z) = \sum_{j=0}^{m} c_j z^j, \qquad m = 0, 1, \ldots, \tag{1.7}$$

form a Cauchy sequence of complex numbers. The limit of this Cauchy sequence is $f(z)$. The only properties of the absolute value needed to prove these results are

(1.8) $|a + b| \leq |a| + |b|, \qquad |\alpha a| = \alpha |a|, \qquad |ab| \leq |a| \cdot |b|,$

for $a, b \in \mathfrak{C}$, $\alpha \geq 0$. Since the matrix norm $\|A\|$ shares these properties we can immediately carry over the fundamental results on power series of complex numbers to power series of matrices. Thus, if $f(z)$ is the power series (1.6) and A is an $n \times n$ matrix we define the matrix power series

(1.9a) $$f(A) = \sum_{j=0}^{\infty} c_j A^j.$$

This series converges to an $n \times n$ matrix provided the partial sums

(1.9b) $$S_m(A) = \sum_{j=0}^{m} c_j A^j, \qquad m = 0, 1, 2, \ldots,$$

form a Cauchy sequence with respect to the matrix norm $\|\cdot\|$. Then

$$f(A) = \lim_{m \to \infty} S_m(A).$$

It follows from the above remarks that there exists a nonegative number r such that (1.8) converges if $\|A\| < r$ and diverges if $\|A\| > r$. For example. the power series defining $\exp A$ corresponds to $r = +\infty$.

If $f_1(z) = \sum a_j z^j, f_2(z) = \sum \mathrm{b}_k \mathrm{z}^k$ have radii of convergence r_1, r_2, respectively, then

$$f_1(z) f_2(z) = (\sum a_j z^j)(\sum b_k z^k) = \sum c_l z^l, \qquad |z| < r,$$

where

$$c_l = \sum_{j=0}^{l} a_j b_{l-j}, \qquad r = \min\{r_1, r_2\}.$$

The proof of this result uses only the properties (1.8) (Ahlfors [1]), hence it carries over immediately to matrix functions $f_1(A) f_2(A)$. For example, the identity $e^{-z} e^z = 1$ for all z leads to the identity

(1.10) $$\exp(-A) \exp(A) = E_n$$

for all $n \times n$ matrices A. Similarly, the identity $(e^z)^m = e^{mz}$, $m = 0, \pm 1, \ldots,$ leads to $[\exp A]^m = \exp(mA)$.

Since $\ln(1 + z) = \sum_{j=1}^{\infty} (-1)^{j+1} z^j / j$ converges for $|z| < 1$ we can define the $n \times n$ matrix

(1.11) $$\ln(E_n + A) = \sum_{j=1}^{\infty} (-1)^{j+1} \frac{A^j}{j}, \qquad \|A\| < 1.$$

Furthermore, the formulas

$$\exp[\ln(1 + z)] = 1 + z, \quad |z| < 1, \qquad \ln(\exp z) = z, \quad |z| < \ln 2,$$

which are proved by power series expansion, immediately lead to

(1.12)
$$\exp[\ln(E_n + A)] = E_n + A, \quad ||A|| < 1; \qquad \ln(\exp A) = A, \quad ||A|| < \ln 2.$$

Let L_n be the space of all $n \times n$ matrices, either real or complex, and let G_n be the group of all nonsingular matrices in L_n.

Theorem 5.3. The exponential mapping $A \longrightarrow \exp A$ transforms L_n into G_n. There exist positive constants $\varepsilon, \delta < 1$ such that the neighborhood $\mathcal{A}$ of the zero matrix in L_n is mapped 1–1 onto the neighborhood $\mathcal{B}$ of the identity matrix E_n in G_n. Here

$$\mathcal{A} = \{A \in L_n: ||A|| < \varepsilon\}, \qquad \mathcal{B} = \{B \in L_n: ||B - E_n|| < \delta\}.$$

Furthermore the mapping $A \longrightarrow \exp A$, $A \in \mathcal{A}$, and its inverse $B \longrightarrow \ln B$, $B \in \mathcal{B}$, are analytic, i.e,. the matrix elements of $\exp A$ and $\ln B$ are analytic functions of the matrix elements of A and B, respectively.

The proof is immediate from identities (1.12). The theorem shows that any matrix $B \in G_n$ in a sufficiently small neighborhood of E_n can be written **uniquely** in the form $B = \exp A$, with A in a sufficiently small neighborhood of Z.

Let t be a parameter, real or complex. We write $A(t)$ to denote an $n \times n$ matrix whose components $A_{ij}(t)$ are functions of t. If each of the matrix components is differentiable we say that the matrix $A(t)$ is differentiable with derivative $\dot{A}(t) = dA(t)/dt$, the matrix with components $\dot{A}_{ij}(t)$. If $A(t)$ and $B(t)$ are differentiable $n \times n$ matrices, the identities

(1.13)
$$\begin{aligned}(d/dt)[\alpha A(t) + \beta B(t)] &= \alpha \dot{A}(t) + \beta \dot{B}(t)\\ (d/dt)[A(t)B(t)] &= \dot{A}(t)B(t) + A(t)\dot{B}(t)\end{aligned}$$

hold, as the reader can easily prove.

If the power series $f(z) = \sum c_j z^j$ has radius of convergence $r > 0$ then $f(z)$ is differentiable and $f'(z)$ is obtained by differentiating the power series for $f(z)$ term by term:

$$f'(z) = df(z)/dz = \sum_{j=1}^{\infty} jc_j z^{j-1}, \qquad |z| < r.$$

Since the proof of this result employs only the properties (1.8) it can easily be extended to compute the derivative of the matrix function $\exp(tA)$, A a constant matrix. Thus,

(1.14)
$$\frac{d}{dt}\exp(tA) = \sum_{j=1}^{\infty} \frac{t^{j-1}}{(j-1)!} A^j = A\exp(tA) = [\exp(tA)]A.$$

We can use this differential formula to derive other interesting properties of the exponential function. Note that the general solution of the matrix differential equation $\dot{A}(t) = Z$ (the zero matrix) is $A(t) = A_0$, where A_0 is any constant matrix. To show the utility of this remark we will use it to find the general solution of the equation $\dot{A}(t) = AA(t)$, where A is a constant matrix. Suppose $A(t)$ is a solution and consider the matrix $B(t) = [\exp(-tA)]A(t)$. From (1.13) and (1.14), we have

$$\dot{B}(t) = -[\exp(-tA)]AA(t) + [\exp(-tA)]AA(t) = Z$$

so $B(t)$ is a constant matrix $A_0 = B(0)$. Thus $A(t) = [\exp(tA)]A_0$. Conversely, it is easy to show that $A(t) = [\exp(tA)]A_0$ satisfies the equation for all matrices A_0.

It is an elementary exercise to show that $B(\exp A) = (\exp A)B$ for any $n \times n$ matrices such that $AB = BA$.

Theorem 5.4. If $AB = BA$ then $\exp(A + B) = (\exp A)(\exp B)$.

Proof. We could obtain this result by explicit power series expansion. A more elegant proof is obtained from a consideration of the matrix function

$$C(t) = \exp[-t(A + B)](\exp tA)(\exp tB).$$

From (1.14) we have

$$\dot{C}(t) = \{\exp[-t(A + B)]\}(-A - B + A + B)(\exp tA) \exp tB = Z.$$

Therefore, $C(t)$ is a constant matrix. Since $C(0) = E_n$ we find $\{\exp[-t(A + B)]\}$ $(\exp tA)$ $\exp tB = E_n$ or $(\exp tA)$ $\exp tB = \exp[t(A + B)]$ for all t. Q.E.D.

If the matrices A and B do not commute then Theorem 5.4 is no longer valid and the simple analogy between the matrix exponential and the usual exponential breaks down. However, from Theorem 5.3, if A and B are sufficiently close to Z then there must exist a unique matrix C sufficiently close to Z such that $(\exp A) \exp B = \exp C$. If A and B commute then $C = A + B$, but in the general case C is a complicated function $C(A, B)$ of A and B. There is an explicit formula for $C(A, B)$ called the Campbell–Baker–Hausdorff formula, which we will derive by means of several lemmas, some of which are useful in their own right.

Let L_n be the n^2-dimensional space of all complex $n \times n$ matrices. For $A \in L_n$ we define the linear transformation Ad A on L_n by

(1.15) $$\text{Ad } A(B) = [A, B],$$

where

(1.16) $$[A, B] = AB - BA$$

is the **commutator bracket.** The matrix of Ad A with respect to a basis for L_n is $n^2 \times n^2$. Clearly Ad A is the zero operator if and only if A commutes with all $B \in L_n$. By $(\text{Ad } A)^m$ we mean the operator

$$(\text{Ad } A)^m(B) = [A, [A, \ldots, [A, B]], \ldots] \quad (m \quad \text{times}).$$

We will frequently write e^A for $\exp A$.

Lemma 5.3. Let $A, B \in L_n$. Then

$$e^A B e^{-A} = (\exp(\text{Ad } A))B = \sum_{j=0}^{\infty} (j!)^{-1}(\text{Ad } A)^j(B)$$
$$= E_n + [A, B] + \tfrac{1}{2}[A, [A, B]] + \cdots.$$

Proof. Set $B(t) = e^{tA}Be^{-tA}$. Clearly $B(0) = B$ and the matrix elements of $B(t)$ are entire functions of t. Thus, there exist matrices $C_j \in L_n$ such that $B(t) = \sum C_j t^j$, where

$$C_0 = B, \qquad C_j = \frac{1}{j!}\frac{d^j}{dt^j} B(t)\bigg|_{t=0}, \qquad j = 1, 2, \ldots.$$

Now

$$\dot{B}(t) = Ae^{tA}Be^{-tA} - e^{tA}Be^{-tA}A = [A, B(t)] = \text{Ad } A(B(t))$$

and by induction on j,

$$d^jB(t)/dt^j = (\text{Ad } A)^j(B(t)).$$

We conclude that $C_j = (j!)^{-1}(\text{Ad } A)^j B$. Q.E.D.

Lemma 5.4. Let $A(t) \in L_n$ such that each of the matrix elements $A_{ij}(t)$ is an analytic function of the parameter t and let

$$f(z) = \frac{e^z - 1}{z} = 1 + \frac{z}{2!} + \frac{z^2}{3!} + \cdots,$$

(an entire function of z). Then

$$e^{A(t)}\, de^{-A(t)}/dt = -f(\text{Ad } A(t))(\dot{A}(t)). \tag{1.17}$$

Recall that we can consider $\text{Ad } A(t)$ as an $n^2 \times n^2$ matrix.

Proof. Set $B(s, t) = e^{sA(t)}\, d(e^{-sA(t)})/dt$. Since $B(s, t)$ is an entire function of s it can be written in the form

$$B(s, t) = \sum_{j=0}^{\infty} C_j(t)s^j, \qquad C_j(t) = \frac{1}{j!}\frac{\partial^j}{\partial s^j} B(s, t)\bigg|_{s=0}.$$

Now

(1.18)

$$\partial B/\partial s = A(t)B(s, t) - B(s, t)A(t) - \dot{A}(t) = (\text{Ad } A(t))B(s, t) - \dot{A}(t).$$

Differentiating this expression successively with respect to s we find

$$\partial^j B(s, t)/\partial s^j = (\text{Ad } A(t))^j B(s, t) - (\text{Ad } A(t))^{j-1}\dot{A}(t), \qquad j = 1, 2, \ldots.$$

Since $B(0, t) = 0$ we have

$$C_0(t) = Z, \qquad C_j(t) = (-1/j!)(\text{Ad } A(t))^{j-1}\dot{A}(t), \qquad j = 1, 2, \ldots .$$

Thus,

$$e^{sA(t)} \frac{de^{-sA(t)}}{dt} = -\sum_{j=1}^{\infty} s^j \frac{(\text{Ad } A(t))^{j-1}}{j!} \dot{A}(t).$$

Setting $s = 1$ we obtain (1.17). Q.E.D.

If $A, B \in L_n$ have sufficiently small norms there exists a unique $C \in L_n$ sufficiently close to Z such that $e^A e^B = e^C$, or $C = \ln(e^A e^B)$, where the logarithm is defined by (1.11). That C must lie sufficiently close to Z for uniqueness follows from the identity $\exp(C + 2\pi i D) = \exp C$, where D is any diagonal matrix with integer coefficients. The logarithm exists provided $\| e^A e^B - E_n \| < 1$. The function

$$g(z) = \frac{\ln z}{z-1} = \sum_{j=0}^{\infty} \frac{(1-z)^j}{j+1} = 1 + \frac{1}{2}(1-z) + \frac{1}{3}(1-z)^2 + \cdots$$

is analytic for $|1 - z| < 1$. Thus the expression

$$g(F) = E_k + \tfrac{1}{2}(E_k - F) + \tfrac{1}{3}(E_k - F)^2 + \cdots$$

defines a matrix for any $k \times k$ matrix F with $\| E_k - F \| < 1$.

Theorem 5.5 (Campbell–Baker–Hausdorff). For A, B in a sufficiently small neighborhood of Z in L_n and $C = \ln(e^A e^B)$, then

$$C = B + \int_0^1 g[\exp(t \text{ Ad } A) \exp(\text{Ad } B)](A)\, dt. \tag{1.19}$$

Indeed,

$$C = A + B + \tfrac{1}{2}[A, B] + \tfrac{1}{12}[A, [A, B]] - \tfrac{1}{12}[B, [B, A]] + \cdots \tag{1.20}$$

and the matrix elements of C are analytic functions of the matrix elements of A and B.

Proof. We will derive a differential equation for $C(t) = \ln(e^{tA} e^B)$, where $0 \le t \le 1$. Now, $e^{tA} e^B = e^{C(t)}$ and from Lemma 5.3,

$$\{\exp[\text{Ad } C(t)]\}H = e^{tA}(e^B H e^{-B})e^{-tA} = [\exp(\text{Ad } tA)](\exp \text{Ad } B)H$$

for any $H \in L_n$. Thus

$$\exp[\text{Ad } C(t)] = [\exp(\text{Ad } tA)] \exp(\text{Ad } B), \tag{1.21}$$

and for A, B such that $\| [\exp(\text{Ad } tA)] \exp(\text{Ad } B) - E_{n^2} \| < 1$ for all $0 \le t \le 1$, then

$$\text{Ad } C(t) = \ln\{[\exp(\text{Ad } tA)] \exp \text{Ad } B\}. \tag{1.22}$$

By definition of $C(t)$,

$$e^{C(t)}\, de^{-C(t)}/dt = -e^{tA}e^{B}e^{-B}e^{-tA}A = -A.$$

Lemma 5.4 implies

$$f(\mathrm{Ad}\, C(t))\dot{C}(t) = A, \tag{1.23}$$

where $f(z) = (e^z - 1)/z$. We can solve this expression for $\dot{C}(t)$ by noting that $f(\ln z)g(z) = 1$ for $|1 - z| < 1$. Thus, $f(\ln F)g(F) = E_{n^2}$ for any $n^2 \times n^2$ matrix F with $\| E_{n^2} - F \| < 1$. In particular, $g(F) = f(\ln F)^{-1}$. Setting $F = [\exp(\mathrm{Ad}\, tA)] \exp(\mathrm{Ad}\, B)$ and using (1.22), we see that (1.23) becomes

$$\dot{C}(t) = g\{[\exp(\mathrm{Ad}\, tA)] \exp(\mathrm{Ad}\, B)\}(A) \tag{1.24}$$

or

$$C(t) = \int_0^t g\{[\exp(\mathrm{Ad}\, tA)] \exp(\mathrm{Ad}\, B)\}(A)\, dt + C_0, \tag{1.25}$$

where C_0 is a constant matrix. Since $C(0) = \ln e^B = B$, $C_0 = B$. Expression (1.19) follows from (1.25) by setting $t = 1$. Expression (1.20) is obtained by writing out the power series expansion of the integrand in (1.19) and integrating term by term. Q.E.D.

The preceding theorem will prove to be of great importance in the theory of Lie groups.

5.2 Local Lie Groups

Roughly speaking, a Lie group is an infinite group whose elements can be parametrized analytically. Thus, any group element g can be denoted $g(\lambda_1, \ldots, \lambda_n)$ in terms of parameters $\lambda_1, \ldots, \lambda_n$. The parameters of the product gg' are analytic functions of the parameters of g and g'. Using the notion of Lie group one can apply both calculus and algebra to group theory and obtain many results which neither discipline alone yields.

We denote by F the field of either the real numbers R or complex numbers $\mathfrak{C}$. Let F_n be the vector space of n-tuples $\mathbf{g} = (g_1, \ldots, g_n)$, $g_i \in F$, and let $\mathbf{e} = (0, \ldots, 0)$ be the zero vector in F_n. Suppose V is an open set in F_n containing $\mathbf{e}$. (We adopt the usual topology for n-tuples. The reader can choose V as an open sphere in n-space.)

Definition. An n-dimensional **local Lie group** G in the neighborhood $V \subseteq F_n$ is determined by a function $\boldsymbol{\varphi}(\mathbf{g}, \mathbf{h})$ with the following properties:

(1) $\boldsymbol{\varphi}(\mathbf{g}, \mathbf{h}) \in F_n$ for all $\mathbf{g}, \mathbf{h} \in V$.
(2) $\boldsymbol{\varphi}(\mathbf{g}, \mathbf{h})$ is an analytic function of each of its $2n$ arguments.
(3) If $\boldsymbol{\varphi}(\mathbf{g}, \mathbf{h}) \in V$ and $\boldsymbol{\varphi}(\mathbf{h}, \mathbf{k}) \in V$ then $\boldsymbol{\varphi}(\boldsymbol{\varphi}(\mathbf{g}, \mathbf{h}), \mathbf{k}) = \boldsymbol{\varphi}(\mathbf{g}, \boldsymbol{\varphi}(\mathbf{h}, \mathbf{k}))$.
(4) $\boldsymbol{\varphi}(\mathbf{e}, \mathbf{g}) = \boldsymbol{\varphi}(\mathbf{g}, \mathbf{e}) = \mathbf{g}$ for all $\mathbf{g} \in V$.

To make the above definition more palatable we write $\boldsymbol{\varphi}(\mathbf{g}, \mathbf{h}) = \mathbf{gh}$. [We interpret $\boldsymbol{\varphi}(\mathbf{g}, \mathbf{h})$ as the product of the group elements $\mathbf{g}$ and $\mathbf{h}$.] Then property (3) is easily recognized as the associative law $(\mathbf{gh})\mathbf{k} = \mathbf{g}(\mathbf{hk})$. Property (4) becomes $\mathbf{eg} = \mathbf{ge} = \mathbf{g}$ and shows that $\mathbf{e}$ is the identity element for group multiplication.

We have not required the existence of an inverse for a group element. However, this fact follows from properties (2)–(4). An inverse for $\mathbf{g} \in V$ would be a group element $\mathbf{x}$ such that $\boldsymbol{\varphi}(\mathbf{g}, \mathbf{x}) = \mathbf{gx} = \mathbf{e}$. Let $\boldsymbol{\varphi}_j(\mathbf{g}, \mathbf{x})$ be the components of the n-tuple $\boldsymbol{\varphi}(\mathbf{g}, \mathbf{x})$. Then from property (4) we have $[\partial\varphi_j/\partial x_i(\mathbf{e}, \mathbf{x})]|_{\mathbf{x}=\mathbf{e}} = \delta_{ij}$. It follows that the Jacobian $\det[\partial\varphi_j/\partial x_i(\mathbf{e}, \mathbf{e})] = 1$. Therefore the inverse function theorem (Apostol [1]) guarantees the existence of a unique solution $\mathbf{x} \in V$ of the equation $\boldsymbol{\varphi}(\mathbf{g}, \mathbf{x}) = \mathbf{e}$ for all $\mathbf{g}$ in some open set V_r, $(\mathbf{e} \in V_r \subseteq V)$. Furthermore the components of $\mathbf{x}$ are analytic functions of the parameters of $\mathbf{g}$. This proves the existence of a right inverse $\mathbf{x}_r$ of $\mathbf{g}$ for $\mathbf{g} \in V_r$. Similarly, we could find a unique left inverse $\mathbf{x}_l$ for each $\mathbf{g} \in V_l$, an open set containing $\mathbf{e}$. Let $V^{-1} = V_r \cap V_l$. Then V^{-1} is an open set containing $\mathbf{e}$ and for each $\mathbf{g} \in V^{-1}$ we have

$$\mathbf{x}_l = \mathbf{x}_l\mathbf{e} = \mathbf{x}_l(\mathbf{g}\mathbf{x}_r) = (\mathbf{x}_l\mathbf{g})\mathbf{x}_r = \mathbf{e}\mathbf{x}_r = \mathbf{x}_r$$

so the left and right inverses coincide. We write $\mathbf{x}_l = \mathbf{x}_r = \mathbf{g}^{-1}$.

By the implicit function theorem for $\mathbf{h} \in V$ the mapping $\mathbf{x} \longrightarrow \mathbf{hx}$, $\mathbf{x} \in U$, is a homeomorphism of a suitably small neighborhood U of $\mathbf{e}$ onto a neighborhood U' of $\mathbf{h}$. Indeed $U' = \mathbf{h}U$.

A local Lie group G is not necessarily a group in the sense of Chapter 1. Indeed the group axioms are satisfied only for elements in a sufficiently small neighborhood of $\mathbf{e}$. It may not even make sense to write $\mathbf{gh}$ for n-tuples $\mathbf{g}, \mathbf{h}$ not both in V. Furthermore an n-tuple $\mathbf{g} \in V$ will not in general have a unique inverse unless $\mathbf{g} \in V^{-1} \subseteq V$. It is useful (and correct) to think of a local Lie group as a neighborhood of the identity element of a (global) group.

If $F = R$, then G is a **real** local Lie group; if $F = \mathfrak{C}$, G is a **complex** group. For the time being we shall develop the theory of real and complex Lie groups simultaneously.

As an example of a local Lie group consider the matrix group $GL(m, \mathfrak{C})$. We can uniquely write any $m \times m$ matrix $\mathbf{g}$ in the form

$$\mathbf{g} = (g_{ij}) = (\delta_{ij} + x_{ij}), \qquad 1 \leq i, j \leq m$$

where δ_{ij} is the Kronecker delta. For all (x_{ij}) in a suitably small neighborhood of the zero matrix, $\det(\delta_{ij} + x_{ij}) \neq 0$, so $\mathbf{g} \in GL(m, \mathfrak{C})$ and we can parametrize the group using the m^2 parameters x_{ij}. It is easy to find a smaller neighborhood V of the zero matrix such that for $(x_{ij}), (x'_{ij}) \in V$ the matrix product $\mathbf{g}(x_{ij})\mathbf{g}'(x'_{ij})$ can be expressed as $\mathbf{g}''(x''_{ij})$, where the parameters x''_{ij} are analytic functions of x_{ij} and x'_{ij}. Finally, the identity element $\mathbf{e}$ of $GL(m, \mathfrak{C})$ corresponds to the parameters $x_{ij} = 0$. Thus, $GL(m, \mathfrak{C})$ is an

m^2-dimensional complex local Lie group in the neighborhood V. Similarly, $GL(m, R)$ is a real local Lie group.

Our definition of $GL(m, \mathfrak{C})$ as a local Lie group depends on the selection of coordinates x_{ij} and neighborhood V. If we adopt different coordinates or choose a different neighborhood then we obtain a different local Lie group. However, we will show that all such local Lie groups are essentially the same, i.e., they are locally isomorphic. For this reason it makes sense to speak of *the* local Lie group $GL(m, \mathfrak{C})$ without explicitly mentioning the choice of local coordinates.

Most of the local Lie groups encountered in applications are groups of matrices. It is convenient, therefore, to introduce the concept of a local linear Lie group. Let W be an open, connected set containing $\mathbf{e}$ in the space F_n of all n-tuples $\mathbf{g} = (g_1, \ldots, g_n)$. (Without loss of generality the reader can assume W is an open sphere with center $\mathbf{e}$.)

Definition. An n-dimensional **local linear Lie group** G is a set of $m \times m$ nonsingular matrices $A(\mathbf{g}) = A(g_1, \ldots, g_n)$, defined for each $\mathbf{g} \in W$, such that

(1) $A(\mathbf{e}) = E_m$ (the identity matrix).
(2) The matrix elements of $A(\mathbf{g})$ are analytic functions of the parameters $g_1, \ldots, g_n$ and the map $\mathbf{g} \longrightarrow A(\mathbf{g})$ is 1–1.
(3) The n matrices $\partial A(g)/\partial g_j$, $j = 1, \ldots, n$, are linearly independent for each $\mathbf{g} \in W$. That is, these matrices span an n-dimensional subspace of the m^2-dimensional space of all $m \times m$ matrices.
(4) There exists a neighborhood W' of $\mathbf{e}$ in F_n, $W' \subseteq W$, with the property that for every pair of n-tuples $\mathbf{g}, \mathbf{h}$ in W' there is an n-tuple $\mathbf{k}$ in W satisfying

$$A(\mathbf{g})A(\mathbf{h}) = A(\mathbf{k}) \tag{2.1}$$

where the operation on the left is matrix multiplication.

Every local linear Lie group G defines a local Lie group. Indeed we can identify the matrix $A(\mathbf{g})$ with the group element $\mathbf{g}$. It follows from (2), (3), and the implicit function theorem that the parameters g_i are analytic functions of the matrix elements of $A(\mathbf{g})$. Then (4) implies that there exists a nonzero neighborhood V of $\mathbf{e}$ such that $\mathbf{k} = \boldsymbol{\varphi}(\mathbf{g}, \mathbf{h})$ for all $\mathbf{g}, \mathbf{h} \in V$ where $\boldsymbol{\varphi}$ is an analytic vector-valued function of its $2n$ arguments and $\mathbf{g}, \mathbf{h}, \mathbf{k}$ are related by (2.1). (Clearly we can assume $V \subseteq W'$.) Thus,

$$A(\mathbf{g})A(\mathbf{h}) = A(\boldsymbol{\varphi}(\mathbf{g}, \mathbf{h})). \tag{2.2}$$

Since matrix multiplication is associative, $\boldsymbol{\varphi}(\mathbf{g}, \mathbf{h})$ satisfies the associative law. Finally, $A(\mathbf{e}) = E_m$, so $\boldsymbol{\varphi}(\mathbf{e}, \mathbf{g}) = \boldsymbol{\varphi}(\mathbf{g}, \mathbf{e}) = \mathbf{g}$ for all $\mathbf{g} \in V$. This proves that an n-dimensional local linear Lie group is an n-dimensional local Lie group.

It is now apparent that $GL(m, \mathfrak{C})$ is an m^2-dimensional local linear Lie group. However, $GL(m, \mathfrak{C})$ is not only a local Lie group, it is also a group in the abstract sense of Section 1.1, i.e., it is a global group. This remark leads us to the notion of a Lie group. A (global) **Lie group** G is (1) an abstract group and (2) an analytic manifold such that (3) group multiplication and group inversion are analytic with respect to the manifold structure. Unfortunately, it would take many pages to clarify this definition, particularly property (2). Rather than embark on such a topological digression, we refer the interested reader to Helgason [1] or Hausner and Schwartz [1], and merely show how one can construct a (global) **linear Lie group** from a local linear Lie group. Since the vast majority of Lie groups occurring in physics are linear Lie groups this simplification of the theory is worthwhile.

Let G be a local linear Lie group consisting of $m \times m$ matrices. We will define the (connected) global linear Lie group $\tilde{G}$ containing G. Algebraically, $\tilde{G}$ is the abstract subgroup of $GL(m, \mathfrak{C})$ generated by the matrices of G. That is, $\tilde{G}$ consists of all possible products of finite sequences of elements in G. In addition, the elements of $\tilde{G}$ can be parametrized analytically. If $B \in \tilde{G}$ we can introduce coordinates in a neighborhood of B by means of the map $g \longrightarrow BA(\mathbf{g})$ where $\mathbf{g}$ ranges over a suitably small neighborhood Z of $\mathbf{e}$ in F_n. In particular, the coordinates of B will be $\mathbf{e} = (0, \ldots, 0)$. Proceeding in this way for each $B \in \tilde{G}$ we can cover $\tilde{G}$ with local coordinate systems or "coordinate patches." The same group element C will have many different sets of coordinates depending on which coordinate patch containing C we happen to consider. Suppose C lies in the intersection of coordinate patches around B_1 and B_2, respectively. Then C will have coordinates $\mathbf{g}_1, \mathbf{g}_2$, respectively, where $C = B_1A(\mathbf{g}_1) = B_2A(\mathbf{g}_2)$. Since

$$A(\mathbf{g}_1) = B_1^{-1}B_2A(\mathbf{g}_2), \qquad A(\mathbf{g}_2) = B_2^{-1}B_1A(\mathbf{g}_1)$$

it follows that in a suitably small neighborhood of $\mathbf{e}$ the coordinates $\mathbf{g}_2$ are analytic functions $\mathbf{g}_2 = \boldsymbol{\rho}(\mathbf{g}_1)$ of the coordinates $\mathbf{g}_1$, $\boldsymbol{\rho}$ is 1–1, and the Jacobian of the coordinate transformation is nonzero (Freudenthal and De Vries [1], Cohn [1]). Thus, every element of $\tilde{G}$ is covered by coordinate neighborhoods and the transformation relating any two overlapping neighborhoods is (a) 1–1, (b) analytic, and (c) has a nonzero Jacobian. This makes $\tilde{G}$ into an analytic manifold. In addition to the coordinate neighborhoods described above we can always add more coordinate neighborhoods to $\tilde{G}$ provided they satisfy conditions (a)–(c) on the overlap with any of the original coordinate systems. We leave it to the reader to show that the Lie group $\tilde{G}$ is **connected.** That is, any two elements A, B in $\tilde{G}$ can be connected by an analytic curve $C(t)$ lying entirely in $\tilde{G}$. Here t is a real parameter, $0 \leq t \leq 1$, $C(0) = A$, $C(1) = B$, and whenever $C(t) \in \tilde{G}$ lies in a coordinate patch, its coordinates $\mathbf{g}(t)$ are analytic functions of t. (It is enough to show that any element A in $\tilde{G}$ can be connected to the identity by an analytic curve in $\tilde{G}$. For, if A is connected to E_m and E_m is connected to B, then the composite curve connects A to B.)

In general an n-dimensional (global) **linear Lie group** K is an abstract matrix group which is also an n-dimensional local linear group. A linear Lie group need not be connected. Clearly $GL(m, \mathfrak{C})$ is an m^2-dimensional Lie group. A single coordinate patch suffices to cover $GL(m, \mathfrak{C})$, but this is not true for other Lie groups such as $O(m)$. Furthermore, we shall show later that $GL(m, \mathfrak{C})$ is connected, but $O(m)$ is not.

5.3 Lie Algebras

Let G be a local Lie group defined in the neighborhood $V \subseteq F_n$. An **analytic curve through the identity** on G is a mapping $t \longrightarrow \mathbf{g}(t) = (g_1(t), \ldots, g_n(t))$ of a neighborhood of $0 \in F$ into V such that $\mathbf{g}(0) = \mathbf{e}$ and the $g_j(t)$ are analytic in t. The **tangent vector** to $\mathbf{g}(t)$ at $\mathbf{e}$ is the vector

$$\alpha = [d\mathbf{g}(t)/dt]\,|_{t=0} = (\dot{g}_1(0), \ldots, \dot{g}_n(0)) \in F_n. \tag{3.1}$$

Every vector in F_n is the tangent vector at $\mathbf{e}$ for some analytic curve. Indeed, the curve

$$\alpha t = (\alpha_1 t, \alpha_2 t, \ldots, \alpha_n t) \tag{3.2}$$

clearly has the tangent vector $\alpha = (\alpha_1, \ldots, \alpha_n)$ at $\mathbf{e}$, so we can identify the set of tangent vectors with F_n. In particular the tangent vectors at $\mathbf{e}$ form an n-dimensional vector space.

For $\mathbf{g}, \mathbf{h} \in V$ we have $\mathbf{gh} = \boldsymbol{\varphi}(\mathbf{g}, \mathbf{h})$, where $\boldsymbol{\varphi}$ is an analytic vector-valued function of its $2n$ arguments. Thus, the components $(\mathbf{gh})_j = \varphi_j(\mathbf{g}, \mathbf{h})$ can be expressed as Taylor series in g_l, h_l about $\mathbf{g} = \mathbf{h} = \mathbf{e}$. Since $\boldsymbol{\varphi}(\mathbf{e}, \mathbf{g}) = \boldsymbol{\varphi}(\mathbf{g}, \mathbf{e}) = \mathbf{g}$ there follows

$$\varphi_j(\mathbf{g}, \mathbf{h}) = g_j + h_j + \sum_{l,s=1}^{n} c_{j,ls} g_l h_s + r_j(\mathbf{g}, \mathbf{h}), \tag{3.3}$$

where r_j consists of terms of order greater than two in g_l, h_l and

$$c_{j,ls} = (\partial^2/\partial g_l\, \partial h_s)\varphi_j(\mathbf{g}, \mathbf{h})\,|_{\mathbf{g}=\mathbf{h}=\mathbf{e}}. \tag{3.4}$$

We now examine the relationship between analytic curves through $\mathbf{e}$ and vector space operations on tangent vectors. Let $\mathbf{g}(t)$, $\mathbf{h}(t)$ be analytic curves through $\mathbf{e}$ with tangent vectors α, β, respectively. Then for any constants $a, b \in F$, $\mathbf{k}(t) = \mathbf{g}(at)\mathbf{h}(bt) = \boldsymbol{\varphi}(\mathbf{g}(at), \mathbf{h}(bt))$ is also an analytic curve through $\mathbf{e}$. The tangent vector $\gamma = \dot{\mathbf{k}}(0)$ can be obtained by differentiating both sides of (3.3): $\gamma = a\alpha + b\beta$. [Note that only the first two terms on the right-hand side of (3.3) contribute to the result.] Here the plus sign $(+)$ refers to vector addition in F_n.

The curve $\mathbf{e}(t) = \mathbf{e}$ obviously has tangent vector $\theta = (0, \ldots, 0)$ at $\mathbf{e}$.

Let α' be the tangent vector at $\mathbf{e}$ of the analytic curve $\mathbf{g}^{-1}(t)$. Since $\mathbf{e}(t) = \mathbf{g}(t)\mathbf{g}^{-1}(t)$ it follows that $\theta = \alpha + \alpha'$, or $\alpha' = -\alpha$.

The preceding results use only the first-order terms in the expansion for $\boldsymbol{\varphi}(\mathbf{g}, \mathbf{h})$. We now introduce an operation on tangent vectors at $\mathbf{e}$ which depends on the second-order terms. With $\mathbf{g}(t)$, $\mathbf{h}(t)$ as given above, we define the **commutator** $[\alpha, \beta]$ of α and β as the tangent vector at $\mathbf{e}$ of the curve

$$\mathbf{k}(t) = \mathbf{g}(\tau)\mathbf{h}(\tau)\mathbf{g}^{-1}(\tau)\mathbf{h}^{-1}(\tau), \qquad t = \tau^2. \tag{3.5}$$

Thus,

$$[\alpha, \beta] = [d/d(\tau^2)]\mathbf{g}(\tau)\mathbf{h}(\tau)\mathbf{g}^{-1}(\tau)\mathbf{h}^{-1}(\tau)\,|_{\tau=0}, \tag{3.6}$$

or more precisely, $[\alpha, \beta]$ is the coefficient of τ^2 in the Taylor series expansion for $\mathbf{k}$. (This definition makes sense even if $\mathbf{k}$ is not an analytic function of τ^2, because the coefficient of τ is zero.)

Theorem 5.6. $[\alpha, \beta]_j = \sum_{l,s=1}^{n} c_j^{ls}\alpha_l\beta_s$, $1 \leq j \leq n$, where $c_j^{ls} = c_{j,ls} - c_{j,sl}$.

Proof. We write $g_j(\tau)$, $h_j(\tau)$, $k_j(t)$ as Taylor series,

$$g_j(\tau) = \alpha_j\tau + b_j\tau^2 + \cdots, \qquad h_j(\tau) = \beta_j\tau + c_j\tau^2 + \cdots,$$
$$k_j(t) = \rho_j\tau + a_j\tau^2 + \cdots,$$

where the omitted terms are of order greater than two. Clearly $[\alpha, \beta]_j = a_j$ if $\rho_j = 0$. We will verify that $\rho_j = 0$. Since $\mathbf{k}(t)\mathbf{h}(\tau)\mathbf{g}(\tau) = \mathbf{g}(\tau)\mathbf{h}(\tau)$ we find by (3.3)

$$(\rho_j + \beta_j + \alpha_j)\tau + [a_j + c_j + b_j + \sum c_{j,ls}(\beta_l\alpha_s + \rho_l\alpha_s + \rho_l\beta_s)]\tau^2$$
$$= (\beta_j + \alpha_j)\tau + (c_j + b_j + \sum c_{j,ls}\alpha_l\beta_s)\tau^2,$$

where only the terms of order less than three are written explicitly. Comparing first-order terms in τ on both sides of the identity, we obtain $\rho_j = 0$. A comparison of second-order terms yields $a_j = \sum (c_{j,ls} - c_{j,sl})\alpha_l\beta_s$. Q.E.D.

The c_j^{ls} are called the **structure constants** of G. A knowledge of the structure constants yields information about the second-order terms in the expansion of $\boldsymbol{\varphi}(\mathbf{g}, \mathbf{h})$. It is, however, a remarkable fact that one can essentially determine the function $\boldsymbol{\varphi}(\mathbf{g}, \mathbf{h})$ from the structure constants, i.e., all information conveyed by the local Lie group G is already contained in the c_j^{ls}. For example, if G is abelian ($\mathbf{gh} = \mathbf{hg}$ for all $\mathbf{g}, \mathbf{h} \in V$) then it follows from (3.6) that $[\alpha, \beta] = \theta$ for all tangent vectors α, β at $\mathbf{e}$, and the structure constants are all zero. Conversely, we shall show that if the structure constants are zero then G is necessarily abelian.

The basic properties of the commutator are given by the following theorem.

Theorem 5.7. Let $\alpha, \beta, \gamma \in F_n$ and $a, b \in F$.

(1) $[\alpha, \beta] = -[\beta, \alpha]$.
(2) $[a\alpha + b\beta, \gamma] = a[\alpha, \gamma] + b[\beta, \gamma]$,
(3) $[[\alpha, \beta], \gamma] + [[\gamma, \alpha], \beta] + [[\beta, \gamma], \alpha] = \theta$ (Jacobi equality).

Proof. Properties (1) and (2) follow directly from the expression for the commutator in Theorem 5.6. The Jacobi equality is an expression of the associative law $(\mathbf{gh})\mathbf{k} = \mathbf{g}(\mathbf{hk})$ for local Lie groups. It is an identity for the structure constants which is obtained by substituting (3.3) into the associative law and equating terms of the same order. Q.E.D.

Definition. The **Lie algebra** $L(G)$ of a local Lie group G is the set of all tangent vectors at $\mathbf{e}$ equipped with the operations of scalar multiplication, vector addition, and commutator product.

Definition. An **abstract Lie algebra** $\mathcal{G}$ over F is a vector space over F together with a product $[\alpha, \beta] \in \mathcal{G}$ defined for all $\alpha, \beta \in \mathcal{G}$ such that for all $\alpha, \beta, \gamma \in \mathcal{G}$ and $a, b \in F$ (θ is the zero vector):

(1) $[\alpha, \beta] = -[\beta, \alpha]$.
(2) $[a\alpha + b\beta, \gamma] = a[\alpha, \gamma] + b[\beta, \gamma]$.
(3) $[[\alpha, \beta], \gamma] + [[\gamma, \alpha], \beta] + [[\beta, \gamma], \alpha] = \theta$.

Property (1) expresses the skew-symmetry of the commutator, while (1) and (2) imply that $[\alpha, \beta]$ is bilinear. The identity (3) is again called the Jacobi equality. We will always assume that $\mathcal{G}$ is a finite-dimensional vector space.

Clearly, $L(G)$ is an abstract Lie algebra. It is not clear (but true) that any abstract Lie algebra is in fact the Lie algebra of some local Lie group. The exact meaning of this statement will be clarified later.

The significance of our results concerning Lie algebras and commutators becomes much clearer when we consider linear Lie groups. Let G be an n-dimensional local linear Lie group of $m \times m$ matrices and let $A(t) = A(\mathbf{g}(t))$, $A(0) = E_m$, be an analytic curve through the identity. We can identify the tangent vector α at $\mathbf{e}$ with the matrix

$$(3.7)\qquad \mathfrak{a} = \frac{d}{dt} A(\mathbf{g}(t))\Big|_{t=0} = \sum_{j=1}^{n} \frac{\partial}{\partial g_j} A(\mathbf{g})\Big|_{\mathbf{g}=\mathbf{e}} \dot{g}_j(0) = \sum_{j=1}^{n} \mathfrak{C}_j \alpha_j, \qquad \mathfrak{C}_j = \frac{\partial}{\partial g_j} A(\mathbf{g})\Big|_{\mathbf{g}=\mathbf{e}}.$$

Thus, we identify the tangent space at $\mathbf{e}$ with the space of all $m \times m$ matrices of the form (3.7). From the definition of linear Lie groups, the set $\{\mathfrak{C}_j\}$ is linearly independent and spans an n-dimensional subspace of the space of all $m \times m$ matrices. Since the vectors α are arbitrary we see again that the tangent space is n-dimensional.

Many of the properties which were obtained for general local Lie groups are much simpler to verify for local linear Lie groups. For example, let $A(t)$, $B(t)$ be analytic curves in G with tangent matrices $\mathcal{A}$, $\mathcal{B}$ at the identity. Since

$$(d/dt)(A(t)B(t)) = \dot{A}(t)B(t) + A(t)\dot{B}(t) \tag{3.8}$$

and $A(0) = B(0) = E_m$, it follows that $A(t)B(t)$ is an analytic curve with tangent matrix $\mathcal{A} + \mathcal{B}$ at E_m.

With $A(t)$, $B(t)$ as above we define the **commutator** of the tangent matrices $\mathcal{A}$, $\mathcal{B}$ by

$$[\mathcal{A}, \mathcal{B}] = (d/dt)[A(\tau)B(\tau)A^{-1}(\tau)B^{-1}(\tau)]_{t=0}, \tag{3.9}$$

where $t = \tau^2$, i.e,. $[\mathcal{A}, \mathcal{B}]$ is the tangent matrix at the identity of the curve $C(t) = A(\tau)B(\tau)A^{-1}(\tau)B^{-1}(\tau)$.

Theorem 5.8. $[\mathcal{A}, \mathcal{B}] = \mathcal{A}\mathcal{B} - \mathcal{B}\mathcal{A} \in L(G)$.

Proof. We copy the proof of Theorem 5.6 taking advantage of the simplifications afforded by matrix groups. Expanding $A(\tau)$, $B(\tau)$, and $C(t)$ as Taylor series in τ we find

$$A(\tau) = E_m + \mathcal{A}\tau + \mathcal{A}'\tau^2 + \cdots, \qquad B(\tau) = E_m + \mathcal{B}\tau + \mathcal{B}'\tau^2 + \cdots,$$
$$C(t) = E_m + \mathcal{C}\tau + \mathcal{C}'\tau^2 + \cdots.$$

We will show that $\mathcal{C} = Z$, so (3.9) is well-defined and $\mathcal{C}' = [\mathcal{A}, \mathcal{B}]$. Since $C(t)B(\tau)A(\tau) = A(\tau)B(\tau)$ there follows the identity

$$E_m + (\mathcal{A} + \mathcal{B} + \mathcal{C})\tau + (\mathcal{A}' + \mathcal{B}' + \mathcal{C}' + \mathcal{B}\mathcal{A} + \mathcal{C}\mathcal{B} + \mathcal{C}\mathcal{A})\tau^2 + \cdots$$
$$= E_m + (\mathcal{A} + \mathcal{B})\tau + (\mathcal{A}' + \mathcal{B}' + \mathcal{A}\mathcal{B})\tau^2 + \cdots.$$

Equating terms in τ and τ^2 we find $\mathcal{C} = Z$ and $\mathcal{C}' = \mathcal{A}\mathcal{B} - \mathcal{B}\mathcal{A}$. Q.E.D.

Clearly, $L(G)$ cannot be an arbitrary subspace of matrices, for we must have $[\mathcal{A}, \mathcal{B}] \in L(G)$ for all $\mathcal{A}, \mathcal{B} \in L(G)$. This requirement greatly restricts the possibilities for matrix Lie algebras.

Using Theorem 5.6 and (3.7), we can relate the structure constants c_j^{ls} of G to the matrix commutator:

$$[\mathcal{C}_l, \mathcal{C}_s] = \sum_{j=1}^{n} c_j^{ls}\mathcal{C}_j, \qquad 1 \le l, s \le n. \tag{3.10}$$

It follows directly from Theorem 5.7 that the matrix commutator is skew-symmetric, bilinear, and satisfies the Jacobi equality. However, it is worthwhile to verify these properties directly from the definition $[\mathcal{A}, \mathcal{B}] = \mathcal{A}\mathcal{B} - \mathcal{B}\mathcal{A}$. Skew-symmetry and bilinearity are trivial to check. Moreover, as the reader can easily show, the matrix commutator automatically satisfies the Jacobi equality. Thus, a set $\mathcal{G}$ of $m \times m$ matrices which is closed under addition of matrices, scalar multiplication, and matrix commutation is necessarily a Lie algebra. It is not necessary to check the Jacobi equality. (Note: We do not yet know that $\mathcal{G}$ is actually the Lie algebra of some local Lie group.) This fact makes matrix algebras much easier to study than general abstract Lie algebras.

Fortunately, we can restrict ourselves to the study of matrix Lie algebras with no loss of generality. To show this we need the concepts of homomorphism and isomorphism of Lie algebras. Let $\mathcal{G}$, $\mathcal{G}'$ be abstract Lie algebras over F with operations $+$, $[-,-]$ and $+'$, $[-,-]'$, respectively.

Definition. A **homomorphism** from $\mathcal{G}$ to $\mathcal{G}'$ is a map $\tau: \mathcal{G} \longrightarrow \mathcal{G}'$ such that

(1) $\tau(a\alpha + b\beta) = a\tau(\alpha) + b\tau(\beta)$.
(2) $\tau([\alpha, \beta]) = [\tau(\alpha), \tau(\beta)]'$, $\quad a, b \in F, \quad \alpha, \beta \in \mathcal{G}$.

A Lie algebra homomorphism which is a 1–1 map of $\mathcal{G}$ onto $\mathcal{G}'$ is an **isomorphism.** An isomorphism of $\mathcal{G}$ onto $\mathcal{G}$ is an **automorphism.**

Note that a homomorphism τ is a linear mapping of $\mathcal{G}$ into $\mathcal{G}'$ which preserves commutators. We can identify isomorphic Lie algebras $\mathcal{G}$, $\mathcal{G}'$ since they have the same structure as abstract Lie algebras.

Theorem 5.9 (Ado). Every abstract Lie algebra is isomorphic to a matrix Lie algebra.

The proof of Ado's Theorem is difficult and will not be given here (see Jacobson [1]). Moreover, there is no known general method to construct a matrix Lie algebra isomorphic to any given abstract Lie algebra, even though such a matrix algebra must exist. We quote Ado's theorem as moral support. Even though we limit ourselves to matrix Lie algebras in the following sections, we will actually be studying all Lie algebras (up to isomorphism).

Definition. A (local) **analytic homomorphism** of a local Lie group G into a local Lie group G' is a map $\mu: G \longrightarrow G'$, where $\mu(\mathbf{g})$ is defined for $\mathbf{g}$ in a suitably small neighborhood W of $\mathbf{e}$ such that

$$\mu(\mathbf{gh}) = \mu(\mathbf{g})\mu(\mathbf{h}), \qquad \mathbf{g}, \mathbf{h}, \mathbf{gh} \in G, \tag{3.11}$$

and μ is an analytic function of the coordinates of G. The group multiplication on the right-hand side of (3.11) takes place in G'. If μ is a 1–1 homomorphism of G onto a neighborhood of $\mathbf{e}'$ in G' such that the mapping $\mu^{-1}: G' \longrightarrow G$ is analytic, then μ is called an **isomorphism** and G is (locally) isomorphic to G'. An isomorphism of G onto G is an **automorphism.**

It follows from (3.11) that $\mu(\mathbf{e}) = \mathbf{e}'$, where $\mathbf{e}'$ is the identity element of G', and $\mu(\mathbf{g}^{-1}) = \mu(\mathbf{g})^{-1}$ for $\mathbf{g}$ in a suitably small neighborhood of $\mathbf{e}$. Note that μ is purely local. Even if G and G' are (global) Lie groups, it is not necessary that μ be defined on all of G. Furthermore, μ need not be an abstract group homomorphism of G in the sense of Chapter 1, since (3.11) is required to hold only in a neighborhood of $\mathbf{e}$. As usual, we identify isomorphic local Lie groups.

In Section 5.5 it will be shown that two local Lie groups are isomorphic if and only if they have isomorphic Lie algebras.

5.4 The Classical Groups

Here we introduce the **classical groups,** a family of linear Lie groups which is of fundamental importance in physics and geometry. The reader should note the significant role of Theorem 5.3 in the following analysis. Among the classical groups are the following.

$\boldsymbol{G_m = GL(m, \mathfrak{C})}$. We have already seen that G_m is an m^2-dimensional complex Lie group. (If we write the complex coordinates as sums of real and imaginary parts we can also consider G_m as a $2m^2$-dimensional real Lie group.) The Lie algebra of this group consists of all matrices of the form

$$\mathfrak{A} = (d/dt)A(t)|_{t=0}, \qquad A(0) = E_m,$$

where $A(t)$ is an analytic curve in G_m. Clearly $\mathfrak{A}$ is an $m \times m$ matrix. Conversely, if $\mathfrak{A}$ is any $m \times m$ matrix then $A(t) = \exp t\mathfrak{A}$ is an analytic curve in G_m with tangent matrix $\mathfrak{A}$. [See expression (1.14).] Thus, $L(G_m)$ is the m^2-dimensional space of all $m \times m$ matrices. There is an analogous result for the real Lie group $GL(m, R)$.

$\boldsymbol{SL(m) = SL(m, \mathfrak{C})}$. The special linear group

$$SL(m) = \{A \in GL(m, \mathfrak{C}): \det A = 1\}$$

is an $(m^2 - 1)$-dimensional complex Lie group. Indeed, by Theorem 5.3 and Corollary 5.3, every $A \in SL(m)$ sufficiently close to the identity can be written uniquely in the form $A = \exp \mathfrak{A}$, where $\mathfrak{A}$ is an $m \times m$ matrix with $\operatorname{tr} \mathfrak{A} = 0$. It is easy to see that $\mathfrak{A}$ has $m^2 - 1$ independent parameters which can be used as local coordinates for $SL(m)$. With these coordinates $SL(m)$ is a complex linear Lie group. [If we write the matrix elements of $\mathfrak{A}$ in terms of real and imaginary parts we can also consider $SL(m)$ as a $2(m^2 - 1)$-dimensional real Lie group.]

Theorem 5.10. The Lie algebra of $SL(m)$ is the space of all $m \times m$ matrices of trace zero.

Proof. Let $A(t)$, $A(0) = E_m$, be an analytic curve in $SL(m)$. For t sufficiently close to zero we can write $A(t) = \exp \mathfrak{A}(t)$, $\mathfrak{A}(0) = Z$, where $\mathfrak{A}(t)$ is an analytic curve in the space of traceless matrices. Let $\mathcal{B} = \dot{\mathcal{A}}(0)$. Since $d/dt(\operatorname{tr} \mathcal{A}(t)) = \operatorname{tr} \dot{\mathcal{A}}(t)$ we have $\operatorname{tr} \mathcal{B} = 0$. A simple computation shows that $\mathcal{B}$ is the tangent matrix to $A(t)$ at E_m.

Conversely, let $\mathcal{B}$ be an $m \times m$ matrix with $\operatorname{tr} \mathcal{B} = 0$ and consider the analytic curve $A(t) = \exp t\mathcal{B}$. Since $\det A(t) = \exp(\operatorname{tr} t\mathcal{B}) = 1$, the curve $A(t)$ lies in $SL(m)$. Furthermore $A(0) = E_m$, and the tangent matrix to $A(t)$ at E_m is $\dot{A}(0) = \mathcal{B}$, so $\mathcal{B}$ is an element of the Lie algebra. Q.E.D.

There is a similar result for the real Lie groups $SL(m, R)$.

O(m) = O(m, ℭ). Recall that $O(m)$ is the group of all complex matrices A such that $A^t A = E_m$. For each $A \in O(m)$ sufficiently close to E_m we can find a unique matrix $\mathcal{A}$ sufficiently close to Z such that $A = \exp \mathcal{A}$. Since $A \in O(m)$, we have $(\exp \mathcal{A}^t)(\exp \mathcal{A}) = E_m = \exp Z$. For $\mathcal{A}$ sufficiently close to Z this is possible only if $\mathcal{A}^t + \mathcal{A} = Z$, i.e., only if $\mathcal{A}$ is skew-symmetric. Conversely, if $\mathcal{A}$ is skew-symmetric and $A = \exp \mathcal{A}$ then $A^t A = \exp \mathcal{A}^t \cdot \exp \mathcal{A} = E_m$, so $A \in O(m)$. We can use the independent matrix elements of $\mathcal{A}$ as local coordinates for $O(m)$. As the reader can check, there are $m(m-1)/2$ independent matrix elements in an $m \times m$ skew-symmetric matrix, so $O(m)$ is an $m(m-1)/2$-dimensional complex Lie group, [or an $(m^2 - m)$-dimensional real Lie group].

Theorem 5.11. The Lie algebra of $O(m)$ is the space of all $m \times m$ skew-symmetric matrices.

Proof. Let $A(t)$, $A(0) = E_m$, be an analytic curve in $O(m)$ with tangent matrix $\mathcal{A}$ at the identity. Differentiating the expression $A^t(t)A(t) = E_m$ and setting $t = 0$ we find $\mathcal{A}^t + \mathcal{A} = Z$. Thus, the elements of the Lie algebra are skew-symmetric matrices.

Conversely, let $\mathcal{A}$ be a skew-symmetric matrix. Then $A(t) = \exp t\mathcal{A}$ is an analytic curve in $O(m)$ with tangent matrix $\dot{A}(0) = \mathcal{A}$, so $\mathcal{A} \in L(O(m))$. Q.E.D.

There is a similar result for $O(m, R)$.

Note that the elements of $O(m)$ which lie in a small neighborhood of the identity all have determinant $+1$. Indeed if $A = \exp \mathcal{A}$ with $\mathcal{A}^t = -\mathcal{A}$ then $\det A = \exp(\operatorname{tr} \mathcal{A}) = +1$ since the trace of a skew-symmetric matrix is zero. Thus, the Lie algebra of $O(m)$ is exactly the same as the Lie algebra of $SO(m) = SO(m, ℭ)$:

$$SO(m) = \{A \in O(m): \det A = +1\}.$$

We shall see later that $SO(m)$ is a connected Lie group. In particular, every $A \in SO(m)$ can be connected to the identity by an analytic curve $A(t)$,

$A(0) = E_m$, $A(1) = A$, $0 \le t \le 1$, which lies entirely in $SO(m)$. However, $O(m)$ is not connected. If it were, we could find an analytic curve $A(t)$ in $O(m)$ such that $A(0) = E_m$ and $A(1) = -E_m$. Then for m odd, $\det A(t)$ would be an analytic function of t taking only the values ± 1 with $\det A(0) = -\det A(1) = 1$. No such function exists. Thus, $O(m)$ consists of two connected components, the cosets $SO(m)$ and $-E_m \cdot SO(m)$. We leave to the reader a verification that $O(m)$ also has two connected components for m even.

***U*(*m*).** The **unitary group** is the set

$$U(m) = \{A \in GL(m, \mathfrak{C}): \bar{A}^t A = E_m\}$$

of all $m \times m$ unitary matrices. Given $A \in U(m)$ in a sufficiently small neighborhood of E_m we can find a unique matrix $\mathfrak{A}$ in a sufficiently small neighborhood of Z such that $A = \exp \mathfrak{A}$. The requirement $\bar{A}^t A = E_m$ implies $\exp \bar{\mathfrak{A}}^t \exp \mathfrak{A} = E_m$ or $\exp \bar{\mathfrak{A}}^t = \exp(-\mathfrak{A})$. For $\mathfrak{A}$ sufficiently close to Z this implies $\bar{\mathfrak{A}}^t = -\mathfrak{A}$, i.e., $\mathfrak{A}$ is skew-Hermitian. On the other hand if $\mathfrak{A}$ is a skew-Hermitian matrix and $A = \exp \mathfrak{A}$ then $\bar{A}^t A = \exp \bar{\mathfrak{A}} \exp \mathfrak{A} = E_m$, so $A \in U(m)$. Thus we can use the matrix elements of $\mathfrak{A}$ to provide local coordinates in $U(m)$. As the reader can check, an $m \times m$ skew-Hermitian matrix has m^2 *real* independent parameters. (Write the matrix elements of $\mathfrak{A}$ in terms of real and imaginary parts.) Using these parameters as local coordinates we see that $U(m)$ is an m^2-dimensional *real* Lie group. Here $U(m)$ is a real Lie group even though it consists of complex matrices.

Theorem 5.12. The Lie algebra of $U(m)$ is the space of all $m \times m$ skew-Hermitian matrices.

Proof. Analogous to Theorem 5.11.

If $\mathfrak{A}$ is skew-Hermitian then the matrix $\mathfrak{H} = -i\mathfrak{A}$ is Hermitian, i.e., $\bar{\mathfrak{H}}^t = \mathfrak{H}$. Thus, any unitary matrix A sufficiently close to the identity can be expressed $A = \exp(i\mathfrak{H})$, where $\mathfrak{H}$ is Hermitian. It is of considerable interest that such an expression is valid for all unitary matrices A, not just those sufficiently close to the identity.

Theorem 5.13. If $A \in U(m)$ there exists a Hermitian matrix $\mathfrak{H}$ (not unique) such that $A = \exp(i\mathfrak{H})$.

Proof. We know from linear algebra that the eigenvalues ε_j of A take the form $\varepsilon_j = e^{i\theta_j}$, θ_j real, $j = 1, \ldots, m$, and that A can be diagonalized by a unitary similarity transformation. Thus there exists $B \in U(m)$ and a diagonal matrix D with diagonal entries, $e^{i\theta_1}, \ldots, e^{i\theta_m}$, such that $BAB^{-1} = D$ (Finkbeiner [1]). Clearly, $D = \exp(i\mathfrak{D})$, where $\mathfrak{D}$ is the diagonal matrix with matrix elements $\theta_1, \ldots, \theta_m$ down the diagonal. Thus $A = B^{-1}DB = \exp(i\mathcal{H})$, where $\mathcal{H} = B^{-1}\mathfrak{D}B$. The matrix $\mathcal{H}$ is Hermitian because $\bar{\mathcal{H}}^t = \bar{B}^t\bar{\mathfrak{D}}^t(\bar{B}^t)^{-1} = B^{-1}\mathfrak{D}B = \mathcal{H}$. Q.E.D.

We can use the above result to show that $U(m)$ is connected. Indeed, for any $A \in U(m)$ the matrix function $A(t) = \exp(it\mathcal{H})$, $A(1) = A$, $0 \le t \le 1$, defines an analytic curve in $U(m)$ connecting A to E_m.

SU(m). The **special unitary group** is the set

$$SU(m) = \{A \in U(m): \det A = +1\},$$

a subgroup of $U(m)$. Recall that $|\det A| = 1$ for $A \in U(m)$. It is left to the reader to show that $SU(m)$ is an $(m^2 - 1)$-dimensional real Lie group whose Lie algebra consists of all skew-Hermitian matrices with trace zero.

Sp(m). Let J be the $2m \times 2m$ skew-symmetric matrix

$$J = \begin{pmatrix} Z & E_m \\ -E_m & Z \end{pmatrix}. \tag{4.1}$$

Definition. A **symplectic** matrix is a $2m \times 2m$ complex matrix A such that $A^tJA = J$. The symplectic group $Sp(m) = Sp(m, \mathfrak{C})$ is the set of all $2m \times 2m$ symplectic matrices.

Theorem 5.14. *Sp(m) is a matrix group.*

Proof. Clearly the identity matrix belongs to $Sp(m)$. Suppose $A, B \in Sp(m)$. Then $(AB)^tJ(AB) = B^t(A^tJA)B = B^tJB = J$, so $AB \in Sp(m)$. Thus, we need only show that A^{-1} belongs to $Sp(m)$ for each $A \in Sp(m)$. Taking the determinant of both sides of the identity $A^tJA = J$ and using the fact $\det J = 1$, we obtain $(\det A)^2 = 1$, so A and A^t are invertible. Then $(A^{-1})^t JA^{-1} = (A^t)^{-1}(A^tJA)A^{-1} = J$ and $A^{-1} \in Sp(m)$. Q.E.D.

Since $J^{-1} = -J$ the relation $A^tJA = J$ implies $A^{-1} = -JA^tJ$.

For $A \in Sp(m)$ in a sufficiently small neighborhood of E_{2m} we can find a unique matrix $\mathcal{A}$ in a sufficiently small neighborhood of Z such that $A = \exp \mathcal{A}$. Since $A^{-1} = J^{-1}A^tJ$ it follows that $\exp(-\mathcal{A}) = J^{-1}(\exp \mathcal{A}^t)J = \exp(J^{-1}\mathcal{A}^tJ)$, or $-\mathcal{A} = J^{-1}\mathcal{A}^tJ$ for $\mathcal{A}$ in a sufficiently small neighborhood of Z. This result is usually expressed in the form $J\mathcal{A} + \mathcal{A}^tJ = Z$. Conversely, if $A = \exp \mathcal{A}$, where $\mathcal{A}$ satisfies the preceding equality, then it is easy to show that $A \in Sp(m)$. Thus, $Sp(m)$ is a complex Lie group with the independent matrix elements of $\mathcal{A}$ as parameters. As the reader can check, there are

$2m^2 + m$ independent parameters. Note that we can also consider $Sp(m)$ as a $(4m^2 + 2m)$-dimensional real Lie group.

Theorem 5.15. The Lie algebra of $Sp(m)$ is the space of all $2m \times 2m$ matrices $\mathcal{A}$ such that $J\mathcal{A} + \mathcal{A}^t J = Z$.

Proof. Analogous to Theorem 5.11.

In each of the above examples, every element A of the Lie group G sufficiently close to the identity can be expressed uniquely in the form $A = \exp \mathcal{A}$, where $\mathcal{A}$ is an element of the corresponding Lie algebra $L(G)$ sufficiently close to Z. Thus the structure of G as a local Lie group is uniquely determined by $L(G)$. In the next section we shall show that this is a general property of local Lie groups: G is uniquely determined by $L(G)$. Thus, all questions concerning the structure of G can be reduced to purely algebraic questions about the structure of the Lie algebra $L(G)$.

5.5 The Exponential Map of a Lie Algebra

Let G be a local Lie group, not necessarily a matrix group. If $\mathbf{g}, \mathbf{h} \in V$ then

$$\mathbf{gh} = \boldsymbol{\varphi}(\mathbf{g}, \mathbf{h}) = (\varphi_1(\mathbf{g}, \mathbf{h}), \ldots, \varphi_n(\mathbf{g}, \mathbf{h}))$$

is an analytic function of $\mathbf{g}$ and $\mathbf{h}$. For fixed $\mathbf{h}$ we expand this function in a Taylor series about $\mathbf{g} = \mathbf{e}$:

$$\begin{aligned} \varphi_i(\mathbf{g}, \mathbf{h}) &= h_i + \sum_{j=1}^{n} g_j F_{ij}(\mathbf{h}) + \cdots, \\ F_{ij}(\mathbf{h}) &= [\partial\varphi_i(\mathbf{g}, \mathbf{h})/\partial g_j]\,|_{\mathbf{g}=\mathbf{e}}, \end{aligned} \tag{5.1}$$

where the omitted terms are of order two or more in the g_j. By the associative law, $\mathbf{g}(\mathbf{hk}) = (\mathbf{gh})\mathbf{k}$ for $\mathbf{g}, \mathbf{h}, \mathbf{k} \in V$ sufficiently close to $\mathbf{e}$. Expanding each side of this equality about $\mathbf{g} = \mathbf{e}$, we find

$$\varphi_i(\mathbf{g}, \mathbf{hk}) = \varphi_i(\mathbf{h}, \mathbf{k}) + \sum_{j=1}^{n} g_j F_{ij}(\mathbf{hk}) + \cdots, \tag{5.2}$$

$$\begin{aligned} \varphi_i(\mathbf{gh}, \mathbf{k}) &= \varphi_i(\varphi_l(\mathbf{g}, \mathbf{h}), k_l) = \varphi_i(h_l + \sum_{j=1}^{n} g_j F_{lj}(\mathbf{h}) + \cdots, k_l) \\ &= \varphi_i(\mathbf{h}, \mathbf{k}) + \sum_{j,l=1}^{n} [\partial\varphi_i(\mathbf{h}, \mathbf{k})/\partial h_l] F_{lj}(\mathbf{h}) g_j + \cdots. \end{aligned} \tag{5.3}$$

A comparison of first order terms in g_j leads to the identity

$$F_{ij}(\mathbf{hk}) = \sum_{l=1}^{n} [\partial\varphi_i(\mathbf{h}, \mathbf{k})/\partial h_l] F_{lj}(\mathbf{h}) \tag{5.4}$$

for the functions $F_{ij}(\mathbf{h})$.

Let $\mathbf{g}(t)$ be the analytic curve in G defined in some neighborhood W of $0 \in F$ such that $\mathbf{g}(0) = \mathbf{e}$ and

$$\mathbf{g}(s)\mathbf{g}(t) = \boldsymbol{\varphi}(\mathbf{g}(s), \mathbf{g}(t)) = \mathbf{g}(s+t), \qquad s, t, s+t \in W. \tag{5.5}$$

Such a curve is called a **one-parameter subgroup** of G. Suppose $\alpha \in L(G)$ is the tangent vector to $\mathbf{g}(t)$ at $\mathbf{e}$. Differentiating (5.5) with respect to s, setting $s = 0$ and making use of (5.1), we obtain the system of differential equations

$$dg_i(t)/dt = \sum_{j=1}^{n} \alpha_j F_{ij}(\mathbf{g}), \qquad i = 1, \cdots, n. \tag{5.6}$$

Thus, $\mathbf{g}(t)$ is a solution of the differential system (5.6) such that $\mathbf{g}(0) = \mathbf{e}$, i.e., $g_i(0) = 0$.

Now (5.6) is a system of first-order analytic ordinary differential equations. There is a standard existence and uniqueness theorem for such systems which states that for given constants $\mathbf{a} = (a_1, \ldots, a_n)$ in V, Eq. (5.6) have one and only one solution $\mathbf{g}(t)$ with $\mathbf{g}(0) = \mathbf{a}$. This solution is defined and analytic for all $|t| < \varepsilon$, where $\varepsilon > 0$ depends on the F_{ij} but not on α (see Ince [1]).

We derived (5.6) by assuming $\mathbf{g}(t)$ was a one-parameter subgroup with tangent vector α. Now we choose an arbitrary $\alpha \in L(G)$, and look for a solution $\mathbf{g}(t)$ of (5.6) such that $\mathbf{g}(0) = \mathbf{e}$. [Recall that the functions $F_{ij}(\mathbf{g})$ are known.] By the theorem quoted above, there exists a unique solution, defined and analytic for $|t| < \varepsilon$. We write $\mathbf{g}(t) = \exp \alpha t$.

Theorem 5.16. The integral curve $\mathbf{g}(t) = \exp \alpha t$ is a one-parameter subgroup of G with tangent vector α at $\mathbf{e}$. In particular, $\exp \alpha(s+t) = (\exp \alpha s)(\exp \alpha t)$ for suitably small values of $|s|, |t|$.

Proof. We have shown that $\exp \alpha t$ is an analytic curve in G such that $\exp \alpha 0 = \mathbf{e}$. By (5.1), $F_{ij}(\mathbf{e}) = \delta_{ij}$. Setting $t = 0$ in (5.6) we find $\dot{g}_i(0) = \alpha_i$, $1 \le i \le n$, so α is the tangent vector to $\exp \alpha t$ at $\mathbf{e}$. Let $\mathbf{h}(t) = \mathbf{g}(s+t) = \exp \alpha(s+t)$ for fixed s sufficiently close to zero. Then

$$\dot{h}_i(t) = \frac{dg_i(s+t)}{dt} = \sum_{j=1}^{n} \alpha_j F_{ij}(\mathbf{h}(t)), \qquad \mathbf{h}(0) = \mathbf{g}(s), \tag{5.7}$$

as follows easily from (5.6). On the other hand the function $\mathbf{k}(t) = \mathbf{g}(t)\mathbf{g}(s)$ satisfies

$$\begin{aligned} \dot{k}_i(t) &= \frac{d\varphi_i(\mathbf{g}(t), \mathbf{g}(s))}{dt} = \sum_{j,l=1}^{n} \frac{\partial \varphi_i(\mathbf{g}(t), \mathbf{g}(s))}{\partial h_l} \alpha_j F_{lj}(\mathbf{g}(t)) \\ &= \sum_{j=1}^{n} \alpha_j F_{ij}(\mathbf{k}(t)), \qquad \mathbf{k}(0) = \mathbf{g}(s). \end{aligned} \tag{5.8}$$

Here we have made use of (5.4) and (5.6). The expressions $\partial\varphi_i/\partial h_l$ refer to the derivatives of φ_i with respect to its first n arguments. Thus $\mathbf{h}(t)$ and $\mathbf{k}(t)$

satisfy the same differential system and have the same value at $t = 0$. According to the uniqueness theorem these solutions must be equal: $\mathbf{h}(t) = \mathbf{k}(t)$. Therefore, $(\exp \boldsymbol{\alpha}s)(\exp \boldsymbol{\alpha}t) = \exp \boldsymbol{\alpha}(s + t)$. Q.E.D.

This proof shows that for each $\boldsymbol{\alpha} \in L(G)$ there is a unique one-parameter subgroup with tangent vector $\boldsymbol{\alpha}$ at the identity. Furthermore, every one-parameter subgroup (5.5) is of the form $\mathbf{g}(t) = \exp \boldsymbol{\alpha}t$. Our notation suggests the relation

$$\exp (a\boldsymbol{\alpha})t = \exp \boldsymbol{\alpha}(at) \tag{5.9}$$

for $a \in F$, $\boldsymbol{\alpha} \in L(G)$, and sufficiently small values of $|t|$. That is, if $\mathbf{g}_{\boldsymbol{\alpha}}(t) = \exp \boldsymbol{\alpha}t$ then (5.9) asserts $\mathbf{g}_{a\boldsymbol{\alpha}}(t) = \mathbf{g}_{\boldsymbol{\alpha}}(at)$. The reader can check this equality by verifying that both sides of (5.9) are one-parameter subgroups of G with tangent vector $a\boldsymbol{\alpha}$ at $\mathbf{e}$.

It follows that we can extend the definition of $\exp \boldsymbol{\alpha}t$ to *all* $\boldsymbol{\alpha} \in L(G)$ and $t \in F$ in such a way that

$$(\exp \boldsymbol{\alpha}s)(\exp \boldsymbol{\alpha}t) = \exp \boldsymbol{\alpha}(s + t). \tag{5.10}$$

Indeed $\exp \boldsymbol{\alpha}t$ is originally defined for $|t| < \varepsilon$ and all $\boldsymbol{\alpha}$. If $|t| \geq \varepsilon$ we define $\exp \boldsymbol{\alpha}t = \exp a\boldsymbol{\alpha}(t/a)$, where a is any constant such that $|t/a| < \varepsilon$. The reader can verify that this definition is independent of a and leads to an entire function of t satisfying the group property (5.10).

For fixed t, $\mathbf{g}(t) = \exp \boldsymbol{\alpha}t$ is an entire function of $\alpha_1, \ldots, \alpha_n$. To see this we expand $\mathbf{g}(t)$ in a Taylor series about $t = 0$:

$$g_k(t) = \sum_{m=1}^{\infty} \left.\frac{d^m g_k(s)}{ds^m}\right|_{s=0} \frac{t^m}{m!}, \qquad 1 \leq k \leq n.$$

Now $\dot{g}_k(0) = \alpha_k$, and in general $g_k^{(m)}(0)$ is a homogeneous polynomial of order m in $\alpha_1, \ldots, \alpha_n$. [This follows from (5.6), the chain rule, and a simple induction argument.] Since for suitably small fixed $t \neq 0$ the Taylor series converges for all $\alpha_1, \ldots, \alpha_n$ it follows that $\mathbf{g}(t)$ is an entire function of these variables.

Now set $t = 1$ and consider the **exponential map** $\boldsymbol{\alpha} \longrightarrow \exp \boldsymbol{\alpha}(1) = \exp \boldsymbol{\alpha}$. If G is a global Lie group then $\exp \boldsymbol{\alpha}$ maps all of $L(G)$ into G. If G is only local then $\exp \boldsymbol{\alpha}$ maps a neighborhood of $\boldsymbol{\theta}$ in $L(G)$ into G. The n coordinates α_j of $\boldsymbol{\alpha}$ can be used to parametrize a neighborhood of $\mathbf{e}$ in G. Indeed, by (5.1), $F_{ij}(\mathbf{e}) = \delta_{ij}$. It follows from (5.6) that

$$(\partial/\partial\alpha_j)(\exp \boldsymbol{\alpha})_i|_{\boldsymbol{\alpha}=\boldsymbol{\theta}} = \delta_{ij},$$

so the Jacobian of the transformation $\boldsymbol{\alpha} \longrightarrow \exp \boldsymbol{\alpha} = \mathbf{g}$ is nonzero in a neighborhood of $\boldsymbol{\theta} \in L(G)$. Therefore, by the inverse function theorem the exponential map defines an analytic coordinate transformation on some neighborhood of $\mathbf{e}$. We can use a neighborhood of the zero vector in $L(G)$

to parametrize the local Lie group G. The coordinates α are called **canonical coordinates of the first kind** or just **canonical** coordinates (Pontrjagin [1]). In canonical coordinates the one-parameter subgroups are the straight lines αt through θ. Group multiplication in a one-parameter subgroup is given by $\alpha t + \alpha s = \alpha(t + s)$, i.e., by vector addition. Note, however, that it is in general *not* true that $(\exp \alpha)(\exp \beta) = \exp(\alpha + \beta)$ if α and β are linearly independent vectors.

In the special case where G is a local linear Lie group of $m \times m$ matrices the above analysis becomes much more transparent, From (5.5) a one-parameter subgroup of G is an analytic curve $A(t)$ such that

$$A(s)A(t) = A(s + t) \tag{5.11}$$

for sufficiently small values of $|s|, |t|$. Differentiating this expression with respect to s and setting $s = 0$, we obtain the matrix equation

$$dA(t)/dt = \mathcal{A}A(t), \qquad A(0) = E_m, \tag{5.12}$$

where $\mathcal{A} \in L(G)$ is the tangent matrix to $A(t)$ at the identity. Clearly, (5.12) is the analogy of (5.6) for matrix groups. It is a consequence of our general theory that the solutions of (5.12) define the exponential mapping of $L(G)$ into G, as $\mathcal{A}$ ranges over $L(G)$. We have already seen that (5.6) always has a unique solution $\exp \alpha t$ for each $\alpha \in L(G)$. However, since we did not know the functions $F_{ij}(g)$ explicitly we were not able to write down the general form of the solution. In the special case where G is a linear group, however, this difficulty vanishes. Indeed from the remarks following (1.14), (5.12) has the unique solution

$$A(t) = \exp \mathcal{A}t = \sum_{j=0}^{\infty} \mathcal{A}^j t^j / j!, \tag{5.13}$$

so the exponential map for linear Lie groups is the ordinary matrix exponential. Thus the transformation

$$\mathcal{A} \longrightarrow \exp \mathcal{A}, \qquad \mathcal{A} \in L(G),$$

is a 1–1 analytic map of a neighborhood of Z in $L(G)$ onto a neighborhood of E_m in G. We can introduce canonical coordinates in G by choosing a basis $\{\mathcal{B}_j\}$ for $L(G)$. If $\mathcal{A} \in L(G)$, $\mathcal{A} = \sum \alpha_j \mathcal{B}_j$, then the transformation $\alpha \longrightarrow \exp(\sum \alpha_j \mathcal{B}_j)$ defines canonical coordinates in G. Every one-parameter subgroup in G is of the form $\exp \mathcal{A}t$, $\mathcal{A} \in L(G)$.

5.6 Local Homomorphisms and Isomorphisms

In Section 5.3 we defined local analytic homomorphisms of Lie groups and homomorphisms of Lie algebras. Here we study the relationship between these two concepts.

Let G, G' be local Lie groups of dimensions n and n', respectively, with corresponding Lie algebras $L(G)$, $L(G')$. Suppose μ is an analytic homomorphism of G into G'. We will show that μ induces a homomorphism μ^* of $L(G)$ into $L(G')$.

Theorem 5.17. An analytic homomorphism $\mu: G \longrightarrow G'$ induces a Lie algebra homomorphism $\mu^*: L(G) \longrightarrow L(G')$ defined by

$$\mu^*(\alpha) = (d/dt)\mu(\mathbf{g}(t))\big|_{t=0} \tag{6.1}$$

where $\mathbf{g}(t)$ is an analytic curve in G with tangent vector α at $\mathbf{e}$ The μ^* is an isomorphism (automorphism) if μ is an isomorphism (automorphism).

Proof. First, we show that the right-hand side of (6.1) is well-defined, i.e., that it depends only on $\alpha \in L(G)$, not on the particular curve $\mathbf{g}(t)$. Since $\mu(\mathbf{g}) = (\mu_1(\mathbf{g}), \ldots, \mu_{n'}(\mathbf{g}))$ is analytic with $\mu(\mathbf{e}) = \mathbf{e}'$, we have the Taylor series expansion

$$\mu_i(\mathbf{g}) = \sum_{j=1}^{n} \mu_i{}^j g_j + \cdots, \qquad 1 \leq i \leq n',$$

where only the first-order term in g_j has been given explicitly. Now $\mathbf{g}(0) = \mathbf{e}$ and $\dot{\mathbf{g}}(0) = \alpha$, so (6.1) yields

$$\mu_i{}^*(\alpha) = \sum_{j=1}^{n} \mu_i{}^j \alpha_j, \qquad 1 \leq i \leq n' \tag{6.2}$$

where $\mu^*(\alpha) = (\mu_1{}^*(\alpha), \ldots, \mu_{n'}^*(\alpha))$ and the constants $\mu_i{}^j$ are determined completely by μ. Thus (6.1) is well-defined. It is clear from (6.2) that μ^* is linear, i.e.,

$$\mu^*(a\alpha + b\beta) = a\mu^*(\alpha) + b\mu^*(\beta), \qquad \alpha, \beta \in L(G), \quad a, b \in F.$$

If $\alpha \in L(G)$ then by (6.1) and the definition of μ, $\mu(\exp \alpha t)$ is the one-parameter subgroup of G' with tangent vector $\mu^*(\alpha)$. Thus,

$$\mu(\exp \alpha t) = \exp(\mu^*(\alpha)t). \tag{6.3}$$

Suppose μ is an isomorphism of G onto G'. If $\alpha \neq \theta$ then $\mu(\exp \alpha t)$ will be a nontrivial one-parameter subgroup of G', i.e., $\mu(\exp \alpha t) \not\equiv \mathbf{e}'$. If $\mu^*(\alpha) = \theta'$ then by (6.3) we would have $\mu(\exp \alpha t) \equiv \mathbf{e}'$, a contradiction. Therefore μ^* is 1–1. Furthermore, μ^* is onto because for each $\alpha' \in L(G')$ there is a unique one-parameter subgroup $\exp \alpha t$ in G such that $\mu(\exp \alpha t) = \exp \alpha' t$. According to (6.3), $\alpha' = \mu^*(\alpha)$.

Now assume only that μ is a homomorphism. Let $g(t) = \exp(\alpha\tau)\exp(\beta\tau)\exp(-\alpha\tau)\exp(-\beta\tau)$, $t = \tau^2$, with $\alpha, \beta \in L(G)$. Then

$$\mu(\mathbf{g}(t)) = \exp(\mu^*(\alpha)\tau)\exp(\mu^*(\beta)\tau)\exp(-\mu^*(\alpha)\tau)\exp(-\mu^*(\beta)\tau), \tag{6.4}$$

where we have used (6.3) and the homomorphism property of μ. According to (3.6) the tangent vector to $\mathbf{g}(t)$ at $\mathbf{e}$ is $[\alpha, \beta]$. Thus, differentiating both sides of (6.4) with respect to t and setting $t = 0$, we obtain

$$\mu^*([\alpha, \beta]) = [\mu^*(\alpha), \mu^*(\beta)]'$$

where the primed commutator refers to $L(G')$. Q.E.D.

We now prove the converse of this theorem for local **linear** Lie groups.

Theorem 5.18. Let G, G' be local linear Lie groups and $\rho: L(G) \longrightarrow L(G')$ a Lie algebra homomorphism. There exists a unique local analytic homomorphism μ of G into G' such that $\mu^* = \rho$. If ρ is an isomorphism (automorphism) then μ is an isomorphism (automorphism).

Proof. By the preceding theorem, if μ exists then

(6.5) $$\mu(\exp \mathcal{A}) = \exp \rho(\mathcal{A}), \qquad \mathcal{A} \in L(G),$$

where exp is the matrix exponential. Since $\mathcal{A} \longrightarrow \exp \mathcal{A}$ is a 1–1 analytic mapping of a neighborhood of Z in $L(G)$ onto a neighborhood of E_m in G, expression (6.5) uniquely determines μ. Just as in the computation of (6.2), we can show that μ^* as defined by

$$\mu^*(\mathcal{A}) = (d/dt)\mu(A(t))|_{t=0}, \qquad \dot{A}(0) = \mathcal{A},$$

depends only on $\mathcal{A}$, not on the particular analytic curve $A(t)$. Then, a direct computation with $A(t) = \exp t\mathcal{A}$ yields

$$\mu^*(\mathcal{A}) = \frac{d}{dt}\,\mu(\exp t\mathcal{A})\bigg|_{t=0} = \frac{d}{dt}\exp t\rho(\mathcal{A})\bigg|_{t=0} = \rho(\mathcal{A}),$$

so $\mu^* = \rho$.

We now show that μ is a local homomorphism. From Theorem 5.5, for $\mathcal{A}, \mathcal{B} \in L(G)$ sufficiently close to Z we have $(\exp \mathcal{A})(\exp \mathcal{B}) = \exp \mathcal{C}$, where

(6.6) $$\mathcal{C}(\mathcal{A}, \mathcal{B}) = \mathcal{B} + \int_0^1 g(\exp(t \operatorname{Ad} \mathcal{A}) \exp(\operatorname{Ad} \mathcal{B}))\mathcal{A}\, dt.$$

Since the integral on the right-hand side of (6.6) is a convergent sum of commutators of elements in $L(G)$ and $L(G)$ is finite-dimensional, it follows that $\mathcal{C} \in L(G)$. (Prove it!) Furthermore $\mathcal{C}$ is an analytic function of the coordinates of $\mathcal{A}$ and $\mathcal{B}$. By definition, $(\operatorname{Ad} \mathcal{A})\mathcal{B} = [\mathcal{A}, \mathcal{B}] = \mathcal{A}\mathcal{B} - \mathcal{B}\mathcal{A}$. Now ρ is a homomorphism, so

$$\rho((\operatorname{Ad} \mathcal{A})\mathcal{B}) = \rho([\mathcal{A}, \mathcal{B}]) = [\rho(\mathcal{A}), \rho(\mathcal{B})] = (\operatorname{Ad} \rho(\mathcal{A}))\rho(\mathcal{B}).$$

Applying ρ to both sides of (6.6) and expanding the integral in a power series, we find $\rho(\mathcal{C}(\mathcal{A}, \mathcal{B})) = \mathcal{C}(\rho(\mathcal{A}), \rho(\mathcal{B})) \in L(G')$. (Note that ρ can be applied term-by-term in the infinite series.) Thus,

$$\begin{aligned}\mu(\exp \mathcal{A})\mu(\exp \mathcal{B}) &= \exp \rho(\mathcal{A}) \exp \rho(\mathcal{B}) = \exp \mathcal{C}(\rho(\mathcal{A}), \rho(\mathcal{B})) \\ &= \exp \rho(\mathcal{C}(\mathcal{A}, \mathcal{B})) = \mu(\exp \mathcal{C}) = \mu(\exp \mathcal{A} \exp \mathcal{B}),\end{aligned}$$

so μ is a local homomorphism. Q.E.D.

Corollary 5.4. Let G and G' be local Linear Lie groups. Then $L(G)$ is isomorphic to $L(G')$ if and only if G is locally isomorphic to G'.

Even though the above results were proved only for **linear** groups they are actually true for any local Lie groups G and G'. One can show that the Campbell–Baker–Hausdorff formula (1.19) is valid for any local Lie group, not just for matrix groups (see Hausner and Schwartz [1]). Once this formula is established, the proof of Theorem 5.18 for general Lie groups follows almost exactly as we have given it.

We know that every local Lie group uniquely defines a Lie algebra. Now suppose $\mathcal{G}$ is an abstract Lie algebra. Does there exist a local Lie group G such that $L(G) = \mathcal{G}$ [or such that $L(G)$ is isomorphic to $\mathcal{G}$]? We establish the affirmative answer to this question in two steps. First we show rigorously that every matrix Lie algebra $\tilde{\mathcal{G}}$ is the Lie algebra $L(G)$ of a local linear Lie group. Then we note by Ado's theorem that any abstract Lie algebra $\mathcal{G}$ is isomorphic to a matrix Lie algebra $\tilde{\mathcal{G}}$. Thus, $\mathcal{G}$ is isomorphic to the Lie algebra $\tilde{\mathcal{G}} = L(G)$ of the local Lie group G.

Theorem 5.19. Let $\tilde{\mathcal{G}}$ be a matrix Lie algebra. There exists a local linear Lie group G such that $L(G) = \tilde{\mathcal{G}}$.

Proof. The obvious candidate for G is the set of matrices $\exp \mathcal{A}$, where $\mathcal{A}$ runs over $\tilde{\mathcal{G}}$. Let the matrices $\{\mathcal{B}_j : j = 1, \ldots, n\}$ form a basis for $\tilde{\mathcal{G}}$. We introduce coordinates $\alpha_1, \ldots, \alpha_n$ in G by writing $A(\alpha) = \exp(\sum \alpha_j \mathcal{B}_j)$ and letting α run over a sufficiently small neighborhood of θ such that the mapping $\alpha \longrightarrow A(\alpha)$ is 1–1. From the Campbell–Baker–Hausdorff formula (1.19),

$$\exp(\textstyle\sum \alpha_j \mathcal{B}_j) \exp(\sum \beta_k \mathcal{B}_k) = \exp(\sum \gamma_l \mathcal{B}_l),$$

where the γ_l are analytic functions of α_j, β_k for α, β sufficiently close to θ. Thus G is a local linear Lie group. It is easy to verify that $L(G) = \tilde{\mathcal{G}}$. Q.E.D.

Let H be a local Lie group with Lie algebra $L(H)$. By Ado's theorem, $L(H)$ is isomorphic to a matrix Lie algebra $\mathcal{G}$. By Theorem 5.18, $\mathcal{G} = L(G)$, where G is a local linear Lie group. Finally, by the extension of Theorem 5.18 to all local Lie groups, we see that H is isomorphic to G. Thus, every local Lie group is isomorphic to a local linear Lie group! (There do exist global Lie groups which are not globally isomorphic to linear groups.) In the remainder of this book we study only local linear Lie groups, but this restriction leads to no loss of generality.

Theorem 5.18 is valid only for local groups. For example, the one-dimensional real Lie groups

$$R = \{x\colon -\infty < x < +\infty\}, \qquad U(1) = \{e^{ix}\colon x \in R\}$$

have isomorphic Lie algebras, so they are locally isomorphic. Indeed the isomorphism is given by $x \longrightarrow e^{ix}$. However, this isomorphism cannot be extended to a global isomorphism since the distinct elements x and $x + 2\pi$ of R are mapped into the same element of $U(1)$. Note that R is isomorphic to the linear Lie group of matrices

$$\begin{pmatrix} 1 & x \\ 0 & 1 \end{pmatrix}, \qquad x \in R.$$

Since every local Lie group is uniquely determined by its corresponding Lie algebra, to compute all local Lie groups up to isomorphism one need only compute all Lie algebras up to isomorphism. This latter problem is purely algebraic and for low-dimensional Lie algebras at least, it can be solved explicitly (see Jacobson [1]). In general, any problem concerning the structure of a local group G can be reduced to a purely algebraic problem concerning the structure of $L(G)$.

As an example of this relationship recall that if G is commutative then $[\mathcal{A}, \mathcal{B}] = Z$ for all $\mathcal{A}, \mathcal{B} \in L(G)$, i.e., the Lie algebra $L(G)$ is **commutative.** Conversely, if $L(G)$ is commutative it follows immediately from the Campbell–Baker–Hausdorff formula that

$$\exp(\mathcal{A} + \mathcal{B}) = \exp \mathcal{A} \exp \mathcal{B} = \exp \mathcal{B} \exp \mathcal{A}$$

for all $\mathcal{A}, \mathcal{B} \in L(G)$, i.e., G is commutative. Thus G is commutative if and only if $L(G)$ is commutative. An n-dimensional commutative local Lie group is locally isomorphic to F_n.

We now examine some special automorphisms of a local linear Lie group G of $m \times m$ matrices. [Since any local linear Lie group can be uniquely extended to a global group, one can assume without loss of generality that G is a (global) Lie group.] For fixed $B \in G$, the map μ_B,

$$\mu_B(A) = BAB^{-1}, \qquad A \in G,$$

is clearly an automorphism of G, an **inner automorphism.** The set G of all inner automorphisms of G is itself a group since

$$(6.7) \qquad \mu_{B_1B_2}(A) = B_1(B_2AB_2^{-1})B_1^{-1} = \mu_{B_1}\mu_{B_2}(A), \qquad \mu_{E_m}(A) = A.$$

Moreover, $\tilde{G}$ is a linear group, the **adjoint group** of G. Indeed we can think of the elements of $\tilde{G}$ as $m^2 \times m^2$ matrices acting on an m^2-dimensional vector space. The map $\mu: G \longrightarrow \tilde{G}$ given by $A \longrightarrow \mu_A$ defines an analytic homomorphism from G onto $\tilde{G}$.

By Theorem 5.17, for fixed $B \in G$ the inner automorphism μ_B of G induces an automorphism $\mu_B{}^*$ of $L(G)$:

$$\mu_B{}^*(\mathfrak{A}) = (d/dt)\mu_B(\exp t\mathfrak{A})|_{t=0}.$$

Thus,

$$\mu_B{}^*(\mathfrak{A}) = (d/dt)[B(\exp t\mathfrak{A})B^{-1}]|_{t=0} = B\mathfrak{A}B^{-1}. \tag{6.8}$$

It is easy to check that $\mu^*_{B_1B_2} = \mu^*_{B_1}\mu^*_{B_2}$ and $\mu^*_{B^{-1}} = (\mu_B{}^*)^{-1}$, so the operators $\mu_B{}^*$ form a linear group, also called the **adjoint group** $\tilde{G}$. (This makes sense because the two adjoint groups introduced above are locally isomorphic.) We will show that $\tilde{G}$ is a local linear Lie group by explicitly introducing canonical coordinates. Think of $L(G)$ as an n-dimensional vector space on which the $n^2 \times n^2$ matrices $\mu_A{}^*$ are acting. Let $B(t)$ be an analytic curve in G with tangent matrix $\mathfrak{B}$ at the identity. We define the linear operator $\text{Ad}\,\mathfrak{B}: L(G) \longrightarrow L(G)$ as follows:

$$\text{Ad}\,\mathfrak{B}(\mathfrak{A}) = \frac{d}{dt}\,\mu^*_{B(t)}(\mathfrak{A})\bigg|_{t=0} = \frac{d}{dt}\,B(t)\mathfrak{A}B^{-1}(t)\bigg|_{t=0} = [\mathfrak{B}, \mathfrak{A}]. \tag{6.9}$$

The operators $\{\text{Ad}\,\mathfrak{B}: \mathfrak{B} \in L(G)\}$ form a Lie algebra $\widetilde{L(G)}$, the **adjoint Lie algebra,** and the map $\text{Ad}: \mathfrak{B} \longrightarrow \text{Ad}\,\mathfrak{B}$ is a homomorphism of $L(G)$ onto $\widetilde{L(G)}$. Indeed

(6.10)

$$\text{Ad}(a\mathfrak{A} + b\mathfrak{B}) = a\,\text{Ad}\,\mathfrak{A} + b\,\text{Ad}\,\mathfrak{B}, \qquad \text{Ad}([\mathfrak{A}, \mathfrak{B}]) = [\text{Ad}\,\mathfrak{A}, \text{Ad}\,\mathfrak{B}].$$

The kernel of the map Ad, i.e., the set of $\mathfrak{B} \in L(G)$ such that $[\mathfrak{B}, \mathfrak{A}] = Z$ for all $\mathfrak{A} \in L(G)$ is called the **center** of $L(G)$. Clearly, the center is a commutative Lie algebra.

For any $\mathfrak{B} \in L(G)$ we can consider $\text{Ad}\,\mathfrak{B}$ as an $n \times n$ matrix acting on the n-dimensional space $L(G)$. Then by Lemma 5.3 we have

$$(\exp(\text{Ad}\,\mathfrak{B}))\mathfrak{A} = (\exp \mathfrak{B})\mathfrak{A}(\exp \mathfrak{B})^{-1} = B\mathfrak{A}B^{-1}, \tag{6.11}$$

where $B = \exp \mathfrak{B} \in G$, so we can identify $\tilde{G}$ with $\exp \widetilde{L(G)}$ in a neighborhood of the identity matrix E_n. This shows that $\tilde{G}$ is a local linear Lie group with Lie algebra $\widetilde{L(G)}$. If the center of $L(G)$ has dimension n' then $\dim \widetilde{L(G)} = n - n'$ and $\tilde{G}$ is an $(n - n')$-dimensional Lie group.

5.7 Subgroups and Subalgebras

Let G be an n-dimensional local Lie group with elements $\mathbf{g} = (g_1, \ldots, g_n)$ defined in some connected neighborhood $V \in F_n$ of $\mathbf{e} = (0, \ldots, 0)$. Suppose H is an m-dimensional local Lie group over F with elements $\mathbf{h} = (h_1, \ldots, h_m)$ and identity $\mathbf{e}' = (0, \ldots, 0)$.

Definition. H is a **local Lie subgroup** of G if there is an analytic homomorphism $\mu: H \to G$ such that (1) $\mu(\mathbf{h})$ is locally 1–1; (2) the vectors $\{\partial\mu(\mathbf{h})/\partial h_j|_{\mathbf{h}=\mathbf{e}'}, 1 \leq j \leq m\}$ are linearly independent.

We can consider the subgroup H as embedded in G. Indeed the image $\mu(H) \subseteq G$ is itself a local Lie group which is isomorphic to H. It is clear that the tangent vectors at the identity of analytic curves in $\mu(H)$ are contained in $L(G)$. Every such analytic curve with tangent vector β is of the form $\mu(\mathbf{h}(t))$, where $\mathbf{h}(t)$ is an analytic curve through $\mathbf{e}'$ in H. Furthermore, if $\mathbf{h}(t)$ has tangent vector α at $\mathbf{e}'$ then $\beta = \mu^*(\alpha)$ by (6.1). Since μ^* is a homomorphism, for every $\beta = \mu^*(\alpha)$, $\beta' = \mu^*(\alpha')$, we have

$$[\beta, \beta'] = [\mu^*(\alpha), \mu^*(\alpha')] = \mu^*([\alpha, \alpha']).$$

Thus the tangent vectors at $\mathbf{e}$ to analytic curves in $\mu(H)$ form a Lie algebra $L(\mu(H))$ contained in $L(G)$. Furthermore, $L(\mu(H))$ is isomorphic to $L(H)$ since μ^* is 1–1.

Definition. Let $\mathcal{G}$ be a Lie algebra and $\mathcal{K}$ a subset of $\mathcal{G}$. We say $\mathcal{K}$ is a **subalgebra** of $\mathcal{G}$ if $\mathcal{K}$ is itself a Lie algebra under the operations in $\mathcal{G}$, i.e., if $\mathcal{K}$ is a vector subspace which is closed under the commutator operation.

Since $L(H)$ is isomorphic to the subalgebra $L(\mu(H))$, we can consider $L(H)$ as a subalgebra of $L(G)$.

Theorem 5.20. If H is a Lie subgroup of the local Lie group G then $L(H)$ is a Lie subalgebra of $L(G)$.

Theorem 5.21. Let G be a local linear Lie group and let $\mathcal{H}$ be a subalgebra of $L(G)$. There exists a unique local Lie subgroup H of G such that $L(H) = \mathcal{H}$.

Proof. Let $\{\mathcal{B}_j\}$, $1 \leq j \leq n$, be a basis for $L(G)$ such that $\{\mathcal{B}_j\}$, $1 \leq j \leq m$, is a basis for $\mathcal{H}$. (Note that $m \leq n$.) By Theorem 5.18 the map

$$(\alpha_1, \ldots, \alpha_n) \longrightarrow \exp(\sum_{j=1}^{n} \alpha_j \mathcal{B}_j) \in G \tag{7.1}$$

defines canonical coordinates on G. Furthermore, by Theorem 5.19 the map

$$(\alpha_1, \ldots, \alpha_m) \longrightarrow \exp(\sum_{j=1}^{m} \alpha_j \mathcal{B}_j) \tag{7.2}$$

defines coordinates in a local linear Lie group (which we call H) as $(\alpha_1, \ldots, \alpha_m)$ ranges over a suitably small neighborhood of θ in F_m. Clearly, $L(H) = \mathcal{H}$ and H is a local Lie subgroup of G. (Here we are considering H as embedded in G). Indeed, in the canonical coordinate system we have chosen, the elements of G that lie in H are just those with coordinates $(\alpha_1, \ldots, \alpha_m, 0, \ldots, 0)$.

If H' is a Lie subgroup of G with $L(H') = \mathcal{H}$, then H' can be described locally by the exponential mapping (7.2). Thus $H' = H$. Q.E.D.

Even though we have proved the above result only for local linear Lie groups it is valid for all local Lie groups. The preceding theorems establish a 1–1 correspondence between Lie subgroups of G and Lie subalgebras of $L(G)$. Any question involving the structure of local Lie subgroups can be reduced to a purely algebraic problem concerning Lie subalgebras.

As an example we work out the relationship between normal subgroups of G and ideals in $L(G)$.

Definition. A local Lie subgroup H of G is a **normal** subgroup if there exists a neighborhood V of the identity in G such that $\mathbf{ghg}^{-1} \in H$ for all $\mathbf{h} \in H \cap V$ and $\mathbf{g} \in G \cap V$.

This definition of normal subgroup agrees with that given in Chapter 1, except that here the normality property is only local.

Definition. A subalgebra $\mathcal{H}$ of the Lie algebra $\mathcal{G}$ is an **ideal** if $[\alpha, \beta] \in \mathcal{H}$ for all $\alpha \in \mathcal{G}$, $\beta \in \mathcal{H}$. We write $[\mathcal{G}, \mathcal{H}] \subseteq \mathcal{H}$.

Theorem 5.22. Let H be a local Lie subgroup of the local linear Lie group G. Then H is normal if and only if $L(H)$ is an ideal.

Proof. Suppose H is a normal subgroup. Given $\mathcal{B} \in L(H)$, we have $A(\exp t\mathcal{B})A^{-1} \in H$ for all $A \in G$ sufficiently close to the identity and all $t \in F$ sufficiently close to zero. Since

$$(d/dt)A(\exp \mathcal{B})A^{-1}|_{t=0} = A\mathcal{B}A^{-1}$$

it follows that $A\mathcal{B}A^{-1} \in L(H)$. Given $\mathcal{A} \in L(G)$, let $A(s) = \exp s\mathcal{A}$. Since $A(s)\mathcal{B}A^{-1}(s)$ is an analytic curve in $L(H)$ and $L(H)$ is finite-dimensional we have

$$[\mathcal{A}, \mathcal{B}] = (d/ds)(\exp(s\mathcal{A})\cdot\mathcal{B}\cdot\exp(-s\mathcal{A}))|_{s=0} \in L(H),$$

so $L(H)$ is an ideal.

Conversely, if $L(H)$ is an ideal and $\mathcal{B} \in L(H)$, $\mathcal{A} \in L(G)$ are chosen in suitably small neighborhoods of Z then $B(t) = \exp \mathcal{A} \exp t\mathcal{B} \exp(-\mathcal{A})$ is a one-parameter subgroup of G with tangent matrix $(\exp \mathcal{A})\mathcal{B}(\exp -\mathcal{A})$ at the identity. However, from Lemma 5.3,

$$(\exp \mathcal{A})\mathcal{B}(\exp -\mathcal{A}) = \sum_{j=0}^{\infty} (\mathrm{Ad}\,\mathcal{A})^j\mathcal{B}/j! \in L(H)$$

since $L(H)$ is an ideal. Thus $B(t) \in H$ for sufficiently small t and H is normal. Q.E.D.

As usual, the above result holds for all local Lie groups, not just linear groups.

5.8 Representations of Lie Groups

We have seen that any local linear Lie group can be uniquely extended to a connected (global) linear Lie group. Here we assume this has been done, so G is not only a local linear Lie group, but also a matrix group in the abstract sense.

Let V be a finite-dimensional vector space over F and let $GL(V)$ be the group of all nonsingular linear transformations of V onto V.

Definition. A **representation** of a linear Lie group G with **representation space** V is an analytic homomorphism $\mathbf{T}: A \longrightarrow \mathbf{T}(A)$ of G into $GL(V)$.

By an **analytic** homomorphism we mean a homomorphism such that the matrix elements $T_{ij}(A)$ with respect to any basis in V are analytic functions of the local coordinates of A in G. (If the matrix elements are analytic with respect to one basis then they will be analytic with respect to every basis.) This definition agrees with that of Section 3.1 except for the analyticity requirement, which is added so that we can take advantage of the Lie structure of G. Note that $GL(V)$ is isomorphic to $GL(m, F)$ where $m = \dim V$ and an isomorphism is defined by choosing a basis for V.

For every $\mathcal{a} \in L(G)$ we define the **infinitesimal operator** $\boldsymbol{\mathcal{a}}$ on V by

$$(8.1) \qquad \boldsymbol{\mathcal{a}} = (d/dt)\mathbf{T}(A(t))\,|_{t=0},$$

where $A(t)$ is an analytic curve in G with tangent matrix $\mathcal{a}$ at the identity. Since $\mathbf{T}$ is an analytic homomorphism, $\boldsymbol{\mathcal{a}}$ depends on $\mathcal{a}$ alone, not on the particular curve $A(t)$. Furthermore, the operators $\boldsymbol{\mathcal{a}}$ form a Lie algebra which is a homomorphic image of $L(G)$.

Definition. A **representation** of a Lie algebra $\mathcal{G}$ with **representation space** V is a map $\boldsymbol{\rho}$ from $\mathcal{G}$ to the space of all linear operators on V such that ($a, b \in F$ and $\alpha, \beta \in \mathcal{G}$):

(1) $\boldsymbol{\rho}(a\alpha + b\beta) = a\boldsymbol{\rho}(\alpha) + b\boldsymbol{\rho}(\beta)$.

(2) $\boldsymbol{\rho}([\alpha, \beta]) = \boldsymbol{\rho}(\alpha)\boldsymbol{\rho}(\beta) - \boldsymbol{\rho}(\beta)\boldsymbol{\rho}(\alpha) = [\boldsymbol{\rho}(\alpha), \boldsymbol{\rho}(\beta)]$.

The operators $\boldsymbol{\mathcal{a}} = \boldsymbol{\rho}(\mathcal{a})$ define a rep of $L(G)$ on V, so every rep of G induces a rep of $L(G)$. Conversely, suppose $\boldsymbol{\rho}$ is a rep of $L(G)$ on V. Set $\boldsymbol{\rho}(\mathcal{a}) = \boldsymbol{\mathcal{a}}$ and define the map $\mathbf{T}$ by

$$(8.2) \qquad \mathbf{T}(\exp \mathcal{a}) = \exp \boldsymbol{\mathcal{a}}, \quad \mathcal{a} \in L(G),$$

where $\mathcal{a}$ ranges over a suitably small neighborhood $\mathcal{V}$ of Z. By Theorem

5.18, **T** is a local analytic homomorphism of G into $GL(V)$. We refer to **T** as a **local rep** of G since it is defined only in some neighborhood of the identity element. If $\boldsymbol{\rho}$ is the Lie algebra rep induced from a (global) rep $\mathbf{T}'$ of G it follows from Theorem 5.18 and (8.2) that $\mathbf{T}(A) = \mathbf{T}'(A)$ for all A in a suitably small neighborhood of the identity. Furthermore, if G is connected then any $A \in G$ can be written as a finite chain

$$A = \exp \mathfrak{a}_1 \exp \mathfrak{a}_2 \cdots \exp \mathfrak{a}_k, \qquad \mathfrak{a}_i \in \mathcal{V},$$

so

$$\mathbf{T}'(A) = \mathbf{T}'(\exp \mathfrak{a}_1) \cdots \mathbf{T}'(\exp \mathfrak{a}_k) = \exp \boldsymbol{\mathfrak{a}}_1 \cdots \exp \boldsymbol{\mathfrak{a}}_k$$

and $\mathbf{T}'$ is **uniquely** determined by $\boldsymbol{\rho}$. Thus, any rep of a connected linear Lie group is uniquely determined by its associated Lie algebra rep. If G is not connected, e.g., $G = O(m, R)$, then only the action of the rep in the connected component containing the identity is determined by the Lie algebra rep.

Warning. If $\boldsymbol{\rho}$ is an arbitrary Lie algebra rep it may not be possible to extend the local rep (8.2) of G to a global rep. For example, let $\{\mathfrak{a}_1\}$ be a basis for the one-dimensional Lie algebra of $U(1) = \{e^{ix}\}$. The choice $\boldsymbol{\rho}(\mathfrak{a}_1) = 1$ defines a one-dimensional rep of $L(U(1))$. From (8.2) we get $\mathbf{T}(e^{ix}) = e^x$. Since $\exp i(x + 2\pi) = \exp ix$ this local rep does not extend globally.

Thus the correspondence between (global) group reps and Lie algebra reps is *not* 1–1. It is possible to make the correspondence 1–1 by considering only **simply connected** Lie groups. However, the study of such groups leads to topological complications beyond the scope of this book. The interested reader can find relevant material in the work of Pontrjagin [1] or Hausner and Schwartz [1].

The definitions of invariant and irred subspaces as well as the Schur lemmas given in Sections 3.1–3.3 carry over immediately to reps of Lie groups and algebras.

Theorem 5.23. Let **T** be a rep of the connected linear Lie group G on the vector space V and let $\boldsymbol{\rho}$ be the associated rep of $L(G)$. Suppose W is a subspace of V. Then, (1) W is **T**-invariant if and only if it is $\boldsymbol{\rho}$-invariant; (2) W is **T**-irred if and only if it is $\boldsymbol{\rho}$-irred.

Proof. A straightforward exercise for the reader.

Thus, the problem of computing all reps (irred reps) **T** of G is equivalent to the purely algebraic problem of computing all reps (irred reps) $\boldsymbol{\rho}$ of $L(G)$ except that we must check every algebra rep $\boldsymbol{\rho}$ to see that it actually determines a global group rep. In the following chapters we shall usually determine the reps $\boldsymbol{\rho}$ first and then obtain the group reps **T** by exponentiation.

5.9 Local Transformation Groups

A knowledge of abstract group theory is not sufficient for the application of group-theoretic methods to physics. In physical problems the groups usually occur as transformation groups acting on some mathematical structure. Here we study the important case of a local Lie group acting on a coordinate manifold. The following definitions and theorems will be stated for complex groups acting on complex manifolds, but the results for real groups acting on real or complex manifolds are completely analogous.

Let U be an open connected set in $\mathfrak{C}_m$. Any $\mathbf{x} \in U$ can be designated by its coordinates $\mathbf{x} = (x_1, \ldots, x_m)$, $x_i \in \mathfrak{C}$. Let G be an n-dimensional local Lie group defined in some connected neighborhood of $\mathbf{e} = (0, \ldots, 0)$ in $\mathfrak{C}_n$. Finally let $\mathbf{Q}$ be a mapping which associates to each pair $(\mathbf{x}, g)$, $\mathbf{x} \in U$, $g \in G$, an element $\mathbf{Q}(\mathbf{x}, g)$ in $\mathfrak{C}_m$. We write $\mathbf{Q}(\mathbf{x}, g) = \mathbf{x}g \in \mathfrak{C}_m$.

Definition. G acts on the manifold U as a **local Lie transformation group** if $\mathbf{Q}$ satisfies the following properties:

(1) $\mathbf{x}g$ is analytic in the $n + m$ coordinates of $\mathbf{x}$ and g;
(2) $\mathbf{x}\mathbf{e} = \mathbf{x}$, all $\mathbf{x} \in U$;
(3) if $\mathbf{x}g \in U$ then $(\mathbf{x}g)h = \mathbf{x}(gh)$, $\quad g, h, gh \in G$.

With the exception of the identity $\mathbf{e}$ the elements of G are printed in lightface type to distinguish them from elements of U. Conditions (2) and (3) express the transformation group property of G. Condition (1) is necessary if we are to make use of the analyticity of G. Note that the group operates on the right ($\mathbf{x}g$) rather than the left ($g\mathbf{x}$) as in Chapter 1. The reason for this notational change will become apparent later.

If $\mathbf{x} \in U$ and g is in a sufficiently small neighborhood of $\mathbf{e}$, properties (2) and (3) imply $(\mathbf{x}g)g^{-1} = \mathbf{x}\mathbf{e} = \mathbf{x}$. Thus the map $\mathbf{x} \longrightarrow \mathbf{x}g$ is locally analytic and 1–1 for fixed g, and the inverse mapping is also analytic.

Let $\exp \alpha t$, $\alpha \in L(G)$, be a one-parameter subgroup of G. For fixed $\mathbf{x}^0 \in U$ we call the curve $\mathbf{x}(t) = \mathbf{x}^0 \exp \alpha t \in U$ the **trajectory** of $\mathbf{x}^0$ under $\exp \alpha t$. [Clearly, $\mathbf{x}(t)$ is defined for sufficiently small values of $|t|$.] The vector-valued function $\mathbf{x}(t)$ can be expanded in a Taylor series in t about $t = 0$:

$$x_i(t) = x_i^{\,0} + t(\partial Q_i/\partial t)(\mathbf{x}^0, \exp \alpha t)\,|_{t=0} + \cdots$$

or

$$x_i(t) = x_i^{\,0} + t \sum_{j=1}^{n} P_{ij}(\mathbf{x}^0)\alpha_j + \cdots, \qquad 1 \le i \le m, \tag{9.1}$$

where

$$P_{ij}(\mathbf{x}^0) = (\partial Q_i/\partial g_j)(\mathbf{x}^0, g)\,|_{g=\mathbf{e}} \tag{9.2}$$

and $\mathbf{Q} = (Q_1, \ldots, Q_m)$ Thus, $\dot{\mathbf{x}}(0)$ is tangent to the trajectory of $\exp \alpha t$ through $\mathbf{x}^0$.

If $f(\mathbf{x})$ is analytic in a neighborhood of $\mathbf{x}^0$ we can define analytic functions $[(\exp \alpha t)f](\mathbf{x}) = f(\mathbf{x} \exp \alpha t)$ for $\alpha \in L(G)$ and suitably small values of $|t|$. More generally, let $\mathfrak{A}_{\mathbf{x}^0}$ be the space of all functions f analytic in a neighborhood of $\mathbf{x}^0$, where the neighborhood is allowed to vary with the function. We define operators $\mathbf{T}(g): \mathfrak{A}_{\mathbf{x}^0} \longrightarrow \mathfrak{A}_{\mathbf{x}^0}$ by

$$[\mathbf{T}(g)f](\mathbf{x}) = f(\mathbf{x}g), \qquad \mathbf{x} \in U, \quad g \in G. \tag{9.3}$$

For a given $f \in \mathfrak{A}_{\mathbf{x}^0}$ the right-hand side will be well-defined for $\mathbf{x}$ suitably close to $\mathbf{x}^0$ and g suitably close to $\mathbf{e}$. We say that (9.3) holds **locally.** Since G is a local transformation group, we have

$$[\mathbf{T}(g_1 g_2)f](\mathbf{x}) = f(\mathbf{x}(g_1 g_2)) = \{\mathbf{T}(g_1)[\mathbf{T}(g_2)f]\}(\mathbf{x}), \tag{9.4}$$

so $\mathbf{T}(g_1 g_2) = \mathbf{T}(g_1)\mathbf{T}(g_2)$ for $g_1, g_2 \in G$. Again the homomorphism property (9.4) holds locally. (If we let G act on U to the left, $\mathbf{x} \longrightarrow g\mathbf{x}$, then to obtain the homomorphism property for $\mathbf{T}$-operators we have to set $[\mathbf{T}(g)f](\mathbf{x}) = f(g^{-1}\mathbf{x})$. To avoid the inverse we have written $\mathbf{x} \longrightarrow \mathbf{x}g$.)

The operators $\mathbf{T}(g)$ define a local rep of G on the infinite-dimensional vector space $\mathfrak{A}_{\mathbf{x}^0}$. In analogy with (8.1) we can define infinitesimal operators corresponding to this rep.

Definition. The **Lie derivative** $L_\alpha f$ of an analytic function $f \in \mathfrak{A}_{\mathbf{x}^0}$ is

$$L_\alpha f(\mathbf{x}) = (d/dt) f(\mathbf{x}g(t))|_{t=0},$$

where $g(t)$ is an analytic curve in G with tangent vector α at $\mathbf{e}$.

A direct computation yields

$$L_\alpha f(\mathbf{x}) = \sum_{i=1}^{m} \sum_{j=1}^{n} (\partial f(\mathbf{x})/\partial x_i) P_{ij}(\mathbf{x}) \alpha_j,$$

where $P_{ij}(\mathbf{x})$ is given by (9.2). In other words, we have

$$L_\alpha = \sum_{i=1}^{m} \sum_{j=1}^{n} P_{ij}(\mathbf{x}) \alpha_j (\partial/\partial x_i). \tag{9.5}$$

It is clear from this expression that L_α depends only on α, not on the particular analytic curve $g(t)$.

Example 1. Let G be the real line R, a one-dimensional real Lie group. For $a \in R$ the mapping $x \longrightarrow x + a$ for all $x \in R$ defines an action of R as a transformation group on itself. Then $[\mathbf{T}(a)f](x) = f(x + a)$. The constant 1 forms a basis for the one-dimensional Lie algebra $L(R)$. Since $g(t) = t$

is an analytic curve in R with tangent vector 1 we have

$$L_1 f(x) = (d/dt) f(x+t)|_{t=0} = df(x)/dx,$$

or $L_1 = d/dx$. If $\alpha = \alpha \cdot 1 \in L(R)$ for $\alpha \in R$ then $L_\alpha = \alpha\, d/dx$.

Example 2. The proper Euclidean group in the plane $E^+(2)$ has the matrix realization

$$g(x_1, x_2, \varphi) = \begin{pmatrix} \cos\varphi & -\sin\varphi & x_1 \\ \sin\varphi & \cos\varphi & x_2 \\ 0 & 0 & 1 \end{pmatrix}, \qquad x_1, x_2, \varphi \in R \tag{9.6}$$

where φ is determined up to a multiple of 2π. The group multiplication rule is

$$g(x_1, x_2, \varphi) g(x_1', x_2', \varphi') = g(x_1' \cos\varphi - x_2' \sin\varphi + x_1, \; x_1' \sin\varphi + x_2' \cos\varphi + x_2, \varphi + \varphi'). \tag{9.7}$$

Thus $\mathbf{a} = (x_1, x_2)$ and

$$\mathbf{O} = \begin{pmatrix} \cos\varphi & -\sin\varphi \\ \sin\varphi & \cos\varphi \end{pmatrix}$$

where $\mathbf{a}$ and $\mathbf{O}$ are analogous to (2.4), Section 2.2. We can consider $E^+(2)$ as a transformation group in the plane R_2, where the group action is

$$\begin{aligned} (y_1, y_2) \longrightarrow ((y_1 - x_1)\cos\varphi + (y_2 - x_2)\sin\varphi,& \\ -(y_1 - x_1)\sin\varphi + (y_2 - x_2)\cos\varphi),& \\ \mathbf{y} = (y_1, y_2) \in R_2.& \end{aligned} \tag{9.8}$$

In terms of the action of $E^+(2)$ as a transformation group defined as in Section 2.2, this reads

$$\mathbf{y} \longrightarrow \{\mathbf{a}, \mathbf{O}\}^{-1}\mathbf{y} = \{-\mathbf{O}^{-1}\mathbf{a}, \mathbf{O}^{-1}\}\mathbf{y}.$$

It follows that $E^+(2)$ is a two-dimensional real Lie group which acts as a Lie transformation group on R_2. As the reader can check, $L(E^+(2))$ is the Lie algebra of all matrices

$$\mathcal{J}(a, b, c) = \begin{pmatrix} 0 & -c & a \\ c & 0 & b \\ 0 & 0 & 0 \end{pmatrix}, \qquad a, b, c \in R. \tag{9.9}$$

A basis is provided by the matrices $\mathcal{J}(1, 0, 0) = \mathcal{J}_1$, $\mathcal{J}(0, 1, 0) = \mathcal{J}_2$, $\mathcal{J}(0, 0, 1) = \mathcal{J}_3$, with commutation relations

$$[\mathcal{J}_1, \mathcal{J}_2] = Z, \qquad [\mathcal{J}_3, \mathcal{J}_1] = \mathcal{J}_2, \qquad [\mathcal{J}_3, \mathcal{J}_2] = -\mathcal{J}_1.$$

Note that $\exp \mathcal{J}_1 t = g(t, 0, 0)$, $\exp \mathcal{J}_2 t = g(0, t, 0)$, $\exp \mathcal{J}_3 t = g(0, 0, t)$.

The Lie derivatives are

$$L_k f(\mathbf{y}) = (d/dt) f(\mathbf{y} \exp \mathcal{J}_k t)|_{t=0}, \qquad k = 1, 2, 3,$$

or

(9.10) $$L_1 = -\partial/\partial y_1, \qquad L_2 = -\partial/\partial y_2, \qquad L_3 = y_2(\partial/\partial y_1) - y_1(\partial/\partial y_2).$$

The Lie derivative corresponding to the Lie algebra element $\mathcal{J}(a, b, c)$ is just $aL_1 + bL_2 + cL_3$. Note that

$$[L_1, L_2]f(\mathbf{y}) = 0, \quad [L_3, L_1]f(\mathbf{y}) = L_2 f(\mathbf{y}), \quad [L_3, L_2]f(\mathbf{y}) = -L_1 f(\mathbf{y}),$$

where $[L_\alpha, L_\beta] = L_\alpha L_\beta - L_\beta L_\alpha$ is the **commutator** of the Lie derivatives L_α, L_β. Thus the L_k satisfy the same commutations relations as the generators of $L(E^+(2))$. In particular the Lie derivatives themselves form a Lie algebra.

Returning to the study of a general transformation group G, we let $\mathbf{x}^0 \in U$. Then for $\alpha \in L(G)$ and sufficiently small values of $|t|, |s|$, $\mathbf{x}(t + s) = \mathbf{x}^0 \cdot \exp \alpha(t + s) = (\mathbf{x}^0 \exp \alpha t) \exp \alpha s = \mathbf{Q}(\mathbf{x}(t), \exp \alpha s)$. Differentiating this expression with respect to s and setting $s = 0$, we find

(9.11) $$(d/dt)x_i(t) = \sum_{j=1}^{n} P_{ij}(\mathbf{x}(t))\alpha_j, \qquad 1 \leq i \leq n, \qquad \mathbf{x}(0) = \mathbf{x}^0,$$

where we have used (9.1) and (9.2). From (9.5) this system can be rewritten as $\dot{x}_i(t) = L_\alpha x_i$, or

(9.12) $$d\mathbf{x}/dt = L_\alpha \mathbf{x}, \qquad \mathbf{x}(0) = \mathbf{x}^0.$$

According to the fundamental existence and uniqueness theorem for differential equations, system (9.12) has a unique solution $\mathbf{x}(t)$. Clearly, this solution is $\mathbf{x}(t) = \mathbf{x}^0 \exp \alpha t$. Since every $g \in G$ sufficiently close to $\mathbf{e}$ lies on a one-parameter subgroup $\exp \alpha t$, the action of G on U is uniquely determined by the solutions of (9.12). Thus a knowledge of the Lie derivatives L_α determines the action of G.

Theorem 5.24. The unique solution of (9.12) is the trajectory $\mathbf{x}(t) = \mathbf{x}^0 \cdot \exp \alpha t$. The Lie derivatives L_α uniquely determine the local Lie transformation group G.

If $f \in \mathfrak{A}_{\mathbf{x}^0}$ and $\mathbf{x}(t) = \mathbf{x}^0 \exp \alpha t$ then $f(\mathbf{x}(t))$ satisfies the differential equation

$$\frac{d}{dt} f(\mathbf{x}(t)) = \sum_{i=1}^{m} \frac{\partial f}{\partial x_i} \frac{\partial x_i}{\partial t} = \sum_{i=1}^{m} \sum_{j=1}^{n} \frac{\partial f}{\partial x_i} P_{ij}(\mathbf{x}(t))\alpha_j,$$

or

(9.13) $$\frac{d}{dt} f(\mathbf{x}(t)) = L_\alpha f(\mathbf{x}(t), \qquad f(\mathbf{x}(0)) = f(\mathbf{x}^0).$$

Again we conclude from the existence and uniqueness theorem that (9.13) has a unique solution $f(\mathbf{x}(t)) \equiv f(\mathbf{x}^0 \exp \alpha t)$ analytic in t. Since $L_\alpha f(\mathbf{x}(t))$ is itself an analytic function of $\mathbf{x}(t)$ it follows that $(d/dt)(L_\alpha f(\mathbf{x}(t))) = L_\alpha(L_\alpha f(\mathbf{x}(t))) = L_\alpha{}^2 f(\mathbf{x}(t))$. Similarly

$$(d/dt)^j f(\mathbf{x}(t)) = L_\alpha{}^j f(\mathbf{x}(t)), \qquad j = 1, 2, \ldots .$$

Now $f(\mathbf{x}(t))$ has a Taylor series expansion in t about $t = 0$:

$$f(\mathbf{x}(t)) = \sum_{j=0}^{\infty} \left(\frac{d}{ds}\right)^j f(\mathbf{x}(s))\bigg|_{s=0} \frac{t^j}{j!} = \sum_{j=0}^{\infty} (L_\alpha{}^j f)(\mathbf{x}^0)\frac{t^j}{j!}$$
$$= (\exp tL_\alpha) f(\mathbf{x}^0),$$

where $\exp tL_\alpha$ is defined by its formal power series expansion.

Theorem 5.25. $f(\mathbf{x} \exp \alpha t) = (\exp tL_\alpha) f(\mathbf{x}) = \sum_{j=0}^{\infty} (t^j/j!) L_\alpha{}^j f(\mathbf{x})$.

Corollary 5.5. $\mathbf{x} \exp \alpha t = (\exp tL_\alpha)\mathbf{x}$.

We apply these results to the Lie derivative $L_1 = d/dx$ of Example 1. Equation (9.12) becomes

$$dx/dt = 1, \qquad x(0) = x^0,$$

or $x(t) = x^0 + t$, so the action of R as a transformation group on itself is just $x \longrightarrow x + t$. Furthermore, by Theorem 5.25, if f is analytic near x then

$$f(x \exp 1t) = f(x + t) = \left(\exp t\frac{d}{dx}\right) f(x) = \sum_{j=0}^{\infty} \frac{t^j}{j!} f^{(j)}(x),$$

which is the usual Taylor series expansion.

In Examples 1 and 2 the Lie derivatives themselves form a Lie algebra. In particular the commutator $[L_\alpha, L_\beta]$ is again a Lie derivative. We show that this is true in general.

Theorem 5.26. The set of all Lie derivatives of a local Lie transformation group G forms a Lie algebra which is a homomorphic image of $L(G)$. In fact, (1) $L_{(a\alpha + b\beta)} = aL_\alpha + bL_\beta$; (2) $L_{[\alpha,\beta]} = L_\alpha L_\beta - L_\beta L_\alpha = [L_\alpha, L_\beta]$ for all $a, b \in \mathfrak{C}$ and $\alpha, \beta \in L(G)$.

Proof. Property (1) follows from expression (9.5) for L_α. Property (2) is a little more complicated. Using Corollary 5.5 we can copy the proof of the analogous result for local linear Lie groups. Let $g(t) = (\exp \alpha\tau)(\exp \beta\tau)(\exp -\alpha\tau)(\exp -\beta\tau)$, where $t = \tau^2$. Then $g(t)$ is a curve in G with tangent vector $[\alpha, \beta]$ at $\mathbf{e}$. Thus, for any $f \in \mathfrak{A}_\mathbf{x}$ we have

(9.14) $$L_{[\alpha,\beta]} f(\mathbf{x}) = (d/dt) f(\mathbf{x}g(t))\,|_{t=0}.$$

However,

$$f(\mathbf{x}g(t)) = \{(\exp \tau L_\alpha)(\exp \tau L_\beta)(\exp -\tau L_\alpha)(\exp -\tau L_\beta)\}f(\mathbf{x})$$
$$= (1 + \tau^2[L_\alpha, L_\beta] + \cdots)f(\mathbf{x}).$$

It follows from this result and (9.14) that $L_{[\alpha,\beta]} = [L_\alpha, L_\beta]$. Q.E.D.

The reader should verify explicitly from (9.5) that $[L_\alpha, L_\beta]$ is a first-order differential operator even though $L_\alpha L_\beta$ and $L_\beta L_\alpha$ are second-order operators.

Two differential operators L_1, L_2 are **equal** on U, $L_1 = L_2$, if $L_1 f \equiv L_2 f$ for all analytic functions f on U. A set $\{L_1, \ldots, L_k\}$ of differential operators is **linearly dependent** on U if there exist constants $a_j \in \mathfrak{C}$ not all zero such that $\sum a_j L_j f \equiv 0$ for all analytic functions f on U. If the set $\{L_j\}$ is not linearly dependent then it is **linearly independent.**

With these definitions, we can discuss the structure of the Lie algebra $\mathfrak{L}(G)$ formed by the Lie derivatives $\{L_\alpha\}$. The map $\alpha \longrightarrow L_\alpha$ is a homomorphism of $L(G)$ onto $\mathfrak{L}(G)$. If this mapping is an isomorphism, i.e., if dim $L(G) =$ dim $\mathfrak{L}(G)$, we say G acts **effectively** as a transformation group.

Every local Lie group G acts effectively as a transformation group on itself: $g \longrightarrow g_0 g$, $g_0, g \in G$. Indeed it follows from Section 5.2 that the Lie derivatives corresponding to this action are

$$(9.15) \qquad L_\alpha = \sum_{i=1}^{n} \sum_{j=1}^{n} R_{ij}(\mathbf{g})\alpha_j \frac{\partial}{\partial g_i}, \qquad R_{ij}(\mathbf{g}) = \left.\frac{\partial \varphi_i(\mathbf{g}, \mathbf{h})}{\partial h_j}\right|_{\mathbf{h}=\mathbf{e}}.$$

Since $R_{ij}(\mathbf{e}) = \delta_{ij}$ the map $\alpha \longrightarrow L_\alpha$ is an isomorphism. We can now use Theorem 5.25 to gain some insight into the exponential mapping:

$$(9.16) \qquad \mathbf{g} \exp \alpha t = (\exp tL_\alpha)\mathbf{g} = \sum_{j=0}^{\infty} (t^j/j!) L_\alpha{}^j \mathbf{g}, \qquad \mathbf{g} \in G.$$

Suppose G is a local Lie transformation group which does not act effectively, i.e., the map $\alpha \longrightarrow L_\alpha$ is not 1–1. Let

$$L' = \{\alpha \in L(G): L_\alpha \equiv 0\}.$$

Then L' is a nontrivial subspace of $L(G)$. Furthermore, if $\alpha \in L'$ and $\beta \in L(G)$, then

$$L_{[\alpha,\beta]} = [L_\alpha, L_\beta] = [0, L_\beta] = 0,$$

so $[\alpha, \beta] \in L'$. This shows that L' is an ideal in $L(G)$. Thus if G does not act effectively then $L(G)$ contains a proper ideal. If a local Lie group G has a Lie algebra $L(G)$ with no proper ideals then G must always act effectively.

The following result is of basic importance for applications of Lie theory. It states that *any* Lie algebra of differential operators is the algebra of Lie derivatives of a local Lie transformation group.

Theorem 5.27. Let

$$L_j = \sum_{i=1}^{m} P_{ij}(\mathbf{x})(\partial/\partial x_i), \qquad j = 1, \cdots, n, \tag{9.17}$$

be n linearly independent differential operators defined and analytic in a connected open set $U \subset \mathfrak{C}_m$. If there exist constants c^l_{jk} such that

$$[L_j, L_k] = L_j L_k - L_k L_j = \sum_{l=1}^{n} c^l_{jk} L_l, \qquad 1 \leq j, k \leq n, \tag{9.18}$$

then the n-dimensional Lie algebra $\mathcal{G}$ generated by the L_j is the algebra of Lie derivatives of a local Lie transformation group G acting effectively on U. The action of G on $\mathbf{x}^0 \in U$ is obtained by solving the equations

$$d\mathbf{x}/dt = \left(\sum_{j=1}^{n} \alpha_j L_j\right)\mathbf{x}, \qquad \mathbf{x}(0) = \mathbf{x}^0. \tag{9.19}$$

Indeed the unique solution of these equations is $\mathbf{x}^0 \exp \alpha t = \mathbf{Q}(\mathbf{x}^0, \exp \alpha t)$.

Proof. By (9.18), the L_j form a basis for an n-dimensional Lie algebra $\mathcal{G}$. Any element α in $\mathcal{G}$ can be written uniquely as $L_\alpha = \sum \alpha_j L_j$, where $\alpha = (\alpha_1, \ldots, \alpha_n)$. (Note: The Jacobi equality is automatically satisfied by any set of linear operators, so $\mathcal{G}$ is indeed a Lie algebra.) By Theorems 5.9 and 5.19, $\mathcal{G}$ is the Lie algebra of some local Lie group G, unique up to isomorphism. We label the elements of G by means of the exponential mapping: $g = \exp \alpha$, $\alpha \in \mathcal{G}$, i.e., we use canonical coordinates in G.

By Theorem 5.24, the action $\mathbf{x}(g) = \mathbf{x} \exp \alpha = \mathbf{Q}(\mathbf{x}, \exp \alpha)$ of G on U must be given by the solutions of (9.19). We need only verify that G is a transformation group on U, i.e., that $(\mathbf{x}g)h = \mathbf{x}(gh)$, where $g = \exp \alpha$ and $h = \exp \beta$.

Let

$$\begin{aligned} \mathbf{y}(t) &= (\mathbf{x}g)h(t) = \mathbf{Q}(\mathbf{x}g, h(t)) = \exp tL_\beta(\mathbf{x}g) \\ \mathbf{z}(t) &= \mathbf{x}(gh(t)) = \mathbf{Q}(\mathbf{x}, gh(t)), \end{aligned} \tag{9.20}$$

where $h(t) = \exp \beta t$ and $(\mathbf{x}k)_i = Q_i(x_1, \ldots, x_m; k_1, \ldots, k_n.)$ We will show that $\mathbf{y}(t) \equiv \mathbf{z}(t)$. The proof of this fact is not trivial and will be accomplished through a chain of lemmas.

Lemma 5.5. Let G be an n-dimensional local Lie transformation group acting on U. Denote the action of G on U by $\mathbf{x}g = \mathbf{Q}(\mathbf{x}, g)$. Then the function $\mathbf{Q}(\mathbf{x}, g)$ satisfies the identity

$$\sum_{j=1}^{n} \frac{\partial Q_i(\mathbf{x}, g)}{\partial g_j} R_{js}(g) = P_{is}(\mathbf{x}g) = P_{is}(\mathbf{Q}), \qquad 1 \leq i \leq m, \quad 1 \leq s \leq n, \tag{9.21}$$

where

$$P_{is}(\mathbf{x}) = \left.\frac{\partial Q_i(\mathbf{x}, h)}{\partial h_s}\right|_{h=\mathbf{e}}, \qquad R_{js}(g) = \left.\frac{\partial \varphi_j(g, k)}{\partial k_s}\right|_{k=\mathbf{e}}.$$

Proof. This result follows from the identity $Q_i(\mathbf{x}, g \exp \alpha t) = Q_i(\mathbf{x}g, \exp \alpha t)$ by differentiating with respect to t and setting $t = 0$ exactly as in the derivation of (5.4) and (5.6). Q.E.D.

Since $R_{js}(\mathbf{e}) = \delta_{is}$, the matrix $(R_{js}(g))$ is invertible for all g in a suitably small neighborhood of $\mathbf{e}$. Denoting the inverse matrix by $S(g)$, we can write (9.21) in the form

$$\partial Q_i(\mathbf{x}, g)/\partial g_j = \sum_{k=1}^{n} P_{ik}(Q) S_{kj}(g), \qquad 1 \le i \le m, \quad 1 \le j \le n. \tag{9.22}$$

Lemma 5.6. Let $P(\mathbf{x})$ be an analytic $m \times n$ matrix of rank n and $S(g) = R^{-1}(g)$ an analytic $n \times n$ matrix with the properties $S(\mathbf{e}) = E_n$ and

(9.23)

$$\sum_{i=1}^{m} \left(P_{ij} \frac{\partial P_{lk}}{\partial x_i} - P_{ik} \frac{\partial P_{lj}}{\partial x_i} \right) = \sum_{r=1}^{n} c^r_{jk} P_{lr}, \qquad 1 \le l \le m, \quad 1 \le j, k \le n,$$

$$\sum_{i=1}^{n} \left(R_{ij} \frac{\partial R_{lk}}{\partial g_i} - R_{ik} \frac{\partial R_{lj}}{\partial g_i} \right) = \sum_{r=1}^{n} c^r_{jk} R_{lr}, \qquad 1 \le j, k, l \le n, \tag{9.24}$$

where the c^r_{jk} are constants. Then (9.22), regarded as a system of equations for $Q_i(\mathbf{x}, g)$, has a unique solution satisfying the initial condition $\mathbf{Q}(\mathbf{x}, \mathbf{e}) = \mathbf{x}$.

Proof. According to the fundamental existence and uniqueness theorem for first-order systems of partial differential equations, the system (9.22) has a unique solution $\mathbf{Q}$ such that $\mathbf{Q}(\mathbf{x}, \mathbf{e}) = \mathbf{x}$ provided the **integrability conditions** $\partial_{g_j} \partial_{g_l} Q_i = \partial_{g_l} \partial_{g_j} Q_i$, or

$$\sum_{q=1}^{m} \sum_{r,s=1}^{n} \left(P_{qr} \frac{\partial P_{is}}{\partial Q_q} - P_{qs} \frac{\partial P_{ir}}{\partial Q_q} \right) S_{rl} S_{sj} = \sum_{s=1}^{n} P_{is} \left(\frac{\partial S_{sl}}{\partial g_j} - \frac{\partial S_{sj}}{\partial g_l} \right) \tag{9.25}$$

are satisfied (Cohen [1], Pontrjagin [1]). It follows from (9.23) and the assumption rank $P = n$ that the integrability conditions become

$$\sum_{r,s=1}^{n} c^k_{rs} S_{rl} S_{sj} = \frac{\partial S_{kl}}{\partial g_j} - \frac{\partial S_{kj}}{\partial g_l}. \tag{9.26}$$

Now $RS = E_n$ (matrix multiplication), so

$$\frac{\partial R}{\partial g_j} S + R \frac{\partial S}{\partial g_j} = Z \qquad \text{or} \qquad \frac{\partial R}{\partial g_j} = -R \frac{\partial S}{\partial g_j} R.$$

Substituting this identity into (9.24), we find (9.24) is equivalent to (9.26). Therefore, the integrability conditions (9.25) are satisfied. Q.E.D.

Now we return to the proof of the theorem. If the function $\mathbf{Q}(\mathbf{x}, g)$ defines the action of a local transformation group on U then it satisfies the system (9.22). Thus, we will construct $\mathbf{Q}(\mathbf{x}, g)$ by requiring that it be a solution of

(9.22) satisfying the initial condition $\mathbf{Q}(\mathbf{x}, \mathbf{e}) = \mathbf{x}$. The matrix functions P and $S = R^{-1}$ are already determined since P is given by (9.17) and R is obtained from G. Furthermore, it follows from (9.15) and (9.18) that P and R satisfy conditions (9.23) and (9.24), where the c^r_{jk} are the structure constants of G. By Lemma 5.6 there is a unique solution $\mathbf{Q}(\mathbf{x}, g)$ of (9.22). This is the same solution we would get by solving (9.19), Indeed $\mathbf{x}(t) = \mathbf{Q}(\mathbf{x}^0, \exp \alpha t)$ satisfies the conditions $\mathbf{x}(0) = \mathbf{x}^0$ and

$$\dot{x}_i(t) = \sum_{jl} \frac{\partial Q_i}{\partial g_j} R_{jl}(\exp \alpha t)\alpha_l = \sum_{sjl} P_{is}(\mathbf{x}(t))S_{sj}(\exp \alpha t)R_{jl}(\exp \alpha t)\alpha_l$$
$$= \sum_{l=1}^{n} P_{il}(\mathbf{x}(t))\alpha_l, \qquad 1 \le i \le m.$$

At this point we can verify the equality $\mathbf{y}(t) = \mathbf{z}(t)$ in (9.20). Indeed

$$\dot{y}_i(t) = L_\beta y_i = \sum_{j=1}^{n} P_{ij}(\mathbf{y})\beta_j$$

$$\dot{z}_i(t) = \sum_{qlj} \frac{\partial Q_i(\mathbf{x}, gh(t))}{\partial k_q} \frac{\partial \varphi_q(g, h(t))}{\partial h_l} R_{lj}(h(t))\beta_j$$

$$= \sum_{sqj} P_{is}(\mathbf{z})S_{sq}(gh(t))R_{qj}(gh(t))\beta_j = \sum_{j=1}^{n} P_{ij}(\mathbf{z})\beta_j$$

and $\mathbf{y}(0) = \mathbf{z}(0) = \mathbf{x}^0 \exp \alpha$. [We have used the identity

$$\sum_{q=1}^{n} \frac{\partial \varphi_q(g, h)}{\partial h_l} R_{lj}(h) = R_{qj}(gh),$$

whose derivation is analogous to (5.4).] Since $\mathbf{y}(t)$ and $\mathbf{z}(t)$ satisfy the same system of equations and the same initial conditions, we have $\mathbf{y}(t) \equiv \mathbf{z}(t)$. Setting $t = 1$ we conclude that $\mathbf{x}^0(\exp \alpha \exp \beta) = (\mathbf{x}^0 \exp \alpha) \exp \beta$. Finally the reader can verify from Theorem 5.25 that $\mathbf{Q}(\mathbf{x}, g)$ is analytic in its $m + n$ arguments. Q.E.D.

Thus, the study of local Lie transformation groups is completely equivalent to the study of Lie algebras of differential operators. We shall need a slight extension of the above result: the equivalence of local multiplier reps and Lie algebras of generalized Lie derivatives.

Let L_j, $1 \le j \le n$, be linearly independent differential operators defined on the neighborhood $U \subset \mathfrak{C}_{m+1}$ and such that $[L_j, L_k] = \sum c^l_{jk} L_l$ for $i, k = 1, \ldots, n$, where the c^l_{jk} are constants, Suppose also that

$$[\partial/\partial x_{m+1}, L_j] = 0, \qquad 1 \le j \le n,$$

i.e., $(\partial/\partial x_{m+1})(P_{ij}(\mathbf{x})) \equiv 0$ for $1 \le j \le n$, $1 \le i \le m + 1$, where

$$L_j = \sum_{i=1}^{m+1} P_{ij}(\mathbf{x})(\partial/\partial x_i).$$

We emphasize the distinction between x_{m+1} and the remaining coefficients of

$\mathbf{x}$ by writing $\mathbf{x} = (\mathbf{x}', w)$, where $\mathbf{x}' = (x_1, \ldots, x_m)$, $w = x_{m+1}$. Theorem 5.27 states that the L_j generate a Lie algebra which is the algebra of Lie derivatives of a local group G acting on U. The action of G is given by integration of the differential system

(9.27)

$$\dot{x}_i(t) = \sum_{j=1}^{n} P_{ij}(\mathbf{x}'(t))\alpha_j, \qquad \dot{w}(t) = \sum_{j=1}^{n} P_{m+1,j}(\mathbf{x}'(t))\alpha_j, \qquad \mathbf{x}(0) = \mathbf{x}.$$

We write $P_{ij}(\mathbf{x}) = P_{ij}(\mathbf{x}')$ since the P_{ij} are independent of $\mathbf{w}$. The right-hand sides of Eq. (9.27) are independent of w, so the solution takes the form $\mathbf{x}(t) = \mathbf{x} \exp \alpha t = (\mathbf{x}' \exp \alpha t, q(\mathbf{x}', \exp \alpha t) + w)$, where q is an analytic scalar-valued function of its $m + n$ arguments. The action of G on U is completely determined by the one-parameter subgroups

(9.28) $$\mathbf{x}g = (\mathbf{x}'g, q(\mathbf{x}', g) + w), \qquad g \in G.$$

If $\mathbf{x}g_1 \in U$ and $g_2 \in G$, then $(\mathbf{x}g_1)g_2 = \mathbf{x}(g_1 g_2)$, which leads to

$$(\mathbf{x}'g_1)g_2 = \mathbf{x}'(g_1 g_2), \qquad q(\mathbf{x}', g_1 g_2) = q(\mathbf{x}', g_1) + q(\mathbf{x}'g_1, g_2),$$

or

(9.29) $$\nu(\mathbf{x}', g_1 g_2) = \nu(\mathbf{x}', g_1)\nu(\mathbf{x}'g_1, g_2),$$

where $\nu(\mathbf{x}', g) = \exp q(\mathbf{x}, g)$. If $g = \mathbf{e}$ then $q(\mathbf{x}', \mathbf{e}) = 0$ and $\nu(\mathbf{x}', \mathbf{e}) = 1$ for all $x \in U$.

Let $\mathfrak{A}_w$ be the set of all functions $f(\mathbf{x})$ on $\mathfrak{C}_{m+1}$, analytic in a neighborhood of $\mathbf{x}^0 \in U$ and such that $f(\mathbf{x}) = \mathbf{e}^w h(\mathbf{x}')$, i.e., $e^{-w}f(\mathbf{x})$ is independent of w. Without loss of generality we can assume $\mathbf{x}^0 = \boldsymbol{\theta}$. Given $g \in G$, we define operators $\mathbf{T}_w{}^g$ on $\mathfrak{A}_w$ by

$$\mathbf{T}_w{}^g f(\mathbf{x}) = f(\mathbf{x}g), \qquad f \in \mathfrak{A}_w.$$

By (9.28), $\mathbf{T}_w{}^g f \in \mathfrak{A}_w$ and these operators define a local rep of G on $\mathfrak{A}_w$.

Let $\mathfrak{A}$ be the set of all functions $f(\mathbf{x}')$ on $\mathfrak{C}_m$ analytic in some neighborhood of $\boldsymbol{\theta}' = (0, \ldots, 0)$. There is a 1–1 mapping ρ of $\mathfrak{A}$ onto $\mathfrak{A}_w$ given by $\rho: f(\mathbf{x}') \longrightarrow e^w f(\mathbf{x}')$. Since this mapping is invertible, the operators $\mathbf{T}_w{}^g$ on $\mathfrak{A}_w$ induce operators $\mathbf{T}^g = \rho^{-1}\mathbf{T}_w{}^g\rho$ on $\mathfrak{A}$ with the properties

(9.30)
$$\begin{aligned} &(1) \quad [\mathbf{T}^g f](\mathbf{x}') = \nu(\mathbf{x}', g) f(\mathbf{x}'g) \\ &(2) \quad [\mathbf{T}^e f](\mathbf{x}') = f(\mathbf{x}') \\ &(3) \quad [\mathbf{T}^{g_1 g_2} f](\mathbf{x}') = [\mathbf{T}^{g_1}(\mathbf{T}^{g_2} f)](\mathbf{x}') \end{aligned}$$

valid for all $f \in \mathfrak{A}$, $\mathbf{x}' \in \mathfrak{C}_m$, and $g_1, g_2 \in G$ such that both sides of expressions (9.30) make sense. The $\mathbf{T}^g$ define a local rep of G on $\mathfrak{A}$.

Definition. Let G be a local Lie transformation group acting on a neighborhood $U \subset \mathfrak{C}_m$, $\mathbf{x}^0 \in U$, and let $\mathfrak{A}$ be the set of all functions analytic in a neighborhood of $\mathbf{x}^0$. A **(local) multiplier representation** $\mathbf{T}$ of G on $\mathfrak{A}$ with

multiplier ν consists of a mapping $\mathbf{T}(g)$ of $\mathfrak{A}$ onto $\mathfrak{A}$ defined for $g \in G$ and $f \in \mathfrak{A}$ by

$$[\mathbf{T}(g)f](\mathbf{x}) = \nu(\mathbf{x}, g)f(\mathbf{x}g).$$

Here $\nu(\mathbf{x}, g)$ is a scalar-valued function analytic in $\mathbf{x}$ and g such that (1) $\nu(\mathbf{x}, \mathbf{e}) = 1$ and (2) $\nu(\mathbf{x}, g_1 g_2) = \nu(\mathbf{x}, g_1)\nu(\mathbf{x}g_1, g_2)$.

Property (2) is equivalent to (3) of expressions (9.30). The mappings (9.30) define a multiplier rep and every multiplier rep can be so obtained. Indeed, if $\mathbf{T}$ is a multiplier rep of G on $U \subset \mathfrak{C}_m$ we can define an action of G as a transformation group on a neighborhood of $\mathfrak{C}_{m+1}$ by $(\mathbf{x}, w) \longrightarrow (\mathbf{x}g, w + \ln[\nu(\mathbf{x}, g)])$, where $\mathbf{x} \in U$ and $w \in \mathfrak{C}$. The mapping $\mathbf{T}^g$, (9.30), induced by this action is just $\mathbf{T}(g)$.

Let $\mathbf{T}$ be a multiplier rep of $G, f \in \mathfrak{A}$, and $\alpha \in L(G)$.

Definition. The **generalized Lie derivative** $D_\alpha f$ of f under the one-parameter group $\exp \alpha t$ is the analytic function

$$D_\alpha f(\mathbf{x}) = (d/dt)[\mathbf{T}(\exp \alpha t)f](\mathbf{x})\,|_{t=0}. \tag{9.31}$$

For $\nu \equiv 1$ the generalized Lie derivative becomes the ordinary Lie derivative L_α.

Direct computation from (9.31) yields

$$D_\alpha f(\mathbf{x}) = \sum_{j=1}^{n} \sum_{i=1}^{m} P_{ij}(\mathbf{x})\alpha_j \frac{\partial f}{\partial x_i} + \sum_{j=1}^{n} \alpha_j P_j(\mathbf{x}) f(\mathbf{x}), \tag{9.32}$$

where the analytic functions $P_{ij}(\mathbf{x})$ are defined by (9.2) and

$$\sum_{j=1}^{n} \alpha_j P_j(\mathbf{x}) = (d/dt)\nu(\mathbf{x}, \exp \alpha t)\,|_{t=0}.$$

Multiplier reps are a particular type of ordinary local reps, so Theorems 5.24–5.27 have immediate analogies for multiplier reps and generalized Lie derivatives. The elementary proofs of the following results are left to the reader.

Theorem 5.28. The generalized Lie derivatives of a local multiplier rep form a Lie algebra under the operations of addition of derivatives and Lie bracket

$$[D_\alpha, D_\beta] = D_\alpha D_\beta - D_\beta D_\alpha.$$

This algebra is a homomorphic image of $L(G)$:

$$D_{a\alpha + b\beta} = aD_\alpha + bD_\beta, \qquad D_{[\alpha, \beta]} = [D_\alpha, D_\beta].$$

Definition. A Lie group G acts **effectively** in a local multiplier rep $\mathbf{T}$ if $L(G)$ is isomorphic to the algebra of generalized Lie derivatives.

Theorem 5.29. A local multiplier rep is completely determined by its generalized Lie derivatives.

Theorem 5.30. $[\mathbf{T}(\exp \alpha t)f](\mathbf{x}) = \sum_{j=0}^{\infty} (t^j/j!)D_\alpha{}^j f(\mathbf{x}) = \exp(tD_\alpha)f(\mathbf{x})$.

Theorem 5.31. Let

$$D_j = \sum_{i=1}^{m} P_{ij}(\mathbf{x})(\partial/\partial x_i) + P_j(\mathbf{x}), \qquad j = 1, \cdots, n,$$

be n linearly independent differential operators defined and analytic in an open set $U \subset \mathfrak{C}_m$. If there exist constants c^l_{jk} such that

$$[D_j, D_k] = \sum_{l=1}^{n} c^l_{jk} D_l, \qquad 1 \le j, k \le n,$$

then the D_j form a basis for a Lie algebra which is the algebra of generalized Lie derivatives of an effective local multiplier rep $\mathbf{T}$. The action of the group G is obtained by integration of the equations

$$\text{(9.33)} \qquad \dot{x}_i(t) = \sum_{j=1}^{n} P_{ij}(\mathbf{x}(t))\alpha_j, \qquad (d/dt)\ln \nu(\mathbf{x}^0, \exp \alpha t) = \sum_{j=1}^{n} \alpha_j P_j(\mathbf{x}(t)),$$

where $\mathbf{x}(0) = \mathbf{x}^0$, $\nu(\mathbf{x}^0, \mathbf{e}) = 1$, $\mathbf{x}(t) = \mathbf{x}^0 \exp \alpha t$, and $1 \le i \le m$.

5.10 Examples of Transformation Groups

Consider the group $SL(2) = SL(2, \mathfrak{C})$ of all 2×2 complex matrices

$$\text{(10.1)} \qquad g = \begin{pmatrix} a & b \\ c & d \end{pmatrix},$$

with $\det g = 1$. As we showed in Section 5.4, the Lie algebra $sl(2)$ of $SL(2)$ is the three-dimensional space of all 2×2 complex matrices α with $\operatorname{tr} \alpha = 0$:

$$\text{(10.2)} \qquad \alpha = \begin{pmatrix} \alpha_1 & \alpha_2 \\ \alpha_3 & -\alpha_1 \end{pmatrix}.$$

The elements

$$\text{(10.3)} \qquad \mathcal{J}^+ = \begin{pmatrix} 0 & -1 \\ 0 & 0 \end{pmatrix}, \qquad \mathcal{J}^- = \begin{pmatrix} 0 & 0 \\ -1 & 0 \end{pmatrix}, \qquad \mathcal{J}^3 = \begin{pmatrix} \frac{1}{2} & 0 \\ 0 & -\frac{1}{2} \end{pmatrix}$$

define a basis for $sl(2)$ with commutation relations

$$\text{(10.4)} \qquad [\mathcal{J}^3, \mathcal{J}^\pm] = \pm \mathcal{J}^\pm, \qquad [\mathcal{J}^+, \mathcal{J}^-] = 2\mathcal{J}^3.$$

An explicit computation shows that $\mathcal{J}^\pm$, $\mathcal{J}^3$ generate one-parameter subgroups

(10.5)

$$\exp a\mathcal{J}^3 = \begin{pmatrix} e^{a/2} & 0 \\ 0 & e^{-a/2} \end{pmatrix}, \quad \exp b\mathcal{J}^+ = \begin{pmatrix} 1 & -b \\ 0 & 1 \end{pmatrix}, \quad \exp c\mathcal{J}^- = \begin{pmatrix} 1 & 0 \\ -c & 1 \end{pmatrix}.$$

In terms of canonical coordinates we can write every $g \in SL(2)$ uniquely in the form

$$g = \exp(a\mathcal{J}^3 + b\mathcal{J}^+ + c\mathcal{J}^-)$$

for g sufficiently close to the identity. However, we shall adopt another method of parametrizing $SL(2)$. Note that

(10.6)

$$(\exp b'\mathcal{J}^+)(\exp c'\mathcal{J}^-)\exp \tau'\mathcal{J}^3 = \begin{pmatrix} (\exp \frac{1}{2}\tau')(1 + b'c') & -b' \exp -\frac{1}{2}\tau' \\ -c' \exp \frac{1}{2}\tau' & \exp -\frac{1}{2}\tau' \end{pmatrix}.$$

Thus, if $g \in SL(2)$, (10.1), is sufficiently close to the identity we can write it uniquely in the form

(10.7) $$g = (\exp b'\mathcal{J}^+)(\exp c'\mathcal{J}^-)\exp \tau'\mathcal{J}^3,$$

where $\exp \frac{1}{2}\tau' = d^{-1}$, $b' = -b/d$, and $c' = -cd$.

Consider the differential operators

(10.8) $$J^+ = -2uz + z^2(d/dz), \qquad J^- = -d/dz, \qquad J^3 = -u + z(d/dz)$$

acting on a neighborhood of $0 \in \mathfrak{C}$. Here $2u$ is a complex number. These operators satisfy the commutation relations

(10.9) $$[J^3, J^\pm] = \pm J^\pm, \qquad [J^+, J^-] = 2J^3.$$

Comparing these relations with (10.4), we see that the J-operators generate a Lie algebra isomorphic to $sl(2)$. Hence the J-operators are generalized Lie derivatives corresponding to a local multiplier rep of $SL(2)$ on $\mathfrak{C}$. We will use Theorem 5.31 to compute this multiplier rep.

According to (9.33) the action of the one-parameter subgroup $\exp \tau\mathcal{J}^3$ on $\mathfrak{C}$ is obtained by integration of the equations

(10.10)

$$\frac{dz}{d\tau} = z, \qquad \frac{d}{d\tau}\ln \nu(z^0, \exp \tau\mathcal{J}^3) = -u, \qquad z(0) = z^0, \quad \nu(z^0, \mathbf{e}) = 1.$$

The solution is

$$z(\tau) = z^0 e^\tau, \qquad \nu(z^0, \exp \tau\mathcal{J}^3) = e^{-u\tau},$$

so if f is analytic in a neighborhood of $z^0 \in \mathfrak{C}$ then

(10.11) $$[\mathbf{T}(\exp \tau\mathcal{J})f](z^0) = e^{-u\tau}f(z^0 e^\tau).$$

To compute the action of $\exp b\mathcal{J}^+$ we must integrate the equations

(10.12)

$$\frac{dz}{db} = z^2, \qquad \frac{d}{db}\ln \nu(z^0, \exp b\mathcal{J}^+) = -2uz, \qquad z(0) = z^0, \quad \nu(z^0, \mathbf{e}) = 1.$$

The solution is

$$z(b) = z^0(1 - bz^0)^{-1}, \qquad \nu(z^0, \exp b\mathcal{J}^+) = (1 - bz^0)^{2u}, \qquad |bz^0| < 1.$$

Thus,

$$[\mathbf{T}(\exp b\mathcal{J}^+)f](z^0) = (1 - bz^0)^{2u} f\left(\frac{z^0}{1 - bz^0}\right), \qquad |bz^0| < 1. \tag{10.13}$$

A similar computation yields

$$[\mathbf{T}(\exp c\mathcal{J}^-)f](z^0) = f(z^0 - c). \tag{10.14}$$

Note. This action of $SL(2)$ is purely local. Unless $2u$ is a nonnegative integer, it is not possible to extend our local multiplier rep to a global rep of $SL(2)$. To give a precise meaning to Eqs. (10.11), (10.13), and (10.14) it would be necessary to state explicitly the values of τ, b, c, and z^0 and the functions f for which the equations are defined. However, these values are easily determined by inspection once the domain of f is given, so ordinarily we will not bother to list them. Similar remarks hold for all computations in this book involving local transformation groups.

We can now determine the local multiplier rep $\mathbf{T}(g)$ defined by the generalized Lie derivatives (10.8). Expressing g by (10.7) we have

$$\begin{aligned}[\mathbf{T}(g)f](z) &= [\mathbf{T}(\exp b'\mathcal{J}^+)\mathbf{T}(\exp c'\mathcal{J}^-)\mathbf{T}(\exp \tau'\mathcal{J}^3)f](z) \\ &= (\exp -u\tau')(1 - b'z)^{2u} f\left(\frac{z(\exp\tau')(1 + c'b') - c'\exp\tau'}{1 - b'z}\right) \\ &= (bz + d)^{2u} f\left(\frac{az + c}{bz + d}\right).\end{aligned}$$

Thus, if $g \in SL(2)$, (10.1), then $zg = (az + c)/(bz + d)$, $\nu(z, g) = (bz + d)^{2u}$, and

$$[\mathbf{T}(g)f](z) = (bz + d)^{2u} f\left(\frac{az + c}{bz + d}\right) \tag{10.15}$$

for g in a sufficiently small neighborhood of $\mathbf{e}$. The operators (10.8) are the generalized Lie derivatives of (10.15). For example, by (10.5),

(10.16)

$$J^+ f(z) = \frac{d}{db}\mathbf{T}(\exp b\mathcal{J}^+)f(z)\Big|_{b=0} = -2uzf(z) + z^2\frac{d}{dz}f(z).$$

Our general theory shows that the operators (10.15) define a local multiplier rep of $SL(2)$: $\mathbf{T}(g_1)\mathbf{T}(g_2) = \mathbf{T}(g_1 g_2)$ for g_1, g_2 in a sufficiently small neighborhood of $\mathbf{e}$, a fact which the reader can also verify directly.

There is one case in which this local rep can actually be extended to a global group rep of $SL(2)$. Suppose $2u$ is a nonnegative integer and let $\mathcal{V}^{(u)}$ be the $(2u + 1)$-dimensional vector space consisting of all polynomials

$$f(z) = \sum_{j=0}^{2u} c_j z^j.$$

Now $\mathcal{V}^{(u)}$ is invariant under the operators $\mathbf{T}(g)$ for *all* $g \in SL(2)$ since by (10.15) each such operator maps a polynomial of order $2u$ into a polynomial of order $2u$. Furthermore, it is easy to check that on $\mathcal{V}^{(u)}$ the group property is valid for *all* $g \in SL(2)$. Indeed, if $g, g' \in SL(2)$ then

$$\text{(10.17)}\quad \mathbf{T}(g)[\mathbf{T}(g')f](z) = [\mathbf{T}(gg')f](z)$$
$$= [(b'a + bd')z + (b'c + dd')]^{2u} f\left(\frac{(a'a + c'b)z + (a'c + c'd)}{(b'a + d'b)z + (b'c + d'd)}\right),$$

where

$$gg' = \begin{pmatrix} a & b \\ c & d \end{pmatrix}\begin{pmatrix} a' & b' \\ c' & d' \end{pmatrix} = \begin{pmatrix} aa' + bc', & ab' + bd' \\ ca' + dc', & cb' + dd' \end{pmatrix}.$$

Thus, we have obtained a class of finite-dimensional reps $\mathbf{D}^{(u)}$, $2u = 0, 1, 2, \ldots$, of the global group $SL(2)$, i.e., reps in the sense of Section 5.8.

Let us show that the reps $\mathbf{D}^{(u)}$ are irred. By Theorem 5.23 it is enough to show that the induced Lie algebra reps (which we also denote $\mathbf{D}^{(u)}$) are irred. Choose the natural basis $h_j(z) = z^j, j = 0, 1, \ldots, 2u$, for $\mathcal{V}^{(u)}$. The action of the generalized Lie derivatives on this basis is

(10.18)

$$J^3 h_j = \left(-u + z\frac{d}{dz}\right)z^j = (j - u)h_j, \qquad J^- h_j = -\frac{dz^j}{dz} = -jh_{j-1},$$

$$J^+ h_j = \left(-2uz + z^2\frac{d}{dz}\right)z^j = (j - 2u)h_{j+1}.$$

This action defines the Lie algebra rep $\mathbf{D}^{(u)}$ of $sl(2)$. We will use one of the Schur lemmas to prove that $\mathbf{D}^{(u)}$ is irred. (Note the remarks at the end of Section 5.8). Let $\mathbf{A}$ be a linear operator on $\mathcal{V}^{(u)}$ which commutes with $J^{\pm}$, J^3. Then $\mathbf{A}$ must commute with all of the J-operators in the rep $\mathbf{D}^{(u)}$. Now $J^3\mathbf{A}h_j = \mathbf{A}J^3h_j = (j - u)\mathbf{A}h_j$, so $\mathbf{A}h_j$ is a multiple of h_j: $\mathbf{A}h_j = \alpha_j h_j$, $\alpha_j \in \mathfrak{C}$. (Each eigenvalue $u - j$ of J^3 has multiplicity one.) Furthermore,

$$J^+\mathbf{A}h_j = (j - 2u)\alpha_j h_{j+1}, \qquad \mathbf{A}J^+h_j = (j - 2u)\alpha_{j+1}h_{j+1},$$

so $\alpha_j = \alpha_{j+1}$, $0 \leq j \leq 2u - 1$. Setting $\alpha_0 = \alpha$ we obtain $\mathbf{A}h_j = \alpha h_j$ or $\mathbf{A} = \alpha\mathbf{E}$, where $\mathbf{E}$ is the identity operator. Therefore, $\mathbf{D}^{(u)}$ is irred. We will see later that the $\{\mathbf{D}^{(u)}\}$ constitute a complete set of nonequivalent irred reps of the complex Lie group $SL(2)$.

Now that we have constructed the reps $\mathbf{D}^{(u)}$ let us examine their relationship to special function theory. The **matrix elements** $D_{lk}(g)$ of $\mathbf{T}(g)$ with respect to the basis $\{h_j\}$ are defined by

$$\text{(10.19)}\quad [\mathbf{T}(g)h_k](z) = \sum_{l=0}^{2u} D_{lk}(g)h_l(z), \qquad g \in SL(2), \quad 0 \leq k \leq 2u,$$

or explicitly,

$$(az + c)^k (bz + d)^{2u-k} = \sum_{l=0}^{2u} D_{lk}(g) z^l. \tag{10.20}$$

Since $\mathbf{T}(g_1 g_2) = \mathbf{T}(g_1)\mathbf{T}(g_2)$ the matrix elements satisfy the **addition theorems**

$$D_{lk}(g_1 g_2) = \sum_{j=0}^{2u} D_{lj}(g_1) D_{jk}(g_2), \qquad l, k = 0, 1, \ldots, 2u. \tag{10.21}$$

Formula (10.20) is a **generating function** for the matrix elements. Using the binomial theorem to expand the left-hand side of this expression we find that the matrix elements are given by

(10.22)

$$D_{lk}(g) = \begin{cases} \dfrac{a^l d^{2u-k} c^{k-l} k!}{l!(k-l)!} \, {}_2F_1\left(-l, -2u-k; k-l+1; \dfrac{bc}{ad}\right), \\ \qquad\qquad 2u \geq k \geq l \geq 0, \\ \dfrac{a^k d^{2u-l} b^{l-k} (2u-k)!}{(2u-l)!(l-k)!} \, {}_2F_1\left(-k, -2u+1; l-k+1; \dfrac{bc}{ad}\right), \\ \qquad\qquad 2u \geq l \geq k \geq 0, \end{cases}$$

where the ${}_2F_1$ are hypergeometric polynomials. (See the Symbol Index.) Relations (10.21) yield addition theorems for the ${}_2F_1$ which are not easy to prove directly. We shall show that this relation between group reps and special functions occurs frequently: Many special functions appear as matrix elements of Lie group reps and the group property leads to addition theorems obeyed by the functions.

We return to the case where $2u$ is an arbitrary complex number, not a nonnegative integer. Then (10.15) and (10.17) make sense only for group elements in a small neighborhood of the identity and functions f analytic in a neighborhood of $z = 0$. Since $2u$ is not a nonnegative integer, the multiplier $\nu(z, g) = (bz + d)^{2u}$ is defined by its power series expansion in z about $z = 0$. This series converges only if $|bz/d| < 1$. Furthermore if f is analytic near zero then $f(zg)$ is analytic near zero only if $(az + c)/(bz + d)$ is in the domain of f and $|az/c| < 1$ and $|bz/d| < 1$. Let $\mathfrak{a}$ be the space of all functions f analytic in a neighborhood of zero. There is now no finite-dimensional subspace of $\mathfrak{a}$ which is invariant under the operators $\mathbf{T}(g)$, (10.15). However, $\mathfrak{a}$ itself is invariant since if $f \in \mathfrak{a}$ we have $\mathbf{T}(g)f \in \mathfrak{a}$ for g is sufficiently close to the identity. Since the elements of $\mathfrak{a}$ are just those functions with convergent power series expansions about $z = 0$, the action of $\mathbf{T}(g)$ on $\mathfrak{a}$ can be determined from a knowledge of $\mathbf{T}(g)$ on the basis functions $h_l(z) = z^l$, $l = 0, 1, 2, \ldots$. We define the matrix elements $B_{lk}(g)$ by

$$[\mathbf{T}(g)h_k](z) = \sum_{l=0}^{\infty} B_{lk}(g) h_l(z), \qquad k = 0, 1, 2, \ldots, \tag{10.23}$$

or

$$(az + c)^k(bz + d)^{2u-k} = \sum_{l=0}^{\infty} B_{lk}(g)z^l. \tag{10.24}$$

Since **T** is a multiplier rep we have

$$B_{lk}(g_1 g_2) = \sum_{j=0}^{\infty} B_{lj}(g_1)B_{jk}(g_2), \qquad l, k = 0, 1, 2, \ldots, \tag{10.25}$$

for g_1, g_2 in a sufficiently small neighborhood of **e**. Thus, (10.24) is a generating function and (10.25) is an addition theorem for the special functions $B_{lk}(g)$. Computing the coefficient of z^l in (10.24) we find

(10.26)

$$B_{lk}(g) = \begin{cases} \dfrac{a^l d^{2u-k} c^{k-l} k!}{l!(k-l)!} \, {}_2F_1(-l, -2u+k; k-l+1; bc/ad), & k \geq l \geq 0, \\[2ex] \dfrac{a^k d^{2u-l} b^{l-k} \Gamma(2u-k+1)}{\Gamma(2u-l+1)(l-k)!} \, {}_2F_1(-k, -2u+l; l-k+1; bc/ad), \\ & l \geq k \geq 0, \end{cases}$$

where $\Gamma(z)$ is the gamma function (see the Symbol Index). Such local group reps are of importance in special function theory, as we shall see.

There are exactly three local transformation groups on the line. The corresponding algebras of differential operators are as follows:

(10.27)

(1) $L_1 = d/dz$, one-dimensional,

(2) $L_1 = d/dz$, $L_2 = z\, d/dz$, $[L_1, L_2] = L_1$, two-dimensional

(3) $L_- = d/dz$, $L_3 = z\, d/dz$, $L_+ = z^2\, d/dz$, three-dimensional

$[L_3, L_\pm] = \pm L_\pm$, $[L_+, L_-] = 2L_3$.

Any local group on the (real or complex) line can be transformed to one of these three by an analytic change of variable. Note that the operators (3) define the action of $SL(2)$ as a transformation group on the line. Any one-parameter group can always be expressed in the form (1) by a change of variable. For a proof of this classification see Lie [1] or Campbell [1].

The classification of local multiplier reps on the line is more complicated. Here it is necessary to classify the Lie algebras of generalized Lie derivatives in one complex variable, e.g., (10.8). There are an infinite number of such algebras. They are listed in the work of Miller [1].

Lie [1] has classified all local transformation groups in the plane. Again there are an infinite number of such groups. Some results on multiplier reps in the plane are given by Miller [1].

Problems

5.1 Let B be an $n \times n$ matrix with distinct eigenvalues $\lambda_1, \ldots, \lambda_n$. Show that Ad B acting on the space of all $n \times n$ complex matrices has the n^2 eigenvalues $\lambda_i - \lambda_j$, $1 \leq i, j \leq n$.

5.2 Verify directly that the matrix commutator satisfies the Jacobi identity.

5.3 Verify that the set of all real matrices

$$\begin{pmatrix} 1 & a & b \\ 0 & c & d \\ 0 & 0 & 1 \end{pmatrix}, \quad c > 0,$$

forms a Lie group, and compute the associated Lie algebra. In particular, determine the commutation relations.

5.4 Show that the following spaces of $m \times m$ matrices form Lie algebras: (a) all upper triangular matrices; (b) all upper triangular matrices with trace zero; (c) all upper triangular matrices with diagonal elements zero. Determine the corresponding Lie groups.

5.5 Compute all real Lie algebras of dimensions one and two (identifying isomorphic algebras). For each algebra compute the associated local linear Lie group.

5.6 Repeat Problem 5.5 for all complex Lie algebras of dimension three (see Jacobson [1, Chapter 1]).

5.7 Show that the element

$$\begin{pmatrix} a & 0 \\ 0 & -a^{-1} \end{pmatrix}, \qquad a > 0, \quad a \neq 1,$$

of $GL(2, R)$ cannot be expressed as $\exp \mathcal{A}$ for any $\mathcal{A} \in gl(2, R)$. Thus the exponential mapping may not cover a Lie group.

5.8 Let $\mu: G \longrightarrow G'$ be an analytic homomorphism of local Lie groups. Show that, in terms of canonical coordinates in G and G', μ is a linear mapping.

5.9 Let G be a linear Lie group with Lie algebra $\mathcal{G}$. Let $\mathcal{G}'$ be the subspace of $\mathcal{G}$ spanned by all elements of the form $[\mathcal{A}, \mathcal{B}]$, $\mathcal{A}, \mathcal{B} \in \mathcal{G}$. Show that $\mathcal{G}'$ is a Lie algebra, the **derived algebra** of $\mathcal{G}$, and that $\mathcal{G}' = L(G_c)$, where G_c is the commutator subgroup of G.

5.10 Verify that $L_1 = z\partial_y + y\partial_z$, $L_2 = x\partial_z + z\partial_x$, and $L_3 = y\partial_x - x\partial_y$, x, y, z real, generate the algebra of Lie derivatives of a local Lie transformation group G. Compute the action of G on R_3.

5.11 For $j = 1, 2$ let $\mathbf{T}_j$ be an analytic rep of the Lie group G on the vector space V_j with associated Lie algebra rep ρ_j. Show that the Lie algebra rep corresponding to $\mathbf{T}_1 \otimes \mathbf{T}_2$ takes the form $\boldsymbol{\rho}_1(\boldsymbol{\alpha}) \otimes \mathbf{E}_2 + \mathbf{E}_1 \otimes \boldsymbol{\rho}_2(\boldsymbol{\alpha})$ on $V_1 \otimes V_2$ where $\mathbf{E}_j$ is the identity operator on V_j and $\boldsymbol{\alpha} \in L(G)$.

5.12 Let G be a local transformation group acting on a neighborhood of $\mathbf{x}^0 \in F_m$. A function f analytic near $\mathbf{x}^0$ is an **invariant** of G if $f(\mathbf{x}g) = f(\mathbf{x})$ for each $\mathbf{x}$ sufficiently close to $\mathbf{x}^0$ and $g \in G$. Prove: A function f is an invariant of G if and only if $Lf(\mathbf{x}) = 0$ for all Lie derivatives L of G.

5.13 Compute the invariants of $E^+(2)$, (9.10), and of the group in Problem 5.10.

Chapter 6

Compact Lie Groups

6.1 Invariant Measures on Lie Groups

Let G be a real n-dimensional global Lie group of $m \times m$ matrices. A function $f(B)$ on G is **continuous** at $B \in G$ if it is a continuous function of the parameters $(g_1, \ldots, g_n)$ in a local coordinate system for G at B. Clearly if f is continuous with respect to one local coordinate system at B it is continuous with respect to all coordinate systems. If f is continuous at every $B \in G$ then it is a **continuous function** on G. We shall show how to define an infinitesimal volume element dA in G with respect to which the associated integral over the group is left-invariant, i.e.,

$$\int_G f(BA)\, dA = \int_G f(A)\, dA, \qquad B \in G, \tag{1.1}$$

where f is any continuous function on G such that either of the integrals converges. In terms of local coordinates $\mathbf{g} = (g_1, \ldots, g_n)$ at A,

$$dA = w(\mathbf{g})\, dg_1 \cdots dg_n = w(\mathbf{g})\, d\mathbf{g}, \tag{1.2}$$

where the continuous function w is called a **weight function.** If $\mathbf{k} = (k_1, \ldots, k_n)$ is another set of local coordinates at A then,

$$dA = \tilde{w}(\mathbf{k})\, dk_1 \cdots dk_n, \qquad \tilde{w}(\mathbf{k}) = w(\mathbf{g}(\mathbf{k}))\,|\det(\partial g_i/\partial k_j)|,$$

where the determinant is the Jacobian of the coordinate transformation. (For a precise definition of integrals on manifolds see Spivak [1].)

Two examples of such left-invariant measures are well known. We can identify R with the group of 2×2 matrices

$$A_x = \begin{pmatrix} 1 & x \\ 0 & 1 \end{pmatrix}, \qquad x \in R. \tag{1.3}$$

The continuous functions on this group are just the continuous functions $f(x)$ on the real line. Here, dx is a left-invariant measure. Indeed by a simple change of variable we have

$$\int_{-\infty}^{+\infty} f(y + x)\, dx = \int_{-\infty}^{\infty} f(x)\, dx, \qquad y \in R.$$

where f is any continuous function on R such that the integrals converge. Since R is abelian, dx is also right-invariant.

Consider the group $U(1) = \{e^{i\theta}\}$. The continuous functions on $U(1)$ can be written $f(\theta)$, where f is continuous for $0 \le \theta \le 2\pi$ and periodic with period 2π. The measure $d\theta$ is left-invariant (right-invariant) since

$$\int_0^{2\pi} f(\varphi + \theta)\, d\theta = \int_0^{2\pi} f(\theta)\, d\theta.$$

We now show how to construct a left-invariant measure for the n-dimensional real linear Lie group G. Let $\{\mathcal{C}_j,\ 1 \le j \le n\}$ be a basis for $L(G)$. We can introduce an inner product on $L(G)$ with respect to which this basis is ON. Associate the n-tuple $\alpha = (\alpha_1, \ldots, \alpha_n)$ with $\sum \alpha_j \mathcal{C}_j \in L(G)$. Now n linearly independent vectors $\alpha^{(1)}, \ldots, \alpha^{(n)}$ in $L(G)$ generate a parallelepiped in $L(G)$ with volume

$$V = |\det(\alpha_j^{(i)})| > 0. \tag{1.4}$$

Expression (1.4) defines volume in the tangent space of the identity element.

Let $A(t)$ be an analytic curve in G such that $A(0) = A$. We call $\dot{A}(0) = \tilde{\mathcal{a}}$ the **tangent matrix** to $A(t)$ at A. The set of all matrices $\tilde{\mathcal{a}}$ as $A(t)$ runs over all analytic curves through A forms a vector space T_A called the **tangent space at** A. If $A(t)$ is an analytic curve through A then $A^{-1}A(t)$ is an analytic curve through E_m. Thus $(d/dt)[A^{-1}A(t)]|_{t=0} = \mathcal{a} \in L(G)$, or $A^{-1}\tilde{\mathcal{a}} = \mathcal{a}$. Conversely, if $\mathcal{a} \in L(G)$ then $A(t) = A \exp t\mathcal{a}$ is an analytic curve through A with tangent matrix $\tilde{\mathcal{a}} = A\mathcal{a}$ at A. Thus every tangent matrix $\tilde{\mathcal{a}}$ at A can be written uniquely as

$$\tilde{\mathcal{a}} = A\mathcal{a}, \qquad \mathcal{a} \in L(G). \tag{1.5}$$

Let $\mathbf{g} = (g_1, \ldots, g_n)$ be local coordinates at A. Without loss of generality we can assume $A(\mathbf{e}) = A$. The matrix functions $A(0, \ldots, 0, g_j, \ldots, 0)$ are analytic curves through A, so $\partial A/\partial g_j(\mathbf{e}) = \tilde{\mathcal{a}}_j \in T_A$. We will define the volume V_A of the parallelepiped in T_A generated by the n tangent vectors $\tilde{\mathcal{a}}_j$. We do this by mapping T_A back to $T_{E_m} = L(G)$. According to (1.5) there exist matrices $\mathfrak{A}_j \in L(G)$ such that

$$A^{-1}\tilde{\mathfrak{A}}_j = \mathfrak{A}_j, \qquad j = 1, \ldots, n.$$

Writing $\mathfrak{A}_j = \sum \alpha_k^{(j)}\mathfrak{C}_k$, we define the volume of the parallelepiped in T_A as the volume of its image in $L(G)$:

$$V_A(\mathbf{g}) = |\det(\alpha_k^{(j)})| > 0.$$

By construction, our volume element is left-invariant. Indeed if $B \in G$ then

$$(BA)^{-1}(\partial/\partial g_j)[BA(\mathbf{g})]|_{\mathbf{g}=\mathbf{e}} = A^{-1}B^{-1}B\tilde{\mathfrak{A}}_j = A^{-1}\tilde{\mathfrak{A}}_j = \mathfrak{A}_j,$$

so $V_{BA} = V_A$. We define the measure d_lA on G by

(1.6) $$d_lA = V_A(\mathbf{g})\, dg_1 \cdots dg_n.$$

Expression (1.6) is actually independent of local coordinates. If $\mathbf{k} = (k_1, \ldots, k_n)$ is another local coordinate system at A then

$$A^{-1}\frac{\partial A}{\partial k_l} = \sum_j A^{-1}\tilde{\mathfrak{A}}_j\frac{\partial g_j}{\partial k_l} = \sum_{j,s}\frac{\partial g_j}{\partial k_l}\alpha_s^{(j)}\mathfrak{C}_s,$$

so

$$V_A(\mathbf{k}) = \left|\det\left(\sum_j \frac{\partial g_j}{\partial k_l}\alpha_s^{(j)}\right)\right| = \left|\det\left(\frac{\partial g_j}{\partial k_l}\right)\right| \cdot |\det(\alpha_s^{(j)})|.$$

Thus,

(1.7) $$\begin{aligned} V_A(\mathbf{k})\, dk_1 \cdots dk_n &= V_A(\mathbf{g})\,|\det(\partial g_j/\partial k_l)|\, dk_1 \cdots dk_n \\ &= V_A(\mathbf{g})\, dg_1 \cdots dg_n. \end{aligned}$$

We have shown that the integral

$$\int_G f(A)\, d_lA = \int_G f(g_1, \ldots, g_n)V_A(\mathbf{g})\, d\mathbf{g}$$

is well-defined provided it converges. Furthermore,

(1.8) $$\begin{aligned} \int_G f(BA)\, d_lA &= \int_G f(BA(\mathbf{g}))V_A(\mathbf{g})\, d\mathbf{g} = \int_G f(BA(\mathbf{g}))V_{BA}(\mathbf{g})\, d\mathbf{g} \\ &= \int_G f(A)V_A(\mathbf{g})\, d\mathbf{g} = \int_G f(A)\, d_lA, \end{aligned}$$

where the third equality follows from the fact that BA runs over G if A does.

By an analogous procedure one can also define a right-invariant measure in G. Indeed the tangent matrices at A can be written uniquely as $\tilde{\mathfrak{A}} = \mathfrak{B}A$, $\mathfrak{B} \in L(G)$. Writing

$$(\partial A/\partial g_j)A^{-1} = \mathfrak{B}_j = \sum \beta_k^{(j)}\mathfrak{C}_k,$$

we define

(1.9) $$W_A(\mathbf{g}) = |\det(\beta_k^{(j)}|, \qquad d_rA = W_A(\mathbf{g})\, dg_1 \cdots dg_n.$$

The reader can verify that d_rA is a right-invariant measure on G.

Since $A(A^{-1}\partial A/\partial g_j)A^{-1} = (\partial A/\partial g_j)A^{-1}$, we have

$$W_A(\mathbf{g}) = |\det \tilde{A}|\cdot V_A(\mathbf{g}), \tag{1.10}$$

where $\tilde{A}$ is the automorphism $\mathfrak{a} \longrightarrow A\mathfrak{a}A^{-1}$ of $L(G)$. Thus, if $\det \tilde{A} = 1$ for all $A \in G$ then $d_l A \doteq d_r A$ and there exists a two-sided invariant measure on G. In the next section we find sufficient conditions for the existence of a two-sided invariant measure.

It can be shown that a much larger class of groups (the locally compact topological groups) possesses left-invariant (right-invariant) measures. Furthermore, the left-invariant (right-invariant) measure of a group is unique up to a constant factor. That is, if dA and δA are left-invariant measures on G then there exists a constant $c > 0$ such that $dA = c\,\delta A$ (Naimark [1], Pontrjagin [1]).

To illustrate our construction, consider the matrix group (1.3). The matrix $A_1 - E_2$ is a basis for the one-dimensional Lie algebra. Let $A_x \in R$. Then

$$A_x^{-1}\frac{\partial A_x}{\partial x} = \begin{pmatrix} 1 & -x \\ 0 & 1 \end{pmatrix}\begin{pmatrix} 0 & 1 \\ 0 & 0 \end{pmatrix} = \begin{pmatrix} 0 & 1 \\ 0 & 0 \end{pmatrix} = A_1 - E_2$$

and $V_A(x) = 1$. Thus, $d_l A = dx$. Similarly $d_r A = dx$.

Now consider $GL(m, R)$. We can choose as parameters for A the m^2 matrix elements A_{ij}. The matrix $\partial A/\partial A_{ij}$ has a one in the ith row and jth column, and zeros every place else. A straightforward computation shows that the $m^2 \times m^2$ matrix $(\alpha_l^{(s)})$ looks like

$$(\alpha_l^{(s)}) = \begin{pmatrix} A^{-1} & & & Z \\ & A^{-1} & & \\ & & \ddots & \\ Z & & & A^{-1} \end{pmatrix}$$

if we suitably rearrange rows and columns. (This rearrangement does not affect the value of $|\det(\alpha_l^{(s)})|$.) Thus,

$$V_A = |\det(\alpha_l^{(s)})| = |\det A|^{-m}$$

and

$$d_l A = |\det A|^{-m} \prod_{j,k=1}^{m} dA_{jk} \tag{1.11}$$

It is obvious from the symmetrical form of (1.11) that $d_r A = d_l A$.

As a final example consider the real group

$$G = \left\{A = \begin{pmatrix} e^a & b \\ 0 & 1 \end{pmatrix},\quad a, b \in R\right\}.$$

Clearly, G acts as a transformation group on the real line: $x \longrightarrow e^a x + b$. The matrices

$$\mathcal{C}_1 = \begin{pmatrix} 1 & 0 \\ 0 & 0 \end{pmatrix}, \qquad \mathcal{C}_2 = \begin{pmatrix} 0 & 1 \\ 0 & 0 \end{pmatrix}$$

form a basis for $L(G)$. Now

$$A^{-1}\frac{\partial A}{\partial a} = \begin{pmatrix} e^{-a} & -e^{-a}b \\ 0 & 1 \end{pmatrix}\begin{pmatrix} e^{a} & 0 \\ 0 & 0 \end{pmatrix} = \begin{pmatrix} 1 & 0 \\ 0 & 0 \end{pmatrix} = \mathcal{C}_1,$$

$$A^{-1}\frac{\partial A}{\partial b} = \begin{pmatrix} e^{-a} & -e^{-a}b \\ 0 & 1 \end{pmatrix}\begin{pmatrix} 0 & 1 \\ 0 & 0 \end{pmatrix} = \begin{pmatrix} 0 & e^{-a} \\ 0 & 0 \end{pmatrix} = e^{-a}\mathcal{C}_2.$$

Thus,

$$V_A(a, b) = \left| \det\begin{pmatrix} 1 & 0 \\ 0 & e^{-a} \end{pmatrix} \right| = e^{-a}, \qquad d_l A = e^{-a}\, da\, db. \tag{1.12}$$

On the other hand

$$\frac{\partial A}{\partial a}A^{-1} = \begin{pmatrix} 1 & -b \\ 0 & 0 \end{pmatrix} = \mathcal{C}_1 - b\mathcal{C}_2, \qquad \frac{\partial A}{\partial b}A^{-1} = \begin{pmatrix} 0 & 1 \\ 0 & 0 \end{pmatrix} = \mathcal{C}_2$$

so

$$W_A(a, b) = \left| \det\begin{pmatrix} 1 & -b \\ 0 & 1 \end{pmatrix} \right| = 1, \qquad d_r A = da\, db. \tag{1.13}$$

The right and left-invariant measures of G are distinct.

6.2 Compact Linear Lie Groups

In Section 5.1 we defined the norm $\|A\|$ of an $m \times m$ matrix A and saw that every Cauchy sequence in the norm $\{A^{(j)}\}$ converges to a unique matrix A. Furthermore, $A_{ik} = \lim_{j\to\infty} A_{ik}^{(j)}$, where $A = (A_{ik})$. Indeed $\{A^{(j)}\}$ is a Cauchy sequence of matrices if and only if each of the sequences of matrix elements $\{A_{ik}^{(j)}\}$, $1 \le i, k \le m$, is Cauchy.

The following result is easy to prove from these remarks.

Lemma 6.1. Let $\{A^{(j)}\}$ and $\{B^{(j)}\}$ be Cauchy sequences of $m \times m$ matrices with limits A and B. Then $\{A^{(j)}B^{(j)}\}$ is a Cauchy sequence with $\lim_{j\to\infty} A^{(j)}B^{(j)} = AB$. Furthermore, if $A^{(j)}$ is nonsingular for all j and A is nonsingular then $\{A^{(j)-1}\}$ is a Cauchy sequence with limit A^{-1}.

In particular, multiplication and inversion in a linear Lie group are continuous with respect to the norm.

A set of $m \times m$ matrices U is **bounded** if there exists a constant $M > 0$ such that $\|A\| \leq M$ for all $A \in U$. Thus, U is bounded if and only if there exists a constant $K > 0$ such that $|A_{ik}| \leq K$ for $1 \leq i, k \leq m$ and all $A \in U$. (Prove it.) The set U is **closed** provided every Cauchy sequence in U converges to an element of U.

A subset S of the real line is **compact** if each countable sequence $\{a_j\}$, $a_i \in S$, contains a subsequence converging to a point in S. Here, S is compact if and only if it is a closed, bounded subset of R (Rudin [1]).

Definition. A (global) group of $m \times m$ matrices is **compact** if it is a bounded, closed subset of the set L_m of all $m \times m$ matrices.

A group G is closed provided every Cauchy sequence $\{A^{(j)}\}$ in G converges to an element of G.

The classical groups $O(m, R)$, $SO(m, R)$, $U(m)$, $SU(m)$, and $USp(m)$ are compact. We verify this fact only for $O(m, R)$ since the other proofs are similar.

If $A \in O(m, R)$ then $A^tA = E_m$, or

$$\sum_{i=1}^{m} A_{il}A_{ik} = \delta_{lk}.$$

Setting $l = k$, we obtain $\sum_i (A_{ik})^2 = 1$, so $|A_{ik}| \leq 1$ for all i, k. Thus, the matrix elements of A are bounded. Let $\{A^{(j)}\}$ be a Cauchy sequence in $O(m, R)$ with limit A. Then

$$E_m = \lim_{j\to\infty} (A^{(j)})^t A^{(j)} = A^tA,$$

So $A \in O(m, R)$ and $O(m, R)$ is compact.

Suppose G is a real, compact, linear Lie group of dimension n. It follows from the Heine–Borel Theorem (Rudin [1]) that the group manifold of G can be covered by a finite number of bounded coordinate patches. Thus, for any continuous function $f(A)$ on G, the integral

$$\int_G f(A)\, d_lA = \int_G f(A)V_A(\mathbf{g})\, d\mathbf{g} \tag{2.1}$$

will converge (since the domain of integration is bounded.) In particular the integral

$$V_G = \int_G 1\, d_lA, \tag{2.2}$$

called the **volume** of G, converges. If G is not compact the integrals (2.1) and (2.2) may not converge. Indeed, if $G = R$, the real line, then (2.2) diverges.

The above remarks also hold for the right-invariant measure d_rA. Moreover, we can show $d_lA = d_rA$ for compact groups.

Theorem 6.1. If G is a compact linear Lie group then $d_lA = d_rA$.

Proof. By (1.10), $d_r A = |\det \tilde{A}|\, d_l A$, where $\tilde{A}$ is the inner automorphism $\mathcal{A} \longrightarrow A\mathcal{A}A^{-1}$ of $L(G)$. We can think of $\tilde{A}$ as an $m^2 \times m^2$ matrix rep of G. Since G is compact the matrices $A, A^{-1} \in G$ are uniformly bounded. Thus the matrices $\tilde{A}$ are bounded and there exists a constant $M > 0$ such that $|\det \tilde{A}| \leq M$ for all $A \in G$. Now fix A and suppose $|\det \tilde{A}| = s > 1$. Then

$$|\det \tilde{A}^j| = |\det \tilde{A}|^j = s^j, \qquad j = 1, 2, \ldots .$$

Choosing j sufficiently large we get $s^j > M$, which is impossible. Thus $s \leq 1$. If $s < 1$ then

$$|\det \tilde{A}^{-1}| = |\det \tilde{A}|^{-1} = s^{-1} > 1$$

which is impossible. Therefore $s = 1$ for all $A \in G$ and $d_l A = d_r A$. Q.E.D.

For G compact we write $dA = d_l A = d_r A$, where the measure dA is both left- and right-invariant.

Using the invariant measure for compact groups, we can mimic the proofs of most of the results for finite groups obtained in Sections 3.1–3.3. In particular, we will show that any finite-dimensional rep of a compact group can be decomposed into a direct sum of irred reps and we will obtain orthogonality relations for the matrix elements and characters of irred reps.

For finite groups K these results were proved using the average of a function over K. If f is a function on K then the **average** of f over K is

$$\mathcal{A}v(f(k)) = [1/n(K)] \sum_{k \in K} f(k). \tag{2.3}$$

If $h \in K$ then

$$\mathcal{A}v(f(hk)) = \mathcal{A}v(f(kh)) = \mathcal{A}v(f(k)). \tag{2.4}$$

Furthermore,

$$\mathcal{A}v(a_1 f_1(k) + a_2 f_2(k)) = a_1 \mathcal{A}v(f_1(k)) + a_2 \mathcal{A}v(f_2(k)), \qquad \mathcal{A}v(1) = 1. \tag{2.5}$$

Properties (2.4) and (2.5) are sufficient to prove most of the fundamental results on the reps of finite groups. Now let G be a compact linear Lie group and let f be a continuous function on G. We define

$$\mathcal{A}v(f(A)) = (1/V_G) \int_G f(A)\, dA = \int_G f(A)\, \delta A \tag{2.6}$$

where dA is the invariant measure on G, $V_G = \int_G 1\, dA$ is the **volume** of G, and $\delta A = V_G^{-1}\, dA$ is the **normalized** invariant measure. Then

$$\begin{aligned} &\mathcal{A}v(f(BA)) = \int_G f(BA)\, \delta A = \int_G f(A)\, \delta A = \mathcal{A}v(f(A)), \\ &\mathcal{A}v(f(AB)) = \mathcal{A}v(f(A)), \qquad \mathcal{A}v(1) = \int_G \delta A = 1, \qquad B \in G, \end{aligned} \tag{2.7}$$

since δA is both left- and right-invariant. Thus, $\mathcal{A}v(f(A))$ also satisfies properties (2.4) and (2.5).

We now study the **continuous** reps of G, i.e., reps $\mathbf{T}$ such that the operators $\mathbf{T}(A)$ are continuous functions of the group parameters of $A \in G$.

Theorem 6.2. Let $\mathbf{T}$ be a continuous rep of the compact linear Lie group G on the finite-dimensional inner product space V. Then $\mathbf{T}$ is equivalent to a unitary rep on V.

Proof. Let $\langle -, - \rangle$ be the inner product on V. We define an inner product $(-, -)$ on V with respect to which $\mathbf{T}$ is unitary. For $\mathbf{u}, \mathbf{v} \in V$ define

$$(2.8) \qquad (\mathbf{u}, \mathbf{v}) = \int_G \langle \mathbf{T}(A)\mathbf{u}, \mathbf{T}(A)\mathbf{v} \rangle \, \delta A = \mathcal{A}v[\langle \mathbf{T}(A)\mathbf{u}, \mathbf{T}(A)\mathbf{v} \rangle].$$

(The integral converges since the integrand is continuous and the domain of integration is finite.) It is straightforward to check that $(-, -)$ is an inner product. In particular the positive-definite property follows from the fact that the weight function is strictly positive. Now

$$\begin{aligned}(\mathbf{T}(B)\mathbf{u}, \mathbf{T}(B)\mathbf{v}) &= \mathcal{A}v[\langle \mathbf{T}(AB)\mathbf{u}, \mathbf{T}(AB)\mathbf{v} \rangle] \\ &= \mathcal{A}v[\langle \mathbf{T}(A)\mathbf{u}, \mathbf{T}(A)\mathbf{v} \rangle] = (\mathbf{u}, \mathbf{v}),\end{aligned}$$

so $\mathbf{T}$ is unitary with respect to $(-, -)$. The remainder of the proof is identical with that of Theorem 3.1. Q.E.D.

The theorem shows that we can restrict ourselves to the study of unitary reps $\mathbf{T}$ with no loss of generality.

Theorem 6.3. If $\mathbf{T}$ is a unitary rep of G on V and W is an invariant subspace of V then $W^{\perp}$ is also an invariant subspace under $\mathbf{T}$.

Theorem 6.4. Every finite-dimensional, continuous, unitary rep of a compact linear Lie group can be decomposed into a direct sum of irred unitary reps.

The proofs of these theorems are identical with the corresponding proofs for finite groups.

Let $\{\mathbf{T}^{(\mu)}\}$ be a complete set of nonequivalent unitary irred reps of G, labeled by the parameter μ. (Here we consider only reps of G on *complex* vector spaces.) Initially we have no way of telling how many distinct values μ can take. (It will turn out that μ takes on a countably infinite number of values, so that we can choose $\mu = 1, 2, \ldots$.) We introduce an ON basis in each rep space $V^{(\mu)}$ to obtain a unitary $n_\mu \times n_\mu$ matrix rep $T^{(\mu)}$ of G.

Now we mimic the construction of the orthogonality relations for finite groups. Given the matrix reps $T^{(\mu)}, T^{(\nu)}$, choose an arbitrary $n_\mu \times n_\nu$ matrix

C and form the $n_\mu \times n_\nu$ matrix

$$D = \mathfrak{Av}[T^{(\mu)}(A)CT^{(\nu)}(A^{-1})] = \int_G T^{(\mu)}(A)CT^{(\nu)}(A^{-1})\,\delta A. \tag{2.9}$$

Just as in the corresponding construction for finite groups, one can easily verify that

$$T^{(\mu)}(B)D = DT^{(\nu)}(B) \tag{2.10}$$

for all $B \in G$. Recall that the Schur lemmas are valid for finite-dimensional reps of all groups, not just finite groups. Thus if $\mu \neq \nu$, i.e., $T^{(\mu)}$ not equivalent to $T^{(\nu)}$, then $D = Z$. If $\mu = \nu$ then $D = \lambda E_{n_\mu}$ for some $\lambda \in \mathfrak{C}$.

$$D(C, \mu, \nu) = \lambda(\mu, C)\,\delta_{\mu\nu} E_{n_\mu}.$$

Letting C run over all $n_\mu \times n_\nu$ matrices, we obtain the independent identities

$$\int_G T_{il}^{(\mu)}(A) T_{ks}^{(\nu)}(A^{-1})\,\delta A = \lambda(\mu, l, k)\,\delta_{\mu\nu}\,\delta_{is}, \tag{2.11}$$

for the matrix elements $T_{il}^{(\mu)}(A)$. To evaluate λ we set $\nu = \mu$ and $s = i$ and sum on i:

$$\sum_{i=1}^{n_\mu} \lambda = n_\mu \lambda = \int_G \sum_{i=1}^{n_\mu} T_{ki}^{(\mu)}(A^{-1}) T_{il}^{(\mu)}(A)\,\delta A = \delta_{kl}.$$

Therefore $\lambda = \delta_{kl}/n_\mu$. Since the matrices $T^{(\mu)}(A)$ are unitary, (2.11) becomes

$$\int_G T_{il}^{(\mu)}(A)\overline{T_{sk}^{(\nu)}(A)}\,\delta A = (\delta_{is}/n_\mu)\,\delta_{lk}\,\delta_{\mu\nu}, \qquad 1 \le i, l \le n_\mu, \quad 1 \le s, k \le n_\nu. \tag{2.12}$$

These are the **orthogonality relations** for matrix elements of irred reps of G.

In the case of finite groups K we were able to relate the orthogonality relations to an inner product on the group ring R_K. We can consider R_K as the space of all functions $f(k)$ on K. Then

$$\langle f_1, f_2 \rangle = [1/n(K)] \sum_{k \in K} f_1(k)\overline{f_2(k)}$$

defines an inner product on R_K with respect to which the functions $\{n_\mu^{1/2} T_{il}^{(\mu)}(k)\}$ form an ON basis. We extend this idea to compact linear Lie groups G as follows: Let $L_2(G)$ be the space of all functions on G which are (Lebesgue) square-integrable:

$$L_2(G) = \left\{ f(A) \colon \int_G |f(A)|^2\,\delta A < \infty \right\}. \tag{2.13}$$

With respect to the inner product

$$\langle f_1, f_2 \rangle = \int_G f_1(A)\overline{f_2(A)}\,\delta A, \tag{2.14}$$

$L_2(G)$ is a Hilbert space (see the Appendix). Note that every continuous function on G belongs to $L_2(G)$. Let

$$\varphi_{ij}^{(\mu)}(A) = n_\mu^{1/2} T_{ij}^{(\mu)}(A). \tag{2.15}$$

It follows from (2.12) and (2.14) that $\{\varphi_{ij}^{(\mu)}\}$, where $1 \le i, j \le n_\mu$ and μ ranges over all equivalence classes of irred reps, forms an ON set in $L_2(G)$.

For finite groups we know that the set $\{\varphi_{ij}^{(\mu)}\}$ is an ON **basis** for the group ring and every function f on the group can be written as a unique linear combination of these basis functions. Similarly one can show that for G compact the set $\{\varphi_{ij}^{(\mu)}\}$ is an ON basis for $L_2(G)$. Thus, every $f \in L_2(G)$ can be expanded uniquely in the (generalized) **Fourier series**

$$f(A) \sim \sum_{\mu=1}^{\infty} \sum_{i,k=1}^{n_\mu} c_{ik}^{\mu} \varphi_{ik}^{(\mu)}(A), \tag{2.16}$$

where

$$c_{ik}^{\mu} = \langle f, \varphi_{ik}^{(\mu)} \rangle. \tag{2.17}$$

Furthermore, we have the **Parseval equality**

$$\langle f, f \rangle = \sum_{\mu=1}^{\infty} \sum_{i,k=1}^{n_\mu} | c_{ik}^{\mu} |^2.$$

[We use $\sim$ rather than $=$ in (2.16) to denote that f and $\sum c_{ij}^{\mu} \varphi_{ij}^{(\mu)}$ are the same Hilbert space vector. We do *not* claim that the two sides of the equality are necessarily pointwise equal.]

We illustrate this result, the celebrated **Peter–Weyl theorem,** for an important example, the circle group $U(1)$.

Lemma 6.2. Let G be an abelian group (not necessarily a Lie group) and let $\mathbf{T}$ be a finite-dimensional irred rep of G on a complex vector space V. Then $\mathbf{T}$ is one-dimensional.

Proof. Suppose $\mathbf{T}$ is irred on V and dim $V > 1$. There must exist a $g \in G$ such that $\mathbf{T}(g)$ is not a multiple of the identity operator on V, for otherwise V would be reducible. Let λ be a eigenvalue of $\mathbf{T}(g)$ and let C_λ be the eigenspace

$$C_\lambda = \{\mathbf{v} \in V \colon \mathbf{T}(g)\mathbf{v} = \lambda \mathbf{v}\}.$$

Clearly C_λ is a proper subspace of V. If $h \in G$ and $\mathbf{w} \in C_\lambda$ then

$$\mathbf{T}(g)(\mathbf{T}(h)\mathbf{w}) = \mathbf{T}(h)(\mathbf{T}(g)\mathbf{w}) = \lambda(\mathbf{T}(h)\mathbf{w})$$

since G is abelian, so C_λ is invariant under the operator $\mathbf{T}(h)$. Therefore, $\mathbf{T}$ is reducible. Impossible! Q.E.D.

Although all complex irred reps of an abelian group are one-dimensional, the lemma is false for real reps (see Problem 6.8).

The circle group $U(1) = \{e^{i\theta}\}$ is compact and abelian. Hence its irred matrix reps are continuous functions $\chi(\theta)$ such that

$$\chi(\theta_1 + \theta_2) = \chi(\theta_1)\chi(\theta_2), \qquad \theta_1, \theta_2 \in R, \tag{2.18}$$

and $\chi(\theta + 2\pi) = \chi(\theta)$. The functional equation (2.18) has only the solutions $\chi(\theta) = e^{a\theta}$ and the periodicity of χ implies $a = im$, where m is an integer. Therefore, there are an infinite number of irreducible unitary representations of $U(1)$:

$$\chi_m(\theta) = e^{im\theta}, \qquad m = 0, \pm 1, \pm 2, \ldots .$$

The invariant measure on $U(1)$ is $d\theta$. The space $L_2(U(1))$ is just the space $L_2[0, 2\pi]$ consisting of all functions $f(\theta)$ with period 2π such that $\int_0^{2\pi} |f(\theta)|^2 \, d\theta < \infty$. By the Peter–Weyl theorem the functions $\{e^{im\theta}\}$ form an ON basis for $L_2[0, 2\pi]$. Every $f \in L_2[0, 2\pi]$ can be expressed uniquely in the form

$$f(\theta) \sim \sum_{m=-\infty}^{\infty} c_m e^{im\theta}, \qquad c_m = (1/2\pi) \int_0^{2\pi} f(\theta) e^{-im\theta} \, d\theta. \tag{2.19}$$

Furthermore,

$$(1/2\pi) \int_0^{2\pi} |f(\theta)|^2 \, d\theta = \sum_{m=-\infty}^{\infty} |c_m|^2. \tag{2.20}$$

Here (2.19) is the well-known Fourier series expansion of a periodic function and (2.20) is Parseval's equality. It is clear from this example that the Peter–Weyl theorem is a group-theoretic generalization of classical Fourier series analysis. Furthermore, we see that the classical theory has a group-theoretic structure.

Theorem 6.5 (Peter–Weyl). If G is a compact linear Lie group, the set $\{\varphi_{ij}^{(\mu)}\}$ is an ON basis for $L_2(G)$.

The proof of this theorem depends heavily on facts about symmetric completely continuous operators in Hilbert space and will not be given here. For the details see Chevalley [1] or Naimark [1].

Corollary 6.1. A compact linear Lie group G has a countably infinite (not finite) number of equivalence classes of irred reps $\{\mathbf{T}^{(\mu)}\}$. Thus, we can label the reps so that $\mu = 1, 2, \ldots$.

Proof. The functions $\{\varphi_{jk}^{(\mu)}\}$ form an ON basis for $L_2(G)$. Since $L_2(G)$ is a separable, infinite-dimensional Hilbert space there are a countably infinite number of basis vectors (Helwig [1]). Q.E.D.

Corollary 6.2. Let G, H be compact linear Lie groups with equivalence classes of irred reps $\{\mathbf{T}^{(\mu)}\}$, $\{\mathbf{U}^{(\nu)}\}$, respectively, Then $\{\mathbf{T}^{(\mu)} \otimes \mathbf{U}^{(\nu)}\}$ is the complete set of equivalence classes of irred reps for the compact group $G \times H$.

The proof, which is left to the reader, consists in showing that the functions $(n_\mu n_\nu)^{1/2} T_{ij}^{(\mu)}(A) U_{kl}^{(\nu)}(B)$ form an ON basis for $L_2(G \times H)$.

The Peter–Weyl theorem refers to continuous reps of the real compact Lie group G, while the Lie-theoretic methods of Chapter 5 apply only to analytic reps. The possibility arises that there may be continuous reps of G which are not analytic. For such reps, Lie-algebraic methods make no sense. Fortunately, the following result eliminates this possibility.

Theorem 6.6. Let $\mathbf{T}$ be a finite-dimensional continuous rep of the real compact linear Lie group G on the inner product space V. Then $\mathbf{T}$ is analytic (with respect to suitable coordinates for G) (see Naimark [2]).

6.3 Group Characters and Representations

The theory of characters for compact Lie groups is almost identical with the character theory for finite groups presented in Section 3.4. The principal difference is that the sum over a finite group is replaced by an integral.

Let $\mathbf{T}$ be a rep of the compact linear Lie group G on the m-dimensional vector space V. With respect to a fixed basis in V the operators $\mathbf{T}(A)$ define a matrix rep $T(A)$. The **character** of $\mathbf{T}$ is the function

$$\chi(A) = \operatorname{tr} T(A).$$

Since $\operatorname{tr}(ST(A)S^{-1}) = \operatorname{tr} T(A)$ the character is independent of basis in V and equivalent reps have the same character. The character of an irred rep is **simple,** while the character of a reducible rep is **compound.** If $\mathbf{T}$ is unitary then its corresponding character satisfies the relation $\overline{\chi(A)} = \chi(A^{-1})$. However, every rep is equivalent to a unitary rep, so the preceding identity is satisfied by all characters. Every character is a continuous function on G.

Let $\{\mathbf{T}^{(\mu)}\}$ be a complete set of nonequivalent unitary irred reps of G and let $\{\chi^{(\mu)}\}$ be the corresponding simple characters. The orthogonality relations (2.12) for matrix elements imply the following orthogonality relations for characters:

$$(3.1) \qquad (\chi^{(\mu)}, \chi^{(\nu)}) = \int_G \chi^{(\mu)}(A)\overline{\chi^{(\nu)}(A)}\,\delta A = \delta_{\mu\nu}, \qquad \mu, \nu = 1, 2, \ldots .$$

The proof is identical with that for finite groups.

Let $\mathbf{T}$ be a finite-dimensional unitary rep of G with character χ. By Theorem 6.4 we can decompose $\mathbf{T}$ into a direct sum of irred reps,

$$(3.2) \qquad \mathbf{T} = \sum_{\mu=1}^{\infty} \oplus\, a_\mu \mathbf{T}^{(\mu)}.$$

Here the integer a_μ denotes the multiplicity of $\mathbf{T}^{(\mu)}$ in $\mathbf{T}$. Only a finite number of the $\{a_\mu\}$ are nonzero. We shall show that the multiplicities a_μ are uniquely determined by $\mathbf{T}$, i.e., they are independent of the method by which $\mathbf{T}$ is decomposed into irred reps. From (3.2), the character of **T** can be expressed in the form

$$\chi = \sum_{\mu=1}^{\infty} a_\mu \chi^{(\mu)}. \tag{3.3}$$

According to the orthogonality relations

$$(\chi, \chi^{(\nu)}) = \sum_{\mu=1}^{\infty} a_\mu (\chi^{(\mu)}, \chi^{(\nu)}) = a_\nu, \qquad \nu = 1, 2, \ldots \tag{3.4}$$

Since (3.4) is independent of basis, the multiplicities a_μ must be uniquely determined.

Theorem 6.7. Let **T** be a rep of G with character χ. The multiplicity a_μ of $\mathbf{T}^{(\mu)}$ in **T** is given by $(\chi, \chi^{(\mu)}) = a_\mu$. Two reps with the same character are equivalent.

Corollary 6.3. The rep **T** is irred if and only if $(\chi, \chi) = 1$.

Example. The simple characters of the circle group $U(1)$ are just $\chi^{(n)}(\theta) = e^{in\theta}$, $n = 0, \pm 1, \ldots$. (Here it is more convenient to let the index of the irred reps run over all integers rather than over the nonnegative integers.) The orthogonality relations are

$$(\chi^{(n)}, \chi^{(m)}) = (1/2\pi) \int_0^{2\pi} e^{i(n-m)\theta}\, d\theta = \delta_{nm}.$$

In Section 3.7 we used the method of projection operators to explicitly decompose a rep into a direct sum of irred reps. These methods carry over to compact Lie groups virtually unchanged. Thus we present the results without detailed proof.

Let **T** be a unitary rep of the compact linear Lie group G on the inner product space V. Corresponding to the decomposition

$$\mathbf{T} = \sum_{\mu=1}^{\infty} \oplus\, a_\mu \mathbf{T}^{(\mu)}$$

of **T** there is a decomposition

$$V = \sum_{\mu=1}^{\infty} \oplus\, V^{(\mu)}, \qquad V^{(\mu)} = \sum_{i=1}^{a_\mu} \oplus\, V_i^{(\mu)}, \tag{3.5}$$

where $\mathbf{T} | V_i^{(\mu)}$ is equivalent to $\mathbf{T}^{(\mu)}$. These spaces $V_i^{(\mu)}$ are not uniquely determined. Define the linear operators $\mathbf{P}_\mu$ on V by

$$\mathbf{P}_\mu = n_\mu \int_G \overline{\chi^{(\mu)}(A)}\mathbf{T}(A)\, \delta A, \qquad \mu = 1, 2, \ldots. \tag{3.6}$$

To make sense of (3.6) choose a basis $\{\mathbf{v}_j\}$ for V with respect to which $T(A)$

is the matrix of $\mathbf{T}(A)$. Then $\mathbf{P}_\mu$ is the operator on V whose matrix is

$$P_\mu = n_\mu \int_G \overline{\chi^{(\mu)}(A)} T(A)\, \delta A.$$

It follows from (3.6) that $\mathbf{T}(B)\mathbf{P}_\mu = \mathbf{P}_\mu \mathbf{T}(B)$ for all $B \in G$. Furthermore, $\mathbf{P}_\mu{}^2 = \mathbf{P}_\mu$ and $\mathbf{P}_\mu{}^* = \mathbf{P}_\mu$. Thus, $\mathbf{P}_\mu$ is a self-adjoint projection operator on V.

Let $\{\mathbf{T}^{(\mu)}\}$ be a complete set of nonequivalent irred unitary matrix reps of G and choose a basis $\{\mathbf{v}_{ij}^{(\nu)}\}$ in each subspace $V_i^{(\nu)}$ such that

$$\text{(3.7)} \qquad \mathbf{T}(A)\mathbf{v}_{ij}^{(\nu)} = \sum_{k=1}^{n_\nu} T_{kj}^{(\nu)}(A)\mathbf{v}_{ik}^{(\nu)}, \qquad 1 \leq j \leq n_\nu .$$

Then, just as in (7.12). Section 3.7, one can prove

$$\text{(3.8)} \qquad \mathbf{P}_\mu \mathbf{v}_{ij}^{(\nu)} = \delta_{\mu\nu}\mathbf{v}_{ij}^{(\nu)}, \qquad \mu, \nu = 1, 2, \ldots, \quad 1 \leq i \leq a_\nu, \quad 1 \leq j \leq n_\nu .$$

Thus $\mathbf{P}_\mu$ projects onto the invariant subspace $V^{(\mu)}$. Since the definition of $\mathbf{P}_\mu$ is basis-independent, $V^{(\mu)}$ is uniquely determined. To find the $V_i^{(\mu)}$ we define operators

$$\text{(3.9)} \qquad \mathbf{P}_\mu^{lk} = n_\mu \int_G \bar{T}_{lk}^{(\mu)}(A)\mathbf{T}(A)\, \delta A, \qquad 1 \leq l, k \leq n_\mu ,$$

which are easily shown to have the properties

$$\text{(3.10)} \qquad \mathbf{P}_\mu^{lk}\mathbf{v}_{ij}^{(\mu)} = \delta_{\mu\nu}\, \delta_{jk}\, \mathbf{v}_{il}^{(\nu)},$$

$$\text{(3.11)} \qquad \mathbf{P}_\mu^{lk}\mathbf{P}_{\mu'}^{l'k'} = \delta_{\mu\mu'}\, \delta_{kl'}\, \mathbf{P}_\mu^{lk'}, \qquad (\mathbf{P}_\mu^{lk})^* = \mathbf{P}_\mu^{kl}, \qquad \mathbf{P}_\mu = \sum_{k=1}^{n_\mu} \mathbf{P}_\mu^{kk}.$$

Thus $\mathbf{P}_\mu^{kk}$ is the self-adjoint projection operator on the a_μ-dimensional space $W_k^{(\mu)}$ spanned by the ON basis vectors $\{\mathbf{v}_{ik}^{(\mu)} : 1 \leq i \leq a_\mu\}$. The remaining details for the construction of the spaces $V_i^{(\mu)}$ are identical with those for finite groups.

We now extend the concept of group rep from finite-dimensional inner product spaces to Hilbert spaces. Let $\mathcal{H}$ be a Hilbert space and G a (global) linear Lie group of $m \times m$ matrices.

Definition. A (bounded) **representation** $\mathbf{T}$ of G on $\mathcal{H}$ is a correspondence which assigns to each $A \in G$ a bounded linear operator $\mathbf{T}(A)$ on $\mathcal{H}$ such that

$$\text{(3.12)} \qquad \mathbf{T}(A)\mathbf{T}(B) = \mathbf{T}(AB), \qquad \mathbf{T}(E_m) = \mathbf{E},$$

where $A, B \in G$ and $\mathbf{E}$ is the identity operator on $\mathcal{H}$.

Note that $\mathbf{T}(A)$ is invertible and $\mathbf{T}(A)^{-1} = \mathbf{T}(A^{-1})$. The rep $\mathbf{T}$ is **irreducible** if $\mathcal{H}$ contains no proper **closed** subspace (closed in the norm) which is invariant under $\mathbf{T}$. Otherwise $\mathbf{T}$ is **reducible.** Every finite-dimensional subspace of a Hilbert space is closed. (Prove it.) Thus for finite-dimensional reps the above definition of irreducibility coincides with that given in Chapter 3.

Suppose $\mathcal{W}$ is an invariant subspace of $\mathcal{H}$. Since $\mathbf{T}(A)$ is bounded, the closure $\overline{\mathcal{W}}$ is invariant under $\mathbf{T}(A)$ for all $A \in G$. (Prove it.) Thus $\overline{\mathcal{W}}$ is also an invariant subspace of $\mathcal{H}$. Since we can always close an invariant subspace, we restrict ourselves to closed invariant subspaces in the definition of irreducibility.

A rep $\mathbf{T}$ is **unitary** if each operator $\mathbf{T}(A)$ is unitary for all A, and **continuous** if $\langle \mathbf{T}(A)\mathbf{v}, \mathbf{w}\rangle$ is a continuous function of A for each $\mathbf{v}, \mathbf{w} \in \mathcal{H}$. Here $\langle -, -\rangle$ is the inner product on $\mathcal{H}$. Unless otherwise stated, we consider only continuous reps.

For G compact we can carry over many of our results for finite-dimensional unitary reps to Hilbert space reps. Let $\mathbf{T}$ be a unitary rep of G on the separable Hilbert space $\mathcal{H}$. We define operators $\mathbf{P}_\mu$, $\mathbf{P}_\mu^{lk}$ on $\mathcal{H}$ by

$$(3.13) \qquad \mathbf{P}_\mu = n_\mu \int_G \bar{\chi}^{(\mu)}(A)\mathbf{T}(A)\,\delta(A), \qquad \mathbf{P}_\mu^{lk} = n_\mu \int_G \bar{T}_{lk}^{(\mu)}(A)\mathbf{T}(A)\,\delta A.$$

To make sense of these expressions choose an ON basis $\{\mathbf{v}_i\}$ for $\mathcal{H}$ and let $T(A)$ be the (possibly infinite) matrix corresponding to $\mathbf{T}(A)$:

$$(3.14) \qquad \mathbf{T}(A)\mathbf{v}_i = \sum_{j=1}^{\infty} T_{ji}(A)\mathbf{v}_j, \qquad i = 1, 2, \ldots .$$

Since $\mathbf{T}$ is continuous the matrix elements $T_{ji}(A)$ are continuous functions on G. By $\mathbf{P}_\mu$ we mean the linear operator on $\mathcal{H}$ whose matrix with respect to $\{\mathbf{v}_i\}$ is

$$P_\mu = n_\mu \int_G \bar{\chi}^{(\mu)}(A)T(A)\,\delta A.$$

There is a similar definition for $\mathbf{P}_\mu^{lk}$. It can be shown that the properties (3.11) which were valid for $\mathcal{H}$ finite-dimensional are true in general. In fact we have the following result.

Theorem 6.8. A unitary rep $\mathbf{T}$ of a compact Lie group G on $\mathcal{H}$ can be decomposed into a direct sum of unitary irred reps $\mathbf{T}^{(\mu)}$: $\mathbf{T} \cong \sum_{\mu=1}^{\infty} \oplus\, a_\mu \mathbf{T}^{(\mu)}$. Indeed there exist mutually orthogonal subspaces $\mathcal{V}_\mu{}^m$, $m = 1, \ldots, a_\mu$, of $\mathcal{H}$ such that $\mathcal{H} = \sum_{m,\mu} \oplus\, \mathcal{V}_\mu{}^m$ and $\mathbf{T} \mid \mathcal{V}_\mu{}^m \cong \mathbf{T}^{(\mu)}$. The $\mathcal{V}_\mu{}^m$ are not unique but the multiplicity $a_\mu = 0, 1, 2, \ldots, \infty$ is unique, as is the space $\mathcal{H}_\mu = \sum_m \oplus\, \mathcal{V}_\mu{}^m$. If $\dim \mathcal{H}_\mu = h_\mu$ is finite then $a_\mu = h_\mu / n_\mu$.

The proof of this theorem makes use of the Peter–Weyl theorem (Naimark [2] or Talman [1]). The proof is constructive in the sense that one can use the method discussed following Eq. (7.18), Section 3.7, to explicitly decompose $\mathcal{H}$. The only difference is that the multiplicity a_μ may be countably infinite.

Problems

6.1 Compute the invariant measure on $U(2)$.

6.2 Compute the left- and right-invariant measures on $SL(n, R)$.

6.3 Prove: If G is a compact linear Lie group then $d(A^{-1}) = dA$, i.e., $\int_G f(A^{-1})\, dA = \int_G f(B)\, dB$. [Hint: Show that $V_{A^{-1}}(\mathbf{g}) = |\det(-\tilde{A})|\, V_A(\mathbf{g})$, where $\tilde{A}$ is the automorphism $\mathcal{a} \longrightarrow A\mathcal{a}A^{-1}$ of $L(G)$, and use the proof of Theorem 6.1.]

6.4 Prove that the identity rep is contained in the tensor product $\mathbf{T}_1 \otimes \mathbf{T}_2$ of two irred reps of a compact Lie group G if and only if $\mathbf{T}_1 \cong \mathbf{T}_2$.

6.5 Prove Corollary 6.3.

6.6 Let G be a compact Lie group with simple characters $\{\chi^{(\mu)}(A)\}$. Show that the $\{\chi^{(\mu)}(A)\}$, suitably renormalized, form an ON basis for the subspace of $L_2(G)$ consisting of all functions constant on conjugacy classes.

6.7 Prove relations (3.11) directly from the definition (3.9) of the $\mathbf{P}_\mu^{lk}$. Do not use the auxiliary relations (3.10).

6.8 Construct a real irred two-dimensional rep of the circle group $U(1)$.

6.9 Show how to decompose any real finite-dimensional rep of $U(1)$ as a direct sum of real irred reps.

Chapter 7

The Rotation Group and Its Representations

7.1 The Groups $SO(3)$ and $SU(2)$

The rotation group $SO(3)$ is of fundamental importance in modern physical theories. Many physical systems admit $SO(3)$ as a symmetry group, a fact which is related to the conservation of angular momentum for such systems. Moreover, the theory of spin and isotopic spin of particles is intimately related to the rep theory of $SO(3)$ and its locally isomorphic companion $SU(2)$. The theory of hypergeometric functions is associated with the study of the Lie algebra of $SO(3)$. Finally, a knowledge of the rep theory of the rotation group and its Lie algebra is indispensible for an understanding of the more complicated rep theory of the classical groups.

Recall that $SO(3) = SO(3, R)$ is the group of all 3×3 real matrices such that $A^t A = E_3$ and $\det A = +1$ (see Section 2.1). This is the natural realization of $SO(3)$ as a transformation group on R_3. We have shown that $SO(3)$ is a three-parameter Lie group whose Lie algebra $so(3)$ consists of all 3×3 real matrices $\mathcal{A}$ such that $\mathcal{A}^t = -\mathcal{A}$. As a convenient basis for $so(3)$ we choose three tangent matrices to the one-parameter groups of rotations about the x, y, and z axes, respectively. The rotations about the z axis are

$$\begin{pmatrix} \cos\varphi & -\sin\varphi & 0 \\ \sin\varphi & \cos\varphi & 0 \\ 0 & 0 & 1 \end{pmatrix}. \tag{1.1}$$

This is a one-parameter subgroup of $SO(3)$ with tangent matrix

$$\mathfrak{L}_3 = \begin{pmatrix} 0 & -1 & 0 \\ 1 & 0 & 0 \\ 0 & 0 & 0 \end{pmatrix} \tag{1.2}$$

at the identity. Similarly

$$\mathfrak{L}_1 = \begin{pmatrix} 0 & 0 & 0 \\ 0 & 0 & -1 \\ 0 & 1 & 0 \end{pmatrix}, \qquad \mathfrak{L}_2 = \begin{pmatrix} 0 & 0 & 1 \\ 0 & 0 & 0 \\ -1 & 0 & 0 \end{pmatrix} \tag{1.3}$$

are tangent matrices to one-parameter subgroups of rotations about the x and y axes, respectively. We have

$$\exp \varphi\mathfrak{L}_1 = \begin{pmatrix} 1 & 0 & 0 \\ 0 & \cos\varphi & -\sin\varphi \\ 0 & \sin\varphi & \cos\varphi \end{pmatrix}, \qquad \exp \varphi\mathfrak{L}_2 = \begin{pmatrix} \cos\varphi & 0 & \sin\varphi \\ 0 & 1 & 0 \\ -\sin\varphi & 0 & \cos\varphi \end{pmatrix}. \tag{1.4}$$

Since these three tangent matrices are linearly independent, they form a basis for $so(3)$. As the reader can easily verify, the commutation relations of the basis vectors are

$$[\mathfrak{L}_1, \mathfrak{L}_2] = \mathfrak{L}_3, \qquad [\mathfrak{L}_3, \mathfrak{L}_1] = \mathfrak{L}_2, \qquad [\mathfrak{L}_2, \mathfrak{L}_3] = \mathfrak{L}_1. \tag{1.5}$$

In Section 5.4 we showed that $SU(2)$ was also a three-parameter real Lie group. As the reader can easily verify, every $A \in SU(2)$ can be written in the form

$$A = \begin{pmatrix} \alpha & \beta \\ -\bar{\beta} & \bar{\alpha} \end{pmatrix}, \tag{1.6}$$

where $|\alpha|^2 + |\beta|^2 = 1$. If $A, A_1, A_2 \in SU(2)$ then

$$A^{-1} = \bar{A}^t = \begin{pmatrix} \bar{\alpha} & -\beta \\ \bar{\beta} & \alpha \end{pmatrix}, \qquad A_1A_2 = \begin{pmatrix} \alpha_1\alpha_2 - \beta_1\bar{\beta}_2, & \alpha_1\beta_2 + \beta_1\bar{\alpha}_2 \\ -\bar{\beta}_1\alpha_2 - \bar{\alpha}_1\bar{\beta}_2, & -\bar{\beta}_1\beta_2 + \bar{\alpha}_1\bar{\alpha}_2 \end{pmatrix}.$$

The Lie algebra $su(2) = L(SU(2))$ consists of all 2×2 complex skew-Hermitian matrices $\mathfrak{A}$ of trace zero:

$$\mathfrak{A} = \begin{pmatrix} ix_3, & -x_2 + ix_1 \\ x_2 + ix_1, & -ix_3 \end{pmatrix}, \qquad x_j \in R. \tag{1.7}$$

As a basis for $su(2)$ we choose the elements

$$\mathfrak{J}_1 = \begin{pmatrix} 0 & i/2 \\ i/2 & 0 \end{pmatrix}, \qquad \mathfrak{J}_2 = \begin{pmatrix} 0 & -1/2 \\ 1/2 & 0 \end{pmatrix}, \qquad \mathfrak{J}_3 = \begin{pmatrix} i/2 & 0 \\ 0 & -i/2 \end{pmatrix}. \tag{1.8}$$

A direct computation shows that these matrices satisfy the commutation relations (1.5). Thus $so(3)$ and $su(2)$ are isomorphic Lie algebras, so $SO(3)$ and $SU(2)$ are locally isomorphic Lie groups. However, this isomorphism is not global.

To exhibit explicitly the relation between $SO(3)$ and $SU(2)$, consider the adjoint rep of $SU(2)$ on its Lie algebra:

$$\mathfrak{A} \longrightarrow \mathfrak{B} = A\mathfrak{A}A^{-1} \in su(2), \qquad \mathfrak{A} \in su(2), \quad A \in SU(2). \tag{1.9}$$

(See Section 5.6.)

Now $\det \mathfrak{B} = \det(A\mathfrak{A}A^{-1}) = \det \mathfrak{A}$. Therefore, writing

$$\mathfrak{B} = \begin{pmatrix} iy_3, & -y_2 + iy_1 \\ y_2 + iy_1, & -iy_3 \end{pmatrix} \tag{1.10}$$

we find

$$y_1^2 + y_2^2 + y_3^2 = \det \mathfrak{B} = \det \mathfrak{A} = x_1^2 + x_2^2 + x_3^2. \tag{1.11}$$

According to (1.9) the y_j are linear combinations of the x_k:

$$y_j = \sum_{k=1}^{3} R(A)_{jk} x_k, \qquad j = 1, 2, 3. \tag{1.12}$$

Since (1.9) defines a rep of $SU(2)$ the 3×3 matrices $R(A)$ satisfy $R(AB) = R(A)R(B)$ for all $A, B \in SU(2)$. Moreover, from (1.11) and (1.12), $R(A)^t R(A) = E_3$, i.e., $R(A) \in O(3)$. The rep $A \longrightarrow R(A)$ is continuous and $SU(2)$ is connected. Thus $\det R(A)$ is a continuous function of A, and since $R(E_2) = E_3$, we conclude that $\det R(A) = +1$ for all $A \in SU(2)$. We have shown that $R(A) \in SO(3)$ and $A \longrightarrow R(A)$ is a homomorphism of $SU(2)$ into $SO(3)$.

We now verify that this homomorphism covers $SO(3)$. Let $R \in SO(3)$ and set $y_j = \sum R_{jk} x_k$. Defining $\mathfrak{A}, \mathfrak{B} \in su(2)$ by (1.7) and (1.10) we find $\operatorname{tr} \mathfrak{A} = \operatorname{tr} \mathfrak{B} = 0$, $\det \mathfrak{A} = \det \mathfrak{B} = x_1^2 + x_2^2 + x_3^2 = q^2$, so the Hermitian matrices $i\mathfrak{A}$ and $i\mathfrak{B}$ have the same eigenvalues, $\pm iq$. Therefore, $i\mathfrak{A}$ and $i\mathfrak{B}$ are similar and there exists a unitary matrix B such that $\mathfrak{B} = B\mathfrak{A}B^{-1}$. Now $|\det B| = 1$ for B unitary, so $B = e^{i\theta}A$, where $e^{2i\theta} = \det B$ and $A \in SU(2)$. Thus $\mathfrak{B} = A\mathfrak{A}A^{-1}$, so $R = R(A)$ and the homomorphism $A \longrightarrow R(A)$ maps $SU(2)$ onto $SO(3)$. Finally, the relation

$$(-A)\mathfrak{A}(-A)^{-1} = A\mathfrak{A}A^{-1}$$

shows that $R(A) = R(-A)$, so two elements of $SU(2)$ map onto a single element of $SO(3)$. Note: The matrix $-A \in SU(2)$ if $A \in SU(2)$.

The reader can check that $R(A) = E_3$ if and only if $A = \pm E_2$. Thus, $SO(3)$ is isomorphic to the factor group $SU(2)/\{\pm E_2\}$. Exactly two elements of $SU(2)$ map onto one element of $SO(3)$. (Since $-A$ is far from E_2 when A is close to E_2 it is clear that this map is locally an isomorphism.)

Writing $\alpha = a + ib$, $\beta = c + id$, $a, b, c, d \in R$, in (1.6) we see that the only restriction on these four real parameters is $a^2 + b^2 + c^2 + d^2 = 1$.

Topologically, $SU(2)$ is homeomorphic to the unit sphere S_4 in four-dimensional space. If $A \in SU(2)$ is a point on this sphere then $-A$ is the point on the other end of the diameter of S_4 passing through A. Topologically, $SO(3)$ is homeomorphic to the projective space obtained by identifying opposite ends of each diameter in S_4. We say that $SU(2)$ is a **covering group** of $SO(3)$ and that it covers $SO(3)$ twice. (For a more geometrical derivation of the relationship between $SU(2)$ and $SO(3)$ see Gel'fand et al, [1].)

The **Euler angles** (φ, θ, ψ) form a convenient coordinate system for $SU(2)$. Consider the product

$$
\begin{aligned}
(1.13)\qquad A(\varphi, \theta, \psi) &= (\exp \varphi \mathcal{J}_3)(\exp \theta \mathcal{J}_1)(\exp \psi \mathcal{J}_3) \\
&= \begin{pmatrix} e^{i\varphi/2} & 0 \\ 0 & e^{-i\varphi/2} \end{pmatrix} \begin{pmatrix} \cos \frac{1}{2}\theta & i \sin \frac{1}{2}\theta \\ i \sin \frac{1}{2}\theta & \cos \frac{1}{2}\theta \end{pmatrix} \begin{pmatrix} e^{i\psi/2} & 0 \\ 0 & e^{-i\psi/2} \end{pmatrix} \\
&= \begin{pmatrix} \alpha & \beta \\ -\bar{\beta} & \bar{\alpha} \end{pmatrix} \\
&= \begin{pmatrix} e^{i(\varphi+\psi)/2} \cos \frac{1}{2}\theta & ie^{i(\varphi-\psi)/2} \sin \frac{1}{2}\theta \\ ie^{i(\psi-\varphi)/2} \sin \frac{1}{2}\theta & e^{-i(\varphi+\psi)/2} \cos \frac{1}{2}\theta \end{pmatrix}.
\end{aligned}
$$

It follows that any $A \in SU(2)$ is determined by Euler angles (φ, θ, ψ), where

$$(1.14)\qquad |\alpha| = \cos \tfrac{1}{2}\theta, \qquad \arg \alpha = \tfrac{1}{2}(\varphi + \psi), \qquad \arg \beta = \tfrac{1}{2}(\varphi - \psi + \pi),$$

$$
\begin{aligned}
(1.15)\qquad & \cos \tfrac{1}{2}\theta = |\alpha|, \qquad \sin \tfrac{1}{2}\theta = |\beta|, \qquad \varphi = \arg \alpha + \arg \beta - \tfrac{1}{2}\pi, \\
& \psi = \arg \alpha - \arg \beta + \tfrac{1}{2}\pi, \qquad |\alpha\beta| \neq 0.
\end{aligned}
$$

If we restrict the Euler angles to the domain

$$(1.16)\qquad 0 \le \varphi < 2\pi, \qquad 0 \le \theta \le \pi, \qquad -2\pi \le \psi < 2\pi,$$

then for $|\alpha\beta| \neq 0$, (φ, θ, ψ) are uniquely determined. (Recall that the argument of a complex number is determined only up to an integer multiple of 2π.) However, if $|\alpha\beta| = 0$, an infinite number of Euler angles describe the same group element. The Euler angles are coordinates on the sphere S_4 somewhat analogous to the coordinates latitude and longitude on the sphere S_3 in three-space. All points on S_3 have unique values of latitude and longitude except the poles, where the longitude becomes indeterminant. The Euler angles are still very useful despite this drawback because the set on which they are indeterminant has lower dimension than three. Thus, if we integrate a function over $SU(2)$ using the invariant measure, the behavior of the function on this set will have no effect on the integral.

Clearly, the Euler angles of the product of two group elements can be expressed as analytic functions of the Euler angles of the factors. The results are given by expressions (2.16).

The invariant measure on $SU(2)$ can be computed directly from the formulas of Section 6.1. Let $A(\varphi, \theta, \psi) \in SU(2)$. Then

$$A^{-1}\frac{\partial A}{\partial \varphi} = \begin{pmatrix} \frac{1}{2}i\cos\theta & -\frac{1}{2}e^{-i\psi}\sin\theta \\ \frac{1}{2}e^{i\psi}\sin\theta & -\frac{1}{2}i\cos\theta \end{pmatrix} = (\sin\psi\sin\theta)\mathcal{J}_1 + (\cos\psi\sin\theta)\mathcal{J}_2 + (\cos\theta)\mathcal{J}_3$$

$$A^{-1}\frac{\partial A}{\partial \theta} = \begin{pmatrix} 0 & \frac{1}{2}ie^{-i\psi} \\ \frac{1}{2}ie^{i\psi} & 0 \end{pmatrix} = (\cos\psi)\mathcal{J}_1 - (\sin\psi)\mathcal{J}_2$$

$$A^{-1}\frac{\partial A}{\partial \psi} = \begin{pmatrix} \frac{1}{2}i & 0 \\ 0 & -\frac{1}{2}i \end{pmatrix} = \mathcal{J}_3.$$

Thus

$$V_A(\varphi, \theta, \psi) = \left| \det \begin{pmatrix} \sin\psi\sin\theta & \cos\psi\sin\theta & \cos\theta \\ \cos\psi & -\sin\psi & 0 \\ 0 & 0 & 1 \end{pmatrix} \right| = \sin\theta$$

and

(1.17) $$dA = \sin\theta\, d\varphi\, d\theta\, d\psi, \qquad 0 \le \varphi < 2\pi, \ \ 0 \le \theta \le \pi, \ \ -2\pi \le \psi < 2\pi.$$

Since $SU(2)$ is compact, dA is both left- and right-invariant. The volume of the group is

(1.18) $$V = \int_{SU(2)} dA = \int_{-2\pi}^{2\pi} d\psi \int_0^{2\pi} d\varphi \int_0^{\pi} \sin\theta\, d\theta = 16\pi^2.$$

Note that the Euler angles φ, ψ are indeterminant only for $\theta = 0, \pi$ and these points make no contribution to the integral.

Now that we have successfully parameterized $SU(2)$ we use the homomorphism $A \longrightarrow R(A)$ to parametrize $SO(3)$. The one-parameter group $\exp t\mathfrak{J}_1$ in $SU(2)$ maps onto the one-parameter group $R(\exp t\mathfrak{J}_1)$ in $SO(3)$. Thus R induces a Lie algebra isomorphism which maps $\mathfrak{J}_1$ to $\mathfrak{L}_1' = (d/dt)R(\exp t\mathfrak{J}_1)|_{t=0}$. By direct computation from (1.9) and (1.12) we see that $\mathfrak{L}_1' = \mathfrak{L}_1$. Similarly, $\mathfrak{J}_2$ maps to $\mathfrak{L}_2$ and $\mathfrak{J}_3$ maps to $\mathfrak{L}_3$.

Thus

(1.19) $$\begin{aligned} R(A) &= R(\exp\varphi\mathfrak{J}_3)R(\exp\theta\mathfrak{J}_1)R(\exp\psi\mathfrak{J}_3) \\ &= (\exp\varphi\mathfrak{L}_3)(\exp\theta\mathfrak{L}_1)(\exp\psi\mathfrak{L}_3), \end{aligned}$$

or from (1.1) and (1.4),

(1.20)

$$R(A) = \begin{pmatrix} \cos\varphi\cos\psi - \sin\varphi\sin\psi\cos\theta, & -\cos\varphi\sin\psi - \sin\varphi\cos\psi\cos\theta, & \sin\varphi\sin\theta \\ \sin\varphi\cos\psi + \cos\varphi\sin\psi\cos\theta, & -\sin\varphi\sin\psi + \cos\varphi\cos\psi\cos\theta, & -\cos\varphi\sin\theta \\ \sin\psi\sin\theta, & \cos\psi\sin\theta, & \cos\theta \end{pmatrix}.$$

Since $R(A) = R(-A)$, two different sets of Euler angles determine the same rotation matrix. Indeed it is easy to check from (1.20) that $R(A(\varphi, \theta, \psi)) = R(A(\varphi, \theta, \psi \pm 2\pi))$. Thus, to uniquely associate a rotation matrix $R(\varphi, \theta, \psi)$ with each set of Euler angles it is enough to restrict the angles to the domain

$$(1.21) \qquad 0 \leq \varphi < 2\pi, \qquad 0 \leq \theta \leq \pi, \qquad 0 \leq \psi < 2\pi,$$

i.e., ψ now runs over a domain of 2π rather than 4π radians. In the cases $\theta = 0, \pi$ only the sum $\varphi + \psi$ is determined by R, but this exceptional set is of lower dimension than three.

Since $SO(3)$ and $SU(2)$ are locally isomorphic, the invariant measure on $SO(3)$ must be given by (1.17), again except that the domain of the variables φ, θ, ψ is given by (1.21) rather than by (1.16). Thus the volume of $SO(3)$ is $8\pi^2$, half that of $SU(2)$.

Let $\mathbf{T}$ be a rep of $SO(3)$ by operators $\mathbf{T}(R)$. Then the operators $\mathbf{T}'(A) = \mathbf{T}(R(A))$, $A \in SU(2)$, define a rep of $SU(2)$ such that $\mathbf{T}'(-A) = \mathbf{T}'(A)$. Conversely, if $\mathbf{S}$ is a rep of $SU(2)$ such that $\mathbf{S}(-A) = \mathbf{S}(A)$ for all $A \in SU(2)$ then the operators $\mathbf{S}'(R(A)) = \mathbf{S}(A)$ define a rep of $SO(3)$. Thus, there is a 1–1 relationship between reps of $SO(3)$ and those reps $\mathbf{S}$ of $SU(2)$ such that $\mathbf{S}(-A) = \mathbf{S}(A)$, i.e., such that $\mathbf{S}(-E_2)$ is the identity operator.

Since $SU(2)$ and $SO(3)$ are compact groups, the problem of constructing all reps of these groups reduces to the problem of constructing all finite-dimensional unitary irred reps. Suppose $\mathbf{S}$ is a unitary irred rep of $SU(2)$ on an m-dimensional vector space. Now $-E_2 \in SU(2)$ commutes with all $A \in SU(2)$, so $\mathbf{S}(-E_2)$ commutes with all operators $\mathbf{S}(A)$. But $\mathbf{S}$ is irred, so by the Schur lemmas, $\mathbf{S}(-E_2) = \alpha\mathbf{E}$, where $\mathbf{E}$ is the identity operator. Since $(-E_2)^2 = E_2$ we have $\alpha^2 = 1$, or $\alpha = \pm 1$. Thus, $\mathbf{S}(-E_2) = \pm\mathbf{E}$. If the plus sign occurs then $\mathbf{S}$ is called **integral** and it defines an irred rep of $SO(3)$. However, if the minus sign occurs then $\mathbf{S}$ does not define a single-valued rep of $SO(3)$. [It is frequently stated that $\mathbf{S}$ defines a **double-valued** rep of $SO(3)$, i.e., two operators are associated with a single group element.] We shall call these reps **half-integral.**

In quantum mechanics the half-integral reps of $SU(2)$ appear even though one is initially concerned only with the rotation group $SO(3)$. The reason for this is that the states of a quantum mechanical system are given by rays in Hilbert space rather than by vectors. Thus the vectors $e^{i\gamma}\mathbf{v}$, $0 \leq \gamma < 2\pi$, all correspond to the same state for fixed $\mathbf{v}$ in the Hilbert space $\mathcal{H}$. A rotation $\mathbf{R}$ of 2π radians about the z axis will transform this state into itself. However, $\mathbf{Rv}$ need not be $\mathbf{v}$. In fact if $\mathbf{Rv} = e^{i\gamma}\mathbf{v}$ then the state will be mapped into itself. It is possible to show that for any action of $SO(3)$ as a continuous transformation group on the states of $\mathcal{H}$ we can always choose the state vectors $\mathbf{v}$ so γ is either 0 or π (Wigner [1]). The case $\gamma = \pi$ actually occurs, e.g., the electron wave functions, so we are led to consider double-valued reps of $SO(3)$.

Let $\mathcal{G}$ be a real n-dimensional matrix Lie algebra. The **complexification** $\mathcal{G}_c$ of $\mathcal{G}$ is the complex n-dimensional Lie algebra consisting of all complex linear combinations of elements in the real algebra $\mathcal{G}$. It is easy to check that isomorphic real Lie algebras have isomorphic complexifications. Let $\mathcal{K}$ be a complex n-dimensional matrix Lie algebra. A subset $\mathcal{K}_r$ is a **real form** of $\mathcal{K}$ if $\mathcal{K}_r$ is a real n-dimensional Lie algebra. A given complex Lie algebra may have several nonisomorphic real forms. If $\mathcal{K}_r$ is a real form of $\mathcal{K}$, then $(\mathcal{K}_r)_c$ is an n-dimensional complex Lie algebra and $(\mathcal{K}_r)_c \subseteq \mathcal{K}$. Since $\mathcal{K}$ is n-dimensional, $(\mathcal{K}_r)_c = \mathcal{K}$. Conversely, if $\mathcal{G}_c$ is the complexification of $\mathcal{G}$ it is obvious that $\mathcal{G}$ is a real form of $\mathcal{G}_c$.

Now $sl(2) = sl(2, \mathfrak{C})$ is the complexification of $su(2)$. Indeed, if we set

(1.22) $$\mathcal{J}^{\pm} = \pm\mathcal{J}_2 + i\mathcal{J}_1, \qquad \mathcal{J}^3 = -i\mathcal{J}_3,$$

where $i = \sqrt{-1}$ and the $\mathcal{J}_k$ are given by (1.8), we find that $\mathcal{J}^{\pm}$, $\mathcal{J}^3$ form a basis for a three-dimensional complex Lie algebra with commutation relations

(1.23) $$[\mathcal{J}^3, \mathcal{J}^{\pm}] = \pm\mathcal{J}^{\pm}, \qquad [\mathcal{J}^+, \mathcal{J}^-] = 2\mathcal{J}^3.$$

Comparing these relations with (10.9), Section 5.10, we see $sl(2) \cong (su(2))_c$. Furthermore, $su(2)$ is a real form of $sl(2)$.

It is clear from these remarks that any rep $\mathbf{T}$ of $su(2)$ on a complex vector space V induces a rep of $sl(2)$. Indeed, $\mathbf{T}(\mathcal{J}^{\pm}) = \pm\mathbf{T}(\mathcal{J}_2) + i\mathbf{T}(\mathcal{J}_1)$, $\mathbf{T}(\mathcal{J}^3) = -i\mathbf{T}(\mathcal{J}_3)$. Conversely, any rep of $sl(2)$ on V induces a rep of $su(2)$ by restriction. One of these reps is irred if and only if the other is irred.

Thus, to find the finite-dimensional irred reps of $su(2)$ it is enough to compute the finite-dimensional irred reps of $sl(2)$ and restrict these reps to $su(2)$. Then the results can be exponentiated to obtain irred reps of $SU(2)$.

7.2 Irreducible Representations of *SU*(2)

In Section 5.10 we constructed a family of finite-dimensional irred reps of $SL(2)$. The rep $\mathbf{D}^{(u)}$, $2u = 0, 1, 2, \ldots,$ is defined by operators

(2.1) $$[\mathbf{T}(A)f](z) = (bz + d)^{2u} f\left(\frac{az + c}{bz + d}\right),$$

$$A = \begin{pmatrix} a & b \\ c & d \end{pmatrix} \in SL(2), \quad f \in \mathcal{V}^{(u)},$$

acting on the $(2u + 1)$-dimensional space of polynomials of order $2u$. The corresponding rep of $sl(2)$ is given by

(2.2) $$J^3 h_j = (j - u)h_j, \qquad J^+ h_j = (j - 2u)h_{j+1}, \qquad J^- h_j = -jh_{j-1},$$

where $h_j(z) = z^j$, $0 \le j \le 2u$, is a basis for $\mathcal{V}^{(u)}$. By the remarks at the end of the preceding section, (2.2) also defines an irred rep of $su(2)$. We need only express the Lie derivatives J_k, $k = 1, 2, 3$, corresponding to $\mathcal{J}_k$ in terms of $J^\pm$, J^3 and use (2.2) to compute the action of J_k on a basis for $\mathcal{V}^{(u)}$. In particular,

(2.3) $$J^\pm = \pm J_2 + iJ_1, \qquad J^3 = -iJ_3.$$

We now exponentiate each rep of $su(2)$ to see if it defines a **global** irred rep of $SU(2)$.

If we consider $SL(2)$ as a real Lie group of dimension six then $SU(2)$ is a connected Lie subgroup. Thus, to obtain the group reps of $SU(2)$ induced by the reps of $su(2)$ we restrict the operators $\mathbf{T}(A)$, (2.1), to $A \in SU(2)$, (1.6):

(2.4) $$[\mathbf{T}(A)f](z) = (\beta z + \bar{\alpha})^{2u} f\left(\frac{\alpha z - \bar{\beta}}{\beta z + \bar{\alpha}}\right), \qquad f \in \mathcal{V}^{(u)}.$$

We shall again denote these $(2u + 1)$-dimensional reps of $SU(2)$ by the symbol $\mathbf{D}^{(u)}$. The $\mathbf{D}^{(u)}$ are irred because their associated Lie algebra reps are irred.

Note that

(2.5) $$[\mathbf{T}(-E_2)f](z) = (-1)^{2u} f(z),$$

or $\mathbf{T}(-E_2) = (-1)^{2u}\mathbf{E}$. Thus, for $u = 0, 1, 2, \ldots$ the reps are **integral** and define irred reps of $SO(3)$. On the other hand, for $u = \frac{1}{2}, \frac{3}{2}, \ldots$ the $\mathbf{D}^{(u)}$ are **half-integral** and yield double-valued reps of $SO(3)$. We shall show later that the $\mathbf{D}^{(u)}$ constitute all the irred reps of $SU(2)$ and the $\mathbf{D}^{(u)}$ for u an integer constitute all the irred reps of $SO(3)$.

Since $SU(2)$ is compact there must exist an inner product $(-, -)$ on $\mathcal{V}^{(u)}$ with respect to which $\mathbf{D}^{(u)}$ is unitary. Thus,

(2.6) $$(\mathbf{T}(A)f, \mathbf{T}(A)h) = (f, h), \qquad A \in SU(2)$$

for all $f, h \in \mathcal{V}^{(u)}$. Let $\exp tJ_k = \mathbf{T}(\exp t\mathcal{J}_k)$, where the $\mathcal{J}_k$ form a basis for $su(2)$. Substituting into (2.6), differentiating with respect to t, and setting $t = 0$, we find

(2.7) $$(J_k f, h) = -(f, J_k h), \qquad k = 1, 2, 3,$$

i.e., $J_k{}^* = -J_k$. Thus the operators J_k are skew-Hermitian. Stated another way, the operators iJ_k are Hermitian, $i = \sqrt{-1}$. It follows from (2.3) that $(J^+)^* = J^-$, $(J^-)^* = J^+$, and $(J^3)^* = J^3$.

The relations

$$(J^3 h_j, h_k) = (h_j, J^3 h_k), \qquad (J^+ h_j, h_k) = (h_j, J^- h_k)$$

together with (2.2) imply

(2.8) $$(h_j, h_k) = 0, \qquad j \neq k,$$

(2.9) $$(2u - j)\|h_{j+1}\|^2 = (j + 1)\|h_j\|^2, \qquad j = 0, 1, \ldots, 2u - 1.$$

Thus the basis vectors $h_j(z) = z^j$ are mutually orthogonal. Expression (2.9) shows the relationship between the norms of the basis vectors. We can normalize the inner product by choosing $\|h_0\|$ arbitrarily. Then (2.9) will fix the remaining norms. We now choose an ON basis $\{f_m\}$ for $\mathcal{V}^{(u)}$. The basis vectors will be labeled by the eigenvalue $m = j - u$ of f_m, with respect to J^3, rather then the parameter j. Normalizing $h_0(z) = 1$ by $\|h_0\|^2 = (2u)!$ we obtain the relation $\|h_j\|^2 = (2u - j)!\,j!$. Therefore, the vectors

(2.10)
$$f_m(z) = \frac{(-1)^j h_j(z)}{[(2u-j)!\,j!]^{1/2}} = \frac{(-z)^{u+m}}{[(u-m)!\,(u+m)!]^{1/2}},$$
$$m = -u, -u+1, \ldots, u-1, u,$$

form an ON basis for $\mathcal{V}^{(u)}$. It follows from (2.2) that

(2.11)
$$J^3 f_m = m f_m, \qquad J^{\pm} f_m = [(u \pm m + 1)(u \mp m)]^{1/2} f_{m\pm 1}.$$

The matrix elements of the rep $\mathbf{D}^{(u)}$ with respect to the ON basis $\{f_m\}$ are

$$T^u_{nm}(A) = (\mathbf{T}(A)f_m, f_n)$$

or

$$[\mathbf{T}(A)f_m](z) = \sum_{n=-u}^{u} T^u_{nm}(A) f_n(z), \qquad -u \le m \le u.$$

Thus,

(2.12)
$$g(A, z) = \frac{(\beta z + \bar{\alpha})^{u-m}(\alpha z - \bar{\beta})^{u+m}}{[(u-m)!\,(u+m)!]^{1/2}} = \sum_{n=-u}^{u} T^u_{nm}(A) \frac{(-1)^{n-m} z^{u+n}}{[(u-n)!\,(u+n)!]^{1/2}}.$$

Equating powers of z on both sides of this expression, we obtain

(2.13)
$$T^u_{nm}(A) = \left[\frac{(u+m)!\,(u-n)!}{(u+n)!\,(u-m)!}\right]^{1/2} \frac{\alpha^{u+n}\bar{\alpha}^{u-n}\bar{\beta}^{m-n}}{\Gamma(m-n+1)}$$
$$\times\, {}_2F_1\left(-u-n, m-u; m-n+1; -\left|\frac{\beta}{\alpha}\right|^2\right).$$

In terms of the Euler angles (1.13) this reads

(2.14)
$$T^u_{nm}(\varphi, \theta, \psi) = i^{n-m}\left[\frac{(u+m)!\,(u-n)!}{(u+n)!\,(u-m)!}\right]^{1/2} \frac{e^{i(n\varphi+m\psi)}(\sin\theta)^{m-n}(1+\cos\theta)^{u+n-m}}{2^u\Gamma(m-n+1)}$$
$$\times\, {}_2F_1\left(-u-n, m-u; m-n+1; \frac{\cos\theta - 1}{\cos\theta + 1}\right)$$
$$= i^{n-m}\left[\frac{(u+m)!\,(u-n)!}{(u+n)!\,(u-m)!}\right]^{1/2} e^{i(n\varphi+m\psi)} P_u^{-n,m}(\cos\theta)$$

(see the Symbol Index). By suitably manipulating these formulas we could obtain many other expressions for the matrix elements. Note the simple dependence of T^u_{nm} on φ and ψ. The group property

(2.15)
$$T^u_{nm}(A_1A_2) = \sum_{j=-u}^{u} T^u_{nj}(A_1)T^u_{jm}(A_2)$$

defines an addition theorem obeyed by the matrix elements. To apply the addition theorem when the $T^u_{nm}(A)$ are parametrized by the Euler angles it is necessary to compute the Euler angles (φ, θ, ψ) of a product $A(\varphi, \theta, \psi) = A_1(\varphi_1, \theta_1, \psi_1)A_2(\varphi_2, \theta_2, \psi_2)$. A straightforward though tedious computation yields

$$\begin{aligned} \cos\theta &= \cos\theta_1 \cos\theta_2 - \sin\theta_1 \sin\theta_2 \cos(\varphi_2 + \psi_1), \\ e^{i\varphi} &= (e^{i\varphi_1}/\sin\theta)(\sin\theta_1 \cos\theta_2 + \cos\theta_1 \sin\theta_2 \cos(\varphi_2 + \psi_1) \\ &\quad + i \sin\theta_2 \sin(\varphi_2 + \psi_1)), \\ e^{i(\varphi+\psi)/2} &= (e^{i(\varphi_1+\psi_1)/2}/\cos\tfrac{1}{2}\theta)(\cos\tfrac{1}{2}\theta_1 \cos\tfrac{1}{2}\theta_2\, e^{i(\varphi_2+\psi_1)/2} \\ &\quad - \sin\tfrac{1}{2}\theta_1 \sin\tfrac{1}{2}\theta_2\, e^{-i(\varphi_2+\psi_2)/2}), \end{aligned} \tag{2.16}$$

and the addition theorems are obtained by substituting (2.14) and (2.16) into (2.15). The unitary property of the operators $\mathbf{T}(A)$ implies

$$T^u_{nm}(A^{-1}) = \overline{T^u_{mn}(A)}, \tag{2.17}$$

or in Euler angles,

$$(-1)^{m-n} P_u^{-n,m}(\cos\theta) = \frac{(u+n)!\,(u-m)!}{(u-n)!\,(u+m)!} P_u^{-m,n}(\cos\theta).$$

Also, $|T^u_{nm}(A)| \leq 1$ or

$$|P_u^{-n,m}(\cos\theta)| \leq \left[\frac{(u+n)!\,(u-m)!}{(u+m)!\,(u-n)!}\right]^{1/2}, \qquad 0 \leq \theta \leq \pi.$$

We can obtain an integral expression for the matrix elements by setting $z = e^{i\gamma}$ in (2.12), multiplying by $e^{-i(u+n)\gamma}$, and integrating both sides of the resulting expression from 0 to 2π:

$$\begin{aligned} T^u_{nm}(\varphi, \theta, \psi) &= \frac{(-1)^{n-m}}{2\pi}\left[\frac{(u-n)!\,(u+n)!}{(u-m)!\,(u+m)!}\right]^{1/2} \\ &\quad \times \int_0^{2\pi} \left(i \sin\frac{\theta}{2} e^{i\gamma} + \cos\frac{\theta}{2}\right)^{u-m} \\ &\quad \times \left(\cos\frac{\theta}{2} e^{i\gamma} + i\sin\frac{\theta}{2}\right)^{u+m} e^{-i\gamma(u+n)}\, d\gamma. \end{aligned} \tag{2.18}$$

The matrix elements $T^l_{0m}(\varphi, \theta, \psi)$, l, m, integers, are proportional to the spherical harmonics $Y_l^m(\theta, \psi)$. Indeed

$$T^l_{0m}(\varphi, \theta, \psi) = i^m\left(\frac{4\pi}{2l+1}\right)^{1/2} Y_l^m(\theta, \psi) = i^m\left[\frac{(l-m)!}{(l+m)!}\right]^{1/2} P_l^m(\cos\theta)e^{im\psi}, \tag{2.19}$$

where the $P_l^m(\cos\theta)$ are the associated Legendre functions. Moreover,

$$T^l_{00}(\varphi, \theta, \psi) = P_l(\cos\theta), \tag{2.20}$$

where $P_l(\cos\theta)$ is the lth Legendre polynomial.

According to the general theory of Section 6.2, the matrix elements $T^u_{nm}(A)$ satisfy the orthogonality relations

$$\int_{SU(2)} T^{u_1}_{n_1m_1}(A)\overline{T^{u_2}_{n_2m_2}(A)}\, dA = \frac{16\pi^2}{2u_1+1}\,\delta_{n_1n_2}\,\delta_{m_1m_2}\,\delta_{u_1u_2}. \tag{2.21}$$

Thus,

$$\int_{-2\pi}^{2\pi} d\psi \int_0^{2\pi} d\varphi \int_0^{\pi} d\theta\, T^{u_1}_{n_1m_1}(\varphi,\theta,\psi)\overline{T^{u_2}_{n_2m_2}(\varphi,\theta,\psi)}\sin\theta = \frac{16\pi^2}{2u_1+1}\,\delta_{n_1n_2}\,\delta_{m_1m_2}\,\delta_{u_1u_2}.$$

The ψ and φ integrations are trivial, while the θ integration gives

$$\int_0^{\pi} P_u^{n,m}(\cos\theta)P_v^{n,m}(\cos\theta)\sin\theta\, d\theta = \frac{2}{2u+1}\frac{(u-n)!\,(u-m)!}{(u+n)!\,(u+m)!}\,\delta_{uv}.$$

For $n = m = 0$ these are the orthogonality relations for the Legendre polynomials. Note: By definition,

$$P_u^{0,-m}(\cos\theta) = P_u{}^m(\cos\theta), \qquad P_u^{0,0}(\cos\theta) = P_u(\cos\theta), \tag{2.22}$$

where $P_u{}^m$, P_u are Legendre functions. At this point we know only that the functions $\{(2u+1)^{1/2}T^u_{nm}(A)\}$ form an ON set in $L_2(SU(2))$, but later we will show that they form a basis, i.e., the $\mathbf{D}^{(u)}$ constitute a complete set of irred reps of $SU(2)$.

We now compute the character $\chi^{(u)}(A)$ of $\mathbf{D}^{(u)}$. By definition,

$$\chi^{(u)}(A) = \sum_{m=-u}^{u} T^u_{mm}(A). \tag{2.23}$$

This expression is too complicated to compute easily. On the other hand we know $\mathbf{T}^{(u)}(BAB^{-1}) = \mathbf{T}^{(u)}(A)$ for all $A, B \in SU(2)$. From elementary matrix theory, every $A \in SU(2)$ can be diagonalized by a unitary similarity transformation. Indeed, there exists a number τ, $-2\pi \le \tau < 2\pi$, and a $B \in SU(2)$ such that

$$BAB^{-1} = \begin{pmatrix} e^{i\tau/2} & 0 \\ 0 & e^{-i\tau/2} \end{pmatrix}.$$

Therefore, the conjugacy classes in $SU(2)$ are labeled by the parameter τ. Passing from $SU(2)$ to $SO(3)$ by the usual homomorphism we see that A represents a rotation through angle τ about a fixed axis. [In $SO(3)$, two rotations about distinct axes are conjugate if and only if they have the same rotation angle.]

We have shown that A is conjugate to the group element C with Euler parameters $(0, 0, \tau)$, or $\alpha = e^{i\tau/2}$, $\beta = 0$. By (2.12), $T^u_{mm}(C) = e^{im\tau}$. Thus,

$$\chi^{(u)}(A) = \sum_{m=-u}^{u} e^{im\tau} = \frac{e^{i(u+1)\tau} - e^{-iu\tau}}{e^{i\tau}-1} = \frac{\sin[(u+\frac{1}{2})\tau]}{\sin(\tau/2)}, \tag{2.24}$$

where we have used the formula for the sum of a geometric series. It is not difficult to express $\chi^{(u)}(A)$ directly in terms of the parameters of A, but the expression is not very enlightening. For $u = l = 0, 1, 2, \ldots$ the formula $\chi^{(l)}(R(A)) = \sin[(l + \frac{1}{2})\tau]/\sin(\tau/2)$ gives the character of the rep $\mathbf{D}^{(l)}$ of $SO(3)$ where $R(A)$ is a rotation through the angle τ about a fixed axis. In this case $\tau \pm 2\pi$ yield the same value as τ.

Let $\mathbf{D}^{(u)}, \mathbf{D}^{(v)}$ be irred reps of $SU(2)$ and consider the tensor product $\mathbf{D}^{(u)} \otimes \mathbf{D}^{(v)}$. This rep is $(2u + 1)(2v + 1)$-dimensional and its character is $\chi^{(u)} \otimes \chi^{(v)}(A) = \chi^{(u)}(A)\chi^{(v)}(A)$. We can determine the decomposition of $\mathbf{D}^{(u)} \otimes \mathbf{D}^{(v)}$ into a direct sum of irred reps of $SU(2)$ by expressing $\chi^{(u)} \otimes \chi^{(v)}$ as a sum of simple characters. Now

$$\chi^{(u)} \otimes \chi^{(v)}(A) = \sum_{m=-u}^{u} \sum_{n=-v}^{v} e^{i(m+n)\tau} = \sum_{w=u-v}^{u+v} \sum_{k=-w}^{w} e^{ik\tau} = \sum_{w=|u-v|}^{u+v} \chi^{(w)}(A),$$

where we have assumed $u \geq v$. [Note: The term $e^{ik\tau}$ occurs $\min(u + v + 1 - |k|, 2v + 1)$ times in the above expansion.] In general

$$\chi^{(u)} \otimes \chi^{(v)}(A) = \sum_{w=|u-v|}^{u+v} \chi^{(w)}(A). \tag{2.25}$$

Therefore,

$$\mathbf{D}^{(u)} \otimes \mathbf{D}^{(v)} \cong \mathbf{D}^{(u+v)} \oplus \mathbf{D}^{(u+v-1)} \oplus \cdots \oplus \mathbf{D}^{(|u-v|)}. \tag{2.26}$$

This expression is known as the **Clebsch–Gordan series**. Note that each irred rep which occurs on the right-hand side of (2.26) has multiplicity one. Thus, the decomposition of the rep space into irred subspaces is unique and independent of basis. In Section 7.7 we discuss this decomposition in detail.

7.3 Irreducible Representations of *sl*(2)

In Section 7.1, we showed that $sl(2)$ is the complexification of the real Lie algebra $su(2) \cong so(3)$. Therefore, there is a 1–1 relationship between irred reps of $sl(2)$ and irred reps of $su(2)$. To determine all finite-dimensional irred reps ρ of these Lie algebras it is enough to classify (up to isomorphism) all finite-dimensional complex vector spaces V and operators $J^{\pm}, J^3$ on V satisfying the commutation relations

$$[J^3, J^{\pm}] = \pm J^{\pm}, \qquad [J^+, J^-] = 2J^3, \tag{3.1}$$

such that V is irred under the J-operators. Here $J^3 = -iJ_3$, $J^{\pm} = \pm J_2 + iJ_1$, and $J_k = \rho(\mathcal{J}_k)$. The operators (3.1) will prove to be much more convenient for computations than the J_k.

The following computation should be familiar to those readers who have studied quantum mechanics. A good understanding of this procedure is essential since similar methods will be used to construct the irred reps of all the classical groups.

Let ρ be a finite-dimensional irred rep of $sl(2)$ on V. As the reader can verify, the **Casimir operator**

$$C = -(J_1)^2 - (J_2)^2 - (J_3)^2 = J^+J^- + J^3J^3 - J^3 \tag{3.2}$$

commutes with $J^\pm, J^3$. By the Schur lemmas, C must be a multiple of the identity operator on V, $C = \lambda\mathbf{E}$.

Let $\mathbf{h}_q \in V$ be an eigenvector of J^3 with eigenvalue q: $J^3\mathbf{h}_q = q\mathbf{h}_q$. Now $[J^3, J^+]\mathbf{h}_q = J^+\mathbf{h}_q$, or $J^3(J^+\mathbf{h}_q) = (q+1)J^+\mathbf{h}_q$. Thus, either $J^+\mathbf{h}_q = \boldsymbol{\theta}$ or $J^+\mathbf{h}_q$ is an eigenvector of J^3 with eigenvalue $q + 1$. Similarly the commutation relation $[J^3, J^-] = -J^-$ implies that $J^-\mathbf{h}_q = \boldsymbol{\theta}$ or $J^-\mathbf{h}_q$ is an eigenvector of J^3 with eigenvalue $q - 1$. By a simple induction argument

$$J^3(J^+)^k\mathbf{h}_q = (q + k)(J^+)^k\mathbf{h}_q, \qquad J^3(J^-)^k\mathbf{h}_q = (q - k)(J^-)^k\mathbf{h}_q, \qquad k = 0, 1, \ldots.$$

Since V is finite-dimensional there exists an integer $r \geq 0$ such that $(J^+)^r\mathbf{h}_q \neq \boldsymbol{\theta}$ and $(J^+)^{r+1}\mathbf{h}_q = \boldsymbol{\theta}$. Set $(J^+)^r\mathbf{h}_q = \mathbf{f}_u$, where $u = q + r$. Then $J^3\mathbf{f}_u = u\mathbf{f}_u$. Similarly there is an integer $s \geq 0$ such that $(J^-)^s\mathbf{f}_u \neq \boldsymbol{\theta}$, $(J^-)^{s+1}\mathbf{f}_u = \boldsymbol{\theta}$. We will show that the eigenvectors $\mathbf{f}_m$, $m = u, u - 1, \ldots, u - s$, where $\mathbf{f}_m = (J^-)^{u-m}\mathbf{f}_u$ form a basis for V.

Now $C\mathbf{f}_u = \lambda\mathbf{f}_u$. On the other hand, by (3.1) and (3.2),

$$C\mathbf{f}_u = (J^-J^+ + J^3J^3 + J^3)\mathbf{f}_u = J^-J^+\mathbf{f}_u + u(u + 1)\mathbf{f}_u.$$

Since $J^+\mathbf{f}_u = \boldsymbol{\theta}$ we obtain $\lambda = u(u + 1)$. Applying C to $\mathbf{f}_{u-s}$ we find

$$C\mathbf{f}_{u-s} = u(u + 1)\mathbf{f}_{u-s} = (J^+J^- + J^3J^3 - J^3)\mathbf{f}_{u-s} = (u - s)(u - s - 1)\mathbf{f}_{u-s},$$

since $J^-\mathbf{f}_{u-s} = \boldsymbol{\theta}$. Thus, $u(u + 1) = (u - s)(u - s - 1)$ or $s = 2u$. It follows that $2u$ is a nonnegative integer. Since $J^3\mathbf{f}_m = m\mathbf{f}_m$, $-u \leq m \leq u$, we obtain

$$C\mathbf{f}_m = u(u + 1)\mathbf{f}_m = (J^+J^- + J^3J^3 - J^3)\mathbf{f}_m = J^+\mathbf{f}_{m-1} + m(m - 1)\mathbf{f}_m,$$

or $J^+\mathbf{f}_m = (u - m)(u + m + 1)\mathbf{f}_{m+1}$, $u - 1 \geq m \geq -u$. We have shown that the $(2u + 1)$-dimensional subspace of V spanned by the $\{\mathbf{f}_m\}$ is invariant and irred under ρ. Since ρ is irred, this subspace must be V itself. The rep ρ is now completely determined:

$$J^3\mathbf{f}_m = m\mathbf{f}_m, \qquad J^-\mathbf{f}_m = \mathbf{f}_{m-1}, \qquad J^+\mathbf{f}_m = (u - m)(u + m + 1)\mathbf{f}_{m+1}, \qquad -u \leq m \leq u. \tag{3.3}$$

(On the right-hand sides of these expressions we adopt the convention: $\mathbf{f}_m = \boldsymbol{\theta}$ if m is not an eigenvalue of J^3.) Conversely, if $2u$ is a nonnegative integer then the operators $J^\pm, J^3$ defined by (3.3) determine an irred rep $\mathbf{D}^{(u)}$ of $sl(2)$. If $u \neq v$ then $\mathbf{D}^{(u)}$ is not equivalent to $\mathbf{D}^{(v)}$ since the two reps have different dimensions.

The rep $\mathbf{D}^{(u)}$ uniquely determines and is determined by the eigenvalues $-u, \ldots, +u$ of J^3. However, the basis vectors are not uniquely determined.

If $\{\gamma_m: -u \le m \le u\}$ is a set of nonzero complex constants then the eigenvectors $\{\mathbf{f}_m' = \gamma_m \mathbf{f}_m\}$ also form a basis for V. If the constants are chosen such that $\gamma_{m+1}/\gamma_m = [(u+m+1)(u-m)]^{1/2}$, $-u \le m \le u-1$, then relations (3.3) become

$$J^3\mathbf{f}_m = m\mathbf{f}_m, \quad J^{\pm}\mathbf{f}_m = [(u \mp m)(u \pm m + 1)]^{1/2}\mathbf{f}_{m\pm 1}, \qquad (3.4)$$
$$C\mathbf{f}_m = u(u+1)\mathbf{f}_m,$$

where we have omitted the prime on $\mathbf{f}_m'$. Note that expressions (3.4) and (2.11) are identical. Thus the reps $\mathbf{D}^{(u)}$, $2u = 0, 1, 2, \ldots$, of $SU(2)$ constructed in the preceding section constitute all the bounded irred reps of $SU(2)$, up to equivalence. [Furthermore, the reps of $SL(2)$ constructed in Section 5.10 constitute all finite-dimensional irred reps of $SL(2)$ as a complex Lie group.]

Another useful basis for V is obtained by setting $\gamma_{m+1}/\gamma_m = -(u+m+1)$. Relations (3.3) become

$$J^3\mathbf{f}_m = m\mathbf{f}_m, \qquad J^{\pm}\mathbf{f}_m = (-u \pm m)\mathbf{f}_{m\pm 1}, \qquad C\mathbf{f}_m = u(u+1)\mathbf{f}_m. \qquad (3.5)$$

Although we have confined ourselves to a search for finite-dimensional reps, expressions (3.5) can also be used to construct infinite-dimensional irred reps of $sl(2)$. (Here we mean V is infinite-dimensional in the algebraic sense. We do not consider V as a Hilbert space.) Indeed if $2u$ is a complex number, not a nonnegative integer, and V is a vector space generated by the vectors $\{\mathbf{f}_m\}$, $m = -u, -u+1, -u+2, \ldots$, then expressions (3.5) define an irred rep $\uparrow_u$ of $sl(2)$ on V, as the reader can verify. Since $J^-\mathbf{f}_{-u} = \boldsymbol{\theta}$ the operator J^3 has a lowest eigenvalue $-u$, i.e., an eigenvalue whose real part is least. However, J^3 has no highest eigenvalue. The rep $\uparrow_u$ is said to be **bounded below**. The reps $\mathbf{D}^{(u)}$ are bounded both below and above. Using similar techniques one can use expressions (3.5) to construct infinite-dimensional reps which are bounded above but not below or which are bounded neither above nor below. A systematic study of such reps is undertaken by Miller [1].

We have already seen the infinite-dimensional reps $\uparrow_u$. In Section 5.10 we constructed the local multiplier rep

$$[\mathbf{T}(A)f](z) = (bz+d)^{2u} f\left(\frac{az+c}{bz+d}\right), \qquad A \in SL(2) \qquad (3.6)$$

of $SL(2)$ on the space $\mathfrak{A}$ of all functions analytic in a neighborhood of $z = 0$. Here $2u$ is not a nonnegative integer. As a basis for $\mathfrak{A}$ we choose the functions $h_j(z) = z^j$, $j = 0, 1, \ldots$. The Lie derivatives associated with (3.6) are easily computed to be

$$J^+ = -2uz + z^2(d/dz), \qquad J^- = -d/dz, \qquad J^3 = -u + z(d/dz). \qquad (3.7)$$

Setting $f_m(z) = h_j(z) = z^j$, where $m + u = j$, we find

$$
\begin{aligned}
&J^+ f_m = (-2uz + z^2\, d/dz)z^{m+u} = (m - u)f_{m+1},\\
(3.8)\qquad &J^- f_m = -dz^{m+u}/dz = -(m + u)f_{m-1}, \qquad J^- f_{-u} = 0,\\
&J^3 f_m = (-u + z\, d/dz)z^{m+u} = mf_m,\\
&\qquad\qquad m = -u, -u + 1, -u + 2, \ldots .
\end{aligned}
$$

Thus the local multiplier rep (3.6) induces the irred rep $\uparrow_u$ of $sl(2)$. Conversely, the infinite-dimensional rep $\uparrow_u$ induces the local multiplier rep (3.6), which we will also call $\uparrow_u$. Note that the group rep $\uparrow_u$ is purely local.

7.4 Expansion Theorems for Functions on $SU(2)$

We have shown that the $(2u + 1)$-dimensional reps $\mathbf{D}^{(u)}$, $2u = 0, 1, 2, \ldots$, constitute a complete set of nonequivalent irred unitary reps of $SU(2)$. Thus, by the Peter–Weyl theorem, the functions $\varphi^u_{nm}(\varphi, \theta, \psi) = (2u + 1)^{1/2} T^u_{nm}(\varphi, \theta, \psi)$, $-u \le m,\ n \le u$, $2u = 0, 1, \ldots$, constitute an ON basis for $L_2(SU(2))$. [Here we use Euler coordinates on $SU(2)$ for the matrix elements (2.14).] The matrix elements satisfy orthogonality relations (2.21). Furthermore, if $f \in L_2(SU(2))$ then

$$
(4.1)\qquad f(\varphi, \theta, \psi) \sim \sum_{2u=0}^{\infty} \sum_{n,m=-u}^{u} a^u_{nm} \varphi^u_{nm}(\varphi, \theta, \psi),
$$

where

(4.2)

$$
a^u_{nm} = (f, \varphi^u_{nm}) = \frac{1}{16\pi^2} \int_{-2\pi}^{2\pi} d\psi \int_0^{2\pi} d\varphi \int_0^{\pi} d\theta\, f(\varphi, \theta, \psi)\overline{\varphi^u_{nm}(\varphi, \theta, \psi)} \sin\theta .
$$

The Parseval equality reads

$$
(4.3)\qquad (f, f) = \sum_{2u=0}^{\infty} \sum_{m,n=-u}^{u} |a^u_{nm}|^2 .
$$

With simple modifications these results apply to functions in $L_2(SO(3))$. The modifications are (1) u takes only integral values, (2) the volume of $SO(3)$ is $8\pi^2$ rather than $16\pi^2$, and (3) the variable ψ runs over the range $0 \le \psi < 2\pi$ rather than $-2\pi \le \psi < 2\pi$.

Some particular cases of (4.1) are of special interest. Suppose $f(\theta, \psi) \in L_2(SO(3))$ is independent of the variable φ. If we think of (θ, ψ) as latitude and longtitude, we can consider f as a function on the unit sphere S_3, square-integrable with respect to the area measure on S_3. Since the φ-dependence of $\varphi^u_{nm}(\varphi, \theta, \psi)$ is $e^{in\varphi}$, it follows from (4.2) that $a^u_{nm} = 0$ unless $n = 0$. The only possible nonzero coefficients are a^u_{0m}, where $u = l = 0, 1, 2, \ldots .$

By (2.19)

(4.4) $$\varphi^l_{0m}(\varphi, \theta, \psi) = (4\pi)^{1/2}\, Y_l^m(\theta, \psi),$$

where Y_l^m is a spherical harmonic. Thus,

(4.5) $$f(\theta, \psi) \sim \sum_{l=0}^{\infty} \sum_{m=-l}^{l} c_m{}^l Y_l^m(\theta, \psi),$$

where

(4.6) $$c_m{}^l = \int_0^{2\pi} d\psi \int_0^{\pi} d\theta\, f(\theta, \psi)\overline{Y_l^m(\theta, \psi)} \sin\theta, \qquad (Y_l^m, Y_{l'}^{m'}) = \delta_{ll'}\,\delta_{mm'}.$$

This is the expansion of a function on the sphere as a linear combination of spherical harmonics. As usual, (4.5) converges in the norm of $L_2(SO(3))$, not necessarily pointwise.

If $f(\theta) \in L_2(SO(3))$ is a function of θ alone then the coefficients a^u_{nm} are zero unless $n = m = 0$. From (2.20),

(4.7) $$\varphi^l_{00}(\varphi, \theta, \psi) = (2l + 1)^{1/2} P_l(\cos\theta), \qquad l = 0, 1, 2, \ldots,$$

where

(4.8)
$$P_l(x) = {}_2F_1\Big(l + 1, -l; 1, \frac{1 - x}{2}\Big) = 2^{-l}(1 + x)^l\,{}_2F_1\Big(-l, -l; 1; \frac{x - 1}{x + 1}\Big)$$

is a Legendre polynomial of order l. The coefficient of x^l in the expansion of $P_l(x)$ is nonzero and $P_l(1) = 1$. The expansion of $f(\theta)$ becomes

(4.9)
$$f(\theta) \sim \sum_{l=0}^{\infty} c_l P_l(\cos\theta), \qquad c_l = \tfrac{1}{2}(2l + 1) \int_0^{\pi} f(\theta) P_l(\cos\theta) \sin\theta\, d\theta,$$
$$\int_0^{\pi} P_l(\cos\theta) P_k(\cos\theta) \sin\theta\, d\theta = 2\delta_{kl}/(2l + 1).$$

Expressions (4.9) can be simplified by introduction of the new variable $x = \cos\theta$, $0 \le \theta \le \pi$.

The reader can construct some examples of the above expansions by considering the generating function (2.12) and the addition theorem (2.15). Other examples can be obtained by manipulation of the integral expression (2.18) for the matrix elements. If $n = m = 0$, $u = l$, (2.18) becomes

(4.10) $$P_l(\cos\theta) = (1/2\pi) \int_0^{2\pi} (\cos\theta + i \sin\theta \cos\gamma)^l\, d\gamma.$$

Setting $z = e^{i\gamma}$, we can write this last equation as a contour integral

(4.11) $$P_l(\cos\theta) = \frac{1}{2\pi i} \oint \Big[\cos\theta + \frac{i}{2} \sin\theta\, (z + z^{-1})\Big]^l \frac{dz}{z},$$

where the contour is a simple closed curve surrounding the origin. The change of variable $z = [t - \cos\theta + (t^2 - 2t\cos\theta + 1)^{1/2}]/(i \sin\theta)$ transforms (4.11) into

$$P_l(\cos\theta) = \frac{1}{2\pi i} \oint \frac{t^l\, dt}{(t^2 - 2t\cos\theta + 1)^{1/2}},$$

where the contour can be chosen as the circle $|t| = r > 1$. Setting $s = t^{-1}$, we find

$$(4.12) \qquad P_l(\cos\theta) = \frac{1}{2\pi i} \oint \frac{s^{-l-1}\, ds}{(s^2 - 2s\cos\theta + 1)^{1/2}}, \qquad |s| = r^{-1}.$$

The analytic function $(s^2 - 2s\cos\theta + 1)^{-1/2} = \sum c_n s^n$ possesses a power series expansion convergent for $|s| < 1$. It follows from (4.12) and the Cauchy integral theorem that $c_n = P_n(\cos\theta)$:

$$(4.13) \qquad h(s, x) = (s^2 - 2sx + 1)^{-1/2} = \sum_{n=0}^{\infty} s^n P_n(x), \qquad -1 \le x \le 1.$$

One can check that $h(s, \cos\theta) \in L_2(SO(3))$ for $|s| < 1$, so this is an example of the expansion (4.9). This generating function is often used to define the Legendre polynomials. Let $P_n'(x) = (d/dx)P_n(x)$.

Theorem 7.1.

(a) $P_n(1) = 1$;
(b) $P_n(-1) = (-1)^n$;
(c) $(2n + 1)xP_n(x) = (n + 1)P_{n+1}(x) - nP_{n-1}(x)$;
(d) $(1 - x^2)P_n'(x) + nxP_n(x) = nP_{n-1}(x)$;
(e) $(1 - x^2)P_n'(x) - (n + 1)xP_n(x) = -(n + 1)P_{n+1}(x)$;
(f) $[(1 - x^2)P_n'(x)]' + n(n + 1)P_n(x) = 0$, $n = 0, 1, 2, \ldots$.

Proof. These results follow from (4.13). (a) $h(s, 1) = (1 - s)^{-1} = \sum s^n$. (b) $h(s, -1) = (1 + s)^{-1} = \sum (-s)^n$. (c) $(s^2 - 2sx + 1)\, \partial h/\partial s = (s - x)h$. Now compare coefficients of s^n on both sides of this equality. (d) Follows from the identity $(1 - x^2)(\partial h/\partial x) + xs(\partial h/\partial s) = s^2(\partial h/\partial s) + sh$. (e) Follows from the identity $(1 - x^2)(\partial h/\partial x) - xs(\partial h/\partial s) - xh = -\partial h/\partial s$. (f) An easy consequence of (d) and (e). Q.E.D.

Any identity we can obtain for the generating function implies an identity for the Legendre polynomials. Thus, the identity $s\, \partial h/\partial s = (x - s)\, \partial h/\partial x$ implies

$$(4.14) \qquad nP_n(x) = xP_n'(x) - P_{n-1}'(x).$$

Identities such as (c)–(e) which relate different Legendre polynomials are called **recurrence formulas**. The differential equation (f) is the **Legendre equation**. Here we have derived these results by manipulation of the generating function $h(s, x)$, but we shall see that all these identities, including the generating function, have a simple group-theoretic interpretation.

7.5 New Realizations of the Irreducible Representations

From an abstract point of view we have completely classified the irred reps of $SU(2)$ and $SO(3)$. We have obtained simple realizations or models of these reps in which the underlying vector spaces consist of polynomials in one complex variable. In actual physical or geometrical systems, however, the group action may appear far different from that in our models. In other words, even though two group reps are abstractly equivalent they may appear physically or geometrically quite different. For this reason it is useful to survey some of the distinct realizations of the reps $\mathbf{D}^{(u)}$ which appear in mathematical physics.

For our first model we consider the natural action of $SO(3)$ as a transformation group on R_3:

(5.1) $$\mathbf{x} \longrightarrow A^{-1}\mathbf{x}, \qquad A \in SO(3), \qquad \mathbf{x} = (x, y, z) \in R_3.$$

(The inverse is necessary to conform to the definition of a Lie transformation group as given in Section 5.9.) Using the basis $\mathcal{L}_1, \mathcal{L}_2, \mathcal{L}_3$ for $so(3)$ as defined by (1.1)–(1.3) and computing the corresponding Lie derivatives we find

(5.2) $$L_1 = z\frac{\partial}{\partial y} - y\frac{\partial}{\partial z}, \qquad L_2 = x\frac{\partial}{\partial z} - z\frac{\partial}{\partial x}, \qquad L_3 = y\frac{\partial}{\partial x} - x\frac{\partial}{\partial y}.$$

As guaranteed by the general theory, these Lie derivatives satisfy the commutation relations

(5.3) $$[L_1, L_2] = L_3, \qquad [L_2, L_3] = L_1, \qquad [L_3, L_1] = L_2$$

and generate a Lie algebra isomorphic to $so(3)$. The Lie derivatives (5.2) are essentially the angular momentum operators of quantum mechanics. We shall construct models of the reps $\mathbf{D}^{(u)}$ where the action of the group and Lie algebra is given by (5.1) and (5.2), and the underlying vector space consists of functions on R_3.

First of all we define operators

(5.4) $$L^{\pm} = \mp L_2 + iL_1, \qquad L^3 = iL_3,$$

which satisfy the commutation relations (1.23) and form a basis for the complex Lie algebra $sl(2)$. [Note: These operators are not identical with (2.3). Nevertheless they satisfy the same commutation relations:

$$[L^3, L^{\pm}] = \pm L^{\pm}, \qquad [L^+, L^-] = 2L^3.$$

The choice (5.4) is more convenient for the computation to follow.] The action (5.1) of $SO(3)$ on R_3 is not transitive. In particular $x^2 + y^2 + z^2$ is invariant under the group. Any sphere of radius r and center at $\mathbf{0}$ is mapped into itself. To exploit this property we introduce spherical coordinates r, θ, φ:

$$x = r \sin\theta \cos\varphi, \qquad y = r \sin\theta \sin\varphi, \qquad z = r\cos\theta, \tag{5.5}$$
$$r \geq 0, \quad 0 \leq \theta \leq \pi, \quad 0 \leq \varphi < 2\pi.$$

Then the L-operators become

$$L^{\pm} = e^{\pm i\varphi}\left(\pm\frac{\partial}{\partial\theta} + i\cot\theta\,\frac{\partial}{\partial\varphi}\right), \qquad L^3 = -i\frac{\partial}{\partial\varphi}, \tag{5.6}$$

independent of r. We now look for realizations of $\mathbf{D}^{(u)}$ such that the basis space $\mathcal{V}^{(u)}$ is a space of analytic functions of θ, φ and the operators $L^{\pm}, L^3$ are given by (5.6). According to expressions (3.4) we must find basis functions $f_m(\theta, \varphi) = Y_u^m(\theta, \varphi)$ for $\mathcal{V}^{(u)}$ such that

$$\begin{aligned} L^3 Y_u^m &= mY_u^m, \qquad L^{\pm}Y_u^m = [(u \mp m)(u \pm m + 1)]^{1/2} Y_u^{m\pm 1}, \\ CY_u^m &= (L^+L^- + L^3L^3 - L^3)Y_u^m = u(u+1)Y_u^m. \end{aligned} \tag{5.7}$$

Since $L^3 = -i\partial/\partial\varphi$ we have

$$-i\,\partial Y_u^m/\partial\varphi = mY_u^m, \qquad Y_u^m(\theta, \varphi) = Q_u^m(\theta)e^{im\varphi},$$

where $Q_u^m(\theta)$ is yet to be determined. The equation $L^+Y_u^u = 0$ becomes

$$(d/d\theta)Q_u^u - u\cot\theta\, Q_u^u = 0,$$

whose solution is

$$Q_u^u = c_u \sin^u\theta = c_u(1 - \cos^2\theta)^{u/2},$$

where c_u is an arbitrary nonzero constant. We can now use the "lowering operator" L^- to obtain the functions Q_u^m recursively from Q_u^u:

$$-(d/d\theta)Q_u^{m+1} - (m+1)(\cot\theta)Q_u^{m+1} = [(u+m+1)(u-m)]^{1/2}Q_u^m. \tag{5.8}$$

A straightforward induction argument and (5.8) yield the explicit expressions

$$Q_u^m(\theta) = c_u\left[\frac{(u+m)!}{(2u)!\,(u-m)!}\right]^{1/2}(1-\cos^2\theta)^{-m/2}\frac{d^{u-m}(1-\cos^2\theta)^u}{d(\cos\theta)^{u-m}}, \tag{5.9}$$
$$-u \leq m \leq u.$$

The equation $L^- Y_u^{-u} = 0$ applied to (5.9) yields the condition

$$-\frac{dQ_u^{-u}}{d\theta} + u(\cot\theta)Q_u^{-u} = \frac{c_u}{(2u)!}(1-\cos^2\theta)^{(u+1)/2}\frac{d^{2u+1}(1-\cos^2\theta)^u}{d(\cos\theta)^{2u+1}} = 0.$$

This condition can be satisfied only if $u = l$ is an integer. For $u = \frac{1}{2}, \frac{3}{2}, \ldots,$ our construction fails. This is not surprising since the angular momentum operators (5.2) were obtained from an action of $SO(3)$ as a transformation group. For $u = l$, however, we have found a highest weight vector Y_l^l and a lowest weight vector Y_l^{-l}. By copying the construction of the reps $\mathbf{D}^{(l)}$ in

Section 7.3, the reader can check that the functions Y_l^m satisfy all the relations (5.7):

$$\pm \frac{d}{d\theta} Q_l^m - m \cot\theta\, Q_l^m = [(l \mp m)(l \pm m + 1)]^{1/2} Q_l^{m\pm 1},$$

(5.10)

$$\frac{1}{\sin\theta}\frac{d}{d\theta}\left(\sin\theta \frac{d}{d\theta} Q_l^m\right) + \left[l(l+1) - \frac{m^2}{\sin^2\theta}\right] Q_l^m = 0, \quad -l \le m \le l,$$

where the last expression is obtained by writing $CY_l^m = l(l+1)Y_l^m$ in terms of differential operators.

The constant c_l is usually fixed by the requirement

$$\int_0^{2\pi}\int_0^{\pi} |Y_l^l(\theta, \varphi)|^2 \sin\theta \, d\theta \, d\varphi = 1$$

or

(5.11) $$c_l = \frac{(-1)^l}{2^l l!}\left[\frac{(2l+1)!}{4\pi}\right]^{1/2},$$

where the phase factor $(-1)^l$ is introduced to conform to convention.

The basis functions $Y_l^m(\theta, \varphi)$ are just the spherical harmonics. To show this explicitly we obtain some new expressions for the matrix elements $T^u_{nm}(A)$ derived in Section 7.2. From (2.12). $T^u_{nm}(A)$ is, to within a constant factor, the coefficient of z^{u-n} in the Taylor series expansion of $g(A, z)$. Thus,

$$T^u_{nm}(A) = (-1)^{n-m}\left[\frac{(u-n)!}{(u+n)!}\right]^{1/2} \frac{d^{u+n} g(A, z)}{dz^{u+n}}\bigg|_{z=0}.$$

In terms of the functions $P_u^{n,m}(\cos\theta)$, (2.14), this reads

(5.12)

$$P_u^{-n,m}(\cos\theta) = \frac{i^{n-m}}{(u+m)!}\frac{d^{u+n}}{dz^{u+n}} \times \left[\left(iz \sin\frac{\theta}{2} + \cos\frac{\theta}{2}\right)^{u-m}\left(z\cos\frac{\theta}{2} + i\sin\frac{\theta}{2}\right)^{u+m}\right]_{z=0}.$$

Setting $y = (iz \sin\theta + \cos\theta - 1)/2$, we find $dy = \frac{1}{2} i \sin\theta \, dz$ and

(5.13) $$P_u^{-n,m}(\cos\theta) = \frac{(-1)^{u+n}}{2^u(u+m)!}(1-\cos\theta)^{(n-m)/2}(1+\cos\theta)^{(n+m)/2} \times \frac{d^{u+n}[(1-\cos\theta)^{u+m}(1+\cos\theta)^{u-m}]}{d(\cos\theta)^{u+n}}.$$

In particular, from (2.14), (2.16), (2.19), and (2.22) we obtain the expressions

(5.14) $$P_l^n(\cos\theta) = \frac{(l+n)!\,(-1)^l}{(l-n)!\,2^l l!}(1-\cos^2\theta)^{-n/2}\frac{d^{l-n}(1-\cos^2\theta)^l}{d(\cos\theta)^{l-n}}$$

for the associated Legendre functions and

(5.15) $$Y_l^m(\theta, \varphi) = \left[\frac{(2l+1)(l-m)!}{4\pi(l+m)!}\right]^{1/2} P_l^m(\cos\theta) e^{im\varphi}, \qquad -l \le m \le l,$$

for the spherical harmonics. This last expression agrees with (5.9) and (5.11), so the basis functions for the realization (5.7) are just spherical harmonics.

We have already seen that special functions appear in Lie theory as matrix elements of group reps. The above example shows that they also appear as basis functions in the underlying vector space of a group rep.

Now that we have found realizations for the reps $\mathbf{D}^{(l)}$ of $so(3)$ we can determine the action of $SO(3)$ on these realizations. Indeed $SO(3)$ acts on R_3 according to (5.1). It is not difficult to show that the resulting identity is

$$T^l_{0n}(AB) = \sum_{m=-l}^{l} T^l_{mn}(B)T^l_{0m}(A), \tag{5.16}$$

a special case of (2.15). Recall that

$$T^l_{0m}(A(\varphi, \theta, \psi)) = i^m\left(\frac{4\pi}{2l+1}\right)^{1/2} Y_l^m(\theta, \varphi).$$

Since $r^2 = x^2 + y^2 + z^2$ is invariant under the action of $SO(3)$ the set $\{f(r)Y_l^m(\theta, \varphi)\colon -l \leq m \leq l\}$ forms a basis for a realization of the irred rep $\mathbf{D}^{(l)}$. Here $f(r)$ is an arbitrary nonzero function. It follows that $L_2(R_3)$, the Hilbert space of all Lebesgue square-integrable functions on R_3, decomposes into a direct sum of irred reps $\mathbf{D}^{(l)}$, each $\mathbf{D}^{(l)}$ with infinite multiplicity.

An important special case of these considerations is the space $\mathcal{W}^l$ of all homogeneous polynomials $u(x, y, z)$ with degree l in x, y, z which satisfy Laplace's equation:

$$\nabla^2 u = \frac{\partial^2 u}{\partial x^2} + \frac{\partial^2 u}{\partial y^2} + \frac{\partial^2 u}{\partial z^2} = 0. \tag{5.17}$$

It is easy to show that under the action (5.1) of $SO(3)$ any solution of Laplace's equation is mapped into another solution. Furthermore, any homogeneous polynomial of degree l is mapped into another homogeneous polynomial of degree l. Thus, $\mathcal{W}^l$ is a finite-dimensional space invariant under the action of $SO(3)$. We shall decompose $\mathcal{W}^l$ into a direct sum of irred subspaces. Introducing the change of variable $\xi = x + iy$, $\eta = x - iy$, we see that every $u \in \mathcal{W}^l$ can be written uniquely in the form

$$u = \sum a_{nm}\xi^n\eta^m z^{l-n-m},$$

where

$$4(\partial^2 u/\partial\xi\,\partial\eta) + (\partial^2 u/\partial z^2) = 0$$

and n, m run over all nonnegative integers such that $0 \leq n + m \leq l$. Thus

$$\sum_{n,m} a_{nm}[4nm\xi^{n-1}\eta^{m-1}z^{l-n-m} + (l-n-m)(l-n-m-1)\xi^n\eta^m z^{l-n-m-2}] = 0,$$

or

$$4(n+1)(m+1)a_{n+1,m+1} + (l-m-n)(l-m-n-1)a_{n,m} = 0. \tag{5.18}$$

It follows from this expression that once values are prescribed for the $2l + 1$ independent constants $a_{0,m}$, $0 \leq m \leq l$, and $a_{n,0}$, $1 \leq n \leq l$, the remaining constants are uniquely determined. Thus $\mathcal{W}^l$ is $(2l + 1)$-dimensional. It is clear that the polynomial

$$r^l Y_l^l(\theta, \varphi) = \left[\frac{(2l + 1)(2l)!}{4\pi}\right]^{1/2} \frac{(-1)^l \xi^l}{2^l l!}$$

belongs to $\mathcal{W}^l$. Now $\mathcal{W}^l$ is invariant under the operators $L^\pm$, L^3, expressions (5.6). From (5.7) we see that the $2l + 1$ linearly independent functions $r^l Y_l^m(\theta, \varphi)$, $-l \leq m \leq l$, all lie in $\mathcal{W}^l$. Since $\mathcal{W}^l$ is $(2l + 1)$-dimensional it transforms irreducibly under the rep $\mathbf{D}^{(l)}$.

A well-known model of the rep $\mathbf{D}^{(u)}$ of $SL(2)$ is defined on the $(2u + 1)$-dimensional space $\mathcal{P}^u$ of homogeneous polynomials of degree $2u$ in the complex variables z_1, z_2. The group action is

$$(z_1, z_2) \longrightarrow (z_1, z_2)A = (z_1, z_2)\begin{pmatrix} \alpha & \beta \\ \gamma & \delta \end{pmatrix}, \qquad A \in SL(2). \tag{5.19}$$

Thus,

$$[\mathbf{T}(A)p](z_1, z_2) = p(\alpha z_1 + \gamma z_2, \beta z_1 + \delta z_2), \qquad p \in \mathcal{P}^u. \tag{5.20}$$

To see the connection between this expression and our previous models, set $w = z_1/z_2$. Then any $p \in \mathcal{P}^u$ can be written uniquely as $p(z_1, z_2) = z_2^{2u} p(w, 1)$, where $p(w, 1) = h(w)$ is a polynomial in w of order at most $2u$. We can factor z_2^{2u} from both sides of (5.20) to obtain the result

$$[\mathbf{T}(A)h](w) = (\beta w + \delta)^{2u} h\left(\frac{\alpha w + \gamma}{\beta w + \delta}\right). \tag{5.21}$$

This expression is identical with the model (2.1) of $\mathbf{D}^{(u)}$. Restricting (5.20) to the subgroup $SU(2)$, we get a model of the rep $\mathbf{D}^{(u)}$ for this subgroup.

We have seen (5.20) before. Indeed, if we let V be the two-dimensional space $V = \{az_1 + bz_2 : a, b \in \mathcal{C}\}$ then $\mathcal{P}^u$ can be identified with the $(2u + 1)$-dimensional subspace of completely symmetric tensors in $V^{\otimes 2u}$. This subspace is determined by the Young frame $[2u]$, i.e., the frame with one row and $2u$ columns. The action (5.20) of $SL(2)$ on this subspace is induced by the action (5.19) of $SL(2)$ on V. In Section 4.3 we showed that $[2u]$ determined an irred rep of $GL(2)$. Now we see that the restriction of this rep to $SL(2)$ and then to $SU(2)$ remains irred. The other irred reps $[f_1, f_2]$, $f_1 \geq f_2$, of $GL(2)$ also restrict to irred reps of $SL(2)$. However, as we shall show later, on restriction to $SL(2)$ we have the equivalences $[f_1, f_2] \cong [f_1 - f_2, 0]$, so the frames $[f_1] = [f_1, 0]$, $f_1 = 0, 1, 2, \ldots,$ exhaust the irred reps of $SL(2)$. In Chapter 9 we will study the irred reps of $SL(n)$ and $SU(n)$, and demonstrate the relationship between these reps and Young diagrams.

For our next example we construct a model of the infinite-dimensional local rep $\uparrow_{-1/2}$ of $SL(2)$. Consider the operators

$$(5.22)\qquad J^{\pm} = t^{\pm 1}\left((x^2-1)\frac{\partial}{\partial x} \pm xt\frac{\partial}{\partial t} + \frac{x}{2}\right), \qquad J^3 = t\frac{\partial}{\partial t},$$

acting on a space of analytic functions of x and t. These operators satisfy the commutation relations

$$[J^3, J^{\pm}] = \pm J^{\pm}, \qquad [J^+, J^-] = 2J^3$$

of $sl(2)$. In order to construct a model of $\uparrow_{-1/2}$ we must find functions $f_k(x, t) = g_k(x)t^{k+(1/2)}$, $k = 0, 1, 2, \ldots$, such that

$$(5.23)\qquad J^3 f_k = (k+\tfrac{1}{2})f_k, \qquad J^+ f_k = (k+1)f_{k+1}, \qquad J^- f_k = -kf_{k-1},$$

[see (3.6) and (3.8)]. It follows that the functions $g_k(x)$ satisfy the recurrence relations

$$(5.24)\qquad \begin{aligned} (x^2-1)g_k' + (k+1)xg_k &= (k+1)g_{k+1}, \\ (x^2-1)g_k' - kxg_k &= -kg_{k-1}. \end{aligned}$$

Furthermore the relation $(J^+J^- + J^3J^3 - J^3)f_k = -\frac{1}{4}f_k$ implies that the $g_k(x)$ satisfy the second-order differential equation

$$(5.25)\qquad [(x^2-1)(d^2/dx^2) + 2x(d/dx) - k(k+1)]g_k(x) = 0, \qquad k = 0, 1, 2, \ldots.$$

Expressions (5.24) determine the $g_k(x)$ up to a multiplicative constant. Indeed the relation $J^-f_0 = 0$ implies $g_0'(x) = 0$, or $g_0(x) = c$. If we set $c = 1$ we can uniquely determine the remaining $g_k(x)$ from the first of the recurrence formulas (5.24). The second recurrence formula and the differential equation (5.25) are consequences of the commutation relations and do not have to be verified explicitly for the $g_k(x)$. Rather then determine the $g_k(x)$ recursively we compare our recurrence formulas with Theorem 7.1 to obtain

$$(5.26)\qquad g_k(x) = P_k(x), \qquad f_k(x, t) = P_k(x)t^{k+(1/2)}.$$

Thus the Legendre polynomials define a model of $\uparrow_{-1/2}$. The operators (5.22) determine a local Lie multiplier rep $\mathbf{T}$ of $SL(2)$. In particular,

$$\mathbf{T}(\exp \alpha\mathcal{J}^3)f(x, t) = f(x, te^{\alpha})$$

$$\mathbf{T}(\exp \beta\mathcal{J}^{\pm})f(x, t) = Q_{\pm}^{-1/4} f\left(\frac{x - \beta t^{\pm 1}}{Q_{\pm}^{1/2}}, tQ_{\pm}^{\mp 1/2}\right),$$

$$Q_{\pm} = \beta^2 t^{\pm 2} - 2\beta x t^{\pm 1} + 1.$$

Just as in (10.22), Section 5.10, we could use these results to compute $\mathbf{T}(A)$ for any $A \in SL(2)$. However, we shall not do this here. The matrix elements $B_{lk}(A)$, (10.26), of the operators $\mathbf{T}(A)$ with respect to the basis f_k are model-independent. That is, they are completely determined by the relations (5.23) and are independent of our particular realization of this rep. We have

(5.27)

$$\mathbf{T}(A)f_k = \sum_{l=0}^{\infty} B_{lk}(A)f_l, \qquad k = 0, 1, 2, \ldots, \qquad A = \begin{pmatrix} a & b \\ c & d \end{pmatrix} \in SL(2).$$

For certain group elements A the functions $B_{lk}(A)$ are very simple. For example,

$$B_{lk}(\exp(-b\mathcal{J}^+)) = \begin{cases} (-b)^{l-k} l!/k!\,(l-k)!, & l \geq k, \\ 0, & l < k, \end{cases} \tag{5.28}$$

$$B_{lk}(\exp(-c\mathcal{J}^-)) = \begin{cases} c^{k-l} k!/l!\,(k-l)!, & k \geq l, \\ 0, & k < l. \end{cases} \tag{5.29}$$

[Note: The reader can obtain these results directly from relations (5.23).] Substituting (5.26) and (5.28) into (5.27) and simplifying, we obtain

(5.30)

$$(1 + b^2 - 2bx)^{-(k+1)/2} P_k\left(\frac{x - b}{(1 + b^2 - 2bx)^{1/2}}\right) = \sum_{l=0}^{\infty} b^l \binom{l+k}{l} P_{k+l}(x),$$

where

$$\binom{n}{m}$$

is the binomial coefficient. This expression makes sense for $|b| < |x \pm (x^2 - 1)^{1/2}|$. For $k = 0$, (5.30) reduces to the standard generating function

$$(1 + b^2 - 2bx)^{-1/2} = \sum_{l=0}^{\infty} b^l P_l(x).$$

Similarly, by substituting (5.26) and (5.29) into (5.27) we obtain

$$(1 + c^2 + 2cx)^{k/2} P_k\left(\frac{x + c}{(1 + c^2 + 2cx)^{1/2}}\right) = \sum_{l=0}^{k} \binom{k}{l} c^l P_l(x). \tag{5.31}$$

The point of this example is that identities such as (5.30) and (5.31) have a group-theoretic interpretation. Using the same operators (5.22) we could construct models of each of the irred reps $\uparrow_u$. The basis functions are essentially the Gegenbauer polynomials $C_k^{-u}(x)$ and our method yields generating functions and relations for the $C_k^{-u}(x)$.

Another interesting model of $\uparrow_u$ is obtained from a consideration of the operators

(5.32)

$$J^+ = t\left(z\frac{\partial}{\partial z} + t\frac{\partial}{\partial t} - z - u\right), \qquad J^- = t^{-1}\left(z\frac{\partial}{\partial z} - t\frac{\partial}{\partial t} - u\right), \qquad J^3 = t\frac{\partial}{\partial t},$$

acting on a space of analytic functions of the complex variables z, t. As the reader can easily verify, these operators satisfy the commutation relations of $sl(2)$. To construct a realization of $\uparrow_u$ we must find functions $f_{k-u}(z, t) = g_k(z)t^{-u+k}$ such that

$$(5.33)\qquad \begin{aligned} J^3 f_m &= m f_m, \qquad J^{\pm} f_m = (-u \pm m) f_{m\pm 1}, \\ C f_m &= (J^+J^- + J^3J^3 - J^3) f_m = u(u+1) f_m. \end{aligned}$$

Thus the special functions $g_k(z)$ satisfy

$$(5.34)\qquad z g_k' + (k - 2u - z) g_k = (k - 2u) g_{k+1}, \qquad z g_k' - k g_k = -k g_{k-1},$$

$$(5.35)\qquad z g_k'' - (2u + z) g_k' + k g_k = 0, \qquad k = 0, 1, 2, \ldots .$$

The functions $g_k(z)$ are determined to within a multiplicative constant by these relations. Indeed the relation $J^- f_u = 0$ implies $g_0' = 0$ or $g_0(z) = c$. Setting $c = 1$ we can then uniquely determine all of the $g_k(z)$ recursively from the first formula (5.34). The solutions are

$$(5.36)\qquad g_k(z) = \frac{\Gamma(-2u)k!}{\Gamma(k - 2u)} L_k^{(-2u-1)}(z), \qquad k = 0, 1, 2, \ldots,$$

where $L_k^{(\alpha)}(z)$ is a generalized Laguerre polynomial of order k and $\Gamma(z)$ is the gamma function (see the Symbol Index). Recall that $2u \neq 0, 1, \ldots$. The function $L_k(z) = L_k^{(0)}(z)$ is an (ordinary) Laguerre polynomial. The $L_k^{(-2u-1)}(z)$ satisfy the Laguerre differential equation (5.35).

A direct computation shows that the operators (5.32) determine a local multiplier rep $\mathbf{T}$ of $SL(2)$ given by

(5.37)

$$\mathbf{T}(A)f(z, t) = (d + bt)^u (a + c/t)^u \exp\left(\frac{bzt}{d + bt}\right) \times f\left(\frac{zt}{(at + c)(bt + d)}, \frac{at + c}{bt + d}\right), \qquad \left|\frac{c}{at}\right| < 1, \quad \left|\frac{bt}{d}\right| < 1.$$

The matrix elements $B_{lk}(A)$ of the $\mathbf{T}(A)$ with respect to the basis $f_{k-u}(z, t)$ are given by (10.26), Section 5.10. Substituting these expressions into

$$\mathbf{T}(A) f_{k-u} = \sum_{l=0}^{\infty} B_{lk}(A) f_{l-u}$$

and simplifying, we obtain identities for the Laguerre polynomials. For example, from (5.28) there follows

$$(5.38)\qquad (1 - b)^{2u-k} \exp\left(\frac{-bz}{1 - b}\right) L_k^{(-2u-1)}\left(\frac{z}{1 - b}\right) = \sum_{l=0}^{\infty} \binom{l + k}{l} b^l L_{k+l}^{(-2u-1)}(z), \qquad |b| < 1.$$

For $k = 0$, $L_0^{(\alpha)}(z) = 1$ and this expression simplifies to a well-known generating function for the Laguerre polynomials:

$$(5.39)\qquad (1-b)^{2u}\exp\left(\frac{-bz}{1-b}\right)=\sum_{l=0}^{\infty} b^l L_l^{(-2u-1)}(z),\qquad |b|<1.$$

Similarly, the matrix elements (5.29) yield the identity

$$(1+c)^k L_k^{(-2u-1)}\left(\frac{z}{1+c}\right)=\sum_{l=0}^{k}\binom{k-2u-1}{l} c^l L_{k-l}^{(-2u-1)}(z).$$

It is shown by Vilenkin [1] and Miller [1] that all hypergeometric and confluent hypergeometric functions can be obtained as basis functions in models of irred reps of $sl(2)$. Furthermore, in the work of Miller [1] it is shown how to derive such models in a systematic fashion.

7.6 Applications to Physics

Here we present a few of the many applications of the rep theory of $SO(3)$ and $SU(2)$ to problems in mathematical physics. In Section 3.8 we studied the relationship between symmetry and perturbation theory in quantum mechanics. Though our discussion was limited to finite symmetry groups it carries over without change to compact Lie symmetry groups.

Recall that the Hamiltonian $\mathbf{H}$ of a nonrelativistic quantum mechanical system containing k particles with masses $m_1, \ldots, m_k$ is

$$(6.1)\qquad \mathbf{H}=\sum_{j=1}^{k}(-1/2m_j)\,\Delta_j+V(\mathbf{x}_1,\ldots,\mathbf{x}_k),$$

where $V(\mathbf{x}_1, \ldots, \mathbf{x}_k)$ is the potential function and $\mathbf{x}_j \in R_3$ designates the coordinates of the jth particle. (We are using units in which $\hbar = 1$.) The Hilbert space $\mathcal{H}$ consists of all Lebesgue square-integrable functions $\Psi(\mathbf{x}_1, \ldots, \mathbf{x}_k)$,

$$\|\Psi\|^2=\int_{R_{3^k}}|\Psi(\mathbf{x}_1,\ldots,\mathbf{x}_k)|^2\,d\mathbf{x}<\infty,\qquad d\mathbf{x}=d^3_{\mathbf{x}_1}\cdots d^3_{\mathbf{x}_k}.$$

The inner product on $\mathcal{H}$ is

$$(\Psi,\Phi)=\int_{R_{3^k}}\Psi(\mathbf{x}_1,\ldots,\mathbf{x}_k)\bar{\Phi}(\mathbf{x}_1,\ldots,\mathbf{x}_k)\,d\mathbf{x}.$$

We can define a unitary rep $\mathbf{T}$ of $SO(3)$ on $\mathcal{H}$ by

$$(6.2)\qquad [\mathbf{T}(A)\Psi](\mathbf{x}_1,\ldots,\mathbf{x}_k)=\Psi(A^{-1}\mathbf{x}_1,\ldots,A^{-1}\mathbf{x}_k),\qquad A\in SO(3),$$

It is an elementary computation to verify $\mathbf{T}(AB) = \mathbf{T}(A)\mathbf{T}(B)$ and $(\mathbf{T}(A)\Psi, \mathbf{T}(A)\Phi) = (\Psi, \Phi)$ for all $A, B \in SO(3)$ and $\Psi, \Phi \in \mathcal{H}$.

Now $SO(3)$ is a symmetry group of $\mathbf{H}$ provided $\mathbf{T}(A)\mathbf{H} = \mathbf{H}\mathbf{T}(A)$ for all $A \in SO(3)$, i.e., provided $V(A\mathbf{x}_1, \ldots, A\mathbf{x}_k) = V(\mathbf{x}_1, \ldots, \mathbf{x}_k)$. If $SO(3)$

is a symmetry group and λ is an eigenvalue of $\mathbf{H}$ then the eigenspace

$$W_\lambda = \{\Psi \in \mathcal{H} : \mathbf{H}\Psi = \lambda\Psi\} \tag{6.3}$$

is invariant under $\mathbf{T}$. By the results of Section 3.7, we can decompose W_λ into a direct sum of subspaces irred under $\mathbf{T}$:

$$W_\lambda = \sum_{l=0}^{\infty} \sum_{i=1}^{a_l} \oplus W_i^{(l)}.$$

Here $\mathbf{T} \mid W_i^{(l)}$ is equivalent to the irred rep $\mathbf{D}^{(l)}$ and a_l is the multiplicity of $\mathbf{D}^{(l)}$ in $\mathbf{T}$. For simplicity we assume $\dim W_\lambda < \infty$, though this assumption could be removed with a little care. Then, only a finite number of the a_l are nonzero. Furthermore, if $SO(3)$ is a maximal symmetry group and there is no accidental degeneracy then only one of the a_l is nonzero.

The most important (and common) case in which $SO(3)$ appears as a symmetry group is the one where the potential takes the form

$$V = V(\|\mathbf{x}_i - \mathbf{x}_j\|, \|\mathbf{x}_l\|). \tag{6.4}$$

That is, V depends only on the mutual distances between particles and/or their distances from a common point. A special case is $V(\mathbf{x}) = V(\|\mathbf{x}\|)$, a single-particle, radially symmetric potential. These potentials admit the larger symmetry group $O(3)$. Indeed $V(A\mathbf{x}_1, \ldots, A\mathbf{x}_k) = V(\mathbf{x}_1, \ldots, \mathbf{x}_k)$ for all $A \in O(3)$.

The irred reps of the compact group $O(3)$ can easily be obtained from those of $SO(3)$. Indeed $SO(3)$ is a normal subgroup of index two in $O(3)$. The left coset decomposition of $O(3)$ is

$$O(3) = \{SO(3), I \cdot SO(3)\},$$

where the inversion $I = -E_3$. Let $\mathbf{D}$ be an irred unitary rep of $O(3)$. Since I commutes with all elements of $O(3)$, $\mathbf{D}(I)$ must be a multiple $\alpha\mathbf{E}$ of the identity operator. But $\mathbf{D}(I)^2 = \mathbf{D}(I^2) = \mathbf{D}(E_3) = \mathbf{E}$, so $\alpha = \pm 1$. Since $\mathbf{D}$ is irred and $\mathbf{D}(I) = \pm\mathbf{E}$ it follows that $\mathbf{D} \mid SO(3)$ is still irred. Therefore $\mathbf{D} \mid SO(3) \cong \mathbf{D}^{(l)}$, $l = 0, 1, 2, \ldots$. We conclude that there are two families $\mathbf{D}_+^{(l)}$, $\mathbf{D}_-^{(l)}$ of irred reps of $O(3)$. Their definitions are

$$\mathbf{D}_+^{(l)}(IA) = \mathbf{D}_+^{(l)}(A) = \mathbf{D}^{(l)}(A), \tag{6.5}$$

$$\mathbf{D}_-^{(l)}(IA) = -\mathbf{D}_-^{(l)}(A) = -\mathbf{D}^{(l)}(A), \qquad A \in SO(3). \tag{6.6}$$

The $\mathbf{D}_+^{(l)}$ are called **positive** reps and the $\mathbf{D}_-^{(l)}$ **negative** reps. Here $\dim \mathbf{D}_\pm^{(l)} = 2l + 1$.

Returning to the study of a system with potential (6.4), we see that each irred subspace $W_i^{(l)}$ of W_λ will transform according to $\mathbf{D}_\pm^{(l)}$. In a one-particle system with central potential $V(\|\mathbf{x}\|)$ we can say more. The space $W_i^{(l)}$ consists of functions $\Psi(\mathbf{x}) = \Psi(x, y, z)$ transforming irreducibly under $\mathbf{D}_\pm^{(l)}$, hence under the rep $\mathbf{D}^{(l)}$ of $SO(3)$. Thus, we can find a basis for $W_i^{(l)}$

of the form $f_m(\mathbf{x}) = j_l(r) Y_l^m(\theta, \varphi)$, $-l \leq m \leq l$, where the Y_l^m are spherical harmonics and r, θ, φ are spherical coordinates. The inversion I maps $\mathbf{x}$ to $-\mathbf{x}$, or in terms of spherical coordinates, (r, θ, φ) to $(r, \pi - \theta, \pi + \varphi)$. In our study of Laplace's equation (5.17) we showed that $r^l Y_l^m(\theta, \varphi)$ is a homogeneous polynomial of degree l in x, y, and z. Thus, under inversion $Y_l^m(\theta, \varphi) \longrightarrow Y_l^m(\pi - \theta, \pi + \varphi) = (-1)^l Y_l^m(\theta, \varphi)$. If l is even then $W_l^{(l)}$ transforms under the representation $\mathbf{D}_+^{(l)}$ of $O(3)$; if l is odd then $W_l^{(l)}$ transforms under $\mathbf{D}_-^{(l)}$. The sign of $(-1)^l$ is sometimes called the **parity** of the rep. In this example the symmetry of the Schrödinger equation under rotations has completely determined the angular dependence of the eigenfunctions. Only the radial dependence $j_l(r)$ remains to be determined from the dynamics of the problem. The well-known separation of variables method applied to the Schrödinger equation yields a second-order ordinary differential equation for $j_l(r)$:

$$(6.7) \qquad \frac{1}{r}\frac{d}{dr}\left[r^2 \frac{d}{dr} j_l(r)\right] + \left[\frac{l(l+1)}{r} + V(r)\right] j_l(r) = \lambda j_l(r).$$

The permissible solutions of this equation are those such that $j_l(r) Y_l^m(\theta, \varphi) \in \mathcal{H}$, i.e., $\int_0^\infty |j_l(r)|^2 r^2\, dr < \infty$. Only for certain values of λ, the eigenvalues, do there exist solutions belonging to $\mathcal{H}$.

The characters of $\mathbf{D}_\pm^{(l)}$ are easily obtained from the characters of the reps $\mathbf{D}^{(l)}$ of $SO(3)$. If R is a rotation through the angle τ about some axis then

$$(6.8) \qquad \chi_\pm^{(l)}(R) = \chi^{(l)}(R) = \{\sin[(l + \tfrac{1}{2})\tau]\}/\sin \tfrac{1}{2}\tau.$$

In the limit as $\tau \longrightarrow 0$ we get $\chi_\pm^{(l)}(E_3) = 2l + 1$. If S is a rotation through the angle τ followed by an inversion, then

$$(6.9) \qquad \chi_+^{(l)}(S) = -\chi_-^{(l)}(S) = \{\sin(l + \tfrac{1}{2})\tau]\}/\sin \tfrac{1}{2}\tau.$$

Suppose the k-particle system with Hamiltonian (6.1) is an atom or molecule. If this system is put into a crystal the new Hamiltonian is

$$(6.10) \qquad \mathbf{H}_1 = \sum_{j=1}^{k} (-1/2m_j)\,\Delta_j + V(\mathbf{x}_1, \ldots, \mathbf{x}_k) + V_1(\mathbf{x}_1, \ldots, \mathbf{x}_k),$$

where V_1 is the potential due to the crystal. Let G be the maximal point symmetry group of this crystal. Note that G is a finite subgroup of $O(3)$. Thus, the symmetry of the system is reduced from $O(3)$ to G under the perturbing potential V_1. If λ is an eigenvalue of $\mathbf{H}$ whose eigenspace transforms according to the $(2l + 1)$-dimensional rep $\mathbf{D}_+^{(l)}$ (or $\mathbf{D}_-^{(l)}$) then under the perturbing potential this degenerate energy level splits into energy levels whose eigenspaces transform according to irred reps of G. We can determine this splitting directly from the simple characters of $O(3)$ and G.

Suppose the eigenspace W_λ of $\mathbf{H}$ transforms according to $\mathbf{D}_+^{(l)}$. Then the restriction of $\mathbf{D}_+^{(l)}$ to the subgroup G splits into a direct sum of irred reps of G:

$$\mathbf{D}_+^{(l)} \mid G \cong \sum_{\mu=1}^{\alpha} \oplus a_\mu \mathbf{T}^{(\mu)}.$$

The character $\psi_l(A)$ of $\mathbf{D}_+^{(l)} \mid G$ is obtained from $\chi_+^{(l)}(A)$ by restricting A to G. Since G is a crystallographic point group it contains only rotations or rotation-inversions through the angles 0, $\pm\pi/3$, $\pm\pi/2$, $\pm 2\pi/3$, and π. Now from (6.8)

$$\chi^{(l)}\left(\pm\frac{2\pi}{n}\right) = \frac{\sin[(2\pi l/n) + (\pi/n)]}{\sin(\pi/n)}, \qquad n = 2, 3, 4, 6. \tag{6.11}$$

For fixed n this expression is periodic in l with period n. Thus, to evaluate ψ_l for any G it is enough to compute (6.11) for $0 \le l \le n$.

For example, suppose $G = O$, the octahedral group. The conjugacy classes are $E, \mathcal{C}_4{}^2, \mathcal{C}_2, \mathcal{C}_4, \mathcal{C}_3$, so O contains only twofold, threefold, and fourfold rotation axes. We compute ψ_l on these conjugacy classes for $0 \le l \le 5$:

l	E	$3\mathcal{C}_4{}^2$	$6\mathcal{C}_2$	$6\mathcal{C}_4$	$8\mathcal{C}_3$
0	1	1	1	1	1
1	3	−1	−1	1	0
2	5	1	1	−1	−1
3	7	−1	−1	−1	1
4	9	1	1	1	0
5	11	−1	−1	1	−1

Indeed, $\chi^{(l)}(E_3) = 2l + 1$, $\chi^{(l)}(\pi) = (-1)^l$, and so on. Using the character table of O, (6.22) of Section 3.6, we can write $\psi_l = \sum a_\mu^{(l)} \chi^{(\mu)}$ and compute the multiplicity $a_\mu^{(l)}$ of $\mathbf{T}^{(\mu)}$ in $\mathbf{D}_+^{(l)} \mid O$. The results are

(6.12)

$$\psi_0 = \chi^{(1)}, \qquad \psi_1 = \chi^{(4)}, \qquad \psi_2 = \chi^{(3)} + \chi^{(5)}, \qquad \psi_3 = \chi^{(2)} + \chi^{(4)} + \chi^{(5)},$$
$$\psi_4 = \chi^{(1)} + \chi^{(3)} + \chi^{(4)} + \chi^{(5)}, \qquad \psi_5 = \chi^{(3)} + 2\chi^{(4)} + \chi^{(5)}.$$

The interpretation of the expansion for ψ_4, for example, is that a ninefold degenerate energy level of $\mathbf{H}$ splits into four energy levels under the perturbation, one of the split levels is nondegenerate, one is twofold degenerate, and two are threefold degenerate. We can continue in this fashion to compute the splitting of an arbitrary $(2l + 1)$-fold energy level under a perturbation with octahedral symmetry. Since O contains only proper rotations the splitting for $\mathbf{D}_-^{(l)} \mid O$ is exactly the same as the splitting for $\mathbf{D}_+^{(l)} \mid O$.

If G contains rotation-inversions the determination of the splitting of the energy levels is analogous to that given above except that the results for $\mathbf{D}_-^{(l)} \mid G$ differ from those for $\mathbf{D}_+^{(l)} \mid G$.

Some Lie subgroups of $O(3)$ are of importance for perturbation theory calculations. Suppose we break the symmetry of a rotationally symmetric system by introducing a perturbing potential which transforms like the z component of an axial vector in xyz space. Then the symmetry group of the perturbed Hamiltonian will be $C_{\infty h} = C_\infty \times \{E, I\}$, where $C_\infty \cong U(1)$ is the group of all rotations about the z axis and $\{E, I\}$ consists of the identity element and the inversion I. (The choice of the z axis is arbitrary. Any other axis of symmetry would do.) As an example of such a perturbation consider an electron in a spherically symmetric field. If a uniform magnetic field parallel to the z axis is applied to this system, the perturbed system has symmetry $C_{\infty h}$. (We are ignoring the spin of the electron. This complication will be considered in Section 7.8).

Since $C_{\infty h}$ is abelian its irred reps are one-dimensional. Furthermore, since $I^2 = E$ and I commutes with all elements of $C_{\infty h}$ it follows that $\mathbf{T}(I) = \pm 1$ for any irred rep $\mathbf{T}$. We already know the irred reps of C_∞. They are denoted by the integer m: $\chi^{(m)}(C(\theta)) = e^{im\theta}$, $m = 0, \pm 1, \ldots$, where $C(\theta)$ is a rotation through the angle θ about the z axis. It follows that the irred reps of $C_{\infty h}$ are $\psi_\pm^{(m)}$, where

$$(6.13) \qquad \psi_\pm^{(m)}(C(\theta)) = e^{im\theta}, \qquad \psi_\pm^{(m)}(C(\theta)I) = \pm e^{im\theta}, \qquad m = 0, \pm 1, \ldots .$$

Suppose the eigenspace W_λ of the unperturbed Hamiltonian transforms according to the irred rep $\mathbf{D}_+^{(l)}$ with character $\chi_+^{(l)}$, (6.8) and (6.9). Now $\mathbf{D}_+^{(l)} \mid C_{\infty h}$ has character $\chi_+^{(l)}(C(\theta)) = \chi_+^{(l)}(C(\theta)I) = \sum_{m=-l}^{l} e^{im\theta} = \sum_{m=-l}^{l} \psi_+^{(m)}(\theta)$. Therefore, under the perturbing potential the degenerate energy level splits completely into $2l + 1$ simple sublevels, each with parity $+1$. Similarly $\mathbf{D}_-^{(l)} \mid C_{\infty h}$ has character $\chi_-^{(l)} \mid C_{\infty h} = \sum_{m=-l}^{l} \psi_-^{(m)}$, so the degenerate energy level splits into $2l + 1$ simple sublevels with parity -1. In the case where the perturbing potential is a magnetic field this splitting of energy levels is called the **Zeeman effect**.

It was shown in Section 2.9 that a molecule whose atoms all lie on a single line L possesses the symmetry group $C_{\infty v}$ consisting of all rotations about L and reflections in all planes in which L lies. Furthermore, if the molecule is also invariant with respect to the reflection σ in a plane perpendicular to L then the symmetry group is $D_{\infty h} = C_{\infty v} \times \{E, \sigma\}$. This occurs if the molecule is symmetric about its center of mass.

If L is the z axis then $C_{\infty v}$ is generated by the rotations $C(\varphi)$ and the reflection σ_v in the xz plane. It is easy to verify that this group has a 2×2 matrix realization

$$C(\varphi) = \begin{pmatrix} e^{i\varphi} & 0 \\ 0 & e^{-i\varphi} \end{pmatrix}, \qquad \sigma_v = \begin{pmatrix} 0 & 1 \\ 1 & 0 \end{pmatrix}.$$

Note that $C(\varphi)\sigma_v = \sigma_v C(-\varphi)$ and $\sigma_v{}^2 = E_2$. The rotations $C(\pm\varphi)$ form a conjugacy class and every reflection is conjugate to σ_v. Let $\mathbf{T}$ be a unitary irred rep of $C_{\infty v}$ on V. Then $\mathbf{T}\,|\,C_\infty$ splits into a direct sum of irred reps $\chi^{(m)}(\varphi) = e^{im\varphi}$ of C_∞. Suppose the nonzero vector $\mathbf{f}_m$ in V transforms according to $\chi^{(m)}$: $\mathbf{T}(C(\varphi))\mathbf{f}_m = e^{im\varphi}\mathbf{f}_m$. Then $\mathbf{T}(C(\varphi))\mathbf{T}(\sigma_v)\mathbf{f}_m = \mathbf{T}(\sigma_v)\mathbf{T}(C(-\varphi))\mathbf{f}_m = e^{-im\varphi}\mathbf{T}(\sigma_v)\mathbf{f}_m$. Thus $\mathbf{T}(\sigma_v)\mathbf{f}_m = \mathbf{f}_{-m}$ is a nonzero vector transforming according to $\chi^{(-m)}$ and $\mathbf{T}(\sigma_v)\mathbf{f}_{-m} = \mathbf{T}^2(\sigma_v)\mathbf{f}_m = \mathbf{f}_m$. Since V is irred under $C_{\infty v}$ it follows that $\mathbf{f}_{\pm m}$ generate V. If $m \neq 0$ then V is two-dimensional and with respect to the basis $\{\mathbf{f}_{\pm m}\}$ we obtain the matrix reps E_m:

$$(6.14)\qquad T(C(\varphi)) = \begin{pmatrix} e^{im\varphi} & 0 \\ 0 & e^{-im\varphi} \end{pmatrix}, \qquad T(\sigma_v) = \begin{pmatrix} 0 & 1 \\ 1 & 0 \end{pmatrix}, \qquad m = 1, 2, 3, \ldots .$$

The characters are $\chi^{(m)}(\varphi) = 2\cos m\varphi$, $\chi^{(m)}(\sigma_v) = 0$. If $m = 0$, then the irreducibility of $\mathbf{T}$ and the property $\mathbf{T}^2(\sigma_v) = \mathbf{E}$ imply V is one-dimensional, $\mathbf{T}(C(\varphi)) = 1$ and $\mathbf{T}(\sigma_v) = \pm 1$. Thus, we get two one-dimensional reps

$$(6.15)\qquad A_1\colon\ \mathbf{T}(C(\varphi)) = 1,\ \ \mathbf{T}(\sigma_v) = 1, \qquad A_2\colon\ \mathbf{T}(C(\varphi)) = 1,\ \ \mathbf{T}(\sigma_v) = -1.$$

The E_m, A_1, A_2 are a complete set of irred reps of $C_{\infty v}$. The character table is

(6.16)

$C_{\infty v}$	$C(\varphi)$	σ_v
A_1	1	1
A_2	1	-1
E_m	$2\cos m\varphi$	0

Now suppose the eigenspace W_λ of a Hamiltonian with spherical symmetry transforms according to $\mathbf{D}_+^{(l)}$, and introduce a perturbing potential with symmetry $C_{\infty v}$. Then the character of $\mathbf{D}_+^{(l)}\,|\,C_{\infty v}$ becomes $\chi(\varphi) = \sin(l + \frac{1}{2})\varphi/\sin(\varphi/2)$ and $\chi(\sigma_v) = \chi(\pi) = (-1)^l$. Clearly,

$$(6.17)\qquad \mathbf{D}_+^{(l)}\,|\,C_{\infty v} = E_l \oplus E_{l-1} \oplus \cdots \oplus E_1 \oplus A_k,$$

where $k = 1$ if l is even and $k = 2$ if l is odd. Thus, the $(2l + 1)$-degenerate eigenvalue λ splits into l eigenvalues with multiplicity two and one single eigenvalue. Similarly if W_λ transforms according to $\mathbf{D}_-^{(l)}$ a simple computation yields

$$(6.18)\qquad \mathbf{D}_-^{(l)}\,|\,C_{\infty v} = E_l \oplus E_{l-1} \oplus \cdots \oplus E_1 \oplus A_j,$$

where $j = 1$ if l is odd and $j = 2$ if l is even. In the case where our system contains only one particle the parity is $(-1)^l$, so we always get the identity rep A_1 in (6.17) and (6.18).

We can achieve $C_{\infty v}$ symmetry by introducing into a spherically symmetric system a perturbing potential which transforms like the z component of a polar vector in xyz space, e.g., an electron in a uniform electric field parallel to the z axis. The splitting of the energy levels due to this perturbation is called the **Stark effect**.

The group $D_{\infty h}$ can be written as the direct product $D_{\infty h} = C_{\infty v} \times \{E, I\}$. Thus, the irred reps can be obtained almost immediately from (6.16). The conjugacy classes are determined by the elements $C(\pm\varphi)$, σ_v, I, $C(\pm\varphi)I$, $\sigma_v I$. If $\mathbf{T}$ is an irred rep of $D_{\infty h}$ then $\mathbf{T}(I) = \pm\mathbf{E}$. The character table is

(6.19)

$D_{\infty h}$	$C(\varphi)$	σ_v	I	$C(\varphi)I$	$\sigma_v I$
A_1^+	1	1	1	1	1
A_1^-	1	1	-1	-1	-1
A_2^+	1	-1	1	1	-1
A_2^-	1	-1	-1	-1	1
E_m^+	$2\cos m\varphi$	0	1	$2\cos m\varphi$	0
E_m^-	$2\cos m\varphi$	0	-1	$-2\cos m\varphi$	0

Here $D_{\infty h}$ has four one-dimensional reps and two infinite families $E_m^{\pm}$ of two-dimensional reps. The determination of the splitting of $\mathbf{D}_{\pm}^{(l)} \mid D_{\infty h}$ is left to the reader.

The use of irred reps of symmetry groups to label the state vectors is of much more importance than perturbation theory alone would indicate. Suppose a quantum mechanical system is in the state $\Psi \in \mathcal{H}$ at time $t = 0$. Then at any other time t the system is in the state $\Psi(t)$, where $\Psi(t)$ is the unique solution of the (time-dependent) Schrödinger equation

$$i\,\partial\Psi(t)/\partial t = \mathbf{H}\Psi(t), \qquad \Psi(0) = \Psi. \tag{6.20}$$

Formally, the solution is $\Psi(t) = [\exp(-it\mathbf{H})]\Psi$. Since $\exp(-it\mathbf{H})$ is a unitary operator, the norm $\|\Psi(t)\|$ is independent of t. To make precise sense out of these statements we would have to employ some sophisticated techniques from functional analysis. (In particular we would need Stone's theorem; see Riesz–Sz.–Nagy [1].) Expression (6.20) is not always well-defined since there exist vectors $\Psi \in \mathcal{H}$ such that $\mathbf{H}\Psi$ has no meaning. Nevertheless, one can show that for the usual Hamiltonians of quantum mechanics there is a dense subspace of $\mathcal{H}$ on which (6.20) does make sense and on which the formal computations to follow can be rigorously justified.

Suppose $\mathbf{T}$ is a unitary rep of $SO(3)$ on $\mathcal{H}$ such that $\mathbf{T}(A)\mathbf{H} = \mathbf{H}\mathbf{T}(A)$ for all $A \in SO(3)$. Let W be an invariant subspace of $\mathcal{H}$ such that $\mathbf{T}\,|\,W$ transforms according to the irred rep $\mathbf{D}^{(l)}$. Then there is an ON basis $\{\Psi_m^{(l)}\}$ for W such that $\mathbf{T}(A)\Psi_m^{(l)} = \sum_n D_{nm}(A)\Psi_n^{(l)}$, where $\{D_{nm}(A)\}$ is a unitary matrix realization of $\mathbf{D}^{(l)}$. Let $\Psi_m^{(l)}(t)$ be solutions of (6.20) such that $\Psi_m^{(l)}(0) = \Psi_m^{(l)}$. Since $\mathbf{H}$ commutes with the operators $\mathbf{T}(A)$ it follows that $\mathbf{T}(A)\Psi_m^{(l)}(t) - \sum D_{nm}(A)\Psi_n^{(l)}(t) = \Phi(t)$ is a solution of (6.20) with initial condition $\Phi(0) \equiv 0$. Thus, $\Phi(t) \equiv 0$ and the vectors $\{\Psi_m^{(l)}(t)\}$ form an ON basis for the rep $\mathbf{D}^{(l)}$ at any time t. We conclude that l and m are good quantum numbers for the system. A state $\Psi_m^{(l)}(t)$ which transforms like the mth basis vector in a realization of $\mathbf{D}^{(l)}$ at one instant of time transforms like the mth basis vector in $\mathbf{D}^{(l)}$ at any time. Physicists refer to this as the conservation of angular momentum. Although this analysis applies only to $SO(3)$, similar results can be easily obtained for any compact symmetry group of $\mathbf{H}$.

It is worthwhile to point out the connection between the time-dependent and time-independent Schrödinger equations. Suppose $\Psi \in \mathcal{H}$ is a nonzero solution of the Schrödinger equation $\mathbf{H}\Psi = \lambda\Psi$. (We assume $\mathbf{H}$ is independent of t.) Then Ψ is an eigenvector of $\mathbf{H}$ with eigenvalue λ. Furthermore, the one-parameter family $\Psi(t) = e^{-it\lambda}\Psi$ is the unique solution of the Schrödinger equation

$$i\,\partial\Psi(t)/\partial t = \mathbf{H}\Psi(t), \qquad \Psi(0) = \Psi.$$

Since the vectors $e^{-it\lambda}\Psi$ belong to the same ray in $\mathcal{H}$ for all t, it follows that any eigenstate of $\mathbf{H}$ remains fixed with passage of time.

As a final application we investigate the quantum mechanical interpretation of the Lie algebra $so(3) \cong su(2)$. As usual we consider the unitary rep $\mathbf{T}$, (6.2), of $SO(3)$ on $\mathcal{H}$. Then $\mathbf{T}$ induces a rep (also called $\mathbf{T}$) of $so(3)$ on $\mathcal{H}$:

$$\mathbf{T}(\mathcal{a}) = (d/dt)\mathbf{T}(\exp t\mathcal{a})|_{t=0}, \qquad \mathcal{a} \in so(3). \tag{6.21}$$

In particular, the operators $\mathbf{T}(\mathcal{L}_j) = \boldsymbol{\mathfrak{L}}_j$ are

$$\begin{aligned} \boldsymbol{\mathfrak{L}}_1 &= \sum_{j=1}^{k}\left(z_j\frac{\partial}{\partial y_j} - y_j\frac{\partial}{\partial z_j}\right), & \boldsymbol{\mathfrak{L}}_2 &= \sum_{j=1}^{k}\left(x_j\frac{\partial}{\partial z_j} - z_j\frac{\partial}{\partial x_j}\right), \\ \boldsymbol{\mathfrak{L}}_3 &= \sum_{j=1}^{k}\left(y_j\frac{\partial}{\partial x_j} - x_j\frac{\partial}{\partial y_j}\right), & \mathbf{x}_j &= (x_j, y_j\, z_j), \end{aligned} \tag{6.22}$$

where the $\mathcal{L}_j$ are given by (1.2) and (1.3). As the reader can easily verify, the $\boldsymbol{\mathfrak{L}}_j$ satisfy the commutation relations (1.5) of $so(3)$ and they form a basis for the Lie algebra of operators $\mathbf{T}(\mathcal{a})$. Proceeding formally by differentiating the identity

$$(\mathbf{T}(\exp t\mathcal{a})\Psi, \mathbf{T}(\exp t\mathcal{a})\Phi) = (\Psi, \Phi), \qquad \Psi, \Phi \in \mathcal{H},$$

with respect to t we obtain

$$(\mathbf{T}(\mathcal{a})\Psi, \Phi) + (\Psi, \mathbf{T}(\mathcal{a})\Phi) = 0 \tag{6.23}$$

at $t = 0$. Thus $\mathbf{T}^*(\mathcal{a}) = -\mathbf{T}(\mathcal{a})$ and the operators $i\mathbf{T}(\mathcal{a})$, $i = \sqrt{-1}$, are **symmetric**. In particular the operators $\mathbf{L}_j = i\boldsymbol{\mathfrak{L}}_j$ are symmetric and satisfy the commutation relations

$$[\mathbf{L}_1, \mathbf{L}_2] = i\mathbf{L}_3, \qquad [\mathbf{L}_3, \mathbf{L}_1] = i\mathbf{L}_2, \qquad [\mathbf{L}_2, \mathbf{L}_3] = i\mathbf{L}_1. \tag{6.24}$$

The $\mathbf{L}_j$ are called the **angular momentum operators**. If the Hamiltonian $\mathbf{H}$ commutes with the operators $\mathbf{T}(A)$, $A \in SO(3)$, then by differentiating the identity $\mathbf{T}(\exp t\mathcal{a})\mathbf{H} = \mathbf{H}\mathbf{T}(\exp t\mathcal{a})$ at $t = 0$ we find $\mathbf{T}(\mathcal{a})\mathbf{H} = \mathbf{H}\mathbf{T}(\mathcal{a})$ for all $\mathcal{a} \in so(3)$. In particular the angular momentum operators commute with

H. Conversely, if the angular momentum operators commute with **H** then the operators $\mathbf{T}(A)$, $A \in SO(3)$, commute with **H**.

Unfortunately the above computations are merely formal. The operators $\mathbf{L}_j$ and **H** are not defined on all of $\mathcal{H}$. For instance if **H** is given formally by (6.1) then $\mathbf{H}\Psi(\mathbf{x})$ makes sense only if $\Psi(\mathbf{x})$ can be differentiated twice in each variable. Furthermore the function $\mathbf{H}\Psi(\mathbf{x})$ must belong to $\mathcal{H}$, i.e., $\|\mathbf{H}\Psi\| < \infty$. Since many functions in $\mathcal{H}$ are not differentiable it is clear that $D_{\mathbf{H}}$ cannot be all of $\mathcal{H}$. The problem of defining explicitly the domain of **H**, or any unbounded operator in quantum mechanics, is outside the scope of this book (see Helwig [1]). It can be shown that each of these operators can be defined on a dense (not necessarily closed) subspace of $\mathcal{H}$. However, the subspace varies with the operator. The angular momentum operators make sense only when applied to differentiable functions $\Psi(\mathbf{x})$ such that $\mathbf{L}_j\Psi(\mathbf{x})$ is square-integrable. Furthermore, the meaning of a commutation relation such as $[\mathbf{L}_1, \mathbf{L}_2] = i\mathbf{L}_3$ is not completely clear since the domains of the left- and right-hand sides may not be the same.

However, it can be shown (Helgason [1, p. 440]) that there exists a dense subspace $\mathfrak{D}$ of $\mathcal{H}$ which is contained in the domains of all the operators **H** and $\mathbf{L}_j$. Furthermore $\mathfrak{D}$ is invariant under the restrictions of $\mathbf{H}, \mathbf{L}_j, \mathbf{T}(A)$ to $\mathfrak{D}$ and has the property that all of the above formal computations are rigorously correct for these restricted operators. Thus, the relation

$$\mathbf{T}(\exp \alpha\mathcal{L}_j)\Psi = \sum_{n=0}^{\infty} \frac{(\alpha\mathcal{L}_j)^n}{n!}\Psi \tag{6.25}$$

is valid for $\Psi \in \mathfrak{D}$. If we accept the fact that $\mathfrak{D}$ exists we can use Lie algebra computations to derive results about infinite-dimensional Lie group reps. Note that the unitary operators $\mathbf{T}(A)$, $A \in SO(3)$, are uniquely determined by the symmetric operators $\mathbf{L}_j$. Indeed $\mathbf{T}(A)$ is uniquely defined on $\mathfrak{D}$ by (6.25). Since $\mathfrak{D}$ is dense in $\mathcal{H}$ and $\mathbf{T}(A)$ is bounded it follows from a standard Hilbert-space argument that $\mathbf{T}(A)$ is uniquely determined on $\mathcal{H}$ (Naimark [2, p. 100]). With these remarks in mind we shall henceforth ignore problems concerning the domains of unbounded operators.

The angular momentum operators can be used to compute the matrix elements of **H** with respect to an ON basis of $\mathcal{H}$. Consider again the unitary rep **T** of $SO(3)$ on $\mathcal{H}$. From the results of Section 6.3 we know that $\mathbf{T} = \sum \oplus a_l\mathbf{D}^{(l)}$, i.e., $\mathcal{H}$ can be decomposed into a direct sum of subspaces irred under **T**. (In general the multiplicities a_l will be infinite.) Thus, there is an ON basis $\{\Psi_{jm}^{(l)}\}$ for $\mathcal{H}$ such that $\mathbf{T}(A)\Psi_{jm}^{(l)} = \sum D_{nm}^{(l)}(A)\Psi_{jn}^{(l)}$ and $1 \leq j \leq a_l$. We have shown in Section 6.3 how such a basis can be constructed without any knowledge of the Hamiltonian **H**.

Since **H** commutes with the $\mathbf{T}(A)$ it also commutes with the operators $\mathbf{L}^{\pm} = \pm\mathcal{L}_2 + i\mathcal{L}_1$ and $\mathbf{L}^3 = -i\mathcal{L}_3$. Here $(\mathbf{L}^+)^* = \mathbf{L}^-$ and $(\mathbf{L}^3)^* = \mathbf{L}^3$.

The ON basis vectors $\mathbf{\Psi}^{(l)}_{jm}$ can be chosen such that

(6.26)

$$\mathbf{L}^3\mathbf{\Psi}^{(l)}_{jm} = m\mathbf{\Psi}^{(l)}_{jm}, \qquad \mathbf{L}^{\pm}\mathbf{\Psi}^{(l)}_{jm} = [(l \pm m + 1)(l \mp m)]^{1/2}\mathbf{\Psi}^{(l)}_{jm\pm 1}$$

$$\mathbf{L} \cdot \mathbf{L}\mathbf{\Psi}^{(l)}_{jm} = l(l+1)\mathbf{\Psi}^{(l)}_{jm}, \qquad \mathbf{L} \cdot \mathbf{L} = \mathbf{L}^+\mathbf{L}^- + \mathbf{L}^3\mathbf{L}^3 - \mathbf{L}^3 = -\sum_{j=1}^{3} \mathfrak{L}_j\mathfrak{L}_j.$$

Now

$$(\mathbf{H}\mathbf{L}^3\mathbf{\Psi}^{(l)}_{jm}, \mathbf{\Psi}^{(l')}_{j'm'}) = (\mathbf{L}^3\mathbf{H}\mathbf{\Psi}^{(l)}_{jm}, \mathbf{\Psi}^{(l')}_{j'm'}) = (\mathbf{H}\mathbf{\Psi}^{(l)}_{jm}, \mathbf{L}^3\mathbf{\Psi}^{(l')}_{j'm'}),$$

so $(m - m')(\mathbf{H}\mathbf{\Psi}^{(l)}_{jm}, \mathbf{\Psi}^{(l')}_{j'm'}) = 0$ and the matrix element is zero unless $m = m'$. Similarly the relation

$$(\mathbf{H}\mathbf{L} \cdot \mathbf{L}\mathbf{\Psi}^{(l)}_{jm}, \mathbf{\Psi}^{(l')}_{j'm'}) = (\mathbf{H}\mathbf{\Psi}^{(l)}_{jm}, \mathbf{L} \cdot \mathbf{L}\mathbf{\Psi}^{(l')}_{j'm'})$$

shows that the matrix element is zero unless $l = l'$. The identity

$$(\mathbf{H}\mathbf{L}^+\mathbf{\Psi}^{(l)}_{jm}, \mathbf{\Psi}^{(l)}_{j'm+1}) = (\mathbf{H}\mathbf{\Psi}^{(l)}_{jm}, \mathbf{L}^-\mathbf{\Psi}^{(l)}_{j'm+1})$$

yields $(\mathbf{H}\mathbf{\Psi}^{(l)}_{jm+1}, \mathbf{\Psi}^{(l)}_{j'm+1}) = (\mathbf{H}\mathbf{\Psi}^{(l)}_{jm}, \mathbf{\Psi}^{(l)}_{j'm})$, i.e., the matrix elements are independent of m. Thus

(6.27) $$(\mathbf{H}\mathbf{\Psi}^{(l)}_{jm}, \mathbf{\Psi}^{(l')}_{j'm'}) = \delta_{ll'}\,\delta_{mm'}\,\lambda(l, j, j'),$$

where $\lambda(l, j, j')$ is independent of m and m'. In Section 7.8 this result will be generalized to obtain information about the matrix elements of operators which do not necessarily commute with the action of $SO(3)$ on $\mathcal{H}$.

7.7 The Clebsch–Gordan Coefficients

In Section 7.2 we derived the Clebsch–Gordan series

(7.1) $$\mathbf{D}^{(u)} \otimes \mathbf{D}^{(v)} = \sum_{w=|u-v|}^{u+v} \oplus\, \mathbf{D}^{(w)}$$

for the tensor product of two irred reps of $SU(2)$. Recall that we also used the symbol $\mathbf{D}^{(u)}$ to denote the $(2u + 1)$-dimensional irred rep of $SL(2)$. Since there is a 1–1 relationship between complex reps of $sl(2)$ and $su(2)$ it follows that expression (7.1) is also valid for $SL(2)$. Furthermore, this same argument shows that any finite-dimensional analytic rep of $SL(2)$ as a complex Lie group can be decomposed into a direct sum of irred reps.

In the following we shall consider (7.1) as a rep of $SL(2)$, but all our results will remain valid on restriction to $SU(2)$. If $\mathbf{D}^{(u)}$, $\mathbf{D}^{(v)}$ are defined on inner product spaces $V^{(u)}$, $V^{(v)}$ then $\mathbf{D}^{(u)} \otimes \mathbf{D}^{(v)}$ is defined on the $(2u + 1)(2v + 1)$-dimensional space $V^{(u)} \otimes V^{(v)}$. As a convenient ON basis for the rep space we choose $\{\mathbf{f}^{(u)}_m \otimes \mathbf{f}^{(v)}_n : -u \le m \le u,\ -v \le n \le v\}$, where $\{\mathbf{f}^{(u)}_m\}$ is a basis for $V^{(u)}$ such that

(7.2) $$J^3\mathbf{f}^{(u)}_m = m\mathbf{f}^{(u)}_m, \qquad J^{\pm}\mathbf{f}^{(u)}_m = [(u \pm m + 1)(u \mp m)]^{1/2}\mathbf{f}^{(u)}_{m\pm 1},$$

and $\{\mathbf{f}_u^{(v)}\}$ is defined similarly. [In the future we will call any ON basis $\{\mathbf{f}_m\}$ satisfying (7.2) a **canonical basis.**] Though $\{\mathbf{f}_m^{(u)} \otimes \mathbf{f}_n^{(v)}\}$ is well adapted to show the tensor product character of our rep, it does not clearly exhibit the decomposition of $\mathbf{D}^{(u)} \otimes \mathbf{D}^{(v)}$ into irred reps. From the right-hand side of (7.1) it follows that $V^{(u)} \otimes V^{(v)}$ contains an ON basis of the form

$$\{\mathbf{h}_k^{(w)}: w = u + v, u + v - 1, \ldots, |u - v|, \quad -w \leq k \leq w\}$$

such that

$$(7.3) \qquad J^3\mathbf{h}_k^{(w)} = k\mathbf{h}_k^{(w)}, \qquad J^{\pm}\mathbf{h}_k^{(w)} = [(w \pm k + 1)(w \mp k)]^{1/2}\mathbf{h}_{k\pm 1}^{(w)},$$

where the J-operators are now those determined by the action of $SL(2)$ on $V^{(u)} \otimes V^{(v)}$. For fixed w the vectors $\{\mathbf{h}_h^{(w)}\}$ are determined up to a phase factor by (7.3). They form an ON basis for the invariant subspace which transforms according to $\mathbf{D}^{(w)}$. The two sets of basis vectors are related by the Clebsch–Gordan (CG) coefficients:

$$(7.4) \qquad \mathbf{h}_k^{(w)} = \sum_{m,n} C(u, m; v, n \mid w, k)\mathbf{f}_m^{(u)} \otimes \mathbf{f}_n^{(v)},$$

$$|u - v| \leq w \leq u + v, \quad -w \leq k \leq w,$$

$$(7.5) \qquad C(u, m; v, n \mid w, k) = (\mathbf{h}_k^{(w)}, \mathbf{f}_m^{(u)} \otimes \mathbf{f}_n^{(v)}),$$

where $(-, -)$ is the inner product on $V^{(u)} \otimes V^{(v)}$. Since $\{\mathbf{h}_k^{(w)}\}$ and $\{\mathbf{f}_m^{(u)} \otimes \mathbf{f}_n^{(v)}\}$ are ON, the matrix formed by the CG coefficients is unitary. Indeed from (7.5)

$$(7.6) \qquad \mathbf{f}_m^{(u)} \otimes \mathbf{f}_n^{(v)} = \sum_{w,k} C(w, k \mid u, m; v, n)\mathbf{h}_k^{(w)},$$

where

$$(7.7) \qquad C(w, k \mid u, m; v, n) = \overline{C(u, m; v, n \mid w, k)}.$$

Later we shall see that it is possible to choose $\{\mathbf{h}_k^{(w)}\}$ such that the CG coefficients are real.

The matrix elements of $\mathbf{D}^{(u)} \otimes \mathbf{D}^{(v)}$ with respect to the $\{\mathbf{f}_m^{(u)} \otimes \mathbf{f}_n^{(v)}\}$ basis are $T^u_{mm'}(A)T^v_{nn'}(A)$, where $A \in SL(2)$ and the $T^u_{mm'}(A)$ are the matrix elements of $\mathbf{D}^{(u)}$ with respect to $\{\mathbf{f}_m^{(u)}\}$. On the other hand, the matrix elements with respect to the $\{\mathbf{h}_k^{(w)}\}$ basis are $T^w_{kk'}(A)$. The matrix of $\mathbf{T}^{(u)} \otimes \mathbf{T}^{(v)}(A)$ in one basis is unitary equivalent to the matrix in the other basis. A straightforward computation yields the identity

$$(7.8) \qquad T^u_{mm'}(A)T^v_{nn'}(A) = \sum_{w,k,k'} C(u, m; v, n \mid w, k)C(u, m'; v, n' \mid w, k')T^w_{kk'}(A),$$

expressing the product of two matrix elements as a sum of matrix elements. Since in appropriate parameters $T^u_{mm'}(A)$ is essentially a Jacobi polynomial, (7.8) can be viewed as an identity expanding the product of two Jacobi polynomials as a sum of Jacobi polynomials. If we restrict A to the subgroup

$SU(2)$ then the matrix elements are given by (2.13) and (2.14). Applying the orthogonality relations (2.21), we find

$$\text{(7.9)}\qquad \frac{2w+1}{16\pi^2}\int_{SU(2)} T^u_{mm'}(A)T^v_{nn'}(A)\overline{T^w_{kk'}(A)}\,dA = C(u,m;v,n\,|\,w,k)C(u,m';v,n'\,|\,w,k')$$

This expression can be used to explicitly compute the CG coefficients (Wigner [2]). However, we shall adopt another approach which leads to a generating function for the coefficients and yields an independent proof of (7.1).

Consider the model of $\mathbf{D}^{(u)}$ on the vector space $\mathcal{V}^{(u)}$ with ON basis

$$\text{(7.10)}\qquad f_m(z) = \frac{(-z)^{u+m}}{[(u-m)!\,(u+m)!]^{1/2}}, \qquad -u \le m \le u,$$

[see (2.10)]. The action of $SL(2)$ on $\mathcal{V}^{(u)}$ is

$$[\mathbf{T}(A)f](z) = (bz+d)^{2u} f\left(\frac{az+c}{bz+d}\right), \qquad f \in \mathcal{V}^{(u)}, \quad A \in SL(2).$$

The matrix elements of $\mathbf{T}(A)$ with respect to this basis are

$$\mathbf{T}(A)f_m = \sum_{p=-u}^{u} Q^u_{pm}(A) f_p$$

or

$$Q^u_{pm}(A) = \left[\frac{(u+p)!\,(u-p)!}{(u+m)!\,(u-m)!}\right]^{1/2} D_{u+p,u+m}(A)(-1)^{p-m},$$

where $D_{lk}(A)$ is given by (10.22) of Section 5.10. The matrix elements have the symmetric generating function

$$\text{(7.11)}\qquad (1/[2u]!)[(bz+d)+y(az+c)]^{2u} = \sum_{m,p=-u}^{u} f_m(y)Q^u_{pm}(A)f_p(z).$$

In this model the action of the generalized Lie derivatives $J^{\pm}$, J^3 on the basis is described by (7.2).

We can realize $\mathbf{D}^{(u)} \otimes \mathbf{D}^{(v)}$ on the $(2u+1)(2v+1)$-dimensional space $\mathcal{V}^{(u)} \otimes \mathcal{V}^{(v)}$ with ON basis

$$\text{(7.12)}\qquad f^{(u)}_m \otimes f^{(v)}_n(z,y) = \frac{(-z)^{u+m}(-y)^{v+n}}{[(u+m)!\,(u-m)!\,(v+n)!\,(v-n)!]^{1/2}}, \qquad -u \le m \le u, \quad -v \le n \le v.$$

The action of $SL(2)$ is defined by operators $\mathbf{S}(A)$ such that

$$\text{(7.13)}\qquad [\mathbf{S}(A)f](z,y) = (bz+d)^{2u}(by+d)^{2v} f\left(\frac{az+c}{bz+d}, \frac{ay+c}{by+d}\right)$$

for $f \in \mathcal{V}^{(u)} \otimes \mathcal{V}^{(v)}$.

The generalized Lie derivatives $J^{\pm}, J^3$ corresponding to (7.13) are easily computed to be

(7.14)
$$J^3 = -u - v + z\frac{\partial}{\partial z} + y\frac{\partial}{\partial y},$$
$$J^+ = -2uz - 2vy + z^2\frac{\partial}{\partial z} + y^2\frac{\partial}{\partial y}, \qquad J^- = -\frac{\partial}{\partial z} - \frac{\partial}{\partial y}.$$

We will decompose $\mathcal{V}^{(u)} \otimes \mathcal{V}^{(v)}$ into irred subspaces by explicitly computing the ON bases $\{h_k^{(w)}\}$, (7.3), of these subspaces. The lowest weight vectors $h_{-w}^{(w)}$ satisfy $J^- h_{-w}^{(w)} = 0$, $J^3 h_{-w}^{(w)} = -wh_{-w}^{(w)}$. We will use this property to compute the $h_{-w}^{(w)}$ explicitly and then use relations (7.3) to obtain a set of vectors $\{h_k^{(w)}\}$, $-w \le k \le w$, $|u - v| \le w \le u + v$. By showing that these vectors form an ON basis in $\mathcal{V}^{(u)} \otimes \mathcal{V}^{(v)}$ we can verify (7.1) independently. Moreover, our explicit expressions for the $h_k^{(w)}(z, y)$ will enable us to compute the CG coefficients.

The general solution of $J^- f = -(\partial/\partial z + \partial/\partial y) f(z, y) = 0$ is $f(z, y) = \sum_{s=0}^{2q} a_s (z - y)^s$, where the a_s are arbitrary constants and $q = \min(u, v)$. A basis for the q-dimensional solution space is given by the vectors.

$$h_{-w}^{(w)}(z, y) = N_w (z - y)^{u+v-w}, \qquad |u - v| \le w \le u + v,$$

where the N_w are nonzero constants. Indeed

(7.15)
$$J^3 h_{-w}^{(w)} = -wh_{-w}^{(w)}, \qquad J^- h_{-w}^{(w)} = 0.$$

Let $(-, -)$ be the inner product on $\mathcal{V}^{(u)} \otimes \mathcal{V}^{(v)}$ with respect to which the basis $\{f_m^{(u)} \otimes f_n^{(v)}\}$ is ON. It is easy to check that $(J^3)^* = J^3$ and $(J^+)^* = J^-$ for this inner product. We will choose the constants N_w such that $\| h_{-w}^{(w)} \| = 1$. Thus,

$$\| h_{-w}^{(w)} \|^2 = |N_w|^2 \sum_{j=0}^{u+v-w} (u + v - w)!^2 \frac{(2u - j)!\,(v - u + w + j)!}{j!\,(u + v - w - j)!} = 1.$$

Making use of the identity

(7.16)
$$\sum_{j=0}^{k} \frac{(m + k - j)!\,(n + j)!}{j!\,(k - j)!} = \frac{(m + k)!\,n!}{k!}\, {}_2F_1(-k, n + 1; -m - k; 1)$$
$$= \frac{m!\,n!\,(m + n + k + 1)!}{k!\,(n + m + 1)!}$$

(see Lebedev [1, p. 243] for a proof), we obtain

$$N_w = (-1)^{2v}\left[\frac{(2w + 1)!}{(u + v - w)!\,(u - v + w)!\,(v - u + w)!\,(u + v + w + 1)}\right]^{1/2},$$

where the phase factor has been added to conform to convention.

Now we define vectors

$$h_k^{(w)}(z, y) = \left[\frac{(w-k)!}{(w+k)!\,(2w)!}\right]^{1/2} (J^+)^{w+k} h_{-w}^{(w)}(z, y), \qquad -w \leq k \leq w, \tag{7.17}$$

where J^+ is given by (7.14). It follows immediately that $J^+ h_k^{(w)} = [(w+k+1)(w-k)]^{1/2} h_{k+1}^{(w)}$ in agreement with (7.3). Also, from the proof of relations (3.3) we see that $(J^+)^{2w+1} h_{-w}^{(w)} = J^+ h_w^{(w)} = 0$. Each $h_k^{(w)}(z, y)$ is a homogeneous polynomial of order $u + v + k$ in z and y, and there are a total of $(2u+1)(2v+1)$ such polynomials. We will show that the $\{h_k^{(w)}\}$ form an ON basis for $\mathcal{V}^{(u)} \otimes \mathcal{V}^{(v)}$.

Lemma 7.1. (a) $J^+ h_k^{(w)} = [(w+k+1)(w-k)]^{1/2} h_{k+1}^{(w)}$; (b) $J^- h_k^{(w)} = [(w-k+1)(w+k)]^{1/2} h_{k-1}^{(w)}$; (c) $J^3 h_k^{(w)} = k h_k^{(w)}$.

Proof. Identity (a) follows directly from (7.17). Identity (c) follows from $J^3 h_{-w}^{(w)} = -w h_{-w}^{(w)}$ and the fact that $J^3(J^+ f) = (k+1)(J^+ f)$ if $J^3 f = kf$. We prove (b) by induction on k. Since $J^- h_{-w}^{(w)} = 0$ the equation holds for $k = -w$. Assume (b) is valid for $k \leq l$ where $-w \leq l \leq w$. Then

$$J^- h_{l+1}^{(w)} = [(w+l+1)(w-l)]^{-1/2} J^- J^+ h_l^{(w)}$$

from (a). Since $J^- J^+ = J^+ J^- - 2J^3$ we have

$$J^- J^+ h_l^{(w)} = (J^+ J^- - 2J^3) h_l^{(w)} = (w+l+1)(w-l) h_l^{(w)}$$

by the induction hypotheses. Therefore, (b) follows for $k = l + 1$. Q.E.D.

Thus, for fixed w the vectors $\{h_l^{(w)}\}$ form a basis for a subspace of $\mathcal{V}^{(u)} \otimes \mathcal{V}^{(v)}$ which transforms irreducibly under the rep $\mathbf{D}^{(w)}$. Furthermore, by computations analogous to (2.7)–(2.10) we see that the $\{h_l^{(w)}\}$ are ON. The Casimir operator $C = J^+ J^- + J^3 J^3 - J^3$ is symmetric since J^3 is symmetric and $(J^+ J^- f, g) = (J^- f, J^- g) = (f, J^+ J^- g)$. Since $C h_k^{(w)} = w(w+1) h_k^{(w)}$ we obtain

$$w(w+1)(h_k^{(w)}, h_{k'}^{(w')}) = (C h_k^{(w)}, h_{k'}^{(w')}) = (h_k^{(w)}, C h_{k'}^{(w')}) = w'(w'+1)(h_k^{(w)}, h_{k'}^{(w')}),$$

so $(h_k^{(w)}, h_{k'}^{(w')}) = \delta_{ww'} \delta_{kk'}$, i.e., the $\{h_k^{(w)}\}$ form an ON set. Since the cardinality of this set is equal to the dimension of $\mathcal{V}^{(u)} \otimes \mathcal{V}^{(v)}$ we conclude that $\{h_k^{(w)}\}$ is an ON basis. This proves the validity of the Clebsch–Gordan series (7.1) from a Lie-algebraic viewpoint.

We can use our model to obtain an explicit expression for the coefficients. Since the $\{h_k^{(w)}\}$ for fixed w form a basis for $\mathbf{D}^{(w)}$ we have the identity

$$\mathbf{T}(A) h_m^{(w)} = \sum_{p=-w}^{w} Q_{pm}^{w}(A) h_p^{(w)}, \qquad A \in SL(2), \tag{7.18}$$

where the matrix elements are given by (7.11). In the case where $A = \exp(-b\mathcal{J}^+)$ and $m = -w$, (7.18) is especially easy to evaluate. Indeed $h_{-w}^{(w)}(z, y)$

$= N_w(z-y)^{u+v-w}$ and $Q^w_{p,-w}(A) = (-b)^{p+w}\{(2w)!/[(w+p)!(w-p)!]\}^{1/2}$. Thus

(7.19)
$$N_v(bz+1)^{w+u-v}(by+1)^{w+v-u}(z-y)^{u+v-w}$$
$$= \sum_{p=-w}^{w}\sum_{m,n}\left[\frac{(2w)!}{(w+p)!\,(w-p)!\,(u+m)!\,(u-m)!\,(v+n)!\,(v-n)!}\right]^{1/2}$$
$$\times\, C(u,m;v,n\,|\,w,p)\times(-z)^{u+m}(-y)^{v+n}(-b)^{w+p},$$

where we have used (7.4) and (7.10). Since $h_p^{(w)}(z, y)$ is homogeneous of order $u+v+p$ in z and y it follows that $C(u,m;v,n\,|\,w,p)$ is nonzero only if $m+n=p$.

Expression (7.19) is a generating function for the CG coefficients. We can write this expression in a more symmetric form by choosing $b = x_3^{-1}$ and introducing the **3-*j* coefficients**

$$\begin{pmatrix} j_1 & j_2 & j_3 \\ m_1 & m_2 & m_3 \end{pmatrix} = \frac{(-1)^{j_3-m_3}}{(2j_3+1)^{1/2}}C(j_1,m_1;j_2,m_2\,|\,j_3,-m_3). \tag{7.20}$$

In terms of these quantities, (7.19) becomes

(7.21)
$$(x_3-x_1)^{j_1-j_2+j_3}(x_2-x_3)^{-j_1+j_2+j_3}(x_1-x_2)^{j_1+j_2-j_3}$$
$$\times\,[(j_1+j_2-j_3)!\,(j_1-j_2+j_3)!$$
$$\times\,(-j_1+j_2+j_3)!\,(j_1+j_2+j_3+1)]^{-1/2}$$
$$= \sum_{m_i=-j_i}^{j_i}\left\{x_1^{j_1+m_1}x_2^{j_2+m_2}x_3^{j_3+m_3}\begin{pmatrix} j_1 & j_2 & j_3 \\ m_1 & m_2 & m_3 \end{pmatrix}\right.$$
$$\times\,[(j_1+m_1)!\,(j_2+m_2)!\,(j+m_3)!$$
$$\left.\times\,(j_1-m_1)!\,(j_2-m_2)!\,(j_3-m_3)!]^{-1/2}\right\}$$

(We have set $z=-x_1$, $y=-x_2$ in this expression.) Since the left-hand side is homogeneous of degree $j_1+j_2+j_3$ in x_1, x_2, x_3, so is the right-hand side. Thus, the 3-*j* coefficients are zero unless $m_1+m_2+m_3=0$. Furthermore, it follows from the CG series (7.1) that these coefficients are zero unless $j_1+j_2+j_3$ and j_i+m_i are integers, and $-j_i \le m_i \le j_i$, $i=1,2,3$. The 3-*j* coefficients have a high degree of symmetry, as is evident from (7.21). Indeed the left-hand side of (7.21) is fixed under an even permutation of the integers 1, 2, 3. As a consequence

$$\begin{pmatrix} j_1 & j_2 & j_3 \\ m_1 & m_2 & m_3 \end{pmatrix} = \begin{pmatrix} j_3 & j_1 & j_2 \\ m_3 & m_1 & m_2 \end{pmatrix} = \begin{pmatrix} j_2 & j_3 & j_1 \\ m_2 & m_3 & m_1 \end{pmatrix}. \tag{7.22}$$

The left-hand side changes by a phase factor under an odd permutation:

(7.23)
$$\begin{pmatrix} j_1 & j_2 & j_3 \\ m_1 & m_2 & m_3 \end{pmatrix} = (-1)^{j_1+j_2+j_3}\begin{pmatrix} j_2 & j_1 & j_3 \\ m_2 & m_1 & m_3 \end{pmatrix}$$
$$= (-1)^{j_1+j_2+j_3}\begin{pmatrix} j_1 & j_3 & j_2 \\ m_1 & m_3 & m_2 \end{pmatrix}.$$

If we make the substitution $x_i \longrightarrow x_i^{-1}$ and multiply by $x_1^{2j_1}x_2^{2j_2}x_3^{2j_3}$ the generating function changes by the factor $(-1)^{j_1+j_2+j_3}$. There follows the identity

$$\begin{pmatrix} j_1 & j_2 & j_3 \\ m_1 & m_2 & m_3 \end{pmatrix} = (-1)^{j_1+j_2+j_3}\begin{pmatrix} j_1 & j_2 & j_3 \\ -m_1 & -m_2 & -m_3 \end{pmatrix}. \tag{7.24}$$

If we multiply both sides of (7.21) by

$$\frac{(j_1+j_2+j_3+1)^{1/2}\alpha^{-j_1+j_2+j_3}\beta^{j_1-j_2+j_3}\gamma^{j_1+j_2-j_3}}{[(-j_1+j_2+j_3)!\,(j_1-j_2+j_3)!\,(j_1+j_2-j_3)!]^{1/2}}$$

and sum over all j_i for which (7.21) makes sense, we obtain the new generating function

(7.25)

$$\begin{aligned} &\exp[\alpha(x_2-x_3)+\beta(x_3-x_1)+\gamma(x_1-x_2)] \\ &\quad = \sum_{j_1+j_2+j_3=0}^{\infty}\ \sum_{m_i=-j_i}^{j_i} \{(j_1+j_2+j_3+1)^{1/2}\alpha^{-j_1+j_2+j_3}\beta^{j_1-j_2+j_3}\gamma^{j_1+j_2-j_3} \\ &\qquad \times x_1^{j_1+m_1}x_2^{j_2+m_2}x_3^{j_3+m_3}\begin{pmatrix} j_1 & j_2 & j_3 \\ m_1 & m_2 & m_3 \end{pmatrix} \\ &\qquad \times [(-j_1+j_2+j_3)!\,(j_1-j_2+j_3)!\,(j_1+j_2-j_3)!\,(j_1+m_1)! \\ &\qquad \times (j_2+m_2)!\,(j_3+m_3)!\,(j_1-m_1)!\,(j_2-m_2)!\,(j_3-m_3)!]^{-1/2}\}. \end{aligned}$$

A still higher degree of symmetry can be obtained by making the replacements $x_i \longrightarrow x_i/y_i$, $\alpha \longrightarrow y_2y_3\alpha$, $\beta \longrightarrow y_3y_1\beta$, $\gamma \longrightarrow y_1y_2\gamma$ in (7.25). Then this expression takes the form

$$\begin{aligned} \exp(\det B) = &\sum_{j_1+j_2+j_3=0}^{\infty}\ \sum_{m_i=-j_i}^{j_i} b(j_i, m_i)\alpha^{-j_1+j_2+j_3}\beta^{j_1-j_2+j_3}\gamma^{j_1+j_2-j_3} \\ &\times x_1^{j_1+m_1}y_1^{j_1-m_1}x_2^{j_2+m_2}y_2^{j_2-m_2}x_3^{j_3+m_3}y_3^{j_3-m_3}\begin{pmatrix} j_1 & j_2 & j_3 \\ m_1 & m_2 & m_3 \end{pmatrix}, \end{aligned} \tag{7.26}$$

where $b(j_i, m_i)$ is completely symmetric under a permutation of the integers 1, 2, 3. Here B is the matrix

$$\begin{pmatrix} \alpha & \beta & \gamma \\ x_1 & x_2 & x_3 \\ y_1 & y_2 & y_3 \end{pmatrix}.$$

It is now evident that the symmetries (7.22) and (7.23) correspond to permutations of the columns of B. Under an even permutation $\det B$ remains invariant, while under an odd permutation it changes sign. The identity (7.24) follows from the fact that $\det B$ changes sign under a transposition of the second and third rows of B. Note that a change in sign of $\det B$ is equivalent to multiplication of the right-hand side of (7.26) by $(-1)^{j_1+j_2+j_3}$.

In addition to these symmetries we see that arbitrary permutations of the rows of B lead to new symmetries. Furthermore, since $\det B^{\mathrm{t}} = \det B$ we can obtain a new symmetry by interchanging rows and columns of B. The six column permutations, six row permutations, and the transpose generate a group of $6 \times 6 \times 2 = 72$ symmetries of the 3-j coefficients. This symmetry group was discovered by Regge [1]. The symmetries (7.22)–(7.24) generate a subgroup of order 12.

Substituting (7.20) in the above formulas we can obtain corresponding formulas for the CG coefficients. The most frequently used CG coefficients $C(j_1, m_1; j_2, m_2 \mid j_3, m_3)$ are those for which $j_2 = \frac{1}{2}$ or 1. We can easily compute these special cases from (7.21). For $j_2 = \frac{1}{2}$ the coefficients are zero unless $j_3 = j_1 \pm \frac{1}{2}$ and $m_3 = m_1 + m_2$. The nonzero coefficients are given by

(7.27)

	$m_2 = -\frac{1}{2}$	$m_2 = \frac{1}{2}$
$j_3 = j_1 - \frac{1}{2}$	$\left[\dfrac{j_1 + m_3 + \frac{1}{2}}{2j_1 + 1}\right]^{1/2}$	$-\left[\dfrac{j_1 - m_3 + \frac{1}{2}}{2j_1 + 1}\right]^{1/2}$
$j_3 = j_1 + \frac{1}{2}$	$\left[\dfrac{j_1 - m_3 + \frac{1}{2}}{2j_1 + 1}\right]^{1/2}$	$\left[\dfrac{j_1 + m_3 + \frac{1}{2}}{2j_1 + 1}\right]^{1/2}$.

For $j_2 = 1$ the coefficients are zero unless $j_3 = j_1, j_1 \pm 1$, and $m_3 = m_1 + m_2$. The nonzero coefficients $C(j_1, m_2; 1, m_2 \mid i_3, m_3)$ are

(7.28)

	$m_2 = -1$	$m_2 = 0$	$m_2 = 1$
$j_3 = j_1 - 1$	$\left(\dfrac{(j_1 + m_3 + 1)(j_1 + m_3)}{2j_1(2j_1 + 1)}\right)^{1/2}$	$-\left(\dfrac{(j_1 - m_3)(j_1 + m_3)}{j_1(2j_1 + 1)}\right)^{1/2}$	$\left(\dfrac{(j_1 - m_3)(j_1 - m_3 + 1)}{2j_1(j_1 + 1)}\right)^{1/2}$
$j_3 = j_1$	$\left(\dfrac{(j_1 - m_3)(j_1 + m_3 + 1)}{2j_1(j_1 + 1)}\right)^{1/2}$	$\dfrac{m_3}{[j_1(j_1 + 1)]^{1/2}}$	$-\left(\dfrac{(j_1 + m_3)(j_1 - m_3 + 1)}{2j_1(j_1 + 1)}\right)^{1/2}$
$j_3 = j_1 + 1$	$\left(\dfrac{(j_1 - m_3)(j_1 - m_3 + 1)}{(2j_1 + 1)(2j_1 + 2)}\right)^{1/2}$	$\left(\dfrac{(j_1 - m_3 + 1)(j_1 + m_3 + 1)}{(2j_1 + 1)(j_1 + 1)}\right)^{1/2}$	$\left(\dfrac{(j_1 + m_3)(j_1 + m_3 + 1)}{(2j_1 + 1)(2j_1 + 2)}\right)^{1/2}$.

It is not difficult to obtain an explicit expression for an arbitrary CG coefficient. Indeed one can expand one of the generating functions in powers of the independent variables and equate coefficients of like powers. However, the resulting expressions are very complicated (see Hamermesh [1]). For practical (computer) computations it is usually more convenient to use recurrence relations for the CG coefficients. Such relations can be easily derived by differentiating the generating functions with respect to some of the independent variables (Bargmann [2]).

7.8 Applications of the Clebsch–Gordan Series

We return to the study of a k-particle quantum mechanical system as described in Section 7.6. Suppose the Hamiltonian is given by

$$\mathbf{H} = \mathbf{H}_1 + \mathbf{H}_2 + \cdots + \mathbf{H}_k, \tag{8.1}$$

where

$$\mathbf{H}_j = -(1/2m_j)\,\boldsymbol{\Delta}_j + V_j(\mathbf{x}_j), \qquad 1 \le j \le k, \tag{8.2}$$

i.e., $\mathbf{H}$ is a sum of single-particle Hamiltonians. Furthermore, suppose $V_j(A\mathbf{x}_j) = V_j(\mathbf{x}_j)$ for all $A \in SO(3)$, so that each potential function $V_j(\mathbf{x}_j)$ is invariant under $SO(3)$. This system admits the compact symmetry group $G = SO(3) \times SO(3) \times \cdots \times SO(3)$ (k times). Indeed, we can define a unitary rep $\mathbf{S}$ of G on $\mathcal{H}$ by

$$[\mathbf{S}(A_1, \ldots, A_k)\Psi](\mathbf{x}_1, \ldots, \mathbf{x}_k) = \Psi(A_1^{-1}\mathbf{x}_1, \ldots, A_k^{-1}\mathbf{x}_k), \qquad A_j \in SO(3), \quad \Psi \in \mathcal{H}. \tag{8.3}$$

It is easy to check that these operators commute with $\mathbf{H}$.

From Corollary 6.2 it follows that the irred unitary reps of G are products of k unitary irred reps of $SO(3)$. Indeed, the irred reps of G can be denoted $\mathbf{D}^{(l_1,\ldots,l_k)}$, where

$$\mathbf{D}^{(l_1,\ldots,l_k)}(A_1, \ldots, A_k) = \mathbf{D}^{(l_1)}(A_1) \otimes \cdots \otimes \mathbf{D}^{(l_k)}(A_k) \tag{8.4}$$

and $\mathbf{D}^{(l_j)}$ is an irred rep of $SO(3)$.

Suppose λ is an eigenvalue of $\mathbf{H}$ and W_λ is the corresponding eigenspace. If W_λ transforms irreducibly under G according to $\mathbf{D}^{(l_1,\ldots,l_k)}$, the multiplicity of λ is $q = \dim \mathbf{D}^{(l_1,\ldots,l_k)} = (2l_1 + 1)(2l_2 + 1) \cdots (2l_k + 1)$. The functions $\Psi^{l_1}_{m_1}(\mathbf{x}_1) \cdots \Psi^{l_k}_{m_k}(\mathbf{x}_k)$, $-l_j \le m_j \le l_j$ form an ON basis for W_λ where $\Psi^{l_j}_{m_j}(\mathbf{x}_j)$ for fixed j is a canonical ON basis for the rep $\mathbf{D}^{(l_j)}$ and $\Psi^{l_j}_{m_j}$ is an eigenvector of $\mathbf{H}_j$. As we have seen earlier $\Psi^{l_j}_{m_j}(\mathbf{x}) = h_{l_j}(r)Y^{m_j}_{l_j}(\theta, \varphi)$ in spherical coordinates, so the angular dependence of the wave functions is determined. The radial dependence can be obtained only by solving the Schrödinger equation.

In the above system the k particles do not interact with one another. We now consider an interacting system obtained by adding a perturbing potential V' to $\mathbf{H}$:

$$\mathbf{H}' = \mathbf{H} + V'(\mathbf{x}_1, \ldots, \mathbf{x}_k). \tag{8.5}$$

We further assume that V' is invariant under the action $\mathbf{T}$, (6.2), of $SO(3)$ on $\mathcal{H}$ but not under the action $\mathbf{S}$ of G, i.e., the equality

$$V'(A_1\mathbf{x}_1, \ldots, A_k\mathbf{x}_k) = V'(\mathbf{x}_1, \ldots, \mathbf{x}_k)$$

holds in general only if $A_1 = \cdots = A_k = A \in SO(3)$. Thus the symmetry group of the perturbed Hamiltonian $\mathbf{H}'$ will be the subgroup of G consisting of all diagonal elements $A \times A \times \cdots \times A$. This subgroup is obviously isomorphic to $SO(3)$. To determine the splitting of the eigenvalue λ under the perturbation we need only express $\mathbf{D}^{(l_1,\ldots,l_k)} \mid SO(3)$ as a direct sum of irred reps of $SO(3)$.

From (8.4) it is clear that this restricted rep is isomorphic to the k-fold tensor product

$$\mathbf{D}^{(l_1,\ldots,l_k)} \mid SO(3) \cong \mathbf{D}^{(l_1)} \otimes \mathbf{D}^{(l_2)} \otimes \cdots \otimes \mathbf{D}^{(l_k)}. \tag{8.6}$$

We can use the CG series to decompose (8.6) into a direct sum of irred reps and thereby obtain the splitting of the energy levels. For example we could use the CG series to decompose $\mathbf{D}^{(l_1)} \otimes \mathbf{D}^{(l_2)}$, tensor the resulting irred reps with $\mathbf{D}^{(l_3)}$, and apply the CG series again, etc. (In case W_λ is not irred under G we can decompose W_λ into a direct sum of G-irred reps and proceed as above.)

In the simplest case $k = 2$ and

$$\mathbf{D}^{(l_1,l_2)} \mid SO(3) \cong \mathbf{D}^{(l_1)} \otimes \mathbf{D}^{(l_2)} \cong \mathbf{D}^{(l_1+l_2)} \oplus \mathbf{D}^{(l_1+l_2-1)} \oplus \cdots \oplus \mathbf{D}^{(|l_1-l_2|)}. \tag{8.7}$$

Here the $(2l_1 + 1)(2l_2 + 1)$-degenerate energy level λ splits into $2\min(l_1, l_2) + 1$ levels and the energy level corresponding to $\mathbf{D}^{(l)}$ is $(2l + 1)$-degenerate. We can use the CG coefficients to decompose W_λ into a direct sum of subspaces transforming under the irred reps of $SO(3)$ given by the right-hand side of (8.7). Indeed a canonical basis for the subspace transforming according to $\mathbf{D}^{(l)}$ is given by

$$h_m{}^l(\mathbf{x}_1, \mathbf{x}_2) = \sum_{m_1 m_2} C(l_1, m_1; l_2, m_2 \mid l, m)\Psi^{l_1}_{m_1}(\mathbf{x}_1)\Psi^{l_2}_{m_2}(\mathbf{x}_2), \qquad -l \leq m \leq l. \tag{8.8}$$

As we have shown in Section 7.6, the computation of matrix elements of $\mathbf{H}'$ with respect to the canonical basis $\{h_m{}^l\}$ is relatively simple because $SO(3)$ is a symmetry group of $\mathbf{H}'$. This basis is far superior to $\{\Psi^{l_1}_{m_1}\Psi^{l_2}_{m_2}\}$ since it explicitly exhibits the $SO(3)$ symmetry. The matrix elements of $\mathbf{H}'$ are needed in quantum mechanical perturbation theory to compute the perturbed eigenvalues (Schiff [1], Landau and Lifshitz [2]).

The decomposition (8.7)–(8.8) is also of great importance in the study of time-varying systems. We look for solutions Ψ of the Schrödinger equation

$$i\,\partial\Psi(\mathbf{x}_1, \mathbf{x}_2, t)/\partial t = \mathbf{H}'\Psi(\mathbf{x}_1, \mathbf{x}_2, t), \tag{8.9}$$

where $\Psi(\mathbf{x}_1, \mathbf{x}_2, t) \in \mathcal{H}$ for each t. Suppose the functions $\Psi_m{}^l(\mathbf{x}_1, \mathbf{x}_2, t)$ are solutions of (8.9) such that $\Psi_m{}^l(\mathbf{x}_1, \mathbf{x}_2, 0) = h_m{}^l(\mathbf{x}_1, \mathbf{x}_2)$, expression (8.8). Then at $t = 0$ the $\Psi_m{}^l$, $-l \leq m \leq l$, form a canonical basis for the irred rep $\mathbf{D}^{(l)}$. According to the results of Section 7.6, the functions $\Psi_m{}^l(\mathbf{x}_1, \mathbf{x}_2, t)$ form a canonical basis for $\mathbf{D}^{(l)}$ at *every* time t. In particular

$$\mathbf{L}^3\Psi_m{}^l = m\Psi_m{}^l, \qquad \mathbf{L}\cdot\mathbf{L}\Psi_m{}^l = l(l+1)\Psi_m{}^l \tag{8.10}$$

for all t. Thus the quantum numbers l and m are conserved under the interaction.

To see the physical significance of this analysis we consider an (oversimplified) example. Suppose the perturbing potential is a function of time, V'

$= V'(\mathbf{x}_1, \mathbf{x}_2, t)$, such that for all t, V' is $SO(3)$-invariant but not necessarily G-invariant. Furthermore suppose $V' = 0$ for $t \leq 0$ and $t \geq \tau > 0$, where τ is some fixed time. Thus the perturbing potential acts only in the time interval $(0, \tau)$. At all other times $\mathbf{H}' = \mathbf{H}$.

Let W_λ be the eigenspace of $\mathbf{H}$ corresponding to eigenvalue λ. The space W_λ transforms irreducibly under G:

$$\mathbf{S}\,|\,W_\lambda \cong \mathbf{D}^{(l_1, l_2)} \tag{8.11}$$

and has the ON basis

$$\{\Psi^{l_1}_{m_1}(\mathbf{x}_1)\Psi^{l_2}_{m_2}(\mathbf{x}_2) \colon -l_j \leq m_j \leq l_j\}.$$

Now suppose Ψ is a solution of (8.9) such that $\Psi(\mathbf{x}_1, \mathbf{x}_2, 0) = \Psi^{l_1}_{m_1}(\mathbf{x}_1)\Psi^{l_2}_{m_2}(\mathbf{x}_2) \in W_\lambda$, i.e., the first particle has quantum numbers l_1, m_1 and the second has quantum numbers l_2, m_2. As t increases, the particles begin to interact. We assume the interaction is **elastic**, i.e., we end up with the same two particles and energy is conserved. No particles are created or destroyed by the interaction.

After time $t = \tau$ the particles are again noninteracting. By conservation of energy, $\Psi(\tau)$ must have energy λ. Thus $\Psi(\tau) \in W_\lambda$, or

$$\Psi(\mathbf{x}_1, \mathbf{x}_2, \tau) = \sum_{n_1 n_2} a_{n_1 n_2} \Psi^{l_1}_{n_1}(\mathbf{x}_1)\Psi^{l_2}_{n_2}(\mathbf{x}_2) \tag{8.12}$$

and we can describe the interaction by computing $a_{n_1 n_2}$: $|a_{n_1 n_2}|^2$ is the probability that a system in the state $\Psi^{l_1}_{m_1}\Psi^{l_2}_{m_2}$ at $t = 0$ ends up in the state $\Psi^{l_1}_{n_1}\Psi^{l_2}_{n_2}$ at $t = \tau$. Since $SO(3) \times SO(3)$ is *not* a symmetry group of $\mathbf{H}'$, m_1 and m_2 are not conserved by the interaction. Thus, if particle one starts out in the state $\Psi^{l_1}_{m_1}$ there is no reason to assume that it will end up in this state.

On the other hand, $SO(3)$ *is* a symmetry group of $\mathbf{H}'$. If the system is in the state $h_m{}^l$ at $t = 0$ then it must be in the state $h_m{}^l$ at $t = \tau$. Note that the vectors

$$h_m{}^l(\mathbf{x}_1, \mathbf{x}_2), \qquad |l_1 - l_2| \leq l \leq l_1 + l_2, \qquad -l \leq m \leq l, \tag{8.13}$$

form an ON basis for W_λ. Thus, if $\Psi(0) = h_m{}^l$ then by conservation of angular momentum

$$\Psi(\tau) = b_l h_m{}^l. \tag{8.14}$$

Since $\Psi(\tau)$ is a unit vector we must have $|b_l| = 1$, or $b_l = e^{i\theta_l}$, $0 \leq \theta_l < 2\pi$. Just as in Section 7.6 we can easily show that θ_l is independent of m. The basis $\{h_m{}^l\}$ is clearly more convenient for W_λ than the basis $\{\Psi^{l_1}_{m_1}\Psi^{l_2}_{m_2}\}$. On the strength of conservation of angular momentum alone we have proved that $h_m{}^l$ is merely multiplied by a phase factor $e^{i\theta_l}$. The results of the scattering experiment are determined by the scattering angles θ_l, $|l_1 - l_2| \leq l \leq l_1 + l_2$, which must be computed from the dynamical equations.

Now that we know how the $\{h_m{}^l\}$ transform we can use the CG coefficients to determine how the basis $\{\Psi^{l_1}_{m_1}\Psi^{l_2}_{m_2}\}$ transforms. A straightforward computation yields

$$\Psi^{l_1}_{m_1}\Psi^{l_2}_{m_2} \longrightarrow \sum_{ln_1n_2} C(l_1, m_1; l_2, m_2 \mid l, m)e^{i\theta_l}C(l_1, n_1; l_2, n_2 \mid l, m)\Psi^{l_1}_{n_1}\Psi^{l_2}_{n_2}. \tag{8.15}$$

Thus the probability that a system in the state $\Psi^{l_1}_{m_1}\Psi^{l_2}_{m_2}$ at $t = 0$ will be found in the state $\Psi^{l_1}_{n_1}\Psi^{l_2}_{n_2}$ at $t = \tau$ is

$$\left| \sum_{l=|l_1-l_2|}^{l_1+l_2} C(l_1, m_1; l_2, m_2 \mid l, m)e^{i\theta_l}C(l_1, n_1; l_2, n_2 \mid l, m) \right|^2. \tag{8.16}$$

If the system has $k > 2$ particles a similar but more complicated analysis can be used to decompose W_λ into a direct sum of irred subspaces under $SO(3)$. The principal complications arise from the fact that a given irred rep may occur with multiplicity greater than one. Then there is no unique way to decompose W_λ and it may be necessary to relate the various possible decompositions by Racah coefficients (Liubarskii [1]).

In the preceding discussion we have ignored the possibility of spin. However, for many particles such as the electron, the proton and the neutron, physical observations do not agree with the predictions of our theory. To obtain predictions in agreement with experiment it is necessary to postulate more complicated transformation properties of the particle state functions. Intuitively, one may think of a particle with spin, say an orbital electron in an atom, as a billiard ball spinning about its own axis. In addition to its orbital angular momentum the billiard ball possesses an intrinsic spin angular momentum.

To make the discussion concrete we construct the state space of a single nonrelativistic particle with spin s, $2s = 0, 1, 2, \ldots$. The Hilbert space $\mathcal{H}_s$ consists of vector valued functions

$$\boldsymbol{\Psi}(\mathbf{x}) = \begin{pmatrix} \Psi_s(\mathbf{x}) \\ \Psi_{s-1}(\mathbf{x}) \\ \cdot \\ \cdot \\ \cdot \\ \Psi_{-s}(\mathbf{x}) \end{pmatrix} = \sum_{\mu=-s}^{s} \Psi_\mu(\mathbf{x})e_\mu, \tag{8.17}$$

where e_μ is the column vector with a one in row μ and zeros everywhere else. The vector $\boldsymbol{\Psi}(\mathbf{x}) \in \mathcal{H}_s$ if

$$\int_{R_3} \boldsymbol{\Psi}^t(\mathbf{x})\overline{\boldsymbol{\Psi}}(\mathbf{x})\, d\mathbf{x} = \int \sum_{\mu=-s}^{s} |\Psi_\mu(\mathbf{x})|^2\, d\mathbf{x} < \infty$$

and the inner product is

$$(\boldsymbol{\Theta}, \boldsymbol{\Psi}) = \int_{R_3} \boldsymbol{\Theta}^t(\mathbf{x})\overline{\boldsymbol{\Psi}(\mathbf{x})}\, d\mathbf{x} = \int \sum_{\mu=-s}^{s} \Theta_\mu(\mathbf{x})\overline{\Psi_\mu(\mathbf{x})}\, d\mathbf{x}. \tag{8.18}$$

We define a unitary rep $\mathbf{T}$ of $SU(2)$ on $\mathcal{H}_s$ by

$$[\mathbf{T}(A)\mathbf{\Psi}](\mathbf{x}) = T^s(A)\mathbf{\Psi}(R(A^{-1})\mathbf{x}),$$

or in components

$$(8.19) \qquad [\mathbf{T}(A)\mathbf{\Psi}]_\mu(\mathbf{x}) = \sum_{\nu=-s}^{s} T^s_{\mu\nu}(A)\mathbf{\Psi}_\nu(R(A^{-1})\mathbf{x}), \qquad -s \leq \mu \leq s.$$

Here $R(A) \in SO(3)$ is defined by (1.12) and (1.20) and the matrix elements $T^s_{\mu\nu}(A)$ by (2.14). Since the matrices $T^s(A)$ are unitary and satisfy the homomorphism property $T^s(AB) = T^s(A)T^s(B)$, the operators $\mathbf{T}(A)$ are unitary in $\mathcal{H}_s$ and satisfy $\mathbf{T}(AB) = \mathbf{T}(A)\mathbf{T}(B)$. Any vector-valued function $\mathbf{\Psi}(\mathbf{x})$ which transforms under the action of $SU(2)$ according to (8.19) is called a **spinor field of weight s.** (A spinor field need not belong to $\mathcal{H}_s$.) It follows from Section 7.2 that if s is an integer, $\mathbf{T}$ defines a single-valued rep of $SO(3)$, while if s is half-integral, $\mathbf{T}$ is double-valued on $SO(3)$.

In nonrelativistic quantum mechanics it is postulated that the state vectors of the electron, proton, and neutron transform under rotations as spinor fields of weight $\frac{1}{2}$. There are mesons and baryons with spins 0, 1, and $\frac{3}{2}$. The photon in relativistic quantum mechanics has spin one, while the nuclei of various atoms can have spins greater than 1.

We have postulated that the state vectors of a particle with spin belong to $\mathcal{H}_s$ and transform under rotations of space by (8.19). If s is half-integral this postulate seems ambiguous because $\mathbf{T}$ is a double-valued rep of $SO(3)$. Indeed if $R \in SO(3)$ there exists $A \in SU(2)$ such that $R = R(\pm A)$ and $\mathbf{T}(-A)\mathbf{\Psi} = -\mathbf{T}(A)\mathbf{\Psi}$ for $\mathbf{\Psi} \in \mathcal{H}_s$. However $\pm\mathbf{T}(A)\mathbf{\Psi}$ both define the same state (ray) in $\mathcal{H}_s$, so there is no physical contradiction.

In a manner similar to the above construction we can define state spaces for systems containing several particles. As an example we construct the state space for a system containing two electrons. The Hilbert space $\mathcal{H}_{1/2} \otimes \mathcal{H}_{1/2}$ consists of all tensor-valued functions $\mathbf{\Psi}(\mathbf{x}_1, \mathbf{x}_2)$ with components $\Psi_{\mu_1\mu_2}(\mathbf{x}_1, \mathbf{x}_2)$, $\mu_1, \mu_2 = \pm\frac{1}{2}$, such that

$$\int_{R_3{}^2} \sum_{\mu_1\mu_2=-1/2}^{1/2} |\Psi_{\mu_1\mu_2}(\mathbf{x}_1, \mathbf{x}_2)|^2 \, d\mathbf{x}_1 \, d\mathbf{x}_2 < \infty.$$

The inner product is

$$(8.20) \qquad (\mathbf{\Theta}, \mathbf{\Psi}) = \int_{R_3{}^2} \sum_{\mu_1\mu_2} \Theta_{\mu_1\mu_2}(\mathbf{x}_1, \mathbf{x}_2)\overline{\Psi_{\mu_1\mu_2}(\mathbf{x}_1, \mathbf{x}_2)} \, d\mathbf{x}_1 \, d\mathbf{x}_2.$$

Here the spinor indices and spatial coordinates corresponding to particles one and two are $\mu_1, \mathbf{x}_1$ and $\mu_2, \mathbf{x}_2$, respectively. [Actually, by the **Pauli exclusion principle** the state space is the proper closed subspace of $\mathcal{H}_{1/2} \otimes \mathcal{H}_{1/2}$ consisting of vectors $\mathbf{\Psi}$ such that $\Psi_{\mu_1\mu_2}(\mathbf{x}_1, \mathbf{x}_2) + \Psi_{\mu_2\mu_1}(\mathbf{x}_2, \mathbf{x}_1) \equiv 0$. Thus, not all elements of $\mathcal{H}_{1/2} \otimes \mathcal{H}_{1/2}$ have physical significance (see Section 9.8).] In a similar manner one can construct state spaces for systems containing an arbitrary (finite) number of particles with arbitrary spin. A state vector $\mathbf{\Psi}(\mathbf{x}_1, \ldots, \mathbf{x}_k)$ in a k-particle system has components $\Psi_{\mu_1 \cdots \mu_k}(\mathbf{x}_1, \ldots, \mathbf{x}_k)$. If the jth particle has spin s_j then the index μ_j takes values $-s_j, -s_j + 1, \ldots, s_j$. Under a rotation $R(A)$ the state vector $\mathbf{\Psi}$ is transformed to

$$(8.21)\quad [\mathbf{T}(A)\mathbf{\Psi}]_{\mu_1 \cdots \mu_k}(\mathbf{x}_1, \ldots, \mathbf{x}_k) = \sum_{\nu_j = -s_j}^{s_j} T^{s_1}_{\mu_1 \nu_1}(A) \cdots T^{s_k}_{\mu_k \nu_k}(A) \Psi_{\nu_1 \cdots \nu_k}(R(A^{-1})\mathbf{x}_1, \ldots, R(A^{-1})\mathbf{x}_k).$$

The rep $\mathbf{T}$ is single-valued on $SO(3)$ if an even number of spins s_j are half-integral. Otherwise, $\mathbf{T}$ is double-valued.

The rep $\mathbf{T}$ of $SU(2)$ induces a corresponding rep of $su(2)$ defined by operators

$$\boldsymbol{\mathcal{J}} = (d/dt)\mathbf{T}(\exp t\mathcal{J})|_{t=0}, \qquad \mathcal{J} \in su(2).$$

We choose the basis $\mathcal{J}_1, \mathcal{J}_2, \mathcal{J}_3$, (1.8), and compute the operators $\boldsymbol{\mathcal{J}}_1, \boldsymbol{\mathcal{J}}_2, \boldsymbol{\mathcal{J}}_3$ in the case where $\mathbf{T}$ acts on $\mathcal{H}_s$ according to (8.19):

$$(8.22)\qquad \boldsymbol{\mathcal{J}}_j = \mathcal{S}_j + \mathcal{L}_j, \qquad j = 1, 2, 3,$$

where $\mathcal{S}_j$ is a $(2s + 1) \times (2s + 1)$ matrix

$$\mathcal{S}_j = (d/dt)T^s(\exp t\mathcal{J}_j)|_{t=0}$$

acting on spinor components and $\mathcal{L}_j$ is the differential operator (6.22), $(k = 1)$. [If $s = \frac{1}{2}$ then $T^{1/2}(A) = A$ and $\mathcal{S}_j = \mathcal{J}_j$, see (1.8). These three matrices are called the **Pauli spin matrices.**] The action of the spin matrices on the spinors e_μ is given by

$$(8.23)\qquad \begin{aligned} &\mathbf{S}^3 e_\mu = \mu e_\mu, \qquad \mathbf{S}^\pm e_\mu = [(s \pm \mu + 1)(s \mp \mu)]^{1/2} e_{\mu \pm 1}, \\ &\mathbf{S} \cdot \mathbf{S} e_\mu = (\mathbf{S}_1\mathbf{S}_1 + \mathbf{S}_2\mathbf{S}_2 + \mathbf{S}_3\mathbf{S}_3)e_\mu = s(s+1)e_\mu, \qquad -s \le \mu \le s, \end{aligned}$$

where $\mathbf{S}^\pm = \pm i\mathbf{S}_2 + \mathbf{S}_1 = \mp\mathcal{S}_2 + i\mathcal{S}_1$ and $\mathbf{S}^3 = -\mathbf{S}_3 = -i\mathcal{S}_3$. Since $\mathbf{T}$ is unitary the operators $\mathbf{J}_j = i\boldsymbol{\mathcal{J}}_j$ are symmetric on $\mathcal{H}_s$ and satisfy the usual commutation relations

$$(8.24)\qquad [\mathbf{J}_1, \mathbf{J}_2] = i\mathbf{J}_3, \qquad [\mathbf{J}_3, \mathbf{J}_1] = i\mathbf{J}_2, \qquad [\mathbf{J}_2, \mathbf{J}_3] = i\mathbf{J}_1.$$

[Compare with (6.24).] In quantum theory the $\mathbf{J}_j$ are called **total** angular momentum operators. Here $\mathbf{J}_j = \mathbf{S}_j + \mathbf{L}_j$, where the self-adjoint matrices $\mathbf{S}_j = i\mathcal{S}_j$ are **spin** angular momentum operators and the symmetric operators $\mathbf{L}_j = i\mathcal{L}_j$ are **orbital** angular momentum operators. Note that the $\mathbf{S}_j$ and $\mathbf{L}_h$ operators commute with one another since the first acts on the spinor induces alone, while the second acts on the coordinates $\mathbf{x}$ alone.

In case $\mathbf{T}$ acts on a k-particle state space according to (8.21), an analogous computation yields

$$(8.25)\qquad \mathbf{J}_j = \sum_{c=1}^{k} (\mathbf{S}_j^{(c)} + \mathbf{L}_j^{(c)}), \qquad j = 1, 2, 3,$$

where $\mathbf{S}_j^{(c)}$ is a $(2s_c + 1) \times (2s_c + 1)$ matrix acting on the spinor indices μ_c and $\mathbf{L}_j^{(c)} = i\mathfrak{L}_j^{(c)}$ is the differential operator (6.22) acting on the coordinates $\mathbf{x}_c$. The commutation relations are again (8.24).

To investigate some of the physical consequences of this formalism we consider a system containing a single electron ($s = \frac{1}{2}$). Suppose the Hamiltonian $\mathbf{K}$ on $\mathcal{H}_{1/2}$ takes the form

$$\mathbf{K\Psi} = \begin{pmatrix} \mathbf{H} & 0 \\ 0 & \mathbf{H} \end{pmatrix} \begin{pmatrix} \Psi_{1/2} \\ \Psi_{-1/2} \end{pmatrix}, \qquad \mathbf{\Psi} \in \mathcal{H}_{1/2}, \tag{8.26}$$

where $\mathbf{H} = (-1/2m)\Delta + V(\mathbf{x})$, m is the mass of the electron, and $V(\mathbf{x})$ is rotationally invariant. We are assuming that $\mathbf{K}$ is *spin-independent*, i.e., it does not depend on the spinor index μ. Let λ be an eigenvalue of $\mathbf{H}$ acting on the Hilbert space $\mathcal{H}$ (no spin), and assume that the eigenspace W_λ in $\mathcal{H}$ transforms according to the $(2l + 1)$-dimensional irred rep $\mathbf{D}^{(l)}$ of $SO(3)$. Here the action of $SO(3)$ on $\mathcal{H}$ is given by $\Psi(\mathbf{x}) \longrightarrow \Psi(R^{-1}\mathbf{x})$, $R \in SO(3)$. An ON basis for W_λ is $\{j(r)Y_l^m(\theta, \varphi): -l \le m \le l\}$, where $j(r)$ is determined from the solution of $\mathbf{H}\Psi = \lambda\Psi$. It is obvious from (8.26) that the eigenspace W_λ' of $\mathcal{H}_{1/2}$ corresponding to eigenvalue λ is $2(2l + 1)$-dimensional and has an ON basis

$$j(r)\begin{pmatrix} Y_l^m(\theta, \varphi) \\ 0 \end{pmatrix}, \qquad j(r)\begin{pmatrix} 0 \\ Y_l^m(\theta, \varphi) \end{pmatrix}, \qquad -l \le m \le l. \tag{8.27}$$

Thus the degeneracy of λ is twice that in a spinless theory. It is easy to check that both the spin operators $\mathbf{S}_j$ and the angular momentum operators $\mathbf{L}_j$ commute with $\mathbf{K}$. Thus $\mathbf{K}$ admits the six-dimensional symmetry group $SU(2) \times SU(2)$ obtained by letting $SU(2)$ act on the spin indices and spatial coordinates independently in (8.19). Clearly, W_λ' transforms according to the irred rep $\mathbf{D}^{(1/2,l)}$ of $SU(2) \times SU(2)$.

Now we introduce a spin-dependent perturbing (matrix) potential $\mathbf{V}'$ such that the perturbed Hamiltonian $\mathbf{K}' = \mathbf{K} + \mathbf{V}'$ is still rotationally invariant, i.e., such that $\mathbf{K}'$ commutes with the operators (8.19). Then $\mathbf{K}'$ will no longer commute with all the spin operators $\mathbf{S}_j$ and orbital angular momentum operators $\mathbf{L}_j$, but will still commute with the operators $\mathbf{J}_j = \mathbf{S}_j + \mathbf{L}_j$. The symmetry group of $\mathbf{K}'$ is the diagonal subgroup of $SU(2) \times SU(2)$ consisting of those elements (A, B) such that $A = B$. Clearly, this subgroup is isomorphic to $SU(2)$. Since

$$\mathbf{D}^{(1/2,l)} \mid SU(2) \cong \mathbf{D}^{(1/2)} \otimes \mathbf{D}^{(l)} \cong \begin{cases} \mathbf{D}^{(l+[1/2])} \oplus \mathbf{D}^{(l-[1/2])}, & l = 1, 2, \ldots, \\ \mathbf{D}^{(1/2)}, & l = 0, \end{cases} \tag{8.28}$$

as follows from (8.19) and the CG series, we see that for $l \ge 1$ the perturbation splits the $2(2l + 1)$-degenerate eigenvalue λ into two eigenvalues of degeneracy $2l + 2$ and $2l$, respectively. For $l = 0$ the twofold eigenvalue does not split. These predictions are dramatically different than the corresponding predictions for spinless particles, and their experimental verification provides a justification for the introduction of spinor fields into quantum theory.

We can use the CG coefficients to construct a canonical basis for $\mathcal{H}_{1/2}$ corresponding to the decomposition (8.28). Indeed the vectors

$$h_n^{(l+[1/2])} = j(r) \sum_{\mu m} C(l, m; \tfrac{1}{2}, \mu \,|\, l + \tfrac{1}{2}, n) Y_l^m(\theta, \varphi) e_\mu,$$

$$h_n^{(l-[1/2])} = j(r) \sum_{\mu m} C(l, m; \tfrac{1}{2}, \mu \,|\, l - \tfrac{1}{2}, n) Y_l^m(\theta, \varphi) e_\mu,$$

form canonical bases for $\mathbf{D}^{(l+[1/2])}$ and $\mathbf{D}^{(l-[1/2])}$, respectively. This basis is very important in scattering problems involving spin-dependent forces. In such problems spin and orbital angular momentum are not separately conserved but only total angular momentum. Thus s, μ, l, m are not good quantum numbers and only the eigenvalues of $\mathbf{J}^3$ and $\mathbf{J} \cdot \mathbf{J}$ are conserved.

The decomposition of energy eigenstates of a system containing k particles with spins $s_1, \ldots, s_k$ into eigenstates of total angular momentum is analogous to that above.

7.9 Double-Valued Representations of the Crystallographic Groups

We have seen that in a physical system containing particles with spin it is possible that an energy eigenspace W_λ of the rotationally invariant, spin-dependent Hamiltonian $\mathbf{H}$ transforms under a half-integral irred rep $\mathbf{D}^{(u)}$ of $SU(2)$. For example, from (8.21) and the CG series, the eigenspaces of systems containing an *odd* number of electrons transform under half-integral reps. [Those with an even number of electrons transform under integral (single-valued) reps of $SO(3)$.]

Suppose W_λ is such an eigenspace of $\mathbf{H}$ in the Hilbert space $\mathcal{H}$ corresponding to a k-particle system. Now suppose we embed our system in an infinite crystal with crystallographic point symmetry group G (of the first kind). That is, we add to $\mathbf{H}$ the perturbing potential $V'(\mathbf{x}_1, \ldots, \mathbf{x}_k)$ with symmetry group G:

$$\text{(9.1)} \qquad \mathbf{H}' = \mathbf{H} + V', \qquad V'(R\mathbf{x}_1, \ldots, R\mathbf{x}_k) = V'(\mathbf{x}_1, \ldots, \mathbf{x}_k), \qquad R \in G.$$

We assume V' is spin-independent, i.e., V' is a function and does not affect the spinor indices.

Let G' be the set of all $A \in SU(2)$ such that $R(A) \in G$, where $R(A)$ is defined by (1.20). Since $R(-A) = R(A)$, then $A \in G'$ implies $-A \in G'$. In particular $I = -E_2 \in G'$. Clearly, G' is a group. Since the mapping $A \longrightarrow R(A)$ is 2–1, the order of G' is twice that of G. Furthermore, $\{E_2, I\}$ is a normal subgroup of G such that $G'/\{E_2, I\} \cong G$. According to (9.1), $\mathbf{T}(A)V' = V'\mathbf{T}(A)$ for $A \in G'$ and $\mathbf{T}(A)$ given by (8.21). Thus $\mathbf{T}(A)\mathbf{H}' = \mathbf{H}'\mathbf{T}(A)$

for $A \in G'$ and G' is a symmetry group of $\mathbf{H}'$. If G is the largest point group fixing V', then G' is the largest subgroup of $SU(2)$ which is a symmetry group of $\mathbf{H}'$.

To analyze the splitting of the $(2u+1)$-degenerate energy level λ under the perturbing potential V' we must decompose the restricted rep $\mathbf{D}^{(u)}|G'$ into a direct sum of irred reps of G'. If $R(\varphi, \theta, \psi) \in G$ has Euler coordinates φ, θ, ψ then the corresponding elements of G' are $A(\varphi, \theta, \psi)$, (1.13), and $-A = IA$. Since I commutes with the elements of G' and $I^2 = E_2$ it follows that $\mathbf{Q}(I) = \pm\mathbf{E}$ for any unitary irred rep $\mathbf{Q}$ of G'. If $\mathbf{Q}(I)$ is the identity operator then $\mathbf{Q}(A) = \mathbf{Q}(-A)$ and the $\mathbf{Q}$ induces a single-valued irred rep of the factor group $G'/\{E_2, I\} \cong G$. We say $\mathbf{Q}$ is **integral**. On the other hand, if $\mathbf{Q}(I) = -E$ then $\mathbf{Q}(-A) = -\mathbf{Q}(A)$ and $\mathbf{Q}$ induces a double-valued rep of G. We say $\mathbf{Q}$ is **half-integral**. The relationship between G and G' is analogous to that between $SO(3)$ and $SU(2)$.

If u is an integer then the operator $\mathbf{T}(I)$ corresponding to the rep $\mathbf{D}^{(u)}$ of $SU(2)$ is the identity. Thus, $\mathbf{D}^{(u)}|G'$ splits into a direct sum of integral irred reps of G'. We get the same splitting as by restricting the single-valued rep $\mathbf{D}^{(u)}$ of $SO(3)$ to G.

However, if u is half-integral (which is the case which concerns us here) then $\mathbf{T}(I) = -\mathbf{E}$ and $\mathbf{D}^{(u)}|G'$ splits into a direct sum of half-integral irred reps of G' (double-valued reps of G).

To determine this splitting we must find the character table for G'. This is a straightforward computation. Given G of order n we express its elements in terms of Euler angles and determine the group G' of order $2n$. Then we use the techniques of Section 3.6 to compute the character table. The integral characters are easy to find since there is a 1–1 relationship between reps of G and integral reps of G'. If χ is a simple character of G then the corresponding integral simple character of G' is $\chi'(A) = \chi'(-A) = \chi(R(A))$, $A \in G'$. Thus it only remains to compute the half-integral characters of G'. Complete tables of these characters are presented by Hamermesh [1] and Liubarskii [1]. Here, we present without proof the table of simple half-integral characters for O' where O is the octahedral group.

If $R \in O$ with Euler angles φ, θ, ψ we denote by R^+ the corresponding element in O' with the same Euler angles and set $R^- = -R^+ \in O'$. Now O contains 24 elements in five conjugacy classes: E, $\mathcal{C}_4{}^2(3)$, $\mathcal{C}_2(6)$, $\mathcal{C}_4(6)$, $\mathcal{C}_3(8)$. On the other hand, O' contains 48 elements in eight conjugacy classes: E, I, $\{\mathcal{C}_3{}^+(4), \mathcal{C}_3^{2-}(4)\}$, $\{\mathcal{C}_3^{2+}(4), \mathcal{C}_3{}^-(4)\}$, $\{\mathcal{C}_4{}^+(3), \mathcal{C}_4^{3-}(3)\}$, $\{\mathcal{C}_2{}^+(6), \mathcal{C}_2{}^-(6)\}$, $\{\mathcal{C}_4^{3+}(3), \mathcal{C}_4{}^-(3)\}$, $\{\mathcal{C}_4^{2+}(3), \mathcal{C}_4^{2-}(3)\}$. Thus, O' has eight irred reps of dimensions $n_1, \ldots, n_8$ such that $n_1{}^2 + \cdots + n_8{}^2 = 48$. However, in Section 3.6 we already found five irred reps of O (the integral reps of O') with dimensions 1, 1, 2, 3, 3. Thus there are three half-integral reps of O' with dimensions n_6, n_7, n_8, where $n_6{}^2 + n_7{}^2 + n_8{}^2 = 24$. The only solution with $n_6 \le n_7 \le$

n_8 is $n_6 = n_7 = 2$, $n_8 = 4$. The character table can be shown to be

(9.2)

O'	E	I	$\mathcal{C}_3^{+}(4)$ $\mathcal{C}_3^{2-}(4)$	$\mathcal{C}_3^{2+}(4)$ $\mathcal{C}_3^{-}(4)$	$\mathcal{C}_4^{+}(3)$ $\mathcal{C}_4^{3-}(3)$	$\mathcal{C}_4^{3+}(3)$ $\mathcal{C}_4^{-}(3)$	$\mathcal{C}_4^{2+}(3)$ $\mathcal{C}_4^{2-}(3)$	$\mathcal{C}_2^{+}(6)$ $\mathcal{C}_2^{-}(6)$
$(\chi')^{(6)}$	2	-2	1	-1	$\sqrt{2}$	$-\sqrt{2}$	0	0
$(\chi')^{(7)}$	2	-2	1	-1	$-\sqrt{2}$	$\sqrt{2}$	0	0
$(\chi')^{(8)}$	4	-4	-1	1	0	0	0	0

We can use this table to compute the splitting of a $(2u + 1)$-degenerate eigenvalue λ corresponding to the half-integral rep $\mathbf{D}^{(u)}$ of $SU(2)$ under a perturbation with O' symmetry. If $A \in SU(2)$ is similar to $A(0, 0, \tau)$ then the character $\chi^{(u)}(A) = [\sin(u + \frac{1}{2})\tau]/\sin(\tau/2)$. Moreover, $\chi^{(u)}(AI) = -\chi^{(u)}(A)$. With this information we can easily compute the character of $\mathbf{D}^{(u)} \mid O'$:

(9.3)

$\chi^{(u)}$	E	I	$\mathcal{C}_3^{+}(4)$ $\mathcal{C}_3^{2-}(4)$	$\mathcal{C}_3^{2+}(4)$ $\mathcal{C}_3^{-}(4)$	$\mathcal{C}_4^{+}(3)$ $\mathcal{C}_4^{3-}(3)$	$\mathcal{C}_4^{3+}(3)$ $\mathcal{C}_4^{-}(3)$	$\mathcal{C}_4^{2+}(3)$ $\mathcal{C}_4^{2-}(3)$	$\mathcal{C}_2^{+}(6)$ $\mathcal{C}_2^{-}(6)$
$\chi^{(1/2)}$	2	-2	1	-1	$\sqrt{2}$	$-\sqrt{2}$	0	0
$\chi^{(3/2)}$	4	-4	-1	1	0	0	0	0
$\chi^{(5/2)}$	6	-6	0	0	$-\sqrt{2}$	$\sqrt{2}$	0	0
$\chi^{(7/2)}$	8	-8	1	-1	0	0	0	0
$\chi^{(9/2)}$	10	-10	-1	1	$\sqrt{2}$	$-\sqrt{2}$	0	0

Writing $\chi^{(u)} \mid O'$ as a linear combination of simple characters, we obtain the results:

(9.4)

$$\chi^{(1/2)} \mid O' = (\chi')^{(6)}, \qquad \chi^{(3/2)} \mid O' = (\chi')^{(8)}, \qquad \chi^{(5/2)} \mid O' = (\chi')^{(7)} + (\chi')^{(8)},$$

$$\chi^{(7/2)} \mid O' = (\chi')^{(6)} + (\chi')^{(7)} + (\chi')^{(8)}, \qquad \chi^{(9/2)} \mid O' = (\chi')^{(6)} + 2(\chi')^{(8)}.$$

for $u = \frac{1}{2}, \ldots, \frac{9}{2}$. For example, under the perturbation a sixfold eigenvalue $(u = \frac{5}{2})$ splits into one twofold and one fourfold eigenvalue. Notice that $\chi^{(1/2)} \mid O'$ and $\chi^{(3/2)} \mid O'$ are simple, so twofold and fourfold eigenvalues do not split.

7.10 The Wigner–Eckart Theorem and Its Applications

Let $\mathbf{T}$ be a unitary rep of $SU(2)$ on the Hilbert space $\mathcal{H}$. The mapping $\mathbf{Q} \to \mathbf{T}(A)\mathbf{Q}\mathbf{T}^{-1}(A)$ defines a rep of $SU(2)$ on the space $\mathcal{B}(\mathcal{H})$ of all bounded linear operators $\mathbf{Q}$ on $\mathcal{H}$. We could introduce an inner product on $\mathcal{B}(\mathcal{H})$

with respect to which this rep is unitary and then decompose the rep into a direct sum of irred reps $\mathbf{D}^{(u)}$. Rather than carry out such a decomposition we shall merely investigate the irred subspaces of operators.

Let $\mathcal{W}^{(u)}$ be an irred subspace of $\mathcal{B}(\mathcal{H})$ transforming according to $\mathbf{D}^{(u)}$. Then there exists a canonical basis $\{\mathbf{Q}_m: -u \leq m \leq u\}$ for $\mathcal{W}^{(u)}$ such that

$$\mathbf{T}(A)\mathbf{Q}_m\mathbf{T}^{-1}(A) = \sum_{n=-u}^{u} T^u_{nm}(A)\mathbf{Q}_n. \tag{10.1}$$

Operators with transformation properties (10.1) are called **spherical tensors of rank u**. We shall compute the matrix elements $(\mathbf{Q}_m\mathbf{f}^{u_1}_j, \mathbf{g}^{u_2}_h)$, where $\mathbf{f}^{u_1}_j$ and $\mathbf{g}^{u_2}_h$ belong to canonical ON sets in $\mathcal{H}$ transforming irreducibly under $\mathbf{T}$:

$$\mathbf{T}(A)\mathbf{f}^{u_1}_j = \sum_{k=-u_1}^{u_1} T^{u_1}_{kj}(A)\mathbf{f}^{u_1}_k, \qquad \mathbf{T}(A)\mathbf{g}^{u_2}_h = \sum_{s=-u_2}^{u_2} T^{u_2}_{sh}(A)\mathbf{g}^{u_2}_s. \tag{10.2}$$

Our considerations will also apply to unbounded operators $\mathbf{Q}_m$ on $\mathcal{H}$ provided there is a dense subspace $\mathcal{Z}$ of $\mathcal{H}$ such that (a) the domain of each of the $\mathbf{Q}_m$ contains $\mathcal{Z}$, (b) $\mathcal{Z}$ is invariant under the $\mathbf{T}(A)$, and (c) (10.1) holds on $\mathcal{Z}$.

The group rep (10.1) induces a Lie algebra rep of $su(2)$. Indeed if $J = (d/dt)\mathbf{T}(\exp t\mathcal{J})|_{t=0}$, $\mathcal{J} \in su(2)$, then by setting $A = \exp t\mathcal{J}$ in (10.1) and differentiating with respect to t at $t = 0$ we obtain the Lie algebra rep

$$\mathbf{Q}_m \longrightarrow [J, \mathbf{Q}_m] = J\mathbf{Q}_m - \mathbf{Q}_m J. \tag{10.3}$$

Since the $\mathbf{Q}_m$ form a canonical basis we find

$$[J^3, \mathbf{Q}_m] = m\mathbf{Q}_m, \qquad [J^\pm, \mathbf{Q}_m] = [(u \pm m + 1)(u \mp m)]^{1/2}\mathbf{Q}_{m\pm 1}, \tag{10.4}$$

where

$$J^\pm = \pm J_2 + iJ_1, \qquad J^3 = -iJ_3. \tag{10.5}$$

Spherical tensors appear frequently in quantum mechanics. For example a Hamiltonian $\mathbf{H}$ which commutes with the $\mathbf{T}(A)$ is a spherical tensor of rank zero. As another example we set $\mathcal{H} = L_2(R_3)$ and let $[\mathbf{T}(A)\Psi](\mathbf{x}) = \Psi(A^{-1}\mathbf{x})$ for $A \in SO(3)$, $\Psi \in \mathcal{H}$. Then for fixed integer l the multiplicative operators

$$\mathbf{Q}_m\Psi(r, \theta, \varphi) = r^l Y_l^{-m}(\theta, \varphi)\Psi(r, \theta, \varphi) \tag{10.6}$$

are spherical tensors of rank l. Here, the $Y_l^m(\theta, \varphi)$ are spherical harmonics expressed in spherical coordinates. To verify this we will check the relations (10.4). From (5.4), (5.6), and (10.5) we find

$$J^\pm = e^{\mp i\varphi}\left(\mp\frac{\partial}{\partial\theta} + i\cot\theta\frac{\partial}{\partial\varphi}\right), \qquad J^3 = i\frac{\partial}{\partial\varphi}. \tag{10.7}$$

Furthermore, from (5.7)

$$J^3 Y_l^{-m} = mY_l^{-m}, \qquad J^\pm Y_l^{-m} = [(l \pm m + 1)(l \mp m)]^{1/2} Y_l^{-(m\pm 1)}. \tag{10.8}$$

Since the J operators are differential and $\mathbf{Q}_m = r^l Y_l^{-m}$ is multiplicative we find

$$[J, \mathbf{Q}_m]\Psi = J(r^l Y_l^{-m}\Psi) - r^l Y_l^{-m}(J\Psi) = \{J(r^l Y_l^{-m})\}\Psi. \tag{10.9}$$

Together (10.8) and (10.9) yield (10.4) for $u = l$. [Actually the above result is valid for $\mathbf{Q}_m = f(r)Y_l^{-m}(\theta, \varphi)$, where $f(r)$ is arbitrary.]

Let us consider the special case $l = 1$. The canonical basis vectors are

(10.10)
$$\mathbf{f}_1{}^1 = rY_1^{-1} = \left(\frac{3}{8\pi}\right)^{1/2}(x - iy), \qquad \mathbf{f}_1{}^0 = rY_1{}^0 = \left(\frac{3}{4\pi}\right)^{1/2} z,$$
$$\mathbf{f}_1^{-1} = -\left(\frac{3}{8\pi}\right)^{1/2}(x + iy).$$

Multiplying all vectors by $(4\pi/3)^{1/2}$, we see that the vectors

(10.11) $$(1/\sqrt{2})(x - iy), \quad z, \quad -(1/\sqrt{2})(x + iy)$$

form a canonical basis for $\mathbf{D}^{(1)}$. Note that x, y, z does *not* transform as a canonical basis. Here, the multiplicative operators $\mathbf{Q}_x = x$, $\mathbf{Q}_y = y$, $\mathbf{Q}_z = z$ are the **position operators** of quantum theory.

A similar computation using the same J-operators shows that the differential operators

(10.12) $$\partial_1 = \frac{1}{\sqrt{2}}\left(\frac{\partial}{\partial x} - i\frac{\partial}{\partial y}\right), \qquad \partial_0 = \frac{\partial}{\partial z}, \qquad \partial_{-1} = -\frac{1}{\sqrt{2}}\left(\frac{\partial}{\partial x} + i\frac{\partial}{\partial y}\right)$$

also transform as spherical tensors of rank one. Note that the ∂_j are closely related to the **linear momentum** operators in quantum theory:

$$\mathbf{P}_x = -i\,\partial/\partial x, \qquad \mathbf{P}_y = -i\,\partial/\partial y, \qquad \mathbf{P}_z = -i\,\partial/\partial z.$$

We will compute the matrix elements $(\mathbf{Q}_m\mathbf{f}_j^{u_1}, \mathbf{g}_h^{u_2})$ for a set of spherical tensors of rank u. From (10.1) and (10.2) we obtain

(10.13)
$$(\mathbf{Q}_m\mathbf{f}_j^{u_1}, \mathbf{g}_h^{u_2}) = (\mathbf{T}(A)\mathbf{Q}_m\mathbf{f}_j^{u_1}, \mathbf{T}(A)\mathbf{g}_h^{u_2}) = (\mathbf{T}(A)\mathbf{Q}_m\mathbf{T}^{-1}(A)\mathbf{T}(A)\mathbf{f}_j^{u_1}, \mathbf{T}(A)\mathbf{g}_h^{u_2})$$
$$= \sum_{nks} T^u_{nm}(A)T^{u_1}_{kj}(A)\overline{T^{u_2}_{sh}(A)}(\mathbf{Q}_n\mathbf{f}_k^{u_1}, \mathbf{g}_s^{u_2}).$$

Multiplying the left- and right-hand sides of this equality by dA, integrating over $SU(2)$, and making use of the identity (7.9), we find

(10.14) $$(\mathbf{Q}_m\mathbf{f}_j^{u_1}, \mathbf{g}_h^{u_2}) = C(u, m; u_1, j \mid u_2, h)N,$$
$$N = \frac{1}{2u_2 + 1}\sum_{uks} C(u, n; u_1\, k \mid u_2, s)(\mathbf{Q}_n\mathbf{f}_k^{u_1}, \mathbf{g}_s^{u_2}).$$

Theorem 7.1 (Wigner–Eckart). If $\{\mathbf{Q}_m\}$ is a set of spherical tensors of rank u then (10.14) holds where N depends on u, u_1, u_2 but not on m, j, and h.

The point of this theorem is that the dependence of the matrix element on m, j, and h is completely determined by the CG coefficient. If for fixed u, u_1, and u_2 we are able to compute one of the nonzero matrix elements (10.14) then we can solve for N and (10.14) will tell us the values of all the matrix elements. The constant N is sometimes called a **reduced** matrix element. From the known properties of the CG coefficients we see that the left-hand side of (10.14) will be zero unless $m + j = h$ and $u_2 = |u - u_1|$, $|u - u_1| + 1, \ldots, u + u_1$.

We have stated the Wigner–Eckart theorem for reps of $SU(2)$, but actually it holds for reps of any finite group or compact Lie group G. Indeed if we denote by $\mathbf{T}^{(u)}$ a complete set of nonequivalent irred unitary reps of G then we can define by (10.1) the operators of rank u where now $A \in G$. Expression (10.13) is unaltered by our generalization. We can integrate (10.13) over G with respect to the invariant measure dA if G is a Lie group or sum over the group if G is finite. Similarly, expression (7.9) is valid for G provided the factor $(2w + 1)/16\pi^2$ is replaced by n_w/V_G, where n_w is the dimension of $\mathbf{T}^{(w)}$. In particular we can define CG coefficients for G in analogy with those for $SU(2)$. (There is one possible complication here. It may be that $\mathbf{T}^{(w)}$ occurs more than once in the decomposition of $\mathbf{T}^{(u)} \otimes \mathbf{T}^{(v)}$. In this case the CG coefficients will need an extra parameter to denote which of the $\mathbf{T}^{(w)}$-subspaces is under consideration.)

We can get a better understanding of the Wigner–Eckart theorem by recalling the discussion of invariant tensors in Section 3.8. Expression (10.13) shows that the tensor $\mathbf{a}$ with components $a_{nks} = (\mathbf{Q}_n \mathbf{f}_k^{u_1}, \mathbf{g}_s^{u_2})$ is an invariant in a tensor space transforming under the rep

$$\mathbf{T}^{(u)} \otimes \mathbf{T}^{(u_1)} \otimes \overline{\mathbf{T}^{(u_2)}} \tag{10.15}$$

of G, where $\overline{\mathbf{T}^{(u_2)}}$ is the rep whose matrix elements are $\overline{T_{sh}^{u_2}}(A)$. Since $\mathbf{a}$ is invariant it must transform according to the identity rep $\mathbf{T}^{(0)}$. Let q be the multiplicity of $\mathbf{T}^{(0)}$ in (10.15) and let $V^{(0)}$ be the subspace of invariant tensors in the tensor space V. Then $q = \dim V^{(0)}$ and $\mathbf{a} \in V^{(0)}$ is nonzero only if $q > 0$. Furthermore, exactly q parameters are needed to uniquely determine $\mathbf{a}$. Let $\chi^{(u)}, \chi^{(u_1)}, \chi^{(u_2)}$ be the characters of $\mathbf{T}^{(u)}, \mathbf{T}^{(u_1)}, \mathbf{T}^{(u_2)}$, respectively. Then $\overline{\chi^{(u_2)}}(A)$ is the character of $\bar{\mathbf{T}}^{(u_2)}$. Since the character of $\mathbf{T}^{(0)}$ is $\chi^{(0)}(A) \equiv 1$ and the character of (10.15) is $\chi^{(u)}\chi^{(u_1)}\overline{\chi^{(u_2)}}$ we find from the orthogonality relations that q is given by

$$q = \int_G \chi^{(u)}(A)\chi^{(u_1)}(A)\overline{\chi^{(u_2)}}(A)\,\delta A = \langle \chi^{(u)}\chi^{(u_1)}\overline{\chi^{(u_2)}}, 1\rangle = \langle \chi^{(u)}\chi^{(u_1)}, \chi^{(u_2)}\rangle. \tag{10.16}$$

On the other hand, the right-hand side of (10.16) is just the multiplicity of $\mathbf{T}^{(u_2)}$ in the tensor product $\mathbf{T}^{(u)} \otimes \mathbf{T}^{(u_1)}$. Thus, we can obtain q from a knowledge of the CG series for irred reps of G. In particular, if $\mathbf{T}^{(u_2)}$ does not appear in the CG series for $\mathbf{T}^{(u)} \otimes \mathbf{T}^{(u_1)}$ then $q = 0$.

In the special case where $G = SU(2)$ the series is

$$\mathbf{D}^{(u)} \otimes \mathbf{D}^{(u_1)} \simeq \mathbf{D}^{(u+u_1)} \oplus \mathbf{D}^{(u+u_1-1)} \oplus \cdots \oplus \mathbf{D}^{(|u-u_1|)}, \tag{10.17}$$

so $q = 1$ if $u_2 = u + u_1, \ldots, |u - u_1|$; otherwise $q = 0$. In the cases where $q = 1$ the space of invariant tensors is one-dimensional and can be determined by specifying a single constant N.

We now give some applications of these results to quantum mechanics. Let $\mathcal{H}$ be the usual Hilbert space corresponding to a k-particle system (without spin) and let the action of $SO(3)$ on $\mathcal{H}$ be given by (6.2). Consider the position operators $\mathbf{Q}_{sj} = x_{sj}$, $s = 1, 2, 3$ $[\mathbf{x}_j = (x_{1j}, x_{2j}, x_{3j}) = (x_j, y_j, z_j)]$, of the jth particle. We will compute the matrix elements

$$(10.18)\qquad (\mathbf{Q}_{sj}\Psi^{l_1}_{m_1}, \Psi^{l_2}_{m_2}) = \int_{R_{3^k}} x_{sj}\Psi^{l_1}_{m_1}(\mathbf{x}_1, \ldots, \mathbf{x}_k)\overline{\Psi^{l_2}_{m_2}}(\mathbf{x}_1, \ldots, \mathbf{x}_k)\, d\mathbf{x},$$

where the $\Psi^{l_h}_{m_h}$ transform as canonical basis vectors under the representations $\mathbf{D}^{(l_h)}$ of $SO(3)$. According to (10.11) the operators $\mathbf{Q}^{(1)} = 2^{-1/2}(\mathbf{Q}_{1j} - i\mathbf{Q}_{2j})$, $\mathbf{Q}^{(0)} = \mathbf{Q}_{3j}$, $\mathbf{Q}^{(-1)} = -2^{-1/2}(\mathbf{Q}_{1j} + i\mathbf{Q}_{2j})$ determine a spherical tensor of rank one. We first compute the matrix elements

$$(10.19)\qquad (\mathbf{Q}^{(s)}\Psi^{l_1}_{m_1}, \Psi^{l_2}_{m_2}), \qquad s = 1, 0, -1, \quad -l_h \leq m_h \leq l_h.$$

It is obvious that the matrix elements (10.18) can be determined immediately from (10.19). Since $\mathbf{D}^{(1)} \otimes \mathbf{D}^{(l_1)} \cong \mathbf{D}^{(l_1+1)} \oplus \mathbf{D}^{(l_1)} \oplus \mathbf{D}^{(l_1-1)}$ if $l_1 \geq 1$ and $\mathbf{D}^{(1)} \otimes \mathbf{D}^{(0)} \cong \mathbf{D}^{(1)}$, it follows from our above analysis that for $l_1 \geq 1$ the matrix elements are nonzero only if $l_2 = l_1 + 1, l_1$, or $l_1 - 1$, while for $l_1 = 0$ the matrix elements are zero unless $l_2 = 1$. An explicit expression for the matrix elements is given by (10.14).

If the system contains particles with half-integral spin we can form expressions (10.19) where the l_i take half-integral values. The above analysis is unchanged except for the special case $\mathbf{D}^{(1)} \otimes \mathbf{D}^{(1/2)} \cong \mathbf{D}^{(3/2)} \oplus \mathbf{D}^{(1/2)}$, which implies that for $l_1 = \frac{1}{2}$ the matrix elements are zero unless $l_2 = \frac{3}{2}$ or $\frac{1}{2}$.

Now suppose the group acting on $\mathcal{H}$ (no spins) is $O(3)$. Recall that the irred reps of $O(3)$ are $\mathbf{D}^{(l)}_{\pm}$, where the sign denotes **parity**, (6.5), (6.6). The $\mathbf{Q}^{(s)}$ transform like polar vectors under $O(3)$, hence like $\mathbf{D}^{(1)}_{-}$. It is easy to verify the CG series

$$(10.20)\qquad \mathbf{D}^{(1)}_{-} \otimes \mathbf{D}^{(l)}_{\pm} \cong \begin{cases} \mathbf{D}^{(l+1)}_{\mp} \oplus \mathbf{D}^{(l)}_{\mp} \oplus \mathbf{D}^{(l-1)}_{\mp}, & l \geq 1, \\ \mathbf{D}^{(1)}_{\mp}, & l = 0, \end{cases}$$

The selection rules for the matrix elements follow immediately from (10.20). Again the nonzero matrix elements are given explicitly by (10.14).

We see from these results that (10.19) is always zero if $\Psi^{l_1}_{m_1}$ and $\Psi^{l_2}_{m_2}$ have the same parity. An interesting special case of our analysis occurs for one-particle systems ($k = 1$). In this case $\Psi^{l_i}_{m_i}(\mathbf{x}) = j_{l_i}(r)Y^{m_i}_{l_i}(\theta, \varphi)$, where the $Y^{m_i}_{l_i}(\theta, \varphi)$ are spherical harmonics. Recall that $\{Y_l{}^m\}$ transforms according to $\mathbf{D}^{(l)}_{+}$ if l is even and $\mathbf{D}^{(l)}_{-}$ if l is odd. Thus $(\mathbf{Q}^{(s)}\Psi^{l_1}_{m_1}, \Psi^{l_2}_{m_2})$ is nonzero only if $l_2 = l_1 \pm 1$. Parity considerations have eliminated the possibility $l_2 = l_1$.

In case the system contains particles of half-integral spin we have to perform our analysis using the group $SU(2) \times \{E, I\}$ rather than $O(3)$, but this changes the above results in no essential manner.

In quantum theory the matrix elements (10.14) may have interpretations other than those given here. For example, expressions of the form (10.19) occur in the study of emission and absorption of light by atoms (Liubarskii [1]). In this case these expressions are related to the lowest-order (dipole) approximation of the transition probability from one state to another. Our results stating that only certain special matrix elements are nonzero are called **selection rules** in this theory. Similarly, the quadrapole approximation of quantum perturbation theory corresponds to the approximation of a set of operators by spherical tensors of rank two and use of the Wigner–Eckart theorem to simplify the matrix element computation.

7.11 Spinor Fields and Invariant Equations

The Euclidean group $E^+(3)$ frequently appears as a symmetry group in classical and quantum physics. Suppose for example that $\mathcal{H}$ is the Hilbert space of a k-particle system (Section 7.6). Then the operators $\mathbf{T}(\mathbf{a}, \mathbf{O})$ given by

$$(11.1) \qquad [\mathbf{T}(\mathbf{a}, \mathbf{O})\Psi](\mathbf{x}_1, \ldots, \mathbf{x}_k) = \Psi(\mathbf{O}^{-1}(\mathbf{x}_1 - \mathbf{a}), \ldots, \mathbf{O}^{-1}(\mathbf{x}_k - \mathbf{a})),$$
$$\mathbf{a} \in R_3, \qquad \mathbf{O} \in SO(3), \qquad \Psi \in \mathcal{H},$$

define a unitary rep of $E^+(3)$ on $\mathcal{H}$. Note that the restriction of $\mathbf{T}$ to $SO(3)$ yields the usual action of $SO(3)$ on $\mathcal{H}$, while the restriction of $\mathbf{T}$ to the translation subgroup R_3 yields

$$(11.2) \qquad [\mathbf{T}(\mathbf{a}, \mathbf{E})\Psi](\mathbf{x}_1, \ldots, \mathbf{x}_k) = \Psi(\mathbf{x}_1 - \mathbf{a}, \ldots, \mathbf{x}_k - \mathbf{a}).$$

If $E^+(3)$ is a symmetry group of the system then the $\mathbf{T}$-operators commute with the Hamiltonian $\mathbf{H}$:

$$(11.3) \qquad \mathbf{T}(\mathbf{a}, \mathbf{O})\mathbf{H} = \mathbf{H}\mathbf{T}(\mathbf{a}, \mathbf{O}).$$

For $\mathbf{a} = \boldsymbol{\theta}$ we have seen that (11.3) signifies the conservation of angular momentum. On the other hand, if we set $\mathbf{O} = \mathbf{E}$ in (11.3), differentiate both sides of the equation with respect to a_j, and set $\mathbf{a} = \boldsymbol{\theta}$ we find $\mathbf{P}_j\mathbf{H} = \mathbf{H}\mathbf{P}_j$, where, (10.18),

$$(11.4) \qquad \mathbf{P}_j = -i\Big(\sum_{h=1}^{k} \partial/\partial x_{jh}\Big), \qquad j = 1, 2, 3,$$

is a **linear momentum** operator. Thus, $E^+(3)$ symmetry of a system implies conservation of angular and linear momentum. Conversely, conservation of angular and linear momentum implies $E^+(3)$ symmetry. (In the standard quantum mechanics texts it is shown that conservation of linear momentum implies the Schrödinger wave functions can be factored into two parts. One part describes the motion of the center of mass as a free particle and the other describes the relative motion of the system with respect to the center of mass.)

If the system contains particles with spin the proper symmetry group is $\mathcal{E}^+(3)$, consisting of pairs $\{\mathbf{a}, A\}$, $\mathbf{a} \in R_3$, $A \in SU(2)$, such that

$$\{\mathbf{a}_1, A_1\}\{\mathbf{a}_2, A_2\} = \{\mathbf{a}_1 + R(A_1)\mathbf{a}_2, A_1A_2\}, \tag{11.5}$$

where $R(A_1) \in SO(3)$ is given by (1.20). Here, $\mathcal{E}^+(3)$ and $E^+(3)$ are six-dimensional locally isomorphic groups (they have isomorphic Lie algebras). The map

$$\{\mathbf{a}, A\} \longrightarrow \{\mathbf{a}, R(A)\}$$

is a homomorphism of $\mathcal{E}^+(3)$ onto $E^+(3)$ which covers each element of $E^+(3)$ exactly twice.

The elements of $\mathcal{H}$ are spinor-valued functions $\mathbf{\Psi} = \{\Psi_\mu(\mathbf{x}_1, \ldots, \mathbf{x}_k)\}$, $\mu = 1, \ldots, q$. (If there are several spin indices we combine them into one index of larger domain.) The action of $\mathcal{E}^+(3)$ on $\mathcal{H}$ is

$$\begin{aligned} &[\mathbf{T}(\mathbf{a}, A)\mathbf{\Psi}]_\mu(\mathbf{x}_1, \ldots, \mathbf{x}_k) \\ &\quad = \sum_{\nu=1}^{q} T_{\mu\nu}(A)\Psi_\nu(R(A^{-1})(\mathbf{x}_1 - \mathbf{a}), \ldots, R(A^{-1})(\mathbf{x}_k - \mathbf{a})), \end{aligned} \tag{11.6}$$

where the matrices $T(A)$ define a unitary rep of $SU(2)$, not necessarily irred. It is straightforward to check that $\mathbf{T}$ is a unitary rep of $\mathcal{E}^+(3)$ with respect to the **inner product**

$$(\mathbf{\Psi}, \mathbf{\Phi}) = \int_{R_{3^k}} \sum_{\mu=1}^{q} \Psi_\mu(\mathbf{x}_1, \ldots, \mathbf{x}_k)\bar{\Phi}_\mu(\mathbf{x}_1, \ldots, \mathbf{x}_k)\, d\mathbf{x}.$$

As before, if the $\mathbf{T}(\mathbf{a}, A)$ commute with $\mathbf{H}$ then total angular momentum and linear momentum are conserved.

Although we have been led to expression (11.6) through Hilbert-space considerations, this expression makes sense independent of Hilbert space. In general any spinor-valued function which transforms under $\mathcal{E}^+(3)$ by (11.6) is called a **spinor field**. If the matrices $T(A)$ satisfy $T(A) = T(-A)$ then the operators $\mathbf{T}$ define a single-valued rep of $E^+(3)$. In this case the function $\mathbf{\Psi}$ is usually called a **tensor field**. Tensor fields abound in classical physics. For example the electromagnetic field $E_j(\mathbf{x})$, $j = 1, 2, 3$, transforms under $E^+(3)$ as a tensor field of rank one, i.e., the matrices $T(A)$ define a rep equivalent to $\mathbf{D}^{(1)}$. Similarly, magnetic fields, elasticity tensors, current tensors, and moment-of-intertia tensors all transform as tensor fields under $E^+(3)$. True spinor fields occur primarily in quantum mechanics and relativistic physics. The best known example is the Dirac electron field where $q = 4$ and $T(A)$ defines a rep equivalent to $\mathbf{D}^{(1/2)} \oplus \mathbf{D}^{(1/2)}$.

Let $\Psi_\mu(\mathbf{x}, t)$ be a spinor field transforming according to (11.6) with $k = 1$. Suppose $\Psi_\mu(\mathbf{x}, t)$ describes some physical quantity which is a solution of a system of q linear differential equations

$$\sum_{jshp} B^{jshp}(\mathbf{x}, t) \frac{\partial^{j+s+h+p}\mathbf{\Psi}(\mathbf{x}, t)}{\partial x^j \, \partial y^s \, \partial z^h \, \partial t^p} = \boldsymbol{\theta}, \tag{11.7}$$

where the B^{jshp} are $q \times q$ matrix functions, $\mathbf{\Psi}(\mathbf{x}, t)$ is a $1 \times q$ column vector, and $\boldsymbol{\theta}$ is the zero vector. Assuming the isotropy of space–time we see that Eq. (11.7) can be physically meaningful only if they assume the same form in every cartesian coordinate system: If we replace $\mathbf{x}$ by $\mathbf{x}' = R(A)\mathbf{x} - \mathbf{a}$, t by $t' = t + c$, and $\Psi_\mu(\mathbf{x}, t)$ by $\Psi_\mu'(\mathbf{x}', t') = \sum T_{\mu\nu}(A)\Psi_\nu(\mathbf{x}, t)$ in (11.7), then the resulting system of equations should be equivalent to (11.7), i.e., the primed equations should be linear combinations of the unprimed equations and conversely. We shall classify all such Euclidean invariant equations (under certain restrictions). The dynamical equations of *any* physical theory which admits $\mathcal{E}^+(3)$ as a symmetry group, via the rep (11.6), will be found in our classification. Our analysis will provide a group-theoretic framework within which all Euclidean invariant physical theories can be described and compared.

Note first that Eq. (11.7) are invariant under all translations in space and time if and only if the matrices B^{jshp} are independent of $\mathbf{x}$ and t. Now we dispense with translation invariance and restrict our attention to invariance under the operators $\mathbf{T}(A) = \mathbf{T}(\boldsymbol{\theta}, A)$, which form a rep of $SU(2)$. Furthermore, we can eliminate dependence on t in (11.7) by considering only solutions of the form $\mathbf{\Psi}(\mathbf{x}, t) = \mathbf{\Psi}(\mathbf{x})e^{i\omega t}$. Then $\partial/\partial t$ is replaced by $i\omega$. (This amounts to taking the Fourier transform in t.)

We can always write (11.7) as a system of first-order differential equations by introducing new components $\Psi_\mu(\mathbf{x})$, $\mu > q$. This will be shown later when we consider specific examples. Thus, we can reduce (11.7) to a system of l equations

$$\left(B_1\frac{\partial}{\partial x} + B_2\frac{\partial}{\partial y} + B_3\frac{\partial}{\partial z} + C\right)\mathbf{\Psi}(\mathbf{x}) = \boldsymbol{\theta}, \tag{11.8}$$

where B_1, B_2, B_3, C are constant $l \times r$ matrices, $\mathbf{\Psi}(\mathbf{x}) = (\Psi_\mu(\mathbf{x}))$ is a $1 \times r$ column vector and the action of $SU(2)$ on $\mathbf{\Psi}(\mathbf{x})$ is

$$[\mathbf{T}(A)\Psi]_\mu(\mathbf{x}) = \sum_{\nu=1}^{r} S_{\mu\nu}(A)\Psi_\nu(R(A^{-1})\mathbf{x}), \qquad \mu = 1, \ldots, r. \tag{11.9}$$

Here $r \geq q$ and $S(A)$ is a matrix rep of $SU(2)$.

For the present we assume C is a nonsingular $r \times r$ matrix. Then multiplying (11.8) on the left by C^{-1} we see that this system of equations is equivalent to a system of the form

$$\left(L_1\frac{\partial}{\partial x} + L_2\frac{\partial}{\partial y} + L_3\frac{\partial}{\partial z}\right)\mathbf{\Psi}(\mathbf{x}) = \kappa\mathbf{\Psi}(\mathbf{x}), \tag{11.10}$$

where the L_j are $r \times r$ matrices and $\kappa \neq 0$ is a constant. (We could take $\kappa = -1$ but it is convenient to leave it arbitrary.)

By passing to a new basis if necessary we can assume that the matrices $S(A)$ take the form

$$S(A) = \begin{pmatrix} \overbrace{T^{(0)}(A) \quad \ddots \quad T^{(0)}(A)}^{\alpha_0} & & Z \\ & \ddots & \\ & & \overbrace{T^{(u)}(A) \quad \ddots \quad T^{(u)}(A)}^{\alpha_u} \\ Z & & & \ddots \end{pmatrix}, \tag{11.11}$$

where $T^{(u)}(A)$ is a matrix realization of $\mathbf{D}^{(u)}$ and α_u is the multiplicity of $\mathbf{D}^{(u)}$ in $S(A)$. In other words we have decomposed the action S of $SU(2)$ on the components of $\boldsymbol{\Psi}$ into a direct sum of irred reps. In this new basis we relabel the components of $\boldsymbol{\Psi}$ as Ψ^m_{un}, the component of the mth canonical basis vector in the nth occurrence of $\mathbf{D}^{(u)}$ in (11.11). Here $-u \leq m \leq u$ and $1 \leq n \leq \alpha_u$. In terms of the new basis, the system of equations still takes the form (11.10).

We can express the partial derivatives on the left-hand side of our equations as linear combinations of $\partial_1, \partial_0, \partial_{-1}$, (10.12), which form a canonical basis for $\mathbf{D}^{(1)}$. Thus, the left-hand side is a linear combination of terms $\partial_1\Psi^p_{vn}, \partial_0\Psi^p_{vn}, \partial_{-1}\Psi^p_{vn}$. For fixed v and n, and p ranging over $-v, -v+1, \ldots, v$ these $3(2v+1)$ quantities transform according to $\mathbf{D}^{(1)} \otimes \mathbf{D}^{(v)} \cong \mathbf{D}^{(v+1)} \oplus \mathbf{D}^{(v)} \oplus \mathbf{D}^{(v-1)}$. Thus the new basis functions

$$h^{m'}_{u'vn} = \sum_{pq} C(1, j; v, p \mid u', m')\, \partial_j \Psi^p_{vn}$$

$$u' = \begin{cases} v+1, v, v-1, & \text{if } v \geq 1, \\ \frac{3}{2}, \frac{1}{2}, & \text{if } v = \frac{1}{2}, \\ 1, & \text{if } v = 0, \end{cases} \qquad -u' \leq m' \leq u', \tag{11.12}$$

transform irreducibly under $\mathbf{D}^{(u')}$. Since the CG coefficients are unitary we can express each of the terms $\partial_j\Psi^p_{vn'}$ on the left-hand side in (11.10) as a linear combination of the $h^{m'}_{u'vn'}$ and rewrite (11.10) as

$$\sum_{vn'm'u'} B^{um}_{m'u'vn} h^{m'}_{u'vn'} = \kappa \Psi^m_{un}. \tag{11.13}$$

Consider the subsystem of $2u+1$ equations (11.13) for which u and n are fixed, and $-u \leq m \leq u$. Now $\Psi'^m_{un}(\mathbf{x}') = \sum T^u_{km}(A)\Psi^k_{un}(\mathbf{x})$ and $h'^{m'}_{u'vn'}(\mathbf{x}') = \sum T^{u'}_{jm'}(A) h^j_{u'vn'}(\mathbf{x})$ so this subsystem will be invariant under $SU(2)$ if and only if the left-hand side of the subsystem transforms like a canonical basis for $\mathbf{D}^{(u)}$. From (11.12) we see that any invariant system must take the form

$$\sum_{u'n'} B^{un}_{u'n'} \sum_{pj} C(1, j; u'\ m' \,|\, u, m)\partial_j \Psi^{m'}_{u'n'} = \kappa \Psi^m_{un}, \tag{11.14}$$

where the sum is taken over $u' = u + 1, u, u - 1$ and $n' = 1, \ldots, \alpha_{u'}$. The constants $B^{un}_{u'n'}$ are completely arbitrary and there is one equation for each component Ψ^m_{un} of $\mathbf{\Psi}$. Note that integral values of u are never coupled with half-integral values of u in (11.14). If both values occur, the system breaks up into two independent subsystems, one coupling integral and the other coupling half-integral values.

The case where the matrix C is singular or not square is more complicated. Suppose $C = Z$. Equations (11.14) with $\kappa = 0$ clearly fall under this case and in general all invariant equations take roughly this form. However, it is not easy to decide if two systems of equations are equivalent, i.e., there is no simple canonical form for such equations. For $\kappa \neq 0$ this difficulty does not occur: Two systems of equations for the Ψ^m_{un} are equivalent if and only if the constants $B^{un}_{u'n'}$ agree for the two systems.

If C is a singular matrix or is not square then the system of equations can be put in the general form (11.14) where $\kappa \neq 0$ for some equations and $\kappa = 0$ for others. The number of equations is not necessarily equal to the number of components of $\mathbf{\Psi}$ and there is no simple canonical form. Fortunately, in the equations of mathematical physics it is usually true that C is nonsingular.

Our analysis of invariant equations follows Liubarskii [1]. There is another approach to this theory, due to Gel'fand and Shapiro, which is based on Lie algebras. The Lie-algebraic method is much more complicated than that given above but it extends rather easily to the case where the matrices L_1, L_2, L_3 in (11.10) act on infinite-dimensional spaces (Gel'fand et al. [1], Naimark [2]).

For equations invariant under the full orthogonal group $O(3)$ these results have to be slightly modified. The components of $\mathbf{\Psi}$ are labeled $\Psi^m_{u\pm,n}$ corresponding to the reps $\mathbf{D}^{(u)}_{\pm}$, u an integer. The differential operators $\partial_{\pm 1}, \partial_0$ form a canonical basis for $\mathbf{D}^{(1)}_{-}$. It follows from the identities

$$\mathbf{D}^{(1)}_{-} \otimes \mathbf{D}^{(u)}_{\pm} \cong \mathbf{D}^{(u+1)}_{\mp} \oplus \mathbf{D}^{(u)}_{\mp} \oplus \mathbf{D}^{(u-1)}_{\mp} \tag{11.15}$$

that the invariant equations take the form (11.14) except that the components of $\mathbf{\Psi}$ on the left- and right-hand sides of these equations have opposite parity.

We consider some examples. The simplest $\mathcal{E}^+(3)$-invariant equations are those in which the components of $\mathbf{\Psi}$ transform according to the single irred rep $\mathbf{D}^{(u)}$. Denoting the $2u + 1$ components of $\mathbf{\Psi}$ by $\Psi_u{}^m$ we obtain the system of equations

$$a \sum_{j+m'=m} C(1, j; u, m' \,|\, u, m)\, \partial_j \Psi_u{}^{m'} = \kappa \Psi_u{}^m, \qquad -u \leq m \leq u. \tag{11.16}$$

There is a single arbitrary constant a. This system is not $E(3)$-invariant since the parity of the left-hand side is opposite that on the right. For $E(3)$-invariant equations the action of $O(3)$ on the indices of $\mathbf{\Psi}$ must be reducible.

Consider the manifestly $E(3)$-invariant equation

$$\left(\frac{\partial^2}{\partial x^2} + \frac{\partial^2}{\partial y^2} + \frac{\partial^2}{\partial z^2}\right)V(\mathbf{x}) = \kappa V(\mathbf{x}), \tag{11.17}$$

where $V(\mathbf{x})$ is a scalar $\mathbf{D}_+^{(0)}$. We shall write (11.17) as a system of first-order equations by introducing three new components $V_j(\mathbf{x}) = \partial_j V(\mathbf{x}), j = \pm 1, 0$. Clearly the $V_j(\mathbf{x})$ form a canonical basis for $\mathbf{D}_-^{(1)}$. The system (11.17) is equivalent to

$$-\partial_1 V_{-1} + \partial_0 V_0 - \partial_{-1} V_1 = \kappa V, \qquad \partial_j V = V_j, \qquad j = \pm 1, 0. \tag{11.18}$$

Without loss of generality we can assume $\kappa = 1$. The indices of the column vector (V, V_1, V_0, V_{-1}) transform according to $\mathbf{D}_+^{(0)} \oplus \mathbf{D}_-^{(1)}$. By our theory the most general $E(3)$-invariant system with these transformation properties is

$$a \sum_{j=-1}^{1} C(1, j; 1, -j \,|\, 0, 0)\, \partial_j V_{-j} = V, \qquad bC(1, l; 0, 0 \,|\, 1, l)\, \partial_l V = V_l, \qquad l = 0, \pm 1. \tag{11.19}$$

It follows from the table (7.28) that (11.19) is identical with (11.18) provided $a = -\sqrt{3}$, $b = 1$.

Another important example is given by two of Maxwell's equations for an electromagnetic field in a vacuum:

$$\nabla \times \mathbf{E} + \frac{1}{c}\frac{\partial \mathbf{H}}{\partial t} = \mathbf{0}, \qquad \nabla \times \mathbf{H} - \frac{1}{c}\frac{\partial \mathbf{E}}{\partial t} = \mathbf{0}. \tag{11.20}$$

Here $\mathbf{E}(\mathbf{x}, t) = (E_x, E_y, E_z)$ is a vector field transforming according to the rep $\mathbf{D}_-^{(1)}$ of $O(3)$ and $\mathbf{H}(\mathbf{x}, t)$ is a vector field transforming according to $\mathbf{D}_+^{(1)}$. We are using Gaussian units. If we consider solutions of frequency ω, $\mathbf{E}(\mathbf{x}, t) = \mathbf{E}(x)e^{i\omega t}$, $\mathbf{H}(\mathbf{x}, t) = \mathbf{H}(\mathbf{x})e^{i\omega t}$, then the equations become

$$(ic/\omega)\, \nabla \times \mathbf{E} = \mathbf{H}, \qquad -(ic/\omega)\, \nabla \times \mathbf{H} = \mathbf{E}. \tag{11.21}$$

Expressed in terms of canonical basis vectors $\partial_{\pm 1}, \partial_0, E_{\pm 1} = 2^{-1/2}(\pm E_x - iE_y)$, $E_0 = E_z$, $H_{\pm 1} = 2^{-1/2}(\pm H_x - iH_y)$, and $H_0 = H_z$, (11.21) reads

$$\begin{aligned} (c/\omega)(\partial_0 E_1 - \partial_1 E_0) &= H_1, & -(c/\omega)(\partial_0 H_1 - \partial_1 H_0) &= E_1, \\ (c/\omega)(\partial_{-1} E_1 - \partial_1 E_{-1}) &= H_0, & -(c/\omega)(\partial_{-1} H_1 - \partial_1 H_{-1}) &= E_0, \\ (c/\omega)(\partial_{-1} E_0 - \partial_0 E_{-1}) &= H_{-1}, & -(c/\omega)(\partial_{-1} H_0 - \partial_0 H_{-1}) &= E_{-1}. \end{aligned} \tag{11.22}$$

By our theory the most general $O(3)$-invariant system of equations with C nonsingular and indices transforming according to $\mathbf{D}_-^{(1)} \oplus \mathbf{D}_+^{(1)}$ is

$$\begin{aligned} a \sum_{j=-1}^{1} C(1, j; 1, m - j \,|\, 1, m)\, \partial_j E_{m-j} &= H_m, \\ b \sum_{j=-1}^{1} C(1, j; 1, m - j \,|\, 1, m)\, \partial_j H_{m-j} &= E_m, \qquad m = 1, 0, -1. \end{aligned} \tag{11.23}$$

It follows from (7.28) that (11.22) is the special case of (11.23) such that $a = -\sqrt{2}\, c/\omega$, $b = \sqrt{2}\, c/\omega$. The other two Maxwell equations $\nabla \cdot \mathbf{E} = 0$, $\nabla \cdot \mathbf{H} = 0$, correspond to the case where C is singular.

Problems

7.1 Compute the Clebsch–Gordan coefficients for all tensor products of irred reps of C_{3v}.

7.2 Determine how the energy levels of an $SO(3)$-symmetric quantum mechanical system split under the influence of a perturbation with D_6 symmetry.

7.3 Compute the level splitting of an $O(3)$-symmetric system under a perturbation with D_{3d} symmetry.

7.4 Prove identity (7.16).

7.5 Determine the double-valued irred reps of the point groups C_3 and D_3.

7.6 Compute the double-valued irred reps of D_6.

7.7 Compute the splitting of levels transforming according to double-valued irred reps of $SO(3)$ under a perturbation with D_3 symmetry.

7.8 Show that the prescription $R(\tau, \mathbf{u}) = \exp(\tau \mathbf{u} \cdot \mathbf{L}) \in SO(3)$ defines a system of coordinates on $SO(3)$ and determine the geometrical significance of these coordinates. Here $\mathbf{u}$ is a unit vector and $\mathbf{u} \cdot \mathbf{L} = u_1 L_1 + u_2 L_2 + u_3 L_3$. Compute the invariant measure in $(\tau, \mathbf{u})$ coordinates and verify explicitly that the simple characters of $SO(3)$ form an orthogonal set.

7.9 Consider a spherical tensor of rank one which transforms as a polar vector under $O(3)$. Determine the selection rules for matrix elements of the tensor between states transforming as irred reps of D_{4h}.

7.10 Repeat the previous problem for a tensor transforming as an axial vector under $O(3)$.

Chapter 8

The Lorentz Group and Its Representations

8.1 The Homogeneous Lorentz Group

The homogeneous Lorentz group in four-space $L(4)$ is the set of all 4×4 real matrices Λ such that $\Lambda^t G\Lambda = G$, where

$$G = \begin{pmatrix} 1 & & & 0 \\ & 1 & & \\ & & 1 & \\ 0 & & & -1 \end{pmatrix}. \tag{1.1}$$

It is straightforward to verify that $L(4)$ satisfies the group axioms. In particular, if $\Lambda \in L(4)$ then $\Lambda^{-1} = G\Lambda^t G \in L(4)$. Also, E and G belong to $L(4)$. If $\Lambda \in L(4)$ then so is $-\Lambda$ and $\Lambda^t = G\Lambda^{-1}G$.

If $x = (x_1, \ldots, x_4)$ and $y = (y_1, \ldots, y_4)$ are column four-vectors such that $y = \Lambda x$, $\Lambda \in L(4)$, then

$$y_1^2 + y_2^2 + y_3^2 - y_4^2 = y^t G y = (\Lambda x)^t G(\Lambda x) = x^t(\Lambda^t G\Lambda)x = x^t G x.$$

Thus the form $x^t Gx$ is invariant under the action of Λ. Conversely, if Λ is a 4×4 real matrix such that $(\Lambda x)^t G(\Lambda x) = x^t Gx$ for all real four-vectors x, then $\Lambda \in L(4)$. By the methods of Section 5.4 it is easy to show that $L(4)$ is a linear Lie group with Lie algebra

$$so(3, 1) = \{\mathcal{A} : \mathcal{A}^t = -G\mathcal{A}G\}, \tag{1.2}$$

[see (10.4), Section 9.10]. Note that $G^2 = E$. Any element of $so(3, 1)$ can be written in the form

$$\mathcal{A} = \begin{pmatrix} 0 & -\alpha_3 & \alpha_2 & \beta_1 \\ \alpha_3 & 0 & -\alpha_1 & \beta_2 \\ -\alpha_2 & \alpha_1 & 0 & \beta_3 \\ \beta_1 & \beta_2 & \beta_3 & 0 \end{pmatrix}, \tag{1.3}$$

where the real parameters α_j, β_j are arbitrary. Thus $so(3, 1)$ is six-dimensional and $L(4)$ is a six-parameter Lie group. The exponential mapping $\mathcal{A} \longrightarrow \exp \mathcal{A}$ maps $so(3, 1)$ homeomorphically onto a neighborhood of the identity in $L(4)$.

As a basis for $so(3, 1)$ we choose the matrices $\mathcal{L}_j, j = 1, 2, 3$, defined by setting $\alpha_j = 1$ and all other parameters zero, and the matrices $\mathcal{B}_j, j = 1, 2, 3$, defined by setting $\beta_j = 1$ and all other parameters zero. The commutation relations are

$$\begin{aligned} &[\mathcal{L}_i, \mathcal{L}_j] = \sum_k \epsilon_{ijk}\mathcal{L}_k, \qquad [\mathcal{B}_i, \mathcal{L}_j] = \sum_k \epsilon_{ijk}\mathcal{B}_k \\ &[\mathcal{B}_i, \mathcal{B}_j] = -\sum_k \epsilon_{ijk}\mathcal{L}_k, \qquad 1 \leq i, j \leq 3, \end{aligned} \tag{1.4}$$

where ϵ_{ijk} is the completely skew-symmetric tensor such that $\epsilon_{123} = +1$.

Note that $\mathcal{L}_1, \mathcal{L}_2, \mathcal{L}_3$ form a basis for a subalgebra of $so(3, 1)$ isomorphic to $so(3)$. Furthermore, the matrices

$$\begin{pmatrix} & & & 0 \\ & R & & 0 \\ & & & 0 \\ 0 & 0 & 0 & 1 \end{pmatrix}, \qquad R \in O(3), \tag{1.5}$$

form a Lie subgroup of $L(4)$ isomorphic to $O(3)$. For convenience we identify this subgroup with $O(3)$. The corresponding subalgebra is spanned by the matrices $\mathcal{L}_j$.

The one-parameter subgroups $\exp \varphi\mathcal{L}_j$ all belong to $SO(3)$ [see (1.4), Chapter 7]. On the other hand, a simple computation yields

$$\exp b\mathcal{B}_3 = \begin{pmatrix} 1 & 0 & 0 & 0 \\ 0 & 1 & 0 & 0 \\ 0 & 0 & \cosh b & \sinh b \\ 0 & 0 & \sinh b & \cosh b \end{pmatrix} \in L(4), \tag{1.6}$$

with similar results for $\mathcal{B}_1$ and $\mathcal{B}_2$. Since the matrix elements of (1.6) are not bounded it follows that $L(4)$ is not a compact group.

We will find it convenient to complexify the Lie algebra. A useful basis for the complexified algebra is

$$(1.7)\qquad \begin{aligned} \mathcal{L}^{\pm} &= \pm\mathcal{L}_2 + i\mathcal{L}_1, & \mathcal{L}^3 &= -i\mathcal{L}_3,\\ \mathcal{B}^{\pm} &= \pm\mathcal{B}_2 + i\mathcal{B}_1, & \mathcal{B}^3 &= -i\mathcal{B}_3. \end{aligned}$$

The commutation relations are

(1.8)

$$\begin{aligned} &[\mathcal{L}^+, \mathcal{L}^-] = 2\mathcal{L}^3, \qquad [\mathcal{L}^3, \mathcal{L}^{\pm}] = \pm\mathcal{L}^{\pm}, \qquad [\mathcal{L}^3, \mathcal{B}^{\pm}] = \pm\mathcal{B}^{\pm},\\ &[\mathcal{L}^+, \mathcal{B}^3] = -\mathcal{B}^+, \qquad [\mathcal{L}^-, \mathcal{B}^3] = \mathcal{B}^-, \qquad [\mathcal{L}^+, \mathcal{B}^-] = [\mathcal{B}^+, \mathcal{L}^-] = 2\mathcal{B}^3,\\ &[\mathcal{L}^+, \mathcal{B}^+] = [\mathcal{L}^-, \mathcal{B}^-] = [\mathcal{L}^3, \mathcal{B}^3] = Z,\\ &[\mathcal{B}^3, \mathcal{B}^{\pm}] = \mp\mathcal{L}^{\pm}, \qquad [\mathcal{B}^+, \mathcal{B}^-] = -2\mathcal{L}^3. \end{aligned}$$

Note that $\mathcal{L}^{\pm}$, $\mathcal{L}^3$ form a basis for the subalgebra $sl(2)$ of the complexified Lie algebra.

A third useful basis is obtained by choosing

$$(1.9)\qquad \begin{aligned} \mathcal{C}^{\pm} &= \tfrac{1}{2}(\mathcal{L}^{\pm} + i\mathcal{B}^{\pm}), & \mathcal{D}^{\pm} &= \tfrac{1}{2}(\mathcal{L}^{\pm} - i\mathcal{B}^{\pm}),\\ \mathcal{C}^3 &= \tfrac{1}{2}(\mathcal{L}^3 + i\mathcal{B}^3), & \mathcal{D}^3 &= \tfrac{1}{2}(\mathcal{L}^3 - i\mathcal{B}^3). \end{aligned}$$

Then the commutation relations become

$$(1.10)\qquad \begin{aligned} &[\mathcal{C}^3, \mathcal{C}^{\pm}] = \pm\mathcal{C}^{\pm}, \qquad [\mathcal{C}^+, \mathcal{C}^-] = 2\mathcal{C}^3,\\ &[\mathcal{D}^3, \mathcal{D}^{\pm}] = \pm\mathcal{D}^{\pm}, \qquad [\mathcal{D}^+, \mathcal{D}^-] = 2\mathcal{D}^3, \qquad [\mathcal{C}, \mathcal{D}] = Z, \end{aligned}$$

i.e., any $\mathcal{C}$ matrix commutes with any $\mathcal{D}$ matrix. It follows from (1.10) that $so(3, 1)^c \cong sl(2) \oplus sl(2)$. This result holds only for the complexified Lie algebra. It is *not* true that $so(3, 1)$ is the direct sum of two nontrivial real Lie algebras.

Let us return to an examination of the group $L(4)$. If $\Lambda \in L(4)$ then $\Lambda^t G \Lambda = G$. Taking the determinant of this expression we find $(\det \Lambda)^2 = 1$, or $\det \Lambda = \pm 1$. Both signs are possible since $E, G \in L(4)$, with $\det E = -\det G = 1$.

In terms of components, $\Lambda = (\Lambda_{ik}) \in L(4)$ provided

$$(1.11)\qquad \sum_{h=1}^{4} \Lambda_{hj} G_{hh} \Lambda_{hl} = G_{jl}, \qquad 1 \leq j, l \leq 4.$$

For $j = l = 4$ this reads

$$(1.12)\qquad \sum_{h=1}^{3} \Lambda_{h4}^2 - \Lambda_{44}^2 = -1.$$

(Also $\sum \Lambda_{4h}^2 - \Lambda_{44}^2 = -1$ since $\Lambda^t \in L(4)$. Thus $|\Lambda_{44}| \geq 1$, so $\Lambda_{44} \geq 1$ or $\Lambda_{44} \leq -1$. If $\Lambda_{44} \geq 1$, then Λ is **forward-timelike,** otherwise Λ is **backward-timelike.** Since E is forward-timelike and G is backward-timelike it is clear that both cases occur. The forward-timelike matrices form a subgroup of $L(4)$. Indeed, if Λ and Λ' are forward-timelike then $(\Lambda\Lambda')_{44} = \sum_{j=1}^{3} \Lambda_{4j}\Lambda'_{j4} + \Lambda_{44}\Lambda'_{44} > 0$ since $|\sum \Lambda_{4j}\Lambda'_{j4}| \leq [\sum \Lambda_{4j}^2 \sum \Lambda'^2_{j4}]^{1/2} \leq [(\Lambda_{44}^2 - 1)(\Lambda'^2_{44} - 1)]^{1/2} < \Lambda_{44}\Lambda'_{44}$. Similarly it is easy to check that the inverse $\Lambda^{-1} = G\Lambda^{t}G$ of a forward-timelike transformation is forward-timelike.

Using these results we can separate $L(4)$ into four components:

(1.13)

$$L^{\uparrow +}: \Lambda_{44} \geq 1, \quad \det \Lambda = +1 \qquad L^{\uparrow -}: \Lambda_{44} \geq 1, \quad \det \Lambda = -1,$$

$$L^{\downarrow +}: \Lambda_{44} \leq -1, \quad \det \Lambda = +1 \qquad L^{\downarrow -}: \Lambda_{44} \leq -1, \quad \det \Lambda = -1.$$

Every element of $L(4)$ lies in a unique component. It is easy to show that the components are disconnected in the sense that no analytic curve in $L(4)$ can connect two distinct components. The component $L^{\uparrow +}$ is itself a group, the **proper Lorentz group.** It is clear that $L^{\uparrow +}$ contains the connected component of the identity in $L(4)$.

Lemma 8.1.

(a) $L^{\uparrow -} = SL^{\uparrow +} = L^{\uparrow +}S$, where $S = -G$.

(b) $L^{\downarrow +} = (-E)L^{\uparrow +}$.

(c) $L^{\downarrow -} = GL^{\uparrow +} = L^{\uparrow +}G$.

Proof. (a) Clearly $S = -G \in L^{\uparrow -}$. If $\Lambda \in L^{\uparrow +}$ then $\det(S\Lambda) = \det(\Lambda S) = \det S = -1$ and $(S\Lambda)_{44} = (\Lambda S)_{44} = \Lambda_{44} \geq 1$, so $S\Lambda$ and ΛS belong to $L^{\uparrow -}$. Thus $L^{\uparrow -} \supseteq SL^{\uparrow +}, L^{\uparrow -} \supseteq L^{\uparrow +}S$. Conversely if $\Lambda \in L^{\uparrow -}$ then $S\Lambda$ and ΛS belong to $L^{\uparrow +}$. Setting $S\Lambda = \Lambda_1$, $\Lambda S = \Lambda_2$ and using the relation $S^2 = E$, we obtain $\Lambda = S\Lambda_1 = \Lambda_2 S$. Therefore, $L^{\uparrow -} = SL^{\uparrow +} = L^{\uparrow +}S$. Parts (b) and (c) are proved in the same manner. Q.E.D.

The matrices S, $-E$ and G are of special importance in the theory of $L(4)$. Here S is called **space inversion,** G is **time inversion,** and $-E = SG = GS$ is **total inversion.** We will discuss the physical significance of these names in the next section.

It follows from the lemma that a parametrization of the whole group can be obtained directly from a parametrization of the proper Lorentz group $L^{\uparrow +}$. We can choose local coordinates for $L^{\uparrow +}$ by merely selecting six independent matrix elements. However, the following construction yields a more useful coordinate system.

Lemma 8.2. Let $\Lambda \in L^{\uparrow +}$. Then $\Lambda \in SO(3)$ if and only if $\Lambda_{44} = +1$.

Proof. From (1.12), $\sum_{h=1}^{3} \Lambda_{h4}^2 = \sum_{h=1}^{3} \Lambda_{4h}^2 = \Lambda_{44}^2 - 1$. Since $\Lambda \in L^{\uparrow +}$ we have $\Lambda_{44} \geq 1$. Thus $\Lambda_{h4} = \Lambda_{4h} = 0$, $1 \leq h \leq 3$, if and only if $\Lambda_{44} = 1$. By (1.11), $\Lambda_{44} = 1$ if and only if Λ takes the form (1.5). Q.E.D.

Lemma 8.3. Let $\Lambda, \Lambda' \in L^{\uparrow +}$ and suppose $\Lambda e = \Lambda' e$, where

$$e = \begin{pmatrix} 0 \\ 0 \\ 0 \\ 1 \end{pmatrix}.$$

Then there exists a unique $R \in SO(3)$ such that $\Lambda = \Lambda' R$. Conversely, if $\Lambda' \in L^{\uparrow +}$ and $R \in SO(3)$ such that $\Lambda = \Lambda' R$ then $\Lambda e = \Lambda' e$. We are considering $SO(3)$ as the subgroup of matrices (1.5).

Proof. If $\Lambda e = \Lambda' e$ for $\Lambda, \Lambda' \in L^{\uparrow +}$ then $R = (\Lambda')^{-1}\Lambda \in L^{\uparrow +}$ and $Re = e$. Thus $R_{h4} = 0$ for $1 \leq h \leq 3$ and $R_{44} = 1$. By the preceding lemma, $R \in SO(3)$.

Conversely, if $\Lambda' \in L^{\uparrow +}$ and $R \in SO(3)$ then $Re = e$ and $\Lambda' R e = \Lambda' e$. Q.E.D.

Theorem 8.1. Every $\Lambda \in L^{\uparrow +}$ can be represented in the form

$$\Lambda = R_1(\exp b\mathfrak{B}_3)R_2, \qquad R_1, R_2 \in SO(3).$$

Proof. It is obvious that all elements of the form $R_1(\exp b\mathfrak{B}_3)R_2$ lie in $L^{\uparrow +}$. We will show that such elements exhaust $L^{\uparrow +}$. Suppose $\Lambda \in L^{\uparrow +}$. Then

$$\Lambda e = \Lambda \begin{pmatrix} 0 \\ 0 \\ 0 \\ 1 \end{pmatrix} = \begin{pmatrix} \Lambda_{14} \\ \Lambda_{24} \\ \Lambda_{34} \\ \Lambda_{44} \end{pmatrix}, \qquad \Lambda_{44} \geq 1.$$

If $\Lambda_{44} = 1$ then $\Lambda \in SO(3)$ by Lemma 8.2 and the theorem follows with $b = 0$. If $\Lambda_{44} > 1$ then

$$\Lambda_{14}^2 + \Lambda_{24}^2 + \Lambda_{34}^2 = \Lambda_{44}^2 - 1 = r^2 > 0, \tag{1.14}$$

where we assume $r > 0$. Since $\Lambda_{44}^2 - r^2 = 1$ there exists a unique number $b > 0$ such that $r = \sinh b$, $\Lambda_{44} = \cosh b$. Indeed, $b = \ln[\Lambda_{44} + (\Lambda_{44}^2 - 1)^{1/2}]$. Then (1.6) implies

$$(\exp b\mathfrak{B}_3)e = \begin{pmatrix} 0 \\ 0 \\ r \\ \Lambda_{44} \end{pmatrix}.$$

According to (1.14) there exist spherical coordinates r, θ_1, φ_1, such that

$$\Lambda_{14} = r \sin\theta_1 \cos\varphi_1, \qquad \Lambda_{24} = r \sin\theta_1 \sin\varphi_1, \qquad \Lambda_{34} = r\cos\theta_1.$$

It follows from (1.20), Chapter 7, that the matrix $R_1 = R(\varphi_1 + \pi/2, \theta_1, 0)$ (Euler parameters) satisfies

$$R_1 \begin{pmatrix} 0 \\ 0 \\ r \\ \Lambda_{44} \end{pmatrix} = \begin{pmatrix} \Lambda_{14} \\ \Lambda_{24} \\ \Lambda_{34} \\ \Lambda_{44} \end{pmatrix}.$$

Clearly, $R_1 \exp b\mathfrak{B}_3 \in L^{\uparrow +}$ and $R_1(\exp b\mathfrak{B}_3)e = \Lambda e.$ By Lemma 8.3 there exists a unique $R_2 \in SO(3)$ such that $\Lambda = R_1(\exp b\mathfrak{B}_3)R_2$. Q.E.D.

If Λ is not an element of $SO(3)$ then the preceding factorization is unique. If the Euler parameters of R_2 are $\varphi_2, \theta_2, \psi_2$ we have

$$\Lambda = R_1(\varphi_1 + \tfrac{1}{2}\pi, \theta_1, 0)(\exp b\mathfrak{B}_3)R_2(\varphi_2, \theta_2, \psi_2) \tag{1.15}$$

and the six parameters $\varphi_1, \theta_1, b, \varphi_2, \theta_2, \psi_2$ serve as coordinates for Λ. If $\theta_1, \theta_2 = 0, \pi$ these coordinates are not 1–1. Similarly, if $\Lambda \in SO(3)$ then $b = 0$ and only the product $R_1 R_2$ is prescribed, not the individual factors. However, those points at which the coordinates are not 1–1 form a lower-dimensional manifold on the group and do not affect the invariant measure.

It follows from (1.15) that any $\Lambda \in L^{\uparrow +}$ can be connected to the identity element by an analytic curve lying entirely in $L^{\uparrow +}$. Indeed we can choose the curve $(t\varphi_1, \ldots, t\psi_2)$, $0 \le t \le 1$. Thus $L^{\uparrow +}$ coincides with the connected component containing the identity in $L(4)$. This proves that $L(4)$ consists of four connected components. The Lie algebra yields information only about $L^{\uparrow +}$. To study the other three connected components we make use of Lemma 8.1.

In Section 7.1 we showed that $SU(2)$ was a double covering group of $SO(3)$. There is a similar relationship between $SL(2) = SL(2, \mathfrak{C})$ and $L^{\uparrow +}$. Indeed, $sl(2)$ considered as a six-dimensional *real* Lie algebra is isomorphic to $so(3, 1)$. Thus, the real Lie groups $SL(2)$ and $L^{\uparrow +}$ are locally isomorphic. To show this we recall that $sl(2)$ consists of all 2×2 complex matrices $\mathfrak{A}$ with trace zero:

$$\mathfrak{A} = \begin{pmatrix} z_1 & z_2 \\ z_3 & -z_1 \end{pmatrix}, \qquad z_j \in \mathfrak{C}. \tag{1.16}$$

Writing $z_j = x_j + iy_j$, we see that $sl(2)$ is a six-dimensional Lie algebra over the reals. As a basis for $sl(2)$ we shoose the matrices $\mathcal{J}_1, \mathcal{J}_2, \mathcal{J}_3$ [(1.8), Chapter 7] and

$$\mathfrak{F}_1 = \begin{pmatrix} 0 & -\frac{1}{2} \\ -\frac{1}{2} & 0 \end{pmatrix}, \qquad \mathfrak{F}_2 = \begin{pmatrix} 0 & -\frac{1}{2}i \\ \frac{1}{2}i & 0 \end{pmatrix}, \qquad \mathfrak{F}_3 = \begin{pmatrix} -\frac{1}{2} & 0 \\ 0 & \frac{1}{2} \end{pmatrix}. \tag{1.17}$$

These matrices satisfy the commutation relations (1.4), with $\mathfrak{L}_j$ replaced by $\mathcal{J}_j$ and $\mathfrak{B}_j$ by $\mathfrak{F}_j$.

To explicitly exhibit the global relationship between $SL(2)$ and $L^{\uparrow +}$ we consider the four-dimensional space S of all 2×2 skew-Hermitian matrices $\mathcal{S} = -\bar{\mathcal{S}}^t$. Each such matrix can be uniquely written as

$$\mathcal{S} = \begin{pmatrix} i(x_4 - x_3) & -x_2 + ix_1 \\ x_2 + ix_1 & i(x_4 + x_3) \end{pmatrix}, \qquad x_j \quad \text{real.} \tag{1.18}$$

[These matrices form the Lie algebra of $U(2)$.] The mapping

$$\mathcal{S} \longrightarrow \mathcal{K} = A\mathcal{S}\bar{A}^t, \qquad A = \begin{pmatrix} a & b \\ c & d \end{pmatrix} \in SL(2), \tag{1.19}$$

is a rep of $SL(2)$ on S. Indeed $\bar{\mathcal{K}}^t = A\bar{\mathcal{S}}^t\bar{A}^t = -A\mathcal{S}\bar{A}^t = -\mathcal{K}$, so $\mathcal{K} \in S$. The homomorphism property is just as obvious. Now $\det \mathcal{K} = \det(A\mathcal{S}\bar{A}^t) = \det \mathcal{S}$, so, writing

$$\mathcal{K} = \begin{pmatrix} i(y_4 - y_3) & -y_2 + iy_1 \\ y_2 + iy_1 & i(y_4 + y_3) \end{pmatrix} \tag{1.20}$$

we obtain

$$y_1^2 + y_2^2 + y_3^2 - y_4^2 = \det \mathcal{K} = \det \mathcal{S} = x_1^2 + x_2^2 + x_3^2 - x_4^2. \tag{1.21}$$

From (1.19), the y_j are linear combinations of the x_k:

$$y_i = \sum_{k=1}^{4} L(A)_{jk} x_k, \qquad 1 \leq j \leq 4. \tag{1.22}$$

From (1.21) and the remarks following (1.1) we conclude that $L(A) \in L(4)$. Furthermore, since (1.19) defines a rep of $SL(2)$ we have the group property $L(AB) = L(A)L(B)$, $A, B \in SL(2)$.

The map $A \longrightarrow L(A)$ is continuous in the parameters of A and $SL(2)$ is connected. Therefore, $L(A)$ must lie in $L^{\uparrow +}$, the connected component of the identity in $L(4)$. We have established the existence of a real analytic homomorphism $A \longrightarrow L(A)$ of $SL(2)$ into $L^{\uparrow +}$. Clearly the kernel of this homomorphism is $\{\pm E_2\}$. Thus, $L(A) = L(-A)$ and exactly two elements of $SL(2)$ map onto each element in the range of the homomorphism.

Suppose $A \in SU(2)$, a real subgroup of $SL(2)$. Then $\bar{A}^t = A^{-1}$ and a comparison of (1.9), Chapter 7, with (1.19) shows that $L(A) = R(A) \in SO(3)$, where $R(A)$ is defined by (1.12), Chapter 7 ($y_4 = x_4$). Thus the homomorphism maps the subgroup $SU(2)$ 2–1 onto the subgroup $SO(3)$ of $L^{\uparrow +}$. We will use this result to show that $A \longrightarrow L(A)$ is a homomorphism of $SL(2)$ **onto** $L^{\uparrow +}$.

Let us compute $L(\exp b\mathcal{F}_3)$ where $\mathcal{F}_3 \in sl(2)$ is given by (1.17). Clearly

$$\exp b\mathcal{F}_3 = \begin{pmatrix} e^{-b/2} & 0 \\ 0 & e^{b/2} \end{pmatrix} \tag{1.23}$$

and

$$\begin{pmatrix} e^{-b/2} & 0 \\ 0 & e^{b/2} \end{pmatrix} \begin{pmatrix} i(x_4 - x_3) & -x_2 + ix_1 \\ x_2 + ix_1 & i(x_4 + x_3) \end{pmatrix} \begin{pmatrix} e^{-b/2} & 0 \\ 0 & e^{b/2} \end{pmatrix}$$
$$= \begin{pmatrix} i(y_4 - y_3) & -y_2 + iy_1 \\ y_2 + iy_1 & i(y_4 + y_3) \end{pmatrix},$$

where $y_1 = x_1, y_2 = x_2, y_3 = x_3 \cosh b + x_4 \sinh b$, and $y_4 = x_3 \sinh b + x_4 \cosh b$, so $L(\exp b\mathfrak{F}_3) = \exp b\mathfrak{B}_3$ [expression (1.6)]. Now suppose $\Lambda \in L^{\uparrow +}$ is given by (1.15). If $A_1 = A(\varphi_1 + \pi/2, \theta_1, 0)$ and $A_2 = A(\varphi_2, \theta_2, \psi_2)$ are elements of $SU(2)$ expressed in Euler coordinates, we have

$$L(A_1(\exp b\mathfrak{F}_3)A_2) = L(A_1)L(\exp b\mathfrak{F}_3)L(A_2) = R_1(\exp b\mathfrak{B}_3)R_2 = \Lambda,$$

so the map $A \longrightarrow L(A)$ covers $L^{\uparrow +}$. We can also use the parameters $\varphi_1, \theta_1, b, \varphi_2, \theta_2, \psi_2$ as coordinates on $SL(2)$, where

(1.24)
$$0 \le \varphi_1, \varphi_2 < 2\pi, \qquad 0 \le \theta_1, \theta_2 \le \pi, \qquad 0 \le b, \qquad -2\pi \le \psi_2 < 2\pi.$$

The parameters of $-A$ are the same as those of A except that ψ_2 is replaced by $\psi_2 \pm 2\pi$. On $L^{\uparrow +}$ the parameters range over the same values except that ψ_2 is restricted to $0 \le \psi_2 < 2\pi$.

Since our group homomorphism is locally 1–1 it induces a Lie algebra isomorphism $\mathfrak{a} \longrightarrow L(\mathfrak{a})$ of $sl(2)$ onto $so(3, 1)$. It is straightforward to check that $L(\mathfrak{J}_j) = \mathfrak{L}_j, L(\mathfrak{F}_j) = \mathfrak{B}_j, 1 \le j \le 3$.

If $\mathbf{T}$ is a rep of the proper Lorentz group by operators $\mathbf{T}(\Lambda)$ then the operators $\mathbf{T}'(A) = \mathbf{T}(L(A))$ define a rep of $SL(2)$ such that $\mathbf{T}'(-A) = \mathbf{T}'(A)$. On the other hand, if $\mathbf{S}$ is a rep of $SL(2)$ such that $\mathbf{S}(A) = \mathbf{S}(-A)$ then the operators $\mathbf{S}'(L(A)) = \mathbf{S}(A)$ define a rep of $L^{\uparrow +}$. Thus, there is a 1–1 correspondence between single-valued reps of $L^{\uparrow +}$ and reps $\mathbf{S}$ of $SL(2)$ such that $\mathbf{S}(-E_2)$ is the identity operator.

Since $SL(2)$ and $L(4)$ are not compact, the results of Chapter 6 do not hold for these groups. In particular a finite-dimensional rep of $SL(2)$ is not necessarily equivalent to a unitary rep. For example the matrices $L(A)$, $A \in SL(2)$, define a four-dimensional irred rep of $SL(2)$. Since the matrix elements of $L(A)$ are unbounded this rep cannot be equivalent to a unitary matrix rep.

Furthermore, we shall see that $SL(2)$ has infinite-dimensional unitary irred reps, which is not possible for compact groups. An arbitrary rep of $SL(2)$ cannot necessarily be decomposed into a direct sum of irred reps.

Suppose $\mathbf{S}$ is a finite-dimensional irred rep of $SL(2)$. Since $-E_2$ commutes with all elements of $SL(2)$, the operator $\mathbf{S}(-E_2)$ commutes with all $\mathbf{S}(A)$. By the Schur lemmas, $\mathbf{S}(-E_2) = \alpha\mathbf{E}$, where $\mathbf{E}$ is the identity operator. Furthermore, $[\mathbf{S}(-E_2)]^2 = \mathbf{S}(E_2) = \mathbf{E}$, so $\alpha^2 = 1$ and $\alpha = \pm 1$. Thus,

$\mathbf{S}(-E_2) = \pm\mathbf{E}$. If $\alpha = +1$ then $\mathbf{S}$ defines a single-valued irred rep of $L^{\uparrow +}$. However, if $\alpha = -1$ then $\mathbf{S}(-A) = -\mathbf{S}(A)$ and $\mathbf{S}$ determines a double-valued rep of $L^{\uparrow +}$. These are the only possibilities.

In quantum mechanics the double-valued reps appear naturally for the same reasons that double-valued reps of $SO(3)$ appear. Thus $SL(2)$ is the group to study for quantum mechanical Lorentz invariance.

8.2 The Physical Significance of Lorentz Invariance

We briefly discuss a realization of the Lorentz group which appears in Einstein's special theory of relativity. In this theory space–time is viewed as a four-dimensional real manifold called **Minkowski space.** The elements or points of this space are **events.** In Minkowski space we distinguish a family of coordinate systems called **inertial frames** or **observers.** With respect to an inertial frame the coordinates of an event are denoted $x = (x_1, x_2, x_3, x_4) = (\mathbf{x}, x_4)$, where the cartesian coordinates $\mathbf{x} = (x_1, x_2, x_3)$ are the spatial coordinates of the event and $x_4 = ct$, where t is the time coordinate of the event. Here c is the velocity of light in a vacuum. The points of Minkowski space are swept out as the x_j range over all real numbers.

Let $\mathcal{I}$ be an inertial frame and let p, q be events with coordinates x, y in $\mathcal{I}$. Here x and y are column 4-vectors. We define the squared space–time distance between these two events by

$$(2.1)\qquad \|x - y\|^2 = \sum_{j=1}^{3} (x_j - y_j)^2 - (x_4 - y_4)^2 = (x - y)^t G(x - y),$$

where G is given by (1.1). Now suppose $\mathcal{I}'$ is another coordinate system with respect to which the events p, q have coordinates x', y', respectively. We postulate that $\mathcal{I}'$ is an inertial frame (with respect to $\mathcal{I}$) provided

$$(2.2)\qquad \|x - y\|^2 = \|x' - y'\|^2$$

for all pairs of events p, q, i.e., provided the space–time distance between events is preserved. By a computation analogous to that carried out in Section 2.2 one can show that if $\mathcal{I}'$ is inertial then the relationship between the coordinates of the event p in $\mathcal{I}$ and $\mathcal{I}'$ is

$$(2.3)\qquad x_j' = \sum_{k=1}^{4} \Lambda_{jk} x_k + a_j, \qquad j = 1, \ldots, 4,$$

where $\Lambda \in L(4)$ and $a = (a_1, \ldots, a_4)$ is a real four-tuple. Conversely, if $\mathcal{I}$ is inertial and $\mathcal{I}'$ is a coordinate system related to $\mathcal{I}$ by (2.3) then $\mathcal{I}'$ is inertial. (For a proof that the coordinate transformation must be linear see the work of Rätz [1].)

It is clear from definition (2.2) that the inertial frames form an equivalence class. That is, (a) $\mathcal{I}$ is inertial with respect to $\mathcal{I}$, (b) if $\mathcal{I}'$ is inertial with respect to $\mathscr{I}$ then $\mathscr{I}$ is inertial with respect to $\mathscr{I}'$, and (c) if $\mathscr{I}'$ is inertial with respect to $\mathscr{I}$ and $\mathscr{I}''$ is inertial with respect to $\mathscr{I}'$ then $\mathscr{I}''$ is inertial with respect to $\mathscr{I}$. Once one inertial frame is chosen it is easy to obtain the rest.

Let p be an event with coordinates x, x', x'' in the inertial frames $\mathscr{I}, \mathscr{I}'$ $\mathscr{I}''$. Then the relations between these coordinates are given by

(1) $x_s' = \sum_k \Lambda_{sk} x_k + a_s.$

(2) $x_l'' = \sum_s \Lambda'_{ls} x_s' + a_l'.$

(3) $x_l'' = \sum_k \Lambda''_{lk} x_k + a_k'', \quad \Lambda, \Lambda', \Lambda'' \in L(4).$

From (1) and (2) we have

$$x_l'' = \sum_k (\sum_s \Lambda'_{ls}\Lambda_{sk}) x_k + \sum_s \Lambda'_{ls} a_s + a_l'.$$

A comparison of this expression with (3) yields

$$\Lambda'' = \Lambda'\Lambda, \qquad a'' = \Lambda' a + a'. \tag{2.4}$$

It follows that the set of all pairs $\{a, \Lambda\}$ forms a group with product

$$\{a', \Lambda'\}\{a, \Lambda\} = \{\Lambda' a + a', \Lambda'\Lambda\}, \qquad \Lambda, \Lambda' \in L(4), \quad a, a' \in R_4. \tag{2.5}$$

This is a ten-parameter Lie group called the **Poincaré** or **inhomogeneous Lorentz** group P. There is a 1–1 relationship between inertial frames and elements of P.

In the theory of special relativity it is postulated that the laws of physics must take the same form in any inertial frame. Since the elements of P determine the coordinate changes from one inertial frame to another, this means the dynamical equations of physics must be invariant under the Poincaré group. For differential equations we mean this invariance in the same sense as Euclidean invariance in Section 7.11. From (2.5) the set of all elements $\{\mathbf{b}, R\}$, $R \in O(3)$, $\mathbf{b} = (a_1, a_2, a_3, 0)$, forms a subgroup of P isomorphic to $E(3)$. Thus, Poincaré-invariant equations are automatically Euclidean-invariant. We shall determine the possible Poincaré-invariant equations in Section 8.5.

Let p be an event and consider the set i_p of all inertial frames in which the coordinates of p are (0, 0, 0, 0), i.e., the inertial frames whose origin of coordinates is p. Let us fix a system $\mathscr{I} \in i_p$. Then if $\mathscr{I}' \in i_p$ there is a $\{a, \Lambda\} \in P$ such that the coordinates x in $\mathscr{I}$ and x' in $\mathscr{I}'$ are related by $x_s' = \sum \Lambda_{sk} x_k + a_s$. This equation must hold for $x = x' = (0, 0, 0, 0)$, so a is the zero vector. Similarly if $\mathscr{I}'$ is a coordinate system related to $\mathscr{I}$ by $x_s' = \sum \Lambda_{sk} x_k$, $\Lambda \in L(4)$, then $\mathscr{I}' \in i_p$. Thus there is a 1–1 correspondence between elements of i_p and elements of $L(4)$ a subgroup of P. In the following we restrict ourselves to inertial frames in i_p.

We now investigate the physical significance of Lorentz transformations.

Let x be a column four-vector with spatial components $\mathbf{x} = (x_1, x_2, x_3)$ and time component $x_4 = ct$. Under space inversion $Sx = (-\mathbf{x}, x_4)$, under time inversion $Gx = (\mathbf{x}, -x_4)$, and under total inversion $SGx = -x$, so the meaning of these coordinate transformations is clear. According to Lemma 8.1 we need only determine the physical significance of transformations in $L^{\uparrow +}$. To do this we prove a variant of Theorem 8.1.

Theorem 8.2. Every $\Lambda \in L^{\uparrow +}$ can be represented uniquely in the form $\Lambda = V(\mathbf{b})R$, where $R \in SO(3)$ and

$$V(\mathbf{b}) = \exp(b_1\mathcal{B}_1 + b_2\mathcal{B}_2 + b_3\mathcal{B}_3).$$

The group elements $V(\mathbf{b})$ are called **velocity transformations.**

Proof. By Theorem 8.1.

$$\Lambda = R_1(\varphi_1 + \tfrac{1}{2}\pi, \theta_1, 0)(\exp b\mathcal{B}_3)R_2(\varphi_2, \theta_2, \psi_2),$$

where

(2.6) $\quad \Lambda_{14} = r \sin \theta_1 \cos \varphi_1, \qquad \Lambda_{24} = r \sin \theta_1 \sin \varphi_1, \qquad \Lambda_{34} = r \cos \theta_1$

and $r = \sinh b$, $b \geq 0$. Suppose $r > 0$, in which case this factorization is unique. Now $\Lambda = R_1(\exp b\mathcal{B}_3)R_1^{-1}(R_1R_2)$. The matrices $B(t) = R_1 \times (\exp tb\mathcal{B}_3)R_1^{-1}$ form a one-parameter subgroup of $L^{\uparrow +}$ as t runs over all real numbers, and the tangent matrix at the identity is $bR_1\mathcal{B}_3R_1^{-1}$. A direct computation gives

$$R_1\mathcal{B}_3R_1^{-1} = (\cos \varphi_1 \sin \theta_1)\mathcal{B}_1 + (\sin \varphi_1 \sin \theta_1)\mathcal{B}_2 + (\cos \theta_1)\mathcal{B}_3.$$

Since the tangent matrix completely determines the one-parameter subgroup we have

(2.7) $\quad R_1(\exp tb\mathcal{B}_3)R_1^{-1} = \exp(t[b_1\mathcal{B}_1 + b_2\mathcal{B}_2 + b_3\mathcal{B}_3]) = V(t\mathbf{b}),$

where

(2.8) $\quad \mathbf{b} = (b \cos \varphi_1 \sin \theta_1, b \sin \varphi_1 \sin \theta_1, b \cos \theta_1), \qquad r = \sinh b.$

Setting $t = 1$, we obtain $\Lambda = V(\mathbf{b})R$, where $R = R_1R_2 \in SO(3)$. By construction this factorization is unique if $r > 0$, i.e., if $\Lambda \notin SO(3)$. However, if $r = 0$ then $b = 0$ and $\Lambda \in SO(3)$. In this case $V(\mathbf{b}) = E$ and $\Lambda = R$, so again the factorization is unique. Q.E.D.

Since the $\mathcal{B}_j$ are symmetric matrices it follows that $V(\mathbf{b})$ is a positive-definite symmetric matrix. Thus the product $\Lambda = V(\mathbf{b})R$ is just the well-known **polar decomposition** of a real nonsingular matrix into the product of a positive-definite symmetric matrix and an orthogonal matrix.

Let $\mathcal{I}'$ be the inertial frame related to $\mathcal{I}$ by the velocity transformation $x' = V(\mathbf{b})x$. In frame $\mathcal{I}$ the origin of spatial coordinates at time t has coordinates $x = (0, 0, 0, ct)$. In frame $\mathscr{S}'$ this event has coordinates

$$(2.9)\qquad \begin{aligned} x' &= V(\mathbf{b})x \\ &= ct(\sinh b \sin \theta_1 \cos \varphi_1, \sinh b \sin \theta_1 \sin \varphi_1, \sinh b \cos \theta_1, \cosh b) \\ &= (\mathbf{x}', ct'). \end{aligned}$$

Thus, in $\mathscr{S}'$ the coordinates of the event are related by the equations

$$(2.10)\qquad \mathbf{x}' = \mathbf{v}t', \qquad \mathbf{v} = \frac{c}{b}(\tanh b)\mathbf{b} = \frac{rc}{(1 + r^2)^{1/2}}\hat{\mathbf{b}},$$

where $\hat{\mathbf{b}}$ is a unit three-vector in the direction of $\mathbf{b}$. The spatial origin of coordinates in system $\mathscr{S}$ is moving with uniform velocity $\mathbf{v}$ with respect to the spatial origin of coordinates in $\mathscr{S}'$. Note that

$$\sinh b = \frac{v/c}{(1 - v^2/c^2)^{1/2}}, \qquad \cosh b = \frac{1}{(1 - v^2/c^2)^{1/2}}, \qquad v = \|\mathbf{v}\|.$$

From the definition of $V(\mathbf{b})$ it is easy to show that this velocity transformation leaves invariant any vector $x = (\mathbf{x}, 0)$ such that $\mathbf{x}\cdot\mathbf{v} = 0$, i.e., $\mathbf{x}\cdot\mathbf{b} = 0$. Indeed

$$(b_1\mathscr{B}_1 + b_2\mathscr{B}_2 + b_3\mathscr{B}_3)\begin{pmatrix} \mathbf{x} \\ 0 \end{pmatrix} = \begin{pmatrix} \mathbf{0} \\ \mathbf{b}\cdot\mathbf{x} \end{pmatrix} = \begin{pmatrix} \mathbf{0} \\ 0 \end{pmatrix}.$$

In the special case where the velocity $\mathbf{v}$ is in the direction of the positive z axis then

$$V(\mathbf{b}) = \begin{pmatrix} 1 & 0 & 0 & 0 \\ 0 & 1 & 0 & 0 \\ 0 & 0 & 1/\gamma & v/\gamma c \\ 0 & 0 & v/\gamma c & 1/\gamma \end{pmatrix}, \qquad \gamma = (1 - v^2/c^2)^{1/2},$$

and the coordinate transformation becomes

$$(2.11)\qquad \begin{aligned} &x' = x, \qquad y' = y, \qquad z' = \frac{z + vt}{(1 - v^2/c^2)^{1/2}}, \\ &t' = \frac{t + zv/c^2}{(1 - v^2/c^2)^{1/2}}, \qquad \mathbf{x} = (x, y, z). \end{aligned}$$

Equations (2.11) are the usual Lorentz transformations discussed in textbooks on special relativity. The physical significance of $R \in SO(3)$ is obvious, so a Lorentz transformation $\Lambda = V(\mathbf{b})R$ can be interpreted as a rotation of spatial coordinates followed by a velocity transformation.

Warning. The velocity transformations do not form a subgroup of $L^{\uparrow +}$ because the product of two velocity transformations is not necessary a velocity transformation.

In the above discussion we have given a passive interpretation of Lorentz transformations: The space remains fixed and the observers (inertial frames) transform under $L(4)$. Alternatively, we could adopt the active interpretation: There is one fixed coordinate system and the Lorentz group transforms the points of Minkowski space. In the active interpretation a velocity transformation maps a state in which a particle is at rest into a state where the particle has velocity $\mathbf{v}$.

8.3 Representations of the Lorentz Group

To find the analytic irred reps of $L^{\uparrow +}$ we compute the analytic irred reps $\mathbf{T}$ of $SL(2)$ considered as a real Lie group and determine which of these reps satisfy $\mathbf{T}(-E_2) = \mathbf{E}$. We have already computed the irred reps $\mathbf{D}^{(u)}$ of $SL(2)$ which are analytic functions of the complex group parameters. If $D^{(u)}(A)$ is a matrix realization of $\mathbf{D}^{(u)}$ then the complex conjugate matrices $\overline{D^{(u)}}(A)$ also define an irred rep of $SL(2)$ which is analytic in the real group parameters but not in the complex group parameters (Prove it!) Since any rep equivalent to a complex analytic rep is complex analytic it follows that $\mathbf{D}^{(u)}$ and $\overline{\mathbf{D}^{(u)}}$ are nonequivalent irred reps.

As a convenient basis for the real six-dimensional Lie algebra $sl(2) \cong so(3, 1)$ we choose the matrices $\mathcal{J}_j, \mathcal{F}_j = i\mathcal{J}_j, 1 \leq j \leq 3$, where the $\mathcal{J}_j$ are defined by (1.8), Chapter 7. These matrices satisfy the commutation relations (1.4) with $\mathcal{L}_j$ replaced by $\mathcal{J}_j$ and $\mathcal{B}_j$ by $\mathcal{F}_j$. Now we forget the origin of our basis as a set of matrices and merely consider the abstract Lie algebra $sl(2)$ spanned by linearly independent basis elements $\mathcal{J}_j, \mathcal{F}_j$ with commutation relations (1.4). From (1.9) we see that $sl(2)^c$, the complexification of our real Lie algebra, has a basis $\mathcal{C}_j, \mathcal{D}_k$ with commutation relations

$$(3.1) \qquad [\mathcal{C}_j, \mathcal{D}_k] = 0, \qquad [\mathcal{C}_j, \mathcal{C}_k] = \sum_l \epsilon_{jkl}\mathcal{C}_l, \qquad [\mathcal{D}_j, \mathcal{D}_k] = \sum_l \epsilon_{jkl}\mathcal{D}_l,$$

where $\mathcal{C}_j = (\mathcal{J}_j - i\mathcal{F}_j)/2$, $\mathcal{D}_j = (\mathcal{J}_j + i\mathcal{F}_i)/2$. The Lie algebra of the group $SU(2) \times SU(2) = G$ is another real form of the complex algebra (3.1). Since G is compact we know that its global irred reps are just $\mathbf{D}^{(u,v)} = \mathbf{D}^{(u)} \otimes \mathbf{D}^{(v)}$, $2u, 2v = 0, 1, 2, \ldots$. Therefore, the possible irred finite-dimensional reps of $L(G)$ are just the Lie algebra reps induced by $\mathbf{D}^{(u,v)}$. Since there is a 1–1 correspondence between reps of a complex Lie algebra and reps of any of its real forms we conclude the the irred reps of $sl(2)$ and $sl(2)^c$ are $\mathbf{D}^{(u,v)}$. Indeed if we denote the operators corresponding to such a rep by $C_j = \mathbf{T}(\mathcal{C}_j)$, $D_j = \mathbf{T}(\mathcal{D}_j)$ and set

$$C^{\pm} = \pm C_2 + iC_1, \qquad C^3 = -iC_3, \qquad D^{\pm} = \pm D_2 + iD_1, \qquad D^3 = -iD_3,$$

then there exists a basis $\{f_{mn}^{(u,v)}\}$ for the rep space $\mathcal{V}^{(u,v)}$ corresponding to the

$(2u+1)(2v+1)$-dimensional rep $\mathbf{D}^{(u,v)}$ such that

$$(3.2)\qquad \begin{aligned} C^3 f_{mn}^{(u,v)} &= m f_{mn}^{(u,v)}, & C^{\pm} f_{mn}^{(u,v)} &= [(u \pm m + 1)(u \mp m)]^{1/2} f_{m\pm 1,n}^{(u,v)} \\ D^3 f_{mn}^{(u,v)} &= n f_{mn}^{(u,v)}, & D^{\pm} f_{mn}^{(u,v)} &= [(v \pm n + 1)(v \mp n)]^{1/2} f_{m,n\pm 1}^{(u,v)} \\ -C\cdot C f_{mn}^{(u,v)} &= u(u+1) f_{mn}^{(u,v)}, & -D\cdot D f_{mn}^{(u,v)} &= v(v+1) f_{mn}^{(u,v)}. \end{aligned}$$

[We call a basis satisfying (3.2) **canonical.**] The operators $C\cdot C = C_1C_1 + C_2C_2 + C_3C_3$ and $D\cdot D$ commute with the C_j and D_j, so they must be multiples of the identity operator for any irred rep of $sl(2)$.

Now we will show that the Lie algebra reps $\mathbf{D}^{(u,v)}$ induce global reps of $SL(2)$. To begin we consider the complex analytic rep $\mathbf{D}^{(u)}$ of $SL(2)$ determined in Chapter 7. Clearly, the induced Lie algebra rep has the property $F_j = iJ_j$, $1 \leq j \leq 3$. Thus $C^{\pm} = J^{\pm}$, $C^3 = J^3$, and the D-operators are zero. We conclude that $\mathbf{D}^{(u)}$ is equivalent to the rep $\mathbf{D}^{(u,0)}$. On the other hand the Lie algebra rep induced by $\bar{\mathbf{D}}^{(v)}$ has the property $F_j = -iJ_j$. Hence $D^{\pm} = J^{\pm}$, $D^3 = J^3$, and the C-operators are zero. This shows that $\bar{\mathbf{D}}^{(v)}$ is equivalent to $\mathbf{D}^{(0,v)}$. Similarly, if we compute the Lie algebra rep induced by the group rep $\mathbf{D}^{(u)} \otimes \bar{\mathbf{D}}^{(v)}$ of $SL(2)$ on $\mathcal{V}^{(u)} \otimes \mathcal{V}^{(v)}$ we get exactly the results (3.2), by making the identification $f_{mn}^{(u,v)} = f_m^{(u)} \otimes g_n^{(v)}$, where $\{f_m^{(u)}\}$ and $\{g_n^{(v)}\}$ are canonical bases for $\mathcal{V}^{(u)}$ and $\mathcal{V}^{(v)}$, respectively.

To sum up, we have shown that a complete set of finite-dimensional analytic irred reps of the real Lie group $SL(2)$ is given by $\mathbf{D}^{(u,v)}$, $2u, 2v = 0, 1, 2, \ldots$. The matrix elements of these reps with respect to a suitable (not canonical) basis are

$$(3.3)\qquad \mathbf{T}(A)h_{mn}^{(u,v)} = \sum_{m'=-u}^{u} \sum_{n'=-v}^{v} T_{m'm}^{(u)}(A)\overline{T_{n'n}^{(v)}}(A) h_{m'n'}^{(u,v)}.$$

Note: If the vectors $f_m^{(u)}$ form a canonical basis for $\mathbf{D}^{(u)}$ it is *not* true that the complex conjugate vectors $\bar{f}_m^{(u)}$ form a canonical basis for $\bar{\mathbf{D}}^{(u)}$. To see this, choose a matrix realization of $\mathbf{D}^{(u)}$ so that the $f_m^{(u)}$ are $(2u+1)$-component column vectors. This group rep induces a matrix Lie algebra rep of $sl(2)$. The matrices F_j, J_j satisfy the properties

$$(3.4)\qquad \begin{aligned} & C^3 f_m^{(u)} = m f_m^{(u)}, \qquad C^{\pm} f_m^{(u)} = [(u \pm m + 1)(u \mp m)]^{1/2} f_{m\pm 1}^{(u)}, \\ & D^{\pm} = D^3 = Z, \end{aligned}$$

where the C and D matrices are defined by the expression following (3.1). Now denote the corresponding matrices induced from the complex conjugate matrix rep $\bar{D}^{(u)}$ with stars. Then $J_j{}^* = \bar{J}_j$, $F_j{}^* = \bar{F}_j$, so

$$C_j{}^* = (J_j{}^* - iF_j{}^*)/2 = \bar{D}_j, \qquad D_j{}^* = (J_j{}^* + iF_j{}^*)/2 = \bar{C}_j.$$

Thus

$$C^{*\pm} = C^{*3} = Z, \qquad D^{*\pm} = -\bar{C}^{\mp}, \qquad D^{*3} = -\bar{C}^3.$$

Substituting these results in (3.4), we find

$$C^{*\pm} = C^{*3} = Z, \quad D^{*3}\bar{f}_m^{(u)} = -\overline{C^3 f_m^{(u)}} = -m\bar{f}_m^{(u)},$$
$$D^{*\pm}\bar{f}_m^{(u)} = -\overline{C^{\mp} f_m^{(u)}} = -[(u \mp m + 1)(u \pm m)]^{1/2}\bar{f}_{m\mp 1}^{(u)}.$$

This shows that the vectors $g_m^{(u)} = (-1)^{u-m}\bar{f}_{-m}^{(u)}$ form a canonical basis for $\bar{\mathbf{D}}^{(u)}$.

With respect to a canonical basis the matrix elements of $\bar{\mathbf{D}}^{(u)}$ are

$$\mathbf{T}(A)g_m^{(u)} = \sum_{n=-u}^{u} (-1)^{m-n}\overline{T_{-n,-m}^{(u)}(A)}g_n^{(u)},$$

where the $T_{nm}^{(u)}(A)$ are the matrix elements of $\mathbf{D}^{(u)}$ in a canonical basis. It follows immediately from (3.4) that $\mathbf{T}(-E_2) = (-1)^{2(u+v)}\mathbf{E}$, so $\mathbf{D}^{(u,v)}$ determines a single-valued rep of the Lorentz group if and only if $u + v$ is an integer.

By construction $\mathbf{D}^{(u,v)} \cong \mathbf{D}^{(u)} \otimes \bar{\mathbf{D}}^{(v)}$. Taking the complex conjugate of

$$\mathbf{D}^{(u)} \otimes \mathbf{D}^{(u')} \cong \sum_{w=|u-u'|}^{u+u'} \oplus\, \mathbf{D}^{(w)}$$

we obtain an analogous relation for $\bar{\mathbf{D}}^{(u)} \otimes \bar{\mathbf{D}}^{(u')}$. [Note: Even though we have defined $\bar{\mathbf{D}}^{(u)}$ by taking the complex conjugate of a matrix realization of $\mathbf{D}^{(u)}$ with respect to a fixed basis, it is easy to show that $\bar{\mathbf{D}}^{(u)}$ is basis-independent. Indeed, one merely verifies that two matrix reps $T(A)$, $T'(A)$ are equivalent if and only if $\overline{T(A)}$ and $\overline{T'(A)}$ are equivalent.] Thus

(3.5)
$$\begin{aligned}\mathbf{D}^{(u,v)} \otimes \mathbf{D}^{(u',v')} &\cong (\mathbf{D}^{(u)} \otimes \bar{\mathbf{D}}^{(v)}) \otimes (\mathbf{D}^{(u')} \otimes \bar{\mathbf{D}}^{(v')})\\ &\cong (\mathbf{D}^{(u)} \otimes \mathbf{D}^{(u')}) \otimes (\bar{\mathbf{D}}^{(v)} \otimes \bar{\mathbf{D}}^{(v')})\\ &\cong \sum_{w=|u-u'|}^{u+u'} \sum_{z=|v-v'|}^{v+v'} \oplus\, \mathbf{D}^{(w,z)}\end{aligned}$$

is the CG series for irred reps of $SL(2)$. Note that each irred rep $\mathbf{D}^{(w,z)}$ occurring in the decomposition of $\mathbf{D}^{(u,v)} \otimes \mathbf{D}^{(u',v')}$ has multiplicity one. Therefore, it is easy to project out the subspace $\mathcal{W}^{(w,z)}$ of $\mathcal{V}^{(u,v)} \otimes \mathcal{V}^{(u',v')}$ which transforms irreducibly under $\mathbf{D}^{(w,z)}$. Indeed from (3.3) and the results of Section 7.7 a canonical basis for $\mathcal{W}^{(w,z)}$ is given by the vectors

(3.6)
$$h_{kl}^{(w,z)} = \sum_{mnm'n'} C(u, m; u', m' \mid w, k)C(v, n; v', n' \mid z, l)f_{mn}^{(u,v)} \otimes f_{m'n'}^{(u'v')},$$
$$-w \le k \le w, \quad -z \le l \le z,$$

where the $C(-|-)$ are the CG coefficients (7.21), Chapter 7. The coefficients for the real Lie group $SL(2)$ are products of the coefficients for the complex group $SL(2)$.

If we restrict the rep $\mathbf{D}^{(u,v)}$ of $SL(2)$ to the subgroup $SU(2)$ it decomposes into a direct sum of irred reps of $SU(2)$. To determine the decomposition we note the rep $\bar{\mathbf{D}}^{(u)}$ of $SU(2)$ is equivalent to $\mathbf{D}^{(u)}$. Indeed $\bar{\mathbf{D}}^{(u)}$ is irred and every irred rep of $SU(2)$ is equivalent to some $\mathbf{D}^{(v)}$. Since dim $\bar{\mathbf{D}}^{(u)} = 2u + 1$ we must have $\bar{\mathbf{D}}^{(u)} \cong \mathbf{D}^{(u)}$. Alternatively, the character $\chi^{(u)}$ of $\mathbf{D}^{(u)}$ is real, so $\mathbf{D}^{(u)}$ and $\bar{\mathbf{D}}^{(u)}$ have the same character. Using this result we obtain

$$\mathbf{D}^{(u,v)} \mid SU(2) \cong \mathbf{D}^{(u)} \otimes \mathbf{D}^{(v)} \cong \sum_{w=|u-v|}^{u+v} \oplus \mathbf{D}^{(w)}. \tag{3.7}$$

Note the special cases $\mathbf{D}^{(u,0)} \mid SU(2) \cong \mathbf{D}^{(0,u)} \mid SU(2) \cong \mathbf{D}^{(u)}$.

Although the canonical basis $\{f_{mn}^{(u,v)}\}$ is very convenient for computational purposes, it does not clearly exhibit the decomposition (3.7) when the Lie algebra $sl(2)$ is restricted to $su(2)$. According to (3.7) there exists an ON basis $\{f_k^{(w)} : |u - v| \le w \le u + v, -w \le k \le w\}$ for $\mathcal{V}^{(u,v)}$ such that

$$J^3 f_k^{(w)} = k f_k^{(w)}, \qquad J^{\pm} f_k^{(w)} = [(w \pm k + 1)(w \mp k)]^{1/2} f_{k\pm 1}^{(w)}. \tag{3.8}$$

Recall that the J-operators satisfy the commutation relations of $su(2)$. We could use CG coefficients to express the $\{f_k^{(w)}\}$ basis in terms of the $\{f_{mn}^{(u,v)}\}$ basis. However, it is more instructive to compute the action of the Lie algebra on the $\{f_k^{(w)}\}$ basis directly.

For this purpose we choose the operators $J^{\pm}, J^3, F^{\pm}, F^3$ with commutation relations (1.8), ($J \equiv \mathcal{L}$, $F \equiv \mathcal{B}$), as the generators of our rep. The action of the J-operators on $f_k^{(w)}$ is given by (3.8). To determine the action of the F-operators we note from (1.8) that the operators $\mathbf{Q}_1 = -F^+$, $\mathbf{Q}_0 = \sqrt{2}F^3$, and $\mathbf{Q}_{-1} = F^-$ transform as a spherical tensor of rank one under the action of $SU(2)$. According to the Wigner–Eckart theorem (10.14), Chapter 7,

$$\begin{aligned} (F^{\pm} f_k^{(w)}, f_{k'}^{(w')}) &= \mp N(w, w') C(1, \pm 1; w, k \mid w', k'), \\ (F^3 f_k^{(w)}, f_{k'}^{(w')}) &= 2^{-1/2} N(w, w') C(1, 0; w, k \mid w', k'). \end{aligned} \tag{3.9}$$

In particular, these matrix elements are zero unless $w' = w \pm 1, w$. Explicit expressions for the CG coefficients are given in (7.28), Chapter 7, so we need only compute the constants N. These constants can be obtained from the remaining commutation relations

$$[F^3, F^{\pm}] = \mp J^{\pm}, \qquad [F^+, F^-] = -2J^3. \tag{3.10}$$

It follows from (3.9) that

$$\begin{aligned} F^{\pm} f_k^{(w)} = {} & \pm [(w \mp k)(w \mp k - 1)]^{1/2} A_w f_{k\pm 1}^{(w-1)} \\ & - [(w \pm k + 1)(w \mp k)]^{1/2} B_w f_{k\pm 1}^{(w)} \\ & \pm [(w \pm k + 1)(w \pm k + 2)]^{1/2} C_{w+1} f_{k\pm 1}^{(w+1)}, \end{aligned} \tag{3.11}$$

$$\begin{aligned} F^3 f_k^{(w)} = {} & [(w - k)(w + k)]^{1/2} A_w f_k^{(w-1)} \\ & - k B_w f_k^{(w)} - [(w + k + 1)(w - k + 1)]^{1/2} C_{w+1} f_k^{(w+1)}, \end{aligned} \tag{3.12}$$

where the constants A_w, B_w, C_w depend only on w. We can simplify the above formulas by renormalizing the vectors $f_k^{(w)}$. If we introduce new basis vectors

$f_k'^{(w)} = \alpha_w f_k^{(w)}$, where the α_w are nonzero complex numbers, then Eq. (3.8) will remain unchanged in the primed basis while (3.11) and (3.12) will maintain the same form with A_w, B_w, C_w replaced by

$$(3.13) \qquad A_w' = (\alpha_w/\alpha_{w-1})A_w, \qquad B_w' = B_w, \qquad C_w' = (\alpha_{w-1}/\alpha_w)C_w.$$

The new basis vectors $f_k'^{(w)}$ will be orthogonal but not necessarily of length one. Note the product $A_w'C_w' = A_wC_w$ is invariant under renormalization and must be nonzero for $|u - v| + 1 \leq w \leq u + v$ since $\mathbf{D}^{(u,v)}$ is irred. Thus we can choose the constants α_w so $A_w = C_w$. We will suppose that this is the case in expressions (3.11) and (3.12).

Now we use the commutation relations (3.10) to compute A_w and B_w. Substituting (3.8), (3.11), and (3.12) in $[F^+, F^3]f_k^{(w)} = J^+f_k^{(w)}$ and equating coefficients of $f_{k+1}^{(w')}$ on both sides of the resulting relations, we find

$$(3.14) \qquad [(w + 1)B_w - (w - 1)B_{w-1}]A_w = [(w + 2)B_{w+1} - wB_w]A_{w+1} = 0,$$

$$(3.15) \qquad (2w - 1)A_w^2 - (2w + 3)A_{w+1}^2 - B_w^2 = 1.$$

The other two equations (3.10) lead to the same results. Since $A_w \neq 0$ it follows from (3.14) that

$$B_{w+1} = wB_w/(w + 2), \qquad w = |u - v|, \ldots, u + v.$$

The solution is

$$(3.16) \qquad B_w = B_{w_0}w_0(w_0 + 1)/[w(w + 1)] = iw_0w_1/[w(w + 1)],$$

where $iw_1 = B_{w_0}(w_0 + 1)$, $w_0 = |u - v|$. We will determine the constant w_1 later. Substituting (3.16) into (3.15), we get a recurrence relation for A_w^2:

$$(3.17) \qquad (2w - 1)A_w^2 - (2w + 3)A_{w+1}^2 = 1 + \frac{w_0^2 w_1^2}{w^2(w + 1)^2},$$

$$w = w_0, \ldots, u + v - 1.$$

Since $f_k^{(w_0-1)}$ does not belong to $\mathcal{V}^{(u,v)}$ we must require $A_{w_0} = 0$. With this restriction Eq. (3.17) determine A_w^2. The solution is

$$(3.18) \qquad A_w = \frac{i}{w}\left[\frac{(w^2 - w_0^2)(w^2 - w_1^2)}{4w^2 - 1}\right]^{1/2}.$$

(By choosing the normalization factors α_w appropriately we can always assume $|\arg A_w| \leq \pi/2$.)

To determine w_1 we note from (3.11) that $A_{u+v+1} = 0$ since $f_k^{(u+v+1)}$ does not belong to $\mathcal{V}^{(u,v)}$. Therefore, (3.18) implies $w_1^2 = (u + v + 1)^2$, or $w_1 = \pm(u + v + 1)$. To determine the proper sign we must distinguish between $\mathbf{D}^{(u,v)}$ and $\mathbf{D}^{(v,u)}$.

It follows from (3.2) that $-C\cdot C = u(u + 1)\mathbf{E}$ and $-D\cdot D = v(v + 1)\mathbf{E}$ for the rep $D^{(u,v)}$. If we express the C_j and D_j in terms of the operators (3.11) and (3.12) and use (3.16)and (3.18) we find

$$-C\cdot C = \tfrac{1}{4}(w_0 + w_1 + 1)(w_0 + w_1 + 3)\mathbf{E},$$
$$-D\cdot D = \tfrac{1}{4}(w_1 - w_0 + 1)(w_1 - w_0 + 3)\mathbf{E}. \tag{3.19}$$

Thus, if we allow w_0 to be negative we can make the unique assignment

$$w_0 = u - v, \qquad w_1 = u + v + 1. \tag{3.20}$$

This is permissible since expressions (3.11), (3.12), (3.16), and (3.18) depend only on $w_0{}^2$, $w_1{}^2$, and $w_0 w_1$. In particular the rep defined by the pair (w_0, w_1) is equivalent to the rep $(-w_0, -w_1)$. Here $2w_0$, $2w_1$, and $w_0 + w_1$ are integers with $|w_0| \leq |w_1|$.

Summing up, there is a basis $\{f_k^{(w)}\}$ for the rep space of $\mathbf{D}^{(u,v)}$ such that

$$J^{\pm} f_k^{(w)} = [(w \pm k + 1)(w \mp k)]^{1/2} f_{k\pm 1}^{(w)}, \qquad J^3 f_k^{(w)} = k f_k^{(w)}, \tag{3.21}$$

$$\begin{aligned} F^{\pm} f_k^{(w)} = {} & \pm[(w \mp k)(w \mp k - 1)]^{1/2} A_w f_{k\pm 1}^{(w-1)} \\ & - [(w \pm k + 1)(w \mp k)]^{1/2} B_w f_{k\pm 1}^{(w)} \\ & \pm [(w \pm k + 1)(w \pm k + 2)]^{1/2} A_{w+1} f_{k\pm 1}^{(w+1)}, \end{aligned} \tag{3.22}$$

$$\begin{aligned} F^3 f_k^{(w)} = {} & [w^2 - k^2]^{1/2} A_w f_k^{(w-1)} - k B_w f_k^{(w)} \\ & - [(w+1)^2 - k^2]^{1/2} A_{w+1} f_k^{(w+1)} \end{aligned} \tag{3.23}$$

$$w = |w_0|, |w_0| + 1, \ldots, |w_1|, \qquad -w \leq k \leq w,$$

where

$$B_w = \frac{i w_0 w_1}{w(w+1)}, \qquad A_w = \frac{i}{w}\left[\frac{(w^2 - w_0{}^2)(w^2 - w_1{}^2)}{4w^2 - 1}\right]^{1/2} \tag{3.24}$$

and $w_0 = u - v$, $w_1 = u + v + 1$.

In many respects the basis $\{f_k^{(w)}\}$ is more convenient than the basis $\{f_{mn}^{(u,v)}\}$. This is particularly true in problems where one is interested in the restriction of a rep of $SL(2)$ to the subgroup $SU(2)$.

The noncompact group $SL(2)$ also has bounded infinite-dimensional irred reps. If $\mathbf{T}$ is such a rep, a slight extension of the results of Section 6.3 shows that $\mathbf{T} \mid SU(2)$ decomposes into a direct sum of irred reps of the compact group $SU(2)$:

$$\mathbf{T} \mid SU(2) \cong \sum_{2w=0}^{\infty} \oplus\, a_w \mathbf{D}^{(w)}. \tag{3.25}$$

For the present we assume that the multiplicity a_w of $\mathbf{D}^{(w)}$ is either zero or one, i.e., each rep $\mathbf{D}^{(w)}$ appears at most once in the decomposition. Furthermore, we assume that the usual relationships between the bounded operators $\mathbf{T}(A)$, $A \in SU(2)$, and $J^{\pm}$, J^3, $F^{\pm}$, F^3 hold for these infinite-dimensional reps.

Let I be the set of all w such that $\mathbf{D}^{(w)}$ is contained in the decomposition of $\mathbf{T} \mid SU(2)$. There exists a basis $\{f_k^{(w)} : w \in I, -w \leq k \leq w\}$ for the rep space such that Eq. (3.8) are satisfied. Indeed for fixed w, $\{f_k^{(w)}\}$ is a canonical basis for $\mathbf{D}^{(w)}$. Applying the Wigner–Eckart theorem, we see that the operators $F^{\pm}$, F^3 satisfy Eq. (3.11) and (3.12) exactly as in the finite-dimensional case. Let w_0 be the smallest number in the index set I. It follows from (3.11), (3.12), and the irreducibility of $\mathbf{T}$ that $I = \{w_0 + n : n = 0, 1, 2, \ldots\}$. Thus

$$\mathbf{T} \mid SU(2) \cong \sum_{n=0}^{\infty} \oplus \mathbf{D}^{(w_0+n)}.$$

(There can be no gaps in the sequence of $\mathbf{D}^{(w)}$ since $\mathbf{T}$ is irred. If the sequence is finite then $\mathbf{T}$ is isomorphic to one of the finite-dimensional reps $\mathbf{D}^{(u,v)}$ which we have already classified.)

The computation of relations (3.21)–(3.24) is exactly the same for $\mathbf{T}$ as for finite-dimensional reps. The only difference is that w_1 is no longer an integer or half-integer such that $w_1 = w_0 + k$ for some integer $k \geq 0$; otherwise $\mathbf{T}$ would be finite-dimensional. Thus, w_1 is an arbitrary complex number not satisfying the above requirement. We conclude that the infinite-dimensional irred reps of $SL(2)$ can be labeled by the parameters (w_0, w_1) where $2w_0$ is a nonnegative integer and w_1 is a complex number such that $w_1 \neq w_0 + k$, $k = 0, 1, 2, \ldots$. However, it is not clear that each of these Lie algebra reps can be exponentiated to a global irred rep of $SL(2)$. Naimark [2] proves that there is in fact a global group rep corresponding to each of our Lie algebra reps. Furthermore, Naimark shows that any irred rep $\mathbf{T}$ of $SL(2)$ when restricted to $SU(2)$ contains each $\mathbf{D}^{(u)}$ at most once.

Let us check to see which of our reps are unitary. If $\mathbf{T}$ is unitary a simple computation (which should be familiar to the reader by now) shows that

$$(3.26) \qquad (J^3)^* = J^3, \qquad (J^+)^* = J^-, \qquad (F^3)^* = F^3, \qquad (F^+)^* = F^-.$$

Just as in Section 7.7 we can use the requirements on the J-operators to prove $(f_k^{(w)}, f_{k'}^{(w')}) = 0$ unless $w = w'$ and $k = k'$. Furthermore $\| f_k^{(w)} \| = \| f_j^{(w)} \|$, $-w \leq k, j \leq w$. Since F^3 is symmetric we have

$$(F^3 f_k^{(w)}, f_k^{(w')}) = (f_k^{(w)}, F^3 f_k^{(w')}).$$

Substituting (3.23) into this expression, we find

$$(3.27) \qquad B_w = \bar{B}_w, \qquad A_w \| f_k^{(w-1)} \|^2 = -\bar{A}_w \| f_k^{(w)} \|^2.$$

The relation $(F^+)^* = F^-$ yields no additional constraints. By (3.24), B_w is real if and only if (a) $w_1 = ic$, c real, or (b) $w_0 = 0$.

Writing $A_w = R_w + iI_w$ in terms of real and imaginary parts, we see that the second relation (3.27) implies $R_w = 0$, $\| f_k^{(w)} \| = \| f_k^{(w-1)} \|$. Since all of the basis vectors have the same length we can normalize them so $\| f_k^{(w)} \| = 1$. By (3.24) the requirement $R_w = 0$ is identically satisfied in case (a) since $-w_1{}^2 = c^2 \geq 0$. In case (b) this requirement will be satisfied provided

$(w^2 - w_1^2)/(4w^2 - 1) \geq 0$ for $w = 0, 1, 2, \ldots$. This is possible if and only if $0 \leq w_1^2 \leq 1$. Thus $-1 \leq w_1 \leq 1$.

This discussion shows that the unitary irred reps (w_0, w_1) of $SL(2)$ fall into two classes:

(3.28) The principal series: w_1 pure imaginary.

(3.29) The complementary series: $w_0 = 0$, w_1 real, $|w_1| \leq 1$.

The only finite-dimensional unitary rep is the identity rep $(0, 1) \cong \mathbf{D}^{(0,0)}$.

We now construct models of the corresponding Lie group reps, starting with the finite-dimensional reps $\mathbf{D}^{(u,v)}$. From (2.1), Section 7.2, we know $\mathbf{D}^{(u,0)} \cong \mathbf{D}^{(u)}$, $2u = 0, 1, 2, \ldots$, has a model in terms of operators

$$[\mathbf{T}(A)f](z) = (bz + d)^{2u} f\left(\frac{az + c}{bz + d}\right), \qquad A \in SL(2), \tag{3.30}$$

acting on the $(2u + 1)$-dimensional space $\mathcal{V}^{(u)}$ of polynomials with order $2u$ in z. The vectors $f_m^{(u)} = (-z)^{u+m}/[(u + m)!(u - m)!]^{1/2}$ form a canonical basis.

It follows that $\mathbf{D}^{(0,v)} \cong \bar{\mathbf{D}}^{(v)}$ has the model

$$[\mathbf{T}(A)f](\bar{z}) = (\overline{bz + d})^{2v} f\left(\frac{\bar{a}\bar{z} + \bar{c}}{\bar{b}\bar{z} + \bar{d}}\right) \tag{3.31}$$

on the $(2v + 1)$-dimensional space $\bar{\mathcal{V}}^{(v)}$ of polynomials with order $2v$ in $\bar{z}$. The vectors $g_n^{(v)} = (-1)^{v-n} \bar{f}_n^{(v)} = (\bar{z})^{v-n}/[(v + n)!(v - n)!]^{1/2}$ form a canonical basis.

According to (3.5), $\mathbf{D}^{(u,v)} \cong \mathbf{D}^{(u)} \otimes \bar{\mathbf{D}}^{(v)}$. Thus $\mathbf{D}^{(u,v)}$ has a model defined by operators

$$[\mathbf{T}(A)f](z, \bar{z}) = (bz + d)^{2u}(\overline{bz + d})^{2v} f\left(\frac{az + c}{bz + d}, \frac{\bar{a}\bar{z} + \bar{c}}{\bar{b}\bar{z} + \bar{d}}\right) \tag{3.32}$$

acting on the $(2u + 1)(2v + 1)$-dimensional space of polynomials with order $2u$ in z and $2v$ in $\bar{z}$.

It is clear from (3.32) that $\bar{\mathbf{D}}^{(u,v)} \cong \mathbf{D}^{(v,u)}$. Only the diagonal reps $\mathbf{D}^{(u,v)}$ are equivalent to their own complex conjugates. Such reps are called **real.**

It is easy to find a model of the unitary reps in the principal series (w_0, ic), (3.28). Indeed by comparing the eigenvalues of the invariant operators $C \cdot C$ and $D \cdot D$ in the $\{f_{mn}^{(u,v)}\}$ and $\{f_k^{(w)}\}$ bases we have concluded that $w_0 = u - v$, $w_1 = u + v + 1$. This suggests that the action of (w_0, ic) can be obtained from (3.32) by setting $2u = w_0 + ic - 1$ and $2v = -w_0 + ic - 1$:

$$[\mathbf{T}(A)f](z) = |bz + d|^{-2w_0 + 2ic - 2}(bz + d)^{2w_0} f\left(\frac{az + c}{bz + d}\right). \tag{3.33}$$

Here we regard $f(z) = f(x, y)$ as a function of the two real variables x, y, where $z = x + iy$, and we suppress the argument z. If the operators $\mathbf{T}(A)$

act on the Hilbert space $L_2(R_2)$,

$$(f_1, f_2) = \int_{R_2} f_1(x,y)\overline{f_2(x, y)}\, dx\, dy,$$

then one can show that they define a global irred unitary rep of $SL(2)$ whose induced Lie algebra rep is equivalent to (w_0, ic), $2w_0 = 0, 1, 2, \ldots$, c real (Naimark [2]).

It is just as easy to formally compute the action of the unitary reps from the complementary series. However, in this case there is some difficulty in determining the proper Hilbert space on which the rep acts. Naimark works out the details.

The computation of the infinitesimal generators (generalized Lie derivatives) for all of the above models is straightforward but will not be carried out here. Furthermore, we will now limit ourselves to finite-dimensional reps. The infinite-dimensional unitary reps of the homogeneous Lorentz group seem to be of less importance for physical applications (see, however, Ruhl [1]).

Because of the isomorphism between the Lie algebras $sl(2)$ and $su(2) \oplus su(2)$ we can conclude that any finite-dimensional rep of $SL(2)$ or $L^{\uparrow +}$ can be decomposed into a direct sum of irred reps $\mathbf{D}^{(u,v)}$. (This is false for infinite-dimensional reps.) We shall use this fact to compute the (finite-dimensional) irred reps of the general Lorentz group $L(4)$. We shall also compute the irred reps of the **complete** Lorentz group $L^{\uparrow} = \{SL^{\uparrow +}, L^{\uparrow +}\}$ obtained by adding the space reflection S to the proper Lorentz group.

Let $\mathbf{T}$ be an irred rep of $L^{\uparrow}$ and let $\mathbf{S} = \mathbf{T}(S)$. Since S commutes with all rotations [see (1.5)] it follows that

$$\mathbf{S}J^{\pm}\mathbf{S}^{-1} = J^{\pm}, \qquad \mathbf{S}J^3\mathbf{S}^{-1} = J^3, \qquad \mathbf{S} = \mathbf{S}^{-1}.$$

On the other hand, by (1.3)

$$\mathbf{S}B^{\pm}\mathbf{S}^{-1} = -B^{\pm}, \qquad \mathbf{S}B^3\mathbf{S}^{-1} = -B^3.$$

In terms of the C- and D-operators [(1.9)] these results become

$$\mathbf{S}C^{\pm}\mathbf{S}^{-1} = D^{\pm}, \qquad \mathbf{S}C^3\mathbf{S}^{-1} = D^3. \tag{3.34}$$

Suppose the rep $\mathbf{D}^{(u,v)}$ is contained in $\mathbf{T}\,|\,L^{\uparrow +}$. Then there exist vectors $\{f_{mn}^{(u,v)}: -u \leq m \leq u,\ -v \leq n \leq v\}$ spanning a subspace $\mathcal{V}^{(u,v)}$ of the rep space $\mathcal{V}$ which transform under the C- and D-operators according to (3.2). Define vectors $g_{nm}^{(v,u)} \in \mathcal{V}$ by $g_{nm}^{(v,u)} = \mathbf{S}f_{mn}^{(u,v)}$. Then by (3.34) and (3.2)

$$C^3 g_{nm}^{(v,u)} = n g_{nm}^{(v,u)}, \qquad C^{\pm} g_{nm}^{(v,u)} = [(v \pm n + 1)(v \mp n)]^{1/2} g_{n\pm 1,m}^{(v,u)}, \tag{3.35}$$

with similar results for the D-operators. Thus the vectors $\{g_{nm}^{(v,u)}\}$ span a subspace $\mathcal{V}^{(v,u)}$ of $\mathcal{V}$ which transforms under $\mathbf{D}^{(v,u)}$. Since $\mathbf{S}^2 = \mathbf{E}$ we have $\mathbf{S}g_{nm}^{(v,u)} = \mathbf{S}^2 f_{mn}^{(u,v)} = f_{mn}^{(u,v)}$. Furthermore, the space $\mathcal{V}^{(u,v)} + \mathcal{V}^{(v,u)}$ is invariant under **T**. Since **T** is irred this space must coincide with $\mathcal{V}$ itself. There are two possibilities depending on whether or not $u = v$. If $u \neq v$ the set $\{f_{mn}^{(u,v)}, g_{kl}^{(v,u)}\}$ is linearly independent and $\mathcal{V} = \mathcal{V}^{(u,v)} \oplus \mathcal{V}^{(v,u)}$. We designate the $2(2u + 1) \times (2v + 1)$-dimensional rep by

$$\mathbf{D}^{(u,v)} \oplus \mathbf{D}^{(v,u)}, \qquad u \neq v. \tag{3.36}$$

Conversely, it is easy to show that each pair of reps of $L^{\uparrow +}$ taking the form (3.36) does define an irred rep of $L^{\uparrow}$.

Now suppose $u = v$ and define new vectors $h_{mn}^{\pm} = f_{mn}^{(u,u)} \pm g_{mn}^{(u,u)}$. The $\{h_{mn}^{+}\}$ span a subspace $\mathcal{V}^{(+)}$, while the $\{h_{mn}^{-}\}$ span $\mathcal{V}^{(-)}$. Here

$$\begin{aligned} C^3 h_{mn}^{\pm} &= m(f_{mn}^{(u,u)} \pm g_{mn}^{(u,u)}) = m h_{mn}^{\pm}, \qquad D^3 h_{mn}^{\pm} = n h_{mn}^{\pm} \\ C^+ h_{mn}^{\pm} &= [(u + m + 1)(u - m)]^{1/2} h_{m+1,n}^{\pm}, \\ D^+ h_{mn}^{\pm} &= [(u + n + 1)(u - n)]^{1/2} h_{mn+1}^{\pm}, \end{aligned} \tag{3.37}$$

with similar results for C^- and D^-. Also,

$$\mathbf{S} h_{mn}^{\pm} = \mathbf{S} f_{mn}^{(u,u)} \pm \mathbf{S} g_{mn}^{(u,u)} = g_{nm}^{(u,u)} \pm f_{mn}^{(u,u)} = \pm h_{nm}^{\pm}. \tag{3.38}$$

As a consequence, both $\mathcal{V}^{(+)}$ and $\mathcal{V}^{(-)}$ are invariant under **T**. Since **T** is irred, either $\mathcal{V} = \mathcal{V}^{(+)}$ and $\mathcal{V}^{(-)} = \{\theta\}$ or $\mathcal{V} = \mathcal{V}^{(-)}$ and $\mathcal{V}^{(+)} = \{\theta\}$. If the first case holds then $h_{mn}^{-} = \theta$, so $f_{mn}^{(u,u)} = g_{mn}^{(u,u)} = \mathbf{S} f_{nm}^{(u,u)}$ and **S** transposes the lower indices of the basis vectors $f_{mn}^{(u,u)}$ for $\mathcal{V}$. We denote the corresponding irred rep by $\mathbf{D}_{+}^{(u,u)}$. If the second case holds then $h_{mn}^{+} = \theta$ and $\mathbf{S} f_{nm}^{(u,u)} = -f_{mn}^{(u,u)}$. We denote this rep by $\mathbf{D}_{-}^{(u,u)}$. Here $\dim \mathbf{D}_{+}^{(u,u)} = \dim \mathbf{D}_{-}^{(u,u)} = (2u + 1)^2$.

Thus the possible irred reps of $L^{\uparrow}$ are $\mathbf{D}^{(u,v)} \oplus \mathbf{D}^{(v,u)}$, $u > v$, and $\mathbf{D}_{+}^{(u,u)}$, $\mathbf{D}_{-}^{(u,u)}$. Only those reps such that $u + v$ is an integer are single-valued on $L^{\uparrow}$. The remaining reps are double-valued on $L^{\uparrow}$ but they are single-valued reps of the group generated by S and $SL(2)$.

We can obtain the irred reps of $L(4)$ by noting that $L(4) = \{L^{\uparrow}, I \cdot L^{\uparrow}\}$, where $I = -E$ is the total inversion operation. Since I commutes with all elements of $L(4)$ and $I^2 = E$ it follows that $\mathbf{T}(I) = \pm \mathbf{E}$ for each irred rep **T**. Thus, to each irred rep of $L^{\uparrow}$ there correspond exactly two reps of $L(4)$. In one rep $\mathbf{T}(I) = \mathbf{E}$ and in the other $\mathbf{T}(I) = -\mathbf{E}$. If $u + v$ is not an integer then these reps are double-valued on $L(4)$ but single-valued on the group generated by S, I, and $SL(2)$.

We mention some of the simplest examples of our reps. The usual 4×4 matrix realization (1.1) of $L^{\uparrow +}$ is equivalent to the real rep $\mathbf{D}^{(1/2,1/2)}$ of $SL(2)$. The usual 2×2 matrix realization is equivalent to $\mathbf{D}^{(1/2,0)}$. The matrices $\bar{A}$, $A \in SL(2)$, define the rep $\mathbf{D}^{(0,1/2)}$. The usual 4×4 realization of $L^{\uparrow}$ is equivalent to $\mathbf{D}_{-}^{(1/2,1/2)}$. A quantity transforming under $L^{\uparrow}$ according to $\mathbf{D}_{+}^{(0,0)}$ is called a **scalar;** one transforming according to $\mathbf{D}_{-}^{(0,0)}$ is a **pseudoscalar. Vectors** and **pseudovectors** transform according to $\mathbf{D}_{-}^{(1/2,1/2)}$ and

$\mathbf{D}_+^{(1/2,1/2)}$, respectively. Finally the usual 4×4 realization of $L(4)$ is equivalent to $\mathbf{D}_-^{(1/2,1/2)}$ with $\mathbf{T}(I) = -\mathbf{E}$.

8.4 Models of the Representations

Let V be a complex two-dimensional vector space with basis $\{\mathbf{v}_1, \mathbf{v}_2\}$. We define a model of the rep $\mathbf{D}^{(1/2,0)}$ of $SL(2)$ on V by

$$\mathbf{A}\mathbf{v}_j = \sum_{l=1}^{2} A_{lj}\mathbf{v}_l, \qquad A = (A_{lj}) \in SL(2). \tag{4.1}$$

The vectors $\mathbf{f}_{1/2}^{(1/2)} = \mathbf{v}_1$ and $\mathbf{f}_{-1/2}^{(1/2)} = \mathbf{v}_2$ form a canonical basis.

Similarly we can define a model of the rep $\mathbf{D}^{(0,1/2)} \cong \bar{\mathbf{D}}^{(1/2,0)}$ on the two-dimensional vector space W with basis $\{\mathbf{w}_1, \mathbf{w}_2\}$:

$$\bar{\mathbf{A}}\mathbf{w}_j = \sum_{l=1}^{2} \bar{A}_{lj}\mathbf{w}_l. \tag{4.2}$$

The vectors $\mathbf{g}_{1/2}^{(1/2)} = \mathbf{w}_2$ and $\mathbf{g}_{-1/2}^{(1/2)} = -\mathbf{w}_1$ form a canonical basis.

Now consider a rep of $SL(2)$ on the $2^{(p+q)}$-dimensional space $V^{\otimes p} \otimes W^{\otimes q}$ defined by

$$\begin{aligned}\mathbf{A}(\mathbf{v}_{\alpha_1} \otimes \cdots \otimes \mathbf{v}_{\alpha_p} \otimes \mathbf{w}_{\beta_1} \otimes \cdots \otimes \mathbf{w}_{\beta_q}) \\ = \sum_{\alpha_j', \beta_k' = 1}^{2} A_{\alpha_1'\alpha_1} \cdots A_{\alpha_p'\alpha_p}\bar{A}_{\beta_1'\beta_1} \cdots \bar{A}_{\beta_q'\beta_q}\mathbf{v}_{\alpha_1'} \otimes \cdots \otimes \mathbf{w}_{\beta_q'}.\end{aligned} \tag{4.3}$$

The elements $\mathbf{a}$ of this space are called **spinors** of rank $p + q$. In terms of the components $a^{\alpha_1 \cdots \alpha_p \beta_1 \cdots \beta_q}$ of $\mathbf{a}$ with respect to the basis $\mathbf{v}_{\alpha_1}, \otimes \cdots \otimes \mathbf{w}_{\beta_q}$ the group action (4.3) reads

$$\mathbf{A}a^{\alpha_1 \cdots \alpha_p \beta_1 \cdots \beta_q} = \sum_{\alpha_j', \beta_k' = 1}^{2} A_{\alpha_1\alpha_1'} \cdots \bar{A}_{\beta_q\beta_q'}a^{\alpha_1' \cdots \alpha_p' \beta_1' \cdots \beta_q'}. \tag{4.4}$$

It is evident that the spinors of rank $p + q$ transform according to the rep $(\mathbf{D}^{(1/2,0)})^{\otimes p} \otimes (\mathbf{D}^{(0,1/2)})^{\otimes q}$. We can use the Clebsch–Gordan series (3.5) repeatedly to decompose this rep into irred reps $\mathbf{D}^{(u,v)}$ but the resulting expression is complicated. However, it is easy to verify that $\mathbf{D}^{(p/2,q/2)}$ is the irred rep of highest weight contained in the reducible rep and its multiplicity is exactly one. We show how to determine the subspace transforming under $\mathbf{D}^{(p/2,q/2)}$.

Let $\mathcal{S}^p$ be the subspace of completely symmetric spinors in $V^{\otimes p}$. The elements of $\mathcal{S}^p$ are symmetric in the spinor indices $a^{\alpha_1 \cdots \alpha_p}$. As shown in Section 4.3, $\dim \mathcal{S}^p = p + 1$ and $\mathbf{a} \in \mathcal{S}^p$ is uniquely determined by the independent components $a^{11 \cdots 1, 22 \cdots 2} = a^{(s)}$, where s is the number of twos and $p - s$ the number of ones, $s = 0, 1, \ldots, p$. Furthermore, $\mathcal{S}^p$ is invariant under the induced action of $SL(2)$ on $V^{\otimes p}$. We have shown earlier that $\mathcal{S}^p$

transforms irreducibly under the rep $[p, 0] = [p]$ of $GL(2)$ [Section 4.3]. We will now show that $\mathcal{S}^p$ remains irred when $GL(2)$ is restricted to the subgroup $SL(2)$. Let $\mathbf{a}^{s_0}$ be the element of $\mathcal{S}^p$ such that $a^{(s_0)} = 1$ and all independent components of $\mathbf{a}^{s_0}$ are zero. Clearly, the tensors $\mathbf{a}^0, \mathbf{a}^1, \ldots, \mathbf{a}^p$ form a basis for $\mathcal{S}^p$. Furthermore a direct computation from (4.4) with $q = 0$ shows

$$(\exp tC^3)\mathbf{a}^{s_0} = (\exp tJ^3)\mathbf{a}^{s_0} = \exp[t(\tfrac{1}{2}p - s_0)]\mathbf{a}^{s_0}.$$

Recall that the D-operators are zero for this rep. It follows that the highest weight vector in $\mathcal{S}^p$ with respect to C^3 has eigenvalue $p/2$. Thus $\mathcal{S}^p$ must contain a subspace transforming according to $\mathbf{D}^{(p/2,0)}$. Since $\dim \mathbf{D}^{(p/2,0)} = p + 1 = \dim \mathcal{S}^p$, $\mathcal{S}^p$ is irred.

An exactly similar argument shows that the subspace $\bar{\mathcal{S}}^q$ of completely symmetric spinors in $W^{\otimes q}$ transforms according to $\mathbf{D}^{(0,q/2)}$. Now the subspace $\mathcal{S}^p \otimes \bar{\mathcal{S}}^q$ of $V^{\otimes p} \otimes W^{\otimes q}$ consists of spinors $a^{\alpha_1 \cdots \alpha_p \beta_1 \cdots \beta_q}$ symmetric in the indices $\alpha_1, \ldots, \alpha_p$ and in the indices $\beta_1, \ldots, \beta_q$ simultaneously. Furthermore, $\mathcal{S}^p \otimes \bar{\mathcal{S}}^q$ transforms under $\mathbf{D}^{(p/2,0)} \otimes \mathbf{D}^{(0,q/2)} \cong \mathbf{D}^{(p/2,q/2)}$. This shows that $\mathcal{S}^p \otimes \bar{\mathcal{S}}^q$ is the subspace of $V^{\otimes p} \otimes W^{\otimes q}$ which carries the rep $\mathbf{D}^{(p/2,q/2)}$. Letting p and q range over all nonnegative integers we can obtain models of all reps $\mathbf{D}^{(u,v)}$ of $SL(2)$.

The use of spinors to provide models of $SL(2)$ reps is very popular in mathematical physics. An extensive spinor calculus has been evolved which enables one to perform operations on spinors to yield new spinors. For example, if $a^{\alpha_1 \cdots \beta_q}$ is a spinor of rank $p + q$ and $b^{\alpha_1' \cdots \beta_q'}$ is a spinor of rank $p' + q'$, then the quantity with components $a^{\alpha_1 \cdots \beta_q} b^{\alpha_1' \cdots \beta_q'}$ transforms as a spinor of rank $(p + p') + (q + q')$. For more details on the spinor calculus see the work of Gel'fand *et al.* [1].

Let V be a four-dimensional real vector space with basis $\{\mathbf{v}_1, \ldots, \mathbf{v}_4\}$ and define a rep of $L^{\uparrow +}$ on V by

$$\text{(4.5)} \qquad \Lambda \mathbf{v}_j = \sum_{l=1}^{4} \Lambda_{lj} \mathbf{v}_l, \qquad \Lambda \in L^{\uparrow +}.$$

This rep is clearly irred; in fact it is equivalent to $\mathbf{D}^{(1/2,1/2)}$. We will verify this explicitly.

If we restrict the rep (4.5) to the subgroup $SO(3)$ then $\mathbf{v}_4$ remains fixed and $\mathbf{v}_1, \mathbf{v}_2, \mathbf{v}_3$ transform under the vector rep $\mathbf{D}^{(1)}$. The only four-dimensional irred reps of $L^{\uparrow +}$ are $\mathbf{D}^{(3/2,0)}$, $\mathbf{D}^{(0,3/2)}$, $\mathbf{D}^{(1/2,1/2)}$ and the first two of these reps remain irred when restricted to $SO(3)$. However,

$$\text{(4.6)} \qquad \mathbf{D}^{(1/2,1/2)} \mid SO(3) \cong \mathbf{D}^{(1)} \oplus \mathbf{D}^{(0)},$$

in agreement with our comments above, so (4.5) defines a rep equivalent to $\mathbf{D}^{(1/2,1/2)}$. One can verify from (1.3) that the vectors

$$\text{(4.7)} \qquad \mathbf{f}_{\pm 1}^{(1)} = (1/\sqrt{2})(\pm \mathbf{v}_1 - i\mathbf{v}_2), \qquad \mathbf{f}_0^{(1)} = \mathbf{v}_3, \qquad \mathbf{f}_0^{(0)} = -i\mathbf{v}_4$$

form a canonical basis which exhibits the decomposition (4.6). Furthermore from (1.9) the vectors

$$\mathbf{f}^{(1/2,1/2)}_{\pm 1/2, \mp 1/2} = \mathbf{v}_3 \pm \mathbf{v}_4, \qquad \mathbf{f}^{(1/2,1/2)}_{\pm 1/2, \pm 1/2} = \pm \mathbf{v}_1 - i\mathbf{v}_2 \tag{4.8}$$

form a canonical basis satisfying relations (3.2).

We can extend the action (4.5) to the space $V^{\otimes n}$. If $\mathbf{a} \in V^{\otimes n}$ with tensor components $a^{i_1 \cdots i_n}$, $1 \leq i_j \leq 4$, then the action of $L^{\uparrow +}$ on $V^{\otimes n}$ is given by

$$\boldsymbol{\Lambda} a^{i_1 \cdots i_n} = \sum_{j_1 \cdots j_n = 1}^{4} \Lambda_{i_1 j_1} \cdots \Lambda_{i_n j_n} a^{j_1 \cdots j_n}. \tag{4.9}$$

Clearly this rep is equivalent to $(\mathbf{D}^{(1/2,1/2)})^{\otimes n}$ and the Clebsch–Gordan series (3.5) can be used to decompose it into irred reps. Note that every irred part of $(\mathbf{D}^{(1/2,1/2)})^{\otimes n}$ is a single-valued rep of $L^{\uparrow +}$ and every single-valued irred rep can be so obtained. The elements of $V^{\otimes n}$ are called **tensors** in distinction to the spinors (4.4) which lead to double-valued reps.

The Lorentz group acts as a natural transformation group on Minkowski space according to the formula

$$x \longrightarrow \boldsymbol{\Lambda}^{-1} x, \qquad \boldsymbol{\Lambda} \in L^{\uparrow +}, \tag{4.10}$$

where $x = (x, y, z, ct)$ is a column four-vector. The Lie derivatives corresponding to this action are

(4.11)

$$L_1 = z\frac{\partial}{\partial y} - y\frac{\partial}{\partial z}, \qquad L_2 = x\frac{\partial}{\partial z} - z\frac{\partial}{\partial x}, \qquad L_3 = y\frac{\partial}{\partial x} - x\frac{\partial}{\partial y},$$

$$B_1 = -t\frac{\partial}{\partial x} - x\frac{\partial}{\partial t}, \qquad B_2 = -t\frac{\partial}{\partial y} - y\frac{\partial}{\partial t}, \qquad B_3 = -t\frac{\partial}{\partial z} - z\frac{\partial}{\partial t}.$$

In these equations and for the computations to follow, we choose units in which $c = 1$.

The components x, y, z, t form a basis for a realization of $\mathbf{D}^{(1/2,1/2)}$ under the action (4.10). Indeed, comparing (4.11) with (3.21)–(3.24) we see that a canonical basis exhibiting the decomposition (4.6) is given by

$$f^{(1)}_{\pm 1} = (1/\sqrt{2})(\pm x - iy), \qquad f^{(1)}_0 = z, \qquad f^{(0)}_0 = it. \tag{4.12}$$

Furthermore, the vectors

$$f^{(1/2,1/2)}_{\pm 1/2, \mp 1/2} = z \mp t, \qquad f^{(1/2,1/2)}_{\pm 1/2, \pm 1/2} = \pm x - iy \tag{4.13}$$

form a canonical basis satisfying relations (3.2).

Another model of $\mathbf{D}^{(1/2,1/2)}$ which will prove useful is obtained by using (4.10) to induce a group rep on the four-dimensional space $\mathfrak{D}$ spanned by the derivatives $\partial/\partial x_j$, $j = 1, \ldots, 4$, where $x = (x_1, \ldots, x_4) = (x, y, z, t)$. Indeed if $x_j' = \sum (\Lambda^{-1})_{jl} x_l$ then $x_l = \sum \Lambda_{lj} x_j'$ and

$$\frac{\partial}{\partial x_j'} = \sum_{l=1}^{4} \frac{\partial x_l}{\partial x_j'} \frac{\partial}{\partial x_l} = \sum_{j=1}^{4} \Lambda_{lj} \frac{\partial}{\partial x_l}.$$

The derivatives

(4.14) $$\partial_{\pm 1/2, \mp 1/2} = \frac{\partial}{\partial z} \pm \frac{\partial}{\partial t}, \qquad \partial_{\pm 1/2, \pm 1/2} = \pm \frac{\partial}{\partial x} - i \frac{\partial}{\partial y}$$

form a canonical basis satisfying relations (3.2) for $u = v = \frac{1}{2}$.

At this point it is convenient to describe the relationship between energy and momentum of a particle in the theory of special relativity. Let $\mathscr{I}$ be an inertial frame with coordinates $x = (x_1, \ldots, x_4) = (x, y, z, t)$. We describe the path of a particle with mass m in this frame using the parametric equations $x_j = h_j(s)$, $1 \leq j \leq 4$, where the parameter s is determined by

(4.15) $$\begin{aligned} ds &= [1 - (dx/dt)^2 - (dy/dt)^2 - (dz/dt)^2]^{1/2} dt \\ &= \pm[dt^2 - dx^2 - dy^2 - dz^2]^{1/2} = (1 - \mathbf{v}\cdot\mathbf{v})^{1/2}\, dt \end{aligned}$$

and $\mathbf{v}$ is the velocity of the particle at time t. [The sign on the right-hand side of (4.15) is the sign of dt.] Since no massive particle can have a velocity as great as the velocity of light ($c = 1$ in this case), ds is always real. The **world time** between two events q_1 and q_2 with coordinates $x^{(1)} = (\mathbf{x}^{(1)}, t^{(1)})$, $x^{(2)} = (\mathbf{x}^{(2)}, t^{(2)})$, $t^{(1)} \neq t^{(2)}$ which lie on the path of the particle is

(4.16) $$s_2 - s_1 = \int_{t^{(1)}}^{t^{(2)}} ds,$$

where the integral is taken along the particle path from $x^{(1)}$ to $x^{(2)}$. Note that s_1 and s_2 are not uniquely determined by (4.16) but only their difference $s_2 - s_1$. The expression $dt^2 - dx^2 - dy^2 - dz^2$ is obviously invariant under the Lorentz group, so the world time between two events q_1, q_2 is the same for *all* inertial frames $\mathscr{I}'$ related to $\mathscr{I}$ by an element of $L^{\uparrow}$. However, if $\mathscr{I}'$ is related to $\mathscr{I}$ by an element of $L^{\downarrow +}$ or $L^{\downarrow -}$ then dt and dt' have opposite signs and $ds = -ds'$. In particular, under time inversion $(dx, dy, dz, dt) \longrightarrow (dx, dy, dz, -dt)$. In this case the magnitude of the world time between two events is conserved but the sign is reversed.

If the particle is moving with uniform velocity $\mathbf{v}$ (with resepct to $\mathscr{I}$) then the frame $\mathscr{I}'$ with spatial axes parallel to the spatial axes of $\mathscr{I}$ and spatial origin of coordinates embedded in the particle is also an inertial frame. In $\mathscr{I}'$ the world time difference between q_1 and q_2 is just the ordinary time interval between the two events as determined by a clock fixed in the particle.

The **momentum** $\boldsymbol{p}$ of the particle is defined as

(4.17) $$\boldsymbol{p} = (m\, dx/ds, m\, dy/ds, m\, dz/ds),$$

where $x(s)$, $y(s)$, $z(s)$ are the spatial coordinates of the particle with respect to $\mathscr{I}$. The **total energy** is given by

(4.18) $$E = (\boldsymbol{p}\cdot\boldsymbol{p} + m^2)^{1/2}$$

and the **four-vector momentum** by

(4.19) $$p = (\boldsymbol{p}, E) = (p_1, p_2, p_3, p_4).$$

Note that

$$p_4^2 - p_3^2 - p_2^2 - p_1^2 = m^2. \tag{4.20}$$

Now ds is Lorentz-invariant and (dx, dy, dz) transforms under the Lorentz group exactly as (x, y, z). Let $x = (x, y, z, t) = (\mathbf{x}, t)$ be the coordinates in $\mathcal{S}$ of an event q such that $t > 0$ and $t^2 - x^2 - y^2 - z^2 = m^2 > 0$. Then we can write

$$x = (\mathbf{x}, t) = (\mathbf{x}, (\mathbf{x}\cdot\mathbf{x} + m^2)^{1/2}). \tag{4.21}$$

Comparing (4.19) and (4.21), we see that both of these vectors must transform in exactly the same manner under $L^\uparrow$. (Here we are assuming that the mass of a particle is the same in all inertial frames.) Since x transforms according to $\mathbf{D}^{(1/2,1/2)}$, so does $\boldsymbol{p}$. In particular the expression for the four-momentum of a particle takes the same form in all inertial systems, as it must in order to be physically meaningful. This shows that expression (4.20) is also Lorentz-invariant. Note, however, that x and $\boldsymbol{p}$ do not transform in the same way under time inversion G. Under G, x goes to $(\mathbf{x}, -t)$ and $\boldsymbol{p}$ goes to $(-\boldsymbol{p}, E)$.

In relativistic quantum physics the states of a one-particle system at time t are given by spinor-valued functions $\boldsymbol{\Psi} = \{\Psi_\mu(x)\}$, $\mu = 1, \ldots, q$, where $x = (x, y, z, t)$. The action of the Poincaré group $\mathcal{P}$ on these state functions is given by

$$[\mathbf{T}(a, A)\Psi]_\mu(x) = \sum_{\nu=1}^{q} T_{\mu\nu}(A)\Psi_\nu(L(A^{-1})(x - a)), \qquad a \in R_4, \quad A \in SL(2), \tag{4.22}$$

where $L(A) \in L^{\uparrow +}$ is given by (1.18)–(1.22) and $T(A)$ is a $q \times q$ matrix rep of $SL(2)$. Here $\mathcal{P}$ is the set of all pairs $\{a, A\}$ with group product

$$\{a_1, A_1\}\{a_2, A_2\} = \{a_1 + L(A_1)a_2, A_1A_2\}. \tag{4.23}$$

The map

$$\{a, A\} \longrightarrow \{a, L(A)\}$$

is a homomorphism of $\mathcal{P}$ onto the ordinary Poincaré group P, (2.5), which covers each element of P exactly twice.

The construction of state functions for relativistic k-particle systems is analogous to that discussed in (11.6), Chapter 7, and is left to the reader. Furthermore, we shall be concerned only with the group-theoretic properties of the transformation (4.22) and shall omit any discussion of Hilbert spaces containing the state vectors $\boldsymbol{\Psi}$. For such a discussion see the work of Schweber [1].

In general, functions $\boldsymbol{\Psi}$ which transform under $\mathcal{P}$ by (4.22) are called **spinor fields.** If $T(A) = T(-A)$ for all $A \in SL(2)$ then (4.22) defines a single-valued rep of P and the functions are called **tensor fields.** Among the important tensor and spinor fields of relativistic physics are the four-momentum and the vector four-potential, $T \cong D^{(1/2,1/2)}$, and the Dirac electron field, $T \cong D^{(1/2,0)} \oplus D^{(0,1/2)}$ (Roman [1], Landau and Lifshitz [3]).

If we set $a = \theta$ in expression (4.22) we obtain a rep of $SL(2)$. The Lie algebra rep of $sl(2)$ induced by this group action takes the form

$$\mathcal{J}_j = \mathcal{S}_j + L_j, \qquad \mathcal{B}_j = \mathcal{K}_j + B_j \qquad j = 1, 2, 3, \tag{4.24}$$

where the Lie derivatives L_j, B_j are given by (4.11) and the matrices

$$\mathcal{S}_j = \frac{d}{dt} T(\exp t\mathcal{J}_j)|_{t=0}, \qquad \mathcal{K}_j = \frac{d}{dt} T(\exp t\mathcal{B}_j)|_{t=0}, \qquad \mathcal{J}_j, \mathcal{B}_j \in sl(2), \tag{4.25}$$

act on the spinor indices of $\mathbf{\Psi}$. Suppose we restrict the group rep $\mathbf{T}(A) = \mathbf{T}(\theta, A)$ to the subgroup $SU(2)$. Then the matrix rep $T(A)$ will decompose into a direct sum of irred reps $D^{(u)}$ of $SU(2)$. The spinor components of $\mathbf{\Psi}$ can always be chosen so that $T(A) \mid SU(2)$ explicitly exhibits this direct sum decomposition:

$$T(A) = \begin{pmatrix} T^{(u_1)}(A) & & & Z \\ & \cdot & & \\ & & \cdot & \\ Z & & \cdot & T^{(u_k)}(A) \end{pmatrix}, \qquad A \in SU(2).$$

Thus, on restriction to $SU(2)$ the field $\mathbf{\Psi}$ transforms as a sum of spinor fields of weights $s = u_1, \ldots, u_k$ with respect to $SU(2)$. This last statement is meant in the sense of (8.19), Chapter 7. The above remarks constitute the relativistic interpretation of spin. If a particle state function transforms according to (4.22) with $T \cong D^{(u,v)}$ in a relativistic theory then the formula

$$\mathbf{D}^{(u,v)} \mid SU(2) \cong \mathbf{D}^{(u+v)} \oplus \mathbf{D}^{(u+v-1)} \oplus \cdots \oplus \mathbf{D}^{(|u-v|)} \tag{4.26}$$

shows that this particle can have spins $s = u + v, u + v - 1, \ldots, |u - v|$. However, there is no known particle with more than one spin. For particles transforming according to $\mathbf{D}^{(u,0)}$ or $\mathbf{D}^{(0,u)}$ this restriction to one spin is achieved automatically: $s = u$. However, for particles which transform according to $\mathbf{D}^{(u,v)}$ with $u, v > 0$ it is necessary to subject the spinor function $\mathbf{\Psi}$ to certain additional constraints which, in a fixed inertial coordinate system, require that all components Ψ_μ of $\mathbf{\Psi}$ are zero except those transforming according to a single rep $\mathbf{D}^{(s)}$ of $SU(2)$.

For example, the photon transforms according to the four-dimensional representation $\mathbf{D}^{(1/2,1/2)}$. Since $\mathbf{D}^{(1/2,1/2)} \mid SU(2) \cong \mathbf{D}^{(1)} \oplus \mathbf{D}^{(0)}$ we would expect the photon to have spins one and zero. However, the system of equations obeyed by the photon includes a supplementary condition which suppresses the component transforming according to $\mathbf{D}^{(0)}$ and we say that the photon has spin one (see the work of Jauch and Rohrlich [1]). The Dirac electron field transforms according to $\mathbf{D}^{(1/2,0)} \oplus \mathbf{D}^{(0,1/2)}$. On restriction to $SU(2)$ we obtain $\mathbf{D}^{(1/2)} \oplus \mathbf{D}^{(1/2)}$, so the electron has the single spin $\frac{1}{2}$, even though $\mathbf{D}^{(1/2)}$ occurs with multiplicity two.

8.5 Lorentz-Invariant Equations

In Section 8.2 we enunciated the basic principle of relativistic physics: The equations and laws of a physical theory must have the same form in any inertial coordinate system. Stated another way, the equations of a physical theory must maintain their form under the action of the Poincaré group. We shall use this principle to classify, under suitable conditions, the possible linear differential equations which can appear in a relativistic theory. Our analysis will be analogous to that for the Euclidean-invariant equations in Section 7.11.

Let $\Psi_\mu(x)$ be a q-component spinor field transforming according to the rule (4.22) under the Poincaré group. We suppose that the components Ψ_μ satisfy a system of q linear partial differential equations in the independent variables $x = (x, y, z, t)$. By introducing new components if necessary we can assume the system takes the form

$$\left(C_1\frac{\partial}{\partial x} + C_2\frac{\partial}{\partial y} + C_3\frac{\partial}{\partial z} + C_4\frac{\partial}{\partial t} + D\right)\mathbf{\Psi}(x) = 0, \tag{5.1}$$

where C_j and D are $q \times q$ matrix functions of x and $\mathbf{\Psi}(x) = (\Psi_\mu(x))$ is a q-component column vector. We will investigate the conditions under which the system (5.1) maintains its form under the action (4.22) of $\mathcal{P}$. First, it is clear that (5.1) is invariant under all translations of coordinates if and only if the matrices C_j and D are constant. Assuming these matrices constant we reduce the problem to one of invariance under $SL(2)$ (the homogeneous Lorentz group):

$$[\mathbf{T}(A)\mathbf{\Psi}]_\mu(x) = \Psi_\mu'(x) = \sum_{\nu=1}^{q} T_{\mu\nu}(A)\Psi_\nu(L(A^{-1})x), \tag{5.2}$$

or

$$\Psi_\mu'(x') = \sum_\nu T_{\mu\nu}(A)\Psi_\nu(x), \qquad x' = L(A)x, \tag{5.3}$$

where $T(A)$ is a $q \times q$ matrix rep of $SL(2)$. Lorentz invariance of (5.1) means exactly that if we replace x by x' and $\Psi_\mu(x)$ by $\Psi_\mu'(x')$ then the resulting system of equations is equivalent to the original system, i.e., the primed equations are linear combinations of the unprimed equations and conversely.

To simplify the discussion we assume D is nonsingular. Then multiplying (5.1) on the left by D^{-1} we obtain the equivalent system

$$\left(L_1\frac{\partial}{\partial x} + L_2\frac{\partial}{\partial y} + L_3\frac{\partial}{\partial z} + L_4\frac{\partial}{\partial t}\right)\mathbf{\Psi} = \kappa\mathbf{\Psi}, \tag{5.4}$$

where κ is a nonzero constant. (We could choose $\kappa = -1$ but it is preferable to leave it arbitrary.)

We may assume without loss of generality that the matrix rep $T(A)$ has already been decomposed into a direct sum of irred reps:

(5.5)

$$T(A) = \begin{pmatrix} \overbrace{D^{(0,0)}(A) \; \cdots \; D^{(0,0)}(A)}^{\alpha_{00}} & & & Z \\ & \ddots & & \\ & & \overbrace{D^{(u,v)}(A) \; \cdots \; D^{(u,v)}(A)}^{\alpha_{uv}} & \\ Z & & & \ddots \end{pmatrix},$$

where $D^{(u,v)}(A)$ is a matrix realization of $\mathbf{D}^{(u,v)}$ and α_{uv} is the multiplicity of $\mathbf{D}^{(u,v)}$ in $T(A)$. We can label the components of $\boldsymbol{\Psi}$ as $\Psi^{uv}_{mn,k}$, the spin component corresponding to the canonical basis vector $f^{(u,v)}_{mn}$ in the kth occurrence of $\mathbf{D}^{(u,v)}$ in (5.5). Here we are using the basis (3.2). Thus,

$$[\mathbf{T}(A)\boldsymbol{\Psi}]^{uv}_{mn,k}(x) = \sum_{m'=-u}^{u} \sum_{n'=-v}^{v} D^{(u,v)}_{mn,m'n'}(A)\Psi^{uv}_{m'n',k}(L(A^{-1})x). \tag{5.6}$$

The partial derivatives on the left-hand side of (5.4) can be expressed as linear combinations of the derivatives $\partial_{\pm 1/2, \pm 1/2}$, (4.14), which form a canonical basis for a realization of $\mathbf{D}^{(1/2,1/2)}$. Thus the left-hand side of (5.4) is a linear combination of terms $\partial_{\pm 1/2, \pm 1/2}\Psi^{uv}_{m'n',k}$. For fixed u, v, and k, and m', n' ranging over $-u \le m' \le u$, $-v \le n' \le v$, these $4(2u+1)(2v+1)$ quantities form a basis for the rep

$$\mathbf{D}^{(1/2,1/2)} \otimes \mathbf{D}^{(u,v)} \cong \mathbf{D}^{(u+1/2,v+1/2)} \oplus \mathbf{D}^{(u+1/2,v-1/2)} \oplus \mathbf{D}^{(u-1/2,v+1/2)} \oplus \mathbf{D}^{(u-1/2,v-1/2)}. \tag{5.7}$$

If either u or v is zero, this expression has an obvious modification. By (5.7) and (3.6), the new basis functions

$$h^{u'v'}_{m'n'}(uv, k) = \sum_{mnjl} C(\tfrac{1}{2}, j; u, m \mid u', m')C(\tfrac{1}{2}, l; v, n \mid v', n')\, \partial_{j,l}\Psi^{uv}_{mn,k} \tag{5.8}$$

transform irreducibly according to $\mathbf{D}^{(u',v')}$. Here $u' = u \pm \frac{1}{2}$ and $v' = v \pm \frac{1}{2}$ for $u, v > 0$. Again the results must be slightly modified if either $u = 0$ or $v = 0$.

Due to the unitarity of the CG coefficients we can uniquely express each of the terms $\partial_{j,l}\Psi^{uv}_{mn,k}$ on the left-hand side of (5.4) as linear combinations of the $h^{u'v'}_{m'n'}(uv, k)$. The resulting system takes the form

$$\sum_{m'n'k_1u_1v_1} A^{uv,u'v',k}_{mn,m'n'}(u_1v_1, k_1)h^{u'v'}_{m'n'}(u_1v_1, k_1) = \kappa\Psi^{uv}_{mn,k}. \tag{5.9}$$

We consider a subsystem of $(2u+1)(2v+1)$ equations (5.9) for which

u, v, k are fixed and $-u \le m \le u$, $-v \le n \le v$. From (5.3) and (5.6) we obtain

$$\Psi'^{uv}_{mn,k}(x') = \sum_{jl} D^{(u,v)}_{mn,jl}(A)\Psi^{uv}_{jl,k}(x), \tag{5.10}$$

$$h'^{u'v'}_{m'n'}(u_1v_1, k_1)(x') = \sum_{j'l'} D^{(u',v')}_{m'n',j'l'}(A)h^{u'v'}_{j'l'}(u_1v_1, k_1)(x). \tag{5.11}$$

If follows that our subsystem will maintain its form under the action of $SL(2)$ if and only if the left-hand side of the subsystem also transforms according to $\mathbf{D}^{(u,v)}$. The necessary and sufficient condition for invariance is that all constants $A^{uv,u'v'}$ are zero except those for which $u = u'$, $v = v'$, $m = m'$, and $n = n'$. Furthermore, the nonzero constants must be independent of the spin indices m and n. Thus, any invariant system takes the form

$$\sum_{u_1v_1k_1} A^{uv,k}_{u_1v_1,k_1} h^{uv}_{mn}(u_1v_1, k_1) = \kappa\Psi^{uv}_{mn,k}, \qquad k = 1, \ldots, \alpha_{uv} \tag{5.12}$$

where h^{uv}_{mn} is given by (5.8) and the pair (u, v) ranges over all irred reps in $T(A)$. The constants $A^{uv,k}_{u_1v_1,k_1}$ are arbitrary and there is one equation for each component of $\mathbf{\Psi}$. We see from this analysis that the component $\Psi^{uv}_{mn,k}$ on the right is coupled with those components $\Psi^{u_1v_1}_{m_1n_1,k_1}$ on the left such that

$$u_1 = \begin{cases} u+\frac{1}{2}, u-\frac{1}{2} & \text{if } u > 0 \\ \frac{1}{2} & \text{if } u = 0, \end{cases} \qquad v_1 = \begin{cases} v+\frac{1}{2}, v-\frac{1}{2} & \text{if } v > 0 \\ \frac{1}{2} & \text{if } v = 0. \end{cases} \tag{5.13}$$

Note that there are no nontrivial invariant equations in which the spinor indices transform according to a single irred rep $\mathbf{D}^{(u,v)}$. With a single $\mathbf{D}^{(u,v)}$ we could not achieve a coupling (5.13).

In case the matrix D in (5.1) is singular or not square the analogous discussion in Section 7.11 is applicable. If $D = Z$ we can construct invariant equations of the form (5.12) with $\kappa = 0$ although the number of such equations need not be equal to the number of components of $\mathbf{\Psi}$. In this case it is possible to construct invariant equations in which the spinor indices transform according to a single irred rep $\mathbf{D}^{(u,v)}$. For an arbitrary singular matrix D one can construct systems of the form (5.12) in which κ is zero for some equations and nonzero for others in the system.

Naimark [2] presents a complicated derivation of results equivalent to Eq. (5.12) based on computations using the Lie algebra of the Lorentz group. His derivation has the useful feature that it generalizes to the case where the matrices L_j in (5.4) are infinite. In this case the infinite-dimensional irred reps of the Lorentz group may appear.

It is worth mentioning that all Lorentz-invariant equations are automatically Euclidean-invariant since $\mathcal{E}^+(3)$ is a subgroup of $\mathcal{P}$. Thus the Lorentz-invariant equations are already contained in the analysis of Section 7.11.

Our results must be modified if we demand invariance under the complete Lorentz group $L^{\uparrow}$ obtained by adding space reflection S to the proper Lorentz group. In Section 4.3 we showed that the irred reps of $L^{\uparrow}$ were

(5.14) $$\mathbf{D}^{(u,v)} \oplus \mathbf{D}^{(v,u)}, \qquad u > v, \qquad \mathbf{S}f_{mn}^{(u,v)} = f_{nm}^{(v,u)},$$

(5.15) $$\mathbf{D}_{\pm}^{(u,u)}, \qquad \mathbf{S}f_{mn}^{(u,u)} = \pm f_{nm}^{(u,u)},$$

where $\mathbf{S}$ is the operator corresponding to space reflection in each rep space. It follows from (4.14) that $\partial_{j,l} \longrightarrow -\partial_{l,j}$ under space reflection, so the $\partial_{j,l}$ form a canonical basis for the rep $\mathbf{D}_{-}^{(1/2,1/2)}$ of $L^{\uparrow}$.

Suppose $\boldsymbol{\Psi} = \{\Psi_\mu(x)\}$ is a spinor field transforming under $L^{\uparrow}$. In addition to the transformation equations (5.2) we have

(5.16) $$[\mathbf{T}(S)\boldsymbol{\Psi}]_\mu(x) = \Psi_\mu{}'(x) = \sum_{\nu=1}^{q} T_{\mu\nu}(S)\Psi_\nu(S^{-1}x),$$

where the matrices $T(A)$, $T(S)$ generate a rep of $L^{\uparrow}$ (possibly double-valued). Here $T(S)^2 = E$. The matrix rep can be decomposed into a direct sum of irred reps of $L^{\uparrow}$. (Prove it!) Thus, each component $\Psi_{mn,k}^{uv}$, $u \neq v$, is associated with a component $\Psi_{nm,k}^{vu}$ such that

(5.17) $$\mathbf{T}(S)\Psi_{mn,k}^{uv}(x') = \Psi_{nm,k}^{vu}(x), \qquad x' = Sx,$$
$$-u \leq m \leq u, \quad -v \leq n \leq v.$$

For $u = v$ there are possible components $\Psi_{mn,k}^{uu+}$ and $\Psi_{mn,k}^{uu-}$ such that

(5.18) $$\mathbf{T}(S)\Psi_{mn,k}^{uu\pm}(x') = \pm\Psi_{nm,k}^{uu\pm}(x).$$

We assume $\boldsymbol{\Psi}$ satisfies the equations (5.4) with $\kappa \neq 0$ and require that this system is $L^{\uparrow}$-invariant. Clearly, the system is $L^{\uparrow +}$ invariant so it can be expressed in the form (5.12). To guarantee $L^{\uparrow}$-invariance we need only determine the requirements on the constants $A_{u_1v_1,k_1}^{uv,k}$ in order that the system of equations remains invariant under space inversion.

Choose one of the equations (5.12) and replace x by $x' = Sx$ and $\Psi_{m'n',k'}^{u'v'}(x)$ by $\mathbf{T}(S)\Psi_{m'n',k'}^{u'v'}(x')$ on both sides of the equation. If $u \neq v$ then the right-hand side becomes $\kappa\Psi_{nm,k}^{uv}$, while the vectors $h_{mn}^{uv}(u_1v_1, k_1)$ become

(5.19)
$$\mathbf{T}(S)h_{mn}^{uv}(u_1v_1, k_1) = \begin{cases} -h_{nm}^{vu}(v_1u_1, k_1) & \text{if} \quad u_1 \neq v_1 \\ \mp h_{mn}^{vu}(u_1\pm, k_1) & \text{if} \quad u_1 = v_1 \text{ and } \Psi_{m_1n_1,k_1}^{u_1v_1} = \Psi_{m_1n_1,k_1}^{u_1u_1\pm}. \end{cases}$$

Here we have used (5.8), (5.17), (5.18), and the fact that $\partial'_{j,l} = -\partial_{l,j}$. If the system is $L^{\uparrow}$-invariant then this transformed equation must be identical with the original equation for the component $\kappa\Psi_{nm,k}^{uv}$. But from (5.19) this is possible if and only if

(5.20) $$A_{u_1v_1,k_1}^{uv,k} = \begin{cases} -A_{v_1u_1,k_1}^{vu,k} & \text{for} \quad u_1 \neq v_1 \\ \mp A_{u_1\pm,k_1}^{vu,k} & \text{for} \quad u_1 = v_1, \quad \text{parity} \quad \pm. \end{cases}$$

If the term on the right-hand side is $\kappa\Psi_{mn,k}^{u,u\pm}$ then under space inversion it is mapped to $\pm\kappa\Psi^{u,u\pm}_{nm,k}$. Again, for $L^{\uparrow}$-invariance the transformed equation must be the same as the original equation for the component $\kappa\Psi^{uu\pm}_{nm,k}$. This is possible if and only if

$$A^{u\pm,k}_{u_1v_1,k_1} = \begin{cases} \mp A^{u\pm,k}_{v_1u_1,k_1} & \text{for } u_1 \neq v_1, \\ \mp A^{u\pm,k}_{u_1+,k_1} & \text{for } u_1 = v_1, \quad \text{positive parity}, \\ \pm A^{u\pm,k}_{u_1-,k_1} & \text{for } u_1 = v_1, \quad \text{negative parity}, \end{cases} \tag{5.21}$$

as can be shown by a proof similar to that of (5.20). Expressions (5.20) and (5.21) are necessary and sufficient for $L^{\uparrow}$-invariance of the system (5.12).

One of the simplest examples of an $L^{\uparrow +}$-invariant equation is the **Klein–Gordon Equation**

$$\left(\frac{\partial^2}{\partial x^2} + \frac{\partial^2}{\partial y^2} + \frac{\partial^2}{\partial z^2} - \frac{\partial^2}{\partial t^2}\right)\varphi(x) = m_0{}^2\varphi(x). \tag{5.22}$$

Here $\varphi(x)$ transforms as a scalar field under $\mathcal{P}$:

$$[\mathbf{T}(a, \Lambda)\varphi](x) = \varphi(\Lambda^{-1}(x - a)). \tag{5.23}$$

In relativistic physics this equation describes fields corresponding to particles with mass m_0. Since the spin index of φ transforms according to $\mathbf{D}^{(0,0)}$ and $\mathbf{D}^{(0,0)} \mid SO(3) \cong \mathbf{D}^{(0)}$, these particles must have spin zero.

To see the connection between the Klein–Gordon equation and elementary particles recall that in classical relativistic physics the relation between momentum and energy of a particle with mass m_0 is

$$E^2 - p_1{}^2 - p_2{}^2 - p_3{}^2 = m_0{}^2, \tag{5.24}$$

[Eq. (4.20)]. In quantum physics we associate the classical momenta and energy with differential operators according to the rule

$$\begin{aligned} p_1 &\longleftrightarrow i\,\partial/\partial x = \mathbf{P}_1, & p_2 &\longleftrightarrow i\,\partial/\partial y = \mathbf{P}_2, \\ p_3 &\longleftrightarrow i\,\partial/\partial z = \mathbf{P}_3, & E &\longleftrightarrow i\,\partial/\partial t = \mathbf{H}. \end{aligned} \tag{5.25}$$

From (5.24) and the usual correspondence principle between classical and quantum physics we see that the state function $\varphi(\mathbf{x}, t)$ describing a particle of mass m_0 satisfies the equation

$$(\mathbf{H}^2 - \mathbf{P}_1{}^2 - \mathbf{P}_2{}^2 - \mathbf{P}_3{}^2)\varphi(\mathbf{x}, t) = m_0{}^2\varphi(\mathbf{x}, t). \tag{5.26}$$

Making the substitutions (5.25), we obtain the Klein–Gordon equation.

Let us write (5.22) in the canonical form (5.12). We introduce four new components $\varphi_{m,n}(x) = \partial_{m,n}\varphi(x)$, $m, n = \pm\frac{1}{2}$, which form a canonical basis for $\mathbf{D}^{(1/2,1/2)} \otimes \mathbf{D}^{(0,0)} \cong \mathbf{D}^{(1/2,1/2)}$. From (4.14), the Klein–Gordon equation is equivalent to the system

$$\begin{gathered} \partial_{m,n}\varphi = \varphi_{m,n}, \qquad m, n = \pm\tfrac{1}{2} \\ \tfrac{1}{2}(-\partial_{1/2,1/2}\varphi_{-1/2,-1/2} + \partial_{1/2,-1/2}\varphi_{-1/2,1/2} + \partial_{-1/2,1/2}\varphi_{1/2,-1/2} \\ - \partial_{-1/2,-1/2}\varphi_{1/2,1/2}) = m_0{}^2\varphi. \end{gathered} \tag{5.27}$$

The indices of the spinor field $\mathbf{\Phi} = (\varphi_{m,n}, \varphi)$ transform according to $\mathbf{D}^{(1/2,1/2)} \oplus \mathbf{D}^{(0,0)}$. Now the most general $L^{\uparrow +}$-invariant system of equations for such a field takes the form

$$(5.28)\qquad \begin{aligned} &a \sum_{jl=-1/2}^{1/2} C(\tfrac{1}{2}, j; 0, 0 \mid \tfrac{1}{2}, m) C(\tfrac{1}{2}, l; 0, 0 \mid \tfrac{1}{2}, n)\, \partial_{j,l}\varphi = \kappa \varphi_{m,n}, \\ &b \sum_{jlm_1n_1=-1/2}^{1/2} C(\tfrac{1}{2}, j; \tfrac{1}{2}, m_1 \mid 0, 0) C(\tfrac{1}{2}, l; \tfrac{1}{2}, n_1 \mid 0, 0)\, \partial_{j,l}\varphi_{m,n} = \kappa\varphi. \end{aligned}$$

According to table (7.27), Section 7.7 these two systems are identical provided $\kappa = 1$, $a = 1$, $b = -1/m_0^2$. [Recall that$C(\frac{1}{2}, j; 0, 0 \mid \frac{1}{2}, m) = \delta_{jm}$.]

Now consider the behavior of (5.22) under space inversion. Under the group $L^{\uparrow}$, φ transforms as $\mathbf{D}_+^{(0,0)}$ or $\mathbf{D}_-^{(0,0)}$, i.e., as a **scalar** or a **pseudoscalar.** Thus $[\mathbf{T}(S)\varphi](x) = \pm\varphi(Sx)$. In either case it is obvious that the Klein–Gordon equation remains invariant under space inversion. However, it is instructive to verify this result for the system (5.28). If φ is a scalar then $\mathbf{\Phi} = (\varphi_{m,n}, \varphi)$ transforms as $\mathbf{D}_-^{(1/2,1/2)} \oplus \mathbf{D}_+^{(0,0)}$. It then follows from (5.21) that the system is $L^{\uparrow}$-invariant. Similarly, if φ is pseudoscalar then $\mathbf{\Phi}$ transforms as $\mathbf{D}_+^{(1/2,1/2)} \oplus \mathbf{D}_-^{(0,0)}$ and (5.28) is $L^{\uparrow}$-invariant.

Note that (5.27) is Lorentz-invariant even if $m_0 = 0$, in which case it corresponds to the system (5.1) with D singular.

We cannot write a nontrivial first-order system of equations for a spinor field transforming as $\mathbf{D}^{(0,0)}$. The next simplest possibility is $\mathbf{D}^{(1/2,0)}$. A particle described by such a spinor field would have spin $\frac{1}{2}$. As we have already remarked, this field cannot satisfy a system of the form (5.12) with $\kappa \neq 0$ since $\mathbf{D}^{(1/2,0)}$ cannot couple with itself. However, for $\kappa = 0$ the relation $\mathbf{D}^{(1/2,1/2)} \otimes \mathbf{D}^{(1/2,0)} \cong \mathbf{D}^{(1,1/2)} \oplus \mathbf{D}^{(0,1/2)}$ suggests the system

$$\sum_{j=-1/2}^{1/2} C(\tfrac{1}{2}, j; \tfrac{1}{2}, -j \mid 0, 0)\, \partial_{j,l}\Psi_{-j} = 0, \qquad l = \pm\tfrac{1}{2},$$

or

$$(5.29)\qquad \partial_{1/2,l}\Psi_{-1/2} - \partial_{-1/2,l}\Psi_{1/2} = 0, \qquad l = \pm\tfrac{1}{2}.$$

The left-hand side of this system transforms as $\mathbf{D}^{(0,1/2)}$. (We reject a system whose left-hand side transforms according to $\mathbf{D}^{(1,1/2)}$ since it would subject the two spinor components to six conditions). Expression (5.29) is the equation of the **two-component neutrino.** This equation cannot possibly be invariant under space reflection because $\mathbf{\Psi}$ transforms as $\mathbf{D}^{(1/2,0)}$. Thus $\mathbf{T}(S)\mathbf{\Psi}$ transforms as $\mathbf{D}^{(0,1/2)}$ and the system does not admit S as a symmetry. It is easy to verify the formulas

$$\left(\frac{\partial^2}{\partial x^2} + \frac{\partial^2}{\partial y^2} + \frac{\partial^2}{\partial z^2} - \frac{\partial^2}{\partial t^2}\right)\Psi_{\pm 1/2}(x) = 0$$

from (5.29), so each component of the field is a solution of the Klein–Gordon equation with $m_0 = 0$. We conclude that the neutrino has spin $\frac{1}{2}$, mass zero, and does not conserve parity, i.e., does not transform according to a rep of the complete Lorentz group.

From experimental results the electron is known to have spin $\frac{1}{2}$, nonzero mass, and to conserve parity in those reactions in which it takes part. Thus we would expect a spinor field $\boldsymbol{\Psi}$ corresponding to an electron to have spin $\frac{1}{2}$ and to satisfy an $L^{\uparrow}$-invariant first-order system. The simplest possibility is that $\boldsymbol{\Psi}$ transforms as $\mathbf{D}^{(1/2,0)} \oplus \mathbf{D}^{(0,1/2)}$ under $L^{\uparrow}$. Then Eq. (5.12) take the form

$$(5.30)\qquad \begin{aligned} &a \sum_{l=-1/2}^{1/2} C(\tfrac{1}{2}, l; \tfrac{1}{2}, -l \mid 0, 0)\, \partial_{m,l} \Psi_{-l}^{-} = \kappa \Psi_m^{+} \\ &b \sum_{j=-1/2}^{1/2} C(\tfrac{1}{2}, j; \tfrac{1}{2}, -j \mid 0, 0)\, d_{j,m} \Psi_{-j}^{+} = \kappa \Psi_m^{-}, \qquad m = \pm\tfrac{1}{2}, \end{aligned}$$

where $\boldsymbol{\Psi} = \{\Psi_m^{+}, \Psi_m^{-}\}$ is a four-component spinor, $\{\Psi_m^{+}\}$ forms a canonical basis for $\mathbf{D}^{(1/2,0)}$, and $\{\Psi_m^{-}\}$ forms a canonical basis for $\mathbf{D}^{(0,1/2)}$. Under space inversion Ψ_m^{+} goes to Ψ_m^{-} and Ψ_m^{-} goes to Ψ_m^{+}. Thus the system (5.30) is invariant under space inversion if and only if $a = -b$. If we choose $a = -b = \sqrt{2}$, (5.30) becomes

$$(5.31)\qquad \begin{aligned} \partial_{m,1/2} \Psi_{-1/2}^{-} - \partial_{m,-1/2} \Psi_{1/2}^{-} &= \kappa \Psi_m^{+}, \\ -\partial_{1/2,m} \Psi_{-1/2}^{+} + \partial_{-1/2,m} \Psi_{1/2}^{+} &= \kappa \Psi_m^{-}, \end{aligned}$$

or in matrix form

$$(5.32)\qquad \left(L_1 \frac{\partial}{\partial x} + L_2 \frac{\partial}{\partial y} + L_3 \frac{\partial}{\partial z} + L_4 \frac{\partial}{\partial t}\right) \boldsymbol{\Psi} = m_0 \boldsymbol{\Psi}, \qquad i\kappa = m_0,$$

where

$$(5.33)\qquad \begin{aligned} &L_1 = i \begin{pmatrix} 0 & 0 & 0 & 1 \\ 0 & 0 & 1 & 0 \\ 0 & -1 & 0 & 0 \\ -1 & 0 & 0 & 0 \end{pmatrix}, \qquad L_2 = i \begin{pmatrix} 0 & 0 & 0 & -i \\ 0 & 0 & i & 0 \\ 0 & i & 0 & 0 \\ -i & 0 & 0 & 0 \end{pmatrix}, \\ &L_3 = i \begin{pmatrix} 0 & 0 & -1 & 0 \\ 0 & 0 & 0 & 1 \\ 1 & 0 & 0 & 0 \\ 0 & -1 & 0 & 0 \end{pmatrix}, \qquad L_4 = i \begin{pmatrix} 0 & 0 & -1 & 0 \\ 0 & 0 & 0 & -1 \\ -1 & 0 & 0 & 0 \\ 0 & -1 & 0 & 0 \end{pmatrix}. \\ &\boldsymbol{\Psi} = \begin{pmatrix} \Psi_{1/2}^{+} \\ \Psi_{-1/2}^{+} \\ \Psi_{1/2}^{-} \\ \Psi_{-1/2}^{-} \end{pmatrix}. \end{aligned}$$

Note that the matrices L_j satisfy the relations

(5.34) $$L_jL_k + L_kL_j = 2G_{jk},$$

Where $G = (G_{jk})$ is the matrix (1.1).

From (5.32)

(5.35) $$\Big(\sum_{j=1}^{4} L_j\, \partial/\partial x_j\Big)^2 \mathbf{\Psi} = m_0{}^2\mathbf{\Psi}.$$

On the other hand, we can use (5.34) to evaluate the left side of this expression:

(5.36) $$\Big(\sum_{j=1}^{4} L_j \frac{\partial}{\partial x_j}\Big)^2 = \sum_{j,k=1}^{4} L_jL_k \frac{\partial^2}{\partial x_j\,\partial x_k} = \frac{\partial^2}{\partial x^2} + \frac{\partial^2}{\partial y^2} + \frac{\partial^2}{\partial z^2} - \frac{\partial^2}{\partial t^2}.$$

Thus, (5.35) becomes

(5.37) $$\Big(\frac{\partial^2}{\partial x^2} + \frac{\partial^2}{\partial y^2} + \frac{\partial^2}{\partial z^2} - \frac{\partial^2}{\partial t^2}\Big)\mathbf{\Psi} = m_0{}^2\mathbf{\Psi},$$

so each component of $\mathbf{\Psi}$ satisfies the Klein–Gordon equation. The system (5.32) is one form of the **Dirac electron equation.** A solution of this equation corresponds to a particle of mass m_0 and spin $\frac{1}{2}$ which conserves parity. We will investigate other possible forms of the Dirac equation in Section 9.6.

Maxwell's equations for an electromagnetic field in a vacuum provide another important example of a Lorentz-invariant system. See the work of Gel'fand *et al.* [1] for the details.

Problems

8.1 Let y be a four-vector such that $y^tGy = -m^2 < 0$ and $y_4 > 0$. (We say y is **forward-timelike.**) Show that there exists a $\Lambda \in L^{\uparrow +}$ such that $x = \Lambda y$ where $x_1 = x_2 = x_3 = 0$, $x_4 = m$.

8.2 Use the polar decomposition to obtain an alternate proof of Theorem 8.2.

8.3 Let $\mathbf{D}^{(u,v)}$ be a finite-dimensional irred rep of the real Lie group $SL(2, \mathfrak{C})$. Express the $\{f_{mn}^{(u,v)}\}$ basis in terms of the $\{f_k^{(w)}\}$ basis (Section 8.3).

8.4 Verify directly that the operators (3.33) define a global irred unitary rep of $SL(2)$ on $L_2(R_2)$ whose induced Lie algebra rep is equivalent to (w_0, ic).

8.5 Decompose the reps $(\mathbf{D}_+^{(1/2,\,1/2)})^{\otimes n}$ and $(\mathbf{D}_-^{(1/2,\,1/2)})^{\otimes n}$ of $L^\uparrow$ into irred reps for $n =$ 2, 3, 4, 5.

8.6 Discuss the Lorentz invariance of Maxwell's equations using the methods of Section 8.5. Include a discussion of invariance under space inversion. (See Landau and Lifshitz [3] for the relativistic transformation properties of Maxwell's equations.)

8.7 Discuss the simplest relativistic equations suitable for describing a particle with spin $\frac{3}{2}$. Which equations are invariant under space inversion?

8.8 Answer Problem 8.7 for particles with spin one.

Chapter 9

Representations of the Classical Groups

9.1 Representations of the General Linear Groups

In Section 4.3 we computed all the tensor irred reps of the general linear groups $GL(m) = GL(m, \mathfrak{C})$. The reps were determined by Young frames. Here we use Lie-algebraic methods to determine all analytic irred finite-dimensional reps of $GL(m)$. A comparison of the Lie-algebraic method with the method based on Young symmetrizers will yield results which are not easily obtainable from either method alone. The Lie-algebraic approach to the rep theory of $GL(m)$ is patterned closely after the corresponding treatment of $SL(2)$ and $SU(2)$ in Section 7.3.

Recall that $GL(m)$ is an m^2-dimensional complex Lie group. Its Lie algebra $gl(m)$ consists of all $m \times m$ complex matrices. The unimodular group $SL(m) = SL(m, \mathfrak{C})$ is an $(m^2 - 1)$-dimensional subgroup of $GL(m)$ with Lie algebra $sl(m)$ consisting of all $m \times m$ complex matrices of trace zero.

As a basis for $gl(m)$ we choose the matrices $\mathcal{E}_{hj}$, $1 \leq h, j \leq m$, where $\mathcal{E}_{hj}$ is the matrix with a one for the entry in row h, column j, and zeros everywhere else. It is easy to verify the commutation relations

$$(1.1) \qquad [\mathcal{E}_{hj}, \mathcal{E}_{kl}] = \delta_{jk}\mathcal{E}_{hl} - \delta_{lh}\mathcal{E}_{kj}.$$

Denote the diagonal elements of the basis by $\mathcal{H}_h = \mathcal{E}_{hh}$, $h = 1, \ldots, m$. The set $\mathfrak{h}_m$ of all diagonal matrices

$$(1.2) \qquad \mathcal{H} = \sum_{j=1}^{m} \lambda_j \mathcal{H}_j = \begin{pmatrix} \lambda_1 & & & & \\ & \lambda_2 & & & \\ & & \cdot & & \\ & & & \cdot & \\ & & & & \lambda_m \end{pmatrix}$$

forms an m-dimensional commutative subalgebra of $gl(m)$. From (1.1)

$$(1.3) \qquad [\mathcal{H}, \mathcal{E}_{kl}] = (\lambda_k - \lambda_l)\mathcal{E}_{kl}, \qquad [\mathcal{H}, \mathcal{H}'] = Z, \qquad \mathcal{H}, \mathcal{H}' \in \mathfrak{h}_m.$$

It is easy to show from (1.3) that $\mathfrak{h}_m$ is a maximal commutative subalgebra of $gl(m)$, i.e., if $\mathfrak{h}'$ is a commutative subalgebra of $gl(m)$ and $\mathfrak{h}' \supseteq \mathfrak{h}_m$ then $\mathfrak{h}' = \mathfrak{h}_m$.

From now on we use the notation $\mathcal{E}_{kl}$ only for $k \neq l$ and reserve $\mathcal{H}$, $\mathcal{H}_j$ to denote elements of $\mathfrak{h}_m$. The mapping $\mathcal{H} \longrightarrow \operatorname{ad} \mathcal{H}$, where

$$(1.4) \qquad \operatorname{ad} \mathcal{H}(\mathcal{A}) = [\mathcal{H}, \mathcal{A}], \qquad \mathcal{A} \in gl(m),$$

defines a rep of $\mathfrak{h}_m$ on $gl(m)$, the **adjoint representation,** as we saw in Section 5.6. According to (1.3) the element $\mathcal{E}_{kl}$ is a simultaneous eigenvector for all operators ad $\mathcal{H}(\lambda_1, \ldots, \lambda_m)$ and corresponds to the eigenvalue $\lambda_k - \lambda_l$. The nonzero elements of $\mathfrak{h}_m$ are eigenvectors corresponding to the eigenvalue zero.

Note that the eigenvalues $\lambda_k - \lambda_l = \alpha(\mathcal{H})$ are linear functionals on the elements $\mathcal{H} = \sum \lambda_j \mathcal{H}_j$ of $\mathfrak{h}_m$. These $m(m-1)$ distinct functionals for $k \neq l$ are called **roots.** The eigenvector $\mathcal{E}_{kl}$ is called the **branch** belonging to the root $\lambda_k - \lambda_l = \alpha$. We will sometimes write $\mathcal{E}_{kl} = \mathcal{E}_\alpha$ to denote this branch. Furthermore, we define $\mathcal{H}_\alpha = \mathcal{H}_k - \mathcal{H}_l$ for $\alpha = \lambda_k - \lambda_l$.

Lemma 9.1.

(a) If α is a root then $-\alpha$ is a root.

(b) $[\mathcal{E}_\alpha, \mathcal{E}_{-\alpha}] = \mathcal{H}_\alpha \neq Z$.

(c) $[\mathcal{E}_\alpha, \mathcal{E}_\beta] = Z$ if $\alpha + \beta$ is not a root and $\alpha \neq -\beta$.

(d) $[\mathcal{E}_\alpha, \mathcal{E}_\beta] = \pm\mathcal{E}_{\alpha+\beta}$ if $\alpha + \beta$ is a root.

(e) $[\mathcal{H}, \mathcal{E}_\alpha] = \alpha(\mathcal{H})\mathcal{E}_\alpha$, $[\mathcal{H}_\alpha, \mathcal{E}_\alpha] = 2\mathcal{E}_\alpha$.

Proof. (a) If $\alpha = \lambda_k - \lambda_l$ is a root then $-\alpha = \lambda_l - \lambda_k$ is a root. (b) $[\mathcal{E}_\alpha, \mathcal{E}_{-\alpha}] = [\mathcal{E}_{kl}, \mathcal{E}_{lk}] = \mathcal{H}_k - \mathcal{H}_l = \mathcal{H}_\alpha$. (c) If $\alpha = \lambda_h - \lambda_j$ and $\beta = \lambda_k - \lambda_l$ and $\alpha + \beta = \lambda_h + \lambda_k - \lambda_j - \lambda_l$ is not a root or zero, then $j \neq k$, $h \neq l$, and $[\mathcal{E}_{hj}, \mathcal{E}_{kl}] = Z$ by (1.1). (d) If $\alpha + \beta$ is a root then from (c) either $j = k$, in which case $[\mathcal{E}_{hj}, \mathcal{E}_{kl}] = \mathcal{E}_{hl} = \mathcal{E}_{\alpha+\beta}$, or $l = h$, in which case $[\mathcal{E}_{hj}, \mathcal{E}_{kl}] = -\mathcal{E}_{kj} = -\mathcal{E}_{\alpha+\beta}$. (e) This follows directly from (1.3). Q.E.D.

Let ρ be a rep of $gl(m)$ by operators $\rho(\mathcal{A})$, $\mathcal{A} \in gl(m)$, on the complex vector space V. Setting $\rho(\mathcal{E}_\alpha) = E_\alpha$, $\rho(\mathcal{H}) = H = \sum \lambda_j H_j$, we obtain the relations

$$(1.5) \qquad [E_\alpha, E_\beta] = \begin{cases} H_\alpha & \text{if } \alpha + \beta = 0, \\ 0 & \text{if } \alpha + \beta \text{ is nonzero and not a root,} \\ \pm E_{\alpha+\beta} & \text{if } \alpha + \beta \text{ is a root.} \end{cases}$$

$$[H, E_\alpha] = \alpha(\mathcal{H})E_\alpha, \qquad [H, H_j] = 0, \qquad 1 \leq j \leq m.$$

where $\alpha = \lambda_k - \lambda_l$ and $\beta = \lambda_h - \lambda_j$ are roots and $[A, B] = AB - BA$ for linear operators A, B on V. To determine the reps of $gl(m)$ it is enough to determine the possible operators $\{E_\alpha, H_j\}$ which satisfy the commutation relations (1.5).

If ρ is a rep of $gl(m)$ on the n-dimensional space V then V has a basis of simultaneous eigenvectors of the operators $H = \rho(\mathcal{H})$, $\mathcal{H} \in \mathscr{h}_m$. A vector $\mathbf{v} \neq \boldsymbol{\theta}$ is a simultaneous eigenvector if there exist constants $c_1, \ldots, c_m$ such that $H_j\mathbf{v} = c_j\mathbf{v}$, $1 \leq j \leq m$. Then

$$H\mathbf{v} = \Lambda(\mathcal{H})\mathbf{v}, \qquad \Lambda = \sum_{j=1}^{m} c_j\lambda_j \tag{1.6}$$

for $H = \sum \lambda_j H_j$. The linear functional $\Lambda(\mathcal{H})$ on $\mathscr{h}_m$ is called a **weight** and $\mathbf{v}$ is a **weight vector.** Before proving the existence of a basis of weight vectors we note that in the special case where $V = gl(m)$ and ρ is the adjoint rep of $gl(m)$ acting on itself, the weights are just the $m(m-1)$ roots $\alpha = \lambda_k - \lambda_l$ plus the zero weight. The weight vectors $\{\mathcal{E}_\alpha, \mathcal{H}_j\}$ form a basis for the rep space.

Lemma 9.2. If ρ is a rep of $gl(m)$ on V then it contains at least one weight.

Proof. Since V is complex the operator $H_1 = \rho(\mathcal{H}_1)$ has a least one eigenvalue c_1. Let W_1 be the nonzero eigenspace of V corresponding to eigenvalue c_1. If $\mathbf{v} \in W_1$ then $H_1(H_j\mathbf{v}) = H_j(H_1\mathbf{v}) = c_1H_j\mathbf{v}$, so $H_j\mathbf{v} \in W_1$ for $2 \leq j \leq m$. Since W_1 is invariant under H_2, $H_2 \mid W_1$ has an eigenvalue c_2. Let $W_2 \subseteq W_1$ be the corresponding nonzero eigenspace. Then $H_k\mathbf{v} = c_k\mathbf{v}$, $k = 1, 2$, for $\mathbf{v} \in W_2$. Continuing in this manner, we finally obtain a nonzero vector $\mathbf{w} \in W_m$ such that $H_j\mathbf{w} = c_j\mathbf{w}$, $1 \leq j \leq m$. Clearly, $\Lambda = \sum \lambda_j c_j$ is a weight. Q.E.D.

Note. The above proof merely demonstrates that a set of commuting operators on a finite-dimensional vector space has a simultaneous eigenvector.

The next result shows that by applying the operator E_α to a weight vector we may be able to generate a new weight.

Lemma 9.3. Let $\mathbf{v}$ be a weight vector with weight Λ. If α is a root and $E_\alpha\mathbf{v} \neq \boldsymbol{\theta}$ then $\Lambda + \alpha$ is a weight with weight vector $E_\alpha\mathbf{v}$.

Proof. If $H\mathbf{v} = \Lambda\mathbf{v}$ then $H(E_\alpha\mathbf{v}) = E_\alpha H\mathbf{v} + \alpha E_\alpha\mathbf{v} = (\Lambda + \alpha)E_\alpha\mathbf{v}$, as follows from the second relation (1.5). Thus $\Lambda + \alpha$ is a weight if $E_\alpha\mathbf{v} \neq \boldsymbol{\theta}$. Q.E.D.

Theorem 9.1. If ρ is an irred rep of $gl(m)$ on V then V contains a basis of weight vectors.

Proof. By Lemma 9.2 there is a weight vector $\mathbf{v} \neq \boldsymbol{\theta}$ in V with weight Λ. Consider the set $\mathcal{W}$ of all vectors of the form

$$\mathbf{v}, \quad E_{\alpha_1}\mathbf{v}, \quad E_{\alpha_1}E_{\alpha_2}\mathbf{v}, \ldots, \quad E_{\alpha_1} \cdots E_{\alpha_k}\mathbf{v}, \ldots,$$

where the α_j run over all roots of $gl(m)$. By Lemma 9.3, each nonzero element $E_{\alpha_1} \cdots E_{\alpha_k}\mathbf{v}$ of $\mathcal{W}$ is a weight vector with weight $\Lambda + \alpha_1 + \cdots + \alpha_k$. Let W be the subspace of V spanned by the elements of $\mathcal{W}$. By construction W is invariant under all operators E_α and H. Hence the nonzero subspace W is invariant under ρ. Since ρ is irred, $W = V$. Now choose a maximal linearly independent set of vectors from $\mathcal{W}$. This set is clearly a basis of weight vectors for V. Q.E.D.

The proof of this theorem is valid only for irred reps. However, we will show later that every finite-dimensional rep of $gl(m)$ can be decomposed into a direct sum of irred reps. Thus, Theorem 9.1 is true for all reps.

Let ρ be an irred rep of $gl(m)$ and let $\{\mathbf{v}_j : j = 1, \ldots, n\}$ be a basis of weight vectors from ρ with weights Λ_j. Then every weight Λ of ρ is one of the Λ_j. Indeed if $\Lambda \neq \Lambda_j$ for any j then there exists an $\mathcal{H} \in \mathscr{h}_m$ such that $\Lambda(\mathcal{H}) \neq \Lambda_j(\mathcal{H})$, $1 \leq j \leq n$. This means that the nonzero eigenspace of $H = \rho(\mathcal{H})$ corresponding to eigenvalue $\Lambda(\mathcal{H})$ is linearly independent of the eigenspaces corresponding to the eigenvalues $\Lambda_j(\mathcal{H})$. However, the latter eigenspaces span V by Theorem 9.1. This is a contradiction, so no such weight Λ exists.

Corollary 9.1. If ρ is an n-dimensional rep of $gl(m)$ there are at most n distinct weights.

Let α be a root. Since the rep ρ has only a finite number of weights there must exist a weight Λ^* such that $\Lambda^* + \alpha$ is not a weight. Let $\mathbf{v}_0$ be a weight vector corresponding to Λ^*, so

$$H\mathbf{v}_0 = \Lambda^*(\mathcal{H})\mathbf{v}_0, \qquad E_\alpha \mathbf{v}_0 = \boldsymbol{\theta} \tag{1.7}$$

by Lemma 9.3. We define a sequence of weight vectors recursively by

$$E_{-\alpha}\mathbf{v}_j = \mathbf{v}_{j+1}, \qquad j = 0, 1, 2, \ldots. \tag{1.8}$$

By Lemma 9.3,

$$H\mathbf{v}_j = (\Lambda^* - j\alpha)(\mathcal{H})\mathbf{v}_j, \tag{1.9}$$

so either $\mathbf{v}_j = \boldsymbol{\theta}$ or $\mathbf{v}_j$ is a weight vector with weight $\Lambda^* - j\alpha$. Since ρ is finite-dimensional there must exist a positive integer q such that $\mathbf{v}_q \neq \boldsymbol{\theta}$ and $\mathbf{v}_{q+1} = \boldsymbol{\theta}$. The $q + 1$ weight vectors $\mathbf{v}_0, \ldots, \mathbf{v}_q$ are called an **$\boldsymbol{\alpha}$-ladder** of **ladder length** q. The corresponding weights $\Lambda^*, \Lambda^* - \alpha, \ldots, \Lambda^* - q\alpha$ also constitute an **$\boldsymbol{\alpha}$-ladder**. According to (1.8) we can move down the ladder by applying the operator $E_{-\alpha}$. On the other hand, using

(1.10) $$[E_\alpha, E_{-\alpha}] = H_\alpha,$$

we can show that application of E_α enables us to move up the α-ladder. For convenience we set $\Lambda^*(\mathcal{H}_\alpha) = \Lambda_\alpha^*$.

Lemma 9.4. $E_\alpha \mathbf{v}_j = r_j \mathbf{v}_{j-1}$, $j = 0, 1, \ldots, q+1$, where $r_j = j\Lambda_\alpha^* - \frac{1}{2}j(j-1)\alpha_\alpha$.

Proof. Induction on j. According to (1.7), $r_0 = 0$. Suppose the lemma is valid for $j \leq k \leq q$. We must verify the result for $j = k + 1$. From (1.10) and the induction hypothesis,

(1.11) $$E_\alpha \mathbf{v}_{k+1} = E_\alpha E_{-\alpha} \mathbf{v}_k = (E_{-\alpha} E_\alpha + H_\alpha)\mathbf{v}_k = (r_k + \Lambda_\alpha^* - k\alpha_\alpha)\mathbf{v}_k$$
$$= r_{k+1}\mathbf{v}_k.$$

Thus $r_{k+1} = r_k + \Lambda_\alpha^* - k\alpha_\alpha = (k+1)\Lambda_\alpha^* - \frac{1}{2}(k+1)k\alpha_\alpha$. Q.E.D.

From the first equality in (1.11) we have $r_{q+1} = 0$, since $E_{-\alpha}\mathbf{v}_q = \boldsymbol{\theta}$. Thus

$$(q+1)\Lambda_\alpha^* - \tfrac{1}{2}(q+1)q\alpha_\alpha = 0.$$

Lemma 9.5. $q = 2\Lambda_\alpha^*/\alpha_\alpha$.

Remark. From the commutation relations $[H_\alpha, E_{\pm\alpha}] = \pm\alpha_\alpha E_{\pm\alpha}$, $[E_\alpha, E_{-\alpha}] = H_\alpha$, it follows that the operators $E_{\pm\alpha}$, H_α form a basis for a subalgebra of $gl(m)$ isomorphic to $sl(2)$. Thus the construction of the α-ladder of weights containing Λ^* is essentially the same as the construction of the irred reps of $sl(2)$ in Section 7.6.

As we have shown earlier, $\alpha_\alpha = \alpha(\mathcal{H}_\alpha) = 2$ for $gl(m)$ since $\alpha = \lambda_k - \lambda_l$ and $\mathcal{H}_\alpha = \mathcal{H}_k - \mathcal{H}_l$. Thus $q = \Lambda_\alpha^*$ for $gl(m)$. However, it is convenient to use the notation α_α because with its use we can verify Lemma 9.5 for other classical groups.

Now let Λ be any weight and consider the linear functionals Λ, $\Lambda + \alpha$, $\Lambda + 2\alpha, \ldots$. There will be a smallest nonegative integer h such that $\Lambda + h\alpha$ is a weight but $\Lambda + (h+1)\alpha$ is not a weight. Then $\Lambda^* = \Lambda + h\alpha$ is a maximal weight in the sense of (1.7) and there exists an α-ladder

(1.12) $$\Lambda^*, \quad \Lambda^* - \alpha, \quad \Lambda^* - 2\alpha, \ldots, \quad \Lambda^* - (2\Lambda_\alpha^*/\alpha_\alpha)\alpha,$$

with ladder length $2\Lambda_\alpha^*/\alpha_\alpha = 2\Lambda_\alpha/\alpha_\alpha + 2h$. Since the ladder length is a nonnegative integer it follows that $2\Lambda_\alpha/\alpha_\alpha$ is an integer. Furthermore, in terms of Λ the α-ladder (1.12) is

(1.13) $$\Lambda + h\alpha, \Lambda + (h-1)\alpha, \ldots, \quad \Lambda - [2(\Lambda_\alpha/\alpha_\alpha) + h]\alpha.$$

The midpoint of this ladder is $\frac{1}{2}\{[\Lambda + h\alpha] + [\Lambda - 2(\Lambda_\alpha/\alpha_\alpha)\alpha - h\alpha]\} = \Lambda - (\Lambda_\alpha/\alpha_\alpha)\alpha$. (The midpoint is a weight on the ladder if and only if the ladder contains an odd number of weights.) Similarly we can find a smallest nonnegative integer k such that $\Lambda - k\alpha = \Lambda^{**}$ is a weight but $\Lambda^{**} - \alpha$ is not a weight. In analogy with Lemma 9.4 it is easy to show that Λ^{**} is the lowest rung on an α-ladder of length $q' = -2\Lambda_\alpha^{**}/\alpha_\alpha = -2\Lambda_\alpha/\alpha_\alpha + 2k$. The midpoint of this ladder is again $\Lambda - (\Lambda_\alpha/\alpha_\alpha)\alpha$. (Prove it!) Since both ladders have the same midpoints and no gaps. they must necessarily coincide. [For example, if the first ladder were longer than the second the weight $\Lambda^* - 2(\Lambda_\alpha{}^*/\alpha_\alpha)\alpha$ would lie lower than Λ^{**}, which is impossible.] Thus there is only one ladder (1.13) and Λ belongs to it. Note that Λ lies a distance $(\Lambda_\alpha/\alpha_\alpha)\alpha$ from the midpoint of the ladder. Hence if we reflect the α-ladder in its midpoint, Λ will be mapped into the functional $\Lambda - 2(\Lambda_\alpha/\alpha_\alpha)\alpha$ which is the same distance from the midpoint but on the opposite side. In particular $\Lambda - 2(\Lambda_\alpha/\alpha_\alpha)\alpha$ is a weight. We have proved the following result.

Theorem 9.2. If Λ is a weight and α is a root then $2\Lambda_\alpha/\alpha_\alpha$ is an integer and $\Lambda - 2(\Lambda_\alpha/\alpha_\alpha)\alpha$ is a weight.

Corollary 9.2. The weights of the form $\Lambda + j\alpha$ belonging to ρ are just those for which $j = -k, -k+1, \ldots, h-1, h$, where $\Lambda^* = \Lambda + h\alpha$ is maximal and $\Lambda^{**} = \Lambda - k\alpha$ is minimal, i.e., there are no gaps in the α-ladder. Here, $k - h = 2\Lambda_\alpha/\alpha_\alpha$.

Proof. Suppose $\Lambda^* = \Lambda + h\alpha$ is the maximal weight constructed in the proof of the theorem and let p be the largest integer such that $(\Lambda^*)' = \Lambda^* + p\alpha$ is a weight. Suppose $p > 0$. Then $(\Lambda^*)'$ is maximal and $S^\alpha(\Lambda^*)' = (\Lambda^*)' - 2[(\Lambda_\alpha{}^*)'/\alpha_\alpha]\alpha = \Lambda^* - 2(\Lambda_\alpha{}^*/\alpha_\alpha)\alpha - p\alpha$ is also a weight which lies lower on the α-ladder than Λ^*, since $\Lambda_\alpha{}^* \geq 0$. According to the proof of the theorem, all functionals on the α-ladder between $S^\alpha(\Lambda^*)'$ and $(\Lambda^*)'$ are weights, so Λ^* is not maximal. This contradiction shows that $p = 0$.

Similarly there is no weight on the α-ladder below $S^\alpha\Lambda^* = \Lambda - k\alpha$. By Lemma 9.5 the length of the α-ladder is $h + k = 2\Lambda_\alpha{}^*/\alpha_\alpha = 2(\Lambda_\alpha/\alpha_\alpha) + 2h$ so $2\Lambda_\alpha/\alpha_\alpha = k - h$. Q.E.D.

The map $S^\alpha\Lambda = \Lambda - 2(\Lambda_\alpha/\alpha_\alpha)\alpha$ of the weights of ρ onto themselves is called a **Weyl reflection.** Each Weyl reflection permutes the weights. Hence the totality of reflections S^α as α runs over the roots generates a group of permutations of the weights, called the **Weyl group.** As we have shown, $S^\alpha\Lambda$ is the reflection of Λ with respect to the midpoint of the α-ladder on which Λ lies. In particular if Λ is the highest weight on the ladder then $S^\alpha\Lambda$ is the lowest weight.

Theorem 9.2 greatly restricts the possible weights $\Lambda = \sum c_j\lambda_j$. The requirement that $2\Lambda_\alpha/\alpha_\alpha$ be an integer for all roots $\alpha = \lambda_k - \lambda_l$ implies $\Lambda_\alpha = c_k - c_l$ is an integer for all $k \neq l$.

If ρ is irred and $\mathbf{v} \in V$ is a weight vector with weight Λ then from Theorem 9.1 the possible weights of ρ are all of the form

$$\Lambda + \alpha_1 + \alpha_2 + \cdots + \alpha_j, \qquad \alpha_1, \ldots, \alpha_j \quad \text{roots.}$$

Thus the difference $\Lambda - \Lambda'$ of any two weights of ρ can be expressed as a sum of roots $\alpha_i = \lambda_k - \lambda_l$.

Definition. A linear functional $b_1\lambda_1 + b_2\lambda_2 + \cdots + b_m\lambda_m$ on $\mathscr{h}_m$ is **real** if all the constants b_j are real. A real functional is **positive** if the first nonzero b_j is positive, reading from left to right. A (possibly complex) linear functional Λ is **greater** than another functional Λ' if $\Lambda - \Lambda'$ is a real positive functional.

Since a sum of roots is always a real functional it follows that the difference of any two weights of ρ is real. Thus the above definition defines a lexicographic ordering of the weights of ρ. We say $\Lambda = \sum c_j\lambda_j$ is greater than $\Lambda' = \sum c_j'\lambda_j$ ($\Lambda > \Lambda'$) provided the first nonzero difference $c_l - c_l'$ is positive. With this total ordering it makes sense to speak of the **highest weight** of ρ. (Note that the roots $\alpha = \lambda_k - \lambda_l$ are positive provided $k < l$. If $\alpha > 0$ then $-\alpha$ is negative. Since the roots and zero are the weights of the adjoint rep, the highest weight of the adjoint rep is $\lambda_1 - \lambda_m$.)

If Λ is the highest weight of the irred rep ρ then $\Lambda + \alpha$ cannot be a weight for any positive root α, since $\Lambda + \alpha > \Lambda$. Thus, if $\mathbf{v}$ is a weight vector with weight Λ then $E_\alpha\mathbf{v} = \mathbf{\theta}$ for all $\alpha > 0$. A basis of weight vectors for V can be selected from the set of all vectors of the form

$$E_{\alpha_1}E_{\alpha_2}\cdots E_{\alpha_k}\mathbf{v}, \qquad k = 1, 2, \ldots. \tag{1.14}$$

If $k = 1$, the vectors $E_{\alpha_1}\mathbf{v}$ are zero unless $\alpha_1 < 0$. For $k = 2$ the vectors $E_{\alpha_1}E_{\alpha_2}\mathbf{v}$ are zero unless $\alpha_2 < 0$. If $\alpha_2 < 0$ and $\alpha_1 > 0$ then

(1) $E_{\alpha_1}E_{\alpha_2}\mathbf{v} = (E_{\alpha_2}E_{\alpha_1} \pm E_{\alpha_1+\alpha_2})\mathbf{v} = \pm E_{\alpha_1+\alpha_2}\mathbf{v}$ if $\alpha_1 + \alpha_2 \neq 0$ is a root.
(2) $E_{\alpha_1}E_{\alpha_2}\mathbf{v} = E_{\alpha_2}E_{\alpha_1}\mathbf{v} = \mathbf{\theta}$ if $\alpha_1 + \alpha_2 \neq 0$ is not a root.
(3) $E_{\alpha_1}E_{-\alpha_1}\mathbf{v} = (E_{-\alpha_1}E_{\alpha_1} + H_{\alpha_1})\mathbf{v} = \Lambda_{\alpha_1}\mathbf{v}$.

We have used the commutation relations (1.5) to derive these results. Proceeding in this way, we see that all weight vectors (1.14) can be written as linear combinations of weight vectors

$$E_{-\beta_1}E_{-\beta_2}\cdots E_{-\beta_l}\mathbf{v}, \qquad \beta_j > 0, \quad j = 1, \ldots, l. \tag{1.15}$$

Furthermore, we can express the vectors (1.14) as linear combinations of the vectors (1.15) by a procedure which depends only on the commutation relations (1.5), not on ρ or V.

Clearly we can choose a subset of the vectors (1.15) as a basis for V. Each such basis vector corresponds to the weight $\Lambda - \sum_{j=1}^{l} \beta_j$ with $\beta_j > 0$. The only possible basis vector (1.15) with highest weight Λ is $\mathbf{v}$ itself. This proves the first statement in the following theorem.

Theorem 9.3. The weight space belonging to the highest weight Λ in an irred rep ρ is one-dimensional. Two irred reps with the same highest weight are equivalent.

Proof. Suppose ρ and ρ' are irred reps of $gl(m)$ on V and V', respectively, with the same highest weight Λ. Let $\mathbf{v}$ and $\mathbf{v}'$ be weight vectors belonging to Λ in V and V'. Weight vectors of the form

$$(1.16) \qquad \mathbf{w} = E_{-\beta_1}E_{-\beta_2}\cdots E_{-\beta_l}\mathbf{v}, \qquad \mathbf{w}' = E'_{-\beta_1}E'_{-\beta_2}\cdots E'_{-\beta_l}\mathbf{v}',$$

$\beta_j > 0$, span V and V'. We define a mapping $\mathbf{S}$ from V to V' by

$$(1.17) \qquad \mathbf{S}\Big(\sum_{k=1}^{p} a_k\mathbf{w}_k\Big) = \sum_{k=1}^{p} a_k\mathbf{w}_k', \qquad p = 1, 2, \ldots, \qquad a_k \in \mathfrak{C},$$

where corresponding vectors $\mathbf{w}_k \in V$, $\mathbf{w}_k' \in V'$ are of the form (1.16) and belong to the same weight. It is not clear that this mapping is well-defined. Assuming this for the present, it follows that $\mathbf{S}$ is a linear mapping of V onto V'. Furthermore, the vectors $E_\alpha\mathbf{w}_k$ and $E_\alpha'\mathbf{w}_k'$ for any root α can be written as linear combinations of corresponding weight vectors $\mathbf{w}_l$ and $\mathbf{w}_l'$ by a procedure based solely on the commutation relations. The expansion coefficients in the primed and unprimed spaces will be the same. Similarly $H_\alpha\mathbf{w}_k = (\Lambda - \sum\beta_j)_\alpha\mathbf{w}_k$, $H_\alpha'\mathbf{w}_k' = (\Lambda - \sum\beta_j)_\alpha\mathbf{w}_k'$ where $\Lambda - \sum\beta_j$ is the weight to which $\mathbf{w}_k$ and $\mathbf{w}_k'$ belong. As a consequence,

$$(1.18) \qquad E_\alpha'\mathbf{S} = \mathbf{S}E_\alpha, \qquad H_\alpha'\mathbf{S} = \mathbf{S}H_\alpha,$$

for all roots α. Since $\mathbf{S}$ is nonzero it follows from (1.18) and the Schur lemmas that ρ and ρ' are equivalent reps.

To finish the proof we must verify that $\mathbf{S}$ is well-defined, i.e., that whenever $\sum_{k=1}^{p} a_k\mathbf{w}_k = \boldsymbol{\theta}$ in V, then $\mathbf{S}(\sum a_k\mathbf{w}_k) = \sum_{k=1}^{p} a_k\mathbf{w}_k' = \boldsymbol{\theta}'$ in V'. Consider the set W' of all vectors $\mathbf{z}' = \sum a_k\mathbf{w}_k'$ such that $\mathbf{z} = \sum a_k\mathbf{w}_k = \boldsymbol{\theta}$ in V. Clearly W' is a subspace of V'. Furthermore, by (1.18), $E_\alpha'\mathbf{z}' = E_\alpha'\mathbf{S}\mathbf{z} = \mathbf{S}E_\alpha\mathbf{z}$, where $E_\alpha\mathbf{z} = \sum a_k E_\alpha\mathbf{w}_k = \boldsymbol{\theta}$, so $E_\alpha'\mathbf{z}' \in W'$ for any root α. Similarly, $H_\alpha'\mathbf{z}' \in W'$. Thus W' is invariant under ρ'. Since ρ' is irred, either $W' = V'$ or $W' = \{\boldsymbol{\theta}\}$. But $\mathbf{v}' \notin W'$. For, if $\sum a_k\mathbf{w}_k = \boldsymbol{\theta}$ and $\sum a_k\mathbf{w}_k' = \mathbf{v}'$, each of the $\mathbf{w}_k'$ with nonzero coefficient a_k must be a multiple $b_k\mathbf{v}'$ of $\mathbf{v}'$ since $\mathbf{v}'$ is a highest weight vector. We can assume that each $\mathbf{w}_k$ is a multiple $b_k\mathbf{v}$ of $\mathbf{v}$ and $\sum a_k\mathbf{w}_k' = \sum a_kb_k\mathbf{v}' = \mathbf{v}'$. Thus $\sum a_kb_k = 1$ and $\sum a_k\mathbf{w}_k = \sum a_kb_k\mathbf{v} = \mathbf{v} \neq \boldsymbol{\theta}$. This contradiction shows that $W' = \{\boldsymbol{\theta}\}$ and $\mathbf{S}$ is well-defined. Q.E.D.

Corollary 9.3. If Λ is a weight such that $\Lambda + \alpha$ is not a weight for all roots $\alpha > 0$ then Λ is the highest weight.

According to this theorem each irred rep ρ is uniquely determined by its highest weight Λ. In particular the highest weights can be used to label the irred reps. Let us determine the possible highest weights

$$\Lambda = c_1\lambda_1 + c_2\lambda_2 + \cdots + c_m\lambda_m.$$

In order that Λ be a weight it is necessary that the differences $c_k - c_l$ be integers for all $k \neq l$. If Λ is a highest weight then $\Lambda + \alpha$ is not a weight for all $\alpha > 0$, so Λ is the maximal weight on each α-ladder containing it. From Lemma 9.5 we have $2\Lambda_\alpha/\alpha_\alpha = \Lambda_\alpha \geq 0$. The positive roots are $\alpha = \lambda_k - \lambda_l$, $k < l$, so $\Lambda_\alpha = c_k - c_l \geq 0$. These are the only restrictions on highest weights if we consider Lie algebra reps alone. However, if we restrict ourselves to reps of $gl(m)$ which extend to **global** reps of $GL(m)$ we get an additional requirement on Λ. Let $\mathcal{H} = \sum \lambda_j \mathcal{H}_j \in \mathscr{h}_m$ and suppose $\mathbf{v}$ is a highest weight vector. Setting $H = \rho(\mathcal{H})$, we have $H\mathbf{v} = \Lambda(\mathcal{H})\mathbf{v}$. Then

$$\exp \mathcal{H} = \begin{pmatrix} e^{\lambda_1} & & & Z \\ & e^{\lambda_2} & & \\ & & \ddots & \\ Z & & & e^{\lambda_m} \end{pmatrix} \in GL(m), \tag{1.19}$$

$$(\exp H)\mathbf{v} = e^{\Lambda(\mathcal{H})}\mathbf{v} = \exp(c_1\lambda_1 + \cdots + c_m\lambda_m)\mathbf{v}. \tag{1.20}$$

It is clear that the addition of any integer multiple of $2\pi i$ to a λ_j leaves the group element $\exp \mathcal{H}$ unchanged. Thus, if the Lie algebra rep induces a global group rep, the addition of an integer multiple of $2\pi i$ to a λ_j must leave (1.20) unchanged. This is possible only if the c_j are integers.

We conclude that the possible highest weights are

$$\Lambda^* = p_1\lambda_1 + p_2\lambda_2 + \cdots + p_m\lambda_m, \qquad p_j \quad \text{integer}, \tag{1.21}$$

$$p_1 \geq p_2 \geq \cdots \geq p_m. \tag{1.22}$$

The corresponding irred reps ρ will be denoted $(p_1, \ldots, p_m)$. We shall show that there exists an irred rep of $GL(m)$ corresponding to each such set of integers p_j.

Since $GL(m)$ is not compact we cannot directly apply the theory of Chapter 6 to show that every rep of $GL(m)$ decomposes into a direct sum of irred reps. However, using a technique (the unitary trick) from Chapter 7 we can relate the reps of $GL(m)$ to those of the (compact) unitary group $U(m)$. The real Lie algebra $u(m)$ of $U(m)$ consists of all $m \times m$ matrices $i\mathcal{B}$, where $\mathcal{B}$ is self-adjoint. Now every $\mathcal{A} \in gl(m)$ can be expressed uniquely in the form

$$\mathcal{A} = \mathcal{B} + i\mathcal{C}, \qquad \mathcal{B} = \tfrac{1}{2}(\mathcal{A} + \bar{\mathcal{A}}^t), \qquad \mathcal{C} = -\tfrac{1}{2}i(\mathcal{A} - \bar{\mathcal{A}}^t),$$

where $\mathcal{B}$ and $\mathcal{C}$ are self-adjoint. Thus $u(m)$ is a real form of $gl(m)$ and there is a 1–1 relationship between complex reps ρ of $gl(m)$ and ρ' of $u(m)$. In particular, ρ' is the restriction of ρ to the subalgebra $u(m)$ and ρ is the extension of ρ' to the complexified algebra $u(m)^c = gl(m)$. The reps ρ and ρ' are simultaneously reducible or irred and a decomposition of one rep into irred components induces a decomposition of the other. From Section 5.8, this 1–1 relationship also holds between the finite-dimensional analytic reps of $GL(m)$ and their restrictions to $U(m)$. Since $U(m)$ is compact we deduce that every finite-dimensional analytic rep of $GL(m)$ [or $gl(m)$] can be decomposed into a direct sum of irred reps (This result is false if the rep is infinite-dimensional or if it is not analytic.)

Now we begin the construction of all analytic irred reps ρ of $GL(m)$. The one-dimensional reps

$$A \longrightarrow (\det A)^p, \qquad A \in GL(m), \qquad p \text{ an integer,} \tag{1.23}$$

are clearly analytic and irred. It follows from Corollary 5.3 that the induced rep of $gl(m)$ is

$$\mathcal{A} \longrightarrow p\,\mathrm{tr}(\mathcal{A}), \qquad \mathcal{A} \in gl(m).$$

Choosing $\mathcal{H} \in \mathscr{h}_m$ as in (1.2) we see that each of these reps has a single weight

$$\Lambda = p(\lambda_1 + \lambda_2 + \cdots + \lambda_m). \tag{1.24}$$

Thus, we have constructed the reps $(p, \ldots, p)$, $p = 0, \pm 1, \ldots$.

Let ρ, ρ' be reps of a Lie algebra $\mathcal{G}$ on the vector spaces V, V', respectively. We define the **tensor product representation** $\rho \otimes \rho'$ of $\mathcal{G}$ on $V \otimes V'$ by

$$\rho \otimes \rho'(\alpha)(\mathbf{v} \otimes \mathbf{v}') = (\rho(\alpha)\mathbf{v}) \otimes \mathbf{v}' + \mathbf{v} \otimes (\rho'(\alpha)\mathbf{v}'), \qquad \alpha \in \mathcal{G}, \quad \mathbf{v} \in V, \quad \mathbf{v}' \in V'. \tag{1.25}$$

It is straightforward to verify that $\rho \otimes \rho'$ is indeed a rep of $\mathcal{G}$. In fact it is just the Lie algebra rep induced by the corresponding tensor product of group reps.

Lemma 9.6. Let ρ, ρ' be reps of $gl(m)$ on V and V'. The weights of $\rho \otimes \rho'$ are all functionals of the form $\Lambda + \Lambda'$, where Λ is a weight of ρ and Λ' is a weight of ρ'.

Proof. Let $\{\mathbf{v}_j\}$ be a basis of weight vectors for V and $\{\mathbf{v}_k'\}$ a basis of weight vectors for V'. Then $H\mathbf{v}_j = \Lambda_j(\mathcal{H})\mathbf{v}_j$ and $H'\mathbf{v}_k' = \Lambda_k'(\mathcal{H})\mathbf{v}_k'$, where Λ_j, Λ_k' are the weights of ρ and ρ', respectively. Choose the vectors $\{\mathbf{v}_j \otimes \mathbf{v}_k'\}$ as a basis for $V \otimes V'$. The action of $\rho \otimes \rho'(\mathcal{H}) = H + H'$ on this basis is

$$\begin{aligned} \rho \otimes \rho'(\mathcal{H})\mathbf{v}_j \otimes \mathbf{v}_k' &= (H\mathbf{v}_j) \otimes \mathbf{v}_k' + \mathbf{v}_j \otimes (H'\mathbf{v}_k') \\ &= (\Lambda_j(\mathcal{H}) + \Lambda_k'(\mathcal{H}))\mathbf{v}_j \otimes \mathbf{v}_k'. \end{aligned} \tag{1.26}$$

Thus $\{\mathbf{v}_j \otimes \mathbf{v}_k'\}$ is a weight basis with weights $\Lambda_j + \Lambda_k'$. By the remarks preceding Corollary 9.1 these are the only weights of $\rho \otimes \rho'$. Q.E.D.

Now suppose ρ is the irred rep $(p_1, \ldots, p_m)$, $p_1 \geq p_2 \geq \cdots \geq p_m$, and ρ' is the irred rep $(p_1', \ldots, p_m')$. Furthermore, suppose both ρ and ρ' induce global irred reps of $GL(m)$ on V and V'. Then $\rho \otimes \rho'$ determines a rep of $gl(m)$ which extends to $GL(m)$. It follows easily from the preceding lemma that the highest weight in $\rho \otimes \rho'$ is

$$\Lambda^* = (p_1 + p_1')\lambda_1 + (p_2 + p_2')\lambda_2 + \cdots + (p_m + p_m')\lambda_m, \tag{1.27}$$

i.e., the sum of the highest weights in ρ and ρ'. Furthermore, the weight space of Λ^* has dimension one. Now $\rho \otimes \rho'$ can be decomposed into irred reps and the weight Λ^* must belong to exactly one of the irred pieces. Since Λ^* is the maximal weight of $\rho \otimes \rho'$ this irred piece must be the rep

$$(p_1 + p_1', \ldots, p_m + p_m'). \tag{1.28}$$

Thus the existence of $(p_1, \ldots, p_m)$ and $(p_1', \ldots, p_m')$ implies the existence of the rep (1.28). We shall use this method to prove the existence of the irred reps $(p_1, \ldots, p_m)$ for all integers $p_1 \geq p_2 \geq \cdots \geq p_m$. Unfortunately our procedure only proves existence. To obtain explicit expressions for the reps we fall must back on Young symmetrizer methods developed in Section 4.3.

Consider the tensor product of $\rho = (p_1, \ldots, p_m)$ and the one-dimensional rep $\rho' = (p, p, \ldots, p)$. In this special case $\rho \otimes \rho' \cong (p_1 + p, \ldots, p_m + p)$ is irred. (Prove it!) The group operators of $\rho \otimes \rho'$ are

$$(\det A)^p \mathbf{T}(A), \tag{1.29}$$

where the $\mathbf{T}(A)$ are the operators of ρ. Thus we can limit ourselves to the construction of reps for which $p_1 \geq p_2 \geq \cdots \geq p_m \geq 0$. The remaining reps can be obtained from (1.29).

In Section 4.3 we determined all tensor irred reps of $GL(m)$. Each such rep was determined by the Young frame $[f_1, f_2, \ldots, f_m]$, $f_1 \geq f_2 \geq \cdots \geq f_m \geq 0$. We will now explore the relationship between these reps and the reps $(p_1, \ldots, p_m)$.

Let us first consider the reps

$$[1^s] = [1, 1, \ldots, 1, 0, \ldots, 0], \qquad s = 1, 2, \ldots, m, \tag{1.30}$$

of $GL(m)$. Here the rep space consists of completely skew-symmetric tensors

$$a^{i_1 i_2 \cdots i_s}, \qquad 1 \leq i_j \leq m. \tag{1.31}$$

This space is $\binom{m}{s}$-dimensional since the $\binom{m}{s}$ independent components for which $i_1 < i_2 < \cdots < i_s$ completely determine the tensors. The group action is

$$[T(A)a]^{i_1\cdots i_s} = \sum_{j_1\cdots j_s=1}^{m} A_{i_1j_1}\cdots A_{i_sj_s}a^{j_1\cdots j_s}, \qquad A\in GL(m). \tag{1.32}$$

As a basis for the rep space we choose the tensors $\mathbf{a}(k_1\cdots k_s)$, $1\le k_1<k_2<\cdots<k_s\le m$. Here $\mathbf{a}(k_1\cdots k_s)$ is the skew-symmetric tensor with component $a^{k_1\cdots k_s}=1$ and all linearly independent components zero. It follows immediately from (1.32) and (1.19) that the $\mathbf{a}(k_1\cdots k_s)$ form a weight basis for the rep. Indeed

$$\mathbf{Ha}(k_1\cdots k_s) = (\lambda_{k_1}+\lambda_{k_2}+\cdots+\lambda_{k_s})\mathbf{a}(k_1\cdots k_s), \tag{1.33}$$

so the $\binom{m}{s}$ weights of $[1^s]$ are

$$\lambda_{k_1}+\lambda_{k_2}+\cdots+\lambda_{k_s}, \qquad 1\le k_1<k_2<\cdots<k_s\le m. \tag{1.34}$$

Each weight has multiplicity one and the highest weight is

$$\lambda_1+\lambda_2+\cdots+\lambda_s, \tag{1.35}$$

so $[1^s]\cong(1,\ldots,1,0,\ldots,0)=(1^s)$.

In the above discussion we have used facts about Young frames to conclude that $[1^s]$ is irred. However, we can give an independent proof of irreducibility based on Theorem 9.2. Let

$$\Lambda = n_1\lambda_1+\cdots+n_k\lambda_k+\cdots+n_l\lambda_l+\cdots+n_m\lambda_m \tag{1.36}$$

be a weight belonging to an irred rep ρ and let $\alpha=\lambda_k-\lambda_l$ be a root. Then $S^\alpha\Lambda=\Lambda-\Lambda_\alpha\alpha$ is also a weight. A simple computation gives

$$\begin{aligned} S^\alpha\Lambda &= \Lambda-(n_k-n_l)(\lambda_k-\lambda_l)\\ &= n_1\lambda_1+\cdots+n_k\lambda_l+\cdots+n_l\lambda_k+\cdots+n_m\lambda_m, \end{aligned} \tag{1.37}$$

i.e., $S^\alpha\Lambda$ is obtained from Λ by interchanging λ_l and λ_k. Thus the group generated by the Weyl reflections is the group S_m of all permutations of $\lambda_1,\ldots,\lambda_m$. If Λ is a weight belonging to ρ then every linear functional obtained from Λ by permuting $\lambda_1,\ldots,\lambda_m$ is also a weight belonging to ρ.

Let us apply this result to show that $[1^s]$ is irred. By (1.35), the highest weight of $[1^s]$ is $\Lambda^*=\lambda_1+\lambda_2+\cdots+\lambda_s+0\lambda_{s+1}+\cdots+0\lambda_m$, so the rep space contains the irred rep (1^s). Now (1^s) must contain the $\binom{m}{s}$ distinct weights (1.35) obtained by applying all permutations of $\lambda_1,\ldots,\lambda_m$ to Λ^*. However, $\dim[1^s]=\binom{m}{s}$ so $[1^s]\cong(1^s)$.

It is now simple to prove the existence of irred reps $(p_1,\ldots,p_m)$ for all integers $p_1\ge p_2\ge\cdots\ge p_m$. Consider the rep

$$[1^1]^{\otimes k_1}\otimes[1^2]^{\otimes k_2}\otimes\cdots\otimes[1^m]^{\otimes k_m}=\rho, \tag{1.38}$$

where $k_1, \ldots, k_m$ are nonnegative integers. Using an obvious generalization of Lemma 9.6 we see that the highest weight of ρ is $\Lambda^* = p_1\lambda_1 + p_2\lambda_2 + \cdots + p_m\lambda_m$, where

$$\begin{aligned} p_1 &= k_1 + k_2 + \cdots + k_m \\ p_2 &= \phantom{k_1 + {}} k_2 + \cdots + k_m \\ &\;\;\vdots \\ p_{m-1} &= \phantom{k_1 + k_2 + \cdots + {}} k_{m-1} + k_m \\ p_m &= \phantom{k_1 + k_2 + \cdots + k_{m-1} + {}} k_m . \end{aligned} \tag{1.39}$$

Clearly $p_1 \geq p_2 \geq \cdots \geq p_m \geq 0$. Since Λ^* occurs in ρ with multiplicity one it follows that the irred rep $(p_1, \ldots, p_m)$ of $GL(m)$ is contained in ρ with multiplicity one. We can obtain all reps $(p_1, \ldots, p_m)$ with $p_m \geq 0$ by choosing the integers $k_m = p_m$, $k_{m-1} = p_{m-1} - p_m, \ldots, k_2 = p_2 - p_3$, $k_1 = p_1 - p_2$. Then using (1.29) we can relax the requirement $p_m \geq 0$.

Let us now determine which rep $(p_1, \ldots, p_m)$ is equivalent to the irred tensor rep with Young frame $[f_1, \ldots, f_m]$. The elements of the rep space of $[f_1, \ldots, f_m]$ are tensors

$$F^B, \qquad B = \begin{matrix} i_1 & \cdots & i_{f_1} \\ j_1 & \cdots & j_{f_2} \\ \vdots & & \\ z_1 & \cdots & z_{f_m}, \end{matrix} \tag{1.40}$$

with $f_1 + f_2 + \cdots + f_m$ indices, each index taking the values one to m. These tensors are defined in terms of a Young symmetrizer by (3.47), Section 4.3. The action of the induced Lie algebra rep on each tensor component F^B is easily shown to be

$$(\mathbf{H}F)^B = (\lambda_{i_1} + \cdots + \lambda_{i_f} + \lambda_{j_1} + \cdots + \lambda_{z_{f_m}})F^B. \tag{1.41}$$

Since the tensors F^B are skew-symmetric with respect to interchange of indices in the same column of B, the highest possible weight vector is the tensor with component $F^{B_0} = 1$ and all linearly independent components zero. where

$$B_0 = \begin{matrix} 1 & 1 \cdots & 1 \\ 2 & 2 \cdots & 2 \\ \vdots & & \\ m & m \cdots & m \end{matrix} \tag{1.42}$$

(see the discussion of F^{B_0} at the end of Section 4.3). Clearly, this weight vector corresponds to the highest weight $\Lambda^* = f_1\lambda_1 + f_2\lambda_2 + \cdots + f_m\lambda_m$. Thus, $[f_1, \ldots, f_m] \cong (f_1, \ldots, f_m)$. This completes the construction of irred reps of the complex Lie group $GL(m)$.

Since $u(m)$ is a real form of $gl(m)$ and $U(m)$ is a connected subgroup of the connected group $GL(m)$ it follows that there is a 1–1 relationship between finite-dimensional analytic reps of $U(m)$ and $GL(m)$. In particular every rep of $GL(m)$ restricts to a rep of $U(m)$ and every rep of $U(m)$ extends to a unique rep of $GL(m)$. One of these reps is irred if and only if the other is. Hence, the irred reps of $U(m)$ are also denoted $[f_1, \ldots, f_m], f_1 \geq f_2 \geq \cdots \geq f_m$, where we allow the f_j to be negative.

The real Lie group $GL(m) = GL(m, \mathfrak{C})$ is $2m^2$-dimensional. As a basis for the real Lie algebra we choose the elements $\mathcal{E}_{kl}$ and $i\mathcal{E}_{kl}$, where $\mathcal{E}_{kl}$ is the $m \times m$ matrix with a one in row k, column l and zeros everywhere else. Let ρ be a complex rep of $gl(m)$ and set $\rho(\mathcal{E}_{kl}) = E_{kl}$, $\rho(i\mathcal{E}_{kl}) = F_{kl}$. If we introduce a new basis $C_{kl} = \frac{1}{2}(E_{kl} - iF_{kl})$, $D_{kl} = \frac{1}{2}(E_{kl} + iF_{kl})$ for the complexified Lie algebra the commutation relations become

$$(1.43)\qquad \begin{aligned} &[C_{kl}, C_{k'l'}] = \delta_{lk'}C_{kl'} - \delta_{l'k}C_{k'l} \\ &[D_{kl}, D_{k'l'}] = \delta_{lk'}D_{kl'} - \delta_{l'k}D_{k'l}, \qquad [C_{kl}, D_{k'l'}] = 0. \end{aligned}$$

Thus, if we denote the real Lie algebra $gl(m)$ by $gl_r(m)$, relations (1.43) show $[gl_r(m)]^c \cong gl(m) \oplus gl(m)$. Hence the irred reps of $[gl_r(m)]^c$ can be expressed as products $\rho' \otimes \rho''$ of irred reps of $gl(m)$.

Suppose the reps ρ', ρ'' of $gl(m)$ induce matrix reps $\rho'(A)$, $\rho''(A)$ of $GL(m)$. Then reasoning exactly as in (3.3), Section 8.3, we see that $\rho' \otimes \rho''$ induces the matrix rep $\rho'(A)\rho''(\bar{A}) = (\rho'(A)_{kl}\rho''(\bar{A})_{k'l'})$ of $GL_r(m)$. Even if ρ' and ρ'' do not induce global reps of $GL(m)$, $\rho' \otimes \rho''$ may induce a global rep. Indeed the reps $A \longrightarrow (\det A)^\alpha$ and $A \longrightarrow (\det \bar{A})^\alpha$ are only local for arbitrary $\alpha \in \mathfrak{C}$. but their product $A \longrightarrow (\det A)^\alpha(\det \bar{A})^\alpha = |\det A|^{2\alpha}$ is global. We conclude that the analytic irred reps of the real Lie group $GL_r(m)$ are

$$(1.44)\qquad |\det A|^c \cdot [f_1, f_2, \ldots, f_m] \otimes [f_1', f_2', \ldots, f_m'],$$

where $f_1 \geq f_2 \geq \cdots \geq f_m, f_1' \geq f_2' \geq \cdots \geq f_m'$, $c \in \mathfrak{C}$.

9.2 Character Formulas

Since $U(m)$ is a compact group we can use the techniques of Chapter 6 to deduce its simple characters and their orthogonality relations. The following results are due essentially to Weyl [3].

The matrices $\mathcal{H}_k$, $k = 1, \ldots, m$, (1.2), form a basis for the abelian subalgebra h_m of $gl(m)$. Similarly the matrices $i\mathcal{H}_k$ form a basis for the real abelian subalgebra $\tilde{\mathscr{h}}_m$ of $u(m)$:

$$\tilde{\mathscr{h}}_m = \left\{ \begin{pmatrix} i\varphi_1 & & Z \\ & \ddots & \\ Z & & i\varphi_m \end{pmatrix} = \sum_{k=1}^{m} i\varphi_k \mathcal{H}_k : \varphi_k \quad \text{real} \right\}. \tag{2.1}$$

Here $\tilde{\mathscr{h}}_m$ is the Lie algebra of the abelian subgroup Λ_m of $U(m)$:

$$\Lambda_m = \left\{ \begin{pmatrix} e^{i\varphi_1} & & Z \\ & \ddots & \\ Z & & e^{i\varphi_m} \end{pmatrix} = \Phi \right\}. \tag{2.2}$$

Since a unitary matrix can be diagonalized by a unitary similarity transformation, every $A \in U(m)$ is unitary similar to a $\Phi \in \Lambda_m$, where the $e^{i\varphi_k}$ are the eigenvalues of A:

$$A = U\Phi U^{-1}, \qquad U \in U(m). \tag{2.3}$$

Thus A is conjugate to an element of Λ_m. Furthermore, distinct $\Phi, \Phi' \in \Lambda_m$ are conjugate if and only if they have the same diagonal elements (in a different order). Thus we can use the parameters $\varphi_1, \ldots, \varphi_m$, $0 \le \varphi_j < 2\pi$ of Φ to denote the conjugacy classes of $U(m)$. Since a character χ is constant on conjugacy classes we can write $\chi = \chi(\varphi_1, \ldots, \varphi_m)$, where χ is a symmetric function of its m arguments.

To compute the simple characters we choose parameters on $U(m)$ as follows: The first m parameters $\varphi_1, \ldots, \varphi_m$ pick out the conjugacy class in which an element lies and the remaining $m^2 - m$ parametrize the elements in a fixed conjugacy class.

Given $A \in U(m)$ we assign the parameters $\varphi_1, \ldots, \varphi_m$ from (2.3). For definiteness we assume $\varphi_1 \le \varphi_2 \le \cdots \le \varphi_m$. We can obtain all elements in the conjugacy class of A by forming $U\Phi U^{-1}$ and letting U run over $U(m)$. This suggests that the elements in the conjugacy class can be parametrized by the matrices U. However, U is described by m^2 real parameters, so we would obtain $m^2 + m$ parameters for $A \in U(m)$. Since $U(m)$ is m^2-dimensional, some of these parameters must be redundant.

The redundancy occurs in the choice of U. Suppose the eigenvalues of Φ are distinct. Then $U\Phi U^{-1} = V\Phi V^{-1}$ for $U, V \in U(m)$ if and only if $U = V\Phi'$ for $\Phi' \in \Lambda_m$. The m parameters of Φ' are redundant so the elements of a conjugacy class are uniquely determined by $m^2 - m$ local parameters whose exact choice need not concern us. If two eigenvalues of Φ are identical then $m + 2$ parameters of U are redundant. (Prove it!) Since Φ has $m - 1$ independent eigenvalues in this case, a matrix A with two eigenvalues identical is determined by $(m^2 - m - 2) + (m - 1) = m^2 - 3$ parameters.

Thus the submanifold of such matrices has three dimensions less than the manifold; it does not affect the invariant measure.

Let us assume that in some neighborhood $\mathcal{W}$ of $A_0 \in U(m)$ we have

$$A = U\Phi U^{-1}, \qquad A \in \mathcal{W}, \tag{2.4}$$

where $U = U(b_1, \ldots, b_{m^2-m})$, $\Phi = \Phi(\varphi_1, \ldots, \varphi_m)$, and the φ_j are distinct. The φ_j and b_j define local coordinates for A. To determine the invariant measure on $U(m)$ with respect to this coordinate system we choose a real analytic curve $A(t)$ in $\mathcal{W}$ and compute $A^{-1}\dot{A} \in u(m)$. The parameters $b_j(t)$, $\varphi_j(t)$ are analytic functions of t, so from (2.4) we obtain

$$A^{-1}\dot{A} = U\Phi^{-1}U^{-1}(\dot{U}\Phi U^{-1} + U\dot{\Phi}U^{-1} - U\Phi U^{-1}\dot{U}U^{-1}), \tag{2.5}$$

or

$$U^{-1}(A^{-1}\dot{A})U = \Phi^{-1}(U^{-1}\dot{U})\Phi - U^{-1}\dot{U} + \Phi^{-1}\dot{\Phi}. \tag{2.6}$$

Here $A^{-1}\dot{A} = \mathcal{A}$, $U^{-1}\dot{U} = \mathcal{U}$, and $\Phi^{-1}\dot{\Phi} = \mathcal{E}$ are elements of $u(m)$, i.e., these matrices are skew-Hermitian. Furthermore, $U^{-1}(A^{-1}\dot{A})U = B^{-1}\dot{B} = \mathcal{B} \in u(m)$, where $B(t) = U^{-1}A(t)U \in U(m)$. Writing (2.6) in terms of matrix elements, we obtain

$$\mathcal{B}_{jk} = [(\epsilon_j/\epsilon_k) - 1]\mathcal{U}_{jk} + i\dot{\varphi}_j\delta_{jk}, \tag{2.7}$$

where $\epsilon_j = e^{i\varphi_j}$. Thus,

$$\mathcal{B}_{jk} = [(\epsilon_j/\epsilon_k) - 1]\mathcal{U}_{jk}, \qquad j \neq k, \qquad \mathcal{B}_{jj} = i\dot{\varphi}_j. \tag{2.8}$$

A convenient basis for $u(m)$ is given by $i\mathcal{H}_k$, $k = 1, \ldots, m$, and $\mathcal{E}_{jk} - \mathcal{E}_{kj}$, $i(\mathcal{E}_{jk} + \mathcal{E}_{kj})$, $k \neq j$, $k, j = 1, \ldots, m$. From Section 6.1, if $U(m)$ has local coordinates $t_1, \ldots, t_{m^2}$ then the invariant measure on $U(m)$ is given (up to a multiplicative constant) by

$$dA = V_A(t_1 \ldots, t_{m^2})\, dt_1 \ldots dt_{m^2},$$

where

$$V_A = |\det(\alpha_j{}^k)|, \qquad A^{-1}\,\partial A/\partial t_j = \sum_k \alpha_j{}^k \mathcal{A}_k \tag{2.9}$$

and $\{\mathcal{A}_k\}$ is the basis for $u(m)$ given above. Furthermore, since U is unitary and $B^{-1}\dot{B} = U^{-1}(A^{-1}\dot{A})U$ it follows from the proof of Theorem 6.1 that $V_A = |\det(\beta_j{}^k)|$, where

$$\mathcal{B} = B^{-1}\,\partial B/\partial t_j = \sum \beta_j{}^k \mathcal{A}_k. \tag{2.10}$$

Using the parameters $\varphi_1, \ldots, \varphi_m, b_1, \ldots, b_{m^2-m}$ as well as expressions (2.8) and (2.10) it is straightforward to verify the formula

$$dA = \prod_{j \neq k} |(\epsilon_j/\epsilon_k) - 1|\, d\omega_u\, d\varphi_1 \cdots d\varphi_m, \tag{2.11}$$

where

$$d\omega_u = W(b_1, \ldots, b_{m^2-m})\, db_1 \cdots db_{m^2-m}$$

depends only on the b-parameters.

Since $|(1-\epsilon_j/\epsilon_k)(1-\epsilon_k/\epsilon_j)| = (\epsilon_j-\epsilon_k)\overline{(\epsilon_j-\epsilon_k)}$, we can write

$$\left|\prod_{j\neq k}[(\epsilon_j/\epsilon_k)-1]\right| = \Delta\bar{\Delta}$$

where

(2.12) $$\Delta(\epsilon_1,\ldots,\epsilon_m) = \prod_{j<k}(\epsilon_j-\epsilon_k)$$

is skew-symmetric in its m arguments.

Consider a continuous function $f(\varphi_1,\ldots,\varphi_m)$ on $U(m)$ which is constant on conjugacy classes. Performing the b-integration, we find

(2.13) $$\begin{aligned}\int_{U(m)} f\,dA &= \int d\omega_u \int f\Delta\bar{\Delta}\,d\varphi_1\cdots d\varphi_m \\ &= \int_0^{2\pi}\cdots\int_0^{2\pi} f(\varphi_1,\ldots,\varphi_m)\,\Delta\bar{\Delta}\,d\varphi_1\cdots d\varphi_m,\end{aligned}$$

where the measure $d\omega_u$ has been normalized so that $\int d\omega_u = 1$. Thus, the inner product on $L_2(U(m))$ of two functions f, g constant on conjugacy classes is

(2.14) $$(f,g) = V^{-1}\int_0^{2\pi}\cdots\int_0^{2\pi} f(\varphi_1,\ldots,\varphi_m)\bar{g}(\varphi_1,\ldots,\varphi_m)\,\Delta\bar{\Delta}\,d\varphi_1\cdots d\varphi_m,$$

where $V = \int_0^{2\pi}\cdots\int_0^{2\pi}\Delta\bar{\Delta}\,d\varphi_1\cdots d\varphi_m$. [Note: the integral in (2.14) gives an answer $m!$ times too big since a conjugacy class is determined by $\varphi_1 \geq \varphi_2 \geq \cdots \geq \varphi_m$ and f and g are symmetric. However, V is also $m!$ times too large, so the factor $m!$ cancels.]

Since the characters of $U(m)$ are constant on conjugacy classes we can use (2.14) to compute the inner product of two characters. Let χ be a simple character and let $\mathbf{v}_1,\ldots,\mathbf{v}_s$ be a basis of weight vectors for the corresponding rep, with weights $\Lambda_1,\ldots,\Lambda_s$, respectively. In terms of this basis the matrix $\rho(\Phi)$ of $\Phi(\varphi_1,\ldots,\varphi_m)$, (2.2), is

(2.15) $$\rho(\Phi) = \begin{pmatrix} e^{i\Lambda_1(\varphi)} & & & \\ & \cdot & & Z \\ & & \cdot & \\ Z & & & \cdot \\ & & & e^{i\Lambda_s(\varphi)}\end{pmatrix}.$$

Here $\exp[i\Lambda_j(\varphi_1,\ldots,\varphi_m)] = \epsilon_1^{p_1}\cdots\epsilon_m^{p_m}$, where $\Lambda_j = p_1\varphi_1 + \cdots + p_m\varphi_m$. By definition,

(2.16) $$\chi(\varphi_1,\ldots,\varphi_m) = \operatorname{tr}\rho(\Phi) = \sum_{p_j} c_{p_1\cdots p_m}\epsilon_1^{p_1}\cdots\epsilon_m^{p_m},$$

where the integer $c_{p_1\cdots p_m}$ is the multiplicity of the corresponding weight in the rep. Since χ is a symmetric function of $\varphi_1,\ldots,\varphi_m$ the integers $c_{p_1\cdots p_m}$ are symmetric in $p_1,\ldots,p_m$. Thus a weight Λ' obtained from Λ by permuting the coefficients $p_1,\ldots,p_m$ has the same multiplicity as Λ. Recalling the definition of the Weyl reflection S^α we find the following result.

Lemma 9.7. If Λ is a weight and $\Lambda' = S^\alpha \Lambda$ for some root α then Λ and Λ' have the same multiplicity.

If χ is the character of the rep $[f_1, \ldots, f_m]$, $f_1 \geq f_2 \geq \cdots \geq f_m$, then the term of highest weight on the right-hand side of (2.16) is $\epsilon_1^{f_1} \cdots \epsilon_m^{f_m}$ and the coefficient of this term is $c_{f_1 \cdots f_m} = 1$.

Now consider the product $\xi = \chi\Delta$. Since χ is symmetric and Δ is skew-symmetric, ξ is a skew-symmetric function of $\epsilon_1, \ldots, \epsilon_m$. Furthermore, ξ is a finite sum

$$\xi = \sum d_{q_1 \cdots q_m} \epsilon_1^{q_1} \cdots \epsilon_m^{q_m}, \qquad \epsilon_j = e^{i\varphi_j}, \tag{2.17}$$

where the $d_{q_1 \cdots q_m}$ are integers. The highest-order term in (2.17) is clearly $1 \cdot \epsilon_1^{f_1+m-1} \epsilon_2^{f_2+m-2} \cdots \epsilon_{m-1}^{f_{m-1}+1} \epsilon_m^{f_m} = \epsilon_1^{l_1} \epsilon_2^{l_2} \cdots \epsilon_m^{l_m}$, where $l_j = f_j + m - j$ and $l_1 > l_2 > \cdots > l_m$. Since ξ is skew-symmetric and contains $\epsilon_1^{l_1} \cdots \epsilon_m^{l_m}$ it must also contain the terms

$$\xi(l_1, \ldots, l_m) = \sum_{s \in S_m} \delta_s \, \epsilon_{s(1)}^{l_1} \cdots \epsilon_{s(m)}^{l_m} = |\epsilon^{l_1}, \ldots, \epsilon^{l_m}|, \tag{2.18}$$

where δ_s is the parity of the permutation $s \in S_m$ and $|\epsilon^{l_1}, \ldots, \epsilon^{l_m}|$ is the determinant of

$$\begin{pmatrix} \epsilon_1^{l_1} & \epsilon_1^{l_2} & \cdots & \epsilon_1^{l_m} \\ \epsilon_2^{l_1} & \epsilon_2^{l_2} & \cdots & \epsilon_2^{l_m} \\ \cdot & & & \\ \cdot & & & \\ \cdot & & & \\ \epsilon_m^{l_1} & \epsilon_m^{l_2} & & \epsilon_m^{l_m} \end{pmatrix}. \tag{2.19}$$

Note that $\Delta(\epsilon_1, \ldots, \epsilon_m) = \xi(m-1, m-2, \ldots, 1, 0)$. If $\xi \neq \xi(l_1, \ldots, l_m)$, then $\xi - \xi(l_1, \ldots, l_m)$ is skew-symmetric and contains a highest-order term $c'\epsilon_1^{l_1'} \cdots \epsilon_m^{l_m'}$, $l_1' > l_2' > \cdots > l_m'$. (Note that the skew-symmetry of ξ guarantees that any term $\epsilon_1^{p_1} \cdots \epsilon_j^{p_j} \cdots \epsilon_k^{p_k} \cdots \epsilon_m^{p_m}$ with $p_j = p_k$ has coefficient zero.) Hence, $\xi - \xi(l_1, \ldots, l_m) = c'\xi(l_1', \ldots, l_m') + \cdots$. Since ξ is a finite sum of terms, this process must end eventually and we obtain

$$\xi = \xi(l_1, \ldots, l_m) + c'\xi(l_1', \ldots, l_m') + c''\xi(l_1'', \ldots, l_m'') + \cdots. \tag{2.20}$$

Now

$$\int_0^{2\pi} \cdots \int_0^{2\pi} \epsilon_1^{p_1} \cdots \epsilon_m^{p_m} \overline{\epsilon_1^{q_1} \cdots \epsilon_m^{q_m}} \, d\varphi_1 \cdots d\varphi_m = (2\pi)^m \delta_{p_1 q_1} \cdots \delta_{p_m q_m}$$

and it follows easily from (2.18) that

$$\begin{aligned} &\int_0^{2\pi} \cdots \int_0^{2\pi} \xi(l_1, \ldots, l_m) \bar{\xi}(l_1', \ldots, l_m') \, d\varphi_1 \cdots d\varphi_m \\ &\qquad = (2\pi)^m m! \, \delta_{l_1 l_1'} \cdots \delta_{l_m l_m'}. \end{aligned} \tag{2.21}$$

In particular

$$V = \int_0^{2\pi} \cdots \int_0^{2\pi} |\xi(m-1, m-2, \ldots, 1, 0)|^2 \, d\varphi_1 \cdots d\varphi_m = (2\pi)^m m!$$

and

$$(2.22) \qquad (\chi, \chi) = V^{-1} \int_0^{2\pi} \cdots \int_0^{2\pi} |\chi\Delta|^2 \, d\varphi_1 \cdots d\varphi_m = 1 + (c')^2 + (c'')^2 + \cdots,$$

where we have used (2.20) and the relation $\xi = \chi\Delta$. Since χ is a simple character, $(\chi, \chi) = 1$ and $c' = c'' = \cdots = 0$.

Theorem 9.4. The character of the irred rep $[f_1, \ldots, f_m]$ of $U(m)$ is

$$(2.23) \qquad \chi^{f_1 \cdots f_m}(\epsilon_1, \ldots, \epsilon_m) = |\epsilon^{l_1}, \ldots, \epsilon^{l_m}| / |\epsilon^{m-1}, \ldots, \epsilon, 1| = \xi/\Delta,$$

where $l_j = f_j + m - j$.

Expression (2.23) makes sense only if no two of the ϵ_j are equal, since otherwise both the numerator and denominator are zero. However, from (2.16) we see that χ is defined and continuous for all $\epsilon_1, \ldots, \epsilon_m$ on the unit circle. Thus, we can compute the character for equal ϵ_j by taking an appropriate limit in expression (2.23).

For example, the dimension $N(f_1, \ldots, f_m)$ of the rep $[f_1, \ldots, f_m]$ is just $\chi^{f_1 \cdots f_m}_{(1, \ldots, 1)}$. To obtain this value of the character we set $\epsilon_j = \epsilon^{m-j}$ in (2.23) and pass to the limit as $\epsilon \longrightarrow 1$. Then

$$(2.24) \qquad N(f_1, \ldots, f_m) = \lim_{\epsilon \to 1} \left\{ \left[\prod_{j<k} (\epsilon^{l_j} - \epsilon^{l_k}) \right] \Big/ \left[\prod_{j<k} (\epsilon^{m-j} - \epsilon^{m-k}) \right] \right\} = \Delta(l_1, \ldots, l_m)/\Delta(m-1, m-2, \ldots, 1, 0)$$

since $\epsilon^{l_j} - \epsilon^{l_k} = e^{i\varphi l_j} - e^{i\varphi l_k} \longrightarrow i(l_j - l_k)\varphi$ as $\varphi \longrightarrow 0$. Note that $N(f_1 + p, \ldots, f_m + p) = N(f_1, \ldots, f_m)$ for any integer p. This is in agreement with the observation

$$[f_1 + p, \ldots, f_m + p] \cong [p^m] \otimes [f_1, \ldots, f_m],$$

where $[p^m]$ is one-dimensional.

If the requirement that $|\epsilon_j| = 1$ is relaxed, expression (2.23) also defines the character of the irred rep $[f_1, \ldots, f_m]$ of $GL(m)$. Indeed, if $A \in GL(m)$ has m distinct eigenvalues $\epsilon_1, \ldots, \epsilon_m$ then there exists $B \in GL(m)$ such that

$$(2.25) \qquad A = B \begin{pmatrix} \epsilon_1 & & & \\ & \cdot & & \\ & & \cdot & \\ & & & \epsilon_m \end{pmatrix} B^{-1} = BD(\epsilon)B^{-1}.$$

Thus the conjugacy class to which A belongs can be described by the parameters ϵ_j. If ρ is a matrix realization of $[f_1, \ldots, f_m]$ then

$$\chi^{f_1\cdots f_m}(\epsilon_1, \ldots, \epsilon_m) = \text{tr } \rho(A) = \text{tr } \rho(D) = \sum_{p_j} c_{p_1\cdots p_m}\epsilon_1^{p_1} \cdots \epsilon_m^{p_m}, \tag{2.26}$$

where $c_{p_1\cdots p_m}$ is the multiplicity of the weight $\Lambda = \sum p_j\lambda_j$. This expression is exactly the same as (2.16) except that the ϵ_j need not have modulus one. Thus, formula (2.23) must be valid for $\chi^{f_1\cdots f_m}(\epsilon_1, \ldots, \epsilon_m)$. If the eigenvalues of A are not all distinct it may not be possible to diagonalize A by similarity transformations. However, we can always find a sequence of matrices $\{A^{(n)} \in GL(m), n = 1, 2, \ldots\}$ such that each $A^{(n)}$ has m distinct eigenvalues $\epsilon_j^{(n)}$, $1 \leq j \leq m$, and $A^{(n)} \longrightarrow A$, $\epsilon_j^{(n)} \longrightarrow \epsilon_j$ as $n \longrightarrow \infty$. Since $\chi^{f_1\cdots f_m}$ is continuous we have $\chi^{f_1\cdots f_m}(A) = \lim_{n\to\infty} \chi^{f_1\cdots f_m}(A^{(n)}) = \lim_{n\to\infty} \chi^{f_1\cdots f_m}(\epsilon_1^{(n)}, \ldots, \epsilon_m^{(n)})$ and the character is given by an appropriate limiting form of (2.23) even if the eigenvalues of A are not distinct.

9.3 The Irreducible Representations of $GL(m, R)$, $SL(m, \mathfrak{C})$, and $SU(m)$

Since $gl(m, R)$ is a real form of $gl(m, \mathfrak{C})$ there is a 1–1 relationship between reps of these two Lie algebras. Every analytic rep of the complex Lie group $GL(m)$ restricts to an analytic rep of the real Lie group $GL(m, R)$. Conversely, an analytic rep of $GL(m, R)$ uniquely extends to an analytic rep of the **complex** Lie group $GL(m)$. Thus, the irred reps of $GL(m, R)$ can be denoted

$$[f_1, \ldots, f_m], \qquad f_1 \geq f_2 \geq \cdots \geq f_m. \tag{3.1}$$

These are the restrictions of the corresponding reps of $GL(m)$ to $GL(m, R)$. Every finite-dimensional analytic rep of $GL(m, R)$ can be decomposed into a direct sum of irred reps.

The rep theory of the group

$$GL(m, R)^+ = \{A \in GL(m, R)\colon \det A > 0\} \tag{3.2}$$

is slightly different because some Lie algebra reps of $gl(m, R)$ induce global reps of $GL(m, R)^+$ which do not extend to $GL(m, R)$. Indeed, the irred reps of $GL(m, R)^+$ are of the form

$$(\det A)^c \otimes [f_1, \ldots, f_m], \qquad c \in \mathfrak{C}, \quad 0 \leq \text{Re } c < 1. \tag{3.3}$$

Next we consider $SL(m) = SL(m, \mathfrak{C})$, where $m \geq 2$. The restriction of the tensor rep $T \cong [f_1, \ldots, f_m], f_1 \geq f_2 \geq \cdots \geq f_m \geq 0$, of $GL(m)$ to $SL(m)$ yields a matrix rep of $SL(m)$ whose matrix elements are homogeneous polynomials of order $f_1 + \cdots + f_m = s$ in the matrix elements of $A \in SL(m)$. If $T' = T \mid SL(m)$ then for any $B \in GL(m)$ we have $T(B) = (\det B)^{s/m}T'(A)$, where $B = (\det B)^{1/m}A$. Note that $\det A = 1$, so $A \in SL(m)$. Since T is irred, it follows from the Schur lemmas that T' is irred. Thus, the tensor reps $[f_1, \ldots, f_m]$ of $SL(m)$ are irred.

However, these reps are no longer pairwise nonequivalent. Indeed we have shown earlier that

(3.4) $$[f_1 + p, \ldots, f_m + p] \cong [p^m] \otimes [f_1, \ldots, f_m]$$

for $GL(m)$, where $[p^m]$ is the rep

(3.5) $$[p^m]: B \longrightarrow (\det B)^p, \qquad B \in GL(m),$$

and p is an integer. Clearly, on restriction to $B \in SL(m)$ we have $[p^m] \cong [0]$. Thus, the reps with signatures $[f_1 + p, \ldots, f_m + p]$ and $[f_1, \ldots, f_m]$ are equivalent for all integers p. By choosing $p = -f_m$ we can always assume that the irred reps take the form $[g_1, \ldots, g_{m-1}, 0]$, $g_1 \geq g_2 \geq \cdots \geq g_{m-1} \geq 0$. We shall adopt a Lie-algebraic approach to verify that these are the only analytic irred reps of the complex Lie group $SL(m)$ and that they are pairwise nonequivalent.

The rep theory of the Lie algebra $sl(m)$ is almost identical to the theory for $gl(m)$ developed in Section 9.1. As we mentioned earlier, $sl(m)$ is the $(m^2 - 1)$-dimensional Lie algebra of all $m \times m$ complex matrices with trace zero. The set of all diagonal matrices

(3.6) $$\mathcal{H} = \sum_{j=1}^{m} \lambda_j \mathcal{H}_j = \begin{pmatrix} \lambda_1 & & & Z \\ & \cdot & & \\ & & \cdot & \\ Z & & & \lambda_m \end{pmatrix}, \qquad \lambda_1 + \cdots + \lambda_m = 0,$$

forms an $(m - 1)$-dimensional maximal commutative subalgebra $\mathscr{h}_{m-1}$ of $sl(m)$. Let $\mathcal{E}_{hj}$ be the $m \times m$ matrix with a one in row h, column j and zeros everywhere else. The $m^2 - 1$ matrices $\mathcal{E}_{hj}$, $1 \leq h, j \leq m$, $h \neq j$, and $\mathcal{H}_j - \mathcal{H}_m$, $j = 1, 2, \ldots, m - 1$, form a basis for $sl(m)$. We easily obtain the commutation relations

(3.7) $$\begin{aligned} &[\mathcal{E}_{hj}, \mathcal{E}_{kl}] = \delta_{jk}\mathcal{E}_{hl} - \delta_{lh}\mathcal{E}_{kj}, && [\mathcal{H}, \mathcal{H}'] = Z, \\ &[\mathcal{H}, \mathcal{E}_{kl}] = (\lambda_k - \lambda_l)\mathcal{E}_{kl}, && \mathcal{H}, \mathcal{H}' \in \mathscr{h}_{m-1}. \end{aligned}$$

Just as in Section 9.1, we consider the adjoint rep $\mathcal{H} \longrightarrow \operatorname{ad} \mathcal{H}$ of $\mathscr{h}_{m-1}$ on $sl(m)$. According to the second of expressions (3.7) the element $\mathcal{E}_{kl}$ is a simultaneous eigenvector for all operators $\operatorname{ad} \mathcal{H}(\lambda_1, \ldots, \lambda_m)$ with eigenvalue $\alpha(\mathcal{H}) = \lambda_k - \lambda_l$. As before we call the linear functional $\alpha(\mathcal{H})$ a **root** and the corresponding eigenvector $\mathcal{E}_{kl} \equiv \mathcal{E}_\alpha$ the **branch** belonging to root α. Set $\mathcal{H}_\alpha = \mathcal{H}_k - \mathcal{H}_l \in \mathscr{h}_{m-1}$ for $\alpha = \lambda_k - \lambda_l$. Then the $\mathcal{E}_\alpha$ and $\mathcal{H}_\alpha$ span $sl(m)$ as α runs over the $m(m - 1)$ distinct roots.

Lemma 9.8.

(a) If α is a root then $-\alpha$ is a root.

(b) $[\mathcal{E}_\alpha, \mathcal{E}_{-\alpha}] = \mathcal{H}_\alpha \neq Z$.

(c) $[\mathcal{E}_\alpha, \mathcal{E}_\beta] = Z$ if $\alpha + \beta$ is not a root and $\alpha \neq -\beta$.
(d) $[\mathcal{E}_\alpha, \mathcal{E}_\beta] = \pm\mathcal{E}_{\alpha+\beta}$ if $\alpha + \beta$ is a root.
(e) $[\mathcal{H}, \mathcal{E}_\alpha] = \alpha(\mathcal{H})\mathcal{E}_\alpha$, $[\mathcal{H}_\alpha, \mathcal{E}_\alpha] = 2\mathcal{E}_\alpha$.

This lemma and its proof are identical with Lemma 9.1. The only difference between these results and the corresponding results for $gl(m)$ is that $\lambda_1 + \cdots + \lambda_m = 0$ in (3.6). We can consider $\lambda_1, \ldots, \lambda_{m-1}$ as independent variables while $\lambda_m = -\lambda_1 - \cdots - \lambda_{m-1}$.

Let ρ be a finite-dimensional rep of $sl(m)$ on the vector space V and set $\rho(\mathcal{H}) = H$, $\rho(\mathcal{E}_\alpha) = E_\alpha$. Then the operators $\{H, E_\alpha\}$ satisfy the commutation relations given above. Furthermore, the proof of Theorem 9.1 shows that V has a basis of weight vectors. The construction of α-ladders of weights and Theorems 9.2 and 9.3 carry over immediately to reps of $sl(m)$. In particular each irred rep ρ is uniquely determined by its highest weight $\Lambda^*(\lambda_1, \ldots, \lambda_{m-1})$ which has multiplicity one. (We write each weight in the unique form $\Lambda = q_1\lambda_1 + \cdots + q_{m-1}\lambda_{m-1}$ and adopt the usual lexicographic ordering.)

Let us determine the possible weights

$$\Lambda(\mathcal{H}) = \sum_{j=1}^{m} p_j\lambda_j = \sum_{j=1}^{m-1} q_j\lambda_j, \qquad q_j = p_j - p_m,$$
$$\mathcal{H} = \sum_{j=1}^{m} \lambda_j\mathcal{H}_j = \sum_{j=1}^{m-1} \lambda_j(\mathcal{H}_j - \mathcal{H}_m) \in \mathscr{h}_{m-1}. \tag{3.8}$$

If Λ is a weight then $2\Lambda_\alpha/\alpha_\alpha$ is an integer for each root $\alpha = \lambda_k - \lambda_l$, where $\Lambda_\alpha = \Lambda(\mathcal{H}_\alpha) = p_k - p_l$, $\alpha_\alpha = 2$. For $1 \leq k, l \leq m - 1$ we find $p_k - p_l = q_k - q_l$ is an integer. However, if $k = 1, \ldots, m-1$, $l = m$ then $p_k - p_m = q_k$ is an integer. Thus, the possible weights take the form $\Lambda = \sum q_j\lambda_j$, where the q_j are integers.

Now supppose $\Lambda^* = \sum q_j\lambda_j$ is a highest weight vector. Then $\Lambda_\alpha^* \geq 0$ for all positive roots α. The positive roots are $\alpha = \lambda_k - \lambda_l$, $1 \leq k < l \leq m - 1$, and $\alpha = \lambda_k - \lambda_m = \lambda_k + \lambda_1 + \cdots + \lambda_{m-1}$. Thus, if Λ^* is maximal then $q_k - q_l \geq 0$ for $k < l$ and $q_k \geq 0$, i.e., $q_1 \geq q_2 \geq \cdots \geq q_{m-1} \geq 0$. We shall show that each such linear form actually *is* the maximal weight of an irred rep ρ of $sl(m)$ which defines a global rep of $SL(m)$.

Consider the irred rep $[f_1, \ldots, f_m]$ of $GL(m)$. Its highest weight is $\Lambda^* = f_1\lambda_1 + \cdots + f_m\lambda_m$. It is characterized by the fact that $\Lambda^* + \alpha$ is not a weight for any root $\alpha > 0$. If we restrict the rep to $SL(m)$ the weights $\Lambda = p_1\lambda_1 + \cdots + p_m\lambda_m$ on $\mathscr{h}_m$ will restrict to weights $\Lambda' = (p_1 - p_m)\lambda_1 + \cdots + (p_{m-1} - p_m)\lambda_{m-1}$ on $\mathscr{h}_{m-1}$. Furthermore, the positive roots α of $gl(m)$ restrict to the positive roots of $sl(m)$. Thus, $(\Lambda')^* + \alpha$ is not a weight for any root $\alpha > 0$ of $sl(m)$ and

$$(\Lambda')^* = (f_1 - f_m)\lambda_1 + \cdots + (f_{m-1} - f_m)\lambda_{m-1} \tag{3.9}$$

is the maximal weight belonging to $[f_1, \ldots, f_m] \mid SL(m)$. Since this rep is irred we see again that

$$[f_1, \ldots, f_m] \cong [f_1 - f_m, \ldots, f_{m-1} - f_m, 0]$$

and $q_1\lambda_1 + \cdots + q_{m-1}\lambda_{m-1}$ for integers $q_1 \geq q_2 \geq \cdots \geq q_{m-1} \geq 0$ is the highest weight of the rep $[q_1, \ldots, q_{m-1}, 0]$.

The Lie algebra $su(m)$ is a real form of $sl(m)$ and there is a 1–1 relationship between analytic reps of the complex group $SL(m)$ and those of $SU(m)$. Since $SU(m)$ is compact we can use the unitary trick to conclude that every rep of $SL(m)$ can be decomposed into a direct sum of irred reps. Furthermore, the irred reps of $SU(m)$ are just the restrictions of the reps $[f_1, \ldots, f_{m-1}, 0]$ to $SU(m)$.

The simple characters $\psi^{f_1 \cdots f_m}$ of $SU(m)$ are the restrictions of the characters (2.23) to $SU(m)$, i.e., $\psi^{f_1 \cdots f_m} = \chi^{f_1 \cdots f_m}$ with $\epsilon_1 \epsilon_2 \cdots \epsilon_m = 1$. It follows from remarks made above that $\psi^{f_1 \cdots f_m} = \psi^{f_1 + p, \ldots, f_m + p}$. For example, we compute the simple characters $\psi^{f_1 f_2}$ of $SU(2)$. Here $\epsilon = \epsilon_1 = \epsilon_2^{-1}$ and

$$\begin{aligned} \psi^{f_1 f_2} &= \begin{vmatrix} \epsilon^{f_1+1} & \epsilon^{f_2} \\ \epsilon^{-f_1-1} & \epsilon^{-f_2} \end{vmatrix} \bigg/ \begin{vmatrix} \epsilon & 1 \\ \epsilon^{-1} & 1 \end{vmatrix} \\ &= (\epsilon^{(f_1 - f_2 + 1)} - \epsilon^{-(f_1 - f_2 + 1)})/(\epsilon - \epsilon^{-1}) \\ &= \{\sin[(f_1 - f_2 + 1)\tau/2]\}/\sin(\tau/2), \end{aligned} \tag{3.10}$$

where $\epsilon = e^{i\tau/2}$. This is in agreement with (2.24), Section 7.2, and shows that

$$[f_1, f_2] \cong [f_1 - f_2, 0] \cong \mathbf{D}^{([f_1 - f_2]/2)}. \tag{3.11}$$

We can also consider $SL(m)$ as a real $2(m^2 - 1)$-dimensional Lie group. In analogy with the theory of the real group $GL(m)$ one can easily show that the analytic irred reps of the real group $SL(m)$ are

$$[f_1, \ldots, f_{m-1}, 0] \otimes \overline{[g_1, \ldots, g_{m-1}, 0]}. \tag{3.12}$$

Finally we note a general method for decomposing reps of the complex group $SL(m)$ into irred parts. Let ρ be an analytic rep of $SL(m)$ such that

$$\rho \cong \alpha_1[f_1, \ldots, f_{m-1}, 0] \oplus \alpha_2[g_1, \ldots, g_{m-1}, 0] \oplus \cdots, \tag{3.13}$$

where the reps on the right-hand side are listed in lexicographic order. Then ρ contains the highest weight $f_1\lambda_1 + \cdots + f_{m-1}\lambda_{m-1}$ with multiplicity α_1. Suppose we have a list of the weights of ρ, each weight listed as many times as its multiplicity in ρ. Remove all the weights corresponding to α_1 copies of $[f_1, \ldots, f_{m-1}, 0]$ from the list. Then the highest weight remaining will be $g_1\lambda_1 + \cdots + g_{m-1}\lambda_{m-1}$ with multiplicity α_2. Next remove all weights corresponding to α_2 copies of $[g_1, \ldots, g_{m-1}, 0]$. We can continue in this manner until all the weights of ρ have been removed. The process is useful when we know the weights of ρ and want to derive the decomposition (3.13).

As an example we rederive the Clebsch–Gordan series for reps of $SL(2)$. The weights of $[f_1, 0]$ are $n\lambda_1 + (f_1 - n)\lambda_2 = (2n - f_1)\lambda_1$, $n = 0, 1, \ldots, f_1$, as we see from (1.41) and $\lambda_1 + \lambda_2 = 0$. By Lemma 9.6, the weights of $\rho \cong [f_1, 0] \otimes [g_1, 0]$ are $(2n + 2p - f_1 - g_1)\lambda_1$, $n = 0, 1, \ldots, f_1$, $p = 0, 1, \ldots, g_1$. Assuming $f_1 \geq g_1$ we see that the weight $2s - f_1 - g_1$ has multiplicity $s + 1$ if $0 \leq s \leq g_1$, $g_1 + 1$ if $g_1 + 1 \leq s \leq f_1$, and $f_1 + g_1 + 1 - s$ if $f_1 + 1 \leq s \leq f_1 + g_1$. The highest weight of this rep is $(f_1 + g_1)\lambda_1$ with multiplicity one. Thus, ρ contains $[f_1 + g_1, 0]$. If we remove the weights of $[f_1 + g_1, 0]$ the highest remaining weight is $(f_1 + g_1 - 2)\lambda_1$ with multiplicity one. Thus ρ contains $[f_1 + g_1 - 2, 0]$. Continuing in this manner we obtain the reps $[f_1 + g_1 - 2k, 0]$, $0 \leq k \leq g_1$, each with multiplicity one. At this point all weights of ρ are used and the process ends. Thus

$$[f_1, 0] \otimes [g_1, 0] \cong \sum_{k=1}^{g_1} \oplus [f_1 + g_1 - 2k, 0], \tag{3.14}$$

which is the Clebsch–Gordan series (2.26), Chapter 7.

9.4 The Symplectic Groups and Their Representations

Recall that the symplectic group $Sp(m)$ consists of all $2m \times 2m$ complex matrices A such that

$$A^{t}JA = J, \tag{4.1}$$

where J is the skew-symmetric matrix (4.1), Section 5.4. (More generally, we can consider $Sp(m)$ as the set of all linear operators $\mathbf{A}$ on a $2m$-dimensional complex vector space V such that $\langle \mathbf{Au}, \mathbf{Av} \rangle = \langle \mathbf{u}, \mathbf{v} \rangle$ for all $\mathbf{u}, \mathbf{v} \in V$, where $\langle -, - \rangle$ is a nonsingular skew-symmetric bilinear form on V: see Weyl [3].)

From Theorem 5.15, the Lie algebra $sp(m)$ of $Sp(m)$ is the space of all $2m \times 2m$ complex matrices $\mathcal{A}$ such that

$$\mathcal{A}^{t} = J\mathcal{A}J. \tag{4.2}$$

Setting

$$\mathcal{A} = \begin{pmatrix} \mathcal{A}_1 & \mathcal{A}_2 \\ \mathcal{A}_3 & \mathcal{A}_4 \end{pmatrix}, \qquad J = \begin{pmatrix} Z & E_m \\ -E_m & Z \end{pmatrix},$$

where the $\mathcal{A}_j$ are $m \times m$ matrices, we find

$$\mathcal{A}_1^{t} = -\mathcal{A}_4, \qquad \mathcal{A}_2^{t} = \mathcal{A}_2, \qquad \mathcal{A}_3^{t} = \mathcal{A}_3.$$

Thus, $sp(m)$ is $(2m^2 + m)$-dimensional. Denoting by $\mathcal{E}_{jk}$ the matrix with a one in row j, column k and zeros everywhere else we obtain the basis

$$\mathcal{E}_{jk} - \mathcal{E}_{k+m,j+m}, \quad \mathcal{E}_{j,k+m} + \mathcal{E}_{k,j+m}, \quad \mathcal{E}_{j+m,k} + \mathcal{E}_{k+m,j}, \quad 1 \leq j, k \leq m. \tag{4.3}$$

The set $\mathscr{h}_m$ of all diagonal matrices

$$\mathcal{H}(\lambda_1, \ldots, \lambda_m) = \sum_{j=1}^{m} \lambda_j \mathcal{H}_j \tag{4.4}$$

forms a maximal abelian subalgebra of $sp(m)$. Here $\mathcal{H}_j = \mathcal{E}_{jj} - \mathcal{E}_{j+m,j+m}$.

The adjoint rep of $\mathscr{h}_m$ on $sp(m)$ is

$$\begin{aligned} [\mathcal{H}, \mathcal{E}_{jk} - \mathcal{E}_{k+m,j+m}] &= (\lambda_j - \lambda_k)(\mathcal{E}_{jk} - \mathcal{E}_{k+m,j+m}) \\ [\mathcal{H}, \mathcal{E}_{j,k+m} + \mathcal{E}_{k,j+m}] &= (\lambda_j + \lambda_k)(\mathcal{E}_{j,k+m} + \mathcal{E}_{k,j+m}) \\ [\mathcal{H}, \mathcal{E}_{j+m,k} + \mathcal{E}_{k+m,j}] &= (-\lambda_j - \lambda_k)(\mathcal{E}_{j+m,k} + \mathcal{E}_{k+m,j}). \end{aligned} \tag{4.5}$$

Thus the roots are $\alpha = \lambda_j - \lambda_k$ $(j \neq k)$ with branches $\mathcal{E}_\alpha = \mathcal{E}_{jk} - \mathcal{E}_{k+m,j+m}$, $\alpha = \lambda_j + \lambda_k$ $(j \leq k)$ with branches $\mathcal{E}_\alpha = \mathcal{E}_{j,k+m} + \mathcal{E}_{k,j+m}$, and $\alpha = -\lambda_j - \lambda_k$ $(j \leq k)$ with branches $\mathcal{E}_\alpha = \mathcal{E}_{j+m,k} + \mathcal{E}_{k+m,j}$. There are $2m^2$ distinct roots

$$\alpha = \pm\lambda_j \pm \lambda_k, \qquad j < k, \qquad \alpha = \pm 2\lambda_j.$$

By tedious computations we can verify the relations

$$\begin{aligned} &[\mathcal{H}, \mathcal{H}'] = Z, \qquad [\mathcal{H}, \mathcal{E}_\alpha] = \alpha(\mathcal{H})\mathcal{E}_\alpha, \\ &[\mathcal{E}_\alpha, \mathcal{E}_\beta] = \begin{cases} Z & \text{if } \alpha + \beta \text{ is nonzero and not a root,} \\ N_{\alpha\beta}\mathcal{E}_{\alpha+\beta} & \text{if } \alpha + \beta \neq 0 \text{ is a root,} \\ \mathcal{H}_\alpha & \text{if } \alpha = -\beta. \end{cases} \end{aligned} \tag{4.6}$$

Here $N_{\alpha\beta}$ is a nonzero constant depending on α and β, and

$$\mathcal{H}_\alpha = \begin{cases} \pm\mathcal{H}_j \pm \mathcal{H}_k & \text{if } \alpha = \pm\lambda_j \pm \lambda_k, \quad j < k, \\ \pm\mathcal{H}_j & \text{if } \alpha = \pm 2\lambda_j. \end{cases} \tag{4.7}$$

It follows from (4.6) that the proofs of Theorems 9.1–9.3 go through virtually unchanged for reps ρ of $sp(m)$. If $\Lambda = \sum p_j\lambda_j$ is a weight then

$$\Lambda_\alpha = \Lambda(\mathcal{H}_\alpha) = \begin{cases} \pm p_j \pm p_k & \text{if } \alpha = \pm\lambda_j \pm \lambda_k, \quad j < k, \\ \pm p_j & \text{if } \alpha = \pm 2\lambda_j. \end{cases} \tag{4.8}$$

In particular,

$$\alpha_\alpha = \alpha(\mathcal{H}_\alpha) = 2 \tag{4.9}$$

for all roots α.

If Λ is a weight then $2\Lambda_\alpha/\alpha_\alpha$ is an integer and $S^\alpha\Lambda = \Lambda - (2\Lambda_\alpha/\alpha_\alpha)\alpha$ is a weight. Writing $\Lambda = \sum p_j\lambda_j$, we see $2\Lambda_\alpha/\alpha_\alpha = \Lambda_\alpha$ and from (4.8) the p_j must be integers. Furthermore, $S^\alpha\Lambda = \sum p_l\lambda_l - (\pm p_j \pm p_k)(\pm\lambda_j \pm \lambda_k)$, so $S^\alpha\Lambda$ is obtained from Λ by interchanging λ_j and λ_k if $\alpha = \lambda_j - \lambda_k$ or by replacing λ_j with $-\lambda_k$ and λ_k with $-\lambda_j$ if $\alpha = \pm(\lambda_j + \lambda_k)$, $j < k$. If $\alpha = \pm 2\lambda_j$ then $S^\alpha\Lambda = \sum p_l\lambda_l - 2p_j\lambda_j$, so $S^\alpha\Lambda$ is obtained from Λ by replacing λ_j with $-\lambda_j$. The Weyl group has order $m!2^m$.

From Theorem 9.3 an irred rep ρ of $sp(m)$ is uniquely determined by its highest weight Λ^* which occurs with multiplicity one. Furthermore, $\Lambda_\alpha^* \geq 0$

for all positive roots α. With the usual lexicographic ordering, the positive roots are $\lambda_j - \lambda_k, \lambda_j + \lambda_k, j < k$, and $2\lambda_j$. From (4.8), Λ^* is a highest weight only if $p_1 \geq p_2 \geq \cdots \geq p_m \geq 0$. Thus, the possible irred reps of $sp(m)$ can be denoted

$$(4.10) \qquad (p_1, \ldots, p_m), \qquad p_1 \geq \cdots \geq p_m \geq 0.$$

We will show that there exists an irred rep corresponding to each signature (4.10). Furthermore, we will show that each Lie algebra rep induces a global group rep of $Sp(m)$.

Since $Sp(m)$ is a subgroup $GL(2m)$, we can construct the rep $[f_1, \ldots, f_{2m}]$ of $Sp(m)$ by restriction from the corresponding rep of $GL(2m)$. However, the restricted reps may not be irred.

First we consider the natural action of $Sp(m)$ on the $2m$-dimensional vector space V, i.e., the rep [1]. The weights of this rep are easily see to be $\pm\lambda_j$, $1 \leq j \leq m$, each with multiplicity one, and the highest weight is λ_1. This rep is irred since, applying the Weyl reflections S^α to λ_1, we get all of the weights $\pm\lambda_j$. Thus, $(1, 0, \ldots, 0) = (1) \cong [1]$. Similarly, it follows from (4.4), (1.19), and (1.33) that the $\binom{2m}{l}$-dimensional rep $[1^l]$, $l = 1, 2, \ldots, m$, has weights

$$(4.11) \qquad \lambda_{i_1} + \lambda_{i_2} + \cdots + \lambda_{i_l}, \qquad 1 \leq i_1 < i_2 < \cdots < i_l \leq 2m,$$

each with multiplicity one. Here $\lambda_{i_j} = -\lambda_k$ if $i_j = m + k$, $k > 0$. The highest weight is clearly

$$(4.12) \qquad \lambda_1 + \lambda_2 + \cdots + \lambda_l.$$

At this point we assume that every finite-dimensional analytic rep of $Sp(m)$ can be decomposed into a direct sum of irred reps. (We will prove this later.) Then from (4.12) there exist irred reps of $Sp(m)$ with signature (1^l), $1 \leq l \leq m$. (Note that we use the reps $[1^l]$ of $GL(2m)$ only for $1 \leq l \leq m$. The reps for $m + 1 \leq l \leq 2m$ have been omitted.)

Now consider the rep

$$(4.13) \qquad [1]^{\otimes k_1} \otimes [1^2]^{\otimes k_2} \otimes \cdots \otimes [1^m]^{\otimes k_m} \cong \rho,$$

where $k_1, \ldots, k_m$ are nonnegative integers. The highest weight of ρ is $p_1\lambda_1 + \cdots + p_m\lambda_m$, where

$$(4.14) \qquad \begin{aligned} p_1 &= k_1 + k_2 + \cdots + k_m \\ p_2 &= \phantom{k_1 + {}} k_2 + \cdots + k_m \\ &\ \ \vdots \\ p_{m-1} &= \phantom{k_1 + k_2 + \cdots + {}} k_{m-1} + k_m \\ p_m &= \phantom{k_1 + k_2 + \cdots + k_{m-1} + {}} k_m, \end{aligned}$$

and this weight occurs with multiplicity one. We can construct the irred rep with signature $(p_1, \ldots, p_m)$ by choosing $k_1 = p_1 - p_2, k_2 = p_2 - p_3, \ldots, k_{m-1} = p_{m-1} - p_m, k_m = p_m$. Thus all our candidates for highest weights actually occur in the irred reps of $Sp(m)$.

Our construction has several gaps. We have not shown that every rep of $Sp(m)$ can be decomposed into a direct sum of irred reps. Second, using Lie-algebraic methods we have computed only reps of the connected component of the identity in $Sp(m)$. If $Sp(m)$ has more than one component [like $O(m)$ or $L(4)$] then there are more irred reps than those we have listed.

If $A \in Sp(m)$ then $A^tJA = J$. Taking the determinant of both sides of this equation, we have $(\det A)^2 = 1$, or $\det A = \pm 1$. If there exist group elements for which $\det A = -1$ then such elements are not in the connected component of the identity. We will show that $Sp(m)$ is connected and $\det A = +1$ always.

The group $USp(m) = Sp(m) \cap U(2m)$ is a subgroup of $Sp(m)$. Furthermore, as the reader can verify, its Lie algebra $usp(m)$ is a real $(2m^2 + m)$-dimensional subalgebra of $sp(m)$. Thus, $usp(m)$ is a real form of $sp(m)$ and there is a 1–1 correspondence between complex reps of these Lie algebras. It follows that the irred reps $(p_1, \ldots, p_m)$ of $Sp(m)$ constructed above, restrict to irred reps of $USp(m)$. We will soon show that $USp(m)$ is connected, so its irred reps are uniquely determined by the irred reps if $usp(m)$. Since $USp(m)$ is compact, every analytic rep of this group or every finite-dimensional complex rep of its Lie algebra decomposes into a direct sum of irred reps. This proves that every rep of $sp(m)$ decomposes into irred reps.

We now examine the structure of $USp(m)$. The elements A of this group are both symplectic and unitary. Thus A preserves the forms

$$y^tJx, \qquad y^t\bar{x} \tag{4.15}$$

simultaneously, where x is the column vector $(x_1, \ldots, x_m, x_1', \ldots, x_m')$. Indeed $(Ay)^tJ(Ax) = y^t(A^tJA)x = y^tJx$ with a similar proof for $y^t\bar{x}$. Since A is unitary, it has $2m$ eigenvalues $\epsilon_1, \ldots, \epsilon_{2m}$ each of modulus one. Let x be an eigenvector of A with eigenvalue ϵ: $Ax = \epsilon x$. Now $(A^{-1})^tJ = JA$ since A is symplectic and $(A^{-1})^t = \bar{A}$ since A is unitary. Thus, $\bar{A}J = JA$ and $JAx = \epsilon Jx = \bar{A}Jx$. Taking the complex conjugate, we obtain

$$A(J\bar{x}) = \bar{\epsilon}(J\bar{x}). \tag{4.16}$$

Thus, if x is an eigenvector of A with eigenvalue ϵ then $J\bar{x}$ is an eigenvector with eigenvalue $\bar{\epsilon}$.

Let $x^{(1)}, \ldots, x^{(l)}$ be an ON basis for the eigenspace $\mathfrak{C}_\epsilon$ of A (usual scalar product $[x^{(j)}]^t\bar{x}^{(k)} = \delta_{jk}$). Then the $J\bar{x}^{(j)}$, $1 \leq j \leq l$, form an ON set in $\mathfrak{C}_{\bar{\epsilon}}$ since $(J\bar{x}^{(j)})^t\overline{(J\bar{x}^{(k)})} = -[\bar{x}^{(j)}]^tJ^2x^{(k)} = [\bar{x}^{(j)}]^tx^{(k)} = \delta_{jk}$. Similarly, an ON basis $\{y^{(j)}\}$ of $\mathfrak{C}_{\bar{\epsilon}}$ is mapped into an ON set $\{J\bar{y}^{(j)}\}$ in $\mathfrak{C}_\epsilon$. It follows that $\dim \mathfrak{C}_\epsilon = \dim \mathfrak{C}_{\bar{\epsilon}} = l$ and $\{J\bar{x}^{(j)}\}$ is an ON basis for $\mathfrak{C}_{\bar{\epsilon}}$.

If ϵ is a complex eigenvalue then $\bar{\epsilon} \neq \epsilon$, $\{x^{(j)}\}$, $1 \leq j \leq l$, is an ON basis of $\mathcal{C}_\epsilon$ and $\{J\bar{x}^{(j)}\}$ is an ON basis of $\mathcal{C}_{\bar{\epsilon}}$. Now suppose ϵ is real, i.e., $\epsilon = \pm 1$. Let $x^{(1)}$ be an element of $\mathcal{C}_\epsilon$ with length one. Then $J\bar{x}^{(1)}$ is a unit vector in $\mathcal{C}_\epsilon$ which is perpendicular to $x^{(1)}$. Indeed $(x^{(1)})^t\overline{(J\bar{x}^{(1)})} = (x^{(1)})^t J x^{(1)} = 0$ since J is skew-symmetric. Thus $\{x^{(1)}, J\bar{x}^{(1)}\}$ is an ON set in $\mathcal{C}_\epsilon$. If this set is not a basis we can find a unit vector $x^{(2)} \in \mathcal{C}_\epsilon$ orthogonal to the above set. A simple computation shows that $J\bar{x}^{(2)}$ is also a unit vector in $\mathcal{C}_\epsilon$ and $\{x^{(j)}, J\bar{x}^{(j)}: j = 1, 2\}$ is an ON set. Continuing in this fashion we eventually obtain an ON basis $\{x^{(j)}, J\bar{x}^{(j)}: j = 1, \ldots, l\}$ for $\mathcal{C}_\epsilon$. In particular, the dimension of $\mathcal{C}_\epsilon$ is even for ϵ real.

Lemma 9.9. The $2m$ eigenvalues of $A \in USp(m)$ occur in pairs: $\epsilon_1, \ldots, \epsilon_m, \bar{\epsilon}_1, \ldots, \bar{\epsilon}_m$. There exists a corresponding ON basis $x^{(1)}, \ldots, x^{(m)}, J\bar{x}^{(1)}, \ldots, J\bar{x}^{(m)}$ of eigenvectors.

Let $\mathbf{f}_j, \mathbf{f}_j'$ be the vectors in V with components $x^{(j)}$, $-J\bar{x}^{(j)}$, respectively, for $1 \leq j \leq m$. Then the $\{\mathbf{f}_j, \mathbf{f}_j'\}$ form an ON basis for V with respect to the usual inner product, and

$$
\begin{aligned}
(x^{(j)})^t J x^{(k)} &= (x^{(j)})^t \overline{(J\bar{x}^{(k)})} = 0 \\
(-J\bar{x}^{(j)})^t J(-J\bar{x}^{(k)}) &= \overline{(x^{(j)})^t J x^{(k)}} = 0 \\
(x^{(j)})^t J(-J\bar{x}^{(k)}) &= (x^{(j)})^t \bar{x}^{(k)} = \delta_{jk}.
\end{aligned}
\tag{4.17}
$$

It follows from this that the $2m \times 2m$ matrix U with columns $(x^{(1)}, \ldots, x^{(m)}, -J\bar{x}^{(1)}, \ldots, -J\bar{x}^{(m)})$ satisfies $U^t J U = J$ and $U^t U = E_{2m}$, i.e., $U \in USp(m)$. With this matrix and relations (4.17) it is straightforward to verify the following result.

Theorem 9.5. If $A \in USp(m)$ then there exists $U \in USp(m)$ such that

$$
U^{-1}AU = D(\epsilon) = \begin{pmatrix} \epsilon_1 & & & & & \\ & \ddots & & & & \\ & & \epsilon_m & & & \\ & & & \bar{\epsilon}_1 & & \\ & & & & \ddots & \\ & & & & & \bar{\epsilon}_m \end{pmatrix} \in USp(m).
$$

This shows that every $A \in USp(m)$ is conjugate in the group to a diagonal matrix $D(\epsilon)$. Since $A = UD(\epsilon)U^{-1}$ it is clear that A can be connected to the identity element by an analytic curve $A(t)$ in $USp(m)$ with $A(1) = A$, $A(0) = E_{2m}$. [Let the eigenvalues $\epsilon(t)$ approach $+1$ as $t \longrightarrow 0$.] Thus $USp(m)$ is connected.

We now sketch a proof of the fact that $Sp(m)$ is connected, leaving many of the details to the reader. As a byproduct we obtain a parametrization of $Sp(m)$.

Any nonsingular matrix A can be written uniquely in the form $A = HU$ where H is a positive-definite Hermitian matrix and U is unitary. (This is the **polar decomposition** of A; see Lancaster [1]). Here $H^2 = AA^*$ and H is the unique positive-definite square root of AA^*.

If $A \in Sp(m)$ then $A^* = \bar{A}^t \in Sp(m)$ and $AA^* \in Sp(m)$. Since the Hermitian matrix $AA^* = H^2$ is positive-definite it can be diagonalized by a unitary similarity transformation. Furthermore, the eigenvalues of H^2 are positive. Let ϵ be an eigenvalue of H^2 with eigenvector x. Now $H^2x = \epsilon x$ and $\bar{H}^{-2}Jx = (H^{-2})^t Jx = JH^2x = \epsilon Jx$ since $H^2 \in Sp(m)$ and $(\bar{H}^2)^t = H^2$. Thus.

$$H^2(J\bar{x}) = \epsilon^{-1}(J\bar{x}) \tag{4.18}$$

and $J\bar{x}$ is an eigenvector of H^2 with eigenvalue ϵ^{-1}. Proceeding almost exactly as in the proof of Theorem 9.5, we can show that the eigenvalues of H^2 take the form $\epsilon_1, \ldots, \epsilon_m, \epsilon_1^{-1}, \ldots, \epsilon_m^{-1}, \epsilon_j > 0$. Furthermore, there exists $W \in USp(m)$ such that

$$H^2 = AA^* = W\begin{pmatrix} \epsilon_1 & & & & & & Z \\ & \ddots & & & & & \\ & & \epsilon_m & & & & \\ & & & \epsilon_1^{-1} & & & \\ & Z & & & \ddots & & \\ & & & & & & \epsilon_m^{-1} \end{pmatrix} W^{-1} \in Sp(m). \tag{4.19}$$

It is clear that the matrix H is given by

$$H = W\begin{pmatrix} \epsilon_1^{1/2} & & & & & & Z \\ & \ddots & & & & & \\ & & \epsilon_m^{1/2} & & & & \\ & & & \epsilon_1^{-1/2} & & & \\ & Z & & & \ddots & & \\ & & & & & & \epsilon_m^{-1/2} \end{pmatrix} W^{-1} = WD(\epsilon)W^{-1}. \tag{4.20}$$

Since $D(\epsilon)$ belongs to $Sp(m)$ we have $H \in Sp(m)$. Hence $U = H^{-1}A \in Sp(m) \cap U(2m) = USp(m)$. so every $A \in Sp(m)$ can be written uniquely in the form $A = HU$, where H is given by (4.20) and $U \in USp(m)$. Since $USp(m)$ is connected, so is $Sp(m)$. In particular det $A = +1$, as follows directly from det A = det H det U and Theorem 9.5.

We have shown that the irred reps $(p_1, \ldots, p_m)$ of $Sp(m)$ are all tensor reps. For an explicit description of the irred tensor spaces see Weyl [2, 3].

The construction of simple characters for $USp(m)$ is analogous to that for $U(2m)$ so we present only the results. From Theorem 9.5 the conjugacy class in which $A \in USp(m)$ lies is determined by the eigenvalues $\epsilon_1, \ldots, \epsilon_m$, $\bar{\epsilon}_1, \ldots, \bar{\epsilon}_m$ $(\bar{\epsilon}_j = \epsilon_j^{-1})$ of A. Writing $\epsilon_j = \exp(i\varphi_j)$, $-\pi \leq \varphi_j < \pi$, we see that a character χ of $USp(m)$ can be written $\chi(\varphi_1, \ldots, \varphi_m)$. Here χ is invariant under any permutation of the φ_j and the transformations $\varphi_k \longrightarrow -\varphi_k$ for any k, i.e., under the Weyl group. In analogy with (2.14), the simple characters satisfy the orthogonality relations

$$(4.21)\qquad \begin{aligned}(\chi^{p_1\cdots p_m}, \chi^{q_1\cdots q_m}) &= \int_{-\pi}^{\pi} \cdots \int_{-\pi}^{\pi} \chi^{p_1\cdots p_m}(\varphi_1, \ldots, \varphi_m) \\ &\quad \times \bar{\chi}^{q_1\cdots q_m}(\varphi_1, \ldots, \varphi_m)\, \Delta\bar{\Delta}\, d\varphi_1 \cdots d\varphi_m \\ &= \delta_{p_1q_1} \cdots \delta_{p_mq_m},\end{aligned}$$

where

$$(4.22)\qquad \Delta(\varphi_1, \ldots, \varphi_m) = \prod_{j=1}^{m} (\epsilon^j - \epsilon^{-j}) \prod_{1\leq j<k\leq m} (\epsilon^j + \epsilon^{-j} - \epsilon^k - \epsilon^{-k}).$$

Now Δ is skew-symmetric with respect to the Weyl group, so $\xi = \chi\Delta$ is skew-symmetric, i.e., ξ changes sign under a transposition $\epsilon_j \longleftrightarrow \epsilon_k$ and under the exchange $\epsilon_j \longleftrightarrow \epsilon_j^{-1}$ for fixed j. Proceeding as in (2.17), we find the possible choices for ξ which give simple χ are

$$(4.23)\qquad \xi(l_1, \ldots, l_m) = |\epsilon^{l_1} - \epsilon^{-l_1}, \ldots, \epsilon^{l_m} - \epsilon^{-l_m}|$$

where the integers l_j satisfy $l_1 > l_2 > \cdots > l_m > 0$, and the determinant $|\cdot|$ is defined by (2.19). In particular

$$\xi(m, m-1, \ldots, 2, 1) = \Delta.$$

The ratio $\chi = \xi/\Delta$ is a finite sum of terms $\epsilon_1^{j_1} \cdots \epsilon_m^{j_m}$. The term with highest weight is $\epsilon_1^{l_1-m}\epsilon_2^{l_2-m+1} \cdots \epsilon_m^{l_m-1}$ and it occurs with multiplicity one. Thus $p_j = l_j - m + j - 1$.

Theorem 9.6.

$$\chi^{p_1\cdots, p_m} = \frac{|\epsilon^{l_1} - \epsilon^{-l_1}, \ldots, \epsilon^{l_m} - \epsilon^{-l_m}|}{|\epsilon^m - \epsilon^{-m}, \ldots, \epsilon - \epsilon^{-1}|}, \qquad l_j = p_j + m - j + 1,$$
$$1 \leq j \leq m.$$

Corollary 9.4. The dimension of $(p_1, \ldots, p_m)$ is

$$N(p_1, \ldots, p_m) = P(l_1, \ldots, l_m)/P(m, m-1, \ldots, 1),$$
$$P(l_1, \ldots, l_m) = \prod_{1\leq j\leq m} l_j \prod_{1\leq j<k\leq m} (l_j^2 - l_k^2).$$

We can use the characters of $USp(m)$ to determine how a symmetry class of tensors $[f_1, \ldots, f_{2m}]$ of $GL(2m)$ decomposes into irred reps $(p_1, \ldots, p_m)$ when $GL(2m)$ is restricted to $Sp(m)$ or $USp(m)$. One need only restrict the character $\chi^{f_1, \ldots, f_{2m}}$ of $U(2m)$, (2.23), to $USp(m)$ and then write it as a sum of simple characters of $USp(m)$. See Weyl [3] for more details.

The reader can check that the analytic irred reps of $Sp(m, R)$ are just the restrictions of the reps $(p_1, \ldots, p_m)$ of $Sp(m)$ to $Sp(m, R)$. Furthermore, the analytic irred reps of the real $2(2m^2 + m)$-dimensional Lie group $Sp(m, \mathfrak{C})$ are $(p_1, \ldots, p_m) \otimes \overline{(p_1', \ldots, p_m')}$.

9.5 The Orthogonal Groups and Their Representations

The usual realization of the complex orthogonal group $O(m, \mathfrak{C})$ is the set of all $m \times m$ complex matrices A such that

$$A^t A = E_m. \tag{5.1}$$

However, we can also consider $O(m, \mathfrak{C})$ as the set of all linear operators $\mathbf{A}$ on an m-dimensional complex vector space V such that $(\mathbf{Au}, \mathbf{Av}) = (\mathbf{u}, \mathbf{v})$ for all $\mathbf{u}, \mathbf{v} \in V$, where $(-, -)$ is a nondegenerate symmetric bilinear form on V. There always exists a basis $\mathbf{f}_1, \ldots, \mathbf{f}_m$ for V such that

$$(\mathbf{f}_j, \mathbf{f}_k) = (\mathbf{f}_k, \mathbf{f}_j) = \delta_{jk}, \qquad 1 \le j, k \le m, \tag{5.2}$$

(see the book of Cullen [1]). Writing $\mathbf{u} = \sum u_j \mathbf{f}_j$, $\mathbf{v} = \sum v_k \mathbf{f}_k$ we find

$$(\mathbf{u}, \mathbf{v}) = \sum_{j=1}^{m} u_j v_j.$$

Furthermore, if A is the matrix of $\mathbf{A}$ in the $\mathbf{f}$-basis a simple computation yields $\mathbf{A} \in O(m, \mathfrak{C})$ if and only if $A^t A = E_m$.

The realization (5.1) is not very convenient for a study of the irred reps of $O(m, \mathfrak{C})$ via the Lie algebra route, for the Lie algebra $o(m)$ in this realization consists of skew-symmetric matrices. Such matrices have all zeros on the diagonal and this is inconvenient since we have become accustomed to the use of a maximal abelian subalgebra $\mathscr{h}_k$ of diagonal matrices. We get around this difficulty by choosing a new basis for V.

If $m = 2n$, $n = 1, 2, \ldots$, we set

$$\mathbf{e}_j = 2^{-1/2}(\mathbf{f}_j + i\mathbf{f}_{n+j}), \qquad \mathbf{e}_j' = 2^{-1/2}(\mathbf{f}_j - i\mathbf{f}_{n+j}), \qquad 1 \le j \le n. \tag{5.3}$$

Then

$$(\mathbf{e}_j, \mathbf{e}_k) = (\mathbf{e}_j', \mathbf{e}_k') = 0, \qquad (\mathbf{e}_j, \mathbf{e}_k') = \delta_{jk} \tag{5.4}$$

and if $\mathbf{u} = \sum (u_j \mathbf{e}_j + u_j' \mathbf{e}_j')$ and $\mathbf{v} = \sum (v_j \mathbf{e}_j + v_j' \mathbf{e}_j')$ we find

$$(\mathbf{u}, \mathbf{v}) = \sum_{j=1}^{n} (u_j v_j' + u_j' v_j). \tag{5.5}$$

If $m = 2n + 1$, $n = 1, 2, \ldots$, we define $\mathbf{e}_j$, $\mathbf{e}_j'$, $1 \leq j \leq n$, by (5.3) and set $\mathbf{e}_0 = \mathbf{f}_{2n+1}$. Then if $\mathbf{v} = \sum (v_j \mathbf{e}_j + v_j' \mathbf{e}_j') + v_0 \mathbf{e}_0$ we find

$$(\mathbf{u}, \mathbf{v}) = \sum_{j=1}^{n} (u_j v_j' + u_j' v_j) + u_0 v_0 . \tag{5.6}$$

From (5.5) the matrices A of orthogonal transformations $\mathbf{A} \in O(2n, \mathfrak{C})$ with respect to the $\{\mathbf{e}_j, \mathbf{e}_j'\}$ basis are those which satisfy

$$A^{\mathrm{t}} K A = K, \qquad K = \begin{pmatrix} Z_n & E_n \\ E_n & Z_n \end{pmatrix}, \tag{5.7}$$

where E_n is the $n \times n$ identity matrix and Z_n is the $n \times n$ zero matrix.

Similarly the matrices A of operators $\mathbf{A} \in O(2n + 1, \mathfrak{C})$ must satisfy

$$A^{\mathrm{t}} K A = K, \qquad K = \begin{pmatrix} 1 & 0 & \cdots & 0 \\ 0 & & & \\ \cdot & & Z_n & E_n \\ \cdot & & & \\ \cdot & & E_n & Z_n \\ 0 & & & \end{pmatrix}, \tag{5.8}$$

where K is a $(2n + 1) \times (2n + 1)$ matrix. Clearly our new realizations of $O(2n, \mathfrak{C})$ and $O(2n + 1, \mathfrak{C})$ are isomorphic (even unitary equivalent) to the old ones. The justification for handling odd m and even m differently is that the rep theory for these two classes is distinctly different.

In our new realization the Lie algebra $o(m)$ is the space of $m \times m$ complex matrices $\mathcal{A}$ such that

$$\mathcal{A}^{\mathrm{t}} K + K \mathcal{A} = Z_m . \tag{5.9}$$

We consider the case $m = 2n$ first. Writing

$$\mathcal{A} = \begin{pmatrix} \mathcal{A}_1 & \mathcal{A}_2 \\ \mathcal{A}_3 & \mathcal{A}_4 \end{pmatrix},$$

where the $\mathcal{A}_j$ are $n \times n$ matrices, we obtain

$$\mathcal{A}_1^{\mathrm{t}} = -\mathcal{A}_4, \qquad \mathcal{A}_2^{\mathrm{t}} = -\mathcal{A}_2, \qquad \mathcal{A}_3^{\mathrm{t}} = -\mathcal{A}_3 . \tag{5.10}$$

It follows that the dimension of $o(2n)$ is $2n^2 - n$. Denoting by $\mathcal{E}_{jk}$ the matrix with a one in row j, column k and zeros elsewhere, we find the basis

$$\begin{aligned} &\mathcal{E}_{jk} - \mathcal{E}_{k+n,j+n}, && j, k = 1, \ldots, n, \\ &\mathcal{E}_{j+n,k} - \mathcal{E}_{k+n,j}, && \mathcal{E}_{j,k+n} - \mathcal{E}_{k,j+n}, \qquad j \neq k. \end{aligned} \tag{5.11}$$

The set $\mathcal{h}_n$ of all diagonal matrices

$$\mathcal{H}(\lambda_1, \ldots, \lambda_n) = \sum_{j=1}^{n} \lambda_j \mathcal{H}_j \tag{5.12}$$

is a maximal abelian subalgebra of $o(2n)$, where $\mathcal{H}_j = \mathcal{E}_{jj} - \mathcal{E}_{j+n,j+n}$.

A straightforward computation shows that the adjoint rep of $\mathcal{h}_n$ on $o(2n)$ is given by

$$\begin{aligned}[\mathcal{H}, \mathcal{E}_{jk} - \mathcal{E}_{k+n,j+n}] &= (\lambda_j - \lambda_k)(\mathcal{E}_{jk} - \mathcal{E}_{k+n,j+n}) \\ [\mathcal{H}, \mathcal{E}_{j+n,k} - \mathcal{E}_{k+n,j}] &= (-\lambda_j - \lambda_k)(\mathcal{E}_{j+n,k} - \mathcal{E}_{k+n,j}) \\ [\mathcal{H}, \mathcal{E}_{j,k+n} - \mathcal{E}_{k,j+n}] &= (\lambda_j + \lambda_k)(\mathcal{E}_{j,k+n} - \mathcal{E}_{k,j+n}),\end{aligned} \tag{5.13}$$

where $j \neq k$ and $1 \leq j, k \leq n$. Note the close relationship between (4.5) and (5.13). Clearly the roots are $\alpha = \lambda_j - \lambda_k, j \neq k$, $\alpha = -\lambda_j - \lambda_k, j < k$, and $\alpha = \lambda_j + \lambda_k, j < k$. There are $2n(n-1)$ distinct roots:

$$\alpha = \pm\lambda_j \pm \lambda_k, \qquad j < k.$$

By straightforward computation one can verify the formulas (4.6) and (4.7).

If $m = 2n + 1$ then $\mathcal{A} \in o(2n+1)$ takes the form

$$\mathcal{A} = \begin{pmatrix} 0 & -a_2{}^{t} & -a_1{}^{t} \\ a_1 & \mathcal{A}_1 & \mathcal{A}_2 \\ a_2 & \mathcal{A}_3 & \mathcal{A}_4 \end{pmatrix},$$

where a_1 and a_2 are $n \times 1$ matrices and $\mathcal{A}_1, \ldots, \mathcal{A}_4$ satisfy (5.10). Thus, the dimension of $o(2n+1)$ is $2n^2 + n$. If we consider the top row of $\mathcal{A}$ as row zero and the left-hand column as column zero then a basis for $o(2n+1)$ is given by the matrices (5.11) plus the matrices

$$\mathcal{E}_{k0} - \mathcal{E}_{0,k+n}, \qquad \mathcal{E}_{0k} - \mathcal{E}_{k+n,0}, \qquad k = 1, \ldots, n. \tag{5.14}$$

The set $\mathcal{h}_n'$ of diagonal matrices

$$\mathcal{H}(\lambda_1, \ldots, \lambda_n) = \sum_{j=1}^{n} \lambda_j \mathcal{H}_j \tag{5.15}$$

is a maximal abelian subalgebra of $o(2n+1)$.

The adjoint rep of $\mathcal{h}_n'$ on $o(2n+1)$ is given by expressions (5.13) plus

$$\begin{aligned}[\mathcal{H}, \mathcal{E}_{k0} - \mathcal{E}_{0,k+n}] &= \lambda_k(\mathcal{E}_{k0} - \mathcal{E}_{0,k+n}) \\ [\mathcal{H}, \mathcal{E}_{0k} - \mathcal{E}_{k+n,0}] &= -\lambda_k(\mathcal{E}_{0k} - \mathcal{E}_{k+n,0}).\end{aligned} \tag{5.16}$$

Thus $o(2n+1)$ has the same roots and branches as $o(2n)$ plus the simple roots $\pm\lambda_k$ with corresponding branches. In summary, the roots of $o(2n+1)$ are

$$\pm\lambda_j \pm \lambda_k, \quad j < k, \qquad \pm\lambda_k. \tag{5.17}$$

Again formulas (4.6) and (4.7) can be verified by direct computation.

It follows that the proofs of Theorems 9.1–9.3 apply to reps ρ of $o(2n)$ and $o(2n+1)$. Here

$$\mathcal{H}_\alpha = \begin{cases} \pm\mathcal{H}_j \pm \mathcal{H}_k & \text{if } \alpha = \pm\lambda_j \pm \lambda_k, \quad j < k, \\ \pm\mathcal{H}_j & \text{if } \alpha = \pm\lambda_j. \end{cases} \tag{5.18}$$

For example, if $\alpha = \lambda_j$ then $-\alpha = -\lambda_j$ and

$$[\mathcal{E}_\alpha, \mathcal{E}_{-\alpha}] = \mathcal{E}_{jj} - \mathcal{E}_{j+n,j+n} = \mathcal{H}_j = \mathcal{H}_\alpha .$$

If $\Lambda = \sum p_j\lambda_j$ is a weight of ρ then

$$(5.19) \qquad \Lambda_\alpha = \Lambda(\mathcal{H}_\alpha) = \begin{cases} \pm p_j \pm p_k & \text{if} \quad \alpha = \pm\lambda_j \pm \lambda_k, \quad j<k, \\ \pm p_j & \text{if} \quad \alpha = \pm\lambda_j . \end{cases}$$

In particular $\alpha_\alpha = 2$ if $\alpha = \pm\lambda_j \pm \lambda_k$ and $\alpha_\alpha = 1$ if $\alpha = \pm\lambda_j$. Note that the latter case occurs only for $m = 2n + 1$.

By Theorem 9.2, $2\Lambda_\alpha/\alpha_\alpha$ is an integer for every root α of $o(m)$ and $S^\alpha\Lambda = \Lambda - (2\Lambda_\alpha/\alpha_\alpha)\alpha$ is a weight of ρ. Thus $\pm p_j \pm p_k, j < k$, are integers for all $o(m)$ and $\pm 2p_j$ are integers for $m = 2n + 1$. If Λ is a weight then the p_j are either all integers of all half integers. If ρ is irred then the weights of ρ either have all integer coefficients or all half-integer coefficients, since any weight Λ' of ρ can be obtained from a single weight Λ by adding suitable sums of roots, $\Lambda' = \Lambda + \alpha_1 + \cdots + \alpha_s$. The roots are weights with integer coefficients.

Let $\Lambda^* = \sum p_j\lambda_j$ be the highest weight of ρ. Then $\Lambda_\alpha^* \geq 0$ for all roots $\alpha > 0$. The positive roots of $o(m)$ are $\lambda_j + \lambda_k, \lambda_j - \lambda_k, 1 \leq j < k \leq n$, and λ_j (if m is odd). Thus $p_j \pm p_k \geq 0$ for $j < k$, and $p_j \geq 0$ (if m is odd). The possible highest weights satisfy

$$(5.20) \qquad \begin{aligned} &p_1 \geq p_2 \geq \cdots \geq p_{n-1} \geq |p_n|, && m = 2n \\ &p_1 \geq p_2 \geq \cdots \geq p_{n-1} \geq p_n \geq 0, && m = 2n + 1, \end{aligned}$$

where the p_j are either all integral or all half-integral. In the case $m = 2n$ we have implicitly assumed $n \geq 2$ and omitted the abelian algebra $o(2)$. Note that p_n need not be positive for even m.

We will show that each signature (5.20) does correspond to the highest weight of an irred rep of $o(m)$. However, we will have some difficulty in using these results to determine the irred reps of $O(m, \mathfrak{C})$. First of all, $O(m, \mathfrak{C})$ is not connected. Indeed one can easily prove from the defining relations (5.7) and (5.8) that if $A \in O(m, \mathfrak{C})$ then $(\det A)^2 = 1$ or $\det A = \pm 1$. Furthermore, both signs occur. Thus $O(m, \mathfrak{C})$ has at least two connected components. (We will prove that there are only two.) The Lie algebra $o(m)$ only furnishes us with information about the connected component of the identity $SO(m, \mathfrak{C})$. Therefore, we have to look at more than the Lie algebra to determine the rep theory of $O(m, \mathfrak{C})$. The examples $O(3)$ and $SO(3)$ which we have treated earlier illustrate the problem to be solved here.

A more serious difficulty is that the half-integral Lie algebra reps do not induce global group reps of $SO(m, \mathfrak{C})$. Indeed, the matrix $\mathcal{H} \in \mathfrak{h}_n$, (5.12) or (5.15), exponentiates to (in case $m = 2n + 1$, say)

(5.21) $$e^{\mathfrak{H}} = \begin{pmatrix} 1 & 0 & \cdots & & & & 0 \\ 0 & e^{\lambda_1} & & & & & \\ & & \ddots & & & Z & \\ \vdots & & & e^{\lambda_n} & & & \\ & & & & e^{-\lambda_1} & & \\ & & Z & & & \ddots & \\ 0 & & & & & & e^{-\lambda_n} \end{pmatrix} \in SO(m, \mathfrak{C}).$$

If ρ is irred, then in a weight basis the matrices $\rho(e^{\mathfrak{H}})$ of the induced local group rep take the form

(5.22) $$\rho(e^{\mathfrak{H}}) = \begin{pmatrix} e^{\Lambda_1(\mathfrak{H})} & & Z \\ & \ddots & \\ Z & & e^{\Lambda_q(\mathfrak{H})} \end{pmatrix},$$

where $\Lambda_1, \ldots, \Lambda_q$ are the weights of ρ. If we replace λ_j by $\lambda_j + 2\pi i$ in (5.21) then $e^{\mathfrak{H}}$ remains unchanged. However, if we make these replacements in (5.22) for each j, $\rho(e^{\mathfrak{H}})$ will remain unchanged only if the weights Λ_k all have integer coefficients. If the weights have half-integer coefficients then ρ does not induce a single-valued rep of $SO(m. \mathfrak{C})$. If we replace λ_j by $\lambda_j + 4\pi i$, however, then $\rho(e^{\mathfrak{H}})$ remains unchanged even for half-integral ρ. This suggests that such ρ define double-valued reps of $SO(m, \mathfrak{C})$. We will see that there is a group Spin(m), locally isomorphic to $SO(m, \mathfrak{C})$ and a homomorphism ν: Spin$(m) \longrightarrow SO(m, \mathfrak{C})$ which covers $SO(m, \mathfrak{C})$ exactly twice. The double-valued reps of $SO(m, \mathfrak{C})$ are single-valued reps of Spin(m).

Keeping these difficulties in mind, we return to the construction of the irred reps of $o(m)$. Since $O(m, \mathfrak{C})$ is a subgroup of $GL(m)$ we obtain reps $[f_1, \ldots, f_m]$ of $O(m, \mathfrak{C})$ by restriction of the corresponding irred reps of $GL(m)$. Most of these reps will no longer be irred, however.

The rep belonging to the symmetry class [1] is just the natural action of $O(m, \mathfrak{C})$ on an m-dimensional vector space. We see immediately from (5.12) and (5.15) that the weights of the induced Lie algebra rep are $\pm\lambda_j$, $1 \leq j \leq n$, each with multiplicity one, plus the simple weight zero if m is odd. The highest weight is clearly λ_1, so $p_1 = 1$, $p_k = 0$, $2 \leq k \leq n$. This rep is irred, as the reader can easily check. Thus, $[1] \cong (1)$.

The weights of the Lie algebra reps induced by the symmetry classes $[1^l]$, $1 \leq l \leq m - 1$, are

(5.23) $$\lambda_{i_1} + \lambda_{i_2} + \cdots + \lambda_{i_l}, \qquad 0 \leq i_1 < i_2 < \cdots < i_l \leq m,$$

where $\lambda_0 = 0$ ($m = 2n + 1$) and $\lambda_{i_j} = -\lambda_k$ if $i_j = n + k$, $k > 0$. The highest weight of this rep is

(5.24) $$\lambda_1 + \lambda_2 + \cdots + \lambda_l.$$

Assuming that every rep of $o(m)$ can be decomposed into a direct sum of irred reps (which will be proved later) we have shown the existence of irred reps of $o(m)$ with signature (1^l), i.e., $p_1 = p_2 = \cdots = p_l = 1$, $p_{l+1} = \cdots = p_n = 0$.

Now suppose $m = 2n + 1$. In Section 9.6 we will construct a spinor rep of $o(2n + 1)$ with signature $(\frac{1}{2}^n)$. Then the highest weight of the rep

(5.25) $$(1)^{\otimes k_1} \otimes (1^2)^{\otimes k_2} \otimes \cdots \otimes (1^{n-1})^{\otimes k_{n-1}} \otimes (\tfrac{1}{2}^n)^{\otimes k_n}$$

is $p_1\lambda_1 + \cdots + p_n\lambda_n$ with multiplicity one, where

(5.26) $$p_j = \tfrac{1}{2}k_n + \sum_{h=j}^{n-1} k_j, \qquad 1 \le j \le n-1, \quad p_n = \tfrac{1}{2}k_n.$$

Here the p_j are integral or half-integral depending on whether k_n is even or odd. We can obtain any highest weight (5.20) by choosing the integers k_j such that $k_n = 2p_n$, $k_{n-1} = p_{n-1} - p_n, \ldots, k_2 = p_2 - p_3$, $k_1 = p_1 - p_2$. Thus we can find an irred rep of $o(2n + 1)$ for each set of integers or half-integers p_j satisfying (5.20).

In the next section we will show that $o(2n)$ has irred spinor reps with signatures $(\frac{1}{2}^n)$ and $(\frac{1}{2}^{n-1}, -\frac{1}{2})$. The highest weight of the rep

(5.27) $$(1)^{\otimes k_1} \otimes (1^2)^{\otimes k_2} \otimes \cdots \otimes (1^{n-2})^{\otimes k_{n-2}} \otimes (\tfrac{1}{2}^n)^{\otimes k_{n-1}} \otimes (\tfrac{1}{2}^{n-1}, -\tfrac{1}{2})^{\otimes k_n}$$

is $p_1\lambda_1 + \cdots + p_n\lambda_n$ with multiplicity one, where

(5.28) $$\begin{gathered} p_j = k_j + k_{j+1} + \cdots + k_{n-2} + \tfrac{1}{2}(k_{n-1} + k_n), \\ l = 1, 2, \ldots, n-2, \\ p_{n-1} = \tfrac{1}{2}(k_{n-1} + k_n), \qquad p_n = \tfrac{1}{2}(k_{n-1} - k_n). \end{gathered}$$

The p_j are all integral or half-integral depending on whether $k_{n-1} + k_n$ is even or odd. We can obtain any highest weight (5.20) with the choice

(5.29) $$k_n = p_{n-1} - p_n, \quad k_{n-1} = p_{n-1} + p_n, \quad k_j = p_j - p_{j+1}, \quad j = 1, \ldots, n-2.$$

The group $SO(m, R) = SO(m, \mathfrak{C}) \cap U(m)$ is a compact subgroup of $SO(m, \mathfrak{C})$. In the realization (5.1) of $O(m, \mathfrak{C})$ this is the group of real orthogonal matrices with determinant $+1$. In the realization (5.7) or (5.8) the matrices of $SO(m, R)$ are unitary equivalent to real orthogonal matrices with determinant $+1$ where the unitary equivalence is determined by (5.3). In the following we use the second realization of $SO(m, R)$. It is easy to verify that the real Lie algebra $o(m, R)$ is a real form of $o(m, \mathfrak{C})$.

Theorem 9.7. If $A \in SO(m, R)$ then there exists $U \in SO(m, R)$ such that

$$U^{-1}AU = D(\epsilon) = \begin{pmatrix} 1 & 0 & \cdots & & & & 0 \\ 0 & \epsilon_1 & & & & & \\ 0 & & \ddots & & Z & & \\ \vdots & & & \epsilon_n & & & \\ & & & & \bar{\epsilon}_1 & & \\ & & Z & & & \ddots & \\ 0 & & & & & & \bar{\epsilon}_n \end{pmatrix} \in SO(m, R), \quad |\epsilon_j| = 1.$$

The first row and column occur only for $m = 2n + 1$.

The proof of this theorem is analogous to that of Lemma 9.9 and theorem 9.5, so we omit it. Since $A = UD(\epsilon)U^{-1}$ it is clear, by the same argument as used for $USp(m)$, that A can be connected to E_m by an analytic curve in $SO(m, R)$. Thus, $SO(m, R)$ is connected.

To show that $SO(m, \mathfrak{C})$ is connected we mimic the corresponding proof for $Sp(m)$, leaving the details to the reader. If $A \in SO(m, \mathfrak{C})$ then by the polar decomposition, $A = HU$, Where H is Hermitian and positive-definite, and U is unitary. Now H is the unique positive-definite solution of $H^2 = AA^*$ $\in SO(m, \mathfrak{C})$. Since H^2 is Hermitian it can be diagonalized within $SO(m, \mathfrak{C})$:

$$H^2 = V\begin{pmatrix} 1 & 0 & \cdots & & & & 0 \\ 0 & \epsilon_1 & & & & & \\ 0 & & \ddots & & Z & & \\ \vdots & & & \epsilon_n & & & \\ & & & & \epsilon_1^{-1} & & \\ & & Z & & & \ddots & \\ 0 & & & & & & \epsilon_n^{-1} \end{pmatrix}V^{-1} = VC(\epsilon)V^{-1}, \qquad V \in SO(m, \mathfrak{C}).$$

The eigenvalues ϵ_j are positive, and the first row and column occur only if $m = 2n + 1$. It follows that H is given by

$$H = VC(\epsilon^{1/2})V^{-1} \in SO(m, \mathfrak{C}). \tag{5.30}$$

Thus, $U = H^{-1}A \in SO(m, \mathfrak{C}) \cap U(m) = SO(m, R)$. Since $SO(m, R)$ is connected and H is given by (5.30) we see that $SO(m, \mathfrak{C})$ is connected. Thus any group rep of $SO(m, \mathfrak{C})$ or $SO(m, R)$ is uniquely determined by its induced Lie algebra rep of $o(m)$. Furthermore, since $SO(m, R)$ is compact and $o(m, R)$ is a real form of $o(m)$ every finite-dimensional analytic rep of $SO(m, \mathfrak{C})$ [and of $O(m, \mathfrak{C})$] can be decomposed into a direct sum of irred reps.

We have seen earlier that some of the irred reps $(p_1, \ldots, p_n)$ of $SO(m, \mathfrak{C})$

and $o(m)$ can be obtained from tensor reps. We clearly cannot obtain the (double-valued) half-integral reps in this way, but we can obtain all of the integral reps. In the case $m = 2n + 1$ the rep $[1^n]$ has highest weight $\lambda_1 + \cdots + \lambda_n$ [see (5.24)]. It is easy to verify that the rep $(p_1, \ldots, p_n)$, integral p_j, is contained in the tensor rep

$$(1)^{\otimes k_1} \otimes (1^2)^{\otimes k_2} \otimes \cdots \otimes (1^{n-1})^{\otimes k_{n-1}} \otimes (1^n)^{\otimes k_n} \tag{5.31}$$

exactly once, where $k_n = p_n, k_{n-1} = p_{n-1} - p_n, \ldots, k_1 = p_1 - p_2$. A similar argument shows $(p_1, \ldots, p_n)$ is contained in the symmetry class $[p_1, \ldots, p_n]$ exactly once.

For $m = 2n$ the highest weight of $[1^n]$ is still $\Lambda = \lambda_1 + \cdots + \lambda_n$. From (5.23), $\Lambda' = \lambda_1 + \lambda_2 + \cdots + \lambda_{n-1} - \lambda_n$ is also a weight of $[1^n]$ and occurs with multiplicity one. Furthermore, if α is a positive root $\lambda_j + \lambda_k$ or $\lambda_j - \lambda_k$, $1 \leq j < k \leq n$, then $\Lambda' + \alpha$ is not a weight of $[1^n]$. (This is false for $m = 2n + 1$.) Thus Λ' must be the highest weight of the irred rep $(1^{n-1}, -1)$ contained in $[1^n]$ with multiplicity one. This shows that (1^n) and $(1^{n-1}, -1)$ are tensor reps. Finally the integral rep $(p_1, \ldots, p_n)$ is contained in the tensor rep

$$(1)^{\otimes k_1} \otimes (1^2)^{\otimes k_2} \otimes \cdots \otimes (1^{n-1})^{\otimes k_{n-1}} \otimes (1^{n-1}, -1)^{\otimes k_n} \otimes (1^n)^{\otimes k_{n+1}} \tag{5.32}$$

with multiplicity one, where $k_j = p_j - p_{j+1}$, $1 \leq j \leq n-2$, and

$$\begin{aligned} k_{n-1} &= p_{n-1} - p_n, & k_n &= 0, & k_{n+1} &= p_n & &\text{if} \quad p_n \geq 0 \\ k_{n-1} &= p_{n-1} + p_n, & k_n &= -p_n, & k_{n+1} &= 0 & &\text{if} \quad p_n < 0. \end{aligned} \tag{5.33}$$

By counting dimensions one can show

$$\begin{aligned} &[1^l] \cong (1^l), \quad 1 \leq l \leq n-1, \\ &[1^n] \cong \begin{cases} (1^n) & \text{if} \quad m = 2n+1 \\ (1^n) \oplus (1^{n-1}, -1) & \text{if} \quad m = 2n. \end{cases} \end{aligned} \tag{5.34}$$

We have now determined all single-valued irred reps of $SO(m, \mathfrak{C})$. For m odd it is easy to extend these results to compute the irred reps of $O(m, \mathfrak{C})$. Indeed, if $m = 2n + 1$ then $-E_m \in O(m, \mathfrak{C})$ and $\det(-E_m) = -1$. Thus, $O(m, \mathfrak{C})$ has the coset decomposition $\{SO(m, \mathfrak{C}), -E_m \cdot SO(m, \mathfrak{C})\}$.

Let $\mathbf{T}$ be an irred rep of $O(m, \mathfrak{C})$. Since $-E_m$ commutes with all elements of $O(m, \mathfrak{C})$ it follows from the Schur lemmas that $\mathbf{T}(-E_m) = \lambda \mathbf{E}$. The property $(-E_m)^2 = E_m$ implies $\lambda = \pm 1$. Thus $\mathbf{T} \mid SO(m, \mathfrak{C})$ must still be irred. The irred reps of $O(2n + 1, \mathfrak{C})$ are

$$\begin{aligned} &(p_1, \ldots, p_n)^+, \quad \mathbf{T}(-E_m) = \mathbf{E}; \\ &(p_1, \ldots, p_n)^-, \quad \mathbf{T}(-E_m) = -\mathbf{E}. \end{aligned} \tag{5.35}$$

For reps of $O(2n, \mathfrak{C})$ the situation is somewhat more complicated. In this case $\det(-E_m) = +1$, so $-E_m$ belongs to the connected component of the identity. We choose the element

$$S = \begin{pmatrix} & & & 0 & & & 0 \\ & & & \cdot & & & \cdot \\ & E_{n-1} & & \cdot & & Z_{n-1} & \cdot \\ & & & \cdot & & & \cdot \\ 0 & \cdot & \cdot & 0 & \cdot & \cdot & 1 \\ & & & \cdot & & & \cdot \\ & Z_{n-1} & & \cdot & & E_{n-1} & \cdot \\ & & & \cdot & & & \cdot \\ 0 & \cdot & \cdot & 1 & \cdot & \cdot & 0 \end{pmatrix}, \qquad \det S = -1, \quad S^2 = E_m. \tag{5.36}$$

Clearly, S and $SO(2n, \mathfrak{C})$ generate $O(2n, \mathfrak{C})$. Let $\mathbf{T}$ be an irred rep of $O(2n, \mathfrak{C})$ and set $\mathbf{S} = \mathbf{T}(S)$. Then $\mathbf{T}' = \mathbf{T} \mid SO(2n, \mathfrak{C})$ decomposes into a direct sum of irred reps. Let W be a subspace of the rep space V transforming irreducibly under $\mathbf{T}'$. Then the subspace $\mathbf{S}W$ also transforms irreducibly under $\mathbf{T}'$. Indeed if $\mathbf{w}^* = \mathbf{Sw} \in \mathbf{S}W$ and $A \in SO(2n, \mathfrak{C})$ then $S^{-1}AS \in SO(2n, \mathfrak{C})$ [since $\det(S^{-1}AS) = +1$] and

$$\mathbf{T}(A)\mathbf{w}^* = \mathbf{S}\mathbf{S}^{-1}\mathbf{T}(A)\mathbf{S}\mathbf{w} = \mathbf{S}(\mathbf{T}(S^{-1}AS)\mathbf{w}) \in \mathbf{S}W,$$

so $\mathbf{S}W$ is invariant under $\mathbf{T}'$. If W' is a nonzero invariant subspace of $\mathbf{S}W$ then $\mathbf{S}W'$ is a nonzero invariant subspace of $\mathbf{S}(\mathbf{S}W) = W$. Since W is irred, $\mathbf{S}W' = W$, so $W' = \mathbf{S}W$ and $\mathbf{S}W$ is irred.

The space $W + \mathbf{S}W$ is invariant under $\mathbf{T}$ and nonzero. Thus $W + \mathbf{S}W$ is the entire rep space V.

From (5.12) and (5.36) we have

$$\begin{gathered} \mathcal{H}(\lambda_1, \ldots, \lambda_{n-1}, \lambda_n)S = S\mathcal{H}(\lambda_1, \ldots, \lambda_{n-1}, -\lambda_n), \qquad E_\alpha S = SE_{\tilde{\alpha}} \\ \alpha = p_1\lambda_1 + \cdots + p_{n-1}\lambda_{n-1} + p_n\lambda_n, \\ \tilde{\alpha} = p_1\lambda_1 + \cdots + p_{n-1}\lambda_{n-1} - p_n\lambda_n. \end{gathered} \tag{5.37}$$

If $\mathbf{w}$ is a weight vector in W with weight $\mathbf{\Lambda} = p_1\lambda_1 + \cdots + p_n\lambda_n$ then $\mathbf{Sw}$ is a weight vector in $\mathbf{S}W$ with weight $\mathbf{\Lambda}' = p_1\lambda_1 + \cdots + p_{n-1}\lambda_{n-1} - p_n\lambda_n$. It follows that the weights of W and $\mathbf{S}W$ are related by a change of sign in λ_n. Let $p_1\lambda_1 + \cdots + p_{n-1}\lambda_{n-1} + p_n\lambda_n$ be the highest weight of W, $p_1 \geq \cdots \geq p_{n-1} \geq |p_n|$. It is straightforward to show that $p_1\lambda_1 + \cdots + p_{n-1}\lambda_{n-1} - p_n\lambda_n$ is the highest weight of $\mathbf{S}W$. If $p_n \neq 0$ then W and $\mathbf{S}W$ are linearly independent and

$$\mathbf{T} \cong (p_1, \ldots, p_n) \oplus (p_1, \ldots, p_{n-1}, -p_n). \tag{5.38}$$

Note that W and $\mathbf{S}W$ have the same dimension. If $p_n = 0$ we define $\mathbf{h}^+ = \mathbf{w} + \mathbf{Sw}$, $\mathbf{h}^- = \mathbf{w} - \mathbf{Sw}$, where $\mathbf{w}$ is the highest weight vector of W. Note that $\mathbf{Sw}$ is the highest weight vector of $\mathbf{S}W$ and $\mathbf{w}$, $\mathbf{Sw}$ correspond to the same weight $\mathbf{\Lambda}^* = p_1\lambda_1 + \cdots + p_{n-1}\lambda_{n-1}$. Now $\mathbf{Sh}^\pm = \pm\mathbf{h}^\pm$. It follows from (5.37) and (1.14) that the spaces $H^\pm$ spanned by all vectors of the form $E_{\alpha_1}E_{\alpha_2}\cdots$

$E_{\alpha_k}\mathbf{h}^{\pm}$ are invariant under $\mathbf{T}$. Since $\mathbf{T}$ is irred, either $H^+ = V, H^- = \{\theta\}$ or $H^+ = \{\theta\}, H^- = V$. In the first case we denote the rep by

(5.39)
$$(p_1, \ldots, p_{n-1}, 0)^+, \qquad \mathbf{Sw} = \mathbf{w}, \qquad \mathbf{T} \mid SO(2n, \mathfrak{C}) \cong (p_1, \ldots, p_{n-1}, 0)$$

and in the second case by

(5.40)
$$(p_1, \ldots, p_{n-1}, 0)^-, \qquad \mathbf{Sw} = -\mathbf{w}, \qquad \mathbf{T} \mid SO(2n, \mathfrak{C}) \cong (p_1, \ldots, p_{n-1}, 0).$$

This completes our catalog of irred reps of $O(2n, \mathfrak{C})$.

We have mentioned earlier that the Lie algebra of the compact group $SO(m, R)$ is a real form of the complex Lie algebra $o(m)$. Thus there is a 1–1 relationship between reps of $SO(m, \mathfrak{C})$ and $SO(m, R)$. Using the same techniques as in Section 9.2 we construct the characters for the irred reps $(p_1, \ldots, p_n)$ of $SO(m, R)$. We assume that the p_j are integral, although the results are virtually unchanged for half-integral p_j. Most results will be given without their straight-forward proofs.

From Theorem 9.7 a character χ of $SO(2n + 1, R)$ can be considered as a function $\chi(\varphi_1, \ldots, \varphi_n)$, $\epsilon_j = e^{i\varphi_j}$, $-\pi \leq \varphi_j < \pi$. If $\{\mathbf{v}_1, \ldots, \mathbf{v}_s\}$ is a weight basis for the corresponding rep and the weights are $\Lambda_1, \ldots, \Lambda_s$ then

$$\chi = \operatorname{tr}[e^{iH(\varphi_1, \cdots, \varphi_n)}] = \sum_{j=1}^{s} e^{i\Lambda_j(\varphi)} = \sum_{q_1 \cdots q_n} c_{q_1 \cdots q_n} \epsilon_1^{q_1} \cdots \epsilon_n^{q_n}, \tag{5.41}$$

where $\Lambda_j = q_1\varphi_1 + \cdots + q_n\varphi_n$ and $c_{q_1 \cdots q_n}$ is the multiplicity of this weight. Since χ is defined on conjugacy classes it follows from Theorem 9.7 that we can permute the φ_j or replace any subset of the φ_j by $-\varphi_j$ without changing the value of the character. Thus the character admits a symmetry group of order $2^n n!$. This is just the Weyl group. From (5.41) we see that the weights Λ and $S^\alpha\Lambda$ must have the same multiplicity for each root α.

The inner product on $L_2(O(2n + 1, R))$ for functions constant on conjugacy classes is

$$(f, g) = V^{-1} \int_{-\pi}^{\pi} \cdots \int_{-\pi}^{\pi} f(\varphi_1, \ldots, \varphi_n)\bar{g}(\varphi_1, \ldots, \varphi_n)\, \Delta\bar{\Delta}\, d\varphi_1 \cdots d\varphi_n \tag{5.42}$$

where

$$V = \int_{-\pi}^{\pi} \cdots \int_{-\pi}^{\pi} \Delta\bar{\Delta}\, d\varphi_1 \cdots d\varphi_n$$

and

$$\Delta(\varphi_1, \ldots, \varphi_n) = \prod_{j=1}^{n} (\epsilon_j^{1/2} - \epsilon_j^{-1/2}) \prod_{1 \leq j < k \leq n} (\epsilon_j + \epsilon_j^{-1} - \epsilon_k - \epsilon_k^{-1}). \tag{5.43}$$

Here Δ is skew-symmetric in its arguments and double-valued (due to the occurrence of $\epsilon_j^{1/2}$). Thus if χ is the character of $(p_1, \ldots, p_n)$, the function $\xi = \chi\Delta$ is skew-symmetric and double-valued. Furthermore, the substitution

$\varphi_j \longrightarrow -\varphi_j$ for a single j causes ξ to change sign. Note that ξ is a finite sum of the form (2.17) whose highest-order term is

(5.44) $$1 \cdot \epsilon_1^{p_1+n-(1/2)}\epsilon_2^{p_2+n-(3/2)} \cdots \epsilon_n^{p_n+(1/2)}.$$

Because of the symmetry properties, since the sum for ξ contains (5.44) it must also contain

(5.45) $$\xi(l_1, \ldots, l_n) = |\epsilon^{l_1} - \epsilon^{-l_1}, \ldots, \epsilon^{l_n} - \epsilon^{-l_n}|$$

where $|\cdot|$ is defined similarly to (2.19) and $l_j = p_j + n - j + \frac{1}{2}$. Here, $\Delta(\epsilon_1, \ldots, \epsilon_n) = \xi(n - \frac{1}{2}, n - \frac{3}{2}, \ldots, \frac{1}{2})$.

Continuing in this fashion we obtain the expansion $\xi = \xi(l_1, \ldots, l_n) + c'\xi(l_1', \ldots, l_n') + \cdots$. However, the requirement $(\chi, \chi) = 1$ implies $1 = 1 + |c'|^2 + \cdots$, so $c' = \cdots = 0$ and $\xi = \xi(l_1, \ldots, l_n)$.

Theorem 9.8. The character of the rep $(p_1, \ldots, p_n)$ of $SO(2n + 1, R)$ is

$$\chi^{p_1 \cdots p_n}(\epsilon_1, \ldots, \epsilon_n) = \frac{|\epsilon^{l_1} - \epsilon^{-l_1}, \ldots, \epsilon^{l_n} - \epsilon^{-l_n}|}{|\epsilon^{l_1^0} - \epsilon^{-l_1^0}, \ldots, \epsilon^{l_n^0} - \epsilon^{-l_n^0}|},$$

where $l_j = p_j + n - j + \frac{1}{2}$, $l_j^0 = n - j + \frac{1}{2}$. The dimension of this rep is

$$N(p_1, \ldots, p_n) = Q(l_1, \ldots, l_n)/Q(l_1^0, \ldots, l_n^0),$$

where

$$Q(l_1, \ldots, l_n) = \prod_{1 \le j < k \le n} (l_j^2 - l_k^2) \prod_{j=1}^{n} l_j.$$

If $m = 2n$ then (5.42) holds with

(5.46) $$\Delta = \prod_{1 \le j < k \le n} (\epsilon_j + \epsilon_j^{-1} - \epsilon_k - \epsilon_k^{-1}).$$

Here Δ is skew-symmetric and single-valued. If χ is the character of $(p_1, \ldots, p_n)$ then $\xi = \chi\Delta$ is skew-symmetric and invariant under an *even* number of sign changes $\varphi_j \longrightarrow -\varphi_j$. Indeed from Theorem 9.7 it follows that we can perform an arbitrary permutation of $\varphi_1, \ldots, \varphi_n$ in $D(e^{i\varphi})$ and still get a diagonal matrix in the same conjugacy class of $SO(2n, R)$. Furthermore, if we make the replacement $\varphi_j \longrightarrow -\varphi_j$ for an even number of angles we stay in the same conjugacy class. However, as the reader can check, the diagonal matrix obtained from an odd number of replacements $\varphi_j \longrightarrow -\varphi_j$ is not conjugate to $D(e^{i\varphi})$ in $SO(2n, R)$, although it is conjugate in $O(2n, R)$. We see that Lemma 9.7 holds also for $o(2n)$ since from (5.19) the above symmetries generate the Weyl group of order $2^{n-1}n!$.

The term of highest weight belonging to ξ is

$$1 \cdot \epsilon_1^{p_1+n-1}\epsilon_2^{p_2+n-2} \cdots \epsilon_n^{p_n}.$$

The symmetry requirements and the condition $(\chi, \chi) = 1$ yield

$$2\xi = |\epsilon^{l_1} + \epsilon^{-l_1}, \ldots, \epsilon^{l_n} + \epsilon^{-l_n}| + |\epsilon^{l_1} - \epsilon^{-l_1}, \ldots, \epsilon^{l_n} - \epsilon^{-l_n}|.$$

Theorem 9.9. The character of the rep $(p_1, \ldots, p_n)$ of $SO(2n, R)$ is

$$\chi^{p_1\cdots p_n}(\epsilon_1, \ldots, \epsilon_n) = \frac{|\epsilon^{l_1} + \epsilon^{-l_1}, \ldots, \epsilon^{l_n} + \epsilon^{-l_n}| + |\epsilon^{l_1} - \epsilon^{-l_1}, \ldots, \epsilon^{l_n} - \epsilon^{-l_n}|}{|\epsilon^{l_1^0} + \epsilon^{-l_1^0}, \ldots, \epsilon^{l_n^0} + \epsilon^{-l_n^0}|}$$

where $l_j = p_j + n - j, l_j^0 = n - j$. The dimension is

$$N(p_1, \ldots, p_n) = \frac{R(l_1, \ldots, l_n)}{R(l_1^0, \ldots, l_n^0)}, \qquad R(l_1, \ldots, l_n) = \prod_{1 \le j < k \le n} (l_j^2 - l_k^2).$$

Note that the reps $(p_1, \ldots, p_{n-1}, \pm p_n)$ have the same dimension. This is also a consequence of the remark following (5.38).

For more detailed proofs of the above theorems see the work of Boerner [1] or Weyl [3].

9.6 Dirac Matrices and the Spin Representations of the Orthogonal Groups

In an attempt to formulate a relativistic theory describing electrons Dirac considered systems of equations of the form

$$\left(L_1\frac{\partial}{\partial x} + L_2\frac{\partial}{\partial y} + L_3\frac{\partial}{\partial z} + L_4\frac{\partial}{\partial t}\right)\mathbf{\Psi} = \kappa\mathbf{\Psi}, \tag{6.1}$$

where the L_j are square matrices, κ is a constant, and $\mathbf{\Psi}$ is a spinor field. He required that this system be compatible with the Klein–Gordon equation (5.37), Section 8.5. Using the same notation as in Section 8.5, we see from (6.1) that

$$\left(\sum_{j=1}^{4} L_j\frac{\partial}{\partial x_j}\right)^2\mathbf{\Psi} = \kappa^2\mathbf{\Psi}. \tag{6.2}$$

If we choose $\kappa = m_0$, the mass of the electron, then (6.2) becomes

$$\sum_{j,k=1}^{4} L_jL_k\frac{\partial^2\mathbf{\Psi}}{\partial x_j\,\partial x_k} = m_0^2\mathbf{\Psi}. \tag{6.3}$$

This is equivalent to the Klein–Gordon equation

$$\left(\frac{\partial^2}{\partial x_1^2} + \frac{\partial^2}{\partial x_2^2} + \frac{\partial^2}{\partial x_3^2} - \frac{\partial^2}{\partial x_4^2}\right)\mathbf{\Psi} = m_0^2\mathbf{\Psi}$$

provided

$$L_jL_k + L_kL_j = 2G_{jk}, \qquad 1 \le j, k \le 4, \tag{6.4}$$

where G is given by (1.1), Section 8.1. We have already seen that the 4×4 matrices (5.33) Section 8.5, are solutions of (6.4). There are many other solutions and we shall compute all of them.

For convenience we modify (6.4) by defining matrices α_j such that

$$\alpha_j = L_j, \quad j = 1, 2, 3, \qquad \alpha_4 = iL_4, \qquad i = \sqrt{-1}.$$

Then the relations become

$$\alpha_j\alpha_k + \alpha_k\alpha_j = 2\delta_{jk}, \tag{6.5}$$

i.e.,

$$\alpha_j^2 = 1, \qquad \alpha_j\alpha_k = -\alpha_k\alpha_j \qquad \text{if} \quad j \neq k. \tag{6.6}$$

Finally we generalize our problem by allowing the indices j and k to run from 1 to m. The associative algebra generated by $\alpha_1, \ldots, \alpha_m$ is called the **Clifford algebra** C_m. We shall not study C_m but rather the multiplicative group G_m generated by ± 1 and $\alpha_1, \ldots, \alpha_m$. A general element of G_m has the form

$$\pm\alpha_{j_1}\alpha_{j_2} \cdots \alpha_{j_q}. \tag{6.7}$$

Making use of (6.6), we can always write this element in the standard form

$$\pm\alpha_{l_1}\alpha_{l_2} \cdots \alpha_{l_s}, \qquad l_1 < l_2 < \cdots < l_s, \quad s \leq m. \tag{6.8}$$

Indeed the relation $\alpha_j\alpha_k = -\alpha_k\alpha_j$, $k \neq j$, permits a reordering of the factors of (6.7) in the normal form (6.8). If two factors in (6.7) are equal one simply reorders the terms such that the two factors are adjacent and then uses $\alpha_j^2 = 1$ to reduce the length of the group element by two factors. Note that no element in standard form can have more than m factors since then two factors would have to be equal. Furthermore, the only elements with m factors in standard form are $\pm\alpha_1\alpha_2 \cdots \alpha_m$. Thus the distinct elements in G_m are

$$1, \quad \alpha_{j_1}, \quad \alpha_{j_1}\alpha_{j_2}, \ldots, \quad \alpha_1\alpha_2 \cdots \alpha_m; \qquad 1 \leq j_1 < j_2 < \cdots < j_{m-1}. \tag{6.9}$$

and their negatives. The order of the group is

$$2\left[\binom{m}{0} + \binom{m}{1} + \cdots + \binom{m}{m-1} + \binom{m}{m}\right] = 2(1+1)^m = 2^{m+1}.$$

It is obvious that every matrix rep of (6.5) determines a matrix rep of G_m. Furthermore every matrix rep T of G_m such that $T(1) = E$ and $T(-1) = -E$, where E is the identity matrix, determines a solution of (6.5). Thus our problem reduces to the determination of the irred reps $\mathbf{T}$ of G_m such that $\mathbf{T}(\pm 1) = \pm\mathbf{E}$. Every rep with this property will be a direct sum of such irred reps.

The number of irred reps is equal to the number of conjugacy classes in G_m. It is straightforward to check that for m even, the classes $\{+1\}$, $\{-1\}$ contain one element each, while the remaining classes are of the form $\{\pm\alpha_{j_1} \cdots \alpha_{j_s}\}$ and contain two elements each. Thus the total number of conjugacy classes is $[(2^{m+1} - 2)/2] + 2 = 2^m + 1$. For odd m the results are the same except that the two elements with m factors each determine a conjugacy class with one element:

$$\{\alpha_1\alpha_2 \cdots \alpha_m\}, \quad \{-\alpha_1\alpha_2 \cdots \alpha_m\}.$$

Thus the total number of conjugacy classes is $2^m + 2$.

The elements ± 1 form a normal subgroup of G_m. Furthermore the factor group $G_m' \cong G_m/\{\pm 1\}$ of order 2^m is abelian since $\alpha_j\alpha_k = -\alpha_k\alpha_j$, and ± 1 correspond to the identity element in G_m'. Thus G_m' has 2^m irred reps $\mathbf{T}_j'$ all one-dimensional. The composed mappings

$$G_m \longrightarrow G_m' \xrightarrow{T_j'} \mathfrak{C}$$

determine 2^m equivalence classes of one-dimensional irred reps of G_m.

Consider the case where $m = 2n$ is even. Here G_m has $2^m + 1$ irred reps of which 2^m are one-dimensional. Since the sum of the squares of the dimensions of the irred reps equals $2^{m+1} = N_G$ we have

$$2^{2n} + q^2 = 2^{2n+1} \qquad \text{or} \qquad q = 2^n = 2^{m/2},$$

where q is the dimension of the remaining irred rep $\mathbf{T}$. The one-dimensional reps map ± 1 to the identity operator, so they are not acceptable as solutions of (6.5).

We will construct an explicit matrix realization of $\mathbf{T}$. Suppose first that $m = 2$, so $\mathbf{T}$ is two-dimensional. The matrices

$$\alpha_1 = \begin{pmatrix} 0 & 1 \\ 1 & 0 \end{pmatrix}, \qquad \alpha_2 = \begin{pmatrix} 0 & i \\ -i & 0 \end{pmatrix}$$

satisfy the relations $\alpha_j^2 = 1$, $\alpha_1\alpha_2 = -\alpha_2\alpha_1 \neq 0$. Thus, these matrices necessarily determine a two-dimensional irred rep of G_2 equivalent to $\mathbf{T}$.

In the general case for $m = 2n$ we form the $2^n \times 2^n$ matrices

(6.10)

$$\alpha_j = \underbrace{\begin{pmatrix} 1 & 0 \\ 0 & -1 \end{pmatrix} \otimes \cdots \otimes \begin{pmatrix} 1 & 0 \\ 0 & -1 \end{pmatrix}}_{j-1} \otimes \begin{pmatrix} 0 & 1 \\ 1 & 0 \end{pmatrix} \otimes \underbrace{\begin{pmatrix} 1 & 0 \\ 0 & 1 \end{pmatrix} \otimes \cdots \otimes \begin{pmatrix} 1 & 0 \\ 0 & 1 \end{pmatrix}}_{n-j}$$

$$\alpha_{n+j} = \underbrace{\begin{pmatrix} 1 & 0 \\ 0 & -1 \end{pmatrix} \otimes \cdots \otimes \begin{pmatrix} 1 & 0 \\ 0 & -1 \end{pmatrix}}_{j-1} \otimes \begin{pmatrix} 0 & i \\ -i & 0 \end{pmatrix} \otimes \underbrace{\begin{pmatrix} 1 & 0 \\ 0 & 1 \end{pmatrix} \otimes \cdots \otimes \begin{pmatrix} 1 & 0 \\ 0 & 1 \end{pmatrix}}_{n-j}$$

$$1 \leq j \leq n.$$

It is straightforward to check that these matrices satisfy relations (6.5). Since the matrices do not commute they cannot be obtained as a direct sum of one-dimensional reps of G_m. Thus, (6.10) defines a matrix rep equivalent to $\mathbf{T}$. It is easy to verify that $\mathbf{T}$ is a 1–1 rep of G_m.

Remark. Recall that if A, B are $m \times m$, $n \times n$ matrices, respectively, then $C = A \otimes B$ is the $mn \times mn$ matrix with matrix elements

$$C_{jk,ls} = A_{jl}B_{ks}.$$

It is easy to verify the relations

$$(A_1 \otimes B_1)(A_2 \otimes B_2) = A_1 A_2 \otimes B_1 B_2$$

for $m \times m$ matrices A_j and $n \times n$ matrices B_j. Furthermore,

$$c(A \otimes B) = (cA) \otimes B = A \otimes (cB).$$

for any scalar c. These relations are easily extended to $C_1 \otimes \cdots \otimes C_k$.

Using the basic theorems on the rep theory of finite groups we can say more about the matrix realizations of (6.5) for even m. First, every irred matrix realization has dimension $2^{m/2}$ and any two such realizations α_j, α_j' are equivalent, i.e., there exists a nonsingular matrix S such that $\alpha_j' = S^{-1}\alpha_j S$, $1 \leq j \leq m$. In particular, every realization is equivalent to (6.10). Furthermore, if the α_j form a realization and S is nonsingular then the $S^{-1}\alpha_j S$ form an equivalent realization. All reducible realizations are direct sums of copies of the single irred rep.

Taking $m = 4$ we obtain all solutions of (6.4) which lead to the Dirac equation. The only irred realizations are in terms of 4×4 matrices and all such realizations are equivalent. All reducible realizations are given by $4k \times 4k$ matrices, where $k \geq 2$.

The case $m = 2n + 1$ is a little more complicated. There are $2^{2n+1} + 2$ equivalence classes of irred reps. Furthermore, since $G_m/\{1, -1\}$ is abelian of order 2^{2n+1} there are 2^{2n+1} one-dimensional reps. This leaves two irred reps of dimensions q_1 and q_2. Since the sum of the squares of the dimensions equals 2^{2n+2} we have

$$2^{2n+1} + q_1^2 + q_2^2 = 2^{2n+2} \qquad \text{or} \qquad q_1^2 + q_2^2 = 2^{2n+1}.$$

A solution of this diophantine equation is $q_1 = q_2 = 2^n$. Indeed, we can exhibit two nonequivalent irred reps of dimension 2^n. For $n = 1$ the matrices

$$\text{(6.11)} \qquad \alpha_1 = \begin{pmatrix} 0 & 1 \\ 1 & 0 \end{pmatrix}, \qquad \alpha_2 = \begin{pmatrix} 0 & i \\ -i & 0 \end{pmatrix}, \qquad \alpha_3 = \begin{pmatrix} 1 & 0 \\ 0 & -1 \end{pmatrix}$$

satisfy (6.5). (These are the **Pauli spin matrices.**) One can use the Schur lemmas to check the irreducibility. A second irred rep is $\alpha_j' = -\alpha_j$, $j = 1, 2, 3$, where the α_j are defined by (6.11). These reps cannot be equivalent because $i\alpha_1\alpha_2\alpha_3 = E_2 = 1$ for (6.11), while $i\alpha_1'\alpha_2'\alpha_3' = -E_2 = -1$. This also shows that neither of these reps is faithful (1–1), although their direct sum is faithful.

For general $m = 2n + 1$ the $2^n \times 2^n$ matrices α_j, α_{n+j}, $1 \leq j \leq n$, defined by (6.10) and the matrix

$$\text{(6.12)} \qquad \alpha_m = \alpha_{2n+1} = \begin{pmatrix} 1 & 0 \\ 0 & -1 \end{pmatrix} \otimes \cdots \otimes \begin{pmatrix} 1 & 0 \\ 0 & -1 \end{pmatrix}$$

satisfy relations (6.5) and, by Schur's lemmas, define an irred rep of G_{2n+1}.

Furthermore, the matrices $\alpha_k' = -\alpha_k$, $1 \le k \le m$, where the α_k are given by (6.10) and (6.12), also define an irred rep of G_{2n+1}. These reps are not equivalent because $i^n\alpha_1 \cdots \alpha_n = E_{2^n}$ for the first representation, while $i^n\alpha_1' \cdots \alpha_n' = -E_{2^n}$ for the second. Neither rep is faithful, but their direct sum is faithful. These results suffice to describe all realizations of relations (6.5) by matrices.

We have related the Clifford algebra C_m to the reps of a finite group G_m. We shall now show that C_m is also related to the reps of a Lie group. Suppose the quantities α_j satisfy the relations

$$\alpha_j\alpha_k + \alpha_k\alpha_j = 2\delta_{jk}, \qquad 1 \le j, k \le m.$$

We define elements α_{jk}, $j \ne k$, $1 \le j, k \le m + 1$, by

$$\alpha_{jk} = \begin{cases} 2\alpha_j\alpha_k & \text{if } j \ne k, \quad 1 \le j, k \le m, \\ \alpha_k & \text{if } j = m + 1, \quad 1 \le k \le m, \\ -\alpha_j & \text{if } k = m + 1, \quad 1 \le j \le m. \end{cases} \tag{6.13}$$

Furthermore, we set $\alpha_{jj} = 0$. It follows that $\alpha_{jk} = -\alpha_{kj}$ and there are $m(m + 1)/2$ independent quantities α_{jk}, $j < k$. The $\{\alpha_{jk}\}$ form a Lie algebra under the commutator bracket

$$[\alpha_{jk}, \alpha_{hl}] = \alpha_{jk}\alpha_{hl} - \alpha_{hl}\alpha_{jk}.$$

Indeed a straightforward computation yields

$$\begin{gathered} [\alpha_{m+1,k}, \alpha_{m+1,l}] = \alpha_{kl}, \qquad [\alpha_{m+1,k}, \alpha_{hl}] = 4\delta_{kh}\alpha_{m+1,l} - 4\delta_{kl}\alpha_{m+1,h}, \\ [\alpha_{jk}, \alpha_{hl}] = 4(\delta_{kh}\alpha_{jl} + \delta_{jl}\alpha_{kh} - \delta_{kl}\alpha_{jh} - \delta_{jh}\alpha_{kl}), \\ 1 \le j, k, h, l \le m. \end{gathered} \tag{6.14}$$

Relations (6.14) are the commutation relations of an $m(m + 1)/2$-dimensional Lie algebra.

Let us compare these results with the commutation relations of the Lie algebra $so(m + 1)$. It will be convenient to consider $so(m + 1)$ as the space of all $(m + 1) \times (m + 1)$ skew-symmetric matrices. Then a basis for $so(m + 1)$ is provided by the matrices $\mathcal{A}_{jk} = \mathcal{E}_{jk} - \mathcal{E}_{kj}$, $1 \le j < k \le m + 1$. Taking account of the rules $\mathcal{A}_{jk} = -\mathcal{A}_{kj}$ for all j, k we easily derive the commutation relations

(6.15)

$$[\mathcal{A}_{jk}, \mathcal{A}_{hl}] = \delta_{kh}\mathcal{A}_{jl} + \delta_{jl}\mathcal{A}_{kh} - \delta_{kl}\mathcal{A}_{jh} - \delta_{jk}\mathcal{A}_{kl}, \qquad 1 \le j, k, h, l \le m + 1.$$

Setting $\mathcal{A}_{jk} = \frac{1}{4}\alpha_{jk}$ and $\mathcal{A}_{j,m+1} = -\frac{1}{2}i\alpha_{j,m+1} = \frac{1}{2}i\alpha_j$ we see that relations (6.14) and (6.15) coincide. Thus, the $\{\alpha_{jk}\}$ span a Lie algebra isomorphic to $so(m + 1)$, so any matrix realization of (6.5) determines a rep of $so(m + 1)$ via the relations (6.14). Furthermore, the nontrivial irred reps of G_m computed above determine irred reps of $so(m + 1)$. We can determine which irred reps of $so(m + 1)$ we have obtained by computing the highest weights.

The maximal abelian subalgebra $\mathscr{h}_s$ of $so(m+1)$ obtained from (5.12) by the change of basis (5.3) can be chosen to take the form

$$\begin{pmatrix} & & & i\lambda_1 & & & 0 \\ & Z_s & & & \ddots & & \vdots \\ & & & & & i\lambda_s & \vdots \\ -i\lambda_1 & & & & & & \vdots \\ & \ddots & & & Z_s & & \\ & & -i\lambda_s & & & & \\ 0 & & & \cdots & & & 0 \end{pmatrix}, \qquad m+1 = 2s+1, \tag{6.16}$$

in the space of skew-symmetric matrices, where the last row and column are missing if $m + 1 = 2s$. A basis for $\mathscr{h}_s$ is given by $\mathcal{H}_j = i\mathcal{A}_{j,s+j} = i\mathcal{E}_{j,s+j} - i\mathcal{E}_{s+j,j}$, $1 \leq j \leq s$.

Now consider the rep of G_{2n} or $so(2n+1)$ given by (6,10). The corresponding operators H_j are

$$H_j = \tfrac{1}{4} i\alpha_{j,n+j} = \tfrac{1}{2} i\alpha_j\alpha_{n+j}, \qquad 1 \leq j \leq n,$$

or

$$H_j = \tfrac{1}{2}\underbrace{E_2 \otimes \cdots \otimes E_2}_{j-1} \otimes \begin{pmatrix} 1 & 0 \\ 0 & -1 \end{pmatrix} \otimes \underbrace{E_2 \otimes \cdots \otimes E_2}_{n-j}. \tag{6.17}$$

The eigenvalues of the H_j are $\pm\frac{1}{2}$. In particular, a weight basis for the rep is given by

$$e(k_1 \cdots k_n) = e_{k_1} \otimes e_{k_2} \otimes \cdots \otimes e_{k_n}, \qquad k_l = \pm\tfrac{1}{2}, \tag{6.18}$$

where

$$e_{1/2} = \begin{pmatrix} 1 \\ 0 \end{pmatrix}, \qquad e_{-1/2} = \begin{pmatrix} 0 \\ 1 \end{pmatrix}.$$

Clearly,

$$H_j e(k_1 \cdots k_j \cdots k_n) = k_j e(k_1 \cdots k_j \cdots k_n),$$

and $e(k_1 \cdots k_n)$ is a weight vector with weight $\sum k_j\lambda_j$ since

$$H(\lambda_1, \ldots, \lambda_n) = \sum_{j=1}^{n} \lambda_j H_j.$$

The highest weight is $\frac{1}{2}\sum \lambda_j$ so we have constructed a realization of the fundamental spin rep $(\frac{1}{2}^n)$ of $so(2n+1)$.

Next we consider the rep of $G_{2n-1} = G_{2(n-1)+1}$ or $so(2n)$ determined by (6.10) and (6.12). The operators H_j are

$$H_j = \tfrac{1}{4} i\alpha_{j,(n-1)+j+1}, \quad 1 \leq j \leq n-1, \qquad H_n = i\mathcal{A}_{n,2n} = -\tfrac{1}{2}\alpha_{(n-1)+1}$$

or

$$H_j = \tfrac{1}{2}\underbrace{E_2 \otimes \cdots \otimes E_2}_{j-1} \otimes S_2 \otimes S_2 \otimes \underbrace{E_2 \otimes \cdots \otimes E_2}_{n-j-2}$$

(6.19)

$$H_n = \tfrac{1}{2}S_2 \otimes \underbrace{E_2 \otimes \cdots \otimes E_2}_{n-2}, \qquad S_2 = \begin{pmatrix} 0 & -i \\ i & 0 \end{pmatrix}.$$

Since the matrix S_2 has eigenvalues ± 1 we can perform a unitary similarity transformation on the H_j to obtain

$$H_j = \tfrac{1}{2}E_2 \otimes \cdots \otimes E_2 \otimes \begin{pmatrix} 1 & 0 \\ 0 & -1 \end{pmatrix} \otimes \begin{pmatrix} 1 & 0 \\ 0 & -1 \end{pmatrix} \otimes E_2 \otimes \cdots \otimes E_2$$

$$H_n = \tfrac{1}{2}\begin{pmatrix} 1 & 0 \\ 0 & -1 \end{pmatrix} \otimes E_2 \otimes \cdots \otimes E_2.$$

Again these matrices have eigenvalues $\pm\frac{1}{2}$ and the $e(k_1 \cdots k_{n-1})$, $k_l = \pm\frac{1}{2}$, form a weight basis. We have

(6.20)
$$H_j e(k_1 \cdots k_{n-1}) = 2k_j k_{j+1} e(k_1 \cdots k_{n-1}),$$
$$1 \le j \le n-1, \quad k_n = \tfrac{1}{2},$$
$$H_n e(k_1 \cdots k_{n-1}) = k_1 e(k_1 \cdots k_{n-1}).$$

We get the highest weight vector for $k_1 = \cdots = k_{n-1} = \frac{1}{2}$. Thus the above construction yields a model of the spin rep $(\frac{1}{2}^n)$ of $so(2n)$.

A second model is obtained by making the replacements $\alpha_j \longrightarrow -\alpha_j$, $1 \le j \le n-1$. This leaves the operators H_j, $1 \le j \le n-1$, unchanged but causes H_n to be replaced by $-H_n$. Again we get the highest weight vector for $k_1 = \cdots = k_{n-1} = \frac{1}{2}$, but this time the highest weight is $\frac{1}{2}(\lambda_1 + \cdots + \lambda_{n-1} - \lambda_n)$. Thus we have constructed a model of the spin rep $(\frac{1}{2}^{n-1}, -\frac{1}{2})$.

As we demonstrated in the preceding section, the spin reps of $so(m)$ do not extend to single-valued reps of $SO(m, \mathfrak{C})$. It is shown explicitly by Boerner [1] and Freudenthal and De Vries [1] that these reps exponentiate to single-valued reps of a compact Lie group Spin(m). Here, Spin(m) is locally isomorphic to $SO(m, \mathfrak{C})$ and there is a 2–1 analytic homomorphism of Spin(m) onto $SO(m, \mathfrak{C})$.

9.7 Examples and Applications

We present several examples showing how the classical groups appear in physical theories.

Consider a family of linear operators $\mathbf{a}_j, \mathbf{a}_j^*$, $1 \le j \le m$, on an inner product space V, satisfying the commutation relations

(7.1)
$$[\mathbf{a}_j, \mathbf{a}_k] = [\mathbf{a}_j^*, \mathbf{a}_k^*] = \mathbf{0}, \qquad [\mathbf{a}_j, \mathbf{a}_k^*] = \delta_{jk}\mathbf{E},$$

where $\mathbf{E}$ is the identity operator. [Such operators appear in the method of second quantization in quantum mechanics where they are defined on a Hilbert space $\mathcal{H}$. There $\mathbf{a}_j^*$ is called a **creation operator** for bosons and its adjoint $\mathbf{a}_j$ is called an **annihilation operator.** These operators are closely related to the harmonic oscillator problem in quantum mechanics and will be treated in detail in Chapter 10. Here, we consider only the abstract commutation relations (7.1).]

Now V must be infinite-dimensional for (7.1) to hold. For, if A and B are matrices such that $[A, B] = \lambda E$ then $\operatorname{tr}([A, B]) = 0$ implies $\lambda = 0$. If V consists of analytic functions in m variables $z_1, \ldots, z_m$ then a realization of (7.1) is provided by the assignment

$$\mathbf{a}_j = \partial/\partial z_j, \qquad \mathbf{a}_j^* = z_j. \tag{7.2}$$

It follows directly from the commutation relations (7.1) that the operators $\mathbf{E}_{jk} = \mathbf{a}_j^*\mathbf{a}_k$ satisfy relations

$$[\mathbf{E}_{jk}, \mathbf{E}_{hl}] = \delta_{kh}\mathbf{E}_{jl} - \delta_{jl}\mathbf{E}_{hk}, \qquad 1 \le j, k, h, l \le m, \tag{7.3}$$

in agreement with (1.1). Thus, making the identification $\mathbf{E}_{jk} \leftrightarrow \mathcal{E}_{jk}$ one can easily construct reps of each of the classical Lie algebras in terms of annihilation and creation operators for bosons. Furthermore one can use the models to decompose V into subspaces transforming irreducibly under these reps.

For our next example we consider a family of operators $\mathbf{a}_j$, $\mathbf{a}_j^*$, $1 \le j \le n$, on a finite-dimensional vector space V satisfying the **anticommutation** relations

$$[\mathbf{a}_j, \mathbf{a}_k]_+ = [\mathbf{a}_j^*, \mathbf{a}_k^*]_+ = \mathbf{0}, \qquad [\mathbf{a}_j, \mathbf{a}_k^*]_+ = \delta_{jk}\mathbf{E}, \qquad 1 \le j, k \le n, \tag{7.4}$$

where $[\mathbf{a}, \mathbf{b}]_+ = \mathbf{ab} + \mathbf{ba}$. Setting

$$\mathbf{a}_j = \tfrac{1}{2}(\alpha_j - i\alpha_{n+j}), \qquad \mathbf{a}_j^* = \tfrac{1}{2}(\alpha_j + i\alpha_{n+j}), \tag{7.5}$$

we obtain the relations

$$\alpha_j\alpha_k + \alpha_k\alpha_j = 2\delta_{jk}\mathbf{E}, \qquad 1 \le j, k \le 2n. \tag{7.6}$$

Conversely, if the α_k satisfy (7.6) and $\mathbf{a}_j$, $\mathbf{a}_j^*$ are defined by (7.5) then the anticommutation relations (7.4) hold. Operators satisfying (7.4) are called **annihilation** and **creation operators** for fermions.

It follows from the preceding section that V can be decomposed into a direct sum of irred subspaces under the $\mathbf{a}_j$ and $\mathbf{a}_k^*$, each subspace transforming as the 2^n-dimensional rep $(\frac{1}{2}^n)$ of $so(2n+1)$. A basis is given by (6.18) which we rewrite as

$$|p_1 \cdots p_n\rangle = f_{p_1} \otimes \cdots \otimes f_{p_n}, \qquad p_j = 0, 1, \tag{7.7}$$

where

$$f_0 = \begin{pmatrix} 1 \\ 0 \end{pmatrix}, \qquad f_1 = \begin{pmatrix} 0 \\ 1 \end{pmatrix}.$$

Then from (6.10) and (7.5) we obtain

$$\text{(7.8)}\qquad \begin{aligned} \mathbf{a}_j \,|\, p_1 \cdots p_j \cdots p_n\rangle &= (-1)^{p_1+\cdots+p_{j-1}} p_j \,|\, p_1 \cdots p_j - 1 \cdots p_n\rangle \\ \mathbf{a}_j^* \,|\, p_1 \cdots p_j \cdots p_n\rangle &= (-1)^{p_1+\cdots+p_{j-1}} (1 - p_j) \,|\, p_1 \cdots p_j + 1 \cdots p_n\rangle. \end{aligned}$$

The physical interpretation of these relations will be given in the next section.

It has been found experimentally that for many nuclear interactions (the **strong** interactions) one can consider the proton p and the neutron n as two states of the same particle. In particular the masses of p and n are approximately equal. Of course there is a small mass difference and p has charge $+1$ while n has charge zero. However, the mass and charge differences can be ascribed to so-called electromagnetic and weak interactions, which are considered as perturbations of the charge-independent strong interactions.

In the theory of strong interactions one considers p and n as two states of the nucleon N described by the state space $\mathcal{K} = \mathcal{H}_{1/2} \otimes \mathcal{I}_2$. Here $\mathcal{H}_{1/2}$ is the state space for a particle with spin $\frac{1}{2}$ as described in Sections 7.8 or 8.4 (for a relativistic theory) and $\mathcal{I}_2$ is a two-dimensional space with basis

$$\text{(7.9)}\qquad \mathbf{e}_{1/2} = \mathbf{p} = \begin{pmatrix} 1 \\ 0 \end{pmatrix}, \qquad \mathbf{e}_{-1/2} = \mathbf{n} = \begin{pmatrix} 0 \\ 1 \end{pmatrix},$$

called **isobaric spin space.** A general element of $\mathcal{K}$ can be written uniquely in the form

$$\mathbf{\Psi} = \mathbf{\Psi}_p \otimes \mathbf{p} + \mathbf{\Psi}_n \otimes \mathbf{n},$$

where $\mathbf{\Psi}_p$, $\mathbf{\Psi}_n$ are themselves two-component spinors. Here $\mathbf{\Psi}_p \otimes \mathbf{p}$ is a pure proton state and $\mathbf{\Psi}_n \otimes \mathbf{n}$ is a pure neutron state. In general, $\mathbf{\Psi}$ is a superposition of proton and neutron states. The inner product in $\mathcal{K}$ is given by

$$\text{(7.10)}\qquad \langle \mathbf{\Psi}^{(1)}, \mathbf{\Psi}^{(2)} \rangle_{\mathcal{K}} = \langle \mathbf{\Psi}_p^{(1)}, \mathbf{\Psi}_p^{(2)} \rangle_{\mathcal{H}_{1/2}} + \langle \mathbf{\Psi}_n^{(1)}, \mathbf{\Psi}_n^{(2)} \rangle_{\mathcal{H}_{1/2}}.$$

The state space for k nucleons is

$$\mathcal{K}^{\otimes k} = (\mathcal{H}_{1/2} \otimes \mathcal{I}_2)^{\otimes k} \cong (\mathcal{H}_{1/2})^{\otimes k} \otimes (\mathcal{I}_2)^{\otimes k}.$$

An element $\mathbf{\Psi}$ of $\mathcal{K}^{\otimes k}$ is determined by the spinor

$$\text{(7.11)}\qquad \mathbf{\Psi}_{s_1 \cdots s_k, t_1 \cdots t_k}(\mathbf{x}_1, \ldots, \mathbf{x}_k), \qquad s_j = \pm\tfrac{1}{2}, \quad t_j = \pm\tfrac{1}{2}.$$

Here $\mathbf{x}_1, \ldots, \mathbf{x}_k$ are the position coordinates of the k particles, the s_j are the ordinary spin indices, and the t_j are isobaric spin indices. For example, if all components of $\mathbf{\Psi}$ are zero except that component for which $s_1 = \cdots = s_k = -\frac{1}{2}$ and $t_1 = \cdots = t_k = \frac{1}{2}$, then $\mathbf{\Psi}$ is a state of k protons each with 3-component of ordinary spin equal to $-\frac{1}{2}$.

If this k-particle system interacts with itself we demand as usual that the interaction admit the symmetry $\mathcal{E}_3$, (8.21), Section 7.8, where the group acts on the position vectors $\mathbf{x}_j$ and the spin indices s_k. [In a relativistic theory we demand that the interaction admit $\mathcal{P}$ as a symmetry group, (4.22), Section

8.4.] Furthermore, for strong interactions we also require that the operators $\mathbf{I}(A)$, $A \in SU(2)$, commute with the interaction Hamiltonian, where

(7.12)

$$[\mathbf{I}(A)\Psi]_{s_1\cdots s_k,t_1\cdots t_k}(\mathbf{x}_1,\ldots,\mathbf{x}_k) = \sum_{t_1'\cdots t_k'} A_{t_1t_1'}\cdots A_{t_kt_k'}\Psi_{s_1\cdots s_k,t_1'\cdots t_k'}(\mathbf{x}_1,\ldots,\mathbf{x}_k).$$

Note that $\mathbf{I}(A)$ acts only on the isobaric spin space $\mathcal{I}_2$ and does not affect the position vectors or ordinary spin indices. Thus for a relativistic theory of strong interactions the symmetry group is assumed to be $\mathcal{P} \times SU(2)$. (Actually one also requires invariance under space and time inversion, but we will not discuss this here.) Since we have already examined the implications of invariance with respect to symmetry groups acting on the position vectors and ordinary spin indices, we restrict the following discussion to the rep (7.12) of $SU(2)$ acting on the isobaric spin indices. Under this action $(\mathcal{I}_2)^{\otimes k}$ transforms according to the rep $(\mathbf{D}^{(1/2)})^{\otimes k}$. We can use the CG series, Section 7.7, to decompose this rep into irred components. If the system is in a state transforming as the canonical basis vector $\mathbf{f}_m^{(u)}$ of $\mathbf{D}^{(u)}$ in isobaric spin space at some time t_0, then the isobaric spin invariance of the system implies that at any later time t the system still transforms as $\mathbf{f}_m^{(u)}$. The mathematical analysis which exploits this invariance is similar to that leading up to expression (8.10) in Section 7.8, but the physical interpretation is different. As an example we consider an interacting two-nucleon system such that the result of the interaction is again a two-nucleon system. The isobaric spin space $\mathcal{I}_2 \otimes \mathcal{I}_2$ for this problem transforms as $\mathbf{D}^{(1/2)} \otimes \mathbf{D}^{(1/2)} \cong \mathbf{D}^{(1)} \otimes \mathbf{D}^{(0)}$ under $SU(2)$. The canonical basis vectors are

(7.13)
$$\mathbf{f}_1^{(1)} = \mathbf{p}\otimes\mathbf{p}, \qquad \mathbf{f}_0^{(1)} = 2^{-1/2}(\mathbf{p}\otimes\mathbf{n} + \mathbf{n}\otimes\mathbf{p}), \qquad \mathbf{f}_{-1}^{(1)} = \mathbf{n}\otimes\mathbf{n},$$
$$\mathbf{f}_0^{(0)} = 2^{-1/2}(\mathbf{p}\otimes\mathbf{n} - \mathbf{n}\otimes\mathbf{p}).$$

Since isobaric spin is conserved by the interaction a *p–p* system ends up as a *p–p* system and, similarly, an *n–n* system ends up as an *n–n* system. Furthermore, the transition probabilities between corresponding eigenstates of orbital and spin angular momentum are exactly the same for *p–p* and *n–n* systems. Both of these systems belong to $\mathbf{D}^{(1)}$, while a mixed system belongs to both $\mathbf{D}^{(1)}$ and $\mathbf{D}^{(0)}$.

The Hamiltonian describing strong interactions between the two nucleons commutes with the action of the isospin group. However, if one takes into account the electromagnetic interaction between nucleons the perturbed Hamiltonian no longer commutes with the isospin operators and it becomes possible to distinguish between p and n. Thus the electromagnetic interaction breaks isospin symmetry.

The concept of isospin symmetry can be applied to other elementary particles. As an example we consider a family of eight baryons, all with (ordinary) spin $\frac{1}{2}$ and approximately the same mass. Two of these particles are p and n which transform as a canonical basis for the rep $\mathbf{D}^{(1/2)}$ in isospace. The sigma hyperons $\Sigma^+, \Sigma^0, \Sigma^-$ have charges $+1, 0, -1$, respectively, and transform as a canonical basis $\mathbf{f}_1^{(1)}, \mathbf{f}_0^{(1)}, \mathbf{f}_{-1}^{(1)}$ for $\mathbf{D}^{(1)}$ in isospace. (Thus the Hilbert space describing a single sigma hyperon takes the form $\mathcal{H}_{1/2} \otimes \mathcal{J}_3$, where $\mathcal{J}_3$ is the rep space for $\mathbf{D}^{(1)}$.) The Λ^0 hyperon has charge zero and transforms as a scalar $\mathbf{D}^{(0)}$ under isospin. Finally, the cascade particles Ξ^0, Ξ^- have charges $0, -1$, respectively, and transform as the canonical basis $\mathbf{f}_{1/2}^{(1/2)}, \mathbf{f}_{-1/2}^{(1/2)}$ of $\mathbf{D}^{(1/2)}$.

It is required that the isospin group $SU(2)$ be an invariance group for the strong interactions of these particles. Thus for nucleon–sigma hyperon scattering the isospin space transforms as $\mathbf{D}^{(1/2)} \otimes \mathbf{D}^{(1)} \cong \mathbf{D}^{(3/2)} \oplus \mathbf{D}^{(1/2)}$, and the isospin space has a basis consisting of a canonical basis for $\mathbf{D}^{(3/2)}$ and a canonical basis for $\mathbf{D}^{(1/2)}$. The basis vectors can be determined from the CG series, e.g., $\mathbf{f}_{3/2}^{(3/2)} = \mathbf{p} \otimes \mathbf{\Sigma}^+$. Isospin invariance implies that a system transforming as $\mathbf{f}_m^{(u)}$ before the scattering must transform the same way after the scattering even if particles are created or destroyed by the interaction. This leads to selection rules and conservation laws in analogy with those derived in Section 7.8. Again, electromagnetic effects break the isospin symmetry, so the conservation laws are only approximately correct. For a detailed examination of these rules and their physical significance or validity see the standard textbooks in elementary particle physics.

We now introduce one more concept important in the theory of strong interactions: the **hypercharge** Y. Let $I^\pm$, I^3 be the generators of isospin satisfying relations (1.23), Section 7.1. (We use the notation I rather than J to distinguish isospin from angular momentum.) The eight elementary particles listed above correspond to eigenvectors of I^3 and the eigenvalues of I^3 are conserved by strong interactions. We know that the charge of a quantum mechanical system is conserved under strong interactions, and each of the canonical basis vectors for isospin space listed above has definite charge. Thus we can uniquely define a charge operator Q by

$$Q\mathbf{r} = q\mathbf{r}, \tag{7.14}$$

where q is the charge of the particle $\mathbf{r}$ and $\mathbf{r}$ runs over a canonical basis for isospin space. For example,

$$Q\mathbf{n} = \boldsymbol{\theta}, \qquad Q\mathbf{p} = \mathbf{p} \tag{7.15}$$

for the nucleon pair n and p. We define the **hypercharge** operator Y by

$$Q = I^3 + \tfrac{1}{2}Y \qquad \text{or} \qquad Y = 2(Q - I^3). \tag{7.16}$$

it follows easily from (7.9) and (7.15) that $Y = 1$ for the pair n, p. A similar computation shows that the hypercharge of the Σ-triplet and Λ^0-singlet is

$Y = 0$ and the hypercharge of the cascade doublet is $Y = -1$. Thus there is a close relation between the charge and the third component of isotopic spin, and this exact relation is determined by the hypercharge. Since charge and the eigenvalues of I^3 are additive quantum numbers conserved by strong interactions, hypercharge must also be an additive conserved quantity. (By additive we mean that a two-particle system consisting of particle a with hypercharge Y_a and particle b with hypercharge Y_b has total hypercharge $Y_a + Y_b$. The sum $Y_a + Y_b$ is conserved under strong interactions, though not Y_a or Y_b separately.)

With the success of the concept of isospin in providing some order in elementary particle physics, attempts were made to find a larger symmetry group G for strong interactions which included the isospin group $SU(2)$ as a proper subgroup. The most successful attempt to date is the "eightfold way" of Gell-Man and Ne'eman [1], in which the symmetry group is $SU(3)$.

Before describing this theory we discuss the irred reps of $SU(3)$. These reps can be denoted $[f_1, f_2] = [f_1, f_2, 0]$, where $f_1 \geq f_2 \geq 0$ are integers. The simple characters are given by Theorem 9.4 and the dimensions by

$$N(f_1, f_2) = \tfrac{1}{2}(f_1 - f_2 + 1)(f_1 + 2)(f_2 + 1). \tag{7.17}$$

We list some of the irred reps of low dimension.

(7.18)

	$[f_1, f_2]$	$N(f_1, f_2)$		$[f_1, f_2]$	$N(f_1, f_2)$
1	[0, 0]	1	**8**	[2, 1]	8
3	[1, 0]	3	**10**	[3, 0]	10
$\overline{\mathbf{3}}$	[1, 1]	3	$\overline{\mathbf{10}}$	[3, 3]	10
6	[2, 0]	6	**27**	[4, 2]	27
$\overline{\mathbf{6}}$	[2, 2]	6			

Here, we have adopted the notation in elementary particle physics where a rep is denoted by its dimension. (Using Young symmetrizers the reader should be able to show that $\overline{\mathbf{3}}$, $\overline{\mathbf{6}}$, and $\overline{\mathbf{10}}$ are the complex conjugate reps of **3**, **6**, and **10**, respectively.)

Clearly, **3** is the usual matrix realization of $SU(3)$. The weights are $\lambda_1, \lambda_2, \lambda_3 = -\lambda_1 - \lambda_2$. The rep $\overline{\mathbf{3}}$ is defined by the matrices $\bar{A}$, $A \in SU(3)$. The weights are $-\lambda_1, -\lambda_2, -\lambda_3 = \lambda_1 + \lambda_2$. The rep $\mathbf{8} \cong [2, 1]$ is just the adjoint rep of $SU(3)$ acting on $su(3)$. Indeed, $su(3)$ is eight-dimensional and the weights of the adjoint rep are just the roots of $su(3)$ plus the weight zero with multiplicity two. Thus the nonzero weights are $\pm(\lambda_1 - \lambda_2)$, $\pm(\lambda_1 - \lambda_3) = \pm(2\lambda_1 + \lambda_2)$, $\pm(\lambda_2 - \lambda_3) = \pm(\lambda_1 + 2\lambda_2)$. The highest weight is $2\lambda_1 + \lambda_2$, so the adjoint rep contains [2, 1]. Since dim [2, 1] = 8 it follows that the adjoint rep is irred.

We can identify $SU(2)$ with the subgroup

$$\begin{pmatrix} SU(2) & & \begin{matrix}0\\0\end{matrix} \\ 0 & 0 & 1 \end{pmatrix}$$

of $SU(3)$. Clearly $[f_1, f_2] \mid SU(2)$ decomposes into a direct sum of irred reps. We determine this decomposition for [2, 1]. On restriction to $SU(2)$, we find $\lambda_1 + \lambda_2 = 0 = \lambda_3$. Thus, the weights of $\rho \cong \mathbf{8} \mid SU(2)$ are $\pm 2\lambda_1$, $\pm\lambda_1$ (multiplicity two), and 0 (multiplicity two). The highest weight is $2\lambda_1$, so ρ contains $[2] \cong \mathbf{D}^{(1)}$. (Recall $[2u] \cong \mathbf{D}^{(u)}$.) Removing the weights $\pm 2\lambda_1, 0$ of [2] we see that λ_1 (multiplicity two) is the highest remaining weight. Therefore, ρ contains $[1] \oplus [1]$. Removing the four weights $\pm\lambda_1$, $\pm\lambda_1$ corresponding to these reps we are left with the single weight 0. Thus,

$$\mathbf{8} \mid SU(2) \cong \mathbf{D}^{(1)} \oplus 2\mathbf{D}^{(1/2)} \oplus \mathbf{D}^{(0)} \cong [2] \oplus 2[1] \oplus [0]. \tag{7.19}$$

In the eightfold way the eight baryons introduced above are identified with a basis for **8** which is canonical with respect to the decomposition (7.19). The $\mathbf{\Sigma}$-triplet with hypercharge $Y = 0$ forms a canonical basis for [2], the two doublets with hypercharge $Y = 1$ and $Y = -1$ form canonical bases for the two occurrences of [1], and the $\mathbf{\Lambda}^0$ with hypercharge $Y = 0$ belongs to [0]. The assignment

$$H(\lambda_1, \lambda_2) = \lambda_1 Q + \lambda_2(Y - Q) \tag{7.20}$$

defines the eight baryons as a weight basis for **8**. We shall not be concerned with the exact normalization of this basis.

Now consider these eight baryons as distinct states of a single particle with spin $\frac{1}{2}$. The Hilbert space describing such a single-particle system takes the form $\mathcal{H}_{1/2} \otimes \mathcal{V}_8$, where $\mathcal{V}_8$ is the eight-dimensional rep space for **8** constructed above. For very strong interactions involving this particle it is required that the interaction Hamiltonian commute with the action of $SU(3)$. Since the isospin group $SU(2)$ is identified as a subgroup of $SU(3)$, this requirement implies conservation of isospin, However, $SU(3)$ invariance clearly leads to additional selection rules and conservation laws. These rules can be obtained from the Wigner–Eckart theorem.

Other particle multiplets can be fitted with irred reps of $SU(3)$. In addition to the baryon octet there is an octet of pseudoscalar mesons which also transforms according to **8**. There is also a baryon decouplet transforming according to **10**. For very strong interactions involving a particle from the baryon octet and a particle from the pseudoscalar meson octet the space on which $SU(3)$ acts transforms according to $\mathbf{8} \otimes \mathbf{8}$. To make full use of the $SU(3)$ symmetry it is necessary to decompose $\mathbf{8} \otimes \mathbf{8}$ into irred reps. The result is

$$\mathbf{8} \otimes \mathbf{8} \cong \mathbf{27} \oplus \mathbf{10} \oplus \overline{\mathbf{10}} \oplus \mathit{28} \oplus \mathbf{1}. \tag{7.21}$$

If the system is in one of the irred subspaces on the right-hand side of (7.21) before the interaction, it will lie in the same subspace after the interaction. The analysis is very similar to that given in Section 7.6. The weight vectors corresponding to irred reps on the right-hand side of (7.21) can themselves be considered as particles or "resonances." If such a resonance decays into a baryon and a meson the possible decay modes and their relative probabilities are given by expressing the resonance as a linear combination of weight vectors from the "natural" baryon–meson basis for $\mathbf{8} \otimes \mathbf{8}$ via the Clebsch–Gordan coefficients. The Clebsch–Gordan coefficients for $SU(3)$ have not been computed in the general case. The principal difficulty is that the tensor products may contain an irred rep with multiplicity greater than one, as illustrated by (7.21). However, the CG coefficients have been tabulated for all tensor products of importance in the eightfold way model, such as (7.21) (Dyson [1]).

The eightfold way model is presumed to be exact only for extremely strong interactions. The other possible interactions between particles are considered as perturbations which break the $SU(3)$ symmetry. We can imagine turning on these perturbations in sequence. First we reduce the symmetry from $SU(3)$ to $SU(2)$, the isospin group. This causes the baryon octet to split into a triplet, two doublets, and a singlet via (7.19). At this point we still have isospin symmetry. Now we turn on the electromagnetic and weak interactions to break the $SU(2)$ symmetry. The electromagnetic interactions conserve I^3 and Q, but the weak interactions conserve only Q. If the perturbing interactions are "small" with respect to the very strong interactions then we expect this model to yield experimental predictions which are at least qualitatively correct. Furthermore, we expect to explain the observed mass differences of the particles in the baryon octet in terms of the perturbing interactions. In fact, there is a great deal of experimental evidence validating the predictions of this model and the observed mass differences can be explained rather well by the Gell-Mann–Okubo mass formula (Dyson [1]). The model has no firm theoretical basis and may be disgarded in time, but it is certainly a useful means of classifying elementary particles and a beautiful application of symmetry groups.

The classical groups have been used extensively in atomic spectroscopy and nuclear physics. For a detailed study of these applications see the work of Hamermesh [1], Loebl [1, 2], or Tinkham [1]. In many cases the mathematical content of the application is the determination of multiplicities of irred reps $\mathbf{T}'$ belonging to a subgroup K of G in the restricted rep $\mathbf{T} \mid K$. Here $\mathbf{T}$ is an irred rep of G. The formulas for the multiplicities are called **branching laws.** Throughout this book we have computed branching laws for various groups of physical interest. Some of the most important laws for the classical groups are given by Boerner [1].

The group $SO(4, R)$ is of special importance in physics because of its relationship to the hydrogen atom. For a (spinless) particle in a spherically symmetric field the Hamiltonian takes the form

$$H = -(1/2m)\Delta + V(r), \qquad r = [x^2 + y^2 + z^2]^{1/2}.$$

For most choices of the potential V the (connected) symmetry group of $\mathbf{H}$ is just $SO(3)$. However, outside of the trivial case V a constant, there are two cases where the connected symmetry group of $\mathbf{H}$ is larger than $SO(3)$:

$$V(r) = c/r, \qquad V(r) = cr^2, \qquad c \text{ a constant.} \tag{7.22}$$

The hydrogen atom (a single particle in an attractive Coulomb field) corresponds to such a potential. Using appropriate units we can choose

$$\mathbf{H} = -\tfrac{1}{2}\Delta - (1/r). \tag{7.23}$$

It can be shown that the eigenvalues of $\mathbf{H}$ are all negative (the boundstate energy levels) (Helwig [1]). Let λ be an eigenvalue of $\mathbf{H}$ and $\mathfrak{C}_\lambda$ the corresponding eigenspace. We look for symmetric operators on $\mathfrak{C}_\lambda$ which commute with $\mathbf{H}$. Since r^{-1} is spherically symmetric the angular momentum operators $\mathbf{L} = (L_1, L_2, L_3)$ [(6.24), Section 7.6] commute with $\mathbf{H}$. Furthermore, a tedious computation shows that the operators $\mathbf{A} = (A_1, A_2, A_3)$,

$$\mathbf{A} = \frac{1}{2(-2\lambda)^{1/2}}\left(\mathbf{L} \times \mathbf{p} - \mathbf{p} \times \mathbf{L} + \frac{2\mathbf{r}}{r}\right), \tag{7.24}$$

are also symmetric and commute with $\mathbf{H}$. Here $\mathbf{p} = (-i\partial_x, -i\partial_y, -i\partial_z)$, $\mathbf{r} = (x, y, z)$, and $\mathbf{L} \times \mathbf{P}$ is the cross-product. For the physical significance of $\mathbf{A}$ (the Runge–Lenz vector) see the book by Pollard [1]. Note that $\mathbf{A}$ depends on λ. The six operators $\mathbf{L}$ and $\mathbf{A}$ satisfy the commutation relations

$$\begin{aligned} &[L_j, L_k] = i \sum_l \epsilon_{jkl} L_l, \qquad [L_j, A_k] = i \sum_l \epsilon_{jkl} A_l, \\ &[A_j, A_k] = i \sum_l \epsilon_{jkl} L_l, \qquad 1 \le j, k, l \le 3, \end{aligned} \tag{7.25}$$

where ϵ_{jkl} is the completely skew-symmetric tensor such that $\epsilon_{123} = +1$. These commutation relations are valid only on the domain $\mathfrak{C}_\lambda$. Setting

$$C_j = -\tfrac{1}{2}i(L_j + A_j), \qquad D_j = -\tfrac{1}{2}i(L_j - A_j), \qquad j = 1, 2, 3,$$

we obtain

$$[C_j, C_k] = \sum \epsilon_{jkl} C_l, \qquad [D_j, D_k] = \sum \epsilon_{jkl} D_l, \qquad [C_j, D_k] = 0. \tag{7.26}$$

Clearly the complexification of the Lie algebra generated by the operators (7.25) is isomorphic to $sl(2) \times sl(2)$. In particular, the commutation relations (7.26) and (3.1), Section 8.3, are identical. Thus the irred reps of the Lie algebra (7.25) can be denoted $\mathbf{D}^{(u,v)}$. The reps are defined by (3.2) and (3.3), Section 8.3. In particular, $\mathbf{C}\cdot\mathbf{C} = u(u + 1)$, $\mathbf{D}\cdot\mathbf{D} = v(v + 1)$, and dim $\mathbf{D}^{(u,v)} = (2u + 1)(2v + 1)$. It follows directly from (7.24) that $\mathbf{A}\cdot\mathbf{L} = \mathbf{L}\cdot\mathbf{A}$

$= 0$ and $\mathbf{L}\cdot\mathbf{L} + \mathbf{A}\cdot\mathbf{A} = -[1 + (1/2\lambda)]$. Thus

$$\mathbf{C}\cdot\mathbf{C} = \mathbf{D}\cdot\mathbf{D} = \tfrac{1}{4}(\mathbf{L}\cdot\mathbf{L} + \mathbf{A}\cdot\mathbf{A}) = -\tfrac{1}{4}[1 + (1/2\lambda)] = u(u+1), \tag{7.27}$$

so the possible irred reps on $\mathcal{C}_\lambda$ obtainable from our model (7.24) are $\mathbf{D}^{(u,u)}$, where

$$\lambda = -1/2(2u+1)^2 = -1/2n^2, \qquad n = 2u + 1 = 1, 2, \ldots. \tag{7.28}$$

The possible energy eigenvalues are given by the Balmer series $\lambda_n = -1/2n^2$ and the degeneracy of the eigenvalues λ_n is n^2. Just as in Section 8.3, the restriction of this Lie algebra to the subalgebra $su(2)$ generated by the angular momentum operators L_1, L_2, L_3 yields

$$\mathbf{D}^{(u,u)} \mid su(2) \cong \mathbf{D}^{(u)} \otimes \mathbf{D}^{(u)} \cong \mathbf{D}^{(2u)} \oplus \cdots \oplus \mathbf{D}^{(1)} \oplus \mathbf{D}^{(0)}. \tag{7.29}$$

The eigenspace corresponding to λ_n is not irred under the angular momentum operators but decomposes into a direct sum of the irred reps $\mathbf{D}^{(l)}$, $0 \le l \le n-1$, each rep with multiplicity one. It follows from (7.29) that we can choose n^2 vectors Ψ_{nlm} as a basis for $\mathcal{C}_{\lambda_n}$, where $l = 0, 1, \ldots, n-1$, $-l \le m \le l$.

The skew-Hermitian operators iL_j, iA_j, $1 \le j \le 3$, generate a real Lie algebra isomorphic to $so(4, R)$. Furthermore, the Lie algebra rep $\mathbf{D}^{(u,u)}$ on $\mathcal{C}_{\lambda_n}$ exponentiates to the irred rep $(2u, 0)$ of $SO(4, R)$. Note that $so(4, R)$ and $so(3, 1)$ are both real forms of $sl(2) \oplus sl(2) \cong so(4)$. The action of $SO(4, R)$ on the coordinate space is rather difficult to compute because the A_j are second-order differential operators (Kursunoğlu [1]).

The harmonic oscillator in three dimensions corresponds to the potential $-kr^2$, $k > 0$. The symmetry group of the Hamiltonian is $SU(3)$. We will study this example in Chapter 10. Outside of these two cases no examples of rotationally symmetric Hamiltonians with connected symmetry group larger than $SO(3)$ are known. The high degree of symmetry of these two examples enables one to find the bound-state energy levels and their multiplicities from group theory alone.

In conclusion, we note that the noncompact classical groups also have infinite-dimensional irred reps. These reps can be constructed using Lie algebras and weight vectors (Sherman [1]) or using the method of induced reps (Gel'fand and Naimark [1]).

9.8 The Pauli Exclusion Principle and the Periodic Table

Consider a Hilbert space $\mathcal{H}_s^{\otimes N}$ corresponding to a physical system of N indistinguishable particles (say electrons) with spin s and mass m. A typical Hamiltonian for this system is

$$\mathbf{H} = -\sum_{j=1}^{N} (1/2m)\Delta_j + V(\mathbf{x}_1, \ldots, \mathbf{x}_N) + \mathbf{W}, \tag{8.1}$$

where $\mathbf{x}_j$ is the position coordinate of particle j, V is a potential function, and

$$\mathbf{W} = (W^{\nu_1 \cdots \nu_N}_{\mu_1 \cdots \mu_N}(\mathbf{x}_1, \ldots, \mathbf{x}_N))$$

is a spin-dependent interaction term. The Hilbert space consists of wave functions $\mathbf{\Psi} = \sum \Psi_{\mu_1 \cdots \mu_N}(\mathbf{x}_1, \ldots, \mathbf{x}_N)\mathbf{e}^{\mu_1} \otimes \cdots \otimes \mathbf{e}^{\mu_N}$, where μ_j is the spin index of the jth particle (see Section 7.8) and the $\mathbf{e}^{\mu_j}$ form a canonical basis in spin space. Here

(8.2)

$$(\mathbf{W\Psi})_{\mu_1 \cdots \mu_N}(\mathbf{x}_1, \ldots, \mathbf{x}_N) = \sum_{\nu_j = -s}^{s} W^{\nu_1 \cdots \nu_N}_{\mu_1 \cdots \mu_N}(\mathbf{x}_1, \ldots, \mathbf{x}_N)\Psi_{\nu_1 \cdots \nu_N}(\mathbf{x}_1, \ldots, \mathbf{x}_N).$$

Let σ be a permutation of the integers $1, \ldots, N$ and define the permutation operator $\boldsymbol{\sigma}$ on $\mathcal{H}_s^{\otimes N}$ by

$$(8.3) \quad (\boldsymbol{\sigma}\mathbf{\Psi})_{\mu_1 \cdots \mu_N}(\mathbf{x}_1, \ldots, \mathbf{x}_N) = \Psi_{\mu_{\sigma^{-1}(1)}, \ldots, \mu_{\sigma^{-1}(N)}}(\mathbf{x}_{\sigma^{-1}(1)}, \ldots, \mathbf{x}_{\sigma^{-1}(N)}).$$

Here $\boldsymbol{\sigma}$ is a self-adjoint operator on $\mathcal{H}_s^{\otimes N}$, as can be seen from the inner product (8.20), Section 7.8. The indistinguishability requirement for the N particles means that $\mathbf{H}$ is symmetric in the coordinates of these particles, i.e.,

$$(8.4) \qquad \boldsymbol{\sigma}\mathbf{H} = \mathbf{H}\boldsymbol{\sigma}$$

for all $\sigma \in S_N$. Thus the permutation group S_N is a symmetry group of $\mathbf{H}$. To see the implications of this fact let λ be an eigenvalue of $\mathbf{H}$ and W_λ the corresponding finite-dimensional eigenspace. If $\mathbf{\Psi} \in W_\lambda$ we have $\boldsymbol{\sigma}\mathbf{\Psi} \in W_\lambda$ for each $\sigma \in S_N$. Thus we can decompose W_λ into a direct sum of subspaces, each subspace transforming irreducibly under S_N. This decomposition was studied in Sections 3.7 and 4.2, and the irred reps were labeled by Young frames $\{f_1, \ldots, f_N\}$.

It has been found experimentally that not all eigenvectors $\mathbf{\Psi}$ in W_λ are physically meaningful. In particular, if the spin s is half-integral, the eigenvectors of $\mathbf{H}$ with eigenvalue λ (corresponding to a physical system) can occupy only that subspace of W_λ belonging to the rep $\{1^N\}$ of S_N. Indeed, the only allowed states $\mathbf{\Psi}$ permitted to such a system are completely skew-symmetric:

$$(8.5) \qquad \boldsymbol{\sigma}\mathbf{\Psi} = \delta_\sigma \mathbf{\Psi}.$$

(Here δ_σ is the parity of $\sigma \in S_N$.) Particles with half-integral spin $s = \frac{1}{2}, \frac{3}{2}, \ldots$ are called **fermions.**

If s is an integer, the eigenvectors corresponding to a physical system can occupy only that subspace of W_λ belonging to the completely symmetric rep $\{N\}$. Thus, the allowed eigenfunctions $\mathbf{\Psi}$ must satisfy

$$(8.6) \qquad \boldsymbol{\sigma}\mathbf{\Psi} = \mathbf{\Psi}$$

for all $\sigma \in S_N$. Particles with integral spin are called **bosons.**

All known elementary particles are either fermions or bosons. In the following discussion we consider only fermions.

If the Hamiltonian is a sum of single-particle Hamiltonians

$$\mathbf{H} = \sum_{j=1}^{N} \mathbf{H}_j, \qquad \mathbf{H}_j = -(1/2m)\,\mathbf{\Delta}_j + V(\mathbf{x}_j) \tag{8.7}$$

then the eigenspace W_λ is spanned by tensor products of single-particle eigenfunctions,

$$\mathbf{\Psi}(\mathbf{x}_1, \ldots, \mathbf{x}_N) = \mathbf{\Psi}^{(1)}(\mathbf{x}_1) \otimes \cdots \otimes \mathbf{\Psi}^{(N)}(\mathbf{x}_N), \qquad \mathbf{H}_j\mathbf{\Psi}^{(j)} = \lambda_j \mathbf{\Psi}^{(j)}, \tag{8.8}$$

where $\lambda = \lambda_1 + \cdots + \lambda_N$. If $\mathbf{\Psi} \in W_\lambda$ then $\boldsymbol{\sigma}\mathbf{\Psi} \in W_\lambda$ for each $\sigma \in S_N$. The vectors $\boldsymbol{\sigma}\mathbf{\Psi}$ span a subspace with dimension $N!$ if the $\{\mathbf{\Psi}^{(j)}\}$ are linearly independent. However, only the one-dimensional subspace consisting of skew-symmetric tensors

$$\tilde{\mathbf{\Psi}} = \sum_{\sigma} \delta_\sigma \mathbf{\Psi}^{(1)}(\mathbf{x}_{\sigma(1)}) \otimes \cdots \otimes \mathbf{\Psi}^{(N)}(\mathbf{x}_{\sigma(N)}) \tag{8.9}$$

is physically meaningful. Thus, for a system of N identical noninteracting fermions the skew-symmetry requirement allows us to discard all but one of our $N!$ linearly independent mathematical eigenfunctions. Furthermore, if two of the single-particle eigenfunctions $\mathbf{\Psi}^{(j)}$, $\mathbf{\Psi}^{(k)}$, $j \neq k$, are linearly dependent then $\tilde{\mathbf{\Psi}} \equiv \mathbf{\theta}$ in (8.9) and there are no permissible eigenfunctions. This yields the **Pauli principle:** A system of identical fermions cannot exist in a state in which two of the fermions are in the same single-particle state. The Pauli principle is a special case of (8.5) since it applies only to noninteracting systems.

To show the significance of the Pauli principle we discuss the N-electron atom. (Electrons are fermions since they have spin $\frac{1}{2}$.) By neglecting the motion of the (relatively) heavy nucleus we can consider an atom as a system of N electrons in a Coulomb field centered at $\mathbf{x} = \mathbf{0}$. The Hamiltonian is

$$\mathbf{H} = \mathbf{H}^{(0)} + \mathbf{H}^{(1)} + \mathbf{H}^{(2)}, \tag{8.10}$$

where

$$\mathbf{H}^{(0)} = \sum_{j=1} \mathbf{H}_j, \qquad \mathbf{H}^{(1)} = \sum_{j<k} V(|\mathbf{x}_j - \mathbf{x}_k|). \tag{8.11}$$

Here the $\mathbf{H}_j$ are single-particle Hamiltonians

$$\mathbf{H}_j = -(1/2m)\,\mathbf{\Delta}_j - Nq^2/|\mathbf{x}_j| \tag{8.12}$$

and m, q are the mass and charge of the electron, respectively. The $V(|\mathbf{x}_j - \mathbf{x}_k|)$ is the potential due to the Coulomb interaction between a pair of electrons and $\mathbf{H}^{(2)}$ denotes the interaction between the spin and orbital angular momentum of the electrons. Only $\mathbf{H}^{(2)}$ acts on the spin indices of the state functions.

We present a group-theoretic discussion of this system under the assumption that $\mathbf{H}^{(1)}$ is "small "in comparison to $\mathbf{H}^{(0)}$ and $\mathbf{H}^{(2)}$ is "small" in comparison to $\mathbf{H}^{(1)}$. The perturbation theory built on this assumption is called the **Russell–Saunders** approximation. It leads to useful results for most atoms, particularly the lighter ones.

To a first approximation the system is described by the Hamiltonian $\mathbf{H}^{(0)}$, i.e., the electrons do not interact with each other, but only with the nucleus. Since $\mathbf{H}^{(0)}$ is a sum of single-particle Hamiltonians its eigenfunctions are tensor products of single-particle eigenfunctions. Furthermore, the single-particle Hamiltonian $\mathbf{H}_j$ is exactly that of the hydrogen atom. Omitting spin considerations for a moment, we see from (7.29) that the single-particle eigenfunctions can be labeled $\Psi_{nlm}(\mathbf{x})$, where the **principal** quantum number takes the value $n = 1, 2, \ldots$, the **orbital** quantum number takes the values $l = 0, 1, \ldots, n - 1$, and $m = -l, -l + 1, \ldots, l$. For fixed n and l the $2l + 1$ vectors $\{\Psi_{nlm}\}$ correspond to an energy level nl of the hydrogen atom. The energy levels increase with increasing n. In the nonrelativistic idealized hydrogen atom, levels nl and nl' have the same energy for $l \neq l'$. However, with a more realistic model of the atom it can be shown that this degeneracy is partially removed and the energy levels increase slightly with increasing l. Thus we can label the distinct energy levels by nl and each such level has multiplicity $2l + 1$. In atomic spectroscopy the orbital quantum numbers $l = 0, 1, 2, 3, \ldots$ are denoted $s, p, d, f, g, \ldots$. Hence, the levels of hydrogen in order of increasing energy are $1s, 2s, 2p, 3s, 3p, 4s, \ldots$.

We have not yet taken the electron spin into account. The spin space of the electron is two-dimensional and the single-particle Hamiltonian does not act on the spin indices. Thus, to each $\Psi_{nlm}(\mathbf{x})$ there correspond two eigenvectors $\Psi_{nlm}(\mathbf{x})\mathbf{e}^{1/2}$ and $\Psi_{nlm}(\mathbf{x})\mathbf{e}^{-1/2}$. We conclude that the multiplicity of the level nl is $2(2l + 1)$.

Now we show that even in the rough first-order approximation where the electrons are noninteracting we can get useful qualitative information. In nature most atoms are found in the ground state (lowest energy state) rather than in some excited state. Using our model of the atom with Hamiltonian $\mathbf{H}^{(0)}$ we will explicitly compute the ground state for each N.

For hydrogen, $N = 1$, the single electron must be in a $1s$ state. For the helium atom, $N = 2$, the lowest energy level is obtained by choosing both electrons in a $1s$ state. In the case of lithium, $N = 3$, it is tempting to choose all three electrons in a $1s$ state, but this is forbidden by the Pauli principle since there are only two $1s$ states. Thus, the lowest energy level of lithium is obtained by choosing two electrons in $1s$ states and the third electron in a $2s$ state. For $N = 4$ the ground state is formed by two electrons in $1s$ states and two electrons in $2s$ states. Since there are only two $2s$ states the next electron must be in a $2p$ state to obtain the lowest energy level for $N = 5$. We can continue adding electrons one at a time until the six $2p$ states are filled. This occurs for $N = 10$ (neon). For $N = 11$ (sodium) the added electron must lie in a $3s$ state. We can continue in this manner adding electrons one at a time in the lowest possible eigenstates consistent with the Pauli principle.

It follows from our construction that electrons in an atom fall into **electron shells** labeled by the principal quantum number n. Electrons in the same shell have approximately the same energy, while the energy difference of electrons in different shells is relatively large.

The first five experimentally observed shells ordered in terms of increasing energy are listed in Table 9.1.

TABLE 9.1

Shell number	Electron states	Number of states in filled shell
1	$1s$	2
2	$2s$, $2p$	8
3	$3s$, $3p$	8
4	$4s$, $3d$, $4p$	18
5	$5s$, $4d$, $5p$	18

The observed composition of the first two shells is just as our simple model predicts. We would expect that the third shell would contain the ten $3d$ state as well as the eight $3s$ and $3p$ states. However, it has been found experimentally that in a complex atom the $3d$ states have higher energy than the $4s$ states and fall in the fourth shell. Similarly the $4d$ states lie in the fifth shell. (Our theoretical model ignores the mutual interaction between electrons. However, as the number of atomic electrons increases so does the electron interaction, so we would expect the model to be less accurate for many-electron atoms.)

As we have seen, for helium the first electron shell is filled, and neon contains exactly two filled shells. Similarly, exactly three, four, and five shells are filled for argon, krypton, and xenon, respectively. These atoms with filled shells all correspond to inert gases. On the other hand, the atoms of the alkali metals, lithium, sodium, potassium, and rubidium, consist of filled shells together with one electron in an s state of the next higher shell. Again these elements are observed to have similar chemical properties. Using the above ideas it is possible to divide the known elements into families with similar chemical properties, based on the structure of the electron shells. This theory provides a quantum mechanical derivation of Mendeleev's periodic table of the elements. Indeed the nth row of the periodic table corresponds exactly to the nth electron shell. For more details the reader should consult standard texts in atomic physics.

We now turn to the problem of analyzing the multiplicity structure of the eigenvalues of the Hamiltonian $\mathbf{H}$, (8.10)–(8.12), corresponding to an N-electron atom. Following the usual perturbation theory technique we first consider the Hamiltonian $\mathbf{H}^{(0)}$, a sum of single-particle Hamiltonians. It is clear that S_N [whose action is described by (8.3)] is a symmetry group of $\mathbf{H}^{(0)}$. Furthermore, the group $[SO(3)]^N = SO(3) \times \cdots \times SO(3)$ (N times) defined by (8.3), Section 7.8, is a symmetry group since each single-particle Hamiltonian $\mathbf{H}_j$ separately commutes with the action of $SO(3)$. [Actually $\mathbf{H}_j$ commutes with $O(3)$ but parity conservation contributes little to the following analysis.] Finally, since $\mathbf{H}^{(0)}$ does not act on the spin indices any unitary transformation in spin space commutes with $\mathbf{H}^{(0)}$. The spin space of an N-electron system is 2^N-dimensional, so the spin transformations form a symmetry group isomorphic to $U(2^N)$. By forming all possible products of symmetries corresponding to S_N, $[SO(3)]^N$, and $U(2^N)$ we can generate a larger symmetry group G_0. Unfortunately the structure of G_0 is so complicated that it is not very useful for perturbation theory. (Note that symmetries from S_N and $[SO(3)]^N$ may not commute.) Therefore, we temporarily restrict our attention to the direct product $K = [SO(3)]^N \times U(2^N)$, a proper subgroup of the maximal unitary symmetry group.

Let λ be an eigenvalue of $\mathbf{H}^{(0)}$ and W_λ the corresponding finite-dimensional eigenspace. Since K is compact, W_λ can be decomposed into a direct sum of K-irred subspaces. Each such subspace transforms according to an irred rep of the form

$$D^{(l_1)} \times D^{(l_2)} \times \cdots \times D^{(l_N)} \times [1], \tag{8.13}$$

where $D^{(l)}$ is the $(2l+1)$-dimensional irred rep of $SO(3)$ and $[1]$ is the 2^N-dimensional rep of $U(2^N)$ equivalent to the usual $2^N \times 2^N$ matrix realization of this group. Note that such an irred subspace is of the form

$$V^{(l_1 \cdots l_N)} \otimes Z^{\otimes N}, \tag{8.14}$$

where $Z^{\otimes N}$ is the 2^N-dimensional spin space with basis $\{\mathbf{e}^{\mu_1} \otimes \cdots \otimes \mathbf{e}^{\mu_N}\}$, $V^{(l_1 \cdots l_N)}$ is the $(2l_1 + 1) \cdots (2l_N + 1)$-dimensional space of scalar functions with basis

$$\Psi_{n_1 l_1 m_1}(\mathbf{x}_1) \cdots \Psi_{n_N l_N m_N}(\mathbf{x}_N), \qquad -l_j \leq m_j \leq l_j, \tag{8.15}$$

and Ψ_{nlm} is the hydrogen atom wave function with quantum numbers n, l, m. The group $U(2^N)$ acts on $Z^{\otimes N}$ irreducibly and $[SO(3)]^N$ acts on $V^{(l_1 \cdots l_N)}$ irreducibly. The dimension of this rep is

$$(2l_1 + 1) \cdots (2l_N + 1) 2^N. \tag{8.16}$$

The numbers (8.16) give information concerning the degeneracies of eigenvalues of $\mathbf{H}^{(0)}$, although, since K is not maximal, several irred subspaces correspond to the same eigenvalue.

Next we consider the effect of the perturbing potential $\mathbf{H}^{(1)}$, (8.11). Clearly the action of $A_1 \times \cdots \times A_N \in [SO(3)]^N$ on configuration space commutes with $\mathbf{H}^{(1)}$ if and only if $A_1 = \cdots = A_N$. These "diagonal" elements generate a subgroup of $[SO(3)]^N$ isomorphic to $SO(3)$. The permutation group S_N and $U(2^N)$ acting on the spin indices are still symmetry groups of $\mathbf{H}^{(1)}$. We can generate a much larger symmetry group G_1 by forming all possible finite products of symmetries associated with $SO(3)$, S_N, and $U(2^N)$. However, the symmetries from S_N and $U(2^N)$ do not commute in general so the structure of this group is very complicated. To get more useful results we consider a subgroup of G_1. The set of all matrices

$$\begin{pmatrix} B & & & \\ & B & & Z \\ & & \ddots & \\ & Z & & B \end{pmatrix} \in U(2^N),$$

where $B \in U(2)$, forms a group isomorphic to $U(2)$. Under the action of this group each of the spin indices of $\mathbf{\Psi}$ is transformed identically under the usual two-dimensional realization of $U(2)$. [If we restrict further to the subgroup $SU(2)$ we get the usual action of $SU(2)$ in spin space.] It is easy to verify that the actions of S_N and $U(2)$ mutually commute. Thus, the symmetry group of $\mathbf{H}^{(1)}$ generated by $SO(3)$, S_N, and $U(2)$ is the direct product group

$$K_1 = SO(3) \times S_N \times U(2). \tag{8.17}$$

The irred reps of K_1 are of the form

$$D^{(L)} \times \{f_1, \ldots, f_N\} \times [g_1, g_2], \tag{8.18}$$

where $D^{(L)}$ is the $(2L + 1)$-dimensional rep of $SO(3)$, the Young frame $\{f_1, \ldots, f_N\}$, $f_1 + \cdots + f_N = N$, denotes an irred rep of S_N as determined in Section 4.2, and the Young frame $[g_1, g_2]$ denotes an irred rep of $U(2)$. Thus each eigenspace of the Hamiltonian $\mathbf{H}^{(0)} + \mathbf{H}^{(1)}$ decomposes into a direct sum of subspaces, each transforming according to reps (8.18). However, the only subspaces of physical interest are those for which $\{f_1, \ldots, f_N\} \cong \{1^N\}$. Furthermore, the rep $[g_1, g_2]$ on restriction to $SU(2)$ is equivalent to $[g_1 - g_2, 0] \cong D^{((g_1-g_2)/2)}$. We conclude that each irred subspace of physical interest is of the form

$$D^{(L)} \times \{1^N\} \times [M + S, M - S], \tag{8.19}$$

where M and S are both integral or both half-integral. The dimension of this space is $(2L + 1)(2S + 1)$ and the action of the subgroup $SO(3) \times SU(2)$ on it yields the irred rep $D^{(L)} \times D^{(S)}$. Such a space is called a **term** and designated by the symbol ${}^{2S+1}L$ in spectroscopic notation. Each wave function in the space has orbital angular momentum L and spin angular momentum

S. (Recall that spin and orbital angular momentum are not coupled at this state of our perturbation procedure.)

Our derivation of (8.19) is not constructive since it does not show how to actually compute such reps for a given physical system. We analyze this problem in more detail by considering a configuration in which the N electrons are in states $n_1 l_1, \ldots, n_N l_N$. Ignoring spin for the moment, we see that the space $Y^{(l_1 \cdots l_N)}$ of all possible coordinate wave functions describing this system is invariant under the group $SO(3) \times S_N^{(1)}$, where $SO(3)$ is related to orbital angular momentum and $S_N^{(1)}$ denotes the action of the permutation group on the spatial coordinates $\mathbf{x}_j$ of the wave functions $\Psi(\mathbf{x}_1, \ldots, \mathbf{x}_N)$. Similarly the 2^N-dimensional spin space $Z^{\otimes N}$ is invariant under $U(2) \times S_N^{(2)}$, where $U(2)$ [and its subgroup $SU(2)$] are related to spin angular momentum, and $S_N^{(2)}$ refers to the action of the permutation group on spin indices. Clearly, the elements of $W_\lambda' = Y^{(l_1 \cdots l_N)} \otimes Z^{\otimes N}$ form an eigenspace of the unperturbed Hamiltonian $\mathbf{H}^{(0)}$ corresponding to some eigenvalue λ. (Here, W_λ' may only be a proper subspace of the total eigenspace W_λ.)

We will decompose W_λ' into terms by first decomposing $Y^{(l_1 \cdots l_N)}$ into irred subspaces under $SO(3) \times S_N^{(1)}$ and $Z^{\otimes N}$ into irred subspaces under $U(2) \times S_N^{(2)}$. For $Y^{(l_1 \cdots l_N)}$ the results depend strongly on $l_1, \ldots, l_N$. All we can say in general is that the irred subspaces are of the form

(8.20) $$D^{(L)} \times \{f_1, \ldots, f_N\}, \qquad f_1 + \cdots + f_N = N.$$

Important special cases of this decomposition are treated by Hamermesh [1] and Lomont [1]. For $Z^{\otimes N}$, on the other hand, we can proceed with complete generality. It follows from the results of Section 4.3 and Section 9.1 that $Z^{\otimes N}$ decomposes under $U(2) \times S_N^{(2)}$ into a direct sum of subspaces of the form

(8.21) $$[g_1, g_2] \times \{g_1, g_2\}, \qquad g_1 + g_2 = N.$$

Indeed the $U(2)$-irred subspaces of $Z^{\otimes N}$ consist of symmetry classes of tensors, each class belonging to a two-rowed Young tableau with N boxes. The space (8.21) is that spanned by all symmetry classes of tensors corresponding to a single frame $[g_1, g_2]$. The dimension of this space is $(g_1 - g_2 + 1) \cdot \dim \{g_1, g_2\}$, where $\dim \{g_1, g_2\}$ is given by Theorem 4.2. Note that the multiplicity of the rep $\{g_1, g_2\}$ of $S_N^{(2)}$ in $Z^{\otimes N}$ is $g_1 - g_2 + 1$. Since $[g_1, g_2] \mid SU(2) \cong [g_1 - g_2, 0]$ the action of $SU(2) \times S_N^{(2)}$ on $Z^{\otimes N}$ decomposes into irred reps

(8.22) $$D^{(S)} \times \{\tfrac{1}{2}N + S, \tfrac{1}{2}N - S\}, \qquad D^{(S)} \cong [2S, 0],$$

where $S = 0, 1, \ldots, N/2$ if N is even and $S = \frac{1}{2}, \frac{3}{2}, \ldots, N/2$ if N is odd. Each irred rep occurs with multiplicity one.

Thus, we can decompose $Y^{(l_1 \cdots l_N)} \otimes Z^{\otimes N}$ into a direct sum of irred reps

(8.23) $$D^{(L)} \times D^{(S)} \times \{f_1, \ldots, f_N\} \times \{\tfrac{1}{2}N + S, \tfrac{1}{2}N - S\}$$

under the action of the group

(8.24) $$SO(3) \times SU(2) \times S_N^{(1)} \times S_N^{(2)}.$$

At this point we couple the actions of S_N on the spatial coordinates and the spin indices, i.e., we restrict $S_N^{(1)} \times S_N^{(2)}$ to the diagonal subgroup $\{\sigma \times \sigma: \sigma \in S_N\}$, isomorphic to S_N. Then the action of the symmetry group

$$SO(3) \times SU(2) \times S_N = K_1 \tag{8.25}$$

on (8.23) is

$$D^{(L)} \times D^{(S)} \times (\{f_1, \ldots, f_N\} \otimes \{\tfrac{1}{2}N + S, \tfrac{1}{2}N - S\}). \tag{8.26}$$

The only possible term in (8.26) is ${}^{2S+1}L$ and the multiplicity of this term is equal to the multiplicity of the alternating rep $\{1^N\}$ in the tensor product

$$\{f_1, \ldots, f_N\} \otimes \{\tfrac{1}{2}N + S, \tfrac{1}{2}N - S\}. \tag{8.27}$$

Let $\{g_1, \ldots, g_p\}^t$ be the Young frame obtained by interchanging rows and columns in the Young frame $\{g_1, \ldots, g_p\}$. In the next section we will show that the multiplicity of $\{1^N\}$ in (8.27) is zero unless

$$\{f_1, \ldots, f_N\} = \{\tfrac{1}{2}N + S, \tfrac{1}{2}N - S\}^t = \{2^{(N/2)-S}, 1^S\}, \tag{8.28}$$

in which case the multiplicity is one.

Thus the only symmetry classes (8.20) that satisfy the Pauli principle are those whose Young frames consist of two columns. The space transforming according to (8.26) contains one or zero terms depending on whether or not (8.28) is satisfied. An explicit construction of the states in ${}^{2S+1}L$ requires a knowledge of the Clebsch–Gordan coefficients for $\{1^N\}$ in the tensor product (8.27), but this explicit construction is seldom necessary in physical problems.

To recapitulate, our method of term analysis is to decompose the orbital and spin wave functions separately into symmetry classes. The terms are formed from tensor products of these two classes that satisfy the Pauli principle.

We are now ready to consider the effect of the spin–orbit interaction $\mathbf{H}^{(2)}$. We assume that $\mathbf{H}^{(2)}$ does not commute with the orbital angular momentum operators $\mathbf{L}_k$ or the spin angular momentum operators $\mathbf{S}_k$ but that it does commute with the operators $\mathbf{J}_k = \mathbf{L}_k + \mathbf{S}_k$ of total angular momentum. Thus, $\mathbf{H}^{(2)}$ couples the actions of $SO(3)$ and $SU(2)$ on the orbital and spin indices. (Compare the analogous discussion in Section 7.8.)

Let ${}^{2S+1}L$ be a term corresponding to a given energy level of $\mathbf{H}^{(0)} + \mathbf{H}^{(1)}$. This space has dimension $(2S + 1)(2L + 1)$ and transforms according to the irred rep $D^{(S)} \times D^{(L)} \times \{1^N\}$ of the symmetry group $K_1 = SO(3) \times SU(2) \times S_N$. The symmetry group of $\mathbf{H}^{(2)}$ is the subgroup

$$K_2 = SU(2) \times S_N, \tag{8.29}$$

where the action of $SU(2)$ is defined as in (8.21), Chapter 7. Decomposing the term ${}^{2S+1}L$ into subspaces irred under K_2, we find

$$D^{(S)} \times D^{(L)} \times \{1^N\} \,|\, K_2 \cong \sum_{J=|S-L|}^{S+L} \oplus (D^{(J)} \times \{1^N\}). \tag{8.30}$$

The $(2J+1)$-dimensional subspace transforming according to $D^{(J)} \times \{1^N\}$ is called a **multiplet** and denoted ${}^{2S+1}L_J$. Thus, under a perturbing potential $\mathbf{H}^{(2)}$ the term ${}^{2S+1}L$ splits into $2K+1$ multiplet levels where $K = \min(S, L)$. This completes our analysis of the Hamiltonian $\mathbf{H} = \mathbf{H}^{(0)} + \mathbf{H}^{(1)} + \mathbf{H}^{(2)}$ on the basis of Russell–Saunders perturbation theory. This method is based on the chain of groups $G_0 \supset K_1 \supset K_2$. Although K_2 is uniquely determined as the symmetry group of $\mathbf{H}$ and G_0 as the symmetry group of a noninteracting system, the intermediate group K_1 is a function of our perturbation assumptions. Different perturbation schemes lead to different chains of symmetry groups $G_1 \supset \cdots \supset G_k$. For some examples see the work of Hamermesh [1] or Loebl [1].

As the simplest nontrivial example of the Russell–Saunders scheme we consider a configuration of two electrons in an *np* state. Here $N = 2$, $l_1 = l_2 = 1$, and the space $Y^{(l_1 l_2)}$ is nine-dimensional. Since the spin space $Z^{\otimes 2}$ is four-dimensional there are 36 possible states belonging to this configuration. However, applying the Pauli principle we see that only

$$\binom{6}{2} = 15$$

of these states can be occupied by a physical system. (Note that a single *np* electron can occupy $3 \times 2 = 6$ states.) We will decompose this configuration into terms. Under the action of $SO(3)$, $Y^{(11)}$ transforms according to $D^{(2)} \otimes D^{(1)} \otimes D^{(0)}$. We want to reduce $Y^{(11)}$ into irred subspaces under $SO(3) \times S_2^{(1)}$. To do this we first split up the space into symmetry classes of tensors, i.e., irred subspaces under the action T of $U(3) \times S_2^{(1)}$ [or $GL(3) \times S_2^{(1)}$]. According to the results of Section 4.3,

$$T \cong [2, 0] \times \{2, 0\} \oplus [1^2] \times \{1^2\}. \tag{8.31}$$

Thus, $Y^{(11)}$ is decomposed into spaces of symmetric tensors (dimension six) and skew-symmetric tensors which carry the reps $[2, 0]$ and $[1^2]$ of $U(3)$, respectively. It is easy to check that $[1^2] \mid SO(3) \cong D^{(1)}$, so we must have $[2, 0] \mid SO(3) \cong D^{(0)} \oplus D^{(2)}$. Thus

$$T \mid SO(3) \times S_2^{(1)} \cong (D^{(0)} \times \{2, 0\}) \oplus (D^{(2)} \times \{2, 0\}) \oplus (D^{(1)} \times \{1^2\}). \tag{8.32}$$

According to our general analysis, the action of $SU(2) \times S_2^{(2)}$ on $Z^{\otimes 2}$ leads to

$$(D^{(0)} \times \{1^2\}) \oplus (D^{(1)} \times \{2, 0\}). \tag{8.33}$$

Now $\{1^2\}^t = \{2, 0\}$ and $\{2, 0\}^t = \{1^2\}$, so the possible terms ${}^{2S+1}L$ transforming according to $D^{(L)} \times D^{(S)} \times \{1^2\}$ are ${}^1S$, ${}^1D$, ${}^3P$, each term occurring once (we use spectroscopic notation for the value of L). The total number of states in the three terms is $1 + 5 + 9 = 15$, in agreement with our earlier result. The decomposition of each term into multiplets follows directly from (8.30). For more complicated examples of term analysis see the work of Hamermesh [1], Lomont [1], or Wybourne [1].

9.9 The Group Ring Revisited

Here we derive some theoretical results of great utility in atomic and nuclear physics. Most of these results are related to the group ring of the symmetric group S_n.

We start by proving a theorem used in the previous section. Let $\{\lambda_j\}$, $\{\lambda_j'\}$ be Young frames with n boxes and let $\chi(\sigma)$, $\chi'(\sigma)$ be the corresponding simple characters of S_n. The rep $\{1^n\} \otimes \{\lambda_j\}$ of S_n has character $\psi\chi(\sigma) = \psi(\sigma)\chi(\sigma)$, where $\psi(\sigma) = \delta_\sigma$ and δ_σ is the parity of $\sigma \in S_n$.

Lemma 9.10. The rep $\{1^n\} \otimes \{\lambda_j\}$ is irred.

Proof. We use the results of Section 3.4 on group characters. By the Corollary to Theorem 3.7 a character ρ of S_n is simple if and only if $\langle \rho, \rho \rangle = 1$, where

$$\langle \chi_1, \chi_2 \rangle = (1/n!) \sum_{\sigma \in S_n} \chi_1(\sigma)\overline{\chi_2(\sigma)} \tag{9.1}$$

for characters χ_1, χ_2. An elementary computation yields $\langle \psi\chi, \psi\chi \rangle = \langle \chi, \chi \rangle = 1$ since χ is simple. Therefore, $\psi\chi$ is a simple character. Q.E.D.

Denote the rep $\{1^n\} \otimes \{\lambda_j\}$ by $\{\tilde{\lambda}_j\}$.

Lemma 9.11. The multiplicity m of the alternating rep $\{1^n\}$ in $\{\lambda_j\} \otimes \{\lambda_j'\}$ is one if $\{\lambda_j'\} \cong \{\tilde{\lambda}_j\}$; otherwise $m = 0$.

Proof. We know $m = \langle \chi\chi', \psi \rangle$. Since the characters of S_n are all real, (9.1) implies $\langle \chi\chi', \psi \rangle = \langle \chi', \chi\psi \rangle$. The characters χ' and $\chi\psi$ are simple, so by the orthogonality relations for characters, $\langle \chi', \chi\psi \rangle = 1$ if and only if $\chi' \equiv \chi\psi$. Otherwise $\langle \chi', \chi\psi \rangle = 0$. Q.E.D.

Since the rep $\{\tilde{\lambda}_j\}$ is irred, it must correspond to some Young frame $\{\mu_j\}$ of S_n. To identify this frame we need a few facts relating characters to the structure of the group ring of S_n. These facts turn out to be valid for all finite groups. Thus we consider an arbitrary finite group G with group ring R_G. The following discussion is based on results derived in Section 3.7.

Let W be a subspace of R_G, invariant under the left regular rep $\mathbf{L}$: $\mathbf{L}(g)x = gx$, $g \in G$, $x \in R_G$. W is a left ideal of R_G and there exists an idempotent c in R_G such that $W = R_G c$, $c^2 = c$. Let $\chi(g)$ be the character of $\mathbf{L} \mid W$. Clearly, $\chi(g)$ is uniquely determined once c is known. We shall derive an expression defining $\chi(g)$ in terms of c. [Recall that every simple character $\chi(g)$ can be obtained in this way from some primitive left ideal W.]

Lemma 9.12. $\chi(g) = \sum_{h \in G} c(h^{-1}g^{-1}h)$, where $c = \sum_{h \in G} c(h) \cdot h$.

Proof. Let $c' = e - c$, where e is the identity element of R_G. Then c' is idempotent and $R_G = W \oplus W'$, where $W' = R_G c'$ is a left ideal. By definition, $\chi(g) = \operatorname{tr}(\mathbf{L}(g) \mid W)$, where $\mathbf{L}(g) \mid W$ is the restriction of $\mathbf{L}(g)$ to W. Let $\mathbf{P}$ be the projection operator $\mathbf{P}x = xc$, $x \in R_G$. Then $\mathbf{P}w = w$ for $w \in W$ and $\mathbf{P}w' = \theta$ for $w' \in W'$. Clearly $\mathbf{L}(g)\mathbf{P}x = \mathbf{L}(g)w$ for $x = w + w' \in R_G$. Thus $\chi(g) = \operatorname{tr}(\mathbf{L}(g)\mathbf{P})$, where $\mathbf{L}(g)\mathbf{P}$ is defined on R_G. We compute this trace using the natural basis $\{k : k \in G\}$ for R_G. We have

$$\mathbf{L}(g)\mathbf{P}k = gkc = \sum_{h \in G} c(h) \cdot gkh = \sum_{h \in G} c(k^{-1}g^{-1}h) \cdot h, \tag{9.2}$$

so the $c(k^{-1}g^{-1}h)$ are the matrix elements of $\mathbf{L}(g)\mathbf{P}$ in the natural basis. Summing the diagonal elements we obtain the lemma. Q.E.D.

We apply this result to S_n. Let T be a Young tableau corresponding to the Young frame $\{\lambda_j\}$. Then T defines a primitive idempotent

$$c = \sum_{p,q} \delta_q pq, \tag{9.3}$$

where p runs over all row permutations of T and q runs over all column permutations. Thus, $c = \sum c(s) \cdot s$, where $c(pq) = \delta_q$ and $c(s) = 0$ if s is not a pq. Let $\tilde{T}$ be the tableau obtained from T by interchanging rows and columns (see Fig. 9.1).

1	3	5	6
2	7	8	
4			

T

1	2	4
3	7	
5	8	
6		

$\tilde{T}$

FIGURE 9.1

Clearly, the row permutations p of T are the column permutations $\tilde{q}$ of $\tilde{T}$ and the column permutations q of T are the row permutations $\tilde{p}$ of $\tilde{T}$. Thus the essential idempotent $\tilde{c}$ corresponding to $\tilde{T}$ is

$$\tilde{c} = \sum_{\tilde{p},\tilde{q}} \delta_{\tilde{q}} \tilde{p}\tilde{q} = \sum_{p,q} \delta_p qp = \sum_s \tilde{c}(s) \cdot s. \tag{9.4}$$

If $s = pq$ then $s^{-1} = q^{-1}p^{-1}$, $\delta_{p^{-1}} = \delta_p$, and $\tilde{c}(s^{-1}) = \delta_p$. If s is not a pq then $\tilde{c}(s^{-1}) = 0$. This proves the following:

$$c(s) = \delta_p \delta_q \tilde{c}(s^{-1}) = \delta_s \tilde{c}(s^{-1}) \tag{9.5}$$

for all $s \in S_n$. (Note that $\delta_p \delta_q = \delta_s$ if $s = pq$.)

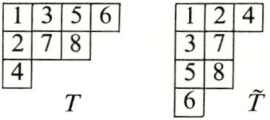

If T is a standard tableau with frame $\{\lambda_j\}$ then $\tilde{T}$ is a standard tableau with frame $\{\tilde{\lambda}_j\}$ and this is a 1–1 relationship between standard tableaux corresponding to these two frames. Hence, by Theorem 4.2 the reps $\{\lambda_j\}$ and $\{\tilde{\lambda}_j\}$ have the same dimension f. By Lemma 4.6 the Young elements $fc/n!$ and $f\tilde{c}/n!$ are generating idempotents for the reps $\{\lambda_j\}$ and $\{\tilde{\lambda}_j\}$, respectively. Applying Lemma 9.12 we obtain

(9.6) $$\chi(s) = \frac{f}{n!} \sum_{t \in S_n} c(t^{-1}s^{-1}t), \qquad \tilde{\chi}(s) = \tilde{\chi}(s^{-1}) = \frac{f}{n!} \sum_t \tilde{c}(t^{-1}st),$$
$$\tilde{c}(t^{-1}st) = \delta_{t^{-1}st} c(t^{-1}s^{-1}t)$$

for the characters of these reps. But $\delta_{t^{-1}st} = \delta_s$, so

(9.7) $$\tilde{\chi}(s) = \delta_s \chi(s).$$

This proves $\{\tilde{\lambda}_j\} \cong \{1^n\} \otimes \{\lambda_j\}$.

Theorem 9.10. The multiplicity of the alternating rep $\{1^n\}$ in $\{\lambda_j\} \otimes \{\mu_j\}$ is zero unless $\{\mu_j\} \cong \{\tilde{\lambda}_j\}$, in which case the multiplicity is one. Here $\{\tilde{\lambda}_j\}$ is the frame obtained from $\{\lambda_j\}$ by interchanging rows and columns.

The frames $\{\lambda_j\}$, $\{\tilde{\lambda}_j\}$ are said to be **conjugate** and their corresponding reps are **conjugate reps.** If $\{\lambda_j\} \cong \{\tilde{\lambda}_j\}$ then $\{\lambda_j\}$ is **self-conjugate.**

Lemma 9.12, relating an idempotent in the group ring to the character of the group rep it generates, is very useful in applied problems. To illustrate this we reexamine the meaning of an induced rep of a finite group G as defined in Section 3.5. If H is a proper subgroup of G we can regard R_H as a subspace (not a subalgebra) of the group ring R_G. Let c be a primitive idempotent in R_H. Then under the action of the left regular rep of H the subspace $R_H c$ of R_H determines an irred rep $\mathbf{T}$ of H with character

(9.8) $$\chi(h) = \sum_{k \in H} c(k^{-1}h^{-1}k), \qquad c = \sum_{h \in H} c(h) \cdot h.$$

All simple characters of H can be so obtained. Now c is also an idempotent in R_G, though not necessarily primitive. Thus, under the action of the left regular rep of G the left ideal $R_G c$ determines a rep $\mathbf{T}'$ of G with character

(9.9) $$\chi'(g) = \sum_{t \in G} \dot{c}(t^{-1}g^{-1}t),$$

where $\dot{c}(t) = c(t)$ if $t \in H$ and $\dot{c}(t) = 0$ if $t \notin H$. Let

$$g_1 H, \quad g_2 H, \ldots, \quad g_m H, \quad n(G) = m \cdot n(H),$$

be the distinct left cosets of H in G, where $g_1 = e$. Then any $t \in G$ can be written uniquely in the form $t = g_j h$ for some $h \in H$. Thus from (9.8),

(9.10) $$\chi'(g) = \sum_{j=1}^{m} \sum_{h \in H} \dot{c}(h^{-1}g_j^{-1}g^{-1}g_j h) = \sum_{j=1}^{m} \dot{\chi}(g_j^{-1}gg_j)$$
$$\dot{\chi}(g) = \begin{cases} \chi(g) & \text{if } \ g \in H \\ 0 & \text{otherwise.} \end{cases}$$

Now $\dot{\chi}(h^{-1}gh) = \dot{\chi}(g)$ for all $h \in H$, so

$$\sum_h \dot{\chi}(h^{-1}g_j^{-1}gg_j h) = n(H)\dot{\chi}(g_j^{-1}gg_j)$$

and

$$\chi'(g) = \frac{1}{n(H)} \sum_{t \in G} \dot{\chi}(tgt^{-1}), \tag{9.11}$$

where $t^{-1} = g_j h$. This expression for $\chi'(g)$ is identical with expression (5.27), Section 3.5, for the induced character χ^G. Thus the rep of G defined by the primitive idempotent c in R_H is equivalent to the induced rep $\mathbf{T}^G$.

Certain induced reps of the symmetric groups are of great importance in atomic physics. We can consider the direct product group $S_n \times S_m$ as a subgroup of S_{n+m}. (If we think of S_{n+m} as the permutation group on $n + m$ letters, then S_n permutes the first n letters alone and S_m permutes the last m letters). The irred reps of $S_n \times S_m$ are of the form $\{\lambda_j\} \times \{\mu_k\}$, where $\{\lambda_j\}$, $\{\mu_k\}$ are irred reps of S_n, S_m, respectively. The rep $\{\lambda_j\} \boxtimes \{\mu_k\}$ of S_{n+m} induced in the sense of Frobenius from $\{\lambda_j\} \times \{\mu_k\}$ is called the **outer product** of $\{\lambda_j\}$ and $\{\mu_k\}$ (Hamermesh [1]). Theorem 3.9 yields the following result.

Lemma 9.13. The multiplicity of the rep $\{\rho_l\}$ of S_{n+m} in the outer product $\{\lambda_j\} \boxtimes \{\mu_k\}$ equals the multiplicity of $\{\lambda_j\} \times \{\mu_k\}$ in the restriction of $\{\rho_l\}$ to the subgroup $S_n \times S_m$.

Simple algorithms have been developed which enable one to decompose any outer product $\{\lambda_j\} \boxtimes \{\mu_k\}$ into a direct sum of irred reps of S_{n+m}. For a discussion of these procedures see the work of Littlewood [1] or Hamermesh [1]. Here we merely show the importance of outer products for quantum mechanics.

Outer products allow us to decompose a tensor product of irred reps of $U(k)$ [or $GL(k)$ or $SU(k)$] into a direct sum of irred reps, i.e., they enable us to determine the Clebsch–Gordan series for $U(k)$. Consider the irred reps $[\lambda_1, \ldots, \lambda_k]$ and $[\mu_1, \ldots, \mu_k]$ of $U(k)$, where $\lambda_1 + \cdots + \lambda_k = n$ and $\mu_1 + \cdots + \mu_k = m$. We can regard the tensor product rep $[\lambda_j] \otimes [\mu_j]$ of $U(k)$ as defined on a subspace W of $V^{\otimes(n+m)} \cong V^{\otimes n} \otimes V^{\otimes m}$, where V is a k-dimensional vector space. Here $W = W_1 \otimes W_2$, W_1 is a symmetry class of tensors in $V^{\otimes n}$ corresponding to a tableau with frame $\{\lambda_j\}$, and W_2 is a symmetry class of tensors in $V^{\otimes m}$ corresponding to a tableau with frame $\{\mu_j\}$. It is a consequence of Theorem 4.11 that $W_1 = \hat{c}_1 V^{\otimes n}$, $W_2 = \hat{c}_2 V^{\otimes m}$, where c_1, c_2 are the Young symmetrizers corresponding to $\{\lambda_j\}$ and $\{\mu_j\}$, respectively. (We can assume that the c_i are primitive idempotents.) Now c_1 belongs to the group ring R_n of S_n and c_2 belongs to the group ring R_m of S_m. Since $S_n \times S_m$ is a subgroup of S_{n+m} we can also regard c_1, c_2 as commuting idempotents in R_{n+m}. Thus, $(c_1 c_2)^2 = c_1{}^2 c_2{}^2 = c_1 c_2$ and $c_1 c_2$ is an idempotent in R_{n+m}. This proves $W = \widehat{c_1 c_2} V^{\otimes(n+m)}$ and associates the $U(k)$-invariant space of tensors W uniquely with the right ideal $\mathcal{I} = c_1 c_2 R_{n+m}$. By Theorem 4.11 again, the decomposition of W into $U(k)$-irred subspaces is equivalent to the decomposition of $\mathscr{I}$ into a direct sum of primitive right ideals in R_{n+m}. We have observed above that $\mathscr{I}$ transforms under the right regular rep of S_{n+m} as the outer product $\{\lambda_j\} \boxtimes \{\mu_j\}$. (Our switch from left regular to right regular rep in no way changes this result.) Thus there is a 1–1 correspondence between irred reps $\{\rho_l\}$ of S_{n+m} occurring in the outer product and $U(k)$-irred subspaces of W transforming according to $[\rho_l]$. (To this assertion we must add the proviso that $\{\rho_l\}$ contain at most k rows since otherwise the tensors in the symmetry class $[\rho_l]$ will all be zero.)

Theorem 9.11. The $U(k)$-irred reps occurring in the tensor product $[\lambda_j] \otimes [\mu_j]$, $\sum \lambda_j = n$, $\sum \mu_j = m$, are of the form $[\rho_j]$ where $\sum \rho_j = n + m$. The multiplicity of $[\rho_j]$ in $[\lambda_j] \otimes [\mu_j]$ equals the multiplicity of the rep $\{\rho_j\}$ of S_{n+m} in the outer product $\{\lambda_j\} \boxtimes \{\mu_j\}$.

Corollary 9.5. The multiplicity of $[\rho_j]$ in $[\lambda_j] \otimes [\mu_j]$ equals the multiplicity of the rep $\{\lambda_j\} \times \{\mu_j\}$ of $S_n \times S_m$ in $\{\rho_j\} \,|\, S_n \times S_m$.

Proof. Immediate from Lemma 9.13.

The above theorem and its corollary demonstrate the importance of the outer product in atomic and elementary particle physics. Use of the outer product enables one to work out the Clebsch–Gordan series for $GL(k)$, $U(k)$, and $SU(k)$ in a straightforward manner. For details concerning the applications see the work of Hamermesh [1] or Wybourne [1].

9.10 Semisimple Lie Algebras

In this chapter we have investigated the rep theory of the classical groups and their Lie algebras. The classical Lie algebras belong to a larger family of Lie algebras called semisimple. Semisimple Lie algebras have been widely studied and there is a vast mathematical literature on their structure and reps. Here we present a number of definitions and results, mostly without proof, to show the relationship between the theory of the classical Lie algebras as presented in this book and the more general theory of semisimple Lie algebras. All of these results are proved in detail in the textbooks by Freudenthal and De Vries [1], Hausner and Schwartz [1], and Jacobson [1]. With the orientation provided here the reader should have no trouble understanding these texts.

Let $\mathcal{G}$ be a Lie algebra. We define a sequence $\{\mathcal{G}^{(n)}\}$ of ideals in $\mathcal{G}$ inductively by $\mathcal{G}^{(1)} = \mathcal{G}$, $\mathcal{G}^{(n+1)} = [\mathcal{G}^{(n)}, \mathcal{G}^{(n)}]$, $n = 1, 2, \ldots$. Clearly, $\mathcal{G}^{(n+1)} \subseteq \mathcal{G}^{(n)}$. We say $\mathcal{G}$ is **solvable** if $\mathcal{G}^{(n)} = \{\theta\}$ for n sufficiently large. We say $\mathcal{G}$ is **semisimple** if it contains no proper solvable ideals and dim $\mathcal{G} > 1$. (An ideal is itself a Lie algebra.) $\mathcal{G}$ is **simple** (dim $\mathcal{G} > 1$) if it contains no proper ideals. Clearly, if $\mathcal{G}$ is simple then it is semisimple. On the other hand, we have the following result.

Theorem 9.12. A Lie algebra $\mathcal{G}$ is semisimple if and only if it can be expressed as a direct sum of simple Lie algebras:

$$\mathcal{G} = \mathcal{G}_1 \oplus \mathcal{G}_2 \oplus \cdots \oplus \mathcal{G}_k, \qquad \mathcal{G}_j \text{ simple.}$$

Simple Lie algebras are the building blocks out of which the semisimple Lie algebras are constructed. Another characterization of semisimple algebras is as follows.

Theorem 9.13 (Cartan's criterion). Let $\gamma_1, \ldots, \gamma_n$ be a basis for $\mathcal{G}$ and define the structure constants c_{ij}^l by

$$[\gamma_i, \gamma_j] = \sum_{l=1}^{n} c_{ij}^l \gamma_l. \tag{10.1}$$

Then $\mathcal{G}$ is semisimple if and only if $\det \mathcal{F} \neq 0$, where $\mathcal{F}$ is the matrix with components $\mathcal{F}_{ik} = \sum_{j,l=1}^{n} c_{ij}^l c_{kl}^j = \mathcal{F}_{ki}$.

It is straightforward to determine which of the Lie algebras of the complex classical groups are simple or semisimple. The algebra $gl(m, \mathfrak{C})$, $m > 1$, is not semisimple, because the set of all multiples of the identity matrix forms a proper solvable ideal. However, the algebras $sl(m, \mathfrak{C})$, $m \geq 2$, are simple, as are $sp(m, \mathfrak{C})$, $m \geq 1$. The one-dimensional algebra $so(2, \mathfrak{C})$ is abelian, but $so(m, \mathfrak{C})$ is simple for $m \geq 3$ with the exception of $so(4, \mathfrak{C}) \cong sl(2, \mathfrak{C}) \oplus sl(2, \mathfrak{C})$, which is semisimple. In the mathematical literature the simple classical Lie algebras are denoted A_m, B_m, C_m, D_m according to the following list

$sl(m+1, \mathfrak{C})$	A_m	$m \geq 1$	$m(m+2)$
$so(2m+1, \mathfrak{C})$	B_m	$m \geq 2$	$m(2m+1)$
$sp(m, \mathfrak{C})$	C_m	$m \geq 3$	$m(2m+1)$
$so(2m, \mathfrak{C})$	D_m	$m \geq 4$	$m(2m-1)$.

The last number in each row is the dimension of the corresponding Lie algebra. The algebras B_1, C_1, C_2, D_3 are also simple, but due to the isomorphisms $B_1 \cong C_1 \cong A_1$, $C_2 \cong B_2$, $D_3 \cong A_3$, they are already included on our list. No two algebras in the above list are isomorphic.

The (complex) simple Lie algebras can be classified up to isomorphism. In addition to the four infinite families $A_m - D_m$ there are exactly five simple algebras E_6, E_7, E_8 of dimensions 52, 78, and 133, respectively, F_4 of dimension 52, and G_2 of dimension 14. Thus, with the exception of these five algebras, the **exceptional** algebras, we have already studied all complex simple Lie algebras. The construction of the exceptional Lie algebras is by no means a trivial matter, but these algebras are rarely used in theoretical physics, so we omit their definition and rep theory. (The exceptional algebra G_2 has been applied in atomic spectroscopy, see the work of Racah [1]).

It is an elementary consequence of the Cartan criterion that any real form of a complex semisimple Lie algebra is itself semisimple. Furthermore, the complexification of any real semisimple algebra is also semisimple. By Theorem 9.12 any real semisimple Lie algebra can be expressed as a direct sum of real simple Lie algebras. The real simple algebras have been classified up to isomorphism. They are of two types. Each real algebra of the first type is obtained by considering a complex simple algebra with dimension n as a real algebra with dimension $2n$. Thus every complex simple algebra A_m – D_m, E_6, E_7, E_8, F_4, G_2 is a real simple algebra of twice the dimension. The real simple algebras of the second type are real forms of the complex simple algebras. The Lie algebras of the groups given in Table 9.2 constitute all algebras of type two which are real forms of the classical complex algebras.

TABLE 9.2[a]

Complex form	Dimension	Real forms
A_m	$m(m+2)$	$SU(m+1)$
	$m \geq 1$	$SU(m+1-q, q), q = 1, \ldots, [(m+1)/2]$
		$SL(m+1, R), m > 1$
		$SU^*(m+1), m+1$ even
B_m	$m(2m+1)$	$SO(2m+1)$
	$m \geq 2$	$SO(2m+1-q, q), q = 1, \ldots, m$
C_m	$m(2m+1)$	$USp(m)$
	$m \geq 3$	$Sp(m-q, q), q = 1, \ldots, [m/2]$
		$Sp(m, R)$
D_m	$m(2m-1)$	$SO(2m)$
	$m \geq 4$	$SO(2m-q, q), q = 1, \ldots, m$
		$SO^*(2m)$

[a]See Eqs. (10.2)–(10.6).

In Table 9.2

(10.2)
$$SU(p, q) = \{A \in SL(p+q, \mathfrak{C}): \bar{A}^t G^p A = G^p\},$$
$$G^p_{ij} = \begin{cases} \delta_{ij}, & 1 \leq i \leq p \\ -\delta_{ij}, & p+1 \leq i \leq p+q; \end{cases}$$

(10.3)
$$SU^*(2n) = \{A \in SL(2n, \mathfrak{C}): AJ = J\bar{A}\}, \qquad J = \begin{pmatrix} Z & E_n \\ -E_n & Z \end{pmatrix};$$

(10.4)
$$SO(p, q) = \{A \in SO(p+q, R): A^t G^p A = G^p\};$$

(10.5)
$$Sp(p, q) = \{A \in Sp(p+q, \mathfrak{C}): \bar{A}^t H^p A = H^p\};$$

$$H^p = \begin{pmatrix} -E_p & & Z & \\ & E_q & & \\ & & -E_p & \\ & Z & & E_q \end{pmatrix};$$

(10.6) $\quad SO^*(2n) = \{A \in SO(2n, \mathfrak{C}) : \bar{A}^t J A = J\}$.

The symbol $[k]$ is the largest integer $\leq k$.

No two algebras in the above list are isomorphic. In addition to the classical algebras of type two there are 17 more algebras which are real forms of the exceptional Lie algebras.

A Lie group G is **locally simple** if it contains no proper normal local Lie subgroups. The commutator subgroup $G^{(1)}$ of G is the group generated by all elements of the form $ghg^{-1}h^{-1}$, $g, h \in G$. Here $G^{(1)}$ is a Lie subgroup of G with Lie algebra $\mathcal{G}^{(1)} = [\mathcal{G}, \mathcal{G}]$. We define groups $G^{(n)}$ inductively by $G^{(n+1)} = (G^{(n)})^{(1)}$. Then $L(G^{(n+1)}) = \mathcal{G}^{(n+1)} = [\mathcal{G}^{(n)}, \mathcal{G}^{(n)}]$. The group G is **solvable** if $G^{(n)} = \{e\}$ for n sufficiently large. A Lie group is **semisimple** if it has no proper solvable normal Lie subgroup.

Theorem 9.14. A Lie group G is locally simple (semisimple, solvable) if and only if $\mathcal{G} = L(G)$ is simple (semisimple, solvable).

According to this result, Theorem 9.12, and our list of simple Lie algebras, the semisimple Lie groups can be classified completely, at least in a neighborhood of the identity. Moreover, use of topological methods enables one to list all global connected semisimple Lie groups.

The real simple Lie algebras $su(m+1)$ $(m \geq 1)$, $so(2m+1)$ $(m \geq 2)$, $usp(m)$ $(m \geq 3)$, and $so(2m)$ $(m \geq 4)$ are called **compact** since the global connected Lie groups associated with these algebras are all compact. The groups associated with all other real classical simple Lie algebras are noncompact. Each simple algebra $A_m - D_m$, E_6, E_7, E_8, F_4, G_2 has exactly one compact real form.

As a final comment on the theory we mention the **Casimir operator** C. Let $\mathcal{G}$ be a simisimple Lie algebra of matrices with basis $\mathcal{A}_1, \ldots, \mathcal{A}_n$ and let T be a finite-dimensional rep of $\mathcal{G}$. If the $\{\mathcal{A}_j\}$ satisfy the commutation relations (10.1) then so do the operators $T_j = T(\mathcal{A}_j)$. The Casimir operator associated with T is

$$C = \sum_{j,k=1}^{n} (\mathcal{F}^{-1})_{jk} T_j T_k, \tag{10.7}$$

where $\mathcal{F}$ is the symmetric nonsingular matrix defined in Theorem 9.13. To demonstrate the significance of C we introduce the **Killing form** for $\mathcal{G}$,

$$(\mathcal{A}, \mathcal{B}) = \mathrm{tr}(\mathrm{Ad}\, \mathcal{A}\, \mathrm{Ad}\, \mathcal{B}), \qquad \mathcal{A}, \mathcal{B} \in \mathcal{G}, \tag{10.8}$$

where $(\mathrm{Ad}\, \mathcal{A})(\mathcal{C}) = [\mathcal{A}, \mathcal{C}]$ is a linear operator on $\mathcal{G}$ (see Section 5.6). It is easy to show that $(\mathcal{A}_j, \mathcal{A}_k) = \mathcal{F}_{jk}$, so by Theorem 9.13 the Killing form of $\mathcal{G}$ is nondegenerate. Furthermore,

$$([\mathcal{A}, \mathcal{B}], \mathcal{C}) + (\mathcal{B}, [\mathcal{A}, \mathcal{C}]) = 0,$$

or, exponentiating,

$$(e^{\mathfrak{A}}\mathfrak{B}e^{-\mathfrak{A}}, e^{\mathfrak{A}}\mathfrak{C}e^{-\mathfrak{A}}) = (\exp(\text{Ad}\,\mathfrak{A})\mathfrak{B}, \exp(\text{Ad}\,\mathfrak{A})\mathfrak{C}) = (\mathfrak{B}, \mathfrak{C}) \tag{10.9}$$

for all $\mathfrak{A}, \mathfrak{B}, \mathfrak{C} \in \mathcal{G}$.

We leave it to the reader to show that C is defined independent of the basis in $\mathcal{G}$, i.e., if we introduce a new basis $\{\mathfrak{B}_j\}$, $\mathfrak{B}_j = \sum h_{lj}\mathfrak{A}_l$, $\mathfrak{F}'_{jk} = (\mathfrak{B}_j, \mathfrak{B}_k)$, and compute $\sum (\mathfrak{F}'^{-1})_{jk} T(\mathfrak{B}_j)T(\mathfrak{B}_k)$ we get C again. Now set $\mathfrak{B}_j = \exp(\text{Ad}\, t\mathfrak{C})\mathfrak{A}_j$, where $\mathfrak{C} \in \mathcal{G}$. The elements $\{\mathfrak{B}_j\}$ form a basis for $\mathcal{G}$ and

$$\mathfrak{F}'_{jk} = (\mathfrak{B}_j, \mathfrak{B}_k) = (\mathfrak{A}_j, \mathfrak{A}_k) = \mathfrak{F}_{jk}, \qquad T(\mathfrak{B}_j) = e^{tT(\mathfrak{C})}T_j e^{-tT(\mathfrak{C})} \tag{10.10}$$

by (10.9). Thus,

$$C = \sum_{j,k} (\mathfrak{F}^{-1})_{jk} e^{tT(\mathfrak{C})} T_j T_k e^{-tT(\mathfrak{C})} = e^{tT(\mathfrak{C})} C e^{-tT(\mathfrak{C})}.$$

Differentiating with respect to t and setting $t = 0$ we find

$$T(\mathfrak{C})C = CT(\mathfrak{C}) \tag{10.11}$$

for all $\mathfrak{C} \in \mathcal{G}$. Thus if T is irred, C must be a multiple of the identity operator, $C = aE$. The value of a is a function of T and can be used to label the rep. We have already observed the utility of Casimir operators for the semisimple algebras $sl(2)$ [(3.2), Section 7.3] and $so(3, 1)$ [(3.2), Section 8.3].

The algebras $A_m - D_m$ are said to be of **rank** m and the irred reps of these algebras are designated by m integers $[\lambda_1, \ldots, \lambda_m]$. It can be shown that corresponding to each simple algebra of rank m one can find m independent invariant operators $C_1 = C, C_2, \ldots, C_m$ such that $T(\mathfrak{A})C_j = C_jT(\mathfrak{A})$ for each rep T of $\mathcal{G}$ and such that the values of the C_j for each irred T completely determine $\lambda_1, \ldots, \lambda_m$ (Racah [2]).

Problems

9.1 Fill in the details of the proof in the text that $Sp(m)$ is connected.

9.2 Prove Theorem 9.7.

9.3 Prove Theorem 9.8 in detail.

9.4 Using weights, decompose the reps $\mathbf{3} \otimes \mathbf{3}$, $\mathbf{3} \otimes \overline{\mathbf{3}}$, $\overline{\mathbf{3}} \otimes \overline{\mathbf{3}}$, and $\mathbf{8} \otimes \mathbf{8}$ of $SU(3)$ into irred reps.

9.5 Consider the electromagnetic interaction as an operator proportional to I^3 which perturbs the strong interactions. Derive the selection rules $I \longrightarrow I, I \pm 1$ for the matrix elements of this operator relating states of different isobaric spin. This theory predicts that to a first approximation, electromagnetic interactions permit transitions only between states whose isobaric spins differ by zero or one.

9.6 Compute the possible terms for the following configurations: (1) two electrons in an *nd* state, (2) three electrons in an *np* state, (3) one electron in an *ns* state and one electron in an *n'p* state.

9.7 Prove: The adjoint rep of a semisimple Lie algebra is faithful. The adjoint rep of a simple Lie algebra is irred.

9.8 Prove: A real compact semisimple Lie algebra has a negative-definite Killing form. (Conversely, a real Lie algebra with negative-definite Killing form is compact, but the verification is more difficult; see the work of Helgason [1, p. 122].)

9.9 Show that the matrix rep

$$g(c) = \begin{pmatrix} 1 & \ln|c| \\ 0 & 1 \end{pmatrix}, \qquad c \neq 0, \quad c \in \mathfrak{C},$$

of $GL(1)$ cannot be expressed as a direct sum of irred reps. Does this example contradict the results of Section 9.1?

9.10 Let $[f_1, \ldots, f_m]$ be an irred rep of $U(m)$ with character $\chi(\epsilon_1, \ldots, \epsilon_m)$ given by Theorem 9.4. We can consider $U(m-1)$ as the subgroup of $U(m)$ such that the character of $[f_1, \ldots, f_m] \mid U(m-1)$ is $\chi(\epsilon_1, \ldots, \epsilon_{m-1}, 1)$. Derive the branching law $[f_1, \ldots, f_m] \mid U(m-1) \cong \sum \oplus [h_1, h_2, \ldots, h_{m-1}]$, where the direct sum is taken over all integers h_j such that $f_1 \geq h_1 \geq f_2 \geq h_2 \geq f_3 \geq \cdots \geq f_{m-1} \geq h_{m-1} \geq f_m$ (see the work of Boerner [1]).

Chapter 10

The Harmonic Oscillator Group

10.1 The Harmonic Oscillator

The two most important nonrelativistic systems whose Schrödinger equations can be completely solved are the hydrogen atom and the harmonic oscillator. We have seen that the tractability of the hydrogen atom is related to its high degree of symmetry and we shall reach similar conclusions for the harmonic oscillator.

We start with a system in one-dimensional space. In suitable units the Hamiltonian for a spinless particle subject to a harmonic oscillator potential is

$$\mathbf{H} = -\frac{1}{2}\frac{d^2}{dx^2} + \frac{x^2}{2}. \tag{1.1}$$

The Hilbert space $\mathcal{H}$ consists of functions $\Psi(x)$ square-integrable on the real line. The inner product is

$$(\Psi, \Phi) = \int_{-\infty}^{\infty} \Psi(x)\overline{\Phi(x)}\, dx. \tag{1.2}$$

Although the eigenvalue problem

$$\mathbf{H}\Psi = \lambda\Psi \tag{1.3}$$

can be solved with special function theory, we can achieve greater insight by adopting a formal Lie-algebraic approach. (Our approach can be made rigorous by careful attention to the domains of definition of **H** and other unbounded operators on $\mathcal{H}$ which we shall define shortly; Helwig [1].) Consider the operators

$$\mathbf{J}^{\pm} = \pm \frac{1}{\sqrt{2}}\left(\frac{d}{dx} \mp x\right) \tag{1.4}$$

on $\mathcal{H}$. It is straightforward to verify the commutation relations

$$[\mathbf{J}^3, \mathbf{J}^{\pm}] = \pm \mathbf{J}^{\pm}, \qquad [\mathbf{J}^+, \mathbf{J}^-] = -\mathbf{E}, \tag{1.5}$$

where $\mathbf{E}$ is the identity operator and

$$\mathbf{J}^3 = \mathbf{H}. \tag{1.6}$$

Furthermore, from the abstract relations (1.5) alone we can check that the operator

$$\mathbf{C} = \mathbf{J}^+\mathbf{J}^- - \mathbf{E}\mathbf{J}^3 = \mathbf{J}^-\mathbf{J}^+ - \mathbf{E} - \mathbf{E}\mathbf{J}^3 \tag{1.7}$$

commutes with $\mathbf{J}^{\pm}$ and $\mathbf{J}^3$. Thus $\mathbf{J}^{\pm}$, $\mathbf{J}^3$ and $\mathbf{E}$ form a basis for a four-dimensional complex Lie algebra $\mathcal{G}$ and $\mathbf{C}$ is an invariant operator for $\mathcal{G}$ (analogous to the Casimir operator for semisimple algebras). Corresponding to an irred rep of $\mathcal{G}$ we expect $\mathbf{C}$ to be a multiple of the identity operator. In fact for the model of $\mathcal{G}$ defined by (1.1) and (1.4) we find

$$\mathbf{C} = -\tfrac{1}{2}\mathbf{E}. \tag{1.8}$$

Under the assumption that $\Psi(x)$, $\Phi(x)$, and their first derivatives vanish as $|x| \longrightarrow \infty$, the formal relations

$$(\Psi, \mathbf{J}^3\Phi) = (\mathbf{J}^3\Psi, \Phi), \qquad (\Psi, \mathbf{J}^+\Phi) = (\mathbf{J}^-\Psi, \Phi) \tag{1.9}$$

can easily be verified. (Integrate by parts.) Let Ψ be a normalized eigenvector of $\mathbf{J}^3 = \mathbf{H}$ with eigenvalue λ. The commutation relations (1.5) imply

$$\mathbf{J}^3(\mathbf{J}^{\pm}\Psi) = (\lambda \pm 1)\mathbf{J}^{\pm}\Psi, \tag{1.10}$$

so $\mathbf{J}^{\pm}$ are raising and lowering operators in the usual sense. Given an eigenvector Ψ with eigenvalue λ we can obtain a ladder of eigenvectors with eigenvalues $\lambda + n$. We assume the vectors $\mathbf{J}^{\pm}\Psi$ still belong to $\mathcal{H}$. Then

$$(\mathbf{J}^+\Psi, \mathbf{J}^+\Psi) = (\mathbf{J}^-\mathbf{J}^+\Psi, \Psi) = (\{\mathbf{J}^3 + \tfrac{1}{2}\mathbf{E}\}\Psi, \Psi) = \lambda + \tfrac{1}{2}. \tag{1.11}$$

Similarly,

$$0 \leq ||\mathbf{J}^-\Psi||^2 = \lambda - \tfrac{1}{2}. \tag{1.12}$$

It follows that $\lambda \geq \frac{1}{2}$, so the eigenvalues of $\mathbf{J}^3$ are bounded below.

If λ_0 is the lowest eigenvalue of $\mathbf{J}^3$ and Ψ_0 is a corresponding eigenvector then $\mathbf{J}^-\Psi_0 = 0$, since $\lambda_0 - 1$ is not an eigenvalue. By (1.12), $\lambda_0 = \frac{1}{2}$. Using (1.4) we can solve this first-order differential equation for Ψ_0 and obtain

$$\Psi_0(x) = \pi^{-1/4}\exp(-x^2/2), \tag{1.13}$$

where the factor $\pi^{-1/4}$ is chosen so $||\Psi_0|| = 1$. From (1.7) and (1.8), $\mathbf{J}^3\Psi_0 = (\mathbf{J}^+\mathbf{J}^- + \frac{1}{2}\mathbf{E})\Psi_0 = \frac{1}{2}\Psi_0$, so $\mathbf{J}^3$ has a lowest eigenvalue $\frac{1}{2}$ with multiplicity one. From (1.10) and (1.11) we can define normalized eigenvectors Ψ_n with eigenvalues $n + \frac{1}{2}$ recursively by

$$\mathbf{J}^+\Psi_n = (n+1)^{1/2}\Psi_{n+1}, \qquad n = 0, 1, 2, \ldots . \tag{1.14}$$

Thus we use the raising operator $\mathbf{J}^+$ to move up the ladder of eigenvalues. According to (1.11), $\|\mathbf{J}^+\Psi_n\|^2 = n + 1 > 0$, so this process never ends. Furthermore, the commutation relations imply the formulas

$$\mathbf{J}^3\Psi_n = (n + \tfrac{1}{2})\Psi_n, \qquad \mathbf{J}^-\Psi_n = n^{1/2}\Psi_{n-1}. \tag{1.15}$$

We have shown that $\mathbf{H}$ has eigenvalues $n + \frac{1}{2}$, $n = 0, 1, \ldots$. It is left to the reader to verify that there are no other eigenvalues and each eigenspace is one-dimensional.

Substituting the operators (1.1) and (1.4) into (1.14) and (1.15), we obtain a second-order differential equation and two recurrence formulas for the special functions $\Psi_n(x)$. We can obtain a generating function from the first-order operator $\mathbf{J}^+$. A simple computation using Theorem 5.31 yields

$$[(\exp \alpha \mathbf{J}^+)\Psi](x) = \exp(-\tfrac{1}{4}\alpha^2 - 2^{-1/2}\alpha x)\Psi(x + 2^{-1/2}\alpha).$$

On the other hand, (1.14) implies

$$[(\exp \alpha \mathbf{J}^+)\Psi_n](x) = \sum_{k=0}^{\infty} \left[\frac{(n+k)!}{n!}\right]^{1/2} \frac{\alpha^k}{k!}\Psi_{n+k}(x).$$

Comparing these equations, we have the identity ($\beta = 2^{-1/2}\alpha$)

$$\left[\exp\left(-\frac{\beta^2}{2} - \beta x\right)\right]\Psi_n(x+\beta) = \sum_{k=0}^{\infty}\left[\frac{2^k(n+k)!}{n!}\right]^{1/2}\frac{\beta^k}{k!}\Psi_{n+k}(x). \tag{1.16}$$

In the special case $n = 0$, (1.13) yields

$$\pi^{-1/4}\exp(-\beta^2 - 2\beta x - \tfrac{1}{2}x^2) = \sum_{k=0}^{\infty} 2^{k/2}\beta^k(k!)^{-1/2}\Psi_k(x), \tag{1.17}$$

a simple generating function for the $\Psi_k(x)$. Comparing this with the well-known generating function

$$\exp(-\beta^2 + 2\beta x) = \sum_{k=0}^{\infty} \beta^k H_k(x)/k!$$

For the Hermite polynomials $H_k(x)$ (Erdélyi *et al.* [2, p. 194]) we obtain

$$\Psi_k(x) = \pi^{-1/4}(k!)^{-1/2}(-1)^k 2^{-k/2}[\exp(-x^2/2)]H_k(x). \tag{1.18}$$

The above series converge for all x and β. Since the $\{\Psi_n(x)\}$ form an ON set in $\mathcal{H}$ we easily obtain the formula

$$\int_{-\infty}^{\infty} H_n(x)H_k(x)\exp(-x^2)\,dx = \pi^{1/2}2^n n!\,\delta_{nk}. \tag{1.19}$$

Our Lie-algebraic analysis of the harmonic oscillator problem has not only determined the eigenvalues of $\mathbf{H}$ but also enabled us to derive the eigenfunctions and a number of their properties in a very simple manner.

The generalized Lie derivatives (1.1) and (1.4) determine the action of a connected four-parameter Lie group G on $\mathcal{H}$ called the **harmonic oscillator group**. Here G is *not* a symmetry group of the Hamiltonian since $\mathbf{J}^{\pm}$ do not commute with $\mathbf{H}$. However, a knowledge of the rep theory of G enables us to determine not only the multiplicities of the eigenvalues but also the eigenvalues themselves. Such a group G is called a **dynamical symmetry group** of the quantum mechanical system. [To be more precise, we are actually interested in the real Lie algebra $\mathcal{G}^r$ generated by the skew-Hermitian operators

$$i\mathbf{J}^3, \quad \mathbf{E}, \quad \mathbf{J}^+ - \mathbf{J}^-, \quad i(\mathbf{J}^+ + \mathbf{J}^-). \tag{1.20}$$

These operators determine a unitary irred rep of the real dynamical symmetry group G^r on $\mathcal{H}$, where $L(G^r) = \mathcal{G}^r$. However, for Lie-algebraic purposes it is more convenient to work with the complexified algebra $\mathcal{G}$ determined by (1.5).]

The above analysis shows that the harmonic oscillator system in one dimension forms a model of an irred rep of $\mathcal{G}$. (The proof of irreducibility is left to the reader.) Another model of this same rep is provided by the annihilation and creation operators for bosons. In this model the annihilation operator $\mathbf{a}$ and the creation operator $\mathbf{a}^*$ act on a Hilbert space $\mathcal{H}$ and satisfy the commutation relations

$$[\mathbf{a}^*, \mathbf{a}] = -\mathbf{E}. \tag{1.21}$$

Furthermore the number-of-particles operator $\mathbf{N} = \mathbf{a}^*\mathbf{a}$ satisfies the commutation relations

$$[\mathbf{N}, \mathbf{a}^*] = \mathbf{a}^*, \qquad [\mathbf{N}, \mathbf{a}] = -\mathbf{a}. \tag{1.22}$$

There is an ON basis $\{|n\rangle, n = 0, 1, 2, \ldots\}$ for $\mathcal{H}$ such that

$$\begin{gathered} \mathbf{a}|n\rangle = n^{1/2}|n-1\rangle, \qquad \mathbf{a}^*|n\rangle = (n+1)^{1/2}|n+1\rangle, \\ \mathbf{N}|n\rangle = n|n\rangle. \end{gathered} \tag{1.23}$$

The eigenstates $|n\rangle$ of $\mathbf{N}$ are considered to be states of n bosons, which explains the names for $\mathbf{a}$, $\mathbf{a}^*$, and $\mathbf{N}$. Clearly, the operators $\mathbf{a}$, $\mathbf{a}^*$, $\mathbf{N}$, and $\mathbf{E}$ generate the Lie algebra $\mathcal{G}$ and expressions (1.23) determine an irred rep of $\mathcal{G}$ equivalent to that of the one-dimensional harmonic oscillator.

The harmonic oscillator in three-space has Hamiltonian

$$\mathbf{H} = -\frac{1}{2}\Delta + \frac{r^2}{2} = \frac{1}{2}\sum_{j=1}^{3}\left(-\frac{d^2}{dx_j^2} + x_j^2\right). \tag{1.24}$$

We can view this system as composed of three one-dimensional noninteracting harmonic oscillators. Thus the eigenfunctons of $\mathbf{H}$ are of the form

$$\Psi_{n_1n_2n_3}(x) = \Psi_{n_1}(x_1)\Psi_{n_2}(x_2)\Psi_{n_3}(x_3) \tag{1.25}$$

with eigenvalue $\lambda_N = n_1 + n_2 + n_3 + \frac{3}{2} = N + \frac{3}{2}$, $n_j = 0, 1, \ldots$. The multiplicity of λ_N is equal to the number of ways we can select nonnegative integers n_j such that $n_1 + n_2 + n_3 = N$. A simple combinatorial argument gives the multiplicity $(N+1)(N+2)/2$. Here $G \times G \times G$ is a dynamical symmetry group for this system, which enables us to compute the eigenvalues and their multiplicities. However, to get a better understanding of the multiplicities it is useful to compute the ordinary symmetry group of **H**. From (1.24) it is clear that the angular momentum operators commute with **H**, so $SO(3)$ [or $SU(2)$] is a symmetry group. However, the dimensions of the reps $D^{(l)}$ do not coincide with the multiplicities $(N+1)(N+2)/2$. This suggests the existence of a larger symmetry group.

To investigate this group we consider the annihilation and creation operators

$$(1.26)\quad \mathbf{a}_j = -2^{-1/2}\left(\frac{d}{dx_j} + x_j\right), \qquad \mathbf{a}_j^* = 2^{-1/2}\left(\frac{d}{dx_j} - x_j\right), \qquad 1 \leq j \leq 3,$$

with commutation relations

$$(1.27)\qquad [\mathbf{a}_j, \mathbf{a}_k^*] = \delta_{jk}\mathbf{E}, \qquad [\mathbf{a}_j, \mathbf{a}_k] = [\mathbf{a}_j^*, \mathbf{a}_k^*] = \mathbf{0}.$$

From (1.24)

$$(1.28)\qquad \mathbf{H} = \tfrac{1}{2}\sum_{j=1}^{3}(\mathbf{a}_j^*\mathbf{a}_j + \mathbf{a}_j\mathbf{a}_j^*) = \sum_{j=1}^{3}\mathbf{a}_j^*\mathbf{a}_j + \tfrac{3}{2}\mathbf{E}.$$

It is easy to verify that the nine operators $\mathbf{E}_{jk} = \mathbf{a}_j^*\mathbf{a}_k$ commute with **H** and satisfy the commutation relations

$$(1.29)\qquad [\mathbf{E}_{jk}, \mathbf{E}_{hl}] = \delta_{kh}\mathbf{E}_{jl} - \delta_{lj}\mathbf{E}_{hk}$$

[see (7.3), Section 9.7]. (Note that the $\mathbf{E}_{jk}$ preserve N, while operators such as $\mathbf{a}_j\mathbf{a}_k$ do not.) Clearly, the $\mathbf{E}_{jk}$ generate a complex Lie algebra isomorphic to $gl(3, \mathfrak{C})$. The skew-Hermitian operators in this algebra form a real Lie algebra isomorphic to $u(3)$. A basis for $u(3)$ is given by

$$(1.30)\quad i(\mathbf{E}_{jk} + \mathbf{E}_{kj}), \quad \mathbf{E}_{jk} - \mathbf{E}_{kj}, \quad k \neq j; \qquad i\mathbf{E}_{jj}, \quad j, k = 1, 2, 3.$$

The angular momentum operators, given by $\mathbf{E}_{jk} - \mathbf{E}_{kj}$, $k \neq j$, generate a subalgebra isomorphic to $so(3)$.

Under the action of the $\mathbf{E}_{jk}$ the eigenspace of **H** corresponding to eigenvalue λ_N is decomposed into a direct sum of $u(3)$-irred subspaces. Using the results of Section 9.1 we can explicitly carry out this decomposition. The highest weight vector is easily shown to be the eigenfunction with $n_1 = N$, $n_2 = n_3 = 0$. Thus, the rep $[N, 0, 0]$ of $u(3)$ occurs exactly once. Moreover, from (2.24), Section 9.2, $\dim[N, 0, 0] = (N+1)(N+2)/2$, which is the multiplicity of λ_N. Thus, the eigenspace of λ_N transforms as $[N, 0, 0]$. As we have seen, the angular momentum operators generate a subalgebra $so(3)$

of $u(3)$. To determine the branching rule for the subalgebra of angular momentum operators we could compute the weight vectors corresponding to the generator $\mathbf{L}^3 = \mathbf{E}_{12} - \mathbf{E}_{21}$ (this is not easy). The results are

$$(1.31)\qquad [N, 0, 0] \mid so(3) \cong \begin{cases} \sum_{j=0}^{N/2} \oplus D^{(N-2j)}, & N \text{ even} \\ \sum_{j=0}^{(N-1)/2} \oplus D^{(N-2j)}, & N \text{ odd.} \end{cases}$$

Thus, for N even the reps $D^{(l)}$ occur with l even. An alternate proof of (1.31) can be obtained from the character formula for $U(3)$. The subgroup $SO(3)$ is embedded in $U(3)$ in the natural way and it is straightforward to expand $\chi^{N00} \mid SO(3)$ as a sum of simple characters of $SO(3)$.

We say that $U(3)$ is the symmetry group of the three-dimensional harmonic oscillator. The global action of $U(3)$ on $\mathcal{H}$ is fairly difficult to determine in this case since the $\mathbf{E}_{jl}$ are second-order partial differential operators to which local Lie theory does not apply. The group action is expressable in terms of integral operators. For details see the work of Bargmann [1] or Miller [1]. These references also give the action of the harmonic oscillator group on $\mathcal{H}$.

10.2 Representations of the Harmonic Oscillator Group

The Lie algebra $\mathcal{G}$ of the complex harmonic oscillator group G is defined by the commutation relations

$$(2.1)\qquad [\mathcal{J}^3, \mathcal{J}^\pm] = \pm\mathcal{J}^\pm, \qquad [\mathcal{J}^+, \mathcal{J}^-] = -\mathcal{E}, \qquad [\mathcal{E}, \mathcal{J}^\pm] = [\mathcal{E}, \mathcal{J}^3] = 0.$$

We present a brief survey of irred reps of $\mathcal{G}$ which occur in physical theories. Let ρ be a rep of $\mathcal{G}$ on a complex vector space V and set

$$(2.2)\qquad J^\pm = \rho(\mathcal{J}^\pm), \qquad J^3 = \rho(\mathcal{J}^3), \qquad I = \rho(\mathcal{E}).$$

These operators satisfy relations (2.1) again.

The faithful irred reps of $\mathcal{G}$ are all infinite-dimensional. Indeed, if ρ is irred and finite-dimensional then I must be a multiple μE of the identity operator on V, since I commutes with $\rho(\alpha)$ for all $\alpha \in \mathcal{G}$. Thus

$$(2.3)\qquad \mathrm{tr}([J^+, J^-]) = \mathrm{tr}(\mu E) = \mu \dim V$$

and $\mu = 0$ because the trace of a commutator is zero. Hence, $I = 0$ and ρ is not faithful. (A rep ρ of a Lie algebra $\mathcal{G}$ is **faithful** if $\rho(\alpha) \neq 0$ for every $\alpha \neq 0$ in $\mathcal{G}$.)

We make no attempt to classify all irred reps of $\mathcal{G}$ and simply examine a few reps of particular importance. An easy way to construct such reps of $\mathcal{G}$ is via realizations in terms of generalized Lie derivatives in one complex variable. [We tried this same approach for $sl(2)$ in Section 5.10.] Clearly, the generalized Lie derivatives

(2.4) $$J^3 = \lambda + z\frac{d}{dz}, \qquad J^+ = \mu z, \qquad J^- = \frac{\xi}{z} + \frac{d}{dz}, \qquad I = \mu,$$

satisfy the commutation relations (2.1) for all constants λ, μ, and ξ.

For an arbitrary rep ρ the operator

(2.5) $$C = J^+J^- - IJ^3$$

commutes with all $\rho(\alpha)$, $\alpha \in \mathcal{G}$, as the reader can check. If ρ is irred we expect that C is a multiple ωE of the identity operator on V. (However, we have not proved this since Theorem 3.5 applies only to finite-dimensional reps.) For our model (2.4) we find $C = \mu(\xi - \lambda)$.

The operators (2.4) determine a local multiplier rep of G. To compute this rep we need an explicit definition of G. Recall that G is only determined locally by $\mathcal{G}$. Among the linear Lie groups with Lie algebra $\mathcal{G}$ we select the one, unique up to isomorphism, which is simply connected.

Definition. The complex **harmonic oscillator group** G consists of all matrices

(2.6) $$g(a, b, c, \tau) = \begin{pmatrix} 1 & ce^{\tau} & a & \tau \\ 0 & e^{\tau} & b & 0 \\ 0 & 0 & 1 & 0 \\ 0 & 0 & 0 & 1 \end{pmatrix}, \qquad a, b, c, \tau \in \mathfrak{C}.$$

Here G is a four-parameter complex linear Lie group. In terms of the parameters,

(2.7)
$$g(a, b, c, \tau)g(a', b', c', \tau') = g(a + a' + cb'e^{\tau}, b + b'e^{\tau}, c + c'e^{-\tau}, \tau + \tau')$$
$$g^{-1}(a, b, c, \tau) = g(bc - a, -be^{-\tau}, -ce^{\tau}, -\tau).$$

The matrices $\mathcal{J}^{\pm}$, $\mathcal{J}^3$, $\mathcal{E}$ defined by

(2.8) $$g(a, b, c, \tau) = (\exp b\mathcal{J}^+)(\exp c\mathcal{J}^-)(\exp \tau\mathcal{J}^3)(\exp a\mathcal{E})$$

form a basis for $\mathcal{G}$ satisfying the commutation relations (2.1).

Let $\mathbf{T}$ be the local multiplier rep of G determined by the generalized Lie derivatives (2.4). The group identity (2.8) implies

(2.9) $$\mathbf{T}(g) = (\exp bJ^+)(\exp cJ^-)(\exp \tau J^3)(\exp aI)$$

for $|b|, |c|, |\tau|, |a|$ sufficiently small. For simplicity we choose $\lambda = -\omega$, $\xi = 0$ in (2.4). Applying local Lie theory to compute the factors of (2.9) and composing the result, we find

(2.10) $$[\mathbf{T}(g)f](z) = \exp[\mu(bz + a) - \omega\tau]f(e^{\tau}z + e^{\tau}c)$$

for $f(z)$ analytic in some neighborhood of $z = 0$. [Compare the analogous computation for $SL(2)$ in Section 5.10.] Since **T** is a local rep we have

(2.11) $$\mathbf{T}(gg')f = \mathbf{T}(g)[\mathbf{T}(g')f]$$

for g and g' in a suitably small neighborhood of the identity. Moreover, if we restrict f to the space $\mathfrak{A}$ of entire functions then (2.10) is defined for all $g \in G$ and $\mathbf{T}(g)f \in \mathfrak{A}$. In this case the identity (2.11) holds for all $g, g' \in G$, as the reader can prove directly from (2.7) and (2.10).

Every $f \in \mathfrak{A}$ has a unique power series expansion

$$f(z) = \sum_{n=0}^{\infty} a_n z^n$$

which converges for all $z \in \mathfrak{C}$. Thus the functions $h_n(z) = z^n$, $n \geq 0$, form a basis for $\mathfrak{A}$. With respect to this basis we define matrix elements $T_{lk}(g)$ by

(2.12) $$[\mathbf{T}(g)h_k](z) = \sum_{l=0}^{\infty} T_{lk}(g)h_l(z), \qquad k = 0, 1, \ldots ;$$

(2.13) $$\{\exp[\mu(bz + a) + (k - \omega)\tau]\}(z + c)^k = \sum_{l=0}^{\infty} T_{lk}(g)z^l.$$

The group property (2.11) yields the addition theorem

(2.14) $$T_{lk}(gg') = \sum_{j=0}^{\infty} T_{lj}(g)T_{jk}(g'), \qquad g, g' \in G.$$

We can obtain the matrix elements explicitly by expanding the left-hand side of (2.13) in a power series and computing the coefficient of z^l:

(2.15) $$T_{lk}(g) = \{\exp[\mu a + (k - \omega)\tau]\}(\mu b)^{l-k} \sum_s \frac{(\mu bc)^s k!}{(l - k + s)!\,(k - s)!\,s!},$$

where the sum is taken over all integers s such that the summand is defined. From Erdélyi *et al.* [1, p. 268] we find

(2.16) $$T_{lk}(g) = \{\exp[\mu a + (k - \omega)\tau]\}c^{k-l}L_l^{(k-l)}(-\mu bc),$$

where $L_m^{(n)}(x)$ is an associated Laguerre polynomial. Substituting this result into (2.13) we obtain the generating function

(2.17) $$e^{-bz}(z + 1)^k = \sum_{l=0}^{\infty} L_l^{(k-l)}(b)z^l.$$

Furthermore, the addition theorem (2.14) yields

(2.18)
$$e^{-cb'}(c + c')^n L_l^{(n)}[(b + b')(c + c')] = \sum_{j=0}^{\infty} c^{j-l}L_l^{(j-l)}[bc'](c)^{l+n-j}L_j^{(l+n-j)}[b'c'],$$

where the integers $l, l + n$ are nonnegative and $b, b', c, c' \in \mathfrak{C}$.

To exhibit the rep $\uparrow_{\omega,\mu}$ of $\mathfrak{G}$ induced by **T** we label the basis vectors in terms of their eigenvalues with respect to J^3:

(2.19) $$f_m(z) = h_n(z) = z^n, \qquad m = n - \omega.$$

Then

(2.20) $J^3 = -\omega + z(d/dz), \qquad J^+ = \mu z, \qquad J^- = d/dz, \qquad I = \mu,$

and direct computation yields

(2.21)
$$J^3 f_m = m f_m, \qquad J^+ f_m = \mu f_{m+1}, \qquad J^- f_m = (\omega + m) f_{m-1},$$
$$I f_m = \mu f_m, \qquad C f_m = \mu\omega f_m,$$

where $m = -\omega + n$, $n = 0, 1, \ldots$, and $\mu, \omega \in \mathfrak{C}$ with $\mu \neq 0$. As the reader can verify, the Lie algebra rep $\uparrow_{\omega,\mu}$ is irred on the infinite-dimensional vector space of all finite linear combinations of the basis vectors $\{f_m\}$.

The rep $\uparrow_{0,1}$ corresponds to the harmonic oscillator problem in one dimension. Indeed, setting $\omega = 0$, $\mu = 1$, and $|n\rangle = f_n(z)(n!)^{-1/2}$ we obtain

(2.22)
$$J^3|n\rangle = n|n\rangle, \qquad J^+|n\rangle = (n+1)^{1/2}|n+1\rangle, \qquad J^-|n\rangle = n^{1/2}|n-1\rangle,$$

in agreement with (1.23). [We have normalized our basis vectors $\{f_m\}$ so no square roots appear in (2.21).]

We have defined a class $\uparrow_{\omega,\mu}$ of irred reps of $\mathcal{G}$ and used a simple model to compute the matrix elements $T_{lk}(g)$ of this rep extended to G. These matrix elements are uniquely determined by expressions (2.21) and are model-independent. Thus, the model of $\uparrow_{0,1}$ provided by the operators (1.4), (1.6), and basis vectors

(2.23) $$f_n(x) = (-1)^n 2^{-n/2}(\exp -\tfrac{1}{2}x^2) H_n(x)$$

must have matrix elements (2.16). From (1.4) we find

(2.24)
$$[\mathbf{T}(g)f](x) = \{\exp[\tfrac{1}{4}(c^2 - b^2 - 2bc) - 2^{-1/2}x(b+c)]\} f(x + (b-c)2^{-1/2})$$

for $g(0, b, c, 0) = (\exp b\mathcal{J}^+) \exp c\mathcal{J}^-$. Substituting (2.16), (2.23), and (2.24) into

(2.25) $$\mathbf{T}(g) f_k = \sum_{l=0}^{\infty} T_{lk}(g) f_l$$

we obtain (after some simplification)

(2.26) $$[\exp(-b^2 - 2bx)] H_k(x + b + c) = \sum_{l=0}^{\infty} c^{k-l} L_l^{(k-l)}(bc) H_l(x).$$

This expression converges for all b, c, and x.

Problems

10.1 Prove (1.31).

10.2 Show that the 4×4 matrices (2.6) define a rep of the harmonic oscillator group which cannot be expressed as a direct sum of irred reps.

10.3 Compute the recurrence relations and differential equation for Hermite polynomials which result on application of the operators $\mathbf{J}^{\pm}$, $\mathbf{J}^3$, $\mathbf{C}$ (Section 10.1) to Ψ_n.

10.4 Show that $E^+(2)$, the proper Euclidean group in the plane, is isomorphic to the group of matrices

$$g(x, y, \theta) = \begin{pmatrix} \cos\theta & -\sin\theta & x \\ \sin\theta & \cos\theta & y \\ 0 & 0 & 1 \end{pmatrix}.$$

Check that these matrices define a rep of $E^+(2)$ which cannot be expressed as a direct sum of irred reps.

10.5 Show that the generalized Lie derivatives

$$J_1 = i\rho\cos\alpha, \qquad J_2 = i\rho\sin\alpha, \qquad J_3 = -\partial/\partial\alpha,$$

ρ a nonzero real constant, $i = \sqrt{-1}$, $0 < \alpha \leq 2\pi \pmod{2\pi}$, span a Lie algebra isomorphic to $L(E^+(2))$. Compute the operators $\mathbf{T}(g)$ of the multiplier rep of $E^+(2)$ determined by the J_k. [Use the coordinates $g[r, \varphi, \theta] = g(x, y, \theta)$, where $x + iy = re^{i\varphi}$, $r \geq 0$, φ real.]

10.6 Verify that the $\mathbf{T}(g)$ computed in the preceding problem define a unitary rep of $E^+(2)$ on the Hilbert space $L_2[0, 2\pi]$ (see Section 6.2). Compute the matrix elements $T_{n,m}(g) = \langle \mathbf{T}(g)f_m, f_n\rangle$ with respect to the ON basis $f_m(\alpha) = e^{im\alpha}$, $m = 0, \pm 1, \pm 2, \ldots$, and show that these elements can be expressed in terms of Bessel functions. What properties of Bessel functions follow from the unitarity of the $\mathbf{T}(g)$ and the group property $\mathbf{T}(g_1)\mathbf{T}(g_2) = \mathbf{T}(g_1 g_2)$?

Appendix

Hilbert Space

We present some basic ideas and definitions from Hilbert space theory. For a more detailed exposition see the work of Korevaar [1] or Naylor and Sell [1]. All vector spaces will be assumed complex, although the facts for real spaces are essentially the same.

Definition. A vector space $\mathcal{V}$ is an **inner product space (pre-Hilbert space)** with inner product $(-, -)$ if $(\mathbf{u}, \mathbf{v}) \in \mathfrak{C}$ for each $\mathbf{u}, \mathbf{v} \in \mathcal{V}$ and

(a) $(\mathbf{u}, \mathbf{v}) = \overline{(\mathbf{v}, \mathbf{u})}$.
(b) $(a_1\mathbf{u}_1 + a_2\mathbf{u}_2, \mathbf{v}) = a_1(\mathbf{u}_1, \mathbf{v}) + a_2(\mathbf{u}_2, \mathbf{v})$, $a_j \in \mathfrak{C}$, $\mathbf{u}_j, \mathbf{v} \in \mathcal{V}$.
(c) $(\mathbf{u}, \mathbf{u}) \geq 0$ and $(\mathbf{u}, \mathbf{u}) = 0$ only if $\mathbf{u} = \boldsymbol{\theta}$.

We define the **length (norm)** of a vector by $\|\mathbf{u}\| = [(\mathbf{u}, \mathbf{u})]^{1/2}$. Clearly, $\|\mathbf{u}\| \geq 0$ and $\|\mathbf{u}\| = 0$ if and only if $\mathbf{u} = \boldsymbol{\theta}$.

Lemma A1 (Schwarz inequality). If $\mathbf{u}, \mathbf{v} \in \mathcal{V}$ then $|(\mathbf{u}, \mathbf{v})| \leq \|\mathbf{u}\| \cdot \|\mathbf{v}\|$. Equality is obtained if and only if $\mathbf{u}$ and $\mathbf{v}$ are linearly dependent.

Lemma A2 (Triangle inequality). If $\mathbf{u}, \mathbf{v} \in \mathcal{V}$ then $\|\mathbf{u} + \mathbf{v}\| \leq \|\mathbf{u}\| + \|\mathbf{v}\|$.

Every finite-dimensional vector space with an inner product is a pre-Hilbert space. We examine some examples of pre-Hilbert spaces which are not finite-dimensional.

Example 1. By l_2 we mean the set of all sequences $\mathbf{x} = (x_1, x_2, \ldots)$ of complex numbers x_i such that $\sum_{i=1}^{\infty} |x_i|^2 < \infty$. Here l_2 is a vector space with operations

(A.1) $$a\mathbf{x} = (ax_1, ax_2, \ldots), \qquad \mathbf{x} + \mathbf{y} = (x_1 + y_1, x_2 + y_2, \ldots),$$

where $\mathbf{x} = (x_1, x_2, \ldots)$, $\mathbf{y} = (y_1, y_2, \ldots)$. Indeed, $\sum_{i=1}^{\infty} |x_i + y_i|^2 \leq \sum_{i=1}^{\infty} (|x_i|^2 + |y_i|^2 + 2|x_i y_i|) \leq 2 \sum_{i=1}^{\infty} (|x_i|^2 + |y_i|^2) < \infty$ for $\mathbf{x}, \mathbf{y} \in l_2$. Here we have used the property $(a - b)^2 = a^2 + b^2 - 2ab \geq 0$ for all real numbers a, b. The space l_2 is a pre-Hilbert space with inner product

(A.2) $$(\mathbf{x}, \mathbf{y}) = \sum_{i=1}^{\infty} x_i \bar{y}_i.$$

Indeed, the series for $(\mathbf{x}, \mathbf{y})$ converges for all $\mathbf{x}, \mathbf{y} \in l_2$.

Example 2. Let $[a, b]$, $a < b$, be a closed interval on the real line and let $C[a, b]$ be the set of all functions $f(x)$ which are defined and continuous on $[a, b]$. Then $C[a, b]$ is a vector space with operations

(A.3) $$\begin{aligned} (\alpha f)(x) &= \alpha f(x), & \alpha &\in \mathbb{C} \\ (f + g)(x) &= f(x) + g(x), & f, g &\in C[a, b]. \end{aligned}$$

The expression

(A.4) $$(f, g) = \int_a^b f(x)\overline{g(x)}\, dx$$

defines an inner product on $C[a, b]$.

Example 3. Let $\mathfrak{M}$ be a closed bounded connected subset of R_m whose boundary is piecewise smooth and let $C(\mathfrak{M})$ be the set of all functions $f(x)$, $x \in R_m$, which are defined and continuous on $\mathfrak{M}$. Using the definitions (A.3) we can make $C(\mathfrak{M})$ into a vector space. Furthermore if $w \in C(\mathfrak{M})$ and $w(x) > 0$ for all $x \in \mathfrak{M}$ then the expression

(A.5) $$(f, g) = \int_{\mathfrak{M}} f(x)\overline{g(x)} w(x)\, dx, \qquad dx = dx_1 \cdots dx_m$$

defines an inner product on $C(\mathfrak{M})$. Here $w(x)$ is a **weight function.**

Example 4. Let G be a compact linear Lie group and let $C(G)$ be the vector space of all continuous functions on G. Then $C(G)$ is a pre-Hilbert space with respect to the inner product

(A.6) $$(f, g) = \int_G f(A)\overline{g(A)}\, \delta A, \qquad f, g \in C(G),$$

where δA is the normalized invariant measure on G.

Example 5. Let $C^2(R_m)$ be the set of all functions $f(x)$ defined and continuous in R_m and such that

$$\int_{R_m} |f(x)|^2\, dx < \infty, \qquad dx = dx_1 \cdots dx_m.$$

(Note that continuous functions on R_m need not be bounded.) Then $C^2(R_m)$ is a vector space under the usual operations (A.3). Indeed, $|2ab| \leq a^2 + b^2$, for a, b real, so

$$(A.7) \qquad 2\int_{R_m} |f(x)g(x)|\,dx \leq \int_{R_m} |f(x)|^2\,dx + \int_{R_m} |g(x)|^2\,dx < \infty$$

for $f, g \in C^2(R_m)$. Thus the integrals $\int_{R_m} f\bar{g}\,dx$ and $\int_{R_m} |fg|\,dx$ converge. This shows that

$$\int_{R_m} |f(x) + g(x)|^2\,dx \leq \int_{R_m} (|f(x)|^2 + 2|f(x)g(x)| + |g(x)|^2)\,dx < \infty$$

and $f + g \in C^2(R_m)$. Furthermore, by (A.7) the expression

$$(A.8) \qquad (f, g) = \int_{R_m} f(x)\overline{g(x)}\,dx$$

defines an inner product on $C^2(R_m)$.

Definition. The pre-Hilbert spaces $\mathcal{V}, \mathcal{W}$ are **isomorphic** (as pre-Hilbert spaces) if there is a vector space isomorphism $\mathbf{T}: \mathcal{V} \to \mathcal{W}$ such that $(\mathbf{v}_1, \mathbf{v}_2) = (\mathbf{T}\mathbf{v}_1, \mathbf{T}\mathbf{v}_2)$ for all $\mathbf{v}_1, \mathbf{v}_2 \in \mathcal{V}$. Here the first inner product belongs to $\mathcal{V}$ and the second to $\mathcal{W}$.

Example. Every m-dimensional pre-Hilbert space $\mathcal{V}_m$ is isomorphic to $\mathfrak{C}_m$. (Choose an ON basis for $\mathcal{V}_m$.)

Let $\{\mathbf{v}_j\}$, $j = 1, 2, \ldots$, be a sequence of vectors in the pre-Hilbert space $\mathcal{V}$. The sequence $\{\mathbf{v}_j\}$ is said to be **Cauchy** if for every $\epsilon > 0$ there exists a positive integer N_ϵ with the property $||\mathbf{v}_k - \mathbf{v}_j|| < \epsilon$ whenever $k, j > N_\epsilon$. A Cauchy sequence **converges** in case there is a $\mathbf{v} \in \mathcal{V}$ such that $\lim_{j\to\infty} ||\mathbf{v} - \mathbf{v}_n|| = 0$. If a Cauchy sequence converges to both $\mathbf{v}$ and $\mathbf{w}$ then $||\mathbf{v} - \mathbf{w}|| = ||\mathbf{v} - \mathbf{v}_j + \mathbf{v}_j - \mathbf{w}|| \leq ||\mathbf{v} - \mathbf{v}_j|| + ||\mathbf{v}_j - \mathbf{w}|| \to 0$ as $j \to \infty$, so $\mathbf{v} = \mathbf{w}$. Thus the limit of a convergent Cauchy sequence is unique. If a sequence $\{\mathbf{v}_j\}$ converges to $\mathbf{v}$, $\lim_{j\to\infty} ||\mathbf{v}_j - \mathbf{v}|| = 0$, then $\{\mathbf{v}_j\}$ is Cauchy. Indeed, $||\mathbf{v}_j - \mathbf{v}_k|| = ||\mathbf{v}_j - \mathbf{v} + \mathbf{v} - \mathbf{v}_k|| \leq ||\mathbf{v}_j - \mathbf{v}|| + ||\mathbf{v} - \mathbf{v}_k|| \to 0$ as $j, k \to \infty$.

Definition. A pre-Hilbert space $\mathcal{H}$ is a **Hilbert space** if every Cauchy sequence $\{\mathbf{v}_j\}$ in $\mathcal{H}$ converges to an element of $\mathcal{H}$.

Example 1. Every finite-dimensional pre-Hilbert space is a Hilbert space. (Prove it!)

Example 2. The space l_2 is a Hilbert space.

Example 3. The pre-Hilbert spaces $C(\mathfrak{M})$, $C(G)$ and $C^2(R_m)$ are not Hilbert spaces. In each case it is easy to construct a Cauchy sequence of continuous functions which do not converge to an element of the pre-Hilbert space.

The inner product in a pre-Hilbert space is continuous in its two arguments.

Lemma A3. Let $\{\mathbf{u}_j\}$, $\{\mathbf{v}_j\}$ be convergent Cauchy sequences in the pre-Hilbert space $\mathcal{V}$. If $\mathbf{v}_j \longrightarrow \mathbf{v}$, $\mathbf{u}_j \longrightarrow \mathbf{u}$ as $j \longrightarrow \infty$ then $\lim_{j\to\infty} (\mathbf{u}_j, \mathbf{v}_j) = (\mathbf{u}, \mathbf{v})$.

In case $\mathbf{u}_j = \mathbf{v}_j$, Lemma A3 yields $\lim_{j\to\infty} \|\mathbf{v}_j\| = \|\mathbf{v}\|$, i.e., the norm is continuous with respect to convergence in $\mathcal{V}$.

Let $\mathcal{S}$ be a subset of the Hilbert space $\mathcal{H}$. The subset $\mathcal{S}$ is **dense** in $\mathcal{H}$ if for every $\mathbf{u} \in \mathcal{H}$ there exists a Cauchy sequence $\{\mathbf{u}_j\}$ in $\mathcal{S}$ such that $\mathbf{u}_j \longrightarrow \mathbf{u}$.

Example. The set of all $\mathbf{x} = (x_1, x_2, \ldots)$ in l_2 with only finitely many nonzero components x_i is dense in l_2.

A subspace $\mathcal{W}$ of the Hilbert space $\mathcal{H}$ is **closed** in $\mathcal{H}$ if every Cauchy sequence in $\mathcal{W}$ converges to an element of $\mathcal{W}$. The **closure** of a possibly nonclosed subspace $\mathcal{W}$ is the smallest closed subspace of $\mathcal{H}$ containing $\mathcal{W}$. (We order the subspaces by inclusion.)

Lemma A4. Let $\bar{\mathcal{W}}$ be the subset of $\mathcal{H}$ consisting of all $\mathbf{u} \in \mathcal{H}$ such that there exists a Cauchy sequence $\{\mathbf{u}_j\}$ in $\mathcal{W}$ with $\mathbf{u}_j \longrightarrow \mathbf{u}$. Then $\bar{\mathcal{W}}$ is the closure of $\mathcal{W}$.

Theorem A1. Let $\mathcal{W}$ be a pre-Hilbert space. Then there exists a Hilbert space $\mathcal{H}$ (unique up to isomorphism) such that $\mathcal{W}$ is dense in $\mathcal{H}$. Indeed $\mathcal{H} = \bar{\mathcal{W}}$.

The proof of this theorem is not obvious since we do not know *a priori* that there exists a Hilbert space containing $\mathcal{W}$ as a subspace. Until $\mathcal{H}$ is constructed the meaning of $\bar{\mathcal{W}}$ is not clear. To construct $\mathcal{H}$ one considers the Cauchy sequences $\{\mathbf{u}_j\}$ in $\mathcal{W}$ which do not converge. Then one adds new elements $\mathbf{u}$ to $\mathcal{W}$ so that $\mathbf{u}_j \longrightarrow \mathbf{u}$. If $\{\mathbf{v}_j\}$ is a Cauchy sequence in $\mathcal{W}$ such that $\|\mathbf{u}_j - \mathbf{v}_j\| \longrightarrow 0$ then also $\mathbf{v}_j \longrightarrow \mathbf{u}$. It can be shown that $\mathcal{W}$ together with the ideal elements $\{\mathbf{u}\}$ forms a Hilbert space $\mathcal{H}$. See books by Korevaar [1] or Helwig [1] for the details.

By Theorem A1 we can always assume we are dealing with a Hilbert space. (If $\mathcal{W}$ is not a Hilbert space we merely close it to obtain the Hilbert space $\bar{\mathcal{W}}$.) This is fortunate because Hilbert spaces have many nice features not shared by pre-Hilbert spaces.

As stated earlier, $C(\mathfrak{M})$ is not a Hilbert space. However, by the preceding theorem $C(\mathfrak{M})$ is dense in a Hilbert space denoted $L_2(\mathfrak{M})$. It can be shown that to each element f in the closure of $C(\mathfrak{M})$ we can associate a function $f(x)$ on $\mathfrak{M}$. Here $f(x)$ is in general not continuous. If $f(x)$, $g(x)$ are in $L_2(\mathfrak{M})$ then there exist Cauchy sequences $\{f_j\}, \{g_j\}$ in $C(\mathfrak{M})$ such that $f_j \longrightarrow f$, $g_j \longrightarrow g$ in the norm. We define the integral $\int_{\mathfrak{M}} f\bar{g}\,dx$ by

$$\text{(A.9)} \qquad \int_{\mathfrak{M}} f(x)g(\bar{x})\,dx = (f, g) = \lim_{j\to\infty}(f_j, g_j) = \lim_{j\to\infty} \int_{\mathfrak{M}} f_j(x)\overline{g_j(x)}\,dx.$$

The integral $\int_{\mathfrak{M}} f\bar{g}\,dx$ is called the **Lebesgue integral** and $L_2(\mathfrak{M})$ is the space of Lebesgue square-integrable functions on $\mathfrak{M}$. If $f(x)$ and $g(x)$ are functions on $\mathfrak{M}$ such that the Riemann integrals $\int_{\mathfrak{M}} |f|^2\,dx$ and $\int_{\mathfrak{M}} |g|^2\,dx$ converge, then $f, g \in L_2(\mathfrak{M})$ and (A.9) is just the ordinary Riemann integral. However, there exist functions in $L_2(\mathfrak{M})$ which are so discontinuous that they are not Riemann square-integrable. The spaces of continuous and of Riemann square-integrable functions on $\mathfrak{M}$ form pre-Hilbert but not Hilbert spaces. However, the closure of each of these pre-Hilbert spaces is the Hilbert space $L_2(\mathfrak{M})$.

Note: Actually the elements of $L_2(\mathfrak{M})$ are not functions but equivalence classes of functions. We say that two Lebesgue square-integrable functions f, g are **equivalent** if $\int_{\mathfrak{M}} |f(x) - g(x)|^2\,dx = 0$. Equivalent functions correspond to the same Hilbert space element. This distinction does not arise on the subspace $C(\mathfrak{M})$ since if f and g are equivalent continuous functions on $\mathfrak{M}$ then $f(x) = g(x)$ for all $x \in \mathfrak{M}$.

The reader unfamiliar with Lebesgue integration need not despair. Since $C(\mathfrak{M})$ is dense in $L_2(\mathfrak{M})$ we will ordinarily be able to restrict our computations to $C(\mathfrak{M})$.

In complete analogy with the above discussion, the closures of the pre-Hilbert spaces $C(G)$ and $C^2(R_m)$ are $L_2(G)$ and $L_2(R_m)$, the Hilbert spaces of Lebesgue square-integrable functions on G and R_m, respectively.

Definition. A Hilbert space $\mathcal{H}$ is **separable** if it contains a countable dense subset $\{\mathbf{u}_1, \mathbf{u}_2, \ldots\}$.

Example. The space l_2 is separable. Consider the subset $\mathcal{S}$ of all $\mathbf{x} = \{x_1, x_2, \ldots\}$ such that each $x_j = a_j + ib_j$, where the real numbers a_j, b_j are rational, and only a finite number of the x_j are nonzero. The set $\mathcal{S}$ is countable and dense in l_2.

It can be shown that $L_2(\mathfrak{M})$, $L_2(G)$, and $L_2(R_m)$ are separable. In fact every Hilbert space studied in this book is separable. Therefore, from now on, "Hilbert space" means "separable Hilbert space."

Let $\mathfrak{M}$ be a subspace of $\mathcal{H}$. Then the set

$$\mathfrak{M}^{\perp} = \{\mathbf{u} \in \mathcal{H} : (\mathbf{u}, \mathbf{v}) = 0 \quad \text{for all} \quad \mathbf{v} \in \mathfrak{M}\}$$

is clearly a subspace of $\mathcal{H}$. Moreover, $\mathfrak{M}^\perp$ is closed in $\mathcal{H}$ since if $\{\mathbf{u}_j\}$ is a Cauchy sequence in $\mathfrak{M}^\perp$ with $\mathbf{u}_j \longrightarrow \mathbf{u}$ and $\mathbf{v} \in \mathfrak{M}$ we have $(\mathbf{u}, \mathbf{v}) = \lim_{j\to\infty} (\mathbf{u}_j, \mathbf{v}) = 0$, so $\mathbf{u} \in \mathfrak{M}^\perp$.

Theorem A2. Let $\mathfrak{M}$ be a **closed** subspace of the Hilbert space $\mathcal{H}$. Then $\mathcal{H} = \mathfrak{M} \oplus \mathfrak{M}^\perp$, i.e., every $\mathbf{u} \in \mathcal{H}$ can be written uniquely in the form $\mathbf{u} = \mathbf{v} + \mathbf{w}$ with $\mathbf{v} \in \mathfrak{M}$, $\mathbf{w} \in \mathfrak{M}^\perp$.

For $\mathcal{H}$ finite-dimensional this theorem can be proved easily by introducing an appropriate ON basis. In the infinite-dimensional case the proof is not so obvious. The theorem is not true unless $\mathfrak{M}$ is closed. (A finite-dimensional subspace of $\mathcal{H}$ is always closed.)

Definition. A countable set $\{\mathbf{u}_1, \mathbf{u}_2, \ldots\}$ in $\mathcal{H}$ is **orthonormal** (ON) if $(\mathbf{u}_j, \mathbf{u}_k) = \delta_{jk}, j, k = 1, 2, \ldots$. For any $\mathbf{u} \in \mathcal{H}$ the numbers $a_j = (\mathbf{u}, \mathbf{u}_j)$ are the **Fourier coefficients** of $\mathbf{u}$ with respect to the set $\{\mathbf{u}_1, \mathbf{u}_2, \ldots\}$.

Definition. Let $\{\mathbf{v}_1, \mathbf{v}_2, \ldots\}$ be a countable set in $\mathcal{H}$. We say $\sum_{j=1}^\infty \mathbf{v}_j$ **converges** in $\mathcal{H}$ if the partial sums $\mathbf{s}_k = \sum_{j=1}^k \mathbf{v}_j, k = 1, 2, \ldots$ form a Cauchy sequence in $\mathcal{H}$. The **sum** $\mathbf{s}$ of the convergent series is the limit of the Cauchy sequence $\{\mathbf{s}_k\}$.

Theorem A3. Let $\{\mathbf{u}_j\}$ be an ON set in $\mathcal{H}$ and let $a_j \in \mathfrak{C}$, $j = 1, 2, \ldots$. Then $\sum_{j=1}^\infty a_j\mathbf{u}_j$ converges in $\mathcal{H}$ if and only if $\sum_{j=1}^\infty |a_j|^2 < \infty$.

This theorem is not true for pre-Hilbert spaces because there the partial sums of $\sum_{j=1}^\infty a_j\mathbf{u}_j$ may form a Cauchy sequence which does not converge.

Definition. An ON sequence $\{\mathbf{u}_j\}$ in $\mathcal{H}$ is an **orthonormal basis (ON basis)** if every $\mathbf{u} \in \mathcal{H}$ can be expressed as $\mathbf{u} = \sum_{j=1}^\infty a_j\mathbf{u}_j$ for some constants $a_j \in \mathfrak{C}$.

If $\{\mathbf{u}_j\}$ is an ON basis then the constants a_j in the expansion of $\mathbf{u} \in \mathcal{H}$ are uniquely determined. Indeed

$$(\mathbf{u}, \mathbf{u}_j) = \lim_{k\to\infty} \left(\sum_{l=1}^k a_l\mathbf{u}_l, \mathbf{u}_j\right) = \lim a_j = a_j,$$

so $a_j = (\mathbf{u}, \mathbf{u}_j)$, the Fourier coefficient of $\mathbf{u}$ with respect to $\mathbf{u}_j$. Thus,

$$\mathbf{u} = \sum_{j=1}^\infty (\mathbf{u}, \mathbf{u}_j)\mathbf{u}_j. \tag{A.10}$$

Furthermore,

$$||\mathbf{u}||^2 = \lim_{k\to\infty} \left|\left| \sum_{j=1}^k (\mathbf{u}, \mathbf{u}_j)\mathbf{u}_j \right|\right|^2 = \sum_{j=1}^\infty |(\mathbf{u}, \mathbf{u}_j)|^2. \tag{A.11}$$

This is the **Parseval equality**.

Theorem A4. An ON sequence $\{\mathbf{u}_j\}$ is an ON basis for $\mathcal{H}$ if and only if the only vector $\mathbf{v} \in \mathcal{H}$ such that $(\mathbf{v}, \mathbf{u}_j) = 0$, $j = 1, 2, \ldots,$ is $\mathbf{v} = \boldsymbol{\theta}$.

Theorem A5. Every separable Hilbert space $\mathcal{H}$ has an ON basis.

Indeed $\mathcal{H}$ has an infinite number of ON bases if $\dim \mathcal{H} > 0$.

Definition. Let $\mathcal{H}_j$, $j = 1, 2, \ldots,$ be Hilbert spaces, where j runs over a finite or countably infinite number of values. Let $\mathcal{H} = \sum_{j=1}^{\infty} \oplus \mathcal{H}_j$ be the set of all sequences

$$\mathbf{x} = (\mathbf{x}_1, \mathbf{x}_2, \ldots, \mathbf{x}_j, \ldots), \qquad \mathbf{x}_j \in \mathcal{H}_j,$$

such that $\sum_{j=1}^{\infty} \|\mathbf{x}_j\|^2 < \infty$, where $\|\mathbf{x}_j\|$ is the norm of $\mathbf{x}_j$ in $\mathcal{H}_j$. Then $\mathcal{H}$ is an inner product space with operations

$$a\mathbf{x} = (a\mathbf{x}_1, \ldots, a\mathbf{x}_j, \ldots), \qquad a_j \in \mathcal{C},$$

$$\mathbf{x} + \mathbf{y} = (\mathbf{x}_1 + \mathbf{y}_1, \ldots, \mathbf{x}_j + \mathbf{y}_j, \ldots), \qquad \mathbf{x}, \mathbf{y} \in \mathcal{H},$$

$$(\mathbf{x}, \mathbf{y}) = \sum_{j=1}^{\infty} (\mathbf{x}_j, \mathbf{y}_j),$$

where $(\mathbf{x}_j, \mathbf{y}_j)$ is the inner product in $\mathcal{H}_j$. Here $\mathcal{H}$ is called the **direct sum** of the Hilbert spaces $\mathcal{H}_1, \ldots, \mathcal{H}_j, \ldots$.

The verification that $\mathcal{H}$ is an inner product space under the above operations is similar to the corresponding proof for l_2. Moreover, by mimicking the completeness proof for l_2 one can show that $\sum \oplus \mathcal{H}_j$ is a Hilbert space.

Let $\mathcal{H}$ be a Hilbert space. A linear operator $\mathbf{T}: \mathcal{H} \longrightarrow \mathcal{H}$ is **bounded** if

$$\|\mathbf{T}\| = \sup_{\|\mathbf{u}\|=1} \|\mathbf{Tu}\| < \infty, \tag{A.12}$$

i.e., if the least upper bound of the set $\{\|\mathbf{Tu}\|: \|\mathbf{u}\| = 1\}$ is finite. If $\mathbf{T}$ is bounded the number $\|\mathbf{T}\|$ is called the **norm** of $\mathbf{T}$. The sum, product, and scalar multiplication of bounded operators are defined exactly as in the finite-dimensional case.

Lemma A5. Let $\mathbf{S}, \mathbf{T}$ be bounded operators on $\mathcal{H}$. Then (1) $\|\mathbf{Tu}\| \leq \|\mathbf{T}\| \cdot \|\mathbf{u}\|$, $\mathbf{u} \in \mathcal{H}$; (2) $\|\mathbf{S} + \mathbf{T}\| \leq \|\mathbf{S}\| + \|\mathbf{T}\|$; (3) $\|\mathbf{ST}\| \leq \|\mathbf{S}\| \cdot \|\mathbf{T}\|$; (4) $\|a\mathbf{T}\| = |a| \cdot \|\mathbf{T}\|$, $a \in \mathcal{C}$. In particular the sum and product of two bounded operators are bounded operators.

The proof of these results is identical with the proof in Section 5.1 of the corresponding results for operators on finite-dimensional spaces.

Lemma A6. If $\mathbf{s} = \sum_{j=1}^{\infty} \mathbf{v}_j$, where $\sum \mathbf{v}_j$ is a convergent series in $\mathcal{H}$ and $\mathbf{T}$ is a bounded operator, then $\mathbf{Ts} = \sum \mathbf{Tv}_j$.

A bounded operator **T** from $\mathcal{H}$ onto $\mathcal{H}$ is **unitary** if $(\mathbf{Tu}, \mathbf{Tv}) = (\mathbf{u}, \mathbf{v})$, all $\mathbf{u}, \mathbf{v} \in \mathcal{H}$, i.e., **T** preserves inner product. The operator **T** is **symmetric** or **self-adjoint** if $(\mathbf{Tu}, \mathbf{v}) = (\mathbf{u}, \mathbf{Tv})$ for all $\mathbf{u}, \mathbf{v} \in \mathcal{H}$.

The **matrix** $T = (T_{kj})$ of a bounded operator **T** with respect to the ON basis $\{\mathbf{u}_j\}$ is defined by

$$\text{(A.13)} \qquad \mathbf{Tu}_j = \sum_{k=1}^{\infty} T_{kj}\mathbf{u}_k, \qquad T_{kj} = (\mathbf{Tu}_j, \mathbf{u}_k), \qquad j, k = 1, 2, \ldots.$$

To the sum of two operators $\mathbf{S} + \mathbf{T}$ corresponds the sum of their matrices and to the product **ST** corresponds the matrix product:

$$\text{(A.14)} \qquad (S + T)_{kj} = S_{kj} + T_{kj}, \qquad (ST)_{kj} = \sum_{l=1}^{\infty} S_{kl}T_{lj}.$$

The **adjoint** $\mathbf{T}^*$ of the bounded operator **T** is defined by the relation

$$\text{(A.15)} \qquad (\mathbf{Tu}, \mathbf{v}) = (\mathbf{u}, \mathbf{T}^*\mathbf{v}), \qquad \text{all} \quad \mathbf{u}, \mathbf{v} \in \mathcal{H}.$$

In particular, if T_{kj} are the matrix elements of **T** with respect to an ON basis $\{\mathbf{u}_j\}$ then the corresponding matrix elements of $\mathbf{T}^*$ are

$$(T^*)_{kj} = \bar{T}_{jk}: \quad \mathbf{T}^*\mathbf{u}_j = \sum_{k=1}^{\infty} \bar{T}_{jk}\mathbf{u}_k.$$

Thus $\mathbf{T}^*$ is a uniquely determined linear operator on $\mathcal{H}$. Moreover, $\mathbf{T}^*$ is bounded and $\|\mathbf{T}\| = \|\mathbf{T}^*\|$. Note that **T** is self-adjoint if and only if $\mathbf{T} = \mathbf{T}^*$.

References

Ahlfors, L.

[1] "Complex Analysis," 2nd ed. McGraw-Hill, New York, 1966.

Apostol, T.

[1] "Mathematical Analysis." Addison-Wesley, Reading, Massachusetts, 1957.

Bargmann, V.

[1] On a Hilbert space of analytic functions and an associated integral transform, I. *Comm. Pure Appl. Math.* **14,** 187–214 (1961).

[2] On the representations of the rotation group. *Rev. Modern Phys.* **34,** 829–845 (1962); also in Biedenharn and Van Dam [1].

[3] "Group Representations in Mathematics and Physics" (ed.). Springer-Verlag, Berlin and New York, 1970.

[4] Zur Theorie des Wasserstoffatoms. *Z. Physik.* **99,** 576–582 (1936).

Biedenharn, L. C., and Van Dam, H. (eds.)

[1] "Quantum Theory of Angular Momentum." Academic Press, New York, 1965.

Billings, A.

[1] "Tensor Properties of Materials." Wiley (Interscience), New York, 1969.

Boerner, H.

[1] "Representations of Groups." North-Holland Publ., Amsterdam, 1969.

Burckhardt, J.

[1] "Die Bewegungsgruppen Der Kristallographie." Verlag Birkhaeuser, Basel, 1947.

Campbell, J.

[1] "Introductory Treatise on Lie's Theory of Finite Continuous Transformation Groups." Oxford Univ. Press (Clarendon), London and New York, 1903.

Chevalley, C.

[1] "Theory of Lie Groups." Princeton Univ. Press, Princeton, New Jersey, 1946.

Cohen, A.

[1] "An Introduction to the Lie Theory of One-parameter Groups." Heath, Indianapolis, Indiana, 1911.

Cohn, P.
[1] "Lie Groups." Cambridge Univ. Press, London and New York, 1957.
Cornwell, J.
[1] "Group Theory and Electronic Energy Bands in Solids." Wiley (Interscience), New York, 1969.
Cotton, F.
[1] "Chemical Applications of Group Theory," 2nd ed. Wiley (Interscience), New York, 1971.
Courant, R., and Hilbert, D.
[1] Methods of Mathematical Physics," Vol. 1. Wiley (Interscience), New York, 1953.
Coxeter, H.
[1] "Introduction to Geometry." Wiley, New York, 1961.
Cullen, C.
[1] "Matrices and Linear Transformations." Addison-Wesley, Reading, Massachusetts, 1966.
Curtis, C., and Reiner, I.
[1] "Representation Theory of Finite Groups and Associative Algebras. Wiley (Interscience), New York, 1962.
Davydov, A.
[1] "Quantum Mechanics," English Transl. Addison-Wesley, Reading, Massachusetts, 1965.
Dettman, J.
[1] "Mathematical Methods in Physics and Engineering," 2nd ed. McGraw-Hill, New York, 1969.
Dyson, F., ed.
[1] "Symmetry Groups in Nuclear and Particle Physics." Benjamin, New York, 1966.
Eisenhart, L.
[1] "Continuous Groups of Transformations." Dover, New York, 1961 (Reprint).
Erdélyi, A., Magnus, W., Oberhettinger, F., and Tricomi, F.
[1] "Higher Transcendental Functions," Vol. I. McGraw-Hill, New York, 1953.
[2] "Higher Transcendental Functions," Vol. II. McGraw-Hill, New York, 1953.
Finkbeiner, D.
[1] "Introduction to Matrices and Linear Transformations." Freeman, San Francisco, California, 1960.
Freudenthal, H., and De Vries, H.
[1] "Linear Lie Groups." Academic Press, New York, 1969.
Gel'fand, I., and Naimark, M.
[1] "Unitary Representations of the Classical Groups." (Trudy Mat. Inst. Steklov. 39), Moskov-Leningrad, 1950. (German Transl., Akademie-Verlag, Berlin, 1957.)
Gel'fand, I., Minlos, R., and Shapiro, Z.
[1] "Representations of the Rotation and Lorentz Groups and Their Applications," English Transl. Macmillan, New York, 1963.
Gell-Man, M., and Ne'eman, Y., eds.
[1] "The Eightfold Way." Benjamin, New York, 1969.
Guggenheimer, H.
[1] "Differential Geometry." McGraw-Hill, New York, 1963.
Hall, M.
[1] "The Theory of Groups." Macmillan, New York, 1959.
Hamermesh, M.
[1] "Group Theory and Its Applications to Physical Problems." Addison-Wesley, Reading, Massachusetts, 1962.

Hausner, M., and Schwartz, J.
[1] "Lie Groups and Lie Algebras." Gordon & Breach, New York, 1968.
Heine, V.
[1] "Group Theory in Quantum Mechanics." Pergamon, Oxford, 1960.
Helgason, S.
[1] "Differential Geometry and Symmetric Spaces." Academic Press, New York, 1962.
Helwig, G.
[1] "Differential Operators of Mathematical Physics." Addison-Wesley, Reading, Massachusetts, 1967.
Henry, N., and Lonsdale, K., eds.
[1] "International Tables for X-Ray Crystallography," Vol. I. Kynock Press, Birmingham, England, 1965.
Herzberg, G.
[1] "Infrared and Raman Spectra." Van Nostrand-Reinhold, Princeton, New Jersey, 1945.
Ince, E.
[1] "Ordinary Differential Equations." Dover, New York, 1956 (Reprint).
Jacobson, N.
[1] "Lie Algebras." Wiley (Interscience), New York, 1962.
Jauch, J., and Rohrlich, F.
[1] "The Theory of Photons and Electrons." Addison-Wesley, Reading, Massachusetts, 1955.
Judd, B.
[1] "Operator Techniques in Atomic Spectroscopy." McGraw-Hill, New York, 1963.
Kato, T.
[1] "Perturbation Theory for Linear Operators." Springer-Verlag, Berlin and New York 1966.
Korevaar, J.
[1] "Mathematical Methods," Vol. I. Academic Press, New York, 1968.
Kursunǒglu, B.
[1] "Modern Quantum Theory." Freeman, San Francisco, California, 1962.
Lancaster, P.
[1] "Theory of Matrices." Academic Press, New York, 1969.
Landau, L., and Lifshitz, E.
[1] "Statistical Physics." Addison-Wesley, Reading, Massachusetts, 1958.
[2] "Quantum Mechanics." Addison-Wesley, Reading, Massachusetts, 1958.
[3] "The Classical Theory of Fields." Addison-Wesley, Reading, Massachusetts, 1959.
Lebedev, N,
[1] "Special Functions and Their Applications," English Transl. Prentice-Hall, Englewood Cliffs, New Jersey, 1965.
Lie, S.
[1] "Theorie der Transformationsgruppen," Vol. III. Unter Mitwirkung von F. Engel, Leipzig, 1893.
[2] "Differential Gleichungen." Chelsea, Bronx, New York, 1967 (Reprint).
Littlewood, D.
[1] "Theory of Group Characters," 2nd ed. Oxford Univ. Press, (Clarendon), London and New York, 1958.
Liubarskii, G.
[1] "The Application of Group Theory in Physics," English Transl. Pergamon, Oxford, 1960.

Loebl, E. ed.
[1] "Group Theory and its Applications," Vol. I. Academic Press, New York, 1968.
[2] "Group Theory and its Applications," Vol. II. Academic Press, New York, 1971.
Lomont, J.
[1] "Applications of Finite Groups." Academic Press, New York, 1959.
Miller, W.
[1] "Lie Theory and Special Functions." Academic Press, New York, 1968.
[2] A branching law for the symplectic groups. *Pacific J. Math.* **16,** 341–346 (1966).
Muller, C.
[1] "Lectures on the Theory of Spherical Harmonics." Springer-Verlag, Berlin and New York, 1966.
Murnaghan, F.
[1] "The Theory of Group Representations." Dover, New York, 1963 (Reprint).
Naimark, M.
[1] "Normed Rings," English Transl. Noordhoff, Groningen, The Netherlands, 1959.
[2] "Linear Representations of the Lorentz Group," English Transl. Macmillan, New York, 1964.
Naylor, A., and Sell, G.
[1] "Linear Operator Theory in Engineering and Science." Holt, New York, 1971.
Nussbaum, A.
[1] "Applied Group Theory for Chemists, Physicists and Engineers." Prentice-Hall, Englewood Cliffs, New Jersey, 1971.
Nye, J.
[1] "Physical Properties of Crystals." Oxford Univ. Press (Clarendon), London and New York, 1957.
Petrashen, M., and Trifonov, E.
[1] "Applications of Group Theory in Quantum Mechanics," English Transl. MIT Press, Cambridge, Massachusetts, 1969.
Pollard, H.
[1] "Introduction to Celestial Mechanics." Prentice-Hall, Englewood Cliffs, New Jersey, 1965.
Pontrjagin, L.
[1] "Topological Groups." Princeton Univ. Press, Princeton, New Jersey, 1958.
Prugovecki, E.
[1] "Quantum Mechanics in Hilbert Space." Academic Press, New York, 1971.
Racah, G.
[1] Theory of complex Spectra IV. *Phys. Rev.* **76,** 1352 (1949).
[2] Group theory and spectroscopy. *Ergeb. Exakt. Naturwiss.* **37,** 28 (1965).
Rätz, J.
[1] On isometries of generalized inner product spaces. *SIAM J. Appl. Math.* **18,** 6–9 (1970).
Regge, T.
[1] Symmetry properties of Clebsch-Gordan's coefficients. *Nuovo Cimento* [10], **10,** 544–545 (1958).
Riesz, F., and Sz.-Nagy, B.
[1] "Functional Analysis," English Transl. Ungar, New York, 1955.
Robinson, G. de B.
[1] "Representation Theory of the Symmetric Group." Univ. of Toronto Press, Toronto, 1961.

Roman, P.
[1] "Theory of Elementary Particles." North-Holland Publ., Amsterdam, 1960.
Rudin, W.
[1] "Principles of Mathematical Analysis." McGraw-Hill, New York, 1964.
Ruhl, W.
[1] "The Lorentz Group and Harmonic Analysis." Bejamin, New York, 1970.
Rutherford, D.
[1] "Substitutional Analysis." Edinburgh Univ. Press, Edinburgh, 1948.
Schiff, L.
[1] "Quantum Mechanics." McGraw-Hill, New York, 1955.
Schweber, S.
[1] "Relativistic Quantum Field Theory." Harper, New York, 1961.
Sherman, T.
[1] A weight theory for unitary representations. *Canad. J. Math.* **18,** 159–168 (1966).
Slater, J.
[1] "Quantum Theory of Atomic Structure," Vols. I, II. McGraw-Hill, New York, 1960.
[2] "Quantum Theory of Molecules and Solids." McGraw-Hill, New York, 1963.
Smirnov, V.
[1] "Linear Algebra and Group Theory," English Transl. McGraw-Hill, New York, 1961.
Spivak, M.
[1] "Calculus on Manifolds." Benjamin, New York, 1965.
Sposito, G.
[1] "An Introduction to Quantum Physics." Wiley, New York, 1970.
Stakgold, I.
[1] "Boundary Value Problems of Mathematical Physics," Vol. I. Macmillan, New York, 1967.
Steele, D.
[1] "Theory of Vibrational Spectroscopy." Saunders, Philadelphia Pennsylvania, 1971.
Sugano, S., Tanabe, Y., and Kamimura, H.
[1] "Multiplets of Transition Metal Ions in Crystals." Academic Press, New York, 1970.
Talman, J.
[1] "Special Functions. A Group Theoretic Approach." Benjamin, New York, 1968.
Tinkham, M.
[1] "Group Theory and Quantum Mechanics." McGraw-Hill, New York, 1964.
Van der Waerden, B.
[1] "Modern Algebra." Ungar, New York, 1949.
[2] "Gruppentheoretische Methode in der Quantenmechanik." Springer-Verlag, Berlin and New York, 1932.
Vilenkin, N.
[1] "Special Functions and the Theory of Group Representations" Amer. Math. Soc. Transl.,Amer. Math. Soc., Providence, Rhode Island, 1968.
Weyl, H.
[1] "Group Theory and Quantum Mechanics." Dover, New York, 1950 (Reprint).
[2] "The Structure and Representation of Continuous Groups" (Lecture Notes). Princeton Univ. Press, Princeton, New Jersey, 1935.
[3] "The Classical Groups." Princeton Univ. Press, Princeton, New Jersey, 1946.
[4] "Symmetry." Princeton Univ. Press, Princeton, New Jersey, 1952.
Whippman, M.
[1] Branching rules for simple Lie groups. *J. Math. Phys.* **6,** 1534–1539 (1965).

Wigner, E.

[1] On unitary representations of the inhomogeneous Lorentz group. *Ann. of Math.* **40,** 149–204 (1939).

[2] "Group Theory and Its Application to the Quantum Mechanics of Atomic Spectra." Academic Press, New York, 1959.

Wybourne, B.

[1] "Symmetry Principles in Atomic Spectroscopy." Wiley (Interscience), New York, 1970.

Yale, P.

[1] "Geometry and Symmetry." Holden-Day, San Francisco, California, 1968.

Symbol Index

This index lists symbols used frequently in the text, followed by the sections in which they first appear.

$\|\alpha\|$	Absolute value, 5.1.
A^*	Adjoint of the operator A, 3.7.
Ad A, Ad α	Adjoint rep operators, 5.1, 5.6.
B_α	Algebra of permutation operators, 4.3.
a, a^*	Annihilation and creation operators, 9.7.
$A(G)$	Automorphism group of G, 1.3.
$\mathcal{J}^+, \mathcal{J}^-, \mathcal{J}^3$ or J^+, J^-, J^3	Basis for a rep of $sl(2)$, 5.10.
$\mathcal{L}_1, \mathcal{L}_2, \mathcal{L}_3$	Basis for $so(3)$, 7.1.
$J_n(z) = [(z/2)^n/n!]\sum_{k=0}^{\infty}[(-z^2/4)^k/(n+1)_k k!]$, $n = 0, 1, \ldots,$	Bessel function, 10.2.
A_α	Bisymmetric transformations, set of, 4.3.
$\mathcal{E}_{kl}, \mathcal{E}_\alpha$	Branch, 9.1.
$\Gamma_n, \Gamma_n{}^m$	Bravais lattices, 2.8.
h_m	Cartan subalgebra of a classical group, 9.1.
$\chi(g), g \in G$	Character of the group G, 3.4.
C_∞, $U(1)$, $SO(2)$	Circle group, 2.3.
CG	Clebsch–Gordan, 3.5.
$(\mu i, \nu j \| \xi s l)$	Clebsch–Gordan (CG) coefficients, 3.5.
$C(u, m; v, n \| w, k)$	Clebsch–Gordan coefficients for $SU(2)$, 7.7.
α_j	Clifford algebra, elements of, 9.6.
$C(T)$	Column permutations of a Young tableau, 4.2.
$[A, B]$, $[\alpha, \beta]$	Commutator bracket, 5.1, 5.3.
$\mathfrak{C}$	Complex numbers, 1.1.
$\bar{A} = (\bar{A}_{ij})$	Complex conjugate matrix, 3.1.
$\mathcal{C}_m{}^n$	Conjugacy class in a point group, 3.6.

$\mathbf{u} \times \mathbf{v}$	Cross product, 3.8.
C_n	Cyclic group of order n, 2.4.
$C_{\infty v}$, $C_{\infty h}$	Symmetry groups, 2.9, 7.6.
$\Delta(\epsilon_1, \ldots, \epsilon_m) = \prod_{j>k} (\epsilon_j - \epsilon_k)$, 9.2.	
$f'(x), \dot{f}(t)$	Derivative of a function, 5.1.
$\dot{A}(t)$	Derivative of a matrix-valued function, 5.1.
$\det A$	Determinant of matrix A, 2.1.
$\lvert \epsilon^{l_1}, \ldots, \epsilon^{l_m} \rvert = \sum_{s \in S_m} \delta_s \epsilon_{s(1)}^{l_1} \cdots \epsilon_{s(m)}^{l_m}$, Determinant, 4.4, 9.2.	
D_n	Dihedral group, 2.4.
$\dim V$	Dimension of vector space V, 3.2.
$G \times H$	Direct product of groups, 1.5.
$\mathbf{T}_1 \oplus \mathbf{T}_2$	Direct sum of reps $\mathbf{T}_1$ and $\mathbf{T}_2$, 3.2, 3.5.
$V \oplus W$	Direct sum of vector spaces V and W, 3.2, 3.5.
$D_{\infty h}$	Symmetry group, 2.9.
$\mathbf{T} \cong \mathbf{T}'$	Equivalence of reps, 3.1.
$E(n)$, $(E^+(n))$	Euclidean group in n-space (proper Euclidean group), 2.2.
$\{a, O\}$	Euclidean group element, 2.2.
(φ, θ, ψ)	Euler angles, 7.1.
$\exp \alpha: \mathcal{G} \longrightarrow G$	Exponential mapping, 5.5, 5.9.
$\exp A = e^A = \sum_{j=0}^{\infty} A^j/j!$	Exponential of a square matrix, 5.1.
G/N	Factor group, 1.2.
$\Gamma(z) = \lim_{n\to\infty} [n!\, n^z/(z)_{n+1}]$, $\Gamma(z+1) = z\Gamma(z)$, $\Gamma(n+1) = n!$, $n = 0, 1, 2, \ldots,$	Gamma function, 7.5.
$GL(n, \mathfrak{C})$, $GL(n, R)$, $GL(n)$	General linear groups, 1.1, 5.4.
G	Group, 1.1.
R_G	Group algebra (ring) of the group G, 3.1, 3.3.
$x = \sum_{g \in G} x(g) \cdot g$	Group algebra element, 3.1.
$\varphi(g, h)$	Group product in local Lie group, 5.2.
$T(g)$, $T(A)$	Group rep matrices, 3.1.
$\mathbf{T}(g)$, $\mathbf{T}(A)$	Group rep operators, 3.1.
$\mathbf{H}$	Hamiltonian operator, 3.8.
$H_n(x) = (-1)^n \exp(x^2)(d^n/dx^n) \exp(-x^2)$, $n = 0, 1, 2, \ldots,$ Hermite polynomials, 10.1.	
$\mathcal{H}$	Hilbert space, 3.8, Appendix.
$\mu: G \longrightarrow G'$	Homomorphism of groups, 1.3.
$\tau: \mathcal{G} \longrightarrow \mathcal{G}'$	Homomorphism of Lie algebras, 5.3.
${}_2F_1(a, b; c; z) = \sum_{n=0}^{\infty} [(a)_n (b)_n/(c)_n](z^n/n!)$, $\lvert z \rvert < 1$, Hypergeometric series (see Pochhammer symbol), 7.2.	
Y	Icosohedral group, 2.4.
$\mathcal{I}$	Ideal, 4.3.
$\mathcal{O}_\alpha$, $\mathcal{R}_\alpha$	Ideals in the group ring of S_α, 4.3.
e	Identity element in a group, 1.1.
E_n	Identity matrix, $n \times n$, 2.1, 3.3.
$\mathbf{E}$	Identity operator, 2.1.
$\mathbf{T}^G$	Induced rep, 3.5.
$\langle \mathbf{u}, \mathbf{v} \rangle$, $(\mathbf{u}, \mathbf{v})$	Inner product, 2.1, Appendix.
$\int f(x)\, dx$	Integral of $f(x)$, 3.1, 3.8, Appendix.
dA, $d_l A$, $d_r A$, δA	Invariant measures on a linear Lie group, 6.1.
$\mathbf{I} = -\mathbf{E}$	Inversion operator, 2.3.
Irred	Irreducible (representation), 3.2.
$\mathbf{T}^{(\mu)}$	Irred rep of group G, indexed by the integer μ, 3.3.

$[f_1, \ldots, f_m]$ Irred reps of $GL(m)$, $U(m)$, and $SL(m)$, 9.1.
$\mathbf{D}^{(u,v)}$ Irred rep of proper homogeneous Lorentz group, 8.3.
$\mathbf{D}_+^{(u,v)}, \mathbf{D}_-^{(u,v)}$ Irred reps of $L^\uparrow$, 8.3.
$\mathbf{D}_+^{(l)}, \mathbf{D}_-^{(l)}$ Irred reps of $O(3)$, 7.6.
$\mathbf{D}^{(u)}$ Irred reps of $SU(2)$, $SO(3, R)$, and $SL(2, \mathfrak{C})$, 7.2, 7.3.
$G \cong H$ Isomorphism of groups, 1.1, 2.4.
G^x Isotropy subgroup of G at x, 1.4.
$\delta_{ij} = 1$ if $i = j$, $= 0$ if $i \neq j$; Kronecker delta, 1.1.
$L_n^{(\alpha)}(z) = [\Gamma(\alpha + n + 1)/\Gamma(\alpha + 1)n!] \sum_{j=0}^{\infty} [(-n)_j/(\alpha + 1)_j](z^j/j!)$, $n = 0, 1, 2, \ldots,$ Laguerre polynomial, generalized, 7.5, 10.2.
Δ Laplacian, 3.1.
$L_2(G)$ Lebesgue square-integrable functions on the group G, 6.2, Appendix.
$L_2(\mathfrak{M})$ Lebesgue square-integrable functions on domain $\mathfrak{M}$, Appendix.
$\mathbf{L}(g)$ Left regular rep, 1.4, 3.1.
$P_u{}^m(\cos\theta) = P_u^{0,-m}(\cos\theta)$, Legendre function (associated), 7.2.
$P_n(\cos\theta) = {}_2F_1(n + 1, -n; 1; \frac{1}{2}(1 - \cos\theta))$, $n = 0, 1, 2, \ldots,$ Legendre polynomial, 7.2.
$\mathcal{G}$ Lie algebra, 5.3.
$L(G)$ Lie algebra of local Lie group G, 5.3.
$so(3, 1)$ Lie algebra of homogeneous Lorentz group, 8.1, 9.10.
L_α, D_α Lie derivatives, 5.9.
$\mathbf{T}: V \longrightarrow V$ Linear transformation, 2.1, 3.1.
$\ln A$ Logarithm of a matrix, 5.1.
$L^\uparrow$ Lorentz group (complete), 8.3.
$L(4)$ Lorentz group (homogeneous), 8.1.
$L^{\uparrow +}$ Lorentz group (proper), 8.1.
$T = (T_{ij})$ Matrix, 2.1.
${}^{2S+1}L_J$ Multiplet, 9.8.
$\nu(\mathbf{x}, g)$ Multiplier, 5.9.
$\|\mathbf{u}\|, \|\mathbf{A}\|$ Norm of a vector (operator), 2.1, 5.1, Appendix.
$N_\mathbf{A}$ Null space of linear operator $\mathbf{A}$, 3.3.
O Octahedral group, 2.4.
$n(G)$ Order of group G, 1.1.
$W^\perp$ Orthogonal complement of subspace W, 3.2.
$O(n, \mathfrak{C})$, $O(n, R)$, $O(n)$ Orthogonal groups, 5.4.
$\{\lambda_j\} \boxtimes \{\mu_k\}$ Outer product of irred reps of S_n and S_m, 9.9.
δ_s Parity of the permutation s, 4.1, 9.2.
$\partial f/\partial x, \partial_x f, f_x$ Partial derivative, 3.1.
$\partial_{\pm 1/2, \pm 1/2}$ Partial derivatives, 8.4.
$\{f_1, \ldots, f_n\}$ Partition or frame, 4.1.
$\begin{pmatrix} 1 & 2 & \cdots & n \\ p_1 & p_2 & \cdots & p_n \end{pmatrix}, (p_i \cdots p_j), \boldsymbol{\sigma}(x)$, Permutations, 1.1, 4.1.
$\perp$ Perpendicular, 2.8.
$h = 2\pi\hbar \sim 1.054 \times 10^{-27}$ erg sec, Planck's constant, 3.1, 3.8.
$(a)_n = a(a + 1) \cdots (a + n - 1) = \Gamma(a + n)/\Gamma(a)$, Pochhammer's symbol (see Bessel, hypergeometric, and Laguerre functions).
P Poincaré group, 8.2.
$\mathcal{P}$ Poincaré group (covering group of), 8.4.
$C_{nh}, C_{nv}, D_{nh}, D_{nd}$,
$O_h = O \cup \mathbf{I}O, S_{2n}, T_d$,
$T_h = T \cup \mathbf{I}T, Y_h = Y \cup \mathbf{I}Y$, Point groups, 2.5.

$\mathbf{P}$	Projection operator, 3.3, 3.7.
$R_{\mathbf{A}}$	Range of linear operator $\mathbf{A}$, 3.3.
R_n	Real Euclidean n-space or group ring of S_n, 2.1, 4.2.
R	Real numbers, 1.1.
$\sigma_{\mathbf{k}}$	Reflection in plane perpendicular to $\mathbf{k}$, 2.1.
Rep	Representation, 3.1.
$\mathbf{T} \mid W$	Restriction of $\mathbf{T}$ to W, 3.2, 3.5.
$\mathbf{R}(g)$	Right regular rep, 3.1.
$\alpha(\mathcal{H})$	Root form, 9.1.
$C_{\mathbf{k}}(\theta)$	Rotation through angle θ about axis $\mathbf{k}$, 2.1.
$S_{\mathbf{k}}(\theta) = \sigma_{\mathbf{k}} C_{\mathbf{k}}(\theta)$,	Rotation-inversion, 2.1.
$R(T)$	Row permutations of a Young tableau, 4.2.
Spin(m)	Simply connected covering group of $SO(m, \mathfrak{C})$, 9.6.
$SL(n, \mathfrak{C})$, $SL(n, R)$, $SL(n)$ [$sl(n, \mathfrak{C})$, $sl(n, R)$, $sl(n)$],	Special linear groups (Lie algebras), 1.1, 5.4.
$SO(n, \mathfrak{C})$, $SO(n, R)$, $SO(n)$ [$so(n, \mathfrak{C})$, $so(n, R)$, $so(n)$],	Special orthogonal groups (Lie algebras), 1.4.
$SU(m)$ [$so(m)$]	Special unitary group (Lie algebra), 5.4.
$Y_l^m(\theta, \varphi) = [(2l+1)(l-m)!/4\pi(l+m)!]^{1/2} P_l^m(\cos\theta) e^{im\varphi}$, $m = -l, -l+1, \ldots, l$, $l = 0, 1, 2, \ldots,$	Spherical harmonic, 7.4.
$P_u^{-m,n}(\cos\theta) = [(\sin\theta)^{m-n}(1+\cos\theta)^{u+n-m}/2^u\Gamma(m-n+1)]_2F_1(-u-n, m-u; m-n+1; [(\cos\theta-1)/(\cos\theta+1)])$,	Spherical function (generalized), 7.2.
S_n, S_α	Symmetric group, 1.1, 4.4.
S_X	Symmetric group on set X, 1.4.
$Sp(m)$	Symplectic group, 5.4, 9.4.
$a_{j_1 \cdots j_n}$	Tensor components, 3.8.
$\mathbf{T}_1 \otimes \mathbf{T}_2$ ($\mathbf{T}^{\otimes n}$)	Tensor product of group reps (n-fold), 3.5 (3.8).
$\mathbf{u} \otimes \mathbf{v}$	Tensor product of vectors, 3.5.
$V \otimes W$ ($V^{\otimes n}$)	Tensor product of vector spaces (n-fold), 3.5 (3.8).
${}^{2S+1}L$	Term, 9.8.
T	Tetrahedral group, 2.4.
$\begin{pmatrix} j_1 & j_2 & j_3 \\ m_1 & m_2 & m_3 \end{pmatrix}$	3-j coefficients, 7.7.
$\operatorname{tr} A = \sum_{i=1}^n A_{ii}$	Trace of $n \times n$ matrix A, 2.1.
$\mathbf{T}_{\mathbf{a}}$	Translation, 2.2.
$T(n)$	Translation group in n-space, 2.2.
A^{t}	Transpose of matrix A, 2.1.
$U(m)$, $USp(2m)$, $SU(m)$	Unitary classical groups, 5.4, 9.4.
$\mathbf{u}$, $\mathbf{v}$, $\mathbf{x}$	Vectors, 2.1, 3.1.
V, W	Vector spaces, 3.1.
V_G	Volume of compact group G, 6.2.
$V(Q)$	Volume of parallelepiped Q, 2.6.
$\Lambda(\mathcal{H})$	Weight, 9.1.
$S^\alpha\Lambda = \Lambda - 2(\Lambda_\alpha/\alpha_\alpha)\alpha$	Weyl reflection, 9.1.
Z	Zero matrix, 3.7.
$\mathbf{Z}$	Zero operator ($\mathbf{Z}\mathbf{v} = \boldsymbol{\theta}$), 3.7.
$\boldsymbol{\theta}$	Zero vector, 2.2.

Index

Numbers in italics refer to the pages on which the complete references are listed.

A

Addition theorems, 203, 231
Adjoint group, 182, 183
Adjoint Lie algebra, 183
Adjoint operator, 92, 414
Ado's theorem, 170
Ahlfors, L., 156, 157, *415*
Analytic curve, 165, 166
Analytic manifold, 165
Angular momentum operators, 239, 254
  orbital, 269
  spin, 269
  total, 269
Apostol, T., 163, *415*
Annihilation operator, 369
Anticommutation relations, 369
Associative algebra, 132
Associative law, 1, 163
Automorphism
  group, 8
  inner, 8, 182
  local analytic, 170
  of groups, 6
  of Lie algebras, 170
Average, 212
Axis of rotation, 18

B

Balmer series, 377
Bargmann, V., 263, 402, *415*
Baryon, 268, 371
Basis, 17
  canonical, 257, 298
  orthonormal, 18, 66, 412
Bessel functions, 406, 421
Biedenharn, L. C., *415*
Billings, A., *415*
Bisymmetric transformations, 131
Boerner, H., 117, 119, 135, 151, 362, 368, 375, 396, *415*
Boson, 369, 378
Bounded below (rep), 235
Bounded operator, 413
Bounded set, 211
Branch, 322, 341, 345, 353
Branching laws, 375
Burckhardt, J., 57, 59, *415*

C

Campbell, J., 204, *415*
Campbell–Baker–Hausdorff formula, 161
Canonical coordinates, 178
Cartan's criterion, 392
Cascade particle, 372
Casimir operator, 234, 260, 394
Cauchy sequence, 153, 210, 409
Cayley's theorem, 12
Cell, primitive, 37
Center (Lie algebra), 183
Centroid, 25
Character, 74, 217,
  compound, 75, 217
  simple, 75, 217
  table, 85
Charge, 372
Chevalley, C., 216, *415*
Circle group, 215
Clebsch–Gordan coefficients, 81, 256
Clebsch–Gordan series, 81, 233
Clifford algebra, 363
Closed set, 211
Closed subspace, 219, 410
Cohen, A., 195, *415*
Cohn, P., *416*
Commutator bracket, 159, 167, 169, 191
Commutator subgroup, 205
Compact group, 206, 211
Compact set, 211
Complexification (Lie algebra), 228
Complementary series, 304
Congruent sets, 20
Conjugate reps, 389
  self, 389
Conjugacy class, 5
Continuous function, 206
Convergent sequence, 409
Convergent series, 412
Convolution product, 72
Cornwell, J., *416*
Coset, 4
Cotton, F., *416*
Courant, R., *416*
Coxeter, H., *416*
Creation operator, 369
Cross product, 102, 376
Crystal, 52
  classes, 51
  ideal, 53
  real, 53
  system, 40
Crystallographic point group, 38
Crystallographic restriction, 39
Cube, 31, 90
Cullen, C., 351, *416*
Curtis, C., 151, *416*
Cycle notation, 3, 117

D

Davydov, A., *416*
Degeneracy, 110, 114
  accidental, 110, 114
Derived algebra, 205
Determinant, 2, 19
Dettman, J., *416*
De Vries, H., 165, 368, 391, *416*
Dihedral groups, 29
Dirac equation, 320, 365
Dirac matrices, 362
Direct product, 13, 216
Direct sum (reps), 68, 79
Direct sum (Hilbert spaces), 413
Dodecahedron, 32
Domain (of homomorphism), 7
Dynamical symmetry group, 400
Dyson, F., 375, *416*

E

Eigenvalue, 18, 70, 109
Eigenvector, 18, 109
Eightfold way, 373
Eisenhart, L., *416*
Elastic interaction, 266
Electron, 268, 319, 362
  shell, 381
Energy, total, 310, 317
Equivalence classes, 5, 52, 71
Erdélyi, A., 399, 404, *416*
Euclidean algorithm, 28
Euclidean group, 20, 190, 278
Euclidean space, 16
Euler angles, 225, 231
Event, 293
Exponential mapping, 175, 177
Exponential matrix, 152

F

Fermion, 369, 378
Finkbeiner, D., 173, *416*
Fourier coefficients, 412
Fourier series, 215
Fourier transform, 280
Frame
  Young, 120
  inertial, 293
  conjugate, 389
Freudenthal, H., 165, 368, 391, *416*
Frobenius reciprocity theorem, 84
Fundamental domain, 37

G

Gamma function, 204, 246, 422
Gaussian units, 283
Gegenbauer polynomials, 245
Gel'fand, I., 282, 308, 320, 377, *416*
Gell-Man, M., 373, *416*
General linear group, 2, 128, 321
Generating function, 203
Glide reflection, 22
Groups, 1
  abelian (commutative), 1
  algebra (ring), 64, 120, 214, 387
  character, 74
  classical, 171
  conjugate, 25
  covering, 225, 290
  crystal, 53
  cyclic, 6
  discrete, 24
  factor, 5
  isomorphic, 6, 25
  lattice, 34
  Lie, 162
  nonsymmorphic, 57
  space, 53, 55
  symmorphic, 56
Guggenheimer, H., *416*

H

Hall, M., 114, *416*
Hamermesh, M., 34, 60, 114, 120, 151, 263, 272, 375, 384, 386, 390, 391, *416*
Hamiltonian operator, 108, 247
Harmonic oscillator, 110, 397
  group, 397, 400, 403
Hausner, M., 165, 181, 187, 391, *417*
Heine, V., *417*
Heine–Borel theorem, 211
Helgason, S., 165, 255, 396, *417*
Helmholtz equation, 64
Helwig, G., 108, 216, 255, 376, 398, 410, *417*
Henry, N., *417*
Hermitian matrix, 173
Hermite polynomials, 399, 422
Herzberg, G., *417*
Hexagon, 15
Hilbert, D., *416*
Hilbert space, 107, 214, 409
  separable, 411
Holohedry, 39
  cubic, 44, 51
  hexagonal, 47, 51
  monoclinic, 50, 51
  orthorhombic, 49, 51
  rhombohedral, 48, 51
  tetragonal, 47, 51
  triclinic, 50, 51
Homomorphism
  analytic, 186
  group, 6
  Lie algebra, 170, 178
  local analytic, 170, 178
Hydrogen atom, 376
Hypercharge, 372
Hypergeometric functions, 203, 222, 422

I

Icosahedral group, 32, 114
Icosahedron, 32
Ideal
  left, 99
  Lie algebra, 185
  minimal, 99, 100
  right, 99
  two-sided, 99
Idempotent, 97
  essential, 120
  generating, 139
  primitive, 98

Identity element
  group, 1, 163
  matrix, 3, 17
  operator, 17, 20
Implicit function theorem, 163, 164
Ince, E., 176, *417*
Indecomposable vector, 79, 129
Index (subgroup), 4
Infinitesimal operator, 186
Inner (scalar) product, 16, 66, 407
  space (pre-Hilbert space), 66, 407
Integrability conditions, 195
Invariant
  equations, 278, 294, 313
  function, 205
  measure, 206
  normalized measure, 212
  subset, 9
  subspace, 67
Inverse
  function theorem, 163, 177
  group, 1, 163
  matrix, 17
  operator, 17
Inversion operator, 17, 26, 288
  space, 288
  time, 288
  total, 288
Isobaric spin space, 370
Isometry, 20
Isomorphism
  group, 6, 182
  inverted, 142
  Lie algebra, 170, 178
  local analytic, 170, 178
  pre-Hilbert space, 409

J

Jacobian, 163, 165, 177, 206
Jacobi equality, 168, 169
Jacobi polynomial, 257
Jacobson, N., 170, 182, 205, 391, *417*
Jauch, J., 312, *417*
Judd, B., *417*

K

Kamimura, H., *419*
Kato, T., 108, 111, *417*
Kernel of homomorphism, 7
Killing form, 394
Klein–Gordon equation, 317, 362
Korevaar, J., 407, 410, *417*
Kronecker delta, 3, 17, 423
Kursunǒglu, B., 377, *417*

L

Ladder (weight), 324, 342
Ladder length, 324
Lagrange's theorem, 5
Laguerre polynomial, 246, 423
Lancaster, P., 349, *417*
Landau, L., 265, 312, 320, *417*
Laplace equation, 242, 249
Lattice, 37
  base-centered, 49, 50
  body-centered, 46, 47, 49
  Bravais, 42, 51
  crystal, 37
  face-centered, 46, 50
  postulate, 53
  primitive, 47, 48, 49
  type, 43
Lebedev, N., 259, *417*
Lebesgue integral, 214, 411
Legendre equation, 238
Legendre function, 231, 241, 423
Legendre polynomial, 231, 237, 244, 423
Lie, S., 204, *417*
Lie algebra, 168
  abstract, 168, 181
  commutative (abelian), 182
  compact, 394
  complexified, 184, 228
  exceptional, 392
  rank, 395
  semisimple, 391
  simple, 392
  solvable, 391
  subalgebra, 184
Lie derivatives, 189
  equal, 193
  generalized, 196, 198
  linearly dependent, 193
  linearly independent, 193
Lie groups, 162, 165
  abelian, 167, 182
  compact, 206

complex, 163, 340
connected, 165
global, 165
linear, 165, 166
local, 162
local linear, 164, 180
locally isomorphic, 164
locally simple, 394
real, 163
semisimple, 394
simply connected, 187
solvable, 394
Lifshitz, E., 265, 312, 320, *417*
Linear functional, 322
real, 327
positive, 327
Linear transformation, 17, 152
bounded, 219, 413
inverse, 17
product, 17
orthogonal, 67
symmetric, 414
unitary, 64, 414
Littlewood, D., 120, 390, *417*
Liubarskii, G., 267, 272, 278, 282, *417*
Loebl, E., 375, 386, *418*
Lomont, J., 384, 386, *418*
Lonsdale, K., *417*
Lorentz group, 284
complete, 305, 316
homogeneous, 285, 313
inhomogeneous, 294
proper, 288, 316

M

Magnetic symmetry group, 60
Magnus, W., 399, 404, *416*
Matrix, 17, 414
elements, 202
Hermitian (self-adjoint), 173
inverse, 17
nonsingular, 17
product, 17
skew-Hermitian, 173
symplectic, 174, 347
transpose, 17
unitary, 66, 173
Maxwell's equations, 283, 320
Meson, 268, 374
Miller, W., 204, 235, 247, 402, *418*
Minkowski space, 293, 309
Minlos, R., 282, 308, 320, *416*
Momentum operators (linear), 275, 278, 310
four vector, 310
Muller, C., *418*
Multiplet, 386
Multiplicity
of an eigenvalue, 110, 264
of a rep, 69, 75
of a weight, 337, 338
Multiplier, 198
rep, 235
Murnaghan, J., 151, *418*

N

Naimark, M., 209, 216, 217, 220, 255, 282, 303, 305, 315, 377, *416, 418*
Naylor, A., 407, *418*
Ne'eman, Y., 373, *416*
Negative reps, 248
Neutrino, 318
Neutron, 268, 370
Norm (vector), 16, 152, 407
Norm (operator), 152, 413
Nucleon, 370
Null space, 69, 92
Nussbaum, A., *418*
Nye, J., *418*

O

Oberhettinger, F., 399, 404, *416*
Octahedral group, 31, 90
Octahedron, 31
Onto mapping, 7
Orbit, 9, 27
Order, of an element, 6
Order, of a group, 1
Orthogonal group, 16, 172, 351
Orthogonality relations, 72, 77, 214
Outer product, 390

P

Parallelepiped, 35, 207
basic, 37
Parity, 124, 249, 277

Parseval equality, 215, 236, 412
Partition, 119
Pauli exclusion principle, 268, 377, 379
Pauli spin matrices, 269, 365
Periodic table, 377
Permutation, 3, 8, 117
  column, 120, 123
  corresponding, 122
  even, 117
  odd, 117
  row, 120, 123
Perturbations, 111, 249, 264
Peter–Weyl theorem, 215, 216
Petraschen, M., and Trifonov, E., *418*
Photon, 268, 312
Planck's constant, 108, 423
Pochammer's symbol, 423
Poincaré group, 294, 311, 313
Point group
  first kind, 26
  second kind, 26, 32
Polar decomposition, 295, 349
Poles, 27
Pollard, H., 376, *418*
Pontrjagin, L., 178, 187, 195, 209, *418*
Position operator, 275, 277
Positive reps, 248
Potential function, 111, 247, 264, 271
Power series, 156
Principal series, 304
Prism, 30, 34
Projection operator, 92, 137, 218
Proton, 268, 370
Prugovecki, E., *418*
Pseudoscalar, 102, 306, 318
Pseudovector, 306
Pyramid, 30, 107

Q

Quantum number, 380

R

Racah, G., 393, 395, *418*
Racah coefficients, 267
Radius of convergence, 156
Range, 7, 69, 92
Rätz, J., 293, *418*
Ray, 227, 254
Real form (Lie algebra), 228, 287
Recurrence formulas, 238, 244
Reduced matrix element, 276
Reflection, 19, 305, 316
Regge, T., 263, *418*
Reiner, I., 151, *416*
Representation, algebra, 132
Representation, group, 61, 219
  alternating, 117
  bounded, 219
  continuous, 213
  dimension, 61
  double-valued, 227, 293, 355
  equivalent, 62
  faithful, 365
  half-integral, 227
  identity, 78, 120
  induced, 83, 389
  integral, 227
  irreducible, 67, 219
  left (right) regular, 64
  local, 187
  local multiplier, 197
  matrix, 61
  real, 62
  reducible, 67, 219
  space, 61, 186
  unitary, 66, 220
Representation, Lie algebra, 186
  adjoint, 183, 322
  faithful, 402
  space, 186
Restricted rep, 67, 79
Rohrlich, F., 312, *417*
Riesz, F., 253, *418*
Robinson, G. de B., 117, *418*
Roman, P., 312, *419*
Root, 322, 341, 345, 353
Rotation, 18, 21, 26
  inversion, 19, 21, 26
Rotation group, 18
  proper, 18, 222
  improper, 18, 286
Rudin, W., 211, *419*
Ruhl, W., 305, *419*
Runge–Lenz vector, 376
Russell–Saunders approximation, 380
Rutherford, D., 117, *419*

S

Scalar, 102, 306, 318
Scattering, 266
Schiff, L., 265, *419*
Schöenflies notation, 34, 60
Schrödinger equation, 65, 109, 253
Schur lemmas, 69, 70, 187
Schwartz, J., 165, 181, 187, 391, *417*
Schweber, S., 311, *419*
Screw displacement, 22
Selection rules, 278
Self-adjoint operator, 92, 219, 414
Sell, G., 407, *418*
Semidirect product, 14, 21
Semigroup, 130
Shapiro, Z., 282, 308, 320, *416*
Sherman, T., 377, *419*
Sigma hyperon, 372
Skew-Hermitian, 173
Skew-symmetric, 105, 286
Slater, J., *419*
Smirnov, V., *419*
Space groups, 53, 55
  enantiomorphic, 59
  nonsymmorphic, 57
  symmorphic, 56
Special linear group, 3, 171, 340
Special orthogonal group, 9, 172, 222, 356
Special relativity, 294, 310
Special unitary group, 174, 222, 340
Spherical function, 424
Spherical harmonics, 231, 237, 241, 424
Spherical tensor, 274, 284, 300
Spin, 267, 312
Spinor, 307
  field, 268, 279, 311
  rep, 356, 362, 367
Spivak, M., 206, *419*
Sposito, G., *419*
Square, 10
Stakgold, I., *419*
Standard tableau, 127
Stark effect, 252
Steele, D., *419*
Strong interactions, 370
Structure constants, 167
Subalgebra, 183, 184
Subgroup, 1, 183
  commutator, 15, 114
  generated, 6
  improper, 1
  isotropy, 11
  local Lie, 184
  normal, 5, 185
  one-parameter, 176
  proper, 1
Sugano, S., *419*
Symmetric group, 3, 8, 116
Symmetric (Hermitian) operator, 108, 254, 414
Symmetric tensor, 105, 129
Symmetry classes (of tensors), 128, 141
Symmetry group, 23, 108
  complete, 109
Symplectic group, 174, 344
Sz.-Nagy, B., 253, *418*

T

Talman, J., 220, *419*
Tanabe, Y., *419*
Tangent matrix, 168, 207
Tangent space, 207
Tangent vector, 166
Tensor, 128, 309
  axial, 102
  contravariant, 101
  field, 279, 311
  polar, 102, 252
  rank of, 101
Tensor product of matrices, 364
Tensor product of reps, 79, 80
Term, 383
Tetrahedral group, 30, 86, 89
Tetrahedron, 30, 90
Three-*j* coefficients, 261
Timelike vector, 287
  backward, 287
  forward, 287, 321
Tinkham, M., 375, *419*
Trace, 19
Trajectory, 188
Transformation group, 8, 199
  effective, 193, 198
  local Lie, 188
  transitive, 9
Translations, 20

Transposition, 117
Tricomi, F., 399, 404, *416*

U

Unitary group, 173, 329
Upper triangular matrix, 154

V

Van Dam, H., *415*
Van der Waerden, 130, *419*
Vector space, 61, 306
  invariant, 67, 187
  irred, 187
Velocity transformation, 295
Vilenkin, N., 247, *419*
Volume of a group, 211, 212

W

Weight, 323, 345, 354
  function, 206, 408
  maximal (highest), 327, 342, 345, 354
  positive, 327, 342, 354
  vector, 323, 342, 367
Weyl, H., 119, 334, 349, 351, 362, *419*
Weyl group, 326, 345, 350, 360
Weyl reflection, 326, 346, 424
Whippman, M., *419*
Wigner, E., 227, 258, *420*
Wigner–Eckart theorem, 273, 275
World time, 310
Wybourne, B., 386, 391, *420*

Y

Yale, P., 34, *420*
Young operator (symmetrizer), 121, 146, 333
Young tableau (diagram), 119, 120, 331

Z

Zeeman effect, 251
Zero operator, 69

6
B 7
C 8
D 9
E 0
F 1
G 2
H 3
I 4
J 5